Engineering Economy and the Decision-Making Process

Joseph C. Hartman

Industrial and Systems Engineering
Lehigh University

Pearson Education International

Vice President and Editorial Director, ECS: *Marcia J. Horton*
Senior Editor: *Holly Stark*
Editorial Assistant: *Nicole Kunzmann*
Executive Managing Editor: *Vince O'Brien*
Managing Editor: *David A. George*
Production Editor: *Scott Disanno*
Director of Creative Services: *Paul Belfanti*
Art Director: *Jayne Conte*
Cover Designer: *Bruce Kenselaar*
Art Editor: *Greg Dulles*
Manufacturing Manager: *Alexis Heydt-Long*
Manufacturing Buyer: *Lisa McDowell*
Senior Marketing Manager: *Tim Galligan*

Pearson Prentice Hall
Pearson Education, Inc.
Upper Saddle River, NJ 07458

Printed in the United States of America

10 9 8 7 6 5 4 3 2

ISBN: 0-13-158702-1

Pearson Education Ltd., *London*
Pearson Education Australia Pty. Ltd., *Sydney*
Pearson Education Singapore, Pte. Ltd.
Pearson Education North Asia Ltd., *Hong Kong*
Pearson Education Canada, Inc., *Toronto*
Pearson Educación de Mexico, S.A. de C.V.
Pearson Education—Japan, *Tokyo*
Pearson Education Malaysia, Pte. Ltd.
Pearson Education, Inc., *Upper Saddle River, New Jersey*

This book is dedicated to my wife, Karen, for her continual and unwavering love and support.

Contents

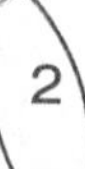

Preface

Consider some of the decisions that engineers have helped make over the past few years:

- Intel announced that it was spending $3.5 billion on a new 300-millimeter wafer fabrication facility (Fab 28) to make microprocessors on 45-nanometer process technology.
- Airbus is reportedly spending $12 billion to design the 550-seat A380 "superjumbo," the largest commercial aircraft ever produced.
- Air France and British Airways recently retired the Concorde, the only supersonic jet ever used in commercial service. The aircraft was developed at a cost of $3 billion in the late 1960s and early 1970s.
- Royal Dutch Shell announced a budget of $19 billion for 2006, of which $10–$11 billion was slated for the development and ramp-up of energy projects and continued exploration.

When one considers the magnitude and impact of these decisions, it becomes clear that *all* engineers must have a fundamental understanding of engineering economy. This subject provides the tools necessary for engineers to economically evaluate their decisions and designs. It also allows engineers to communicate with the "business" world, which is generally more interested in dollars and cents than engineering tolerances and specifications.

Given the importance of engineering economy, there are a number of textbooks available to the student, instructor, and practitioner. So what is different about this text? It is my opinion that *engineering economy* has become synonymous with *financial mathematics*—the process of applying interest rate factors to a cash flow diagram in order to determine some measure of worth (such as present or future worth) that is then used to make a decision. While this topic is of critical importance, it should not overshadow the fact that financial calculations are fairly straightforward when one considers the entire decision-making process. That is, the finances of a project must be *estimated* before it can be analyzed mathematically. This process includes estimating all relevant cash flows and understanding their inputs. Furthermore, the risk and uncertainty inherent in these financial estimates must be taken into serious consideration when the decision is to be made. This book differentiates itself from others by taking the student or practitioner through the *entire* decision-making process. After an introduction to the basics of engineering economy (interest, the time value of money, and equivalence), we begin our journey through the decision-making process, from defining the problem through post implementation analysis, just as one would when building a case for management in order to make a capital investment decision.

This approach is motivated by the "fathers" of engineering economy, including Eugene Grant, John C. L. Fish, and Gerald Thuesen. Studying their texts, written in the early part of the 20th century, one notices that they covered *a lot* of material. In particular, Grant's text covered basic principles (financial mathematics), but also fact-finding, estimation, and judgment, topics that are critical in dealing with the difficulties of making a capital investment decision.

Thus, we elaborate on the topics of estimation, risk, multicriteria, and uncertainty in this text, facilitated by the extensive use of spreadsheets and computer tools. The ultimate goal is that students become better decision makers.

Coverage of Material

The book is broken into five parts as described in detail below.

Part I: Principles of Engineering Economy. Chapters 1 through 4 present material that is fundamental to engineering economics. Basic concepts such as the time value of money, interest, purchasing power, cash flow diagrams, and economic equivalence are explained. Interest formulas are developed for the reader, allowing for the evaluation of cash flows. Examples from a variety of engineering firms or from companies that engage in engineering activities are presented in order to illustrate the concepts.

Part II: Decision-Making Preliminaries. Chapters 5 through 8 lead the reader through the first three steps in the decision-making process: identifying a problem or an opportunity, generating or designing investment alternatives to address the problem or opportunity, and estimating various data associated with each feasible alternative. Chapters 5 and 6 emphasize defining the problem and generating solution alternatives, two concepts not generally associated with engineering economy, but fundamental to the decision-making process. Chapter 7, on cash flow development, provides estimation techniques that are traditional to engineering economy and overviews forecasting. The role of these techniques in developing cash flow estimates for evaluation purposes is explained. This part closes with the introduction of taxes, depreciation, and the development of after-tax cash flows.

Part III: Making the Decision for a Single Project. Now that the investment alternatives have been generated and the cash flows have been estimated, the reader is ready to analyze those alternatives in order to make an informed decision. Chapters 9 through 11 examine the decision of accepting or rejecting a single project. Each chapter builds on the previous one as the content moves from analysis under certainty, to analysis under risk (with both deterministic and probabilistic methods), to the analysis of multicriteria, which includes a consideration of noneconomic factors. Note that this part spends considerable time dealing with the concept of risk with the use of the payback period, project balance, sensitivity analysis, break-even analysis, scenario analysis, and simulation analysis.

Part IV: Making the Decision for Multiple Projects. Chapters 12 through 14 consider the situation of multiple alternatives (either mutually exclusive or independent under constraints). Because many of the analyses used in examining a single project apply directly to this case, we spend a majority of time considering the difficulties having to do with comparing multiple projects with different attributes, be they different service lives or multicriteria. Approaches to solving large-scale problems are also addressed. Mutually exclusive alternatives defined by time—such as a project involving the option to delay or a project with phased expansions—are explicitly considered in Chapter 13.

Part V: Postimplementation Analysis. In Chapters 15 and 16, we stress the fact that the decision-making process does not end with the selection of a project. Rather, as we illustrate through project tracking, this step in the decision-making process ties itself back to the first step for future decisions, as costs are tracked to improve future cost estimates. We also link the concept of tracking to financial and activity-based costing. More importantly, we illustrate how tracking an investment can lead to abandonment, expansion, or replacement decisions. Decisions regarding expansion are covered earlier in the text, but abandonment and replacement decisions are explicitly analyzed in the final chapter.

Key Features of the Text

The fundamental mission of this text is to present engineering economy in the context of a decision-making framework such that the student understands the necessary tools and their application. To motivate the student, the text has the following attributes:

- **The book takes an explicit walk through the entire decision-making process**, from the definition of the problem or opportunity; to the generation of alternatives; to cost estimation and after-tax cash flow development; to analysis under certainty, risk, and uncertainty; to considering options in time; to considering multiple criteria and project acceptance; and, finally, to post-implementation analysis. This process is presented sequentially in Chapters 5 through 16, not just through a listing of steps in an introductory chapter.

- **Examples that stem from real-world applications are presented throughout the book.** We have spent considerable time accumulating information from various sources (e.g. *The Wall Street Journal, Dow Jones Newswires,* and *The Financial Times,*) concerning real-world capital investment decisions that have been made in the past. There are no widgets, there is no pixie dust, and there is not one company named ABC. Examples are derived from problems and opportunities faced by engineering firms, firms that engage in engineering activities, and firms that benefit from engineering activities. The industries represented include the automotive, chemicals, computer hardware, computer software, consumer products, defense and aerospace, energy, industrial goods and services, metals and mining, paper and forest products, pharmaceutical and biotech, semiconductors, telecommunications, transportation, and utility sectors. This diverse set of industries and firms allows the text to cover problems encountered by *all* engineering disciplines. In all, the examples and problems found in this textbook reference nearly 400 articles.
- **The problems at the end of each chapter are divided into four parts: (1) concept questions, (2) drill problems, (3) application problems, and (4) questions preparing the reader for the Fundamentals of Engineering examination.** In the concept section, the student is asked to examine the meaning of topics presented in the chapter. The drill section presents standard questions, which enable the student to practice using the tools presented. The application section presents questions that are motivated by actual decisions that have been made by industrial and government entities. The analyses allow the student to fully understand the decision-making process and its complexity. Each chapter concludes with multiple-choice questions in the same format found on the Fundamentals of Engineering exam, the first step toward achieving a Professional Engineering license. These questions were written such that the problems could be solved with the use of the *Fundamentals of Engineering Supplied-Reference Handbook*, supplied by the National Council of Examiners for Engineering and Surveying (Clemson, SC, www.ncees.org). Solutions to all Fundamentals of Engineering exam problems are given in the appendix. In all, there are nearly 800 problems in this textbook, evenly divided among these four categories of problems.
- **The book introduces the use of computers and spreadsheets and integrates their use into economic analysis.** We introduce spreadsheets to the novice user in Chapter 2 and illustrate how they can aid in engineering-economic analysis. Building upon this foundation, we present advanced material throughout the text, including how to generate graphs and

how to use built-in-functions, Goal Seek, Solver, and Visual Basic for Applications in Excel.

Supplements for the Instructor

A considerable amount of time has been spent creating the examples and application problems in this text in order to help an instructor bring realism to the classroom. To further ease the burden of teaching, the following supplements are provided.

- **Presentation slides in two formats: complete and incomplete.** It is my belief that providing students with slides that are "incomplete" is a wonderful way in which to motivate class participation (and attendance!). That is, the students do not have to spend the entire class writing (and thus not participating). Rather, they only have to fill in key topics that have been removed (such as the final equation of a derivation, the solution to an example, etc.). You, the instructor, are provided with "complete" slides such that the students can "fill in" what is missing during class. (Of course, it is your prerogative to use or not to use the incomplete slides.)
- **Presentation slides with different examples.** Students do not want lectures to simply regurgitate textbook material. While the slides provided cover all topics in the text, no example from the text is repeated in the slides, in order to lend variety to your teaching and to bring more perspectives to the student. These examples continue the theme of bringing realism to the educational experience by referencing real decisions.
- **Presentation slides for a lecture on "Using Engineering Economy in Your Everyday Life."** When I teach the course on which this textbook is based, the final lecture is composed of examples of applications of engineering economy concepts in real life: buying a house, saving for retirement, etc. This is generally well received by students and provides a good wrap-up to the course.
- **Database of summaries of real-world capital investment examples.** Articles from various sources, including *The Financial Times* and *The Wall Street Journal*, are cited, summarized, and provided on the website. The articles are categorized according to topic and are accompanied by short questions. Thus, they can be customized to a target audience and used to enhance discussions in lectures or as the basis of examples or quiz problems. With these summaries being continuously updated and maintained, the instructor will *always* have timely, topical examples at his or her disposal. These are also suitable for case studies.
- **A number of spreadsheet tools are available.** Visual Basic for Applications code (which integrates seamlessly into Microsoft Excel) has been provided to generate drill problems (of the type found at the end of

many chapters). The instructor may use the code to generate problems for additional homework, quizzes, tests, etc. Note that the instructor may alter the code if he or she desires. An after-tax capital-investment analysis tool developed in Visual Basic for Applications in Excel is also available as well as a "live" spreadsheet for generating interest factors in Excel.

- **Solutions to all questions in the textbook are available.** Where applicable, the associated spreadsheets are also available.

These files can be accessed by Instructors only and can be found at: www.prenhall.com/hartman; students will not be granted access to these files via this website.

How to Use This Text

This text is comprehensive in its coverage of engineering economy within a decision-making framework. Due to its numerous topics, it can be adapted according to the needs or desires of the instructor. Table 1 provides a number of suggested uses of the book at the undergraduate level. The table is broken down according to the level of the student, as juniors and seniors have been exposed to more mathematics during their schooling, allowing for a faster coverage of introductory material and leaving more time for advanced topics.

For each student level, classroom hours are suggested, given 20 contact hours (a 2-hour quarter course), 30 contact hours (a 2-hour semester course or 3-hour quarter course), or 45 contact hours (a 3-hour semester course). The number of hours suggested allows the instructor some latitude (e.g., only 41 of the 45 total hours are suggested for 3-hour semester courses).

The text may be used at a more advanced level if the instructor supplements it with case studies (for master's degree students) or journal articles (for Ph.D. students). The author may be contacted directly for suggestions in this area of application.

To the Student

It is hoped that you gain an appreciation for both the difficulty and the importance of the decisions that we are to study. The problems are difficult because they generally involve making an investment today for a better future—increased revenues or benefits or decreased costs that are realized at a later point or points in time—but in a future defined by risk and uncertainty. This textbook outlines an approach to deal with these difficult problems.

After laying a foundation of engineering economy principles, we take steps through the process of making a decision, considering all of the risks

TABLE 0.1 **Suggested outlines for courses, dependent on needs.**

Part	Chapter	Topic	Freshman–Sophomore Contact Hours			Junior–Senior Contact Hours		
			20	30	45	20	30	45
I	1	Engineering Economy and Decision Making	1.0	1.0	1.0	1.0	1.0	1.0
	2	Cash Flows and Time Value of Money	3.0	4.0	4.0	3.0	4.0	4.0
	3	Interest Formulas	4.0	4.0	4.0	3.0	3.0	3.0
	4	Economic Equivalence	2.0	3.0	3.0	2.0	3.0	3.0
II	5	Defining the Problem	Opt	0.5	0.5	Opt	0.5	0.5
	6	Generating Alternatives	Opt	Opt	1.0	Opt	0.5	1.0
	7	Estimating Cash Flows	Opt	0.5	1.0	Opt	1.0	2.0
	8	After-Tax Cash Flows	2.0	3.0	4.0	2.0	3.0	3.0
III	9	Single Project: Deterministic Analysis	3.0	4.0	4.0	3.0	3.0	4.0
	10	Considering Risk	2.0	2.0	4.0	2.0	2.0	4.0
	11	Considering Multiattributes	Opt	Opt	2.0	Opt	Opt	2.0
IV	12	Multiple Projects: Deterministic Analysis	2.0	3.0	5.0	2.0	3.0	5.0
	13	Considering Options in Time	Opt	Opt	2.0	1.0	1.0	2.5
	14	Considering Multicriteria	Opt	Opt	1.0	Opt	Opt	2.0
V	15	Postimplementation Analysis	Opt	1.0	2.0	Opt	1.0	2.0
	16	Abandonment and Replacement	Opt	1.0	2.5	Opt	1.0	2.0
		Total Hours	19	27	41	19	27	41

Note: Opt = optional.

and attributes inherent in real-world decisions. In doing so, we will evaluate problems encountered by *all* engineers from *all* types of industries *all* over the world. It is hoped that the student gains:

- An appreciation of the difficulty, complexity, and risks associated with capital investment decisions.
- Skills with which to analyze problems and the ability to convey the results of the analyses to others.
- Intuition with respect to the time value of money and its importance in decisions with regard to money in general.
- An understanding that engineering economy is a topic for all engineers in all industries with applications throughout the world.

This book puts you, the student, in the "seat" of being a decision maker. This will allow you firsthand to understand the important issues surrounding an investment decision and the difficulties involved in arriving at such a decision.

The Financial Times

We are please to announce a special partnership with *The Financial Times*. For a small additional charge, Prentice Hall offers students a 15-week subscription to *The Financial Times*. Upon adoption of a special package containing the book and the subscription booklet, professors will receive a free one-year subscription. Please contact your Prentice Hall representative for details and ordering information.

The newspaper (and online edition) can be used to enhance classroom discussions of real investment problems. Articles may even be used as the basis for example or homework problems. Feel free to contact the author on how to integrate *The Financial Times* into a course in engineering economy.

Acknowledgments

I would like to acknowledge the many people who have helped make this book possible.

My interest in economics, finance, and investing was instilled at a young age by my father, an accountant and tax attorney. While I found my way to engineering on my own, I thank him for teaching me the basics of money long ago, as it has become the basis for a very enjoyable career.

My interest in engineering economy was born during my graduate studies at Georgia Tech, working with Jack Lohmann, Gunter Sharp, and Gerald Thuesen. It was Jerry Thuesen who gave me my first opportunity to teach while a graduate student (nearly 15 years ago) and who suggested that I write a text on engineering economy. He even helped establish my relationship with Prentice Hall. I owe much of my success in this area to him.

It was also during my graduate studies that I met a number of prominent textbook authors in the area, including Ted Eschenbach, Wolter Fabrycky, Jerome Lavelle, Jim Luxhoj, Don Merino, Chan Park, William Sullivan, and John White. All of these educators have influenced my work over time. I also cannot discount (a little financial humor) my interactions with colleagues in the Engineering Economy Divisions of the American Society for Engineering Education and the Institute of Industrial Engineers, and those involved with *The Engineering Economist*. Discussions over the years with (in no particular order) Kim LaScola Needy, Heather Nachtmann, Janis Terpenny, Hamp Liggett, Marlin Thomas, Jane Fraser, Sarah Ryan, Kevin Dahm, Phil Jones, Hemantha Herath, Hamid Parsaei, Miroslaw Hajdisinski, Tom Boucher, Paul Kauffman, Bill Peterson, Dick Bernhard, Deborah Thurston, Dennis Kulonda, Phillip Ostwald, and John Ristroph have not gone unnoticed!

I would also like to thank the Department of Industrial and Systems Engineering at Lehigh University for supporting me in this endeavor. I have had the pleasure of working with two chairs during my tenure, Louis Martin-Vega and S. David Wu, both of whom strongly encouraged this work. I also thank the School of Management at the University of Edinburgh in Scotland, where I took my sabbatical in the 2003–2004 academic year. I thank Tom Archibald and Kevin Glazebrook for creating a setting that was ideal for work.

The following educators provided valuable input about the text's content at both the proposal and review stages: Phil Farrington, University of Alabama; Huntsville; Richard Bernhard, North Carolina State University; Sergey Sarykalin, University of Florida; Bob White, Western Michigan University; Carter J. Kerk, South Dakota Tech; John Mullen, Clarkson University; and Pete Loucks, Cornell University.

In addition to these reviewers, a number of people helped review specific parts of the manuscript: Robert Storer (Lehigh University) closely reviewed all sections on statistics, probability, and simulation; Raymond Hartman (Triton College) closely reviewed all sections on taxation; and Helen Linderoth (a Georgia Tech graduate) reviewed the *entire* manuscript thoroughly. I also thank Holly Stark, Dorothy Marrero, Scott Disanno, Nicole Kunzmann, and Marcia Horton for support from Prentice Hall.

Of course, I would be remiss if I did not thank the hundreds of undergraduate and graduate students whom I have taught at both Lehigh University and the Georgia Institute of Technology over the years. The lessons I have learned in the classroom have drastically shaped both the way I teach and the layout of this text. Specifically, I would like to thank a number of undergraduate research assistants who, over the years, helped improve educational materials related to engineering economy: Amy Reddington, Alison Totman, Erin Willey, Patricia Kosnik, Alison Kulp, Alison Murphy, Lauren Ross, Jennifer Rudnicki, Ingrid Schafrick, Maura Misiti, Colleen Sullivan, Alexandra Feinstein, and Sara Miller. Special thanks go to Pinar Keles and Huseyin Mac for help in developing solutions. The development of these new materials and this textbook was also sponsored, in part, by a CAREER award from the National Science Foundation (DMI-9984891). This support is gratefully acknowledged.

Joseph C. Hartman
Sotería and George N. Kledaras Endowed Chair
Industrial and Systems Engineering, Lehigh University

Part I

Principles of Engineering Economy

1 Engineering Economy and the Decision-Making Process

(*Courtesy of Airbus North America.*)

Real Decisions: Dueling for Air Supremacy

The market for large commercial aircraft is dominated by Airbus (80% owned by European Aeronautic Defence & Space Co. and 20% by BAE Systems, PLC) and Boeing. In 2003, Airbus delivered more commercial jets to customers than Boeing did for the first time in history.[1] This change only intensified their rivalry in competing for customer orders. Since that time, both companies have also made a number of important capital investment decisions, such as the following:

- In December of 2000, Airbus **launched** the development of the 555-seat A380 "superjumbo" jet at an initial cost of $10.7 billion. The largest plane in commercial history is to enter service in 2006.[2] As of December 13, 2005, Airbus had received 159 firm orders from

[1] Lunsford, J.L., "Bigger Planes, Smaller Planes, Parked Planes," *The Wall Street Journal Online,* February 9, 2004.

[2] Gecker, J., "Airbus officially launches world's largest passenger plane," *Associated Press Newswires,* December 19, 2001.

16 customers for the plane, which carries a list unit price of $280 million.[3]

- In March of 2001, Boeing **abandoned** the development of its 747X, an extended version of its 747 meant to carry 522 passengers and debut in 2005. Boeing supposedly spent millions on developing the plane, but was unable to secure a single order.[4]
- In March of 2001, Boeing announced the **development** of the Sonic Cruiser, a plane designed to travel just under the speed of sound and cut travel times by nearly 20%. The program began with $4 billion in funds originally earmarked for the 747X.[4] In December of 2002, Boeing **abandoned** the program, as the plane's high operating cost did not appeal to potential customers.[5]
- In April of 2004, Boeing **launched** the development of the 7E7 Dreamliner, later dubbed the 787, at a cost estimated between $7 billion and $10 billion. The plane, a fuel-efficient replacement for the 767, but with higher cargo capacity, will enter service in 2008 and carry between 217 and 289 passengers.[6] As of December 13, 2005, Boeing had 354 orders from 26 customers for the plane, which carries a list unit price between $132 million and $150 million.[7]
- In January of 2005, Boeing announced that it would **cease** production of its 100-seat 717 jet in 2006 due to declining orders.[8]
- In October of 2005, Airbus **launched** the development of its A350 jet at the cost of EUR4.35 billion.[9] The plane is expected to enter service in 2010.[10] As of December 30, 2005, Airbus had 170 firm orders from 13 customers for the plane, which carries a list unit price of about $160 million.[11]
- In November of 2005, Boeing **launched** the development of the 450-seat, 747–8, an updated version of its 747 jumbo jet first introduced

---

[3] "UPS signs firm contract for A380," *Press Release,* www.airbus.com, December 13, 2005.

[4] Avery, S., "Boeing to abandon plans for super-jumbo: Major shift in strategy: Mid-sized plane will travel near speed of sound," *Financial Post,* p. C01, March 30, 2001.

[5] Wiggins, J., "Boeing abandons plans for supersonic growth," *The Financial Times,* www.FT.com, December 22, 2005.

[6] Goo, S.K., "Boeing Bets on Dreamliner; Hopes Pinned on Technology for Return to Dominance," *The Washington Post,* p. E01, May 29, 2004.

[7] "Qantas Chooses Boeing 787 Dreamliner," *News Release,* www.boeing.com, December 13, 2005.

[8] Christie, R., "Boeing to Recognize Charges for USAF 767 Tanker Costs and Conclusion of 717 Production," *Dow Jones Newswires,* January 14, 2005.

[9] EUR stands for euro, the monetary unit of the European Union.

[10] Lagrotteria, B., "2nd UPDATE: EADS OKs A350 Launch, US To Continue WTO Case," *Dow Jones Newswires,* October 6, 2005.

[11] "Bangkok Airways selects A350 for new long range services," *Press Release,* www.airbus.com, December 30, 2005.

in 1970. The longer, more fuel-efficient plane with newly designed wings is expected to enter service in 2009. Cargolux and Nippon Cargo Airlines placed firm orders for 18 planes worth about $5 billion at list prices.[12]

Each of these decisions is at the heart of engineering economics and leads to a number of interesting questions:

1. How are the decisions related to engineering economy?
2. Are the decisions typical for a firm that engages in engineering activities? Do the decisions carry risk?
3. How is the design process related to the investment decision-making process?

In addition to answering these questions, after studying this chapter you will be able to:

- Define the role of an engineer and clarify the importance of economic analysis to an engineer's work. (Section 1.1)
- Define engineering economy as a critical competency for any engineer. (Section 1.2)
- Categorize typical engineering economic analyses according to profit-enhancing, cost-control, or public-improvement programs in which expansion, replacement, or abandonment decisions are made. (Section 1.3)
- Define the steps of the engineering economic decision-making process and illustrate its relationship to the engineering design process. (Sections 1.4–1.5)

[12] Lunsford, J.L. and D. Michaels, "Boeing Updates 747 in Battle with Airbus," *The Wall Street Journal,* p. A10, November 16, 2005.

We begin our journey through the process for making engineering investment decisions by using the tools of engineering economy. In this chapter, we introduce the types of problems to be studied, the analyses to be utilized, and the role that you, the engineer, play in this process of determining whether an investment in a given engineering project should be made or not. While engineers are often credited for their analytical and design skills, their "business" skills are frequently overlooked. The tools of engineering economy presented in this text will help bridge this gap between engineering and business that otherwise will prevent the engineer from being successful.

1.1 The Engineer

Engineers are trained to provide answers. Through the rigorous study of science and its application to various problems, an engineer learns to design and implement solutions in his or her field of study. These solutions may be applied to solving problems or opening doors to new opportunities. Consider the design and construction of an airplane, which, as noted at the beginning of this chapter, is a very expensive process that takes input from a number of professionals, including many engineers:

- Aerospace engineers may design the body and wings for optimal lift and minimal wind resistance.
- Chemical engineers may analyze and reduce the corrosiveness of the jet fuel such that tanks do not rupture.
- Civil engineers may reduce the weight of the plane's shell without compromising the structural integrity of the vessel.
- Computer science engineers may develop software for automated takeoff and landing procedures.
- Electrical and computer engineers may design the hardware, circuitry, and sensors used to control the aircraft.
- Environmental engineers may design filters to reduce the pollutants dispersed through the plane's exhaust system.
- Engineering managers may coordinate the activities of the entire team.
- Industrial engineers may design the process to build the airplane (including job and personnel scheduling), as well as locate the facility.
- Material science engineers may develop new composites to reduce the weight of the plane.
- Mechanical engineers may design the turbine engines such that the plane is quieter and more fuel efficient.

To design and build an airplane is a daunting task, as each component and subsystem is complex. The components, either designed and produced in-house or purchased from another vendor, must be precisely and seamlessly integrated into the complete airplane design. This may require numerous decisions throughout the design process. Although the tasks of these engineers may be very different, they have the same ultimate goals during this process:

1. **Technical feasibility.** In order to ensure the implementation of the design, the solutions provided must obey the laws of nature and science.
2. **Technical efficiency.** Because there may be many solutions to a problem, the best technical solution should be sought. Examples include designing solutions that generate the least waste, consume the least energy, perform reliably in adverse conditions, and allow for easy production or maintenance.

While meeting these two objectives would seem to suffice as a valid job description for an engineer, they represent only part of an engineer's responsibilities. In addition to adhering to technical specifications, the engineer must also heed economic considerations:

1. **Economic feasibility.** The solutions provided must not exceed monetary budget limits.
2. **Economic efficiency.** The most economical of the many technical solutions to a problem should be chosen.

Engineers are burdened with satisfying both engineering requirements (the solution must obey the laws of science and nature) *and* economic requirements (the solution must meet budget requirements and achieve the highest return possible). These two requirements are not trivial. Given an infinite sum of money, any person, engineer or not, could design some component of an airplane. However, the engineer who can design the component or system so that it meets specifications under all circumstances at the lowest cost is the engineer who will succeed. This requires trade-offs in analysis, as there may be a conflict between technical and economic efficiency.

Given that a problem exists, it is the role of the engineer to develop a solution to it. In order to ensure that the best possible solution is chosen, a number of solutions should be designed and evaluated. Each of these solutions, although different, must solve the problem at hand. For the aerospace engineer, this may require examining different wing configurations in order to minimize drag and fuel costs. For the mechanical engineer, it may require examining different methods of heat transfer to cool the engines. For the civil engineer, it may require minimizing the amount of material used to support the plane while retaining its structural integrity. For the material science engineer, a number of materials, both natural and synthetic, must be examined. For the chemical engineer, various alternative fuels must be evaluated. For the industrial engineer, different plant layouts and material flow must be analyzed, as must different plant locations.

As these various engineers design different solutions, engineering economic decision analysis provides a method for *all* of them to decide which solution should be accepted and implemented. Engineering economy affords the tools to evaluate these alternative solutions from the perspective of their monetary worth. Provided that all of the solutions designed by the engineers are technically feasible, the tools of engineering economy allow the engineer to implement solutions that are technically *and* economically efficient.

1.2 Engineering Economy

It takes years of study and practice to become a competent engineer—one capable of providing technically feasible and efficient solutions in his or her specific field. This includes studying general subjects, such as mathematics and physics, and topics specific to each engineering discipline, such as control theory for mechanical engineers and linear programming for industrial engineers. In this text, we generally assume that technical solutions are readily available or have already been designed. That is because those solutions are discipline specific.

Every technical solution, whether designed by a mechanical or an electrical engineer, has financial consequences. Among these consequences is the cost to implement and maintain the solution, as well as the revenues or savings achieved from implementing the solution. For example, a civil engineer may consider numerous options to alleviate traffic congestion in a city, including building a bypass road or adding traffic lights with new traffic patterns. The cost to build the bypass road would include acquiring the necessary land and building a road of a specified length and width. Costs to maintain the road over time must also be estimated. The costs for changing the traffic patterns include the possible removal of existing signs and markings and the purchase and installation of additional traffic lights and signals. Maintenance of the traffic signals and the road surface, which must undergo increased traffic, must also be estimated. Once these economic variables have been identified and estimated, the analysis is no longer discipline specific, and the role of engineering economy becomes clear.

Engineering economy is a methodology that provides the necessary tools to describe and evaluate engineering projects according to their economic characteristics. Any engineering project can be defined according to cash inflows and outflows. Outflows include all expenses, while inflows include all savings or revenues derived from implementing a solution. Engineering economics defines how the financial or economic aspects of a project are to be evaluated. When multiple projects are available to solve a problem, the economic consequences of each of the choices can be evaluated and the best alternative selected for implementation.

Note that while the development of the technical solution is discipline specific, the economic evaluation of the financial characteristics of the project is general. That is, engineering economy defines tools to be used by *all* engineers. It is a general subject that transcends disciplines because cash is a measure that is common to all engineers and other professionals. This is why it is a topic on the Fundamentals of Engineering (FE) exam, the first step towards a Professional Engineer (PE) license.

Note also that economic evaluation is the role of an engineer, not just of professionals in finance and accounting departments. Who better to analyze the economic merits of a proposed solution than the engineer who designed the proposal? An engineer should never think that the role of economic analysis resides with those who have business, accounting, finance, or economics degrees. The role of the engineer is to develop technical solutions that are economically viable. Thus, the engineer must evaluate the economic aspects of a project, not just its technical merits.

When an engineer develops a solution to a problem, that solution must eventually be presented to management for evaluation and possible implementation. The engineer's report may consist of justified recommendations and/or a presentation. While management is ultimately responsible for implementation decisions, managers will generally assume that an engineer's solutions are technically feasible. That is, they assume that the solutions presented will solve the problem at hand. After all, that is why engineers are hired. What management will likely focus on is the economic viability of the project. Management will want to know, "Will it pay?" Management will want to know whether implementing *your* engineering solution will make the company or entity more profitable. Answering this question entails evaluating the economic merits of the project and your ability to "speak" intelligently about its financial and economic impact. Engineering economy provides the tools for this evaluation, and this text puts you, the engineering student, in the role of engineer as an economic evaluator.

1.3 Engineering Economic Decisions

Engineers design and develop solutions to problems and also formulate alternatives to take advantage of opportunities. To implement a solution, we generally have to make an investment (spend money) in hopes of increasing profits or achieving some level of savings.

We can categorize the decisions to be analyzed in this text in a number of ways. In general, most investments can be placed into one of the following three categories:

1. **Profit-enhancing programs.** A company may expand production, its product line, or services, in order to increase sales. These ventures generally take considerable investment, with the hope of increasing revenues. Following are some examples:

   - *New-product development.* In 2003, with competitors offering 32- and 64-bit chips, Advanced Micro Devices (AMD) announced the development of Hammer, a family of microprocessor chips designed to efficiently run in both 32-bit and 64-bit programs.[13] AMD was also the first to ship 64-bit processors for the desktop market.[14]

- *New-product acquisition.* Cisco Systems Incorporated, a leading supplier of networking equipment for companies, purchased Linksys systems in 2003 for $500 million in order to expand its reach into the home-networking market.[15] Cisco then continued this expansion by purchasing Scientific-Atlanta, a leading maker of TV set-top boxes for home entertainment, for $6.9 billion in the fall of 2005.[16]
- *Production capacity expansion.* In the summer of 2005, Intel announced plans to build a 300-millimeter wafer fabrication facility in Arizona to produce microprocessors for a range of projects. The factory, dubbed Fab 32, is expected to cost $3 billion and create 1,000 new jobs.[17]
- *Service capacity expansion.* UPS will open 20 new warehouses in China over the next two years (through 2007) at the cost of $500 million. The warehouses will serve both importers and exporters, as well as companies that do not have their own storage facilities.[18]
- *Improved customer service.* In the spring of 2003, T-Mobile USA announced a $300 million investment with Nortel Networks to upgrade T-Mobile's network so that customers could fully utilize the company's portfolio of products, including visual communications, instant messaging, Web browsing, and gaming.[19] T-Mobile followed this announcement in the winter of 2004 with intentions to spend an additional $300 million on Nokia equipment to expand capabilities further.[20]

2. **Cost control programs.** Engineers are often asked to correct errors in systems—errors that cost money. Designing and implementing solutions also costs money, including, at a minimum, the time and effort spent by the engineer in finding a solution to the problem. The expectation is that the engineer will provide a solution that will save more money in the long run. Examples are as follows:

   - *Improving efficiency.* Porsche AG announced a EUR6.7 million expansion of its Leipzig plant, which produces the Cayenne SUV and Carrera

[13] Clark, D., "AMD is Making a 64-Bit Bet on Its Future," *The Wall Street Journal,* June 2, 2003.
[14] Boslet M., "Advanced Micro CEO: Launches Athlon 64 3200+, FX51 Models," *Dow Jones Newswires,* September 23, 2003.
[15] Thurm, S., "Cisco to Buy Home-Networking Leader," *The Wall Street Journal,* April 21, 2003.
[16] Kessler, M., "Cisco Raises the Stakes in Digital Home Entertainment; Company to Buy Set-Top Maker Scientific-Atlanta," *USA Today,* November 21, 2005, p. B3.
[17] Rogow, G., "Intel to Build New 300-mm Wafer Factory in Arizona; $3 B Investment for Future Generation Platform Products," *Dow Jones Newswires,* July 25, 2005.
[18] Souder, E., "UPDATE: UPS Opens Chinese Warehouses for Imports, Exports," *Dow Jones Newswires,* February 9, 2005.
[19] Thomas, S., "Nortel Networks Gets T-Mobile USA Pact," *Dow Jones Newswires,* May 7, 2003.
[20] Wallmeyer, A., "T-Mobile USA Chooses Nokia Equipment for Network Upgrades in a $300 M Pact," *Dow Jones Newswires,* February 11, 2004.

GT. The investment, completed in the fall of 2004, was aimed at improving efficiency, not production capacity, although the production area was increased by about 24%.[21]

- *Streamlining operations.* Bombardier, Inc., sold its consumer products division for C$1.23 billion[22] in the summer of 2003 to focus on small jet aircraft and train production. The consumer unit produced snowmobiles and watercraft.[23]
- *Eliminating waste.* Norampac, Inc., announced the purchase and installation of a wood-residue boiler in the fall of 2004 at its Cabano mill, which produces corrugated products. The C$5.4 million boiler produces 100,000 pounds of steam per hour from wood waste, reducing the mill's reliance on fossil fuels.[24]
- *Reducing liabilities.* In the spring of 2004, American Electric Power Co. (AEP) announced that it would spend $3.5 billion through 2010 on environmental controls for its coal-fired plants. AEP is the largest power generator in the United States, but also the largest emitter of carbon dioxide. AEP said that $1.8 billion would be spent to meet current EPA regulations and the remaining investment would be spent on forthcoming regulations.[25]

3. **Public-improvement programs.** Government entities often make investments, the goal of which is not to increase profits, but rather to increase some measure of public satisfaction. Following are some examples:

   - *Increased public satisfaction.* The U.S. Postal Service set aside $637 million for the fiscal year starting October 1, 2003 (up from $435 million a year earlier) in order to help address the issue of long lines, the number-one complaint of customers.[26]
   - *Increased public safety.* The government expects to pay $7 billion through 2006 to level and remove Rocky Flats, a defense complex used to produce parts for atomic bombs during the Cold War. In addition to removing radioactive waste, The area, located 15 miles northwest of Denver, is to be converted into a wildlife refuge.[27]

---

[21] Reiter, C., "Porsche to Invest EUR6.7M in Leipzig Plant Expansion," *Dow Jones Newswires,* March 18, 2004.

[22] C$ stands for Canadian dollars.

[23] Chipello, C.J., "Bombardier to Sell Subsidiary to Group Led by Bain Capital," *The Wall Street Journal Online,* August 27, 2003.

[24] King, C., "Norampac to Invest C$5.4M to Increase Energy Efficiency," *Dow Jones Newswires,* November 3, 2003.

[25] Kamp, J., "AEP Coal Plants Will Keep Up with Air Quality Rules—CEO," *Dow Jones Newswires,* March 24, 2004.

[26] Brooks, R., "Postal Service to Lift Budget for Construction, in Effort to Speed Delivery, Cut Lines," *The Wall Street Journal Online,* September 11, 2003.

[27] Fialka, J.J., "Nuclear-Site Cleanup Faces First Big Test," *The Wall Street Journal,* February 3, 2003.

- *Improved infrastructure.* The Economic Development Corp. of New York City unveiled plans to upgrade facilities on Manhattan's West Side, as well as to construct two new berths for cruise ships in Brooklyn. A total of $250 million is to be spent over 10 years, starting in 2004.[28]

The preceding categories define the reasons for investing. Either a company wants to grow, through increased production or by offering new products, or the company wants to save money through operational improvements. Thus, an investment is made to increase future revenues or reduce future costs. The tools of engineering economy are designed to look at this trade-off between the status quo and making an investment now for future savings or revenues. Government entities have different focuses, as they are driven, not by profit, but rather by public satisfaction.

The actual decisions that lead to these improvements or increased revenues can be quite different. For example, to enhance profits, a company may consider building a new plant to increase revenues, or it may consider consolidating operations to save money. We classify these decisions into one of three broad categories of (1) expansion, (2) replacement, and (3) abandonment as follows:

1. **Expansion.** This can take many forms, including expanding the production capacity of a current product and expanding into new markets with new products or services. Examples are as follows:

   - *New-product design and development.* Eclipse Aviation aimed to complete testing of its new six-seat corporate jet in 2005 and begin production of the plane in early 2006 in Albuquerque, New Mexico. The plane is expected to cost between $950,000 and $1.2 million per unit, much lower than competitors' price tags. As of the end of 2003, 2,100 orders had been booked through 2008.[29,30]
   - *Expansion of current facility.* Sanyo Electric Co. spent YEN7.5 billion[31] to boost the capacity of solar cells at its plant in Osaka Prefecture. The investment increased capacity in early 2005 from 33 to 103 megawatts per year.[32]
   - *Construction of new facility.* Gillette announced that it would spend EUR120 million on an 80,000-square-meter facility to produce disposable

[28] "NYC Unveils Plan to Improve Ports for Cruise Ships," *Dow Jones Newswires,* February 4, 2004.

[29] "UK's Hampson Signs $380M Mfg Pact with Eclipse Aviation," *Dow Jones Newswires,* December 10, 2003.

[30] "Eclipse Aviation announces sales agreement for 239 airplanes," *Associated Press Newswires,* April 25, 2005.

[31] YEN stands for Japanese yen.

[32] "Sanyo Elec to Boost Solar Cell Output Capacity—Kyodo," *Dow Jones Newswires,* February 12, 2004.

razors and shaving systems in Lodz, Poland. The plant is expected to be fully operational by 2007.[33]

- *Acquiring capacity.* Timken Company acquired Torrington Company from Ingersoll Rand Company for $840 million early in 2003. The purchase resulted in the third-largest producer of bearings in the world.[34]
- *Equipment, process, or technology selection.* The U.S. government has contracted with Space Exploration Technologies Corp. for an estimated $30 million to launch a classified satellite into low orbit by the summer of 2007, using the company's Falcon 9 rocket, a larger version of its least-expensive Falcon 1 rocket. The market for launches in excess of 21,000 pounds has been dominated by Boeing Co. and Lockheed Martin Corp.[35]

2. **Replacement.** A company may want to continue operations or services in some sector, but may want to do so in a more economical manner. This may lead to replacing equipment, changing processes, or changing locations. Following are some examples:

   - *Equipment, process, or technology selection.* Canadian National Railway Co. began taking delivery of 50 Evolution locomotives, a 4400-horsepower engine from General Electric, and 25 SD70 locomotives from Electro-Motive Diesel, Inc. (EMD) in the fall of 2005. The 75 more-fuel-efficient locomotives are to replace 100 older engines. The two engines meet new Environmental Protection Agency emissions regulations and cost around $2 million each.[36]
   - *In-house versus outsourcing.* Auto parts manufacturer Visteon outsourced its computer systems operations, such as data centers and help desks, to IBM. The deal, a 10-year contract signed in 2003, was valued at over $2 billion.[37]

3. **Abandonment.** Although it may not seem like an investment decision, the decision to walk away from a project, such as closing a facility, is a very important economic decision. It also represents the final stage in our decision-making process. Types of abandonment include the following:

   - *Cease production.* In the fall of 2005, with the approval of the U.S. government, IBM agreed to sell its personal computer-manufacturing

[33] McKinnon, J., "Gillette to Build New Manufacturing, Packaging and Warehouse Facility in Poland," *Dow Jones Newswires,* March 16, 2004.

[34] Aeppel, T., and C. Ansberry, "Some Businesses Boost Spending," *The Wall Street Journal,* April 1, 2003.

[35] Pasztor, A., "Space Exploration Gets New Rocket Contract, *The Wall Street Journal,* September 7, 2005.

[36] Scinta, C., "Tales of the Tape: GE on Track with Railroad Technology," *Dow Jones Newswires,* September 26, 2005.

[37] "Visteon and IBM to Partner in Transformation of Visteon's Worldwide IT Services," *IBM Press Release,* www.ibm.com, February 11, 2003.

division to Lenovo of China for $1.75 billion. Lenovo will be allowed to sell PCs under the IBM brand name for five years.[38]

- *Cease product line.* Caterpillar sold the manufacturing and marketing of its Challenger tractor line to AGCO Corporation, a farm-equipment manufacturer, in 2002, in order to focus on its core construction business.[39]
- *Close a facility.* When the telecommunications boom collapsed, Corning responded by mothballing (temporarily closing) or abandoning four fiber-optic-cable production plants by the spring of 2003, leaving only one plant fully operational.[40]
- *Retire equipment.* British Airways and Air France retired their combined fleet of 12 Concordes in 2003, ending the use of the only supersonic plane in commercial service. The research-and-development costs of the planes totalled $3 billion in the late 1960s and early 1970s.[41]

Note that these decisions are generally not made independently. If a company decides to forge into a new area (new-product development), production will generally require additional capacity (expansion). Equipping the facility will also require the evaluation of various production techniques (equipment and process selection). All decisions eventually must face the question of abandonment.

One must realize how important these decisions are and how difficult they can be. Consider the consequences of implementing a solution. An investment is made with hopes of reducing costs or increasing revenues *in the future*. Often, the revenues materialize far in the future. But we live in a dynamic world: Demands are not realized, product testing and research can be unsuccessful, consumers' tastes change, companies alter their focus, world markets open and close, currency valuations fluctuate, and laws are modified. These deviations from our expected investment outcomes define the *risks* of an investment—risks that can lead to lost money. If this occurs too often, a company can go bankrupt.

To illustrate the risks involved with engineering projects, consider an example from the oil industry. In the winter of 2003, Socar, the Azerbaijan state oil company, said that foreign companies had spent more than $500 million searching sectors of the Caspian sea *without finding* recoverable commercial-grade oil. Many engineering projects folded as a result of the (lack of) findings.[42] This is just one example of the risks involved with

[38] Loades-Carter, J., and S. Goff, "The week in technology: IBM gets Lenovo clearance," *Financial Times*, www.FT.com, March 11, 2005.

[39] "Our Company: History of AGCO," www.agcocorp.com.

[40] Berman, D.K., "Corning Tries to Adapt to Changing Times," *The Wall Street Journal,* March 6, 2003.

[41] Michaels, D., "Concorde Flights to End, a Coda for Sound of Speed," *The Wall Street Journal,* April 11, 2003.

[42] Sultanova, A., and A. Raff, "Socar: Foreign Oil Cos. Lost $500M on Offshore Exploration," *Dow Jones Newswires,* December 2, 2003.

engineering projects. An investment is made to achieve future returns or savings, but these are not guaranteed to materialize.

This is why you, the engineer, must analyze the problem in its entirety to determine the best course of action. Doing so requires in-depth research, as alternatives must be identified and the consequences of their implementation must be fully understood. Once estimates have been made, analyses that consider the risks can commence. Only after this investigation is complete can a final decision be made. Note that the process should not be rushed. In February of 2004, resource provider BHP Billiton said that it would conduct a feasibility study to determine whether the company should expand its Western Australian iron ore production capacity. Chief executive Chip Goodyear expected to make a decision concerning the expansion within 12 to 15 months.[43] Note that this was not the time required to *undertake* the expansion, but merely to *study* the expansion. Because these investments are risky, they must be studied extensively and not entered into lightly.

The goal of our studies is to be able to make good decisions in a dynamic and risky environment. Thus, our decision-making process has to be dynamic and the capital-budgeting process, during which all investment alternatives are reviewed, must be continuous. The corporate treasurer may elect to make decisions on a quarterly or annual basis or may evaluate proposals on a rolling basis as needs or problems arise. Regardless of when the decision is made, there are three potential outcomes with respect to an investment proposal:

1. **Invest.** This "accept" or "go" decision releases the necessary funding to undertake a project.
2. **Delay.** This "wait" decision provides time to gather more information about the prospects of an investment or merely for the investment climate to change. On the one hand, the decision to delay may occur before a "go" decision. For example, the groundbreaking for a new facility may not occur until demand in a certain economic sector improves. On the other hand, the decision may occur *after* a "go" decision has been implemented. For example, the structural shell of a facility may be completed, but it will not be equipped and become operational until economic conditions improve.
3. **Do not invest.** This "reject" or "no go" decision eliminates a proposal from further consideration. This decision also encompasses halting funding for a project, as in the abandonment decision just described.

These decisions capture all the options available to a decision maker. Note that they are not independent. A company may decide to invest (go) in a new production line for a facility. However, it will be implemented when demand in the sector rises to a certain level (wait). After production commences, there will come a time when it must be determined whether output should continue (go) or cease (no go) so that funding can be channeled in other directions.

[43] "BHP Billiton: Approves Study for Iron Ore Expansion," *Dow Jones Newswires,* February 13, 2003.

The tools of engineering economy are designed to evaluate the economic implications of each of these decisions. Although we focus on the decisions of profit-making entities, such as the many companies listed up to now, we will also examine decisions from the perspective of a government entity, which has very different goals. We embed our economic analysis in a rigorous decision-making framework in order to ensure that we examine all aspects of our decision and make an educated choice.

1.4 The Decision-Making Process

We have established the fact that engineers solve problems and formulate alternatives to take advantage of opportunities. Their training endows them with the engineering expertise to provide technical solutions. Engineering economy offers the tools required to evaluate engineers' solutions according to their monetary merits. To ensure that the best solution is chosen, these tools are integrated into a decision-making process. Patricia Oerlemans, a spokeswoman for Philips Electronics, the largest European manufacturer of consumer electronics, was describing a new program for evaluating suppliers. When asked how the new requirements for suppliers would be implemented, she said, "We're an engineering company, so everything happens in steps."[44] This is excellent advice for making decisions (whether an engineering firm or not!). The steps in engineering economic analysis are as follows:

1. **Recognition and definition of problem or opportunity.** In order to solve a problem, the problem must be identified, along with the characteristics of its desired solution. The decision-making process cannot commence unless this step occurs.

   In addition to solving problems, engineers provide methods with which to take advantage of opportunities. Whereas problems, such as the failure of a piece of equipment, may abruptly present themselves, investment opportunities are often more subtle. To find these opportunities, one must examine the technical capabilities available and creatively seek additional applications of those capabilities. For example, Intel has applied its technological know-how in personal computer chip technology to communications, such as chips used in cell phones, in hopes of increasing revenues.[45] Alternatively, needs in the market may present themselves and may lead to additional opportunities. Consider the heavy investment in Russia's oil delivery infrastructure, which has been driven by unrest in

---

[44] "Philips to Make All Suppliers Conform to Standards," *Dow Jones Newswires,* March 23, 2004.

[45] Pringle, D., K.J. Delaney, and D. Clark, "PC Industry's Foray into Cellphones is a Siege," *The Wall Street Journal,* February 18, 2003.

the Middle East.[46] As with problems, the opportunities must be rigorously defined before a plan of action can be formulated.

2. **Generation of solution alternatives.** Once a problem or opportunity has been defined, solution alternatives must be identified. Identifying these solutions requires all of the training, creativity, and skill of an engineer, as this is the phase when all engineering solutions are designed for subsequent economic evaluation. A rich collection of possible solutions will make it more likely that an economically viable option will be found. To broaden perspectives and allow various solutions to be explored, engineers generally form teams.

   This process also requires discipline, as one must refrain from passing final judgment on a particular solution. This phase is concerned with generating alternatives, not evaluating them. That is not to say that the design process should accept solutions that do not meet technical specifications. However, judgment must be used such that feasible alternatives are not eliminated too early in this creative process.

   That said, there is a fine line between generating a large number of feasible alternatives and generating the number of alternatives that can be feasibly analyzed with the resources available. While it seems best to generate as many solutions as possible, as this ultimately ensures a good decision, the resources available to analyze the alternatives are limited. Thus, we must produce a *reasonable set* of alternatives. This requires engineering judgment to identify outlandish proposals (those which are likely to fail) and make sure that they are eliminated early, so as not to waste limited resources analyzing them.

3. **Development of feasible alternative cash flows and information gathering.** Once alternatives have been identified, they must be researched. A wide range of factors may be considered in evaluating the economic benefits and costs of engineering proposals, including all expected receipts and disbursements.

   In addition to investigating the associated cash flows, the research process includes gathering relevant information in order to make an informed decision. Risks of a physical and economic nature may be involved and must be evaluated. This includes calculating the probability of success and the likelihood of failure. All other benefits and costs of an alternative, even those which cannot be easily expressed in terms of money, must be identified. For example, the $350 million addition of a fifth runway at Atlanta's Hartsfield airport was supported by the construction of a 5.5-mile conveyor belt to transfer rubble from, and rock to, the site. The conveyor belt was deemed cost effective compared with trucks, but had the additional

[46] Whalen, J., and M. Fackler, "Moscow is Likely to Give the Green Light to Two Major Pipelines," *The Wall Street Journal*, March 7, 2003.

benefit of reducing traffic, emissions, and noise generated from the trucks.[47] The latter is a benefit that is not easily quantified monetarily, but surely influences the decision.

Estimation is generally considered the most difficult step of the decision-making process, as all of the cash flows are expected to occur in an uncertain future. The error can often be reduced, but it can never be eliminated. These estimates, and their uncertainty, are the subject of economic scrutiny in the next decision-making step.

4. **Evaluation of alternatives.** The cash flows defined in the previous step are rigorously analyzed to determine the economic viability of an alternative. The focus here is on examining the *differences* between alternatives, not their similarities. This text offers specific instruction in these methods of analysis under situations of certainty, uncertainty, and risk in environments that are either static or dynamic. In addition to the analysis of economic factors, noneconomic attributes associated with each alternative are also examined. When possible, these are converted into monetary terms.
5. **Selection and implementation of best alternative.** Once all the alternatives have been analyzed economically, the optimal choice is selected. Because noneconomic factors may weigh into the decision, it is the job of the engineer to exercise judgment in making the final decision. This may lead to the implementation of an engineering solution, the delay of an investment until a later date, or the decision to forgo investment completely, if economic success is far from guaranteed.
6. **Postimplementation analysis and evaluation.** Choosing a solution does not complete the decision process. After the solution is implemented, the project must be tracked, and periodic decisions are made as to whether it should continue. We have already mentioned that investment decisions are made concerning an uncertain future. As the future unfolds, the project's direction may need to be adjusted in order to capitalize on opportunities. This may entail accelerating or delaying stages of the project, expanding the project, replacing certain aspects of the project, or abandoning the project altogether. These decisions can be made smartly only if the project is closely watched after a solution is implemented.

   We also track projects in order to provide information for future investment decisions. Estimates required in new economic analyses are usually derived from previous estimates, just as previous designs provide input for new designs in the engineering design process. Thus, close tracking of costs and revenues associated with a project will provide a database of information that can be used to improve the accuracy of future predictions. The result is a link back to steps 1, 2, and 3 of the decision-making process for future decisions.

[47] Harris, N., "Airports Keep Growing Despite the Slowdown," *The Wall Street Journal Online*, October 14, 2002.

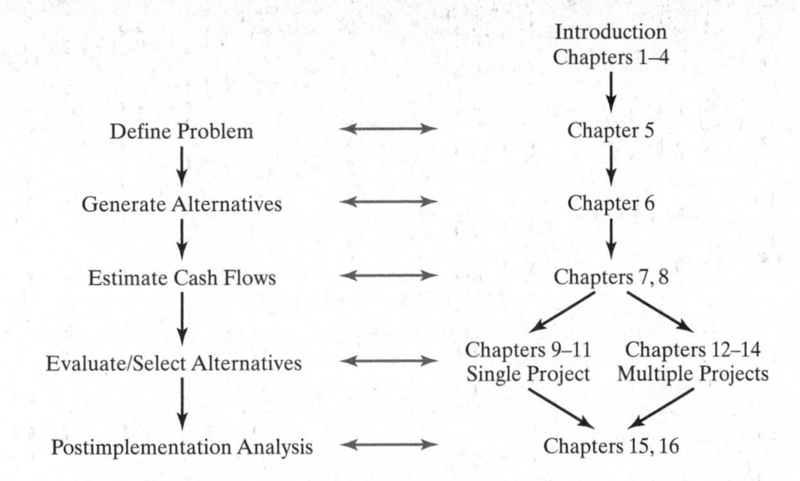

Figure 1.1 Steps in the decision-making process according to this text's outline.

The preceding steps take any decision from its inception, to the implementation of a solution, and, eventually, its phaseout and completion. It can be said that a decision is taken from birth to death in these six steps. To ensure rigorous analysis, each step must be completed. Figure 1.1 illustrates the decision-making process in conjunction with the chapters in this text, as we emphasize each step in its entirety.

1.5 The Engineering Design Process

Engineers examining the steps in the decision-making process may note that it is similar to the engineering design process. The design process is a critical step in developing investment alternatives and clearly falls into step 2 of the decision-making process. For example, consider the options Boeing has in determining what engines to put on its planes. Boeing could buy engines from a vendor, such as General Electric, Rolls-Royce, or Pratt & Whitney (United Technologies), and integrate them into its existing system, or Boeing could design and build its own engines. These are all viable solutions to the problem of providing thrust and power to propel the airplane. If Boeing decides to design and build the engines in-house, then it must go through a separate engineering design process. Likewise, a chosen vendor would follow a similar process. The steps involved are defined in Figure 1.2.

In this text, we are concerned with the overall decision-making process of implementing engineering projects. Clearly, the design process is a critical step, as it defines the alternatives to be evaluated. Because design techniques tend to be discipline specific, we will focus on the ensuing economic analysis, which is common to all engineers. However, it is critical that the engineer understand the role of the design process in the context of the entire implementation process, and it is vitally important that the engineer understand that the

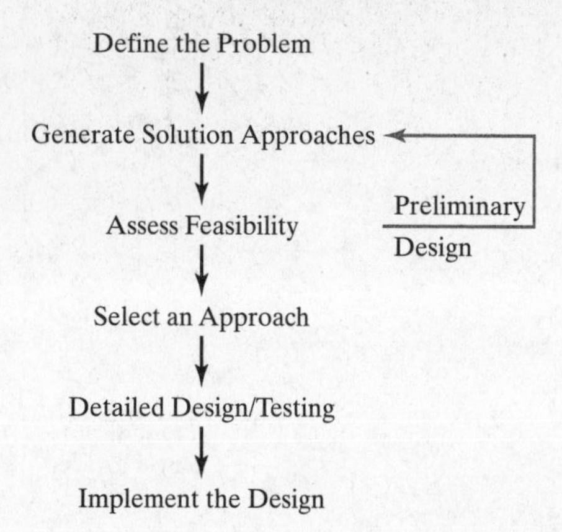

Figure 1.2 Steps in the engineering design process.

economic principles taught in this text are also applicable to the various steps in the design process. For example, when selecting a material or process to be used in a design, an economic analysis should be included with the technical analysis. This economic analysis should follow the same steps of the decision-making process highlighted in the previous section.

To further illustrate how critical it is to integrate engineering economic decision making and engineering design, consider the consequences of a design process that ignores economics. Suppose an engineering design team determines that gold-plated connectors are needed to ensure excellent connectivity in an electronic component. While gold is an excellent choice from an engineering specifications point of view, it may not be the optimal choice considering the economics involved. For example, if the electronics component is a high-volume, low-cost item, the material costs may prohibit sales. In addition, if the application does not require an extremely high level of clarity in the transmission, then a cheaper alternative, such as aluminum, may suffice. Unfortunately, if these *economic* issues are not recognized early in the design process, the consequences can be grave. Suppose the design goes forward and plans are made for production. Equipment may be selected for a production line, and other processes and designs into which the component is to be integrated may be redesigned for the new contacts. If the specification is to be changed from gold to aluminum, all of these dependent decisions may have to change. Clearly, the further the process continues, the more expensive it is to make changes.

The process leading up to production is common to most products and is depicted in Figure 1.3. The cumulative cost of a project grows significantly in the design or development stage. Once the design stage is completed, the project moves to preproduction and, eventually, production phases. The process of readying production can be costly, as it involves plant construction and purchasing equipment. It is not until the production phase that the cumulative cost begins to level out and rewards from the investment are reaped.

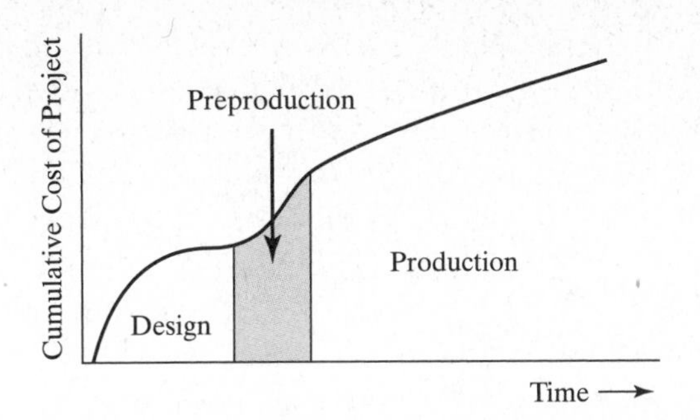

Figure 1.3
Phases of an engineering project, from design to production and eventual disposal.

A general rule states that 80% of the cost endured over the lifetime of a project occurs at the design-and-setup phase. When production finally begins, most of the money to be spent on a product has been spent. This illustrates why investing can be risky: There is much time and money expended before any is returned. It also reiterates why changes in designs are extremely costly and thus the process must always consider the economic ramifications, which, unfortunately, are often realized in the distant future.

1.6 Examining the Real Decision Problems

We return to our introductory example concerning the investment decisions of both Airbus and Boeing over the years and the questions posed:

1. How are the decisions related to engineering economy?

   Engineering economy provides the tools for the two engineering firms to make capital investment decisions. Developing and producing a new airplane requires a significant capital investment while abandoning a new design or terminating production frees up funding for other ventures. Capital investment decisions are not easy, as they are made on the basis of an uncertain future. It is the evaluation of these difficult decisions that lies at the heart of engineering economy.

2. Are these decisions typical for a firm that engages in engineering activities? Do the decisions carry risk?

   Airbus and Boeing have made abandonment, expansion, and replacement decisions over time—investment decisions that are common to most firms. The A380 represents a product expansion for Airbus, while the A350 is a replacement for its A330. Boeing's development of the 787 and 747-8 jets can be viewed as replacements for the 767 and 747-400, respectively, while the company abandoned two programs before launch (the 747X and the Sonic Cruiser) and ended production of the 717. Furthermore, the airlines that purchase these planes are generally expanding their services or replacing older aircraft.

Each of these decisions carry significant risk because of the magnitude of the funding involved (development costs exceeding $10 billion) and the time lag between the investment and revenues from sales (usually three to five years). If the project is abandoned before sales occur, then the investment is lost. If not enough sales are booked, the investment may not achieve a positive return, putting the financial future of the firm at risk.

3. How is the design process related to the investment decision-making process?

As noted in this chapter, these two processes are clearly entwined. Alternative designs must be evaluated before the decision to invest in a project is made. Furthermore, once the project is launched, a number of subsystems must be designed in order to meet project specifications. The development of both the A380 and 787 will utilize new technologies and composite materials. Boeing and Airbus will rely heavily on their engineering expertise to make this possible.

1.7 Key Points

1. Engineers provide answers that either solve problems or offer opportunities for growth.

2. Good solutions are not only *technically* feasible, but *economically* feasible.

3. Although engineers are trained to be *technically* efficient with their solutions, *economic* efficiency determines whether an investment will be successful.

4. Engineering economic decisions generally revolve around enhancing revenues, cutting costs, or improving some other measure of utility, such as customer satisfaction.

5. Economic decisions can generally be categorized as either expansion, replacement, or abandonment decisions.

6. Investment decisions are both important, as they provide avenues of growth for a company, and risky, as their returns are based on an uncertain future.

7. Decision making is a dynamic process, and not all decisions can be made on the basis of a "yes" or "no" answer.

8. Engineers are best suited for making engineering economic decisions because they understand the technical and economic implications of implementing a solution.

9. Engineering economy is a methodology that allows one to evaluate the financial merits of an engineering project.

10. The decision-making process outlines the steps that are necessary for making a decision. These steps include taking a decision from its inception (defining the problem) to its completion (abandonment).

11. The design and decision-making processes are interdependent: The design process generates solutions alternatives for subsequent evaluation in the decision-making process.

12. Design teams must be aware of the economic impact of their decisions, as changes in designs can be extremely costly.

1.8 Further Reading

Ertas, A., and J.C. Jones, *The Engineering Design Process*, 2d ed. John Wiley and Sons, New York, 1996.

Grant, E.L., *Principles of Engineering Economy,* rev. ed. The Ronald Press Company, New York, 1938.

Horenstein, M., *Engineering Design: A Day in the Life of Four Engineers.* Prentice Hall, Upper Saddle River, New Jersey, 1998.

Sullivan, W.G., "A New Paradigm for Engineering Economy," *The Engineering Economist,* 36(3):187–200, 1991.

Morris, W.T., *Engineering Economic Analysis.* Reston Publishing Company, Reston, Virginia, 1976.

1.9 Questions and Problems

1.9.1 Concept Questions

1. What is the role of an engineer in an organization?

2. Why should an engineer be concerned with the economic consequences of his or her decisions?

3. What is the difference between technical efficiency and economic efficiency? Is one more important than the other? Explain.

4. Explain why engineers are well suited to perform economic analysis.

5. What is the goal of a profit-enhancing program? How does this differ from that of a cost-control program? Would you expect the analyses to be the same?

6. How do investments by the government and industry differ? Are there any similarities? Explain.

7. Define expansion, replacement, and abandonment problems. How are they related?
8. When considering an investment, what decisions are generally available to a decision maker? Define each.
9. Why is the decision to delay an investment important?
10. Which step in the decision-making process is most difficult? Defend your answer.
11. How are the decision-making and engineering design processes similar?
12. Is the design process a step in the decision-making process, or vice versa? Explain.

1.9.2 Application Questions

For each of the scenarios that follow, define whether the project is a profit-enhancing, cost-control, or public-improvement program. If possible, define the scenario further as an expansion, replacement, or abandonment decision.

1. Kia Motors, a unit of Hyundai Motor Co., of South Korea announced that it would invest EUR1.1 billion to build an automobile-manufacturing plant in Zilina, Slovakia. The plant is expected to enter operation in 2006 with a capacity of 300,000 units.[48]
2. BP said it would invest $1 billion through 2010 to upgrade process control systems and maintenance procedures at its Texas City refinery, where an explosion in 2005 killed 15 and injured 170. The U.S. Department of Labor uncovered over 300 health and safety violations, resulting in a maximum fine of $21 million.[49]
3. Emerging from its first bankruptcy, US Airways placed orders for 170 new regional jets (with 50–75 seats each), with options on an additional 380 planes to grow the firm's route network. The orders were split between rival manufacturers, with 85 planes from Bombardier and 85 from Embraer. The 170 planes were valued at $4.3 billion, according to list prices.[50]
4. NASA expects to spend $104 billion to return to the moon by 2018 in a "crew exploration vehicle," which is expected to replace the shuttle.[51]
5. The port of Yangshan, 90 miles from Shanghai, opened the first five berths of its shipping facility in December of 2005. This is the first phase of investment, with plans to have as many as 50 deepwater berths by 2020 to provide shipping capacity to the Shanghai region. The Shanghai International Port Group, owned by the Shanghai municipal government, is the sole operator of the berths.[52]

---

[48] "Kia Motors Increases Funds for Slovak Plant to $1.35 Billion," *Dow Jones Newswires,* March 19, 2004.

[49] McNulty, S., "BP to invest Dollars 1bn in Texas plant—OIL & GAS," *The Financial Times,* London edition, p. 21, December 10, 2005.

[50] Carey, S., "US Air Orders 170 Regional Jets in Revamp Plan," *The Wall Street Journal,* May 13, 2003.

[51] Dunn, M., "NASA estimates $104 billion price tag for returning astronauts to moon by 2018," *Associated Press Newswires,* September 19, 2005.

[52] Stanley, B., "China to Open First 5 Berths of Planned Mega-Port," *The Wall Street Journal Asia,* December 9, 2005.

6. Carnival Corp. awarded a $2 billion contract to Italian shipbuilder Fincantieri to build four cruise ships, with deliveries expected in 2008 and 2009. Carnival expects to launch 16 new ships between February 2006 and the fall of 2009.[53]

7. BP PLC agreed to sell older oil fields in the North Sea and Gulf of Mexico to Apache Corporation of Houston in the spring of 2003 for $1.3 billion due to lower profit margins from the fields.[54]

8. Centrica, the United Kingdom's largest electricity and gas supplier, was considering cutting 2000 back-office jobs after installing a computerized billing system at the cost of GBP430 million[55] in the summer of 2005.[56]

9. Honda Motor Co. announced plans to build a new 250,000-square-foot automatic transmission plant to produce 300,000 units annually at the cost of $100 million.[57]

10. DHL Worldwide Express, Inc., announced an investment of $160 million to automate package-sorting processes in the company's Ohio, Pennsylvania, and California facilities.[58]

11. Culp, Inc., agreed to sell its polypropylene yarn-extrusion equipment to a supplier for $1.1 million and close operations in Graham, North Carolina, in October of 2005.[59]

12. Regal Petroleum said it would abandon its RSD-1 exploration well in Romania after it failed to uncover commercial quantities of hydrocarbons.[60]

13. Transcontinental, Inc., replaced seven printing presses with three new Goss presses and automated finishing equipment at the cost of C$53 million in the firm's Owen Sound, Ontario, and Beauceville, Quebec, operations.[61]

14. Apogee Enterprises, Inc., is investing $25 million in a new architectural glass fabrication plant in the southwestern United States in order to meet increasing demand for Viracon glass. Operations are expected to commence in 2007.[62]

15. Department 56, Inc., purchased Lenox, Inc., maker of fine china and giftware, for $196.5 million and announced it would close its Pomona, New Jersey, production

---

[53] Herron, J., "Carnival Corp & Plc Announces Four-Ship Deal with Italian Shipbuilder Fincantieri Worth More than $2B," *Dow Jones Newswires,* December 14, 2005.

[54] "BP Assets in Gulf of Mexico Are Acquired for $509 Million," *Dow Jones Newswires,* March 19, 2003.

[55] GBP stands for Great Britain pounds.

[56] Watkins, M., "Centrica considers cutting 2,000 jobs," *Financial Times,* www.FT.com, July 15, 2005.

[57] Siegel, B., "Honda Confirms to Build Transmission Plant in Georgia as Part of $270 M North American Powertrain Strategy," *Dow Jones Newswires,* November 9, 2004.

[58] Souder, E., "UPDATE: DHL to Spend $160M to Automate Mail-Sorting Ops," *Dow Jones Newswires,* March 24, 2005.

[59] Lam, J., "Culp Announces Reduction of U.S. Yarn Manufacturing Ops," *Dow Jones Newswires,* August 16, 2005.

[60] Garnham, P., and R. Orr, "Small-Caps: Regal Petroleum to Abandon Romanian Well," *The Financial Times,* London Edition, p. 44, November 29, 2005.

[61] Moritsugu, J., "Transcontinental Plans C$53M Capital Investment in State-of-the-Art Equipment," *Dow Jones Newswires,* November 16, 2004.

[62] Rojas, T., "Apogee Announces Plans to Build New Glass Fabrication Plant," *Dow Jones Newswires,* September 13, 2005.

facility at the cost of approximately $7.5 million due to declining demand for ivory china.[63]

16. In the fall of 2005, Grass Valley, a competitor of Sony and Panasonic in the camcorder and recording products market, chose the REV drive portable storage technology from Iomega as its recording medium. This is expected to bring a minimum of $25 to $30 million in annual revenues to Iomega.[64]

17. ATH Resources, one of the United Kingdom's largest coal producers, announced that it would purchase 83 pieces of Caterpillar equipment at the cost of C$56 million from Finning International to replace its existing fleet and provide new equipment. The package included dump trucks, coaling excavators, loading shovels, and bulldozers.[65]

18. Anadarko Petroleum Corp. signed a three-year drilling contract with Dolphin Drilling, Ltd., for the use of a drill ship starting in mid-2007 at the cost of $459 million.[66]

19. In the spring of 2005, BT Group (formerly British Telecom) announced that it would spend GBP10 billion to replace and upgrade its legacy networks to a single Internet-based technology. The new system will allow for the simultaneous transmission of voice, data, and video streaming to any fixed-line connection. It was expected that equipment manufacturers Alcatel of France; Ciena, Cisco, and Lucent of the United States; Ericsson of Sweden; Fujitsu of Japan; Huawei of China; and Siemens of Germany would participate in the extensive project.[67]

20. In the fall of 2005, GlaxoSmithKline, Inc., announced a C$23 million expansion of its facility in Mississauga, Ontario. The 7000-square-foot expansion will enhance both the capacity and the capabilities of the plant, which produces drugs to treat HIV infections, malaria, and allergies.[68]

21. The city of Philadelphia chose Earthlink, Inc., to provide citywide high-speed wireless Internet access. Earthlink would spend between $10 and $15 million to build and operate a 135-square-mile wireless network, with access points posted primarily on public lampposts, for the city of 1.5 million people. Earthlink expects to charge $20 per household per month for service ($10 for low-income families) while providing free access in public places, such as parks. Visitors to the city could purchase hourly, daily, or weekly access.[69]

---

[63] Derpinghaus, T., "Lenox Intends to Consolidate its Fine China Production and Close 31 Retail Stores," *Dow Jones Newswires,* September 13, 2005.

[64] Taylor, P., "Iomega makes video breakthrough," *Financial Times,* www.FT.com, September 12, 2005.

[65] King, C., "Finning Announces C$53 Million Equipment Sale to UK Coal Producer," *Dow Jones Newswires,* August 24, 2005.

[66] Herron, J., "Anadarko Announces Rig Plan to Deliver on Deepwater Strategy," *Dow Jones Newswires,* September 26, 2005.

[67] Odell, M., "BT heads for Pounds 10bn radical overhaul," *Financial Times,* p. 23, April 29, 2005.

[68] King, C., "GlaxoSmithKline Announces Expansion of Mississauga Facility," *Dow Jones Newswires,* September 29, 2005.

[69] Richmond, R., "Earthlink Philly Wi-Fi Win Could Pave Way for More Pacts," *Dow Jones Newswires,* October 4, 2005.

22. Statkraft is spending NOK1.3 billion[70] on a wind farm in Smoela park in northern Norway. The facility will produce 450 gigawatts of electricity annually.[71]

23. Essroc, the fifth-largest cement producer in Northern America, replaced 245 daily dump truck trips between its quarries and plants in Nazareth, Pennsylvania, with a $20 million, 1.7-mile conveyor system in 2003.[72]

24. European steel producer Arcelor opened its new stainless-steel plant in Carinoux in September of 2005. The plant took 26 months to construct at the cost of EUR241 million and will ultimately have an annual capacity of 1 million metric tons of stainless steel.[73]

25. December 9, 2005, marked the last day that a "London bus" (a double-decker, half-cab bus with an old-fashioned bell and conductor) operated on a mainstream route in London. Reasons for the change include the new buses being safer, more accessible to the disabled, more comfortable, and cheaper to operate, as no conductor is needed. The buses are being sold for GBP10,000 each.[74]

26. Consult any media outlet and find three articles which describe investments by firms that engage in engineering activities. Classify the investments as either expansion, replacement, or abandonment.

27. Select the firm of your choice and peruse the press releases available on its website. Identify two articles that describe decisions about capacity (either service or production). Classify these decisions as either expansion, replacement, or abandonment investments.

1.9.3 Fundamentals of Engineering Exam Prep

1. Engineering economics is critical because engineers
   - **(a)** Must design solutions that are economically efficient.
   - **(b)** May interface with accounting and finance departments.
   - **(c)** Require a mechanism to compare alternative solutions.
   - **(d)** All of the above.

2. The best alternative among a set of mutually exclusive alternatives
   - **(a)** Does not meet technical objectives and exceeds budget constraints.
   - **(b)** Meets technical objectives within budget constraints.
   - **(c)** Exceeds technical objectives, but exceeds budget constraints.
   - **(d)** Does not meet technical objectives, but is within budget constraints.

---

[70] NOK stands for Norwegian kroner (singular: krone).

[71] Talley, I., "FOCUS: Statkraft Aims to Build Big Green-Power Empire," *Dow Jones Newswires*, October 11, 2005.

[72] Esack, S., "Essroc Blasts Its Way to a Quieter Nazareth," *The Morning Call*, p. B1, February 7, 2003.

[73] "Arcelor Starts New Belgian Stainless Steel Plant," *Dow Jones Newswires*, September 29, 2005.

[74] Briscoe, S., "End of the road for the traditional London bus," *The Financial Times*, www.FT.com, December 9, 2005.

3. A company generally invests in engineering projects to

(a) Fully expend capital budgets.

(b) Decrease future revenues.

(c) Decrease future costs.

(d) Reduce future capabilities.

4. A government entity performs engineering economy studies when

(a) Passing legislature concerning stem cell research.

(b) Establishing hazardous smog level ratings.

(c) Considering improvements in infrastructure.

(d) Estimating desired police force sizes.

5. Which of the following most closely describes the financial risk associated with an engineering project investment?

(a) Future revenues or cost savings are guaranteed.

(b) The savings or revenues from an investment occur in an uncertain future.

(c) Initial capital investments are often small.

(d) Funding is generally unlimited.

6. Which of the following requires an engineering economic analysis?

(a) Selecting between two technologies for implementation.

(b) Determining when to replace an asset.

(c) Delaying the start of an engineering project for a certain period.

(d) All of the above.

2 Cash Flows and the Time Value of Money

(*Courtesy of Bombardier Inc.*)

Real Decisions: Mover Maintenance

The City of Atlanta awarded Bombardier Transportation a 10-year contract to operate and maintain the automated people-mover system, consisting of 49 vehicles, at Hartsfield–Jackson airport. Bombardier installed the original system in 1980 and has maintained it since that time. Bombardier also extended the system to a new concourse and supplied 24 CX-100 replacement vehicles. The contract is valued at $98 million and includes provisions for two 5-year extensions.[1] This leads to a number of interesting questions:

1. If the $98 million value assumes equal annual payments (starting in 2006) over the 10-year contract without inflation, how can this contract be visualized and analyzed in a spreadsheet,

[1] Moritsugu, J., "Bombardier Gets $98M US Operations, Maintenance Contract with City of Atlanta," *Dow Jones Newswires,* October 4, 2005.

2. Given that Bombardier is a Canadian firm, what is its expected cash flow stream in Canadian dollars?
3. If Bombardier uses a 3.5% quarterly interest rate when analyzing engineering projects, what rate should the company use in analyzing this particular contract?

In addition to answering these questions, after studying this chapter you will be able to:

- Represent an engineering project (investment opportunity) with a cash flow diagram that describes the movement of money in and out of the project over time. (Section 2.1)
- Set up a spreadsheet for cash flow analysis with the use of data, variables, functions, and references. (Section 2.2)
- Explain the concept of the time value of money according to utility, purchasing power, and interest. (Section 2.3)
- Define interest as the means of moving money through time according to an interest rate. (Section 2.4)
- Distinguish between nominal and effective rates and convert the interest rate over the proper period for analysis. (Section 2.5)
- Draw a cash flow diagram that aligns the timing of cash flows and the compounding period of the interest rate. (Section 2.6)
- Define and measure inflation and compute its impact on purchasing power and cash flows. (Sections 2.7)
- Use the inflation rate to convert real-dollar cash flows, which ignore inflation, into current-dollar cash flows, which incorporate inflation effects, and vice versa. (Sections 2.7)
- Compute the impact of exchange rates on cash flows. (Section 2.8)

Chapter 1 highlighted a number of decisions to be analyzed in this text. A common feature among engineering projects that involve these decisions is that they generally require an initial investment which hopefully leads to future savings or additional revenues. Thus, projects are to be described financially over some period of time. In order to analyze a project, we have to develop an understanding of the value of money *over time*, for a sum of money today is clearly not economically equivalent to the same sum of money years from now. That is, if given a choice between the two sums (now or later), we would clearly have a preference. The concept of the *time value of money* describes the change in the value of money over time and is defined on the basis of the concepts of interest and purchasing power, which are introduced in this chapter.

2.1 Cash Flows and Cash Flow Diagrams

To financially analyze engineering projects, we need to model the projects in terms of *cash flows*. Literally, cash flows represent the flow, or movement, of money at some specific time or over some period of time. *Outflows* represent cash that is leaving an account, such as a withdrawal. Outflows are often referred to as *expenses* or *disbursements*. *Inflows* represent cash that is entering an account, such as a deposit. Inflows are often called *revenues* or *receipts*.

The "answers" provided by engineers are defined as engineering projects. They represent investment opportunities for the firm. An engineering project can be viewed as an account with outflows and inflows. The outflows include the costs of implementing and maintaining the project, while the inflows describe the revenues or savings from the project. Cash flow movements can be visually displayed through use of a cash flow diagram, illustrated in the next example.

EXAMPLE 2.1

Cash Flow Diagram

Perryman Co., a titanium producer in Pennsylvania, expanded its operations with the purchase of a $10 million rolling mill in 2005. Annual capacity increased by over 60% to 7 million pounds of titanium ingots, used to produce coiled and bar products.[2] Assume that the new mill was purchased and installed at the beginning of 2005 and runs at peak capacity (4.375 million pounds of output per year) for 10 years. Assume further that a pound of output generates $9 in revenues while costing $3.90 to produce, and maintenance of the equipment is $10 million the first year and grows by $1 million per year. Finally, the mill is to be scrapped at the end of 10 years for $500,000. Draw

[2] Glader, P., "Perryman to Boost Titanium Ops Amid Aerospace Demand," *Dow Jones Newswires,* December 17, 2004.

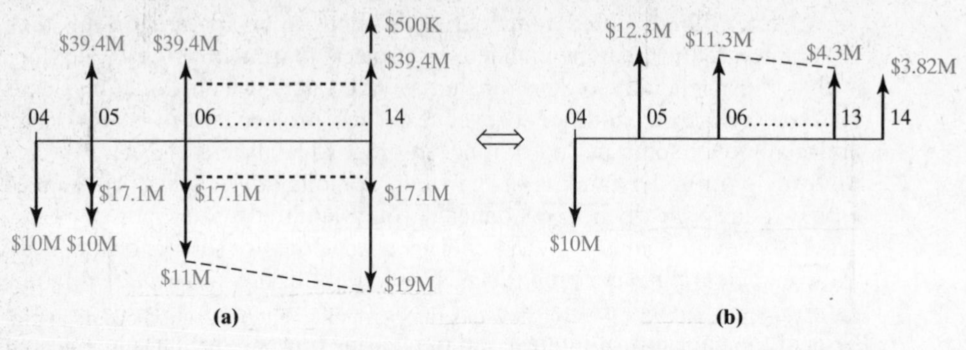

Figure 2.1 Expansion of titanium production represented in a cash flow diagram with (a) individual and (b) net cash flows.

the cash flow diagram for this investment, assuming that all costs and revenues are realized at year-end.

Solution. The cash flow diagrams for individual and net cash flows are given in Figures 2.1(a) and (b), respectively. A net cash flow in a given period is merely the sum of the individual cash flows in that same period.

In Figure 2.1(b), the arrows that point downward from the axis represent outflows. The end of the arrow is labeled with the magnitude of the flow. Conversely, the arrows that point upward from the axis represent inflows, which are also labeled with the magnitude of the flow. The axis itself represents time. Each cash flow emanates from the horizontal line, which signifies time, at the time the cash flow occurs.

Note that the direction of the arrow is critical, as an outflow is a "negative" cash flow. The label on the arrow is merely the magnitude of the flow, whereas the sign of the flow (negative for outflows and positive for inflows) is captured by the arrow. This must be clear, as a negative outflow is an inflow.

The cash flow diagram provides a summary of the inflows and outflows of an engineering project over its horizon. It provides nearly all the information necessary to perform an economic analysis, much like a free-body diagram provides nearly all the information necessary for a civil engineer to analyze forces on a bridge truss or a circuit diagram provides nearly all the information necessary for an electrical engineer to study the flow of current.

A cash flow diagram is defined by two general types of cash flows: discrete and continuous. Discrete cash flows are the most common, representing cash transactions at specific points in time. We assumed discrete cash flows in Example 2.1. In general, a discrete cash flow at time n will be defined as A_n, although values of P, F, and A will also be used to identify cash flows at specific points in time.

Continuous cash flows occur over a period of time and are defined by a *rate* of cash inflow or outflow. One can think of the flows here as being literal, as money moves from one account to another at some rate, like water flowing from one reservoir to another via a pipe. The term $\overline{A}$ identifies a continuous cash flow with a constant rate. A continuous cash flow is depicted in Figure 2.2 in the next example.

EXAMPLE 2.2

Cash Flow Diagram with Continuous Cash Flows

Consider again Example 2.1, in which the ingot producer generated revenues from the sale of titanium. Redraw the cash flow diagram, assuming that production occurs at the *rate* of 4.375 million pounds of titanium ingots *per year*, defining continuous revenues with time. All other cash flows remain discrete.

Solution. The cash flow diagram for individual cash flows is given in Figure 2.2, but it is assumed that production and sales are continuous over time. Here, a block is utilized to show the flow of money over a period of time.

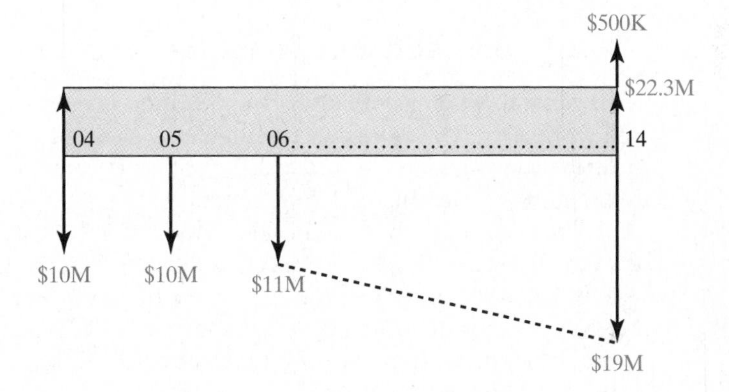

Figure 2.2 Cash flow representation of an investment opportunity with continuous revenues.

Continuous cash flow transactions are actually quite rare, as accounting methods generally require a specific (discrete) transfer of funds. Thus, we will not discuss continuous cash flows at great length in this book, although more information is available on the website. (Approximating cash flow movements as continuous sometimes leads to easier mathematics for economic analysis.)

In general, the importance of the cash flow diagram cannot be understated, as it describes the financial aspects of an engineering project. The correct depiction of a project via its cash flow diagram ensures that the decision maker understands the investment being analyzed. Economic analysis follows strict protocols that assume a correctly drawn cash flow diagram. *Every* time an economic analysis is performed, a cash flow diagram should be drawn.

2.2 Using Spreadsheets to Represent Cash Flows

Cash flow diagrams can also be represented on a computer spreadsheet. There are a number of spreadsheets available to the user for engineering economy applications. In the interest of space, we restrict our discussions to Excel by Microsoft. This decision is not intended to pass judgment on the other programs, but rather, is due strictly to market conditions, as Excel is the best-selling spreadsheet software. As we introduce a number of concepts, we assume that the reader has some familiarity with spreadsheets and computers.

Spreadsheet programs are quite powerful, and their importance in economic analysis has become immeasurable. The ability to copy and reuse information speeds the development process, especially in problems with long horizons, and enables a decision maker to focus on gathering information and the actual decision. Furthermore, spreadsheets allow information to be displayed in a variety of ways with the use of charts and graphs, aiding an engineer who is building a business case.

2.2.1 Spreadsheet Fundamentals

In general, when a spreadsheet application is opened, the user is faced with a *worksheet*. The worksheet resembles a sheet of graph paper, in that lines divide the page into *cells*. A cell's location is defined by its *column* and *row*. Columns are designated by the letters A, B, C, ..., Y, Z, AA, AB, etc. Rows are marked numerically. The numbers and letters are found in the margins of the workspace. Selecting a cell highlights the corresponding column letter and row number so that you can keep track of your location. A *workbook* generally consists of multiple worksheets in the same file. These are usually tabbed at the bottom of the current worksheet. The individual worksheets can be renamed by clicking on the tabs, initially defined as 'Sheet1,' 'Sheet2,' etc.

Highlighting a cell allows the user to provide input to the spreadsheet. Input generally falls into one of five categories,

- **Data.** Data constitute any numerical value used as input for analysis. When input, the value is placed in the cell and can be referenced at any time. Note that once a value is input into a cell, the value will not change (unless the user changes the input to that cell).
- **Variables.** These are the opposite of data, because variables are not fixed, but rather, can change. Defining a variable in a spreadsheet is achieved by referencing another cell. If the cell being referenced changes, then the contents of the cell defined by the reference will change. A cell is referenced with the '=' sign, followed by the cell to be referenced. For example, if we highlight cell C10 and input '=B4', then the value assigned to the C10 cell is equal to the value in the B4 cell. (Note that you can type

'=' and then just click on cell B4, as opposed to typing it.) If the B4 cell contains the data value '4', then the value '4' is assigned to the C10 cell. If the B4 cell contains the value 'Hi', then the value 'Hi' is assigned to the C10 cell. If the B4 cell is empty, then the value '0' is assigned to C10. The important point to understand is that if the referenced cell B4 changes, then the value assigned to C10 will change, because it is a variable. We will discuss referencing shortly.

- **Functions.** Excel has numerous functions to help find various answers. A function is entered by typing '=FUNCTIONNAME(arguments)', where 'FUNCTIONNAME' takes on the name of the function and 'arguments' are the required inputs to the function, separated by commas. Functions may also be input from the 'Insert' pulldown menu at the top of the screen.
- **Combination of inputs.** Clearly, we can input information into a cell that contains variables (a reference to another cell), a function or functions, and data. For example, typing '=B4+ABS(C3)' uses a reference in a function (the absolute-value function) and adds it to another referenced cell.
- **Labels.** The remaining input is labels, which improve the readability of a spreadsheet. Labels include the use of bold, italic, underlined, or colored type or the highlighting of a cell with lines or colors. These do not aid in the computation of solutions but help keep input and output in order.

2.2.2 Spreadsheet Usage

Although a spreadsheet offers the user flexibility in performing an economic analysis, there are some guidelines that should always be followed so that information is not lost and time is not wasted "relearning" a spreadsheet that has not been seen in a while. The following steps should be taken with each spreadsheet:

1. A short title or phrase should identify the problem and the analysis.
2. Data should be labeled and maintained separately from the analysis, and units and measures of time should always be noted. The location of this information will be referred to as the "data center."
3. Cash flow diagrams should be clearly marked with a time scale.
4. Formatting, such as boldface, underlining, color, and currency notation, should be used to its fullest extent (within reason, of course), so that the spreadsheet is readable.

These steps may seem obvious, but, as with other kinds of work, preparing a spreadsheet quickly and improperly will only lead to a time when it will have to be redone. Let us build a spreadsheet model together to illustrate inputs and cell referencing.

EXAMPLE 2.3 *Inputting Data to a Spreadsheet*

In October 2005, Vedanta Resources announced a $2 billion investment in its Jharsuguda smelter in India to produce 400,000 tons of aluminum each year. It was expected to take three years to bring the facility online.[3] Assume that construction takes only one year, with the first year of production reaching 100,000 tons and increasing 50,000 tons per year until reaching the maximum capacity of 400,000 tons. Assume further that production costs are $900 per ton and revenues are $1950 per ton of aluminum. Finally, assume that annual maintenance costs are $2 million and the salvage value of the smelter is $50 million after 15 years of production. Using a spreadsheet, draw the cash flow diagram for this investment.

Solution. Figure 2.3 shows the spreadsheet developed for this investment project. The data center contains the inputs, including the total investment cost, salvage value, interest rate, and production parameters (output, costs, and revenues) over the 15 years of operations, as well as the desired output of "Present Worth" and "IRR" (discussed in later chapters).

The spreadsheet is built by referencing the data center inputs. Cell A18 is highlighted to illustrate the use of a **relative** reference, programmed as

=A17+1.

Thus, the spreadsheet takes the value in A17 and adds 1, resulting in the value '14' being displayed in cell A18, since the value in cell A17 is '13'. To build the spreadsheet quickly, we can input the value '0' in cell A4 and program '=A4+1' in cell A5. This is

	A	B	C	D	E	F
1	Example 2.3: Aluminum Smelter Investment			**Input**		
2				Investment	$2,000,000,000.00	
3	**Period**	**Cash Flow**		Initial Production	100000	tons
4	0	-$2,000,000,000.00		Production Increase	50000	tons
5	1	$103,000,000.00		Max Production	400000	tons
6	2	$155,500,000.00	=-E2	Unit Cost	$900.00	per ton
7	3	$208,000,000.00		Unit Price	$1,950.00	per ton
8	4	$260,500,000.00		Operating Cost	$2,000,000.00	per year
9	5	$313,000,000.00		Salvage Value	$50,000,000.00	
10	6	$365,500,000.00		Interest Rate	8%	per year
11	7	$418,000,000.00		Production Period	15	years
12	8	$418,000,000.00	=$E$3+$E$4*(A10-1)*($E$7-$E$6)-$E$8			
13	9	$418,000,000.00		**Output**		
14	10	$418,000,000.00		Present Worth	$246,345,390.55	
15	11	$418,000,000.00		IRR	9.87%	per year
16	12	$418,000,000.00				
17	13	$418,000,000.00	=$E$5*($E$7-$E$6)-$E$8			
18	14	$418,000,000.00				
19	15	$468,000,000.00				
20		=A17+1				
21						

Figure 2.3 Possible spreadsheet layout for the aluminum smelter investment.

[3] Bream, R., "Vedanta Ready to Invest Dollars 2bn in Aluminum Plant," *Financial Times,* p. 21, October 19, 2005.

defined as a relative reference. Then we can Copy (from the 'Edit' menu) the contents of cell A5 and Paste (again from the 'Edit' menu) it into cell A6. This will result in cell A6 being programmed as '=A5+1'. We can continue to paste down column A, through row 19, to achieve our end result of 15 years of analysis. Note that, alternatively, we can simply highlight the entire column (cells A6 through A19) and Paste in a single step.

Relative referencing is an extremely powerful tool in spreadsheets. When a relative reference is copied and pasted from one cell to another, the reference changes accordingly. In this example, cell A5 was defined by a reference to cell A4. The "distance" between the cell and the reference is thus one row. Hence, when copied to another cell, the reference will be to the cell immediately above the designated cell (one row apart).

If we had defined cell A5 by referencing cell B2, then copying the reference into cell D12 would result in cell D12 referencing cell E9, as the "distance" between D12 and E9 is one column and three rows—the same "distance" between cells A5 and B2.

Now, the cash flow for period zero, cell B4, references the investment cost in the data center as

=-E2.

Note the negative sign in the reference is needed because our data center defines the investment as an absolute value.

Cell B5 represents the first year of operation of the smelter, with 100,000 tons of aluminum produced. This leads to the following net revenues in the first year of operation:

$$100{,}000 \text{ tons} \left(\frac{\$1950}{\text{ton}} - \frac{\$900}{\text{ton}} \right) - \$2\text{M} = \$103 \text{ million.}$$

There are a number of ways in which to program this formula into the cells that correspond to the years of production of the smelter. For example, we could type the following into cell B5 (you do not have to type the cell references, as you can click on the appropriate cell when you are at that point in the equation, but you have to add the '$' manually):

=$E$3*($E$7-$E$6)-$E$8.

The $ in the formula define **absolute** references. When absolute references are copied to another cell, the reference does not change.

Unfortunately, this method of defining cell B5 does not lend itself to copying, as we would have to adjust each individual cell during the increase in production in the ensuing years. (This is known as an arithmetic gradient, which will be discussed extensively in the next chapter.) In order to copy this properly, we can use absolute and relative references simultaneously as

=($E$3+$E$4*(A5-1))*($E$7-$E$6)-$E$8.

The reference to cell A5 is relative and the remaining cell references are absolute. Because the value of A5 is 1, the increase of 50,000 tons (cell E4) is multiplied by zero, leaving production at 100,000 tons (cell E3) in the first year. The benefit of this new approach is that if we Copy this function from B5 and Paste it into B6, the result is

=($E$3+$E$4*(A6-1))*($E$7-$E$6)-$E$8.

Note that the absolute references did not change, but the relative reference changed in that the 5 became 6. If we reference in this manner, then we can copy our programmed cell (B5) down through period eight (cell B11) on the cash flow diagram and be assured that the answer will be correct, since the references are correct. However, we cannot copy beyond that period because the maximum production is reached. To illustrate, cell B10 is highlighted in Figure 2.3.

To determine the remaining cash flows, we can merely change our reference to the maximum production of 400,000 tons (cell E5), such that cells B12 to B19 can be defined as

=$E$5*($E$7-$E$6)-$E$8.

For cell B19, we need to subtract the salvage value defined in cell E9.

This method works fine and is shown on the spreadsheet, but we can in fact be clever in our definitions and have less to change. Note that production starts at 100,000 tons per year and increases by 50,000 tons per year, but cannot exceed 400,000 tons. This condition can be captured in Excel with the MIN function, defined as

MIN(number1,number2, . . .),

where the arguments (number1, number2, etc.) can be numbers or references to cells with numerical values. The function returns the minimum number from this set. Given the period of the cash flow, we can compute the production level by using the gradient, just as we defined cell B5, and compare it with the maximum output, using the MIN function. The value returned by the MIN function can then be used to compute the net revenues. Thus, we can define cell B5 as

=(MIN($E$3+$E$4*(A5-1),$E$5))*($E$7-$E$6)-$E$8.

This function returns the minimum of $100{,}000 + 50{,}000(n - 1)$ and 400,000, depending on the period n. This cell definition can now be copied through cell B19. (The salvage value must also be added to cell B19.)

The ability to reference accurately with both absolute and relative references, together with the use of a data center for input data, leads to powerful spreadsheets. If we decide to change an input, each cell in the spreadsheet that references the changed data updates itself instantaneously. Thus, modeling the spreadsheet in this way allows the user to test a variety of data sets without changing the model. In this example, the investment cost, initial cash flow, production rate, production increase, maximum production, production cost, maintenance cost, salvage value, and interest rate can all be manipulated. Given a change in one of these parameters, the cash flows change, along with the outputs of the present worth and internal rate of return. ("Tweaking" input parameters in this way is generally referred to as sensitivity analysis.)

Thus, the decision maker can examine a number of scenarios without creating new spreadsheets! We will wait to program these output cells until we understand their meaning. We will also continue to introduce spreadsheet capabilities throughout the text.

2.3 The Time Value of Money

The cash flow diagram represents inflows and outflows, either to or from a project, an account, or a person. People who possess money may spend the money as they like. A person may buy a ticket to the theater, while a civil engineering firm may buy new pickup trucks for its engineers. These purchases bring utility, or satisfaction, to the buyers. This is why money has value—because it brings one utility.

However, people or firms that have money may not choose to spend it immediately; they could (1) hold on to it for use at a later date or (2) invest it. These two options are very different, but are central to a fundamental concept in engineering economy known as the **time value of money**.

Simply stated, the time value of money is the reason that a dollar today is not worth a dollar tomorrow. While the example of comparing a dollar from one day to the next may seem trivial, the difference is more pronounced as the time span of comparison is increased, such that a dollar today is not economically equivalent to a dollar one month or one year from now. Under normal conditions, you should prefer the dollar today. Then you could (1) spend it or (2) save it.

If you decide to spend the dollar, then, presumably, you would receive some **utility**. Utility is a measure of satisfaction that can be defined only by the individual. For example, one person may spend the money on a new DVD while another may purchase a candy bar. The fact that each could be equally happy with his or her purchase is described by the fact that they have different utility functions, for they receive satisfaction in different amounts for different things. Often, in engineering applications, engineering firms receive utility by increasing savings or profits. They may also receive utility by reducing the possibility of future losses.

If you decide not to spend the money, then you have some additional options. You could invest the money by putting it the bank, where it would grow at some rate such that there would be more than a dollar at a later point in time. This change is defined by the **interest** that the bank pays you, the investor. Or you could stuff the money in your wallet and let it sit there over time. Although the dollar may not change in appearance, you would notice that you could not buy as much with it because prices have changed. Still, for certain goods, it may be possible that you could actually buy more than previously. However, it is more likely that you would have to settle for less than previously. This phenomenon is known as a change in purchasing power, generally due to a change in price.

These examples provide the reasons that a dollar today is not **economically equivalent** to a dollar tomorrow: interest and purchasing power. When two things are said to be equivalent, one is indifferent to possessing either item. That is, the difference between the choices is zero. We illustrated the idea that a dollar one day is not economically equivalent to a dollar at a later (or earlier) point in time. Thus, the dollars are not economically equivalent. Earlier, we noted that most engineering projects require an initial investment in return for future savings or revenues. Thus, the time value of money plays a critical role in determining whether to make the initial investment, as it defines whether the investment is economically equivalent to future revenues or savings.

2.4 Interest and Interest Rates

Interest is defined as either a cost or a revenue, depending on whether you are a lender or a borrower. If you deposit money into a savings account, the bank pays you, *the depositor,* a fee defined as interest. You receive this payment because the bank is paying you for the use of your money. (Clearly, you cannot spend the money while it is in the bank's possession.) In turn, the bank lends the money to other people or companies through mortgages and loans. In this situation, the roles are reversed, with the bank receiving interest from the borrower. Clearly, the bank must pay you, the depositor, less interest than it receives from its customers, the borrowers, in order to be profitable.

The calculation of the interest payment is identical whether it is a revenue paid to the depositor or a cost paid by the borrower. The value of the interest payment is dependent on the amount of principal involved and the interest rate. The **principal** P is the amount of money being loaned or borrowed and is generally referred to as capital. The **interest rate** i is the percentage of principal charged as a fee over some period for lending money. The interest rate is dependent on a variety of factors, including the prime rate, the profit margin, the risk of loss, and administrative expenses. The prime rate is the rate charged to a bank for borrowed funds. Because the bank often does not have enough depositor money to cover its loans to customers, the bank may also borrow funds. Generally, the bank borrows these funds from the government (via federal reserve banks) and is charged interest at the prime rate. Thus, if it costs a bank 3.5% to provide funds to customers, the bank will charge a rate in excess of this value in order to make a profit. The bank may even inflate the rate if it believes that the loan is risky. The actual calculation of the interest payment is dependent on the terms of the loan.

2.4.1 Simple Interest

Simple interest involves no compounding: The interest paid is directly proportional to the principal. Mathematically, the amount of interest I paid on a loan of principal amount P over N periods is

$$I = PiN, \tag{2.1}$$

where the interest rate i is defined as a percentage per period. Thus, if a loan and its accumulated interest are to be returned to the lender after N periods, then an amount

$$F = P + I = P + PiN = P(1 + iN)$$

is to be returned. One must be sure to match the period to the definition of the interest rate, as that definition is dependent on time. The next example illustrates this concept.

EXAMPLE 2.4

Simple Interest

In 2004, Boeing announced that its new 7E7 Dreamliner (later named the 787), to be produced in 2007, would be sold for up to \$127.5 million.[4] If an upstart airline takes out a loan for that amount and the interest rate is 5.5% per year (without compounding), how much must be paid four years after taking out the loan?

Remember! 5.5% = .055.

Solution. The cash flow diagram for this loan scenario is given in Figure 2.4. The loan is received by the airline and must be repaid at the end of four years.

The interest owed at the end of four years is calculated with Equation (2.1) as follows:

$$I = PiN = (\$127.5 \text{ million})(0.055)(4) = \$28.05 \text{ million}.$$

The total amount to be repaid in the future is

$$F = P + I = \$127.5\text{M} + \$28.05\text{M} = \$155.55 \text{ million}.$$

If the loan is to be repaid at the end of the first quarter after the fourth year, the interest would be

$$I = PiN = (\$127.5\text{M})(0.055)\left(4 + \frac{1}{4}\right) = \$29.8 \text{ million}.$$

The extra quarter results in interest costs rising \$1.75 million.

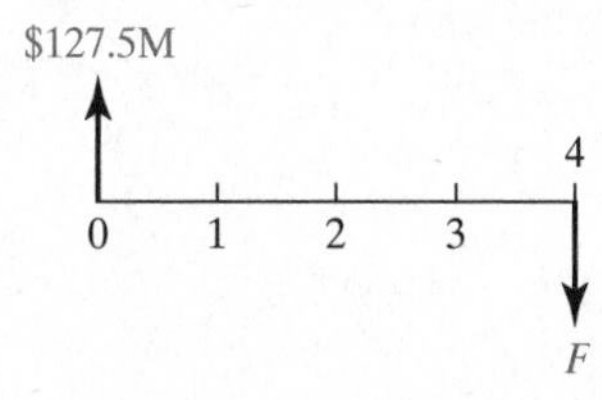

Figure 2.4
Cash flow diagram for received loan and repayment four years later.

[4] Hadhi, A., "U.S. Boeing to Sell New 7E7 Plane for up to US\$127.5M," *Dow Jones Newswires*, February 24, 2004.

Although this example is straightforward, the concept of simple interest is generally not followed in practice. That is, interest is usually paid on the outstanding principal *and* any accrued interest, as discussed in the next section.

2.4.2 Compound Interest

Unlike simple interest, compound interest is paid on the principal amount and on any interest that has previously accrued, or built up, over time. With compound interest, invested funds grow much more quickly than with simple interest. Also, interest costs grow much more rapidly. The interest paid after some period of time is dependent on the principal amount, interest rate, compounding period, and payout schedule. The **compounding period** defines when interest payments are to be made. If interest is not paid, it accrues for further compounding. Let us revisit Example 2.4, but assume that now there is compounding.

EXAMPLE 2.5 ***Compound Interest***

Consider the \$127.5 million purchase cost of the Boeing 787, but the 5.5% interest rate is compounded annually. Determine the total interest paid if the loan must be paid back after four years.

Solution. The cash flow diagram for this transaction is the same as in Figure 2.4, but the calculation of the interest payment is more difficult, as interest is paid on both the principal and any accrued interest. At the end of the first year (the first compounding period), the borrower owes

$$I_1 = Pi = (\$127.5\text{M})(0.055) = \$7.0125 \text{ million.}$$

However, the loan is not to be paid back until after four years, so this interest is not paid, but rather, accrues. After the second period, the amount of

	A	B	C	D	E	F	G
1	Example 2.5: Airplane Loan Analysis				**Input**		
2					Loan Principal	\$127,500,000.00	
3	**Period**	**Interest Accrued**	**Total Owed**		Interest Rate	5.5%	per yr.
4	0	--	\$127,500,000.00		Periods	4	years
5	1	\$7,012,500.00	\$134,512,500.00				
6	2	\$7,398,187.50	\$141,910,687.50		**Output**		
7	3	\$7,805,087.81	\$149,715,775.31		Total Owed	\$157,950,142.95	
8	4	\$8,234,367.64	\$157,950,142.95	=C5+B6			
9			=\$F\$3*C7				
10							

Figure 2.5 Interest accrued and amount owed on loan with a 5.5% rate compounded annually.

interest owed is

$$I_2 = (P + I_1)i = (\underbrace{\$127.5\text{M}}_{\text{Principal}} + \underbrace{\$7.0125\text{M}}_{\text{Accrued Interest}})(0.055) = \$7{,}398{,}187.50.$$

We carry out the calculations to the penny so that you can verify them. The complete schedule is shown in the spreadsheet in Figure 2.5.

This example should illustrate the difference between simple and compound interest, as the amount of interest owed is clearly quite different. In the compound interest example, a total of $30,450,142.95 in interest is owed at the end of four years, in addition to the original purchase amount of $127.5 million. This amount is $2,400,142.95 more than that in the simple interest example.

To analyze this situation mathematically, we can utilize the principle of simple interest. After one *compounding* period ($n = 1$), the amount of interest owed is equivalent to one period of simple interest:

$$I_1 = Pi.$$

The total amount owed at that point is

$$F_1 = P + I_1 = P + Pi = P(1 + i).$$

The cash flow diagram in Figure 2.6 illustrates the payments if the loan is to be repaid after one period. The arc connecting the two periods on the cash flow diagram and labeled with the interest rate emphasizes the fact that the amount of money is growing at the rate i.

We now extend the analysis, assuming that the loan is not repaid after the first period. After the second period, the amount of interest owed is based on the first period's interest and the principal.

With simple interest, we ignored the first period's interest in our calculation. With compound interest, the accrued interest is lumped together with

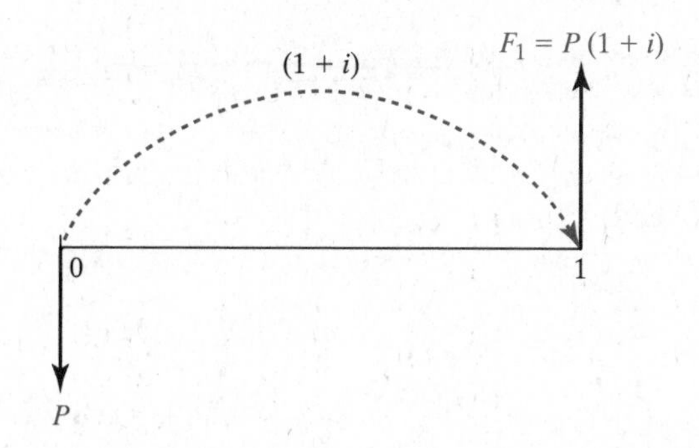

Figure 2.6
Initial loan P and payment F_1 after one period.

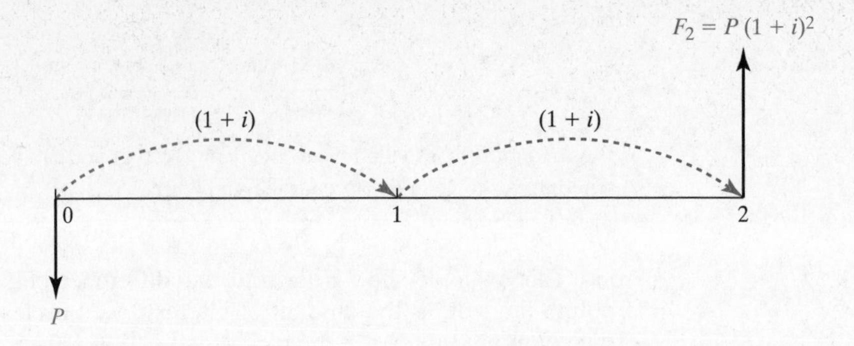

Figure 2.7 Loan repayment after two periods of compound interest.

the principal such that the total amount of interest owed after the second period is

$$I_2 = (P + I_1)i = (P + Pi)i.$$

The value I_1 represents the interest accrued from the first period, which is added to the original principal. The total amount owed after the second period is now

$$F_2 = P + I_1 + I_2 = P + Pi + (P + Pi)i = P + 2Pi + Pi^2 = P(1+i)^2.$$

Figure 2.7 illustrates interest compounding over two periods, resulting in payment F_2 if the loan is paid back after the second period.

To determine the total amount owed after three periods, we could begin with the amount of principal borrowed and then calculate the amount of interest accrued in each period, as we did with the first and second periods. But examining Figure 2.7 for two periods, we see that the total amount owed after the third period is equal to the amount accrued after the second period, with an additional period of interest, or

$$F_3 = F_2 + F_2 i = F_2(1+i) = P(1+i)^2(1+i) = P(1+i)^3.$$

Clearly, a pattern is emerging. We can depict the general case, with the total amount of principal and interest F owed after N periods on a loan of principal P with interest rate i compounded each period n, in Figure 2.8. F is often referred to as a future value of the present amount P, as it includes the effects of compound interest.

For each period that interest accrues, we multiply by $(1 + i)$, resulting in

$$F = P(1+i)^N.$$

Returning to Example 2.5, we easily find the solution:

$$F = P(1+i)^N = \$127.5\text{M}(1+0.055)^4 = \$157{,}950{,}143.00.$$

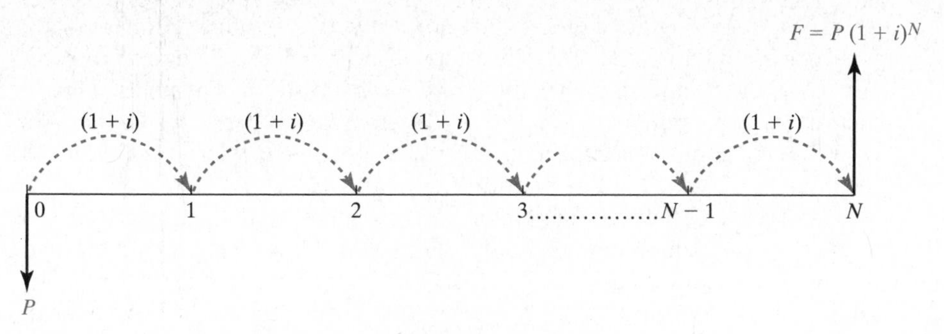

Figure 2.8 Future value of N periods of compound interest.

The interest paid on the loan is

$$I = \$157{,}950{,}143.00 - \$127{,}500{,}000.00 = \$30{,}450{,}143.00.$$

This is the same answer we found with the spreadsheet in Figure 2.5, which computed the periodic interest payments. Note that determining the interest paid and the amount owed requires that the interest rate match the compounding period, as discussed in the next section.

2.5 Nominal and Effective Interest Rates

We have already illustrated the use of cash flow diagrams to represent a series of cash flows that describe an investment or lending opportunity. In the chapters that follow, we illustrate how one can manipulate these cash flows for purposes of evaluation with the use of interest. The interest rate that is used in the calculations depends on the compounding period and the timing of the cash flows for the loan or investment. Thus, we must have an understanding of different types of interest rates and how they may be converted to the proper rate for analysis.

When you investigate interest rates for a loan from a bank, the bank may quote you two numbers. How is this possible? Are you allowed to pick the lower rate for the loan? Although it would be nice to take the lower rate, the two rates quoted are generally equivalent. For various reasons, including marketing aims, loans and investment opportunities are often advertised in terms of nominal interest rates and effective interest rates.

2.5.1 Nominal Interest Rates

Financial institutions generally quote interest rates, termed **nominal interest rates**, in annual terms without the effects of compounding. Nominal interest rates are also termed APRs (annual percentage rates). While these rates are prevalent in advertisements, they do not provide sufficient information for decision-making purposes, since they do not include the effects of compounding and therefore do not fully describe the resulting payments. Thus, a nominal rate must be converted to an **effective interest rate**, which includes the effects of compounding. Define the nominal annual[5] interest rate as r and the number of compounding periods in a year as M.

With these definitions, the nominal rate is generally defined or advertised as r% per year, compounded periodically. This wording is generally shortened to r% compounded periodically. Here, compounding can occur annually, semiannually, quarterly, monthly, weekly, daily, or continuously. For example, 12% compounded monthly is an annual nominal rate. For any discrete compounding period, of which there are M in a year, the value

$$i = \frac{r}{M} \tag{2.2}$$

defines the interest rate per compounding period (defined by M), which is the *effective* interest rate per compounding period. We will discuss continuous compounding shortly. We illustrate the conversion to an effective rate in the next example.

EXAMPLE 2.6

Nominal Rates

First Quantum Minerals, Ltd., received a $30 million credit facility (credit line) from Standard Chartered Bank for its mining operations. The facility carries an interest rate of LIBOR (London Inter-Bank Offered Rate) plus 2.5% and is to be repaid in quarterly payments.[6] Assume that the LIBOR is fixed at 1.37% per year such that the nominal rate is 1.37% + 2.50% = 3.87% per year. Assuming quarterly compounding, find the effective quarterly interest rate.

Always convert nominal rates immediately!

Solution. With quarterly compounding, the interest rate per quarter (i_q) is defined from Equation (2.2) as

$$i_q = \frac{r}{M} = \frac{0.0387}{4} = 0.009675,$$

which translates to 0.97% per quarter.

Similarly, given an interest rate per compounding period, one can convert it to a nominal rate by multiplying it by the number of compounding periods in a year, as in the next example.

[5] Note that a nominal interest rate is not required to be an annual rate. However, because it is most commonly defined as such in practice, we assume it to be an annual interest rate in this text.

[6] Thomas, S., "First Quantum Completes US$30M Loan Facility for Bwana Mkubwa," *Dow Jones Newswires,* November 27, 2002.

EXAMPLE 2.7

Nominal Rates Revisited

Revisit the previous example, but assume that Quantum Minerals' loan had been quoted such that the 2.5% is a quarterly rate and the 3-month LIBOR rate is 0.28%, for a total rate of 2.78% per quarter. What is the nominal rate?

Solution. Solving for the nominal rate r in Equation (2.2), we find that

$$r = i_q M = (0.0278)(4) = 0.1112,$$

or 11.12% compounded quarterly.

As these examples show, the nominal rate is of little use for comparison or analysis purposes, as it does not describe the true interest rate (and thus payments) involved in the transaction. For analysis, effective interest rates are required. Thus, it is suggested that nominal rates be immediately converted to the effective interest rate of the compounding period, which is merely r/M. Once found, this effective interest rate can be converted to the necessary periodic rate for analysis, as illustrated next.

2.5.2 Effective Interest Rates

Effective interest rates, also known as yields on investments or true costs of loans, are much more useful than nominal interest rates, as they incorporate the effects of compounding. We have already shown that r/M is the effective interest rate per compounding period, which is defined by M. Because nominal rates are defined in annual terms, it is useful to define an annual effective interest rate i_a.

Consider a nominal rate that compounds monthly such that

$$i_m = \frac{r}{M}$$

is the effective monthly rate, with $M = 12$. To convert this monthly rate to an annual effective rate, we can use a cash flow diagram. For the monthly rate, the principal compounds 12 times in one year. The question to be answered is "What interest rate allows P to accrue to the amount F in one period?" This situation is illustrated in Figure 2.9, where the small arc represent monthly compounding and the large arcs represents annual compounding.

Examining the figure, we can write down two equivalent equations. First, with the monthly rate, we have

$$F = P(1 + i_m)^{12}.$$

Then, the annual rate yields

$$F = P(1 + i_a).$$

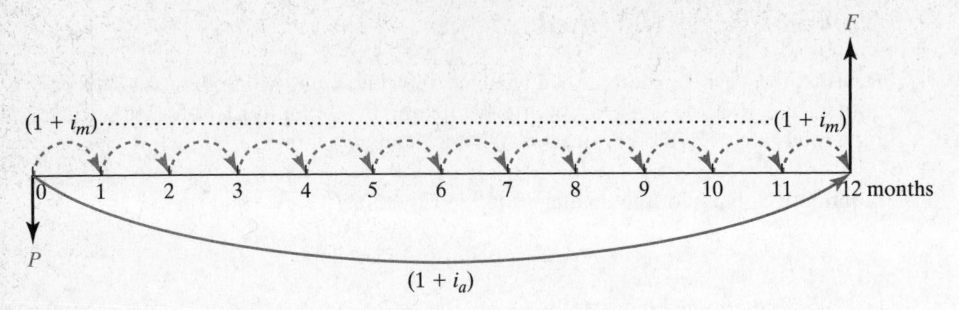

Figure 2.9 Compounding with monthly and annual interest rates.

Equating the F terms and dividing out the P values results in the expression

$$(1 + i_m)^{12} = (1 + i_a)$$

such that

$$i_a = (1 + i_m)^{12} - 1.$$

It should be clear that we can draw the cash flow diagram for any two interest rates and find equivalent expressions. In general, the annual effective interest rate can be derived from a nominal (annual) rate compounded M times in a year as follows:

$$i_a = \left(1 + \frac{r}{M}\right)^M - 1.$$

Conceivably, one may require an effective interest rate over a period different than a year. Assume that the nominal rate is 12% compounded monthly. From what we have learned, we know that this translates to an effective rate of $0.12/12 = 0.01$, or 1% per month. What if we would like to know the semiannual effective rate? This is depicted in Figure 2.10.

The monthly effective rate compounds six times in a half-year, which is equivalent to

$$i_{sa} = (1 + i_m)^6 - 1 = (1 + 0.01)^6 - 1 = 0.0615 = 6.15\% \text{ per six months.}$$

Figure 2.10 Compounding with monthly and semi-annual interest rates.

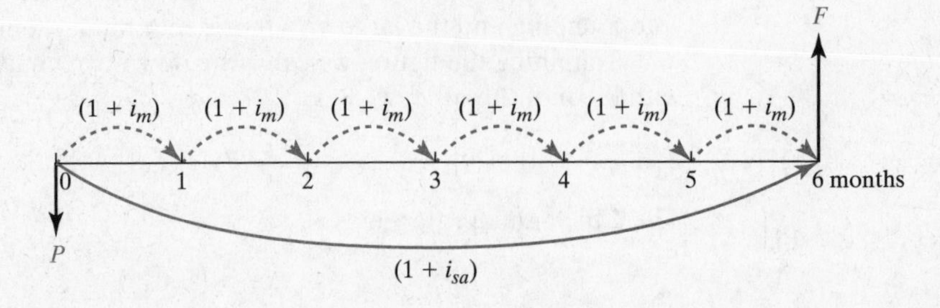

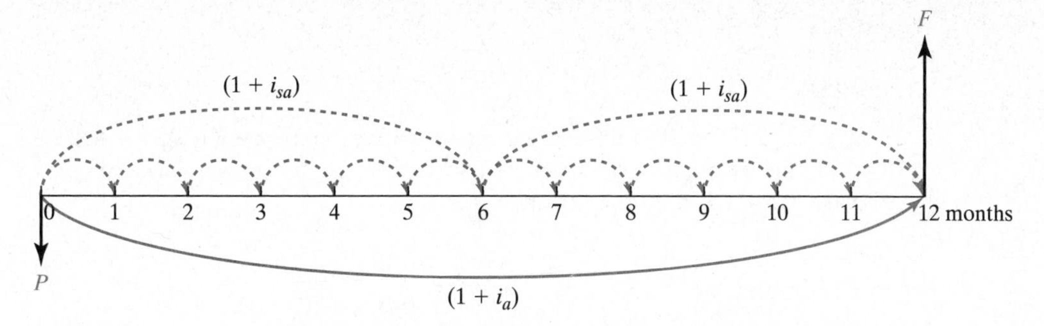

Figure 2.11 Compounding with monthly, semiannual, and annual interest rates.

What if we would like to determine the annual effective rate? We could use the formula, or we could note that the monthly rate compounds 12 times in a year or that the semiannual rate compounds twice in a year. All three cases are depicted in Figure 2.11.

Thus, using the monthly rate, we can equivalently determine i_a as

$$i_a = (1 + i_m)^{12} - 1 = (1 + 0.01)^{12} - 1 = 0.1268 = 12.68\% \text{ per year},$$

or, using the semi-annual rate, as

$$i_a = (1 + i_{sa})^2 - 1 = (1 + 0.0615)^2 - 1 = 0.1268 = 12.68\% \text{ per year}.$$

This method works in a similar manner when one is converting from a rate over a longer period to a shorter one. Assume that you are provided an annual interest rate of 12% compounded annually and you want to find the effective quarterly rate. Figure 2.12 shows the cash flow diagram representing both views.

As before, we equate the two rates as follows:

$$(1 + i_a) = (1 + i_q)^4.$$

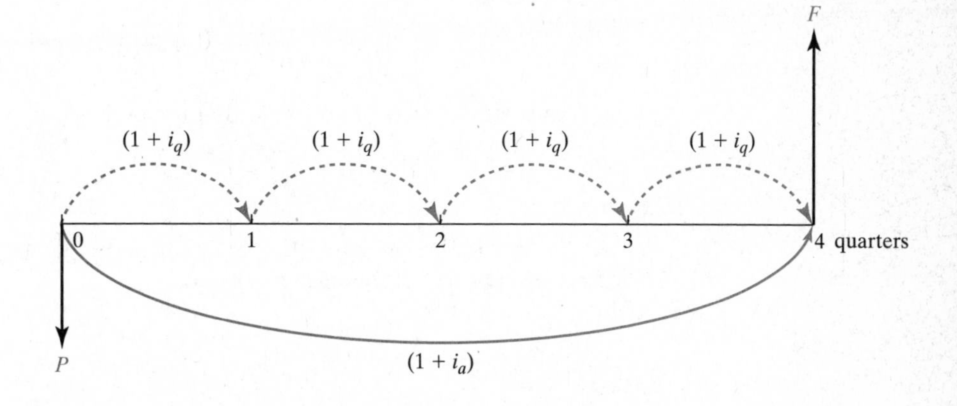

Figure 2.12 Compounding with quarterly and annual interest rates.

Then

$$i_q = (1 + i_a)^{\frac{1}{4}} - 1 = (1 + 0.12)^{\frac{1}{4}} - 1 = 0.0287 = 2.87\% \text{ per quarter.}$$

In essence, an effective interest rate can be converted to an equivalent effective interest rate over *any* period. The cash flow diagrams illustrate the conversions and are summarized in the formula

$$i = \left(1 + \frac{r}{M}\right)^{lM} - 1. \tag{2.3}$$

The interest rate i is the effective interest rate per period l, where l is expressed in terms of years, with the remaining variables as before. Thus, to convert a nominal (annual) interest rate of $r = 13\%$ compounded monthly ($M = 12$) into an effective quarterly interest rate ($l = \frac{1}{4}$), we write

$$i_q = \left(1 + \frac{0.13}{12}\right)^{(\frac{1}{4})(12)} - 1 = (1 + 0.01083)^3 - 1 = 0.0328$$
$$= 3.28\% \text{ per quarter.}$$

Do not be intimidated by this formula. The value (0.13/12) merely converts the nominal rate to an effective monthly rate, while the power term, $(1/4)(12) = 3$, merely compounds the monthly rate three times, since there are three months in a quarter. This is exactly the same calculation that results from drawing the cash flow diagram and analyzing it with monthly and quarterly interest rates.

Equation (2.3) can be used to convert interest rates over discrete periods, such as days, weeks, months, quarters, or years. But interest can, and often is, compounded continuously. How does that change Equation (2.3)? If an interest rate is compounded continuously, then there are an infinite number of compounding periods M over time. Let us concentrate on the effective annual interest rate, such that $l = 1$ and M approaches infinity in Equation (2.3). For continuous compounding,

$$i_a = \lim_{M \to \infty} \left(1 + \frac{r}{M}\right)^M - 1. \tag{2.4}$$

To analyze this equation, we note that the definition of e is

$$e = \lim_{x \to 0} (1 + x)^{1/x} \approx 2.718281828.$$

In Equation (2.4), the value $r/M \to 0$ as $M \to \infty$. Thus, to mimic the definition of e, we can rewrite Equation (2.4) as

$$i_a = \left[\lim_{M \to \infty} \left(1 + \frac{r}{M}\right)^{M/r}\right]^r - 1.$$

By design, the value inside the brackets reduces to e, and we are left with

$$i_a = e^r - 1. \tag{2.5}$$

Thus, given the nominal rate r compounded continuously, the annual effective rate can be found with Equation (2.5). If an effective rate over a different period l is desired, where l is defined as a fraction of a year, the equation reduces to

$$i = e^{lr} - 1.$$

For example, if we require a semiannual rate, then the value of $l = 0.5$ and the effective semiannual rate would be $e^{r/2} - 1$. We illustrate continuous compounding in the next example.

EXAMPLE 2.8

Continuous Compounding

ON Semiconductor Corp. refinanced $369 million of credit in the fall of 2003 at an interest rate of LIBOR plus 325 basis points, or 3.25%.[7] Assume that LIBOR is 1.5% such that the loan has a rate of 4.75% per year, compounded continuously. Find the equivalent effective annual and semiannual interest rates.

Solution. The equivalent annual rate is defined by Equation (2.5):

$$i_a = e^r - 1 = e^{0.0475} - 1 = 0.04865 = 4.87\% \text{ per year.}$$

The equivalent semiannual (six-month) rate is

$$i_{sa} = e^{r/2} - 1 = e^{0.0475/2} - 1 = 0.02403 = 2.4\% \text{ per six months.}$$

Similar calculations follow for other discrete periods.

2.5.3 Comparing Interest Rates

When comparing interest rates, one must use effective rates over a common compounding period. Nominal rates should not be used for comparisons. From the previous sections, the following rules for comparisons apply:

- Convert all nominal rates to the effective interest rate per compounding period by dividing the nominal rate by the number of compounding periods in a year: $i = r/M$.
- Convert an effective rate over a given compounding period to another rate with a compounding period of longer duration through compounding.

[7] Park, J., "ON Semiconductor Refinances $369M of Credit Facilities," *Dow Jones Newswires*, November 26, 2003.

If the rate with the shorter compounding period compounds N times in the longer period, then

$$(1 + i_{\text{longer}}) = (1 + i_{\text{shorter}})^N.$$

- Convert an effective rate over a given compounding period to another rate with a compounding period of shorter duration through discounting. Again, if the rate with the shorter compounding period compounds N times in the longer period, then

$$(1 + i_{\text{shorter}}) = (1 + i_{\text{longer}})^{1/N}.$$

To ensure that the conversion is correct, cash flow diagrams should be drawn. We illustrate this approach from two points of view in the next example.

EXAMPLE 2.9 *Comparing Interest Rates*

Construction equipment can be very expensive. For example, Toromont Industries, Ltd.'s, CAT division sold 33 pieces of Caterpillar equipment, including paving equipment, compactors, wheel loaders, off-highway trucks, motor graders, and excavators, to Lafarge Canada, Inc., for about C$12 million in early 2004.[8] Although no details were given about the purchase in this case, the buyer often has many choices with regard to paying for the equipment. Assume that the dealer is offering financing (through a loan) to purchase the equipment at a rate of 18% compounded monthly. Alternatively, the firm may seek a loan from a local bank, which is offering an APR of 17% compounded continuously. Which is the better rate?

Solution. First, do not get caught into the trap of accepting the bank's rate just because 17% is less than 18%. These are nominal rates and thus cannot be compared. Each rate must be converted to an effective rate over a common period for comparison.

Never compare with r!

First consider the dealer's interest rate of $r = 18\%$ compounded monthly. This translates to an effective monthly rate of

$$i_m = \frac{r}{M} = \frac{0.18}{12} = 0.015 = 1.5\% \text{ per month.}$$

For the bank's rate, we can use Equation(2.5) to convert from a continuously compounded rate to an annual effective rate:

$$i_a = e^r - 1 = e^{0.17} - 1 = 0.1853 = 18.53\% \text{ per year.}$$

Now we must determine whether the monthly effective rate of 1.5% is cheaper than the 18.53% annual effective rate. Let us compare the two rates over an annual period. Since the monthly rate compounds 12 times in a year, the equivalent annual rate is

$$i_a = (1 + 0.015)^{12} - 1 = 0.1956 = 19.56\% \text{ per year.}$$

[8] Tsau, W., "Toromont Announces C$12 million CAT Equipment Order," *Dow Jones Newswires,* February 9, 2004.

The bank offers the engineering firm a cheaper rate. Note that we could have arrived at the same conclusion by converting the bank's annual effective rate to an effective monthly rate:

$$i_m = (1 + 0.1853)^{\frac{1}{12}} - 1 = 0.0143 = 1.43\% \text{ per month.}$$

The decision remains the same, as the bank's offer of 1.43% per month is cheaper than the 1.5% dealer rate.

Notice that each of the preceding analyses comes to the same conclusion that the bank rate is much more affordable. The decision will always be the same, because the analyses are equivalent. Thus, it is best to select a period for analysis that makes computation easy, as the final decision will not be affected by this choice. The key to the analysis is comparing effective interest rates over a similar period of time.

2.6 Cash Flow Timing and the Interest Rate

As we noted in the previous section, two interest rates can be compared only if they are effective interest rates with the same compounding period. Similarly, we can analyze cash flow diagrams only if the compounding period of the interest rate matches the timing of the cash flows. That is, if we have annual cash flows, we need to utilize an annual effective interest rate. If we have a quarterly effective interest rate, then we need to have quarterly cash flows. There are two situations where we may have to adjust either the effective interest rate or the cash flow timing in order to achieve this balance:

1. The interest rate is compounded more frequently than the cash flows occur.
2. The cash flows occur more frequently than the interest rate is compounded.

In the first case, we must adjust the interest rate to match the timing of the cash flows; in the second case, we must adjust the timing of the cash flows. These situations are explained with examples in the next two sections.

2.6.1 Compounding Is More Frequent than Cash Flows

We indirectly addressed this situation in the previous section when we adjusted the effective interest rate for comparison purposes. In this case, the deposits into an account earn interest for a number of periods before new deposits are placed into the account. To ease both analysis and the computations that follow in the next chapter, we can adjust our interest rate to match the timing of the cash flows. We have already detailed the mathematics of this procedure, and we provide another example to illustrate the concept.

EXAMPLE 2.10

Matching the Interest Rate to the Cash Flows

The Tennessee Valley Authority (TVA) signed a number of contracts with Kentucky and Illinois high-sulfur coal providers to guarantee supply for its No. 3 generating unit at its Paradise, Kentucky, power plant. The TVA recently installed "scrubbers" in order to be able to burn the region's high-sulfur coal while meeting federal pollution standards. Among the contracts, Resource Sales, of Slaughters, Kentucky, is to supply 1.2 million tons of coal per year for 20 years. The contract is valued at $803 million.[9] Although coal is delivered on a weekly basis, assume that payments to Resource Sales are made monthly, beginning in January of 2004. The interest rate is 11% per year, compounded continuously.

Solution. According to the contract, TVA will pay $33.46 per ton of coal over the length of the contract. With 100,000 tons of coal delivered each month, this translates to monthly payments of $3.346 million. TVA's cost stream is depicted in the cash flow diagram in Figure 2.13.

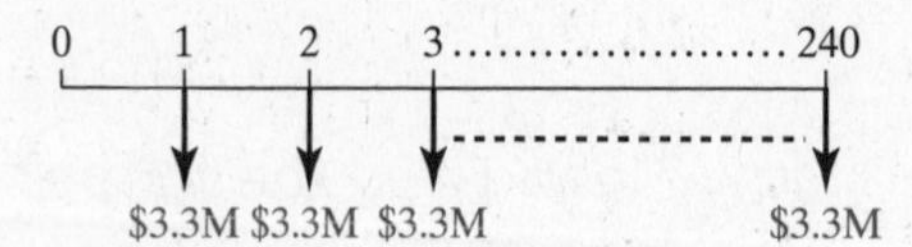

Figure 2.13 Monthly payments for coal deliveries.

Since the interest rate is compounded continuously, the compounding period is more frequent than the cash flows. Thus, we can convert the interest rate to an effective monthly rate to be used in analysis. The monthly rate is

$$i_m = e^{r/12} - 1 = e^{0.11/12} - 1 = 0.0092 = 0.92\% \text{ per month.}$$

This rate can be directly applied to the given cash flow diagram, because the timing of the cash flows and compounding period is the same.

2.6.2 Compounding Is Less Frequent than Cash Flows

It is also possible that the compounding period of the interest rate is greater than the timing of cash flows. For example, if one makes daily deposits into an account that pays quarterly interest, then the cash flow timing is more frequent than the interest rate compounding period. This can result in a number of scenarios:

1. *The timing of any payments made between compounding periods may be ignored.* Thus, for our small example, the daily payments would be lumped together (summed) into a single payment at the end of each quarter.
2. *The payments would earn interest as soon as they are deposited.* In our small example, that would mean that the bank would provide an effective daily

[9] "TVA OKs Contracts Worth up to $3B for Ky Coal Operators," *Dow Jones Newswires,* November 12, 2003.

interest rate which could be used in computation. This is similar to the case where compounding is more frequent than cash flows, as the cash flows are left unchanged and the effective rate is altered.

3. *An average level of the account may be computed over the compounding period.* That value is then used to calculate the interest payment, which is unchanged. This method serves as a compromise between the two methods described previously.

In this textbook, we will assume that the cash flow timing is ignored, so that more frequent payments are lumped into single payments in order to match the timing of the compounding period. Our reasons are straightforward. First, it is rare that the compounding period will be longer than the cash flow timing, as most accounts pay interest rates that are continuously compounded. Second, in the second scenario described, the bank would end up paying more interest than advertised, which is unlikely. If the bank truly wanted to pay interest on a daily basis, then it would have advertised that fact. Finally, although the third scenario is probably the most common, at least for a typical consumer with a savings account, it adds an undesired layer of complication to our analysis. Thus, we will "shift" cash flows in this situation in order to match their timing to the compounding period.

EXAMPLE 2.11

Matching the Cash Flows to the Compounding Period

The TVA also signed a 20-year $1.07 billion contract for 1.5 million tons of coal per year from Alliance Resource Partners.[10] Assume that payments of $4.46 million are made monthly (with the first payment occurring at the end of January 2004), but the interest rate is 12% per year.

Solution. Figure 2.14(a) illustrates the true movement of cash flows under the coal supply contract. However, the cash flows occur every month, while interest is compounded annually. Thus, the monthly cash flows can be aggregated into annual sums of $53.5 million, as given in Figure 2.14(b). This is the convention that we will utilize throughout the textbook unless stated otherwise.

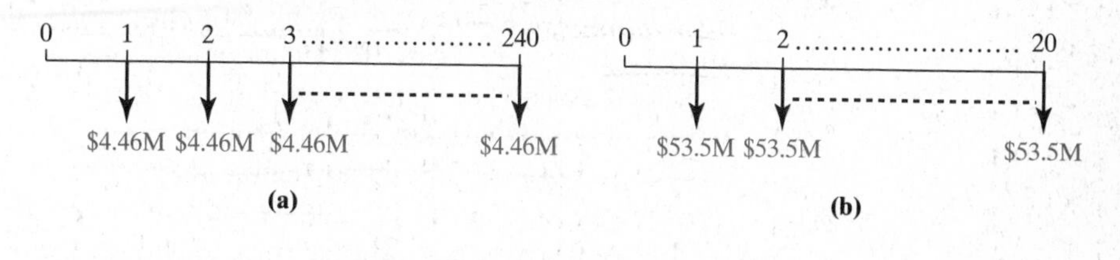

Figure 2.14 (a) Monthly payments for coal deliveries and (b) aggregated annual payments for an annual interest rate.

[10] "TVA OKs Contracts Worth up to $3B for Ky Coal Operators," *Dow Jones Newswires*, November 12, 2003.

Given the two scenarios in Examples 2.10 and 2.11, together with their adjustments, we can assume that cash flows and interest rates can be matched such that the timing and the compounding periods are identical.

2.7 Inflation and Purchasing Power

Recall that we had three options for our money: We could spend it and receive utility, we could invest it and earn interest, or we could hold onto it. If we put the money in our wallet, then after some time, we could pull out the money and it would probably look the same (although maybe a bit rumpled). But is it? Chances are, it is not worth what it used to be. Although there are some products that seem to decrease in price over time, a majority of prices rise with time. This phenomenon is known as **inflation**. The worth or value of your money decreases with inflation because you cannot purchase as much of a commodity with the same amount of money as you could previously. In other words, you have experienced a loss of **purchasing power**.

Inflation, or the rise in prices over time, can occur for various reasons. For example, we have seen drastic increases in the cost of a gallon of gasoline in the last few years in the United States. This is not the case with all commodities, as competition, deregulation, and improved manufacturing efficiencies may actually lead to decreases in the price of a good or commodity (*deflation*) over time. In the next several sections, we examine inflation and its effect on cash flows.

2.7.1 Measuring Inflation

The government provides comprehensive data concerning the change in price of a variety of commodities over time. These numbers are important, because the government has fiscal policies that can curb inflation if it believes that prices are rising too quickly. For example, the government may raise the prime lending interest rate, which *discourages* investment because banks generally follow by *raising* the interest rates they charge for loans. By contrast, a lowering of the rate *encourages* investment because banks generally follow by *lowering* the interest rates they charge for loans.

The measure commonly used to track inflation is the Consumer Price Index, or CPI. This index represents the average cost of goods and services for a typical consumer. The costs include those for food and beverages, housing, transportation, apparel, recreation, education, communication, and personal care. Table 2.1 gives a sampling of CPI values from 1983 through 2005. The data used are the annual CPI values for all urban consumers. This CPI is averaged across United States cities and over the year.

Note that 1987 has a value of 113.6 and 2002 has a value of 179.9. This means that it costs $179.9/113.6 = 1.58$ times more to receive the same goods

TABLE 2.1 **Consumer Price Index for all urban consumers averaged over U.S. cities.**

Year	CPI	Year	CPI
1983	99.6	1994	148.2
1984	103.9	1995	152.4
1985	107.6	1996	156.9
1986	109.6	1997	160.5
1987	113.6	1998	163.0
1988	118.3	1999	166.6
1989	124.0	2000	172.2
1990	130.7	2001	177.1
1991	136.2	2002	179.9
1992	140.3	2003	184.0
1993	144.5	2004	188.9
		2005	195.3

Source: U.S. Department of Labor, Bureau of Labor Statistics, www.bls.gov.

and services in 2002 as it did in 1987. To put it another way, we might say that the purchasing power in 2002 was $113.6/179.9 = .631$ of what it was in 1987.

To calculate the average rate of inflation in any time interval, we merely apply what we have learned about compound interest. The CPI values of 113.6 and 179.9 in years 1987 and 2002 are not cash flows, but we will ignore that fact for one moment. Treat the value of 113.6 as a deposit into an account that earns interest at an annual rate f for 15 years, totaling 179.9 in 2002. With this reasoning and the definition of compound interest, we have

$$113.6(1 + f)^{15} = 179.9.$$

Solving for f, we find the average annual inflation rate between 1987 and 2002 to be 3.11%.

To generalize the definition, the average annual inflation rate over any N periods is

$$\mathrm{CPI}_n(1 + f)^N = \mathrm{CPI}_{n+N}. \tag{2.6}$$

Note that the inflation rate can be calculated in this manner for any index, not just the CPI. Also, if the value of the index is lower at time $n + N$ than at time n, then a period of deflation has occurred and $f < 0$.

Table 2.2 shows Producer Price Index, or PPI, data on construction machinery and semiconductors, respectively. These are measures of the change in selling price, from the perspective of producers, over time. In the table, the construction machinery index has been steadily rising over the past 10 years (inflation) while the semiconductor index has been steadily dropping (deflation). Producer Price Index data are available for a variety of products and can be extremely useful for estimation.

TABLE 2.2

Producer Price Indexes for (1) construction machinery manufacturing and (2) semiconductors and related devices manufacturing.

Year	Construction Machinery PPI	Semiconductor PPI
1993	151.2	141.9
1994	153.8	140.1
1995	157.2	131.8
1996	161.6	122.4
1997	164.4	110.7
1998	167.8	101.7
1999	170.8	97.4
2000	172.7	91.1
2001	173.5	86.8
2002	175.9	83.8
2003	178.3	76.8
2004	183.9	71.7
2005	192.6	69.5

Source: U.S. Department of Labor, Bureau of Labor Statistics, www.bls.gov.

EXAMPLE 2.12

Inflation

Compute the average rate of inflation for construction machinery from 1996 through 2002, using the data in Table 2.2.

Solution. Noting that the index values are 161.6 and 175.9 for the years 1996 and 2002, respectively, we can calculate the inflation rate from Equation (2.6) as

$$161.6(1+f)^6 = 175.9$$

$$\Rightarrow f = (1.0885)^{\frac{1}{6}} - 1 = 0.0142.$$

$f > 0$ is inflation.

This defines an annual inflation rate of 1.42%.

EXAMPLE 2.13

Deflation

Compute the average rate of deflation (negative inflation) for semiconductors and related devices from 1993 through 2002, using the data in Table 2.2.

Solution. The computation for deflation is like that for inflation:

$$141.9(1+f)^9 = 83.8$$

$$\Rightarrow f = (0.5906)^{\frac{1}{9}} - 1 = -0.05684.$$

$f < 0$ is deflation.

Solving for f gives us −5.68%, signaling a period of deflation.

2.7.2 Cash Flows as Real or Current Dollars

The CPI, PPI, and other price indexes exist because we often want to see the change in value of a dollar over time. To determine the worth of something today compared with its worth at an earlier time, we must remove the effects of inflation over the appropriate time span. The resulting values can then be compared more favorably, leading us to the following two definitions:

Current dollars represent the cash flow *at the time* of the transaction. In other words, they are the actual out-of-pocket expenses paid or revenues received. These cash flows *include* the effects of inflation. Current dollars are also known as actual dollars, future dollars, and inflated dollars. We denote a current dollar cash flow during period n as A_n.

Real dollars are current dollars that have been adjusted according to changes in purchasing power defined in some base year. Thus, real dollars *ignore* the effects of inflation. Real dollars are also known as nominal dollars, constant dollars, or deflated dollars. We denote a real-dollar cash flow during period n as A'_n.

The link between real and current dollars is the inflation rate, together with a base period chosen for a point of reference. Consider zero as an arbitrary reference point in time. Then, to convert current dollars in period n to real dollars, we must "remove" n periods of inflation, or

$$A'_n = \frac{A_n}{(1+f)^n}. \tag{2.7}$$

This concept is illustrated in the next example.

EXAMPLE 2.14

Current to Real Dollars

According to the National Association of Colleges and Employers (NACE) winter 2004 survey, the average starting salary for engineering majors (computer, chemical, electrical, mechanical, computer science, industrial, manufacturing, construction, and civil) was $48,000.[11] Assume that an entry-level engineer can expect a starting salary of $50,000 next year. If the engineer receives raises of 4.3% per year and the annual inflation rate is 2.3%, what is the engineer's pay in real dollars over the next five years?

Solution. The cash flow diagram for pay received by the engineer is given in Figure 2.15. Because this is pay that is *actually received* by the engineer, the cash flow diagram is defined by current dollars. Note that the engineer received $50,000 at time 1 (assume end-of-year cash flows), and this value increases by 4.3% per year. Thus, pay received totals $\$50{,}000(1.043)^{n-1}$ in year n.

Current dollars are received or spent!

[11] Sahadi, J., "Most lucrative college degrees," *CNN Money Online,* money.cnn.com, February 5, 2004.

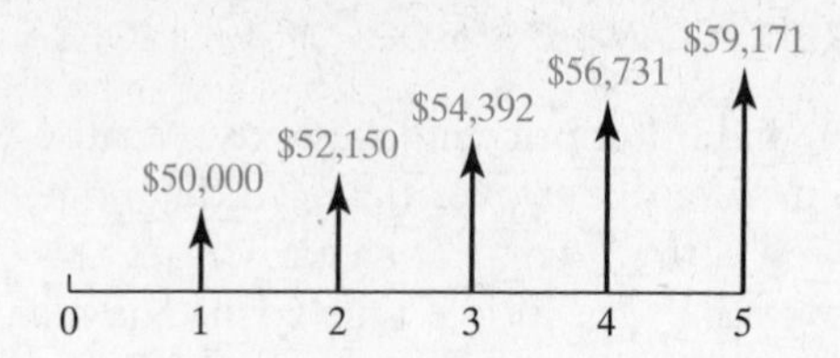

Figure 2.15 Cash flow diagram in current dollars for an engineer's annual salary.

With time zero as the base year, the real-dollar cash flows are computed from Equation (2.7):

$$A_1' = \frac{\$50{,}000}{(1+0.023)^1} = \$48{,}875.86;$$

$$A_2' = \frac{\$50{,}000(1+0.043)}{(1+0.023)^2} = \$49{,}831.40;$$

$$A_3' = \frac{\$50{,}000(1+0.043)^2}{(1+0.023)^3} = \$50{,}805.62;$$

$$A_4' = \frac{\$50{,}000(1+0.043)^3}{(1+0.023)^4} = \$51{,}798.88;$$

$$A_5' = \frac{\$50{,}000(1+0.043)^4}{(1+0.023)^5} = \$52{,}811.57.$$

Thus, the $50,000 received at the end of the first year of work is worth only $48,875.86 in time-zero dollars. Similarly, the $59,170.77 received at the end of year 5 is worth just under $53,000 in time-zero dollars, as inflation erodes one's purchasing power. If the engineer had only received a raise to match the rise in the CPI index each year, then his or her salary would have a constant real-dollar value of $48,875.86 as actual pay only keeps pace with inflation. Note that this conversion is easily performed over a number of years with the use of a spreadsheet.

Sometimes it is difficult to determine whether current or real dollars have been defined. Do not fall into the trap of believing that real dollars are always constant because they do not grow with inflation. Similarly, do not always assume that growing cash flows signal current dollars that have been inflated due to price increases. Rather, determine whether a contract has been agreed upon or some assumption has been made about the future.

EXAMPLE 2.15 *Current to Real Dollars Revisited*

Assume that you purchased a five-year, $10,000 bond from Harley-Davidson (bonds are discussed at great length in Chapter 4) for the price of $9,998.80 in December of 2003 when they were issued. According to the bond, which is a contract, Harley-Davidson will pay you $362.50 per year over the next five years and then the $10,000

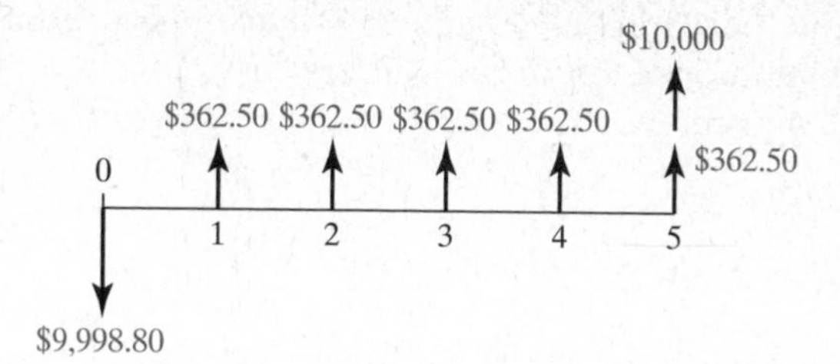

Figure 2.16 Cash flow diagram for bond purchase.

face value at the time the bond matures.[12] The cash flow diagram for these transactions is given in Figure 2.16. Convert the cash flows into real or current dollars, whichever is appropriate, assuming that the annual inflation rate over the period of analysis is 3%.

Solution. Is the cash flow diagram populated with current- or real-dollar cash flows? Since out-of-pocket expenses or in-pocket revenues define current dollars, the diagram is defined by current dollars. The real-dollar cash flow diagram is found by dividing out the appropriate periods of inflation. With time zero as the base year, the real-dollar cash flows are as follows:

$$A'_0 = \frac{-\$9{,}998.80}{(1+0.03)^0} = -\$9{,}998.80;$$

$$A'_1 = \frac{\$362.50}{(1+0.03)^1} = \$351.94;$$

$$A'_2 = \frac{\$362.50}{(1+0.03)^2} = \$341.69;$$

$$A'_3 = \frac{\$362.50}{(1+0.03)^3} = \$331.74;$$

$$A'_4 = \frac{\$362.50}{(1+0.03)^4} = \$322.08;$$

$$A'_5 = \frac{\$10{,}362.50}{(1+0.03)^5} = \$8{,}938.79.$$

Note that these real-dollar cash flows only signify the value of the flows at time zero. They do not represent cash flows actually received.

The conversion of real-dollar cash flows to current-dollar cash flows involves removing the effects of inflation. If we solve for the current-dollar cash flow in Equation (2.7), we see that we now incorporate, or "add" the effects of inflation into the cash flows:

$$A_n = A'_n(1+f)^n.$$

[12] Geressey, K., "Harvey-Davidson $400M 5-Yr 144a Yields 3.62%; Tsys +0.48," *Dow Jones Newswires*, November 18, 2003.

We put the word "add" in quotation marks because that is not the mathematical operation we perform. Rather, we are incorporating inflation just as we dealt with compound interest, as the next example shows.

EXAMPLE 2.16

Real to Current Dollars

Statoil, ASA, of Norway has agreed to supply up to 1.4 billion cubic meters of natural gas per year to utility Essent, NV, of the Netherlands for about five years.[13] Assume that the contract does last five years and exactly 1.4 billion cubic meters are delivered each year for the market price. Assume also that the time-zero (beginning 2004) price is NOK1 per cubic meter. Under the assumption of no inflation, the expenses incurred by Essent are illustrated in Figure 2.17. What are the current-dollar cash flows, assuming that natural-gas prices are expected to grow at an annual rate of 0.5%?

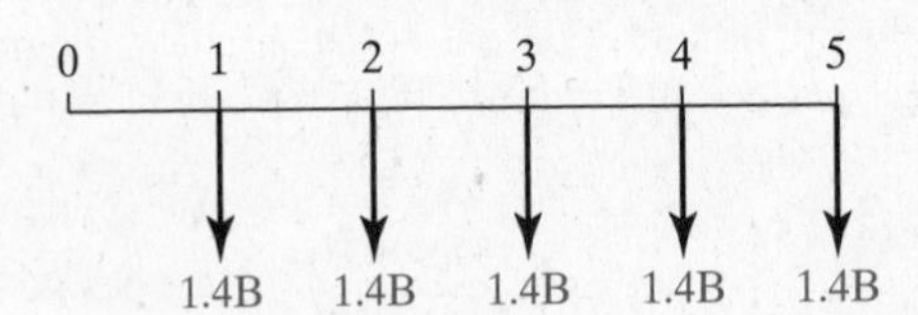

Figure 2.17 Cash flow diagram in real dollars (kroner) for natural-gas deliveries.

Real dollars are fictitious!

Solution. The cash flows in Figure 2.17 do not represent actual out-of-pocket expenses, but rather, real dollars. The current-dollar cash flows are as follows:

$$A'_1 = \text{NOK}1.4\text{B}(1+0.005)^1 = \text{NOK}1.407 \text{ billion;}$$
$$A'_2 = \text{NOK}1.4\text{B}(1+0.005)^2 = \text{NOK}1.414 \text{ billion;}$$
$$A'_3 = \text{NOK}1.4\text{B}(1+0.005)^3 = \text{NOK}1.421 \text{ billion;}$$
$$A'_4 = \text{NOK}1.4\text{B}(1+0.005)^4 = \text{NOK}1.428 \text{ billion;}$$
$$A'_5 = \text{NOK}1.4\text{B}(1+0.005)^5 = \text{NOK}1.435 \text{ billion.}$$

These dollars represent the actual expenditures for the natural gas.

2.7.3 Market and Inflation-Free Interest Rates

We can perform analysis in one of two domains: (1) cash flow diagrams defined by current dollars or (2) cash flow diagrams defined by real dollars. The questions that remain include how these domains are related and which should be utilized in a particular analysis.

[13] Lee, N., "Statoil Seals NOK6.5B Gas Sales Deal with Dutch Co Essent," *Dow Jones Newswires*, February 26, 2004.

Consider the cash flows in a real-dollar cash flow diagram. If these cash flows represent the transaction of placing money into an account and then removing it some number of periods later, then the computation requires the use of an interest rate. For real-dollar cash flows, we require the use of an interest rate that does not include the effects of inflation, as our cash flows do not include those effects. Thus, we require an inflation-free interest rate, which we will define as i', such that

$$F' = P'(1+i')^N.$$

For current dollars, we require an interest rate that *does* incorporate the effects (or expected effects) of inflation. We define this as the market interest rate i, such that

$$F = P(1+i)^N.$$

What is the relationship between i' and i? To define this relationship, we can examine the cash flows of F' and F more closely. We know that they are related according to the inflation rate f as

$$F = F'(1+f)^N.$$

If we substitute our earlier expressions for F' and F, we get

$$P(1+i)^N = P'(1+i')^N(1+f)^N.$$

If time zero is our base period, then P and P' are the same; therefore,

$$(1+i)^N = (1+i')^N(1+f)^N.$$

Examining this equation, we see that the market rate (i) compounds inflation (f) into the inflation-free rate (i'). Assuming that $N = 1$, we can solve for the inflation-free rate:

$$i' = \frac{(1+i)}{(1+f)} - 1. \tag{2.8}$$

Assuming that the inflation rate is constant over the period of study, the inflation-free rate is also constant. Similarly, solving for the market rate i provides the conversion from the inflation-free rate. Note that the market rate is generally known, as banks charge and pay interest on the basis of rates that include the effects of inflation. We illustrate this conversion in the next example.

EXAMPLE 2.17

Market and Inflation-Free Rates

A bank is offering money market accounts that pay effective rates of 4.5% per year. If the annual inflation rate is 1.6%, what is the inflation-free rate for the account?

Solution. We convert the given market rate i to the inflation-free rate with Equation (2.8) as follows:

$$i' = \frac{(1+i)}{(1+f)} - 1 = \frac{(1+.045)}{(1+.016)} - 1 = 0.0285 = 2.85\%.$$

This 2.85% is equivalent to the 4.5% with the effects of inflation removed.

2.7.4 Real- and Current-Dollar Cash Flow Analysis

We have introduced two interest rates (the market and inflation-free rates) and two types of cash flows (current dollars and real dollars). With the definition of two types of cash flows and two types of interest rates, we must be careful as to which combination is used in analysis. We can either analyze current-dollar cash flows with the market rate, since both include the effects of inflation, or we can analyze real-dollar cash flows with the inflation-free rate, since both ignore the effects of inflation. Either analysis leads to the same conclusion, as the methods are differentiated only by inflation. The key point is not to mix real cash flows with market rates or current cash flows with inflation-free rates. (We will return to this point in Chapter 4.)

Unless otherwise specified in this text, it can be assumed that all cash flows are defined by current dollars and the interest rate is the market rate. This is why we chose to define the market interest rate as i, which we also used for our definition of a general interest rate. We will use current cash flows in our analyses because they represent true (not fictitious) out-of-pocket expenses. Also, there are situations, such as when we calculate taxes, in which we must use current-dollar cash flows, as taxes are paid on dollars that are actually exchanged. Furthermore, banks and other lending institutions publish market rates that include the effects of inflation. Thus, to reiterate, one can assume that the cash flows are current dollars and the interest rate is a market rate, unless it is stated otherwise.

2.8 Exchange Rates and Cash Flow Analysis

You have already been exposed to a number of currencies in this text, including euros, kroner, pounds, won, and yen, and will be exposed to many more. We must be aware that we live in a global community and that many companies engage in engineering activities throughout the world. The fact that each country has its own currency adds another layer of complication to our

TABLE 2.3 **Currencies and exchange rates per $1 (USD) ending in 2004 and 2005.**

Country	Currency	Symbol	Exchange Rate, December 31, 2004	Exchange Rate, December 30, 2005
Australia	Dollar	A$	1.2812	1.3620
Brazil	Real	R$	2.6550	2.3340
Canada	Dollar	C$	1.2034	1.1656
China	Yuan	元	8.2765	8.0702
Denmark	Krone	kr	5.4940	6.2985
European Monetary Union	Euro	€	0.7387	0.8445
Hong Kong	Dollar	HK$	7.7723	7.7533
India	Rupee	Rs	43.27	44.95
Japan	Yen	¥	102.68	117.88
Malaysia	Ringgit	RM	3.8000	3.7790
Mexico	Peso	M$	11.154	10.628
New Zealand	Dollar	NZ$	1.3883	1.4609
Norway	Krone	Nkr	6.0794	6.7444
Singapore	Dollar	S$	1.6319	1.6628
South Africa	Rand	R	5.6450	6.3300
South Korea	Won	₩	1035.10	1010.00
Sri Lanka	Rupee	Rs	104.400	102.080
Sweden	Krona	kr	6.6687	7.9370
Switzerland	Franc	Fr	1.1412	1.3148
Taiwan	Dollar	元	31.740	32.800
Thailand	Baht	฿	38.800	40.990
United Kingdom	Pound	£	0.5219	0.5818
Venezuela	Bolivar	Bs	1915.20	2144.60
United States	Dollar	US$	1.0000	1.0000

analyses. The value of one currency relative to another changes constantly and is measured by an exchange rate. Table 2.3 lists a number of exchange rates from December 31, 2004, and December 30, 2005.[14] Specifically, the table defines the number of units of currency that is equivalent to 1 United States dollar at that point in time. For example, it would cost 0.8445 euro to purchase 1 U.S. dollar on December 30, 2005. Euros are the currency of the European Monetary Union, which includes (previous currency in parentheses) Austria (schillings), Belgium (francs), Finland (markkaa), France (francs), Germany (marks), Ireland (pounds), Italy (lire), Luxembourg (francs), Netherlands (guilders), Portugal (escudos), Spain (pesetas), and Greece (drachmas).

The exchange rate allows one to convert the currency of one country into an equivalent amount of currency of another country at the same point in time. Note that this concept is different from that of inflation, which assumes some change in the price level *over* time. However, we include it here, as it is

[14] *Source*: United States Federal Reserve, Foreign Exchange Rates (weekly), Federal Reserve Statistical Release, www.federalreserve.gov/releases/, December 31, 2004.

relevant with respect to converting cash flows. We illustrate in the following example.

EXAMPLE 2.18

Currency Conversion

In October 2003, Motorola signed an agreement with Oman Telecommunications Co. to install 100 GSM stations and expand existing stations in the region of al-Batinah. The deal is worth 11 million rials.[15] Assume that the contract is to be completed over one-and-one-half years with equal payments (totalling 11 million rials) being made quarterly. Assume further that the exchange rate is $1 = OMR0.3840 at time zero, but increases at the rate of 0.125% each quarter. What payments will Motorola receive in U.S. dollars?

Solution. The cash flow diagram for the payments in rials is given in Figure 2.18. To convert to U.S. dollars, the exchange rate must be applied in each period.

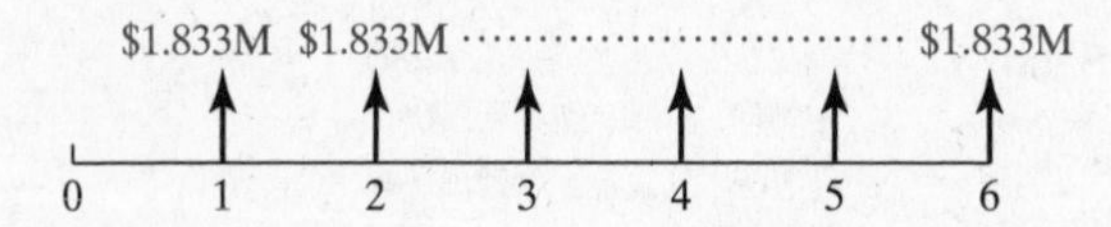

Figure 2.18 Payments for a telecommunications contract in rials.

The exchange rate at the time of the first payment is (0.3840)(1.00125) = 0.3845. Thus, to convert rials to dollars, we have

$$A_1 = \frac{\text{OMR}1.833\text{M}}{(1 + 0.00125)(0.3840)} = \$4.768 \text{ million;}$$

$$A_2 = \frac{\text{OMR}1.833\text{M}}{(1 + 0.00125)^2(0.3840)} = \$4.762 \text{ million;}$$

$$A_3 = \frac{\text{OMR}1.833\text{M}}{(1 + 0.00125)^3(0.3840)} = \$4.756 \text{ million;}$$

$$A_4 = \frac{\text{OMR}1.833\text{M}}{(1 + 0.00125)^4(0.3840)} = \$4.751 \text{ million;}$$

$$A_5 = \frac{\text{OMR}1.833\text{M}}{(1 + 0.00125)^5(0.3840)} = \$4.745 \text{ million;}$$

$$A_6 = \frac{\text{OMR}1.833\text{M}}{(1 + 0.00125)^6(0.3840)} = \$4.739 \text{ million.}$$

Clearly, the value of the payments in U.S. dollars shrinks with time due to the increasing exchange rate.

This example illustrates that the exchange rate can have a drastic effect on the earnings of international companies. In the example, the payments

[15] Fardan, A., "Motorola, Ericsson Sign Telecom Upgrade Contracts in Oman," *Dow Jones Newswires*, October 27, 2003.

made in the currency of the country of origin remained steady, but the value declined in the currency of the company receiving payment.

In this text, we are concerned with making comparisons that are fair. We have introduced the concepts of current and real dollars. Clearly, comparisons between two proposals must both be in either current or real dollars for a fair comparison. That logic extends to foreign currencies, as we cannot make comparisons of projects that are in different currencies. Thus, exchange rates must be utilized for appropriate conversions.

2.9 Examining the Real Decision Problems

We return to our introductory example concerning Bombardier's contract to operate and maintain the automated people-mover system at Atlanta's airport and the questions posed:

1. If the $98 million value assumes equal annual payments (starting in 2006) over the 10-year contract without inflation, how can this contract be visualized and analyzed in a spreadsheet?

2. Given that Bombardier is a Canadian firm, what is its expected cash flow stream in Canadian dollars?

We answer these questions with the use of the spreadsheet given in Figure 2.19. The input data are given in the data center, while columns B and C provide the real- and current-dollar cash flows in U.S. dollars, respectively, and the current Canadian dollars are given in column D. This snapshot assumes a 3.25% annual inflation rate, growing from the end of year 2005 through the project's horizon, and a fixed exchange rate of C$1.1656 per US$1.00 over the horizon. Both the inflation rate and the exchange rate are variables, so each time an input in the data center is changed, the cash flows (and graph) change.

The cash flow diagrams for expected annual payments (current dollars) for both U.S. and Canadian currencies are also shown using a bar chart in Excel. Graphs are an excellent way to display a lot of information succinctly.

A variety of graphs can be created in Excel. We highlight the use of the bar chart here, but generating other graphs in Excel follows similarly.

The easiest way to start the process is to highlight the data to be graphed, cells A3:D13, and click on the "Chart Wizard" toolbar. A dialogue box appears requesting you to select a type of graph. In this example, we select the 'Column' type. (The first of the Column types is the 'Clustered Column.') After this selection, a miniature figure appears with the highlighted data graphed. The graph may initially look strange, but it takes on the correct shape after changing some of the inputs. If the 'Data Range' tab is selected, verify that the 'Rows' option is selected and click on the 'Series' tab. Each series of data, corresponding to a column of inputs, is defined in the scroll box according to

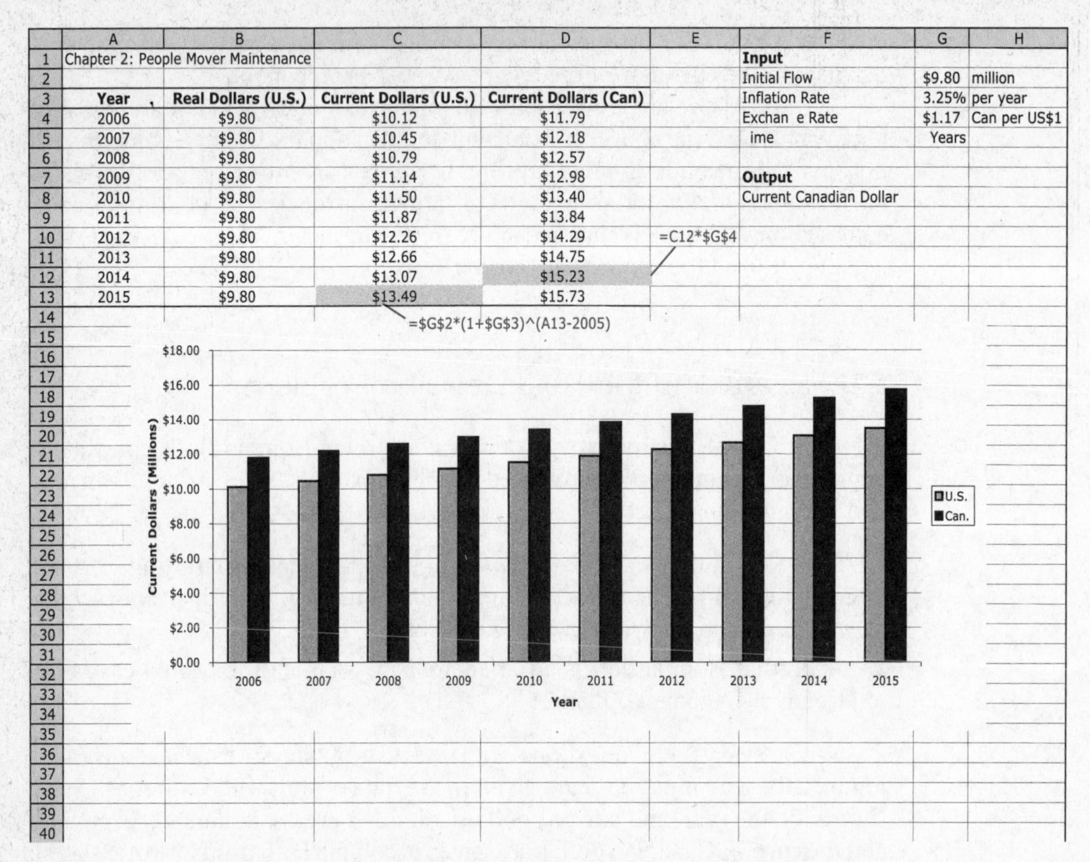

	A	B	C	D	E
1	Chapter 2: People Mover Maintenance				
2					
3	Year	Real Dollars (U.S.)	Current Dollars (U.S.)	Current Dollars (Can)	
4	2006	$9.80	$10.12	$11.79	
5	2007	$9.80	$10.45	$12.18	
6	2008	$9.80	$10.79	$12.57	
7	2009	$9.80	$11.14	$12.98	
8	2010	$9.80	$11.50	$13.40	
9	2011	$9.80	$11.87	$13.84	
10	2012	$9.80	$12.26	$14.29	=C12*$G$4
11	2013	$9.80	$12.66	$14.75	
12	2014	$9.80	$13.07	$15.23	
13	2015	$9.80	$13.49	$15.73	
14			=$G$2*(1+$G$3)^(A13-2005)		

	F	G	H
1	**Input**		
2	Initial Flow	$9.80	million
3	Inflation Rate	3.25%	per year
4	Exchan e Rate	$1.17	Can per US$1
5	ime		Years
6			
7	**Output**		
8	Current Canadian Dollar		

Figure 2.19 Contract cash flows in real and current U.S. dollars and current Canadian dollars.

the headings in row 3. Highlighting a name shows the corresponding data. The name of the data can be changed, as can the y-coordinates (termed 'Values').

Since our first column of data is the input for the x-axis on our graph, we can delete the 'Year' data series. However, for the other three data series, we must input cells A4 through A13 as the 'Category (X) axis labels.' (Excel will generally assume that this is true for all series after you input the first.)

The next page of the dialogue box allows you to manipulate how the graph looks, including adding a title and axes identifiers. Changes can also be made to data labels, grid lines, and the legend. Don't worry if you forget to do something or want to change it later, as you can merely click on the chart and manipulate the appropriate setting at any time.

The final dialogue box asks if you want the chart placed on a separate worksheet or within the one you are working. This is strictly your preference, although it is often useful to have the chart near your data. *Note that if the data used to create the chart changes, the graph will change, too.*

Thus, if generated correctly with variable data, only one graph needs to be created.

Once the chart appears on the worksheet, you can click on any part of it to manipulate it. For example, you can click on the plot or chart area and change the background colors or shades. You can click on the data points and change their appearance. You can click on an axis and change the spread of the data or the labeling of an axis. The number of options is seemingly endless.

Furthermore, when you click on the chart, a new menu entitled 'Chart' appears at the top of the screen. This allows you to go back and change the data being examined or add more data sets to the current chart. There are a number of other functions and formatting options under this menu option. We suggest that you practice making graphs and charts in Excel, as it is a handy function that we will call upon repeatedly throughout the textbook.

We now return to the last question posed in our introductory example:

3. If Bombardier uses a 3.5% quarterly interest rate when analyzing engineering projects, what rate should the company use in analyzing this particular contract?

Because the cash flows occur annually, Bombardier can use an annual interest rate of

$$i_a = (1 + i_q)^4 - 1 = (1 + 0.035)^4 - 1 = 14.75\%$$

for analysis.

2.10 Key Points

1. Cash flows represent the transfer of money into or out of an account. A cash flow diagram is used to illustrate a series of cash flows over time and to financially describe engineering projects.

2. Cash flows can be described as discrete, in which an amount of money is transferred at a single point in time, or continuous, in which money flows to or from an account according to some rate over some period of time.

3. Spreadsheets are a powerful tool for economic analysis. A spreadsheet consists of cells, defined by rows and columns, in which various data and parameters can be input. Inputs and outputs should always be captured in a data center to allow for the reuse of a spreadsheet with different parameters.

4. Data, variables, and functions can be input to a spreadsheet in order to develop cash flow diagrams for analysis. Absolute references copy the

contents of one cell to another cell. Relative references translate data from one cell to another, based on the number of rows and columns between the copied cell and its references.

5. The time value of money describes the change in the worth of money with time. Money can be spent (with the user receiving utility), invested (where it grows), or held (where it is subject to inflation or deflation).

6. Interest is a fee paid for borrowing money or income received for lending money. The amount of interest paid is dependent on the interest rate.

7. Simple interest calculates interest payments solely on the basis of the principal amount. Compound interest calculates payments on the basis of the principal and accrued interest.

8. Nominal interest rates are generally stated as annual interest rates that do not include the effects of compounding. Effective interest rates include the effects of compounding.

9. When comparing interest rates, effective rates over similar periods should be used.

10. An increase in price for a commodity is known as inflation, while a decrease is called deflation. The amount of increase or decrease in prices is defined by the inflation rate. Prices are tracked through various indexes, including the Consumer Price Index and the Producer Price Index.

11. Current dollars are actual out-of-pocket expenses or revenues that include the effects of inflation. Real dollars define cash flows with the effects of inflation removed.

12. The market interest rate includes the effects of inflation, while the inflation-free rate defines the interest rate without the effects of inflation. These rates are mathematically related by the inflation rate.

13. Real dollars must be analyzed with the inflation-free interest rate, while current dollars must be analyzed with the market interest rate.

14. Exchange rates describe the economic conversion of one currency to another at the same point in time. Because the worth of a currency with respect to another currency can change with time, exchanges between currencies are also subject to changes in purchasing power, much like inflation or deflation.

2.11 Further Reading

Denardo, E.V., *The Science of Decision Making: A Problem-Based Approach Using Excel.* John Wiley and Sons, New York, 2002.

Etter, D.M., *Microsoft Excel for Engineers.* Addison-Wesley Publishing Company, Menlo Park, California, 1995.

Fisher, I., *The Theory of Interest: As Determined by Impatience to Spend Income and Opportunity to Invest It.* Augustus M. Kelley, Publishers, New York, 1965.

Fleischer, G.A., *Introduction to Engineering Economy*. PWS Publishing Co., Boston, 1994.

Hartman, J.C., "Inflation and Price Change in Economic Analysis," Chapter 92 (pp. 2394–2405) in *Handbook of Industrial Engineering,* 3d ed., Gavriel Salvendy (ed.), John Wiley and Sons, New York, 2001.

Kuncicky, D.C., *Introduction to Excel.* Prentice Hall, Upper Saddle River, New Jersey, 1999.

Lee, P., and W.G. Sullivan, "Considering Exchange Rate Movements in Economic Evaluation of Foreign Direct Investments," *The Engineering Economist,* 40(2):171–200, 1995.

Park, C.S., *Contemporary Engineering Economics,* 3d ed. Prentice Hall, Upper Saddle River, New Jersey, 2002.

Sullivan, W.G., E.M. Wicks, and J.T. Luxhoj, *Engineering Economy,* 12th ed. Prentice Hall, Upper Saddle River, New Jersey, 2003.

Thuesen, G.J., and W.J. Fabrycky, *Engineering Economy,* 9th ed. Prentice Hall, Upper Saddle River, New Jersey, 2001.

2.12 Questions and Problems

2.12.1 Concept Questions

1. What is a cash flow?
2. Is interest an expense or revenue? Explain.
3. How can cash flow diagrams be useful?
4. What is the time value of money?
5. What is the difference between simple and compound interest?
6. How are nominal and effective rates related? Which rate is more useful? Explain.
7. What types of interest rates can be compared?

8. What is the difference between inflation and deflation? How do they relate to purchasing power?
9. How are price indexes and the inflation rate related?
10. What costs are included in the CPI?
11. What is the difference between real and current dollars? How are they mathematically related?
12. What is the difference between market and inflation-free interest rates? How are they mathematically related?
13. If analyzing real dollars, what interest rate should be used?
14. If using the market interest rate, what cash flows should be analyzed?
15. When reading a problem in this text, what should you assume about the cash flows and interest rates provided?

2.12.2 Drill Problems

1. Draw the net cash flow diagram for the following situations:

   (a) The sum of $100,000 is invested at time zero, resulting in revenues of $50,000 per year and $2000 in annual expenses over the next five years.

   (b) A lease is signed that requires monthly payments of $5000 over the next two years.

   (c) A piece of equipment is purchased for $25,000. Its annual operating and maintenance expenses are $1000 in year 1, increasing $500 per year over the next four years. The asset is sold for $2000 at the end of the fifth year.

   (d) A facility is built over four quarters at the cost of $250,000 per quarter. Net revenues total $40,000 in the first quarter following completion of the facility and are expected to grow by 2.4% per quarter. The facility is closed after five years of operations at the cost of $10,000.

   (e) Annual deposits of $75,000 are made into an account for three years, after which $235,000 is withdrawn.

   (f) A new product is introduced at the cost of $500,000. Revenues for the first year are $100,000, growing by $10,000 per year for the next four years and then declining by $12,000 a year for the next five years. Expenses are $15,000 in the first year and are expected to increase by 3.2% per year over the lifetime of the product. The project has no salvage value.

   (g) A bond that pays $350 every six months over five years is purchased for $10,000 (face value). The face value amount is returned to the purchaser at the end of the life of the bond.

   (h) A loan is taken out for $95,000 and paid back through six annual payments of $18,000.

   (i) A mine is opened at the cost of $15 million at the end of 2005 and generates daily revenues of $100,000 against daily costs of $40,000 for seven years. At the end of the life of the mine, a $1 million remediation fee is paid. Draw the cash flow diagram, with a continuous flow of funds where appropriate.

(j) An asset is purchased for \$2 million, operated for five years at the cost of \$125,000 per year, and sold for \$50,000 at the end of year 5.

2. A \$250,000 simple-interest loan is to be repaid at the end of 1 year. How much is owed, assuming a 7.3% annual rate of interest? Recalculate for 2 years and $3\frac{1}{2}$ years.

3. A \$150,000 simple-interest loan is repaid after four years with a payment of \$175,000. What is the annual interest rate?

4. An \$85,000 loan with a 4.5% annual interest rate is to be repaid in a single payment at the end of five years. Draw a table defining the amount of interest accrued and the amount owed at the end of each year.

5. A \$40,000 loan with a 3.8% semiannual interest rate is to be repaid in one payment at the end of three years. What is the total interest paid on the loan?

6. An interest rate of 6.25% compounded monthly is advertised. Find

(a) The effective monthly interest rate.

(b) The effective quarterly interest rate.

(c) The effective semiannual interest rate.

(d) The effective annual interest rate.

7. An interest rate of 9.5% compounded quarterly is advertised. Find

(a) The effective monthly interest rate.

(b) The effective quarterly interest rate.

(c) The effective semiannual interest rate.

(d) The effective annual interest rate.

8. An interest rate of 8.0% compounded annually is advertised. Find

(a) The effective monthly interest rate.

(b) The effective quarterly interest rate.

(c) The effective semiannual interest rate.

(d) The effective annual interest rate.

9. An interest rate of 7.45% compounded continuously is advertised. Find

(a) The effective daily interest rate.

(b) The effective monthly interest rate.

(c) The effective quarterly interest rate.

(d) The effective semiannual interest rate.

(e) The effective annual interest rate.

10. An interest rate of 3.2% per quarter is advertised. Find the nominal rate (annual).

11. An interest rate of 1.55% per month is advertised. Find the nominal rate (annual).

12. An interest rate of 10.2% per year is advertised. Find the nominal rate (annual).

13. Which of the following represents a cheaper loan, 1.25% per month or 12.0% compounded quarterly?

14. Which of the following represents a better investment, 14.3% compounded semiannually or 2.1% per quarter?

15. Which of the following represents a cheaper loan, 7.35% per year or 8.25% compounded semiannually?

16. Which of the following represents a better investment, 4.35% per quarter or 15.3% compounded continuously?

17. Using Table 2.1, compute the annual inflation rate over the following periods:

(a) Between 1983 and 1987.

(b) Between 1985 and 1992.

(c) Between 1993 and 1999.

(d) Between 2000 and 2001.

(e) Between 2000 and 2002.

18. Using Table 2.2, compute the annual inflation rate with both the construction machinery data and the data on semiconductors and related devices over the following years:

(a) Between 1993 and 1999.

(b) Between 1994 and 2002.

(c) Between 2000 and 2001.

(d) Between 2000 and 2002.

19. Convert the following real-dollar cash flow diagram (in years) to a current-dollar cash flow diagram, assuming an annual inflation rate of 3.75%:

$$A'_0 = -\$50,000;$$
$$A'_1 = \$20,000;$$
$$A'_2 = \$20,000;$$
$$A'_3 = \$20,000;$$
$$A'_4 = \$20,000;$$
$$A'_5 = \$20,000.$$

20. Convert the following current-dollar cash flow diagram (in years) to a real-dollar cash flow diagram, assuming a monthly inflation rate of 0.25%:

$$A_0 = -\$280,000;$$
$$A_1 = \$200,000;$$
$$A_2 = \$300,000;$$
$$A_3 = \$400,000;$$
$$A_4 = \$500,000;$$
$$A_5 = \$100,000.$$

21. Convert the following real-dollar cash flow diagram (in years) to a current-dollar cash flow diagram, assuming a quarterly inflation rate of 1.05% Note that there is no cash flow at time zero nor at time period two:

$$A'_1 = \$120{,}000;$$

$$A'_3 = \$180{,}000;$$

$$A'_4 = \$210{,}000;$$

$$A'_5 = \$250{,}000.$$

22. Convert the following current-dollar cash flow diagram (in years) to a real-dollar cash flow diagram, assuming an annual inflation rate of 2.79%:

$$A_0 = -\$1{,}000{,}000;$$

$$A_1 = \$256{,}975;$$

$$A_2 = \$316{,}974;$$

$$A_3 = \$380{,}120;$$

$$A_4 = \$446{,}543;$$

$$A_5 = \$516{,}377.$$

23. Assume that the cash flows in the previous problem are in Canadian dollars. Convert to Australian dollars, using the conversions listed in Table 2.3 (2004 data). (Assume that the exchange rate is constant over time.)

24. Assume that the cash flows in Problem 22 are in Japanese yen. Convert to euros, using the conversions listed in Table 2.3 (2005 data). (Assume that the exchange rate is constant over time.)

25. If the annual market interest rate is 12.5% and the annual inflation rate is 2.5%, what is the annual inflation-free rate?

26. If the annual inflation-free rate is 7.45% and the annual inflation rate is 3.5%, what is the annual market interest rate?

27. If the annual inflation-free rate is 6.25% and the annual market rate is 9.45%, what is the annual inflation rate?

28. If the annual inflation-free rate is 5.0% and the monthly inflation rate is 0.247%, what is the annual market interest rate?

29. If the monthly inflation-free rate is 1.25% and the monthly market rate is 1.5%, what is the quarterly inflation rate?

30. If the annual market interest rate is 11.0% and the semiannual inflation rate is 1.38%, what is the annual inflation-free rate?

2.12.3 Application Problems

1. Draw the net cash flow diagram for the following situations:

(a) Computer Sciences Corp. was awarded a $150 million contract to provide information technology services for the U.S. Department of the Treasury.[16] Draw the cash flow diagram, assuming a five-year contract with equal sized end-of-year payments.

(b) The Systems Engineering Division of Calian Technology, Ltd., received a C$16 million contract for operations and maintenance services for the Canadian Space Agency's Satellite Operations Directorate.[17] Draw the cash flow diagram, assuming that September 2005 is time zero and equal payments are scheduled monthly from October 2005 through September 2007.

(c) Devon Energy of Oklahoma City, Oklahoma, purchased Ocean Energy for $5.3 billion in 2003, to become the largest independent oil and natural-gas producer in the United States.[18] Assume that Ocean Energy produced 200,000 barrels of oil or oil equivalents per day before the Devon purchase, that oil sells for $40 a barrel, and that the cost to produce a barrel is $14. Consider a 10-year horizon, at which time Devon will sell the fields for $500 million. Assume end-of-year cash flows (250 production days per year) and draw the cash flow diagram for Devon Energy's investment.

(d) Redraw the previous cash flow diagram, assuming continuous cash flows for oil production and sales.

(e) Horizon Air signed an agreement to purchase 12 Q400 turboprop aircraft from Bombardier at the total cost of $294 million.[19] Assume that the jets fly, on average, three legs per day at 60% capacity (the aircraft seats 70 people), with each seat generating a net revenue (revenues minus costs of operation) of $35. Aggregate the net revenues to monthly figures (30 days per month) and generate a five-year cash flow diagram, assuming that the planes are sold at the end of five years for 40% of their initial worth.

(f) Adjust your solution to the previous problem, assuming that filled capacity grows 10% per year (but the same for all months in the same year), starting at 40% in year 1.

(g) Environmental Management Solutions, Inc., signed a five-year contract for C$12 million with Scott Paper, Ltd., for the treatment and disposal of de-inking residual waste from its Crabtree, Quebec, mill.[20] Assume that revenues are spread evenly over five years. Draw end-of-year cash flows in U.S. dollars, assuming a constant exchange rate over time, as given in Table 2.3 (2005 data).

---

[16] Paige, M., "NEWS WRAP: Computer Sciences Wins US Treasury Order," *Dow Jones Newswires,* December 1, 2005.

[17] McKinnon, J., "Calian Gets C$19M in Two Satellite-Related Operations and Maintenance Contracts," *Dow Jones Newswires,* September 23, 2005.

[18] "Devon Energy's holistic strategy for growth redefines the art of getting bigger via mergers and acquisitions," *Oil & Gas Investor*, p. 16, March 1, 2005.

[19] Zachariah, T., "Horizon Air Signs for 12 Bombardier Q400 Airliners," *Dow Jones Newswires,* October 19, 2005.

[20] Tsau, W., "Environ Mgmt Signs 5-Yr Contract with Scott Paper," *Dow Jones Newswires,* August 11, 2005.

(h) Volvo announced that it sold around 400 buses in the Indian market in 2005.[21] Assume that 402 buses are sold in increasing fashion such that six are delivered in January and deliveries increase by five each month thereafter through December. Construct a cash flow diagram, using a spreadsheet that allows you to examine different unit revenues of buses, assuming that revenues are received upon delivery.

2. Crystallex International Corporation announced a $153 million investment to double the capacity of its gold mine at Las Cristinas, Venezuela. The investment will allow 20,000 metric tons of additional ore to be mined daily, yielding 1.27 grams (0.448 ounce) of gold per ton of ore.[22] Assume that the investment is divided over two years, with the first year of production reaching 5,000 tons per day (200 production days per year) and increasing 5,000 tons per day each year until reaching the capacity of 20,000 tons per day. Assume further that production costs are $6.91 per ton and revenues are $300 per ounce of gold. Finally, assume that annual maintenance costs are $200 million and the charge to clean up the mine after 20 years of production is $15 million. Using a spreadsheet, draw the cash flow diagram for this investment.

3. Air France, EDS, and eBay have endorsed the new low-power T1000 and T2000 servers from Sun Microsystems.[23] Assume that eBay purchases 12 servers at the price of $18,000 each with no money down and pays $228,960 at the end of the first year to complete the firm's obligation. What is the effective annual interest rate? What is the effective monthly interest rate? If eBay decides not to make any payments until the end of five years, how much will the company owe, given the same interest rate?

4. Adams County (Wisconsin) Road Commission purchased a $340,000 Bomag MPH454R recycler for pulverizing asphalt pavement and mixing the material with new additives to create a new, structurally superior base. The county shares the machine with a neighboring county to defer the high purchase cost.[24] Assume that a local bank is offering a low interest rate of 5% compounded daily to purchase the machine. If the municipality will pay back the principal plus interest in one lump sum at the end of four years, what is the total interest to be paid? If another local bank has offered a rate of 5.4% compounded semiannually, should the municipality go with the new offer?

5. Taiwan Semiconductor Manufacturing Corporation (TSMC) purchased lithography machinery from Dutch semiconductor equipment maker ASML in the fall of 2005 for EUR54.2 million.[25] If TSMC receives a loan at the cost of 5.75% per year, how much will the company save in interest if it pays off the loan (in one payment)

---

[21] Pfalzer, J., "Volvo Says Should Sell about 400 Coaches in India in 2005," *Dow Jones Newswires,* December 14, 2005.

[22] McKinnon, J., "Crystallex Announces Capital and Operating Costs for Expansion of Las Cristinas to a 40,000 Tpd Operation," *Dow Jones Newswires,* September 26, 2005.

[23] Boslet, M., "Sun Micro Expected to Unveil Two Low Power Servers," *Dow Jones Newswires,* December 6, 2005.

[24] "Bomag's Re-Sized Road Reclaimer Digs In," *Better Roads Magazine,* www.betterroads.com, January 2002.

[25] "Taiwan TSMC Buys Lithography Machinery for 54.2 Mln Euro from Dutch ASML," *Dutch News Digest,* October 6, 2005.

after one year, as opposed to paying it off (again, in one payment) at the end of two years?

6. Assume that a farmer in Europe is looking to upgrade equipment with a new Claas combine harvester at the cost of EUR250,000.[26] A local bank is offering a rate of 8.5% compounded continuously, while an Internet bank is offering an effective monthly rate of 0.75% per month. Strictly on the basis of cost, which loan should the farmer take? Using Table 2.3 (2005 data), what is the cost in U.S. dollars?

7. AltaSteel, Ltd., purchased an eight-stand Danieli roughing train for C$16 million to upgrade its bar mill facilities.[27] Assume that AltaSteel considered two sources of financing: a regional bank offering a rate of 6.5% compounded continuously and a commercial bank offering a rate of 6.6% compounded quarterly. Which loan should have been secured?

8. In late 2005, Hornbeck Offshore Services, Inc., announced that it would sell $75 million of 6.125% bonds due in 2014 to partially fund building a number of new shipping vessels at a total cost of $265 million.[28] Assume that you purchase $10,000 worth of the bonds which pay a coupon payment of $306.25 every six months for nine years, at which time you receive the $10,000 back. Draw both the real- and current-dollar cash flow diagrams, assuming a semi-annual inflation rate of 2.3%.

9. Consider the Devon Energy example in Problem (1c). Assuming an annual inflation rate of 1.2% for oil revenues, 0.5% for costs, and a 1.3% annual inflation rate for the worth of the fields, draw the current-dollar cash flow diagram for the given data.

10. Consider the Horizon Air example in Problem (1e). Assuming a monthly inflation rate of 0.3% for net revenues and a 2.5% annual inflation rate for the worth of the planes, draw the current-dollar cash flow diagram for the given data.

2.12.4 Fundamentals of Engineering Exam Prep

1. Annual revenues from a project are expected to grow at a rate of 3% annually from $100,000 in the first year, while operating and maintenance costs are expected to hold steady at $75,000. The net cash flow in period 3 is most closely

(a) $25,000.

(b) $34,273.

(c) $31,090.

(d) $27,318.

2. Per unit revenues are forecast at $15 per item against per unit costs of $9.50 and fixed costs of $10,000 per period. If 20,000 units are sold each period, what is the periodic net cash flow?

---

26 Milne, R., "Claas reaps benefits of move into tractors," *The Financial Times,* London Edition, p. 23, July 18, 2005.

27 Tsau, W., "Stelco's AltaSteel Announces Bar Mill Expansion," *Dow Jones Newswires,* March 29, 2005.

28 "Hornbeck Offshore Announces $265 M in New Expansion Plans, Provides Update on Current OSV Market Conditions and Other Recent Developments," *Dow Jones Newswires,* September 26, 2005.

(a) \$110,000.

(b) \$100,000.

(c) \$120,000.

(d) \$90,000.

3. An account pays simple interest of 2% per quarter. If \$100 is deposited at time zero, how much is in the account after one year?

(a) \$102.

(b) \$104.

(c) \$106.

(d) \$108.

4. If fixed production costs are \$500,000 per quarter and per unit net revenues are \$125, the minimum number of units sold per period that would result in positive net cash flow is

(a) 4000.

(b) 3500.

(c) Less than 3000.

(d) More than 5000.

5. A supply contract stipulates the delivery of 50,000 parts each year at the market rate. If the cost per part is \$14.50 at time zero and the price is expected to rise 4% per period, the amount paid in period 4 is most closely

(a) \$754,000.

(b) \$848,150.

(c) \$725,000.

(d) \$815,000.

6. In its final year of service, an asset produces \$50,000 in sales against \$30,000 in costs and is salvaged (at the end of the period) for \$25,000. The net cash flow is

(a) \$45,000.

(b) −\$5,000.

(c) \$5,000.

(d) \$75,000.

7. \$20,000 is deposited into an account paying simple interest of 4% per year. The amount in the account after four years is most closely

(a) \$20,000.

(b) \$26,900.

(c) More than \$30,000.

(d) \$23,200.

8. A contract is signed bringing in 150,000 British pounds in year 1 and growing at the rate of 7% per year. If the exchange rate is constant over time at 0.63 pound per U.S. dollar, the equivalent dollars received in year 3 are most closely

(a) $238,000.

(b) $272,600.

(c) $94,500.

(d) $108,193.

9. A nominal annual rate of 12% compounded monthly is the same as

(a) 1% per month.

(b) 4% per quarter.

(c) 12% per year.

(d) 0.08% per month.

10. An interest rate of 2% per month is the same as

(a) 6% per quarter.

(b) A nominal rate of 24% per year compounded monthly.

(c) 24% per year.

(d) None of the above.

11. An effective rate of 0.25% per week is most closely

(a) 13.86% per year.

(b) 13.00% per year.

(c) A nominal rate of 13.86% per year, compounded weekly.

(d) None of the above.

12. An effective rate of 3.5% semiannually is

(a) 7.12% per year.

(b) 7% per year, compounded semiannually.

(c) Both (a) and (b).

(d) None of the above.

13. A bank pays 2.5% per year on its savings account, while a money market account pays 2.4% per year compounded continuously. The bank's rate is

(a) Lower than the money market rate.

(b) Equivalent to the money market rate.

(c) Higher than the money market rate, but by less than 0.10%.

(d) Greater than the money market rate by more than 0.10%

14. A nominal rate of 14% per year compounded annually is the same as

(a) 1% per month.

(b) 7% per six months.

(c) 14.12% per year.

(d) 14% per year.

15. An Internet bank offers loans at 1.5% per quarter, while the local bank advertises 6% per year, compounded monthly. Which of the following is true?

(a) The Internet bank rate is higher.

(b) The rates are equivalent.

(c) The local bank rate is lower.

(d) The local bank rate is equivalent to 1.51% per quarter.

16. A $70,000 bulldozer can be financed at 5% per year. The amount of interest owed at the end of year 1 is

(a) $35,000.

(b) $3500.

(c) $73,500.

(d) $66,500.

17. The price index for raw material was 104.5 in 1995 and 142.6 in 2004. The annual inflation rate over that period was most closely

(a) 3.51%.

(b) 3.65%.

(c) 3.88%.

(d) 4.67%.

18. The price index for raw material was 104.5 in 1995 and 142.6 in 2004. If the material cost $10,000 in 2003 and prices are expected to increase similarly in the future, one could expect the cost in 2007 to be most closely

(a) $11,480.

(b) $10,000.

(c) $13,645.

(d) $10,450.

19. Prices decrease from $1500 per ton in 1999 to $1250 per ton in 2003. The annual inflation rate is most closely

(a) 4.66%.

(b) 20%.

(c) –4.46%.

(d) –5.55%.

20. An effective interest rate of 2.5% per quarter is

(a) 10% per year, compounded quarterly.

(b) 5.06% per six months.

(c) 10.38% per year.

(d) All of the above.

21. A nominal interest rate of 18% per year, compounded quarterly, is most closely

(a) 19.25% per year.

(b) 4.25% per quarter.

(c) 18% per year.

(d) None of the above.

22. A nominal interest rate of 20% per year, compounded continuously, is most closely

(a) 22% per year.

(b) 24% per year.

(c) 11% per six months.

(d) 22.14% per year.

23. A nominal interest rate of 10% per year, compounded monthly, is most closely

(a) 1% per month.

(b) 5.11% per six months.

(c) 10% per year.

(d) 4.98% per six months.

24. Semiconductor lithography equipment cost $5.5 million in 2000. With the semiconductor equipment price index moving from 125.2 in 2000 to 142.4 in 2005, the price of the equipment in 2005 was most closely

(a) $5.89 million.

(b) $5.5 million.

(c) $6.26 million.

(d) None of the above.

25. $2.50 bought 2 gallons of diesel fuel in 2003. If inflation for fuel from 2003 through 2006 has averaged 5.2% per year, how much fuel does $2.50 buy in 2006?

(a) Less than 1.5 gallons.

(b) Between 1.5 and 1.75 gallons.

(c) Between 1.75 and 2 gallons.

(d) More than 2 gallons.

26. Parts purchased in Germany cost 2.30 euros per unit. If the exchange rate is 0.72 euro to the U.S. dollar, the cost in dollars is most closely

(a) $1.66

(b) $3.20

(c) More than $4.00.

(d) Less than $1.50.

27. The inflation-free rate is 2.3% per year and the inflation rate is 2.5% per year. The effective interest rate (market rate) is most closely

(a) 0.2% per year.

(b) 4.80% per year.

(c) 4.86% per year.

(d) 4.2% per year.

28. At time zero, a bond is purchased that pays a $600 coupon payment at the end of every year. If the inflation rate is 3.2% per year, then the time-zero value of the fourth coupon payment is most closely

(a) $581.20.

(b) $600.00.

(c) $564.32.

(d) $528.97.

29. Costs are expected to rise over time at the rate of 4.5%. If costs are $250,000 in time-zero dollars, the costs actually experienced in period 5 are most closely

(a) $261,250.

(b) $250,000.

(c) $285,290.

(d) $311,500.

30. A dealer offers a loan on equipment at 4.5% per year, compounded continuously. This most closely equals

(a) 2.28% per six months.

(b) 4.5% per year.

(c) 4.7% per year.

(d) None of the above.

3 Interest Formulas

(Courtesy of Apple Computer, Inc.)

Real Decisions: Hit Shuffle!

In an effort to meet growing demand for its iPOD, Apple Computer, Inc., announced that it had entered into contracts with five memory chip makers through 2010 for supplies of NAND flash memory chips. The chips are popular for a variety of products, including digital cameras and MP3 players, because they retain data even after power is shut off. Speculation has it that the chips will eventually replace mini hard drives. To ensure a supply of the flash memory chips currently used in its iPOD Nano model, Apple agreed to prepay a total of $1.25 billion: $500 million to Samsung Electronics (South Korea), $250 million to Hynix Semiconductor (South Korea), $250 million to Intel (United States), and $250 million to Micron Technology (United States). Toshiba (Japan) was also announced as a long-term supplier.[1,2] These agreements lead to a number of interesting questions:

1. Given an 18% annual rate of interest, what annual payments at the end of years 2005 through 2010 would have been equivalent to the $1.25 billion prepayment (at the end of 2005)?
2. If Apple makes similar $1.25 billion payments at the end of each year from 2005 through 2010, what is the worth of the contract (payments) at time 2010?
3. What if the payments increase 12% per year due to increasing sales? Is this significant?

In addition to answering these questions, after studying this chapter you will be able to:

- Define the concept of economic equivalence.
- Use formulas, tables, and spreadsheets to develop mathematical relationships, known as interest factors, to convert a cash flow diagram into another equivalent cash flow diagram, assuming discrete cash flows and an interest rate that is compounded over a discrete number of periods. (Sections 3.1–3.4)
- Convert a single cash flow or a series of cash flows into an equivalent future value. (Section 3.1)
- Convert a single cash flow or a series of cash flows into an equivalent present value. (Section 3.2)
- Convert a single cash flow or a series of cash flows into an equivalent series of periodic cash flows. (Section 3.3)
- Utilize multiple interest factors to analyze complicated cash flows. (Section 3.5)
- Establish similar interest factors under the assumption of an interest rate that is continuously compounded. (Section 3.6)

[1] Kawamoto, D., "Apple sews up flash market," *ZDNet UK,* November 22, 2005.

[2] "Korea Samsung, Hynix ink $750-mil in chip sales to Apple," *Platts Commodity News,* November 21, 2005.

Now that we have an understanding of interest, interest rates, and cash flow diagrams, we are prepared to move money through time. Our motivation is for comparison purposes. Recall that we are describing engineering solutions, known as projects, according to their cash flow diagrams. In order to compare two different cash flow diagrams fairly, we must be able to convert each diagram into something similar.

Consider the following problem which many engineering firms face: A technology (a process or a product) is developed and patented. It can either be sold for a single sum of money or be licensed to other companies over time. If the technology is licensed, royalties would be received over some specified time frame. Which option is preferred? How would you answer this question? The key is to convert each option into similar terms such that they can be compared fairly. You were always told never to compare apples with oranges. This is never more true than with monetary transactions.

The interest factors developed in this chapter give you the necessary tools to move money through time and allow you to convert cash flow diagrams into different representations. When we convert one cash flow diagram into another, we say that the two diagrams are **economically equivalent**. This means that we, as decision makers, are indifferent as to which we choose, because the financial consequences are the same. Equivalence depends on three inputs:

1. The **magnitude** of the cash flow(s).
2. The **timing** of the cash flow(s).
3. The **interest rate(s)** over the relevant periods.

Given this information, we can determine whether two cash flows or two series of cash flows are economically equivalent. If they are not equivalent, then we would prefer one over the other. To determine equivalence, we transform the cash flows into common terms such that they may be compared. Methods to transform cash flow diagrams are presented in this chapter.

The first case to be analyzed assumes discrete cash flows and discrete compounding. As defined in Chapter 2, discrete cash flows are cash transactions that occur at a single point in time. These transactions may be as simple as someone paying cash for an item in a store or as complicated as a large bank wire transferring thousands of dollars between its branches. Furthermore, it is assumed that the interest rate used for compounding and discounting is discretely compounded (daily, monthly, quarterly, etc.). The case of continuous compounding will be discussed later in the chapter.

We make the following general assumptions:

1. Interest is compounded per period and the periods designating cash flow transactions are the same as the compounding periods. That is, if the periods are designated in months, then an effective monthly interest rate is assumed. (See Chapter 2 for a discussion of converting interest rates and dealing with the timing of cash flows.)

2. The interest rate does not change over the time frame of the analysis. While this restriction may be relaxed (and we address this issue in later chapters), doing so merely complicates the derivations.
3. Time zero is an arbitrary starting point. That is, the movement of a cash flow five periods in time is the same whether the beginning and ending periods are 0 and 5 or 3 and 8, assuming that the interest rate does not change over any of the periods.
4. P is used to represent cash flows at time zero, while F is used to represent cash flows at time N. Repeated cash flows (series) are denoted with an A and are assumed to cover N periods. Individual cash flows occurring between time zero and time N are designated as A_n, with n designating the period.
5. All cash flows are assumed to occur at the end of the period, unless otherwise specified. Note that the end of one period is equivalent to the beginning of the next.

Under these assumptions, we derive interest factors for discrete cash flows and a discrete compounding period for the interest rate. These tools, which move money efficiently through time, are derived from the principles of interest and interest rates.

3.1 Compound Amount Factors

We first concentrate on moving money forward in time, or compounding. Our goal is to transform a set of cash flows into a single cash flow in the future. We say that the resulting cash flow is economically equivalent to the initial set of cash flows. Specifically, we want to determine the equivalent future value F at time N, assuming a periodic interest rate i, for the following four cash flow diagrams:

1. A single cash flow P at time zero.
2. N consecutive cash flows of equal size A.
3. N consecutive cash flows beginning with size 0 and increasing by G in each consecutive period.
4. N consecutive cash flows beginning with size A_1 and increasing at rate g in each consecutive period.

Mathematically, F is an unknown variable, while the other parameters, P, A, G, A_1, g, i, and N, are all known.

3.1.1 Single-Payment Analysis

The first cash flow diagram described in the preceding list is a single payment P at time zero. The *compound amount factor* defines F, the equivalent value

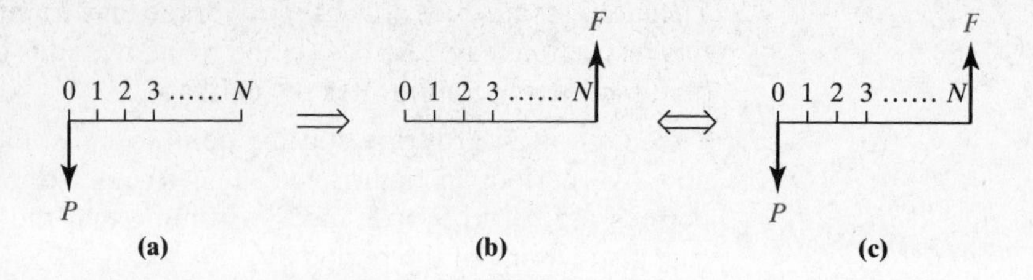

Figure 3.1 (a) Single cash flow P at time zero. (b) Future value F at time N. (c) Cash flow diagram for the compound amount factor for a single payment.

of P at time N, assuming a constant periodic interest rate i. The desired transformation is depicted in Figure 3.1, with (a) referring to the cash flow P at time zero and (b) referring to our desired future cash flow F at time N.

To conserve space and help in our analysis, Figure 3.1(c) illustrates a cash flow diagram with both P and F. In deriving F, this diagram defines an investment scenario in which P is invested in an account that earns periodic interest i for N periods. F is the maximum amount that can be removed from the account at time N. The name "compound amount factor" is derived from the fact that money compounds, or grows, as it moves forward in time, assuming that the interest rate is positive.

In short, we are seeking the answer to the following simple question:

What is F, given P?

Mathematically, F can be calculated in the same manner in which compound interest was analyzed in Chapter 2. If an amount P is invested at rate i, then, after one period,

$$F_1 = P(1+i)$$

is in the account. F_1 includes the principal amount P and the interest Pi earned during the first period.

If we leave F_1 in the account, it earns interest for another period, so that

$$F_2 = F_1(1+i) = P(1+i)^2$$

is available in the account after two periods. Figure 3.2 illustrates interest compounding on the principal amount for N periods, where i acts as a multiplier.

As shown in Chapter 2, the preceding equation for F generalizes such that if P earns interest rate i for N periods, then

$$F = P(1+i)^N. \tag{3.1}$$

The value of $(1+i)^N$ in Equation (3.1) is referred to as the *single-payment compound amount factor*. We often write it in shorthand as

$$\boxed{F = P(1+i)^N = P(F/P, i, N).}$$

Figure 3.2
Periodic compounding of a single payment.

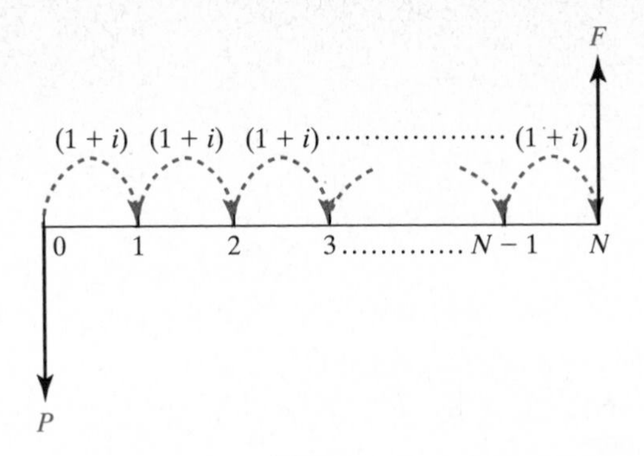

The term $(^{F/P,i,N})$ is also referred to as the *compound amount factor* for a single payment. The shorthand notation provides a way to write down our thoughts quickly and correctly. Treat the F/P notation, stated "F given P," as a common fraction. If we want to find F when P is given, then we multiply P by F/P, leaving us with F, as desired. This is illustrated in the next example.

EXAMPLE 3.1

Compound Amount Factor: Single Payment

Sea Satin Corp. of Greece placed an order with South Korea's Daewoo Shipbuilding and Marine Engineering for one 145,700-cubic-meter liquefied natural gas (LNG) carrier to be delivered by December 31, 2005, for KRW177.2 billion.[3] If the purchase price was paid at the time of the order (assume June 1, 2003), what is the equivalent future value of the cost at the time of delivery, assuming a nominal interest rate of 20% compounded semiannually?

Solution. The cash flow diagram for this example is given in Figure 3.3. The KRW177.2 billion is shown as an outflow at time zero (time 2002.5), with the future value F unknown at time 2005.

Our cash flow diagram is drawn with semiannual periods; thus, we require a semiannual interest rate. From Chapter 2, we know that the semiannual rate for this example is merely r/M, or $20\%/2 = 10\%$.

From the cash flow diagram, we note that P is given and we must calculate F. This condition is quickly captured with our shorthand notation:

$$F = P(^{F/P,i,N}) = \text{KRW177.2B}(^{F/P,10\%,5}).$$

Intuition: $F > P$ if $i > 0$

Having written this equation, we see that all that remains is computing the compound amount factor and finding F. We can use Equation (3.1), defined earlier, as follows:

$$F = P(^{F/P,i,N}) = P(1+i)^N = \text{KRW177.2B}(1+0.10)^5 = \text{KRW 285.38 billion}.$$

[3] Chang, S., "Daewoo Shipbuilding Signs KRW177.2B Deal for LNG Vessel," *Dow Jones Newswires*, September 4, 2003.

Figure 3.3 Equivalent future value F of KRW177.2 billion purchase at time zero.

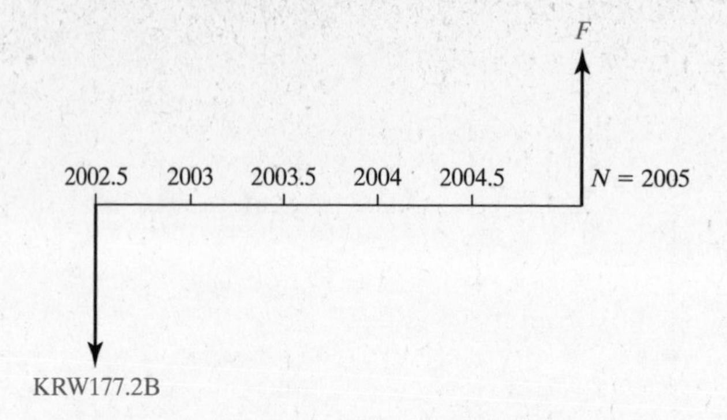

Alternatively, instead of using the formula, one may look up the factor value, using the tables for discrete compounding and discrete cash flows in the appendix. For this example, the 10% factors are given in Table A.16, with the compound amount factor for a single payment located in the first column. Tracing down this column to the $N = 5$ value results in

$$F = P(F/P,i,N) = \text{KRW}177.2\text{B}(\overset{F/P,10\%,5}{1.6105}) = \text{KRW } 285.38 \text{ billion.}$$

Note that the tables contain only four digits of accuracy beyond the decimal point, which can lead to round-off errors when comparing answers found with Equation (3.1). Generally, the error is insignificant for decision-making purposes.

Another solution method is to use a spreadsheet. As noted in Chapter 2, one can either program the formulas directly into the cells or use the built-in functions. To program a cell directly, we can input Equation (3.1) into cell E7 in Figure 3.4 as

$$= \text{B4} * (1 + \text{\$E\$3}) \wedge \text{\$E\$4},$$

making sure to keep the sign correct.

We could have arrived at an equivalent result by programming cell E7 with Excel's FV, or future value, function, defined as

=FV(rate,nper,pmt,pv,type).

	A	B	C	D	E	F
1	Example 3.1: LNG Carrier Payment			**Input**		
2				P	KRW 177.20	billion
3	**Period**	**Cash Flow**		Interest Rate	10%	per six months
4	0	KRW 177.20		Periods	5	(semi-annual)
5	1	--				
6	2	--		**Output**		
7	3	--		F	KRW 285.38	billion
8	4	--				
9	5	--				
10					=B4*(1+E3)^E4	

Figure 3.4 Spreadsheet with Equation (3.1) programmed in cell E7.

In our notation, 'rate' is i, 'nper' is N, and 'pv' is $-P$. The value 'pmt' will be addressed later, and 'type' refers to the timing of cash flows. A value of '1' assumes beginning-of-period flows while a '0', or leaving the argument blank, assumes end-of-period flows.

Thus, we could program cell E7 as

=FV(E3,E4,,–B4).

The additional commas in the function call are required because the input field 'pmt' is not utilized to determine F. Note that Excel returns the negative (−) of the input, so a positive value of F is returned for an input of $-P$. Similar functions exist in Lotus (FVAMOUNT) and Quattro Pro (FVAL). Note that a currency symbol is displayed in the spreadsheet through the 'Format' menu by selecting 'Cell.' If the 'Number' tab is selected, the category 'Currency' can be chosen—leading to a drop down menu of currencies with symbols.

3.1.2 Equal-Payment Series Analysis

Consider the cash flow diagram with N consecutive cash flows of equal size A shown in Figure 3.5(a). The *compound amount factor* in this case defines an equivalent future value F at period N of this cash flow series, as shown in Figure 3.5(b).

As before, we conserve space and combine these diagrams in Figure 3.5(c), which defines an investment scenario in which N consecutive payments of size A are placed into an account earning money at periodic interest rate i. The value F is the maximum amount that can be removed from the account at time N. Note that F is removed immediately after the final payment of A is made.

The question to answer is

What is F, given A?

To determine the value of F, we can simply apply the single-payment analysis from the previous section to each individual value of A and sum the results. This is depicted in Figure 3.6, where each individual cash flow A is compounded forward in time to period N.

Our analysis begins with the final payment at period N. The final payment A does not earn any interest, because all funds are removed immediately

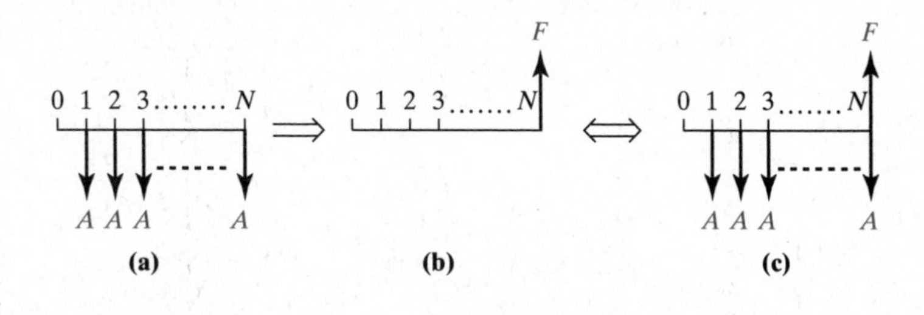

Figure 3.5
(a) Series of cash flows A from period 1 through period N. (b) Future value F at period N. (c) Cash flow diagram for compound amount factor for an equal payment series.

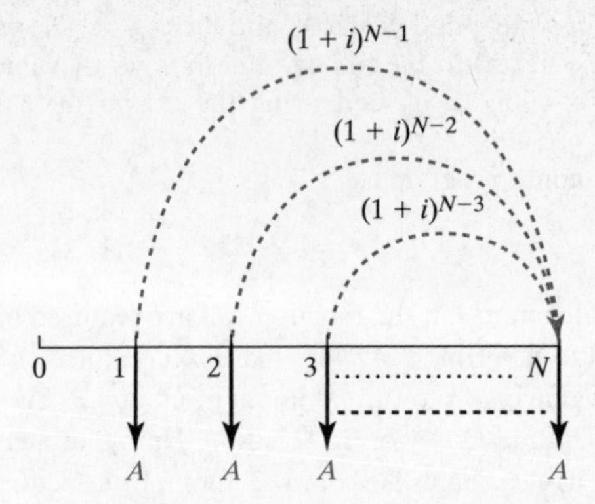

Figure 3.6 Compounding of individual A values to time N.

after it is deposited. The previous payment, at time $N - 1$, earns one period of interest. The payment of A at time $N - 2$ earns two periods of interest. Continuing this logic, the first payment at period 1 earns $N - 1$ periods of interest. Using our single-payment compound amount factors, we can define F by summing each of the individual payments, compounded to period N, as

$$F = \underbrace{A(F/P,i,0)}_{\text{Period } N} + \underbrace{A(F/P,i,1)}_{\text{Period } N-1} + \underbrace{A(F/P,i,2)}_{\text{Period } N-2} + \cdots + \underbrace{A(F/P,i,N-1)}_{\text{Period } 1}.$$

Substituting our single-payment compound amount factors, we have

$$\begin{aligned} F &= A(1+i)^0 + A(1+i) + A(1+i)^2 + \cdots + A(1+i)^{N-1} \\ &= A[1 + (1+i) + (1+i)^2 + \cdots + (1+i)^{N-1}]. \end{aligned}$$

The expression in the brackets is a geometric series[4] which we can write more compactly, so that

$$\begin{aligned} F &= A[1 + (1+i) + (1+i)^2 + \cdots + (1+i)^{N-1}] \\ &= A\left[\frac{1-(1+i)^N}{1-(1+i)}\right] = A\left[\frac{1-(1+i)^N}{-i}\right]. \end{aligned}$$

Thus,

$$F = A\left[\frac{(1+i)^N - 1}{i}\right]. \tag{3.2}$$

The value in the brackets in Equation (3.2) is the *equal-payment series compound amount factor*. As before, this factor can be written in shorthand notation as

$$\boxed{F = A\left[\frac{(1+i)^N - 1}{i}\right] = A(F/A,i,N).}$$

[4] A geometric series is defined as $a + ax + ax^2 + ax^3 + \cdots + ax^{N-1} = a\frac{1-x^N}{1-x}$.

Therefore, if we want to determine F, and A is given, we multiply A by the F/A factor, leaving us with our desired value of F. This is illustrated in the next example.

EXAMPLE 3.2

Compound Amount Factor: Equal-Payment Series

In the fall of 2003, Nortel Networks was selected by Verizon Wireless to supply networking equipment to upgrade voice and data networks in various U.S. cities, including Atlanta, Detroit, and Los Angeles. The network is to be deployed over $3\frac{1}{2}$ years at the cost of \$1 billion.[5] Assume that semiannual payments of equal size are to be made over a four-year period, starting at the beginning of 2004. What is the equivalent future value of this equal-payment series, assuming a semiannual interest rate of 4%?

Solution. The cash flow diagram for this example is given in Figure 3.7. Note that the beginning of year 2004 is also the end of year 2003 and that semiannual cash flows are drawn.

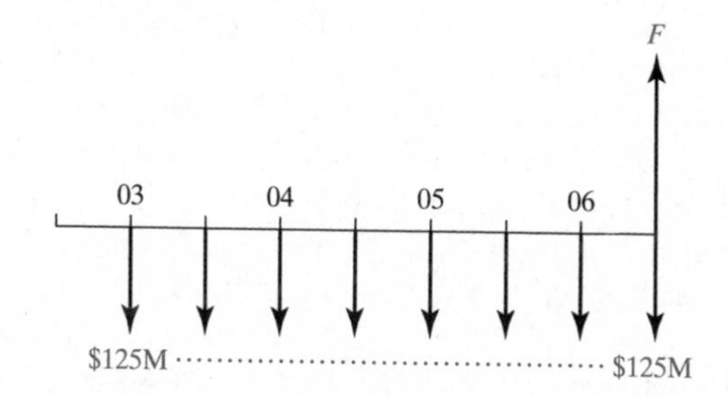

Figure 3.7 Equivalent future value F of \$1 billion equipment purchase spread over four years.

With A given and F unknown, we can describe F in terms of A with our shorthand notation:

$$F = A(^{F/A,i,N}) = \$125\text{M}(^{F/A,4\%,8}).$$

Intuition: $F > AN$ if $i > 0$

Substituting our expression, Equation (3.2), for the compound amount factor, we can determine the amount F as follows:

$$F = A(^{F/A,i,N}) = A\left[\frac{(1+i)^N - 1}{i}\right] = \$125\text{M}\left[\frac{(1+0.04)^8 - 1}{0.04}\right]$$

$$= \$1.152 \text{ billion.}$$

The compound amount factors for an equal-payment series can be found in the second column in the tables in the appendix. For this example, the 4% factors are given in Table A.10. Substituting the appropriate value results in

$$F = A(^{F/A,i,N}) = \$125\text{M}(\overset{F/A,4\%,8}{9.2142}) = \$1.152 \text{ billion.}$$

Note once more that there may be round-off error.

---

[5] "Verizon Wireless Awards Nortel \$1 Billion Contract," *Dow Jones Newswires*, September 3, 2003.

Again, alternatively, We can use the FV function in Excel to calculate F with a spreadsheet. From Example 3.1, recall our definition of FV as

=FV(rate,nper,pmt,pv,type).

To find F, given A, in our notation requires the function call

$$F = \text{FV}(i, N, -A) = \text{FV}(0.04, 8, -125) = 1{,}151.78,$$

in millions of U.S. dollars. Note that the other arguments (pv and type) can be ignored. Lotus and Quattro Pro have similar functions named FV.

3.1.3 Arithmetic Gradient Series Analysis

We now turn our focus to the third type of cash flow diagram, in which a series of cash flows grows or shrinks uniformly by an amount G with the passing of each period. The situation is shown in Figure 3.8, with (a) depicting the growth of the series. This uniform growth is defined as an arithmetic gradient, because we are adding an equal amount G at each period in the series. Note that there is no cash flow at time period 1 with an arithmetic gradient. An initial cash flow of G occurs in period 2, with subsequent cash flows increasing by the amount G through period N. We want to transform this arithmetic gradient series into a single cash flow F at time N, as shown in Figure 3.8(b). Both diagrams are consolidated into one in Figure 3.8(c). Although we have drawn G as a positive value, the results that follow are valid for any value of G.

Here, the question we are trying to answer is

What is F, given G?

Figure 3.8(c) describes an investment scenario in which payments defined by the arithmetic gradient series are placed into an account earning periodic

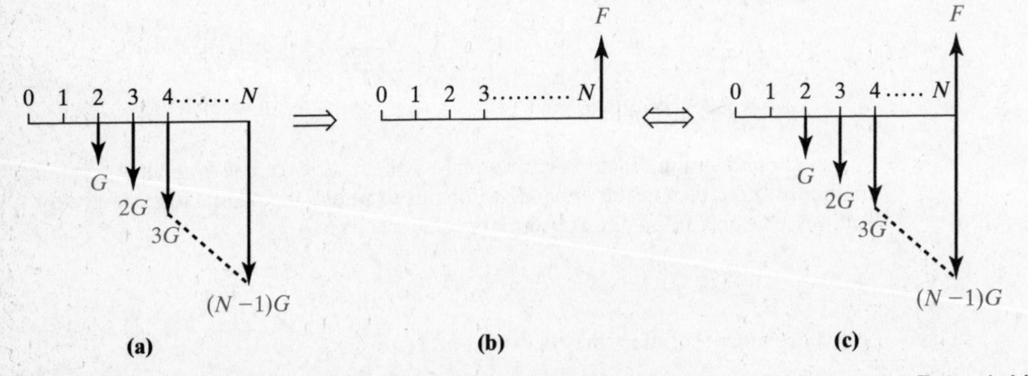

Figure 3.8 (a) Increasing series of cash flows G from period 1 through period N. (b) Future value F at period N. (c) Cash flow diagram for compound amount factor for an arithmetic gradient series.

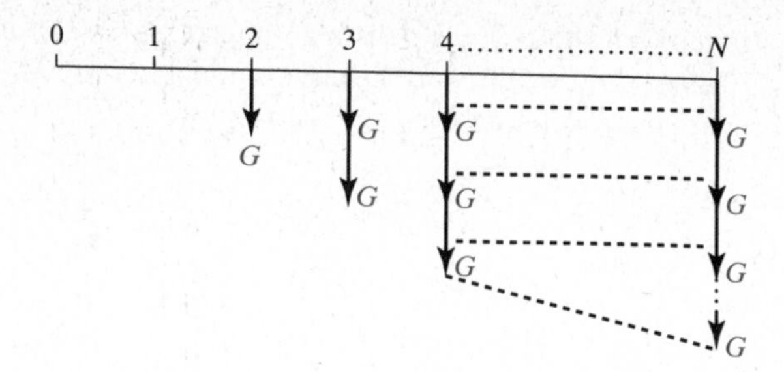

Figure 3.9 Arithmetic gradient series depicted as individual equal payment series of size G.

interest at rate i. The amount F is the maximum amount that can be removed from the account at time N, immediately after the final payment of size $(N-1)G$ is made. F is defined by the *compound amount factor* for an arithmetic gradient series.

In order to derive an expression for F, it helps to redraw the cash flow diagram as a series of equal-payment series of value G, as shown in Figure 3.9. Note that the first payment of G in period 2 repeats in each period through period N. Likewise, the second payment of G in period 3 repeats in each period through period N. These individual series are nothing more than equal payment series, as examined in the previous section. The differences are that the size of the payment is defined as G, not A, and each series runs for a different length of time, between 1 and $N-1$ periods long.

Consider the payment of G at time period 2, which repeats through period N. To bring these $N-1$ cash flows forward to a single payment at time N, we merely use the compound amount factor for the equal-payment series derived in the previous section. Another series of $N-2$ payments of size G begins at time period 3 and repeats through period N, and the same logic applies to each of the series in Figure 3.9. Adding together the cash flows, multiplied by their appropriate compound amount factors for each series, results in our future value F:

$$F = \underbrace{G(^{F/A,i,N-1})}_{\text{Series from period 2}} + \underbrace{G(^{F/A,i,N-2})}_{\text{Series from period 3}} + \cdots + \underbrace{G(^{F/A,i,1})}_{\text{Series from period } N}.$$

Substituting our equal-payment series compound amount factors, we have

$$\begin{aligned}
F &= G\left(\left[\frac{(1+i)^{N-1}-1}{i}\right] + \left[\frac{(1+i)^{N-2}-1}{i}\right] + \cdots + \left[\frac{(1+i)-1}{i}\right]\right) \\
&= \frac{G}{i}\left[(1+i)^{N-1} - 1 + (1+i)^{N-2} - 1 \ldots + (1+i) - 1\right] \\
&= \frac{G}{i}\left[(1+i) + (1+i)^2 + \cdots + (1+i)^{N-1} - (N-1)\right] \\
&= \frac{G}{i}\left[1 + (1+i) + (1+i)^2 + \cdots + (1+i)^{N-1}\right] - \frac{NG}{i}.
\end{aligned}$$

Again, the term in brackets is a geometric series. In fact, it is the same series we found when we derived the compound amount factor for an equal-payment series. Thus, our equation for F simplifies to

$$F = \frac{G}{i}[1 + (1+i) + (1+i)^2 + \cdots + (1+i)^{N-1}] - \frac{NG}{i}$$
$$= \frac{G}{i}\left[\frac{(1+i)^N - 1}{i}\right] - \frac{NG}{i}$$
$$= G\left[\frac{(1+i)^N - 1}{i^2} - \frac{N}{i}\right].$$

With one last simplification of merging the two terms in the brackets, the remaining bracketed term represents the *arithmetic gradient series compound amount factor*, and we obtain

$$F = G\left[\frac{(1+i)^N - Ni - 1}{i^2}\right]. \tag{3.3}$$

The shorthand notation follows as before:

$$\boxed{F = G\left[\frac{(1+i)^N - Ni - 1}{i^2}\right] = G(F/G, i, N).}$$

We illustrate the arithmetic series compound amount factor in Example 3.3.

EXAMPLE 3.3

Compound Amount Factor: Arithmetic Gradient Series

Hamilton Sundstrand Corp., a subsidiary of United Technologies, landed a \$1.3 billion contract to build power systems, which generate and distribute electricity, for up to 500 regional jets for AVIC Commercial Aircraft of Shanghai, China. Test flights for the new ARJ21 jet are set to begin in 2006.[6] Assume that revenues (\$2.6 million per plane) are received when delivered, with 50 expected deliveries in 2006, 100 in 2007, 150 in 2008, and 200 in 2009. What is the equivalent future value of these revenues, assuming an 8% annual interest rate?

Solution. Assuming that 2004 is time zero, the cash flow diagram for this example is given in Figure 3.10. Note that the first cash flow occurs in 2006.

With G given and F unknown, the equation that describes the situation, in our shorthand notation, is

$$F = G(F/G, i, N) = 50 \times \$2.6\text{M}(F/G, 8\%, 5).$$

[6] "Sundstrand Wins \$1.3B Contract with Chinese Aircraft Co.," *Dow Jones Newswires*, September 19, 2003.

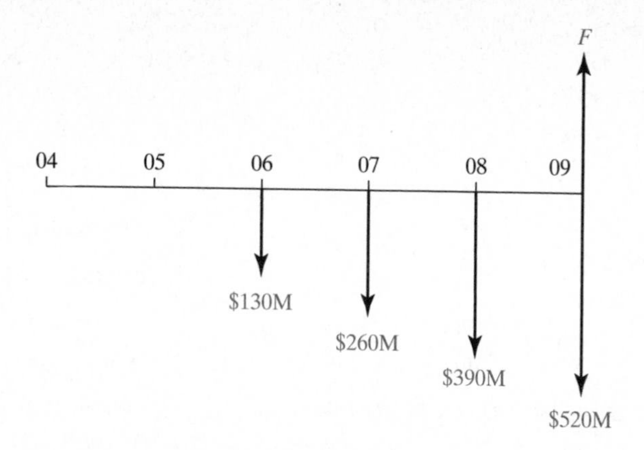

Figure 3.10 Equivalent future value F of increasing revenues of \$130 million over four years.

Intuition: $F > \frac{N(N-1)}{2}G$ if $i > 0$

Substituting our expression from Equation (3.3) for the compound amount factor, we obtain

$$F = G(^{F/G,i,N}) = G\left[\frac{(1+i)^N - Ni - 1}{i^2}\right] = \$130\text{M}\left[\frac{(1+0.08)^5 - (5)(0.08) - 1}{(0.08)^2}\right]$$
$$= \$1.408 \text{ billion.}$$

The compound amount factors for the 8% arithmetic gradient series can be found in column 3 in Table A.14 in the appendix. Substituting the appropriate value results in

$$F = G(^{F/G,i,N}) = \$130\text{M}(\overset{F/G,8\%,5}{10.8325}) = \$1.408 \text{ billion.}$$

There are no spreadsheet functions that calculate future values of arithmetic gradient series. However, we can program Equation (3.3) directly into a spreadsheet, as in cell E7 in Figure 3.11. The cash flow diagram is defined according to the value of G and the period, as shown in cell B5.

	A	B	C	D	E	F	G
1	Example 3.3: Power Systems Purchase			**Input**			
2				G	$130,000,000.00		
3	**Period**	**Cash Flow**		Interest Rate	8%	per year	
4	0	--		Periods	5	years	
5	1	$0.00					
6	2	$130,000,000.00		**Output**			
7	3	$260,000,000.00		F	$1,408,226,560.00		
8	4	$390,000,000.00					
9	5	$520,000,000.00	=$E$2*(A5-1)				
10							
11					=E2*((1+E3)^E4-(E4*E3)-1)/(E3^2)		

Figure 3.11 Spreadsheet solution for finding F, given G.

Note that the arithmetic gradient series analysis assumes no cash flow at time period 1. If a cash flow diagram has a value for period 1 and the ensuing periods follow that of an arithmetic gradient series, we may use the

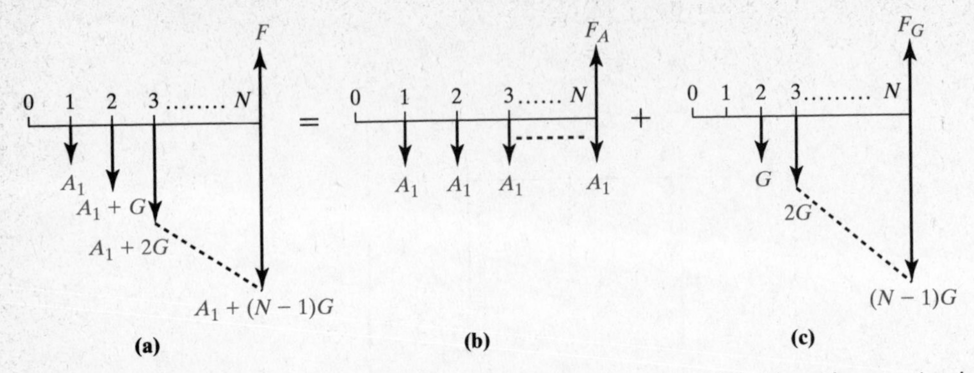

Figure 3.12 (a) Equal-payment and arithmetic gradient series decomposed into (b) equal-payment series and (c) arithmetic gradient series cash flows.

previous analyses to find the desired value of F. Consider, for example, the cash flow diagram in Figure 3.12. Part (a) of the figure shows a cash flow of A_1 at period 1. The cash flow of A_1 increases by an amount G in each period through period N. To analyze the diagram with our formulas, we can separate the A_1-value flows from the gradient cash flows, as shown in Figure 3.12(b) and (c), respectively. Noting that the value of A_1 is merely an equal-payment series, we can now analyze the two parts separately and sum the result. We do this in the next two examples.

EXAMPLE 3.4

Increasing Arithmetic Gradient with Equal Payment Series

Steel Dynamics spent \$75 million to expand its bar mill in Pittsboro, Indiana, in late 2003. The investment included a rolling mill, straightening and stacking equipment, a warehouse, and scrap-handling facilities. The investment was expected to increase annual production capacity at the facility to nearly 600,000 tons of bar products.[7] Assume that the end of 2002 is time zero and production capacity is 400,000 tons for 2003. Assume further that capacity increases 50,000 tons each year thereafter for four years. If a ton of steel bar generates \$250 in revenue, what is the equivalent future value (at the end of 2007) of the *total* revenue from the plant, assuming an annual interest rate of 14%? Assume steel bar is produced at full capacity and all production is sold.

Solution. The revenues from Steel Dynamic's production are depicted in Figure 3.13(a). As noted earlier, the cash flow diagram contains two parts. The first is the equal-payment series, broken out in Figure 3.13(b), and the second is the arithmetic gradient series, shown in part (c).

We first focus our attention on the arithmetic gradient series, or part (c) of Figure 3.13. With G given and F_G unknown (we will use F_G to designate the gradient

[7] Henglein, G., "Steel Dynamics Gets OK to Expand Indiana Mill," *Dow Jones Newswires,* September 2, 2003.

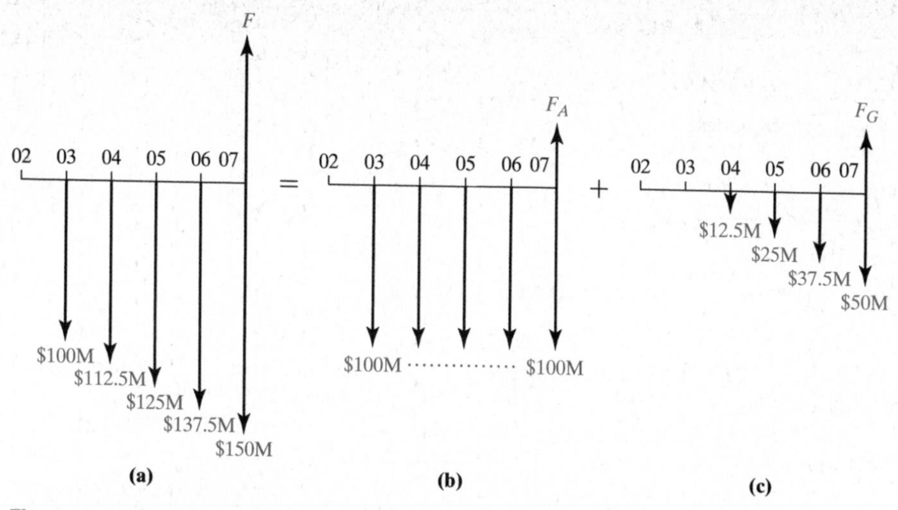

Figure 3.13 Equivalent future value F of (a) equal-payment and arithmetic gradient series decomposed into (b) equal-payment ($100M) series and (c) arithmetic gradient ($12.5M) series cash flows.

portion of our desired value of F), we can write the necessary equation in our shorthand notation:

$$F_G = G(^{F/G,i,N}) = \frac{\$250}{\text{ton}} \times \frac{50{,}000 \text{ tons}}{\text{year}}(^{F/G,14\%,5}).$$

Figure 3.13(b) is an equal-payment series with A given and F_A unknown. We again can write, in our shorthand notation,

$$F_A = A(^{F/A,i,N}) = \frac{\$250}{\text{ton}} \times \frac{400{,}000 \text{ tons}}{\text{year}}(^{F/A,14\%,5}).$$

We can now write our desired solution for F as

$$F = F_G + F_A = \$12.5\text{M}(^{F/G,14\%,5}) + \$100\text{M}(^{F/A,14\%,5}).$$

Substituting the arithmetic gradient series and equal-payment series compound amount factors leaves us with

$$\begin{aligned} F = F_G + F_A &= \$12.5\text{M}(\overset{F/G,14\%,5}{11.5007}) + \$100\text{M}(\overset{F/A,14\%,5}{6.6101}) \\ &= \$143.76\text{M} + \$661.01\text{M} = \$804.77 \text{ million}. \end{aligned}$$

Thus, the five years of revenues are equivalent to $804.77 million at the end of year 2007.

Figure 3.14 presents an alternative approach to finding the value of F with a spreadsheet. The cash flows A_n are defined by the sum of A and $(n-1)G$, as given in column B. In column C, the future value (period 5) of each individual cash flow is defined, using the compound amount factor for a single payment. This leads to the definition of F as the sum of the individual future values in cells C5 through C9. The SUM function in Excel is used to sum the values. Similarly to the MIN function introduced in Chapter 2, the SUM function is defined as

SUM(number1,number2, ...),

	A	B	C	D	E	F	G
1	Example 3.4: Steel Mill Expansion				**Input**		
2					A	$100,000,000.00	
3	**Period**	**Cash Flow**	**Future Value**		G	$12,500,000.00	
4	0	--	--		Interest Rate	14%	per year
5	1	$100,000,000.00	$168,896,016.00		Periods	5	years
6	2	$112,500,000.00	$166,673,700.00				
7	3	$125,000,000.00	$162,450,000.00		**Output**		
8	4	$137,500,000.00	$156,750,000.00		F	$804,769,716.00	
9	5	$150,000,000.00	$150,000,000.00				
10		=$F$2+$F$3*(A8-1)		=B5*(1+$F$4)^($F$5-A5)		=SUM(C5:C9)	
11							

Figure 3.14 Spreadsheet solution using individual future values to find F, given G and A.

where the arguments (number1, number2, etc.) can be numbers or references to cells with numerical values. When selecting a number of consecutive cells as inputs, a colon ':' can be used to define the range. This is illustrated in cell F8 in Figure 3.14.

Earlier, we noted that the arithmetic gradient series compound amount factor can handle positive or negative changes in the cash flows. In the next example, we examine a decreasing arithmetic gradient coupled with an equal-payment series.

EXAMPLE 3.5 *Decreasing Arithmetic Gradient with Equal-Payment Series*

In the fall of 2003, Intel announced that it would build a semiconductor assembly and testing facility in Chengdu, China. The first phase of the project should open in 2006 at the cost of $200 million. Intel expects to invest an additional $175 million in the facility over time.[8] Assume that the end of 2005 is time zero and the phase-one cost of $200 million is paid in 2006. Assume further that the additional investment of $175 million is paid in two payments of $125 million and $50 million in 2007 and 2008, respectively. Assuming an annual interest rate of 15%, what is the equivalent future value (in 2008) of these payments?

Solution. The cash flow diagram depicted in Figure 3.15(a) contains two parts. The equal-payment series is broken out in part (b) and the arithmetic gradient in part (c).

We begin our analysis with the equal-payment series depicted in Figure 3.15(b), with A given as \$200 million and F_A unknown. We write, in our shorthand notation,

$$F_A = A(^{F/A,i,N}) = \$200\text{M}(^{F/A,15\%,3}).$$

In Figure 3.15(c), G is given as −\$75 million and F_G is unknown, so that

$$F_G = G(^{F/G,i,N}) = -\$75\text{M}(^{F/G,15\%,3}).$$

[8] Dean, J., "Intel to Build Plant in Central China," *The Wall Street Journal Online*, August 27, 2003.

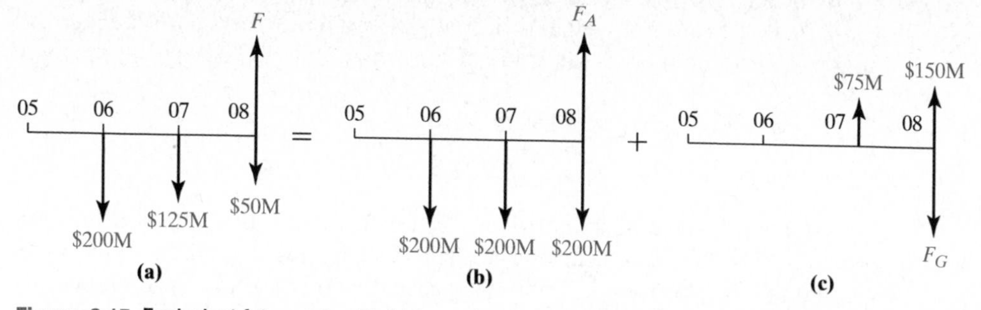

Figure 3.15 Equivalent future value F of (a) equal-payment and arithmetic gradient series decomposed into (b) equal-payment ($200m) series and (c) arithmetic gradient (−$75m) series cash flows.

We can now write our desired solution for F:

$$F = F_A + F_G = \$200\text{M}(^{F/A,15\%,3}) - \$75\text{M}(^{F/G,15\%,3})$$

$$= \$200\text{M}(\overset{F/A,15\%,3}{3.4725}) - \$75\text{M}(\overset{F/G,15\%,3}{3.1500})$$

$$= \$694.50\text{M} - \$236.25\text{M} = \$458.25 \text{ million}.$$

Thus, Intel's investment is worth $458.25 million at the end of year 2008. The spreadsheet approach to this problem follows that of Example 3.4.

3.1.4 Geometric Gradient Series Analysis

Our fourth series depicts cash flows growing or shrinking from an initial value of A_1 at time 1 according to some defined rate g, as depicted in Figure 3.16(a). The *compound amount factor* for a geometric gradient series defines F, which is depicted in Figure 3.16(b). Both cash flow diagrams are combined in Figure 3.16(c), which can again be viewed as an investment scenario. The first

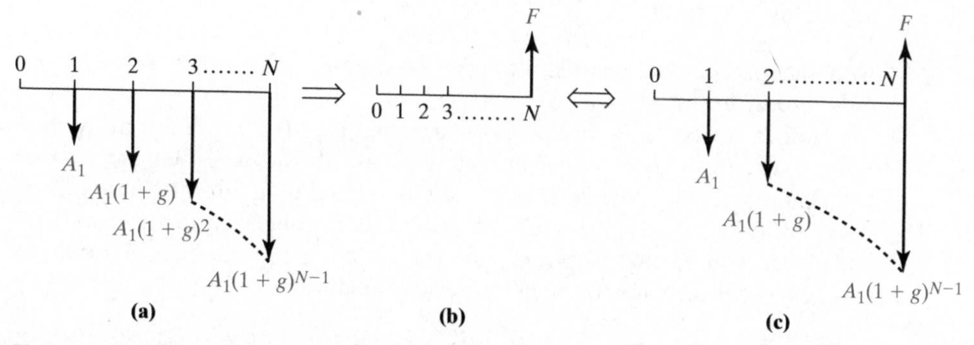

Figure 3.16 (a) Geometric series of cash flows growing from A_1 in period 1 to $A_1(1+g)^{N-1}$ in period N. (b) Future value F at time N. (c) Cash flow diagram for compound amount factor for a geometric gradient series.

payment into the account is of size A_1, and each subsequent payment grows (or shrinks) at rate g through period N. The account earns periodic interest at rate i. The amount F is the value of the account immediately after the final payment.

The question being asked is

What is F, given A_1 and g?

As before, we determine a conversion factor for this series, using our previous knowledge. Specifically, we can use the single-payment compound amount factor to move each individual cash flow forward in time, obtaining

$$F = \underbrace{A_1(^{F/P,i,N-1})}_{\text{Period 1}} + \underbrace{A_1(1+g)(^{F/P,i,N-2})}_{\text{Period 2}} + \underbrace{A_1(1+g)^2(^{F/P,i,N-3})}_{\text{Period 3}} + \cdots$$
$$+ \underbrace{A_1(1+g)^{N-1}(^{F/P,i,0})}_{\text{Period } N}.$$

Substituting our single-payment compound amount factor, we have

$$F = A_1[(1+i)^{N-1} + (1+g)(1+i)^{N-2} + (1+g)^2(1+i)^{N-3} + \cdots$$
$$+ (1+g)^{N-1}(1+i)^0]$$
$$= A_1(1+i)^{N-1}\left[1 + \frac{(1+g)}{(1+i)} + \frac{(1+g)^2}{(1+i)^2} + \cdots + \frac{(1+g)^{N-1}}{(1+i)^{N-1}}\right].$$

Again, the term in brackets represents a geometric series, so that

$$F = A_1(1+i)^{N-1}\left[\frac{1-\left(\frac{1+g}{1+i}\right)^N}{1-\left(\frac{1+g}{1+i}\right)}\right] = A_1(1+i)^{N-1}\left[\frac{\frac{(1+i)^N-(1+g)^N}{(1+i)^N}}{\frac{(1+i)-(1+g)}{(1+i)}}\right].$$

"Cleaning up" terms leaves us with

$$F = A_1\left[\frac{(1+i)^N - (1+g)^N}{i-g}\right],$$

with the bracketed term defining our *geometric gradient series compound amount factor*.

Before we continue, we must realize that i and g are different interest rates. The rate g represents how fast the cash flows are either growing or shrinking over time, while i is the interest rate used in compounding or discounting those cash flows. Clearly, we have a problem in the case where i and g are equal, as our compound amount factor is then undefined. To analyze this case, we refer back to the initial series equation:

$$F = A_1(1+i)^{N-1}\left[1 + \frac{(1+g)}{(1+i)} + \frac{(1+g)^2}{(1+i)^2} + \cdots + \frac{(1+g)^{N-1}}{(1+i)^{N-1}}\right].$$

With $i = g$, each fraction in the series reduces to 1 and the sum of the terms in the brackets is N, so that

$$F = A_1 N(1+i)^{N-1}.$$

Thus, for the *geometric gradient series compound amount factor*, we define the following two cases:

$$F = \begin{cases} A_1\left[\frac{(1+i)^N-(1+g)^N}{i-g}\right] & \text{if } i \neq g \\ A_1 N(1+i)^{N-1} & \text{if } i = g \end{cases}. \tag{3.4}$$

Our shorthand notation is a bit longer for the geometric series, since there are two unknowns, A_1 and g:

$$F = \left\{ \begin{array}{ll} A_1\left[\frac{(1+i)^N-(1+g)^N}{i-g}\right] & i \neq g, \\ A_1 N(1+i)^{N-1} & i = g. \end{array} \right\} = A_1(^{F/A_1,g,i,N}).$$

We illustrate the use of the geometric series compound amount factor in the next two examples.

EXAMPLE 3.6

Compound Amount Factor: Geometric Gradient Series

Haldex AB of Sweden received a five-year contract to supply air disc brakes for medium-sized trucks built by Volvo and Renault. The deal is valued at more than SEK200 million, with deliveries starting in 2006.[9] Assume that the first year's worth of deliveries are valued at SEK40 million and the amount grows at the rate of 3.5% per year. What is the future value (in 2010) of the order, assuming a 12% annual interest rate?

Intuition: $F > A_1 N$ if $i, g > 0$

Solution. The cash flow diagram for this example is given in Figure 3.17. The value of F at time 2010 is

$$F = A_1(^{F/A_1,g,i,N}) = \text{SEK40M}(^{F/A_1,3.5\%,12\%,5}).$$

Substituting our expression, Equation (3.4), for the compound amount factor, we determine the amount F as

$$F = A_1(^{F/A_1,g,i,N}) = A_1\left[\frac{(1+i)^N-(1+g)^N}{i-g}\right] = \text{SEK40M}\left[\frac{(1+0.12)^5-(1+0.035)^5}{0.12-0.035}\right]$$
$$= \text{SEK270.43 million.}$$

[9] "Haldex to Supply Disc Brakes for Volvo Medium Trucks," *Dow Jones Newswires*, November 3, 2003.

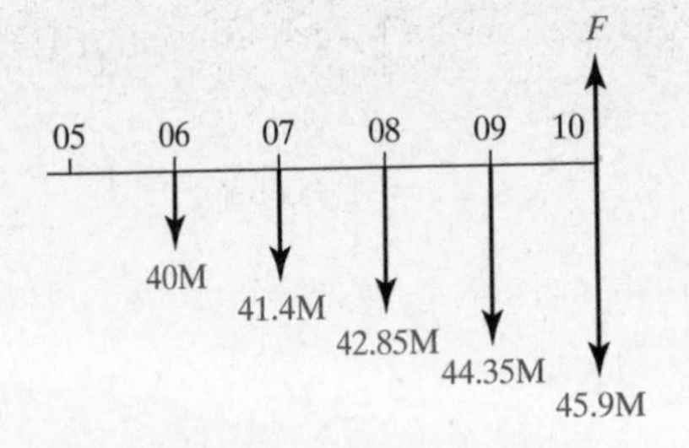

Figure 3.17 Equivalent future value F of revenues from air disc brake supply contract increasing at a rate of 3.5% per year from an initial value of SEK40 million.

	A	B	C	D	E	F	G
1	Example 3.6: Air Disc Brakes Sales				Input		
2					A1	SEK 40,000,000.00	
3	Period	Cash Flow	Future Value		g	3.5%	per year
4	0	--	--		Interest Rate	12%	per year
5	1	SEK 40,000,000.00	SEK 62,940,774.40		Periods	5	years
6	2	SEK 41,400,000.00	SEK 58,164,019.20				
7	3	SEK 42,849,000.00	SEK 53,749,785.60		Output		
8	4	SEK 44,348,715.00	SEK 49,670,560.80		F	SEK 270,426,060.03	
9	5	SEK 45,900,920.03	SEK 45,900,920.03				
10							
11		=$B$5*(1+$F$3)^(A6-1)	=B8*(1+$F$4)^($F$5-A8)			=SUM(C5:C9)	
12							

Figure 3.18 Spreadsheet solution to find F, given A_1 and g.

Note that if our interest rate i had been 3.5%, then we would have solved the following equation, because $i = g$:

$$F = A_1(F/A_1, g, i, N) = \text{SEK40M}(F/A_1, 3.5\%, 3.5\%, 10) = A_1 N(1+i)^{N-1}$$

$$= \text{SEK40M}(5)(1+0.035)^4 = \text{SEK229.50 million}.$$

As with the arithmetic gradient, there is no spreadsheet function that calculates the future value of a geometric gradient series. The spreadsheet in Figure 3.18 determines F from the future values (column C) of the individual cash flows (column B), as illustrated in Example 3.4.

We could have also programmed Equation (3.4) into cell F8, using the logical IF function in Excel, as the equation for F changes with the relationship between i and g. We illustrate this approach in Example 3.17.

EXAMPLE 3.7

Decreasing Geometric Gradient Series

At its peak, the Beatrice oil field in the North Sea off the shores of Scotland produced 50,000 barrels of oil per day for Talisman Energy. After 20 years, daily output has dropped to 5000 barrels.[10] Assuming that revenue is \$40 per barrel, 50,000 × 300 barrels are produced in year 1, and production is decreasing 11% per year for the next 20 years, what is the future value of the revenue stream, assuming a 12% annual rate of interest?

[10] Symon, K., "Wind Farm to Breathe New Life into Oil Field," *Sunday Herald,* Business Section, p. 5, August 31, 2002.

Solution. With production of 50,000 barrels per day over 300 days and revenue of $40 per barrel, the first-year revenues (A_1) total $600 million. Production declines annually such that $g = -11\%$. This defines the cash flow diagram given in Figure 3.19.

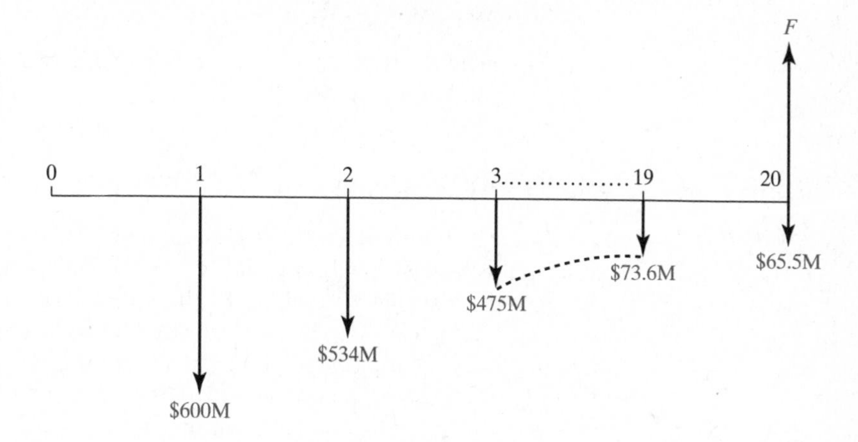

Figure 3.19 Equivalent future value F of revenues from oil output declining at a rate of 11% from initial production of 50,000 barrels daily.

To determine F at period 20, we must solve

$$F = A_1(^{F/A_1,g,i,N}) = \$600\text{M}(^{F/A_1,-11\%,12\%,10}).$$

Substituting our expression, Equation (3.4), for the compound amount factor, we determine the amount F:

$$F = A_1(^{F/A_1,g,i,N}) = A_1\left[\frac{(1+i)^N - (1+g)^N}{i-g}\right] = \$600\text{M}\left[\frac{(1+0.12)^{20} - (1-0.11)^{20}}{0.12+0.11}\right]$$

$$= \$24.91 \text{ billion}.$$

In the case of a decreasing geometric series, it would be quite rare that $i = g$, because that would require a negative interest rate. Although mathematically valid ($i > -1$), this is generally not assumed. The spreadsheet approach follows as previously illustrated.

3.2 Present-Worth Factors

We now turn our attention to moving money back through time, or discounting. Our goal is to transform a set of cash flows into a single cash flow at the present time, or time zero. As before, we say that the resulting cash flow is economically equivalent to the initial set of cash flows. Specifically, we want to determine the equivalent present value P at time zero, assuming a periodic interest rate i, for the following four cash flow diagrams:

1. A single cash flow F at time N.
2. N consecutive cash flows of equal size A.

3. N consecutive cash flows beginning with size 0 and increasing by G in each consecutive period.
4. N consecutive cash flows beginning with size A_1 and increasing at rate g in each consecutive period.

Mathematically, P is an unknown variable, while the other parameters, F, A, G, A_1, g, i, and N, are all known.

3.2.1 Single-Payment Analysis

As with our previous derivations, we begin with the simplest cash flow diagram, which in this case is a single cash flow F at period N, as shown in Figure 3.20(a). The *present-worth factor* defines P, the equivalent value of F at time zero, assuming a constant periodic interest rate i. Our desired cash flow is depicted in Figure 3.20(b), with both cash flows drawn simultaneously on (c). Note that this cash flow diagram is equivalent to that in Figure 3.1, except that F is now assumed to be known and P is unknown.

Figure 3.20(c) defines an investment scenario. The value P is the money required to invest at time zero in an account earning periodic interest rate i such that F can be withdrawn at time N, leaving no money in the account.

Since the value F is known, the following question describes the situation:

What is P, given F?

The term "present-worth factor" comes from the fact that we are describing the future payout F at the present time, or time zero. Interest discounts the future value F to time zero, acting as a divisor, as illustrated in Figure 3.21.

To determine P, we could derive the interest factor much in the same manner as we derived the compound amount factor for a single payment. But the derivation of this interest formula is not necessary, as it is the reciprocal

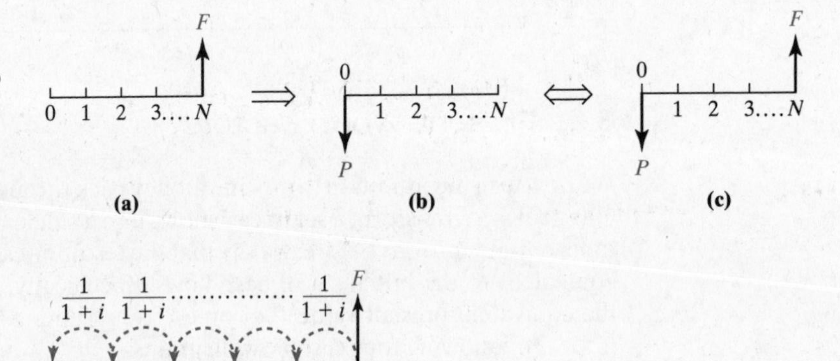

Figure 3.20 (a) Single cash flow F at time N. (b) Present value P at time zero. (c) Cash flow diagram for the present-worth factor for a single payment.

Figure 3.21 Periodic discounting of a single future payment.

of the compound amount factor for a single payment. Solving Equation (3.1) for P, we are left with

$$P = F\frac{1}{(1+i)^N}. \tag{3.5}$$

The value $\frac{1}{(1+i)^N}$ is defined as the *single-payment present-worth factor*. The shorthand notation follows as before:

$$\boxed{P = F\frac{1}{(1+i)^N} = F(^{P/F,i,N}).}$$

The term $(^{P/F,i,N})$ is also referred to as a *present-worth factor* for a single payment. Note that multiplying the future value F by the factor P/F leaves us with the desired value of P. We illustrate the use of the factor in the next example.

EXAMPLE 3.8

Present-Worth Factor: Single Payment

Siemens was chosen to provide 56 three-car diesel trains for the Trans Pennine Express rail franchise in the United Kingdom. The trains, which can travel at speeds of 100 miles per hour, are slated to begin running in 2006. The cost of the trains is EUR360 million.[11] What is the equivalent present value (at the beginning of 2003) of the purchase price if it is paid when the trains begin running (at the beginning of 2006), assuming an 18% annual interest rate?

Solution. The cash flow diagram (end of year flows) is given in Figure 3.22. We want to determine the value of P, with F given as EUR360 million.

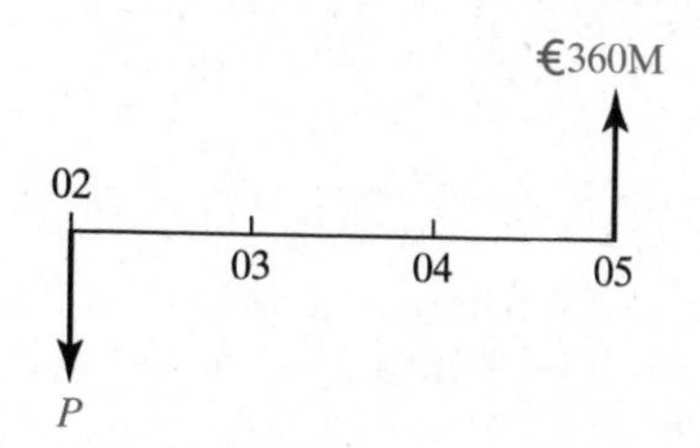

Figure 3.22
Equivalent present value *P* of EUR360 million equipment purchase.

Intuition: $P < F$ if $i > 0$

Again, we use the shorthand notation to compose our thoughts:

$$P = F(^{P/F,i,N}) = \text{EUR360M}(^{P/F,18\%,3}).$$

Substituting the present-worth factor in Equation (3.5), we determine the amount P as follows:

$$P = F(^{P/F,i,N}) = F\frac{1}{(1+i)^N} = \text{EUR360M}\frac{1}{(1+0.18)^3} = \text{EUR219.11 million}.$$

[11] Manzaroli, T., "Siemens Picked as Preferred Bidder for UK Train Order," *Dow Jones Newswires*, August 21, 2003.

The present-worth factors are found in column four of the interest factor tables in the appendix. Table A.24 contains the 18% factors. Tracing down that column to the $N = 3$ value results in

$$P = F(^{P/F,i,N}) = \text{EUR360M}(\overset{P/F,18\%,3}{0.6086}) = \text{EUR219.11 million.}$$

Equation 3.5 can also be programmed directly into a spreadsheet, as in cell E7 in Figure 3.23. Alternatively, the PV function in Excel can be used. The function is defined similarly to the FV function, as

=PV(rate,nper,pmt,fv,type),

Figure 3.23 Spreadsheet with Equation (3.5) programmed.

	A	B	C	D	E	F
1	Example 3.8: Diesel Train Purchase			**Input**		
2				F	€ 360.00	million
3	**Period**	**Cash Flow**		Interest Rate	18%	per year
4	2002	--		Periods	3	years
5	2003	--				
6	2004	--		**Output**		
7	2005	€ 360.00		P	€ 219.11	million
8						
9					=B7/(1+E3)^E4	
10						

differing only by the 'fv' (as opposed to 'pv') argument. In our shorthand notation,

$$P = \text{PV}(i, N, , -F) = \text{PV}(0.18, 3, , -360) = 219.11,$$

in millions of euros. Similar functions exist for Lotus (PVAMOUNT) and Quattro Pro (PVAL).

3.2.2 Equal-Payment Series Analysis

We now consider the equal-payment series described by N consecutive cash flows of size A, as defined in Figure 3.24(a). The *present-worth factor* for this

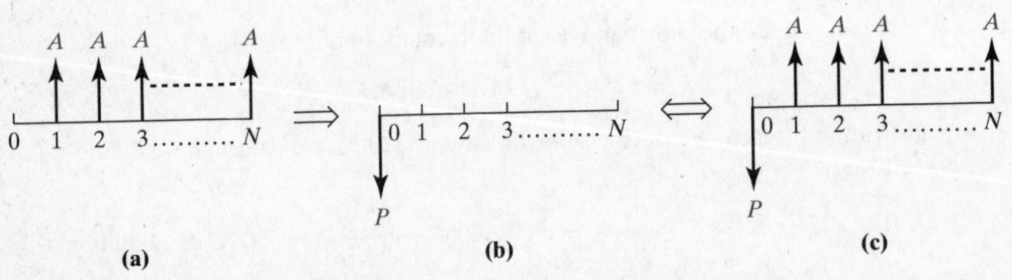

Figure 3.24 (a) Series of cash flows A from period 1 through period N. (b) Present value P at time zero. (c) Cash flow diagram for present-worth factor for an equal-payment series.

series defines the equivalent value P at time zero, as given in Figure 3.24(b), with both cash flows drawn simultaneously in part (c).

Figure 3.24(c) can also be viewed as an investment scenario wherein an amount P is placed into an account earning periodic interest at rate i at time zero such that N consecutive withdrawals of size A can be taken in each successive period, leaving no money in the account after the final withdrawal at time N. An answer to the following question is desired:

What is P, given A?

Examining Figure 3.24, we can use the knowledge gained with the single-payment present-worth factor to compute P. As illustrated in Figure 3.25, we can bring each cash flow A back to time zero by utilizing the present-worth factor that we just derived.

Using this logic, each A value is an individual cash flow in the future. Discounting each cash flow separately gives us

$$P = \underbrace{A(^{P/F,i,1})}_{\text{Period 1}} + \underbrace{A(^{P/F,i,2})}_{\text{Period 2}} + \underbrace{A(^{P/F,i,3})}_{\text{Period 3}} + \cdots + \underbrace{A(^{P/F,i,N})}_{\text{Period } N}.$$

Substituting our single-payment compound amount factors from Equation (3.5), we have

$$P = \frac{A}{(1+i)} + \frac{A}{(1+i)^2} + \frac{A}{(1+i)^3} + \cdots + \frac{A}{(1+i)^N}.$$

Before proceeding with this analysis, we could cleverly note that we have derived an expression for the compound amount factor for an equal-payment series, which we can easily convert to a present amount at time zero. Recall that Equation (3.2) defined the value of F, given A. If we convert the series of A values to a single future value F, we can find the present amount of the future value F, using the present-worth factor for a single payment:

$$P = A\,\underbrace{(^{F/A,i,N})}_{A \Rightarrow F}\,\underbrace{(^{P/F,i,N})}_{F \Rightarrow P}.$$

Mathematically, this can be written as

$$P = A\left[\frac{(1+i)^N - 1}{i}\right]\frac{1}{(1+i)^N}.$$

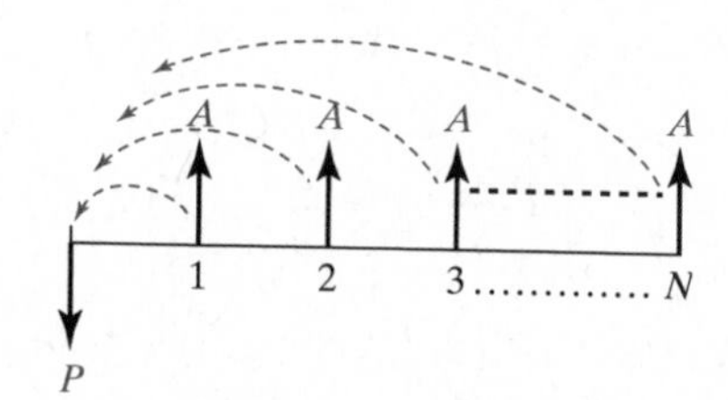

Figure 3.25 Discounting of individual A values to time zero.

Putting the expressions over a common denominator leaves us with

$$P = A\left[\frac{(1+i)^N - 1}{i(1+i)^N}\right]. \tag{3.6}$$

The term in the brackets is defined as the *equal-payment series present-worth factor*. Equivalently,

$$\boxed{P = A\left[\frac{(1+i)^N - 1}{i(1+i)^N}\right] = A(P/A,i,N).}$$

Again, the shorthand notation $(P/A,i,N)$ is also referred to as the *present-worth factor* for an equal-payment series. An example using this factor follows.

EXAMPLE 3.9

Present-Worth Factor: Equal Payment Series

In the summer of 2003, Samsung Electronics announced that it would construct a complex of plants outside of Seoul, Korea, to build flat-panel screens for computers and TVs at the cost of \$17 billion.[12] Assume that the investment is spread equally with annual payments over five years, beginning in January 2004. Given a 15% annual rate of interest, what is the equivalent present value (in January 2003) of this payment series?

Solution. The payments are described by the cash flow in Figure 3.26, with P our unknown variable. (Note that \$17/5 = \$3.4 billion.)

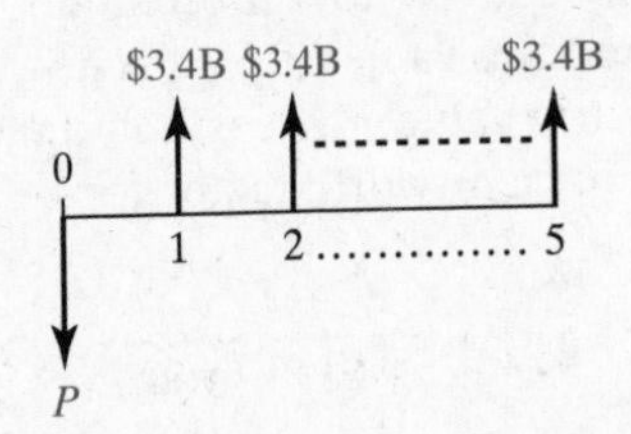

Figure 3.26 Equivalent present value P of \$17 billion plant investment spread over five years.

Intuition: $P < AN$ if $i > 0$

The shorthand notation defines our known and unknown variables:

$$P = A(P/A,i,N) = \$3.4\text{B}(P/A,15\%,5).$$

Using Equation (3.6), we substitute the present-worth factor and calculate P as follows:

$$P = A(P/A,i,N) = A\left[\frac{(1+i)^N - 1}{i(1+i)^N}\right] = \$3.4\text{B}\left[\frac{(1+0.15)^5 - 1}{0.15(1+0.15)^5}\right] = \$11.40 \text{ billion.}$$

---

12 McBride, S., "As the Demand for Flat Screens Surges, Investors Are Tuning In," *The Wall Street Journal Online*, August 11, 2003.

The present-worth factors for an equal payment series are found in column five of the interest factor tables in the appendix. The 15% factors are given in Table A.21. For $N = 5$,

$$P = A(^{P/A,i,N}) = \$3.4\text{B}(\overset{P/A,15\%,5}{3.3522}) = \$11.40 \text{ billion}.$$

As in Example 3.8, we can use the Excel PV function to obtain

$$P = \text{PV}(i, N, -A) = \text{PV}(0.15, 5, -3.4) = 11.397,$$

in billions of U.S. dollars. Similar PV functions exist for Lotus and Quattro Pro.

3.2.3 Arithmetic Gradient Series Analysis

The arithmetic gradient series, with cash flows increasing from size 0 at period 1 to $(N-1)G$ at time N, is given in Figure 3.27(a), with the *present-worth factor* defining the time-zero equivalent P in Figure 3.27(b). As before, the cash flow diagrams are given simultaneously in Figure 3.27(c).

With P the unknown variable, the question to be answered here is

What is P, given G?

The compound amount factor for the arithmetic gradient series is given in Equation (3.3). As with the equal-payment series, we can discount this future value to time zero by using the present-worth factor for a single payment, Equation (3.5), as

$$P = G\,\underbrace{(^{F/G,i,N})}_{G \Rightarrow F}\,\underbrace{(^{P/F,i,N})}_{F \Rightarrow P}.$$

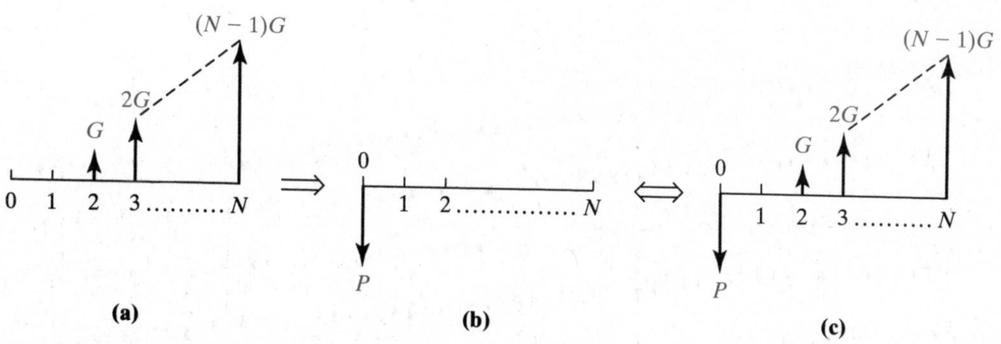

Figure 3.27 (a) Increasing series of cash flows G from period 1 through period N. (b) Present value P at time zero. (c) Cash flow diagram for present-worth factor for an arithmetic gradient series.

Mathematically, we can write this expression as

$$P = G\left[\frac{(1+i)^N - Ni - 1}{i^2}\right]\frac{1}{(1+i)^N}.$$

Putting the terms over a common denominator leaves us with

$$P = G\left[\frac{(1+i)^N - Ni - 1}{i^2(1+i)^N}\right]. \tag{3.7}$$

The term in the brackets is the *present-worth factor* for an arithmetic gradient series. The shorthand notation is

$$\boxed{P = G\left[\frac{(1+i)^N - Ni - 1}{i^2(1+i)^N}\right] = G(^{P/G,i,N}).}$$

The next example illustrates use of this present-worth factor.

EXAMPLE 3.10

Present-Worth Factor: Arithmetic Gradient Series

In the summer of 2003, Evergreen Group, the world's third-largest container-shipping company, based in Taiwan, announced that it would purchase 12 "S-type" vessels from Mitsubishi Heavy Industries in Japan. An "S-type" vessel can carry 6,724 TEUs (20-foot container equivalents) and costs roughly $55 million. The ships are to be delivered between 2005 and 2007.[13] Assume that time zero is 2003, two ships are delivered in 2005, four in 2006, and six in 2007. If payment is made at the time of delivery, what is the equivalent present value of Mitsubishi's revenue, assuming a 5% annual rate of interest?

Solution. The cash flow diagram for this example is given in Figure 3.28. Note that there is no cash flow in 2004.

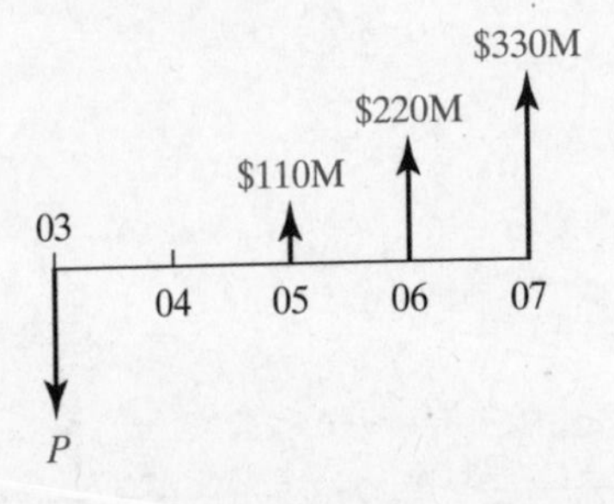

Figure 3.28 Equivalent present value of container vessel sales from 2005 through 2007.

The arithmetic gradient series for this cash flow diagram is defined by $G = 2 \times \$55$ million per year. With G given and P unknown, our solution, in shorthand notation, is

$$P = G(^{P/G,i,N}) = \$110\text{M}(^{P/G,5\%,4}).$$

[13] Dean, J., "Taiwan's Evergreen Plans a $3 Billion Fleet Expansion," *The Wall Street Journal Online,* September 8, 2003.

Intuition:
$P < \frac{N(N-1)}{2}G$ if $i > 0$

Substituting Equation (3.7) for the present-worth factor, we calculate P as follows:

$$P = G(^{P/G,i,N}) = G\left[\frac{(1+i)^N - Ni - 1}{i^2(1+i)^N}\right] = \$110\text{M}\left[\frac{(1+0.05)^4 - (4)(0.05) - 1}{(0.05)^2(1+0.05)^4}\right]$$

$$= \$561.31 \text{ million.}$$

The present-worth factors for the 5% arithmetic gradient series can be found in the sixth column of Table A.11 in the appendix. Substituting the appropriate value for $N = 4$ results in

$$P = G(^{P/G,i,N}) = \$110\text{M}(\overset{P/G,5\%,4}{5.1028}) = \$561.31 \text{ million.}$$

As noted earlier, there are no spreadsheet functions for arithmetic gradient series. However, the use of spreadsheets is similar to that for future value factors, as illustrated in the next example.

As shown in the compound amount factors analysis in Section 3.1.3, arithmetic gradient series are often accompanied by equal-payment series, as there is often a cash flow in period 1. As with the compound amount factor analysis, the cash flow diagram can be decomposed into an equal-payment series and an arithmetic gradient series. An example for this case follows.

EXAMPLE 3.11

Increasing Arithmetic Gradient with Equal-Payment Series

Ryanair, an airline operator based in Ireland, placed an order for 100 Boeing 737-800 jets in the spring of 2003. The list price for the entire order was \$6 billion.[14] Assume that list prices are paid and that payment is made when the planes are delivered, according to the following schedule: Ten planes are to be delivered in 2004, with an increase of 5 planes delivered each ensuing year. That is, 10, 15, 20, 25, and 30 planes are delivered in years 2004 through 2008, respectively. What is the equivalent present value (in 2003) of the revenue Boeing will generate through the sale, assuming a 20% annual rate of interest?

Solution. The cash flow diagram for Ryanair's payments is given in Figure 3.29(a). As we detailed with our compound amount factor analysis, the cash flow diagram contains two parts, with the equal-payment series broken out in part (b) and the arithmetic gradient in part (c).

We first focus our attention on the arithmetic gradient series, or part (c) of Figure 3.29. With G given and P_G unknown (we follow similar notation as before, using P_G for the present value of the gradient portion of our desired value of P), and noting the cost of one airplane is \$60 million, we can write the necessary equation in our shorthand notation:

$$P_G = G(^{P/G,i,N}) = 5 \times \$60\text{M}(^{P/G,20\%,5}).$$

---

[14] Lunsford, J.L., and D. Michaels, "Boeing to Get Order for 100 Planes from Ryanair," *The Wall Street Journal*, January 31, 2003.

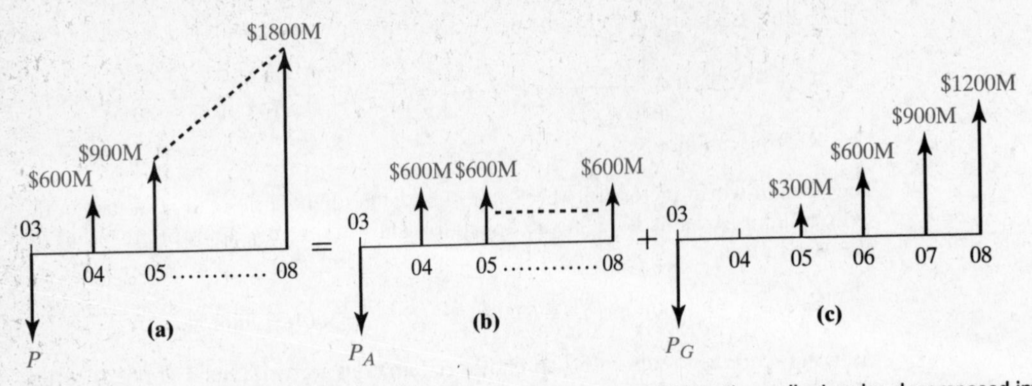

Figure 3.29 (a) Equivalent present value P of an equal-payment and arithmetic gradient series decomposed into (b) equal-payment ($600M) and (c) arithmetic gradient ($300M) series.

	A	B	C	D	E	F	G
1	Example 3.11: Airline Fleet Expansion				**Input**		
2					A	$600,000,000.00	
3	**Period**	**Cash Flow**	**Present Worth**		G	$300,000,000.00	
4	0	--	--		Interest Rate	20%	per year
5	1	$600,000,000.00	$500,000,000.00		Periods	5	years
6	2	$900,000,000.00	$625,000,000.00				
7	3	$1,200,000,000.00	$694,444,444.44		**Output**		
8	4	$1,500,000,000.00	$723,379,629.63		P	$3,266,203,703.70	
9	5	$1,800,000,000.00	$723,379,629.63				
10							
11	=$F$2+$F$3*(A6-1)		=B8/(1+$F$4)^(A8)			=SUM(C5:C9)	
12							

Figure 3.30 Present worth of annual and gradient series cash flows.

Figure 3.29(b) is an equal-payment series with A given and P_A unknown. We again summarize with our shorthand notation to obtain

$$P_A = A(^{P/A,i,N}) = 10 \times \$60\text{M}(^{P/A,20\%,5}).$$

The desired solution is

$$P = P_G + P_A = \$300\text{M}(^{P/G,20\%,5}) + \$600\text{M}(^{P/A,20\%,5}).$$

Substituting the arithmetic gradient series and equal-payment series present-worth factors leaves us with

$$P = P_G + P_A = \$300\text{M}(\overset{P/G,20\%,5}{4.9061}) + \$600\text{M}(\overset{P/A,20\%,5}{2.9906})$$

$$= \$1471.83\text{M} + \$1794.36\text{M} = \$3.27 \text{ billion}.$$

Thus, the time-zero value of the order is $3.27 billion.

The equivalent analysis using a spreadsheet is given in Figure 3.30. As with finding the future value, each individual cash flow is taken to the present in column C and summed in cell F8 to find P.

If the arithmetic gradient series is negative, we merely substitute a negative value in for G, resulting in a subtraction, not an addition, from the equal-payment series factor.

3.2.4 Geometric Gradient Series Analysis

Our final *present-worth factor* transforms a cash flow series beginning with flow A_1 at period 1 that grows (or shrinks) at rate g for N periods, as in Figure 3.31(a), to a single cash flow P, as shown in Figure 3.31(b), at time zero. Both cash flows are depicted in Figure 3.31(c).

We can consider the cash flow diagram in Figure 3.31(c) as an investment scenario in which we require a deposit P at time zero in order to make the given withdrawals and leave nothing in the account after the final withdrawal at time N. The account earns money at the periodic interest rate i. In short, we are answering the question

What is P, given A_1 and g?

Previously, we derived the compound amount factor for a geometric gradient series in Equation (3.4). Again, we can convert this to a present amount by multiplying by the present-worth factor for a single payment, using Equation (3.5):

$$P = A_1 \underbrace{(F/A_1,g,i,N)}_{A_1,\, g \Rightarrow F} \underbrace{(P/F,i,N)}_{F \Rightarrow P}.$$

For the case where $i \neq g$, this can be written as

$$P = A_1 \left[\frac{(1+i)^N - (1+g)^N}{i-g} \right] \frac{1}{(1+i)^N}.$$

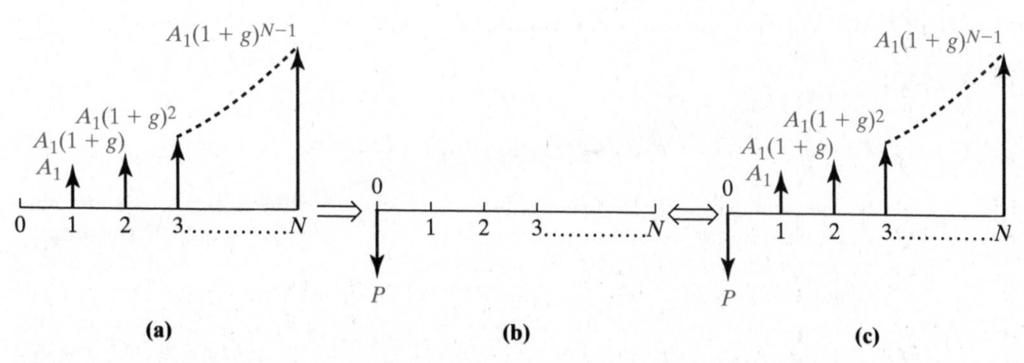

Figure 3.31 (a) Geometric series of cash flows growing from A_1 in period 1 to $A_1(1+g)^{N-1}$ at time period N. (b) Present value P at time zero. (c) Cash flow diagram for present-worth factor for a geometric series.

Aggregating the expressions leaves us with

$$P = A_1\left[\frac{(1+i)^N - (1+g)^N}{(i-g)(1+i)^N}\right]$$

$$= A_1\left[\frac{1 - \frac{(1+g)^N}{(1+i)^N}}{i-g}\right].$$

For the case where $i = g$, we proceed in a similar manner, so that

$$P = A_1(F/A_1,g,i,N)(P/F,i,N)$$

$$= A_1 N(1+i)^{N-1}\frac{1}{(1+i)^N}.$$

Multiplying out the expression leaves us with

$$P = A_1\left[\frac{N}{(1+i)}\right].$$

Thus, for the *geometric gradient series present-worth factor*, we define the following two cases:

$$P = \begin{cases} A_1\left[\frac{1-(1+g)^N(1+i)^{-N}}{i-g}\right] & \text{if } i \neq g \\ A_1\frac{N}{(1+i)} & \text{if } i = g \end{cases} \tag{3.8}$$

As before, our shorthand notation is a bit longer for the geometric series, since there are two unknowns, A_1 and g:

$$\boxed{P = \left\{\begin{array}{ll} A_1\left[\frac{1-(1+g)^N(1+i)^{-N}}{i-g}\right] & i \neq g, \\ A_1\frac{N}{(1+i)} & i = g. \end{array}\right\} = A_1(P/A_1,g,i,N).}$$

We illustrate the use of the geometric series present-worth factor in the next example.

EXAMPLE 3.12

Present-Worth Factor: Geometric Gradient Series

In the summer of 2003, General Electric (GE) Company's Power Systems unit announced that it would provide engineering support for as many as 22 gas turbines and 18 compressors for Nigeria LNG at a main natural gas liquefaction plant on Bonny Island in Rivers State, Nigeria. As of 2003, the GE unit had similar contracts in place at more than 600 sites throughout the world, generating almost \$23 billion in revenue in 2002.[15] Assume that the contracts generate \$23 billion in revenue in 2005 and that

---

15 Jordan, J., "GE to Supply Services for LNG Plants," *Dow Jones Newswires,* August 4, 2003.

revenue continues to increase 12% per year for the next eight years. If the annual interest rate is 19%, what is the present worth (in 2004) of these revenues?

Solution. Figure 3.32 describes the increasing geometric gradient series for this example. Using our shorthand notation, we define P as

$$P = A_1(^{P/A_1,g,i,N}) = \$23\text{B}(^{P/A_1,12\%,19\%,8}).$$

Because $i \neq g$, we calculate P with Equation (3.8) as follows:

$$P = A_1(^{P/A_1,g,i,N}) = A_1\left[\frac{1-(1+g)^N(1+i)^{-N}}{i-g}\right] = \$23\text{B}\left[\frac{1-(1+0.12)^8(1+0.19)^{-8}}{0.19-0.12}\right]$$

$$= \$126.27 \text{ billion.}$$

Intuition:
$P < A_1N$ if $i, g > 0$

If we had defined i to be 12%, then we would have calculated P as

$$P = A_1(^{P/A_1,g,i,N}) = A_1\frac{N}{(1+i)} = \$23\text{B}\frac{8}{(1+0.12)} = \$164.29 \text{ billion.}$$

This results in a much higher value of P, due to the lower interest rate.

The spreadsheet in Figure 3.33 illustrates the individual gradient series cash flows taken back to time zero in column C. These values are then summed in cell F8 to find P.

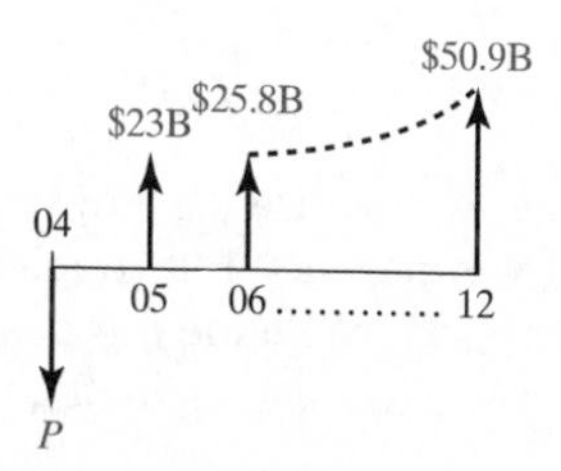

Figure 3.32 Equivalent present value P of increasing revenues from maintenance contracts.

	A	B	C	D	E	F	G
1	Example 3.12: Engineering Support Contracts Revenue				**Input**		
2					A1	$23,000,000,000.00	
3	**Period**	**Cash Flow**	**Present Worth**		g	12%	per year
4	0	--	--		Interest Rate	19%	per year
5	1	$23,000,000,000.00	$19,327,731,092.44		Periods	8	years
6	2	$25,760,000,000.00	$18,190,805,734.06				
7	3	$28,851,200,000.00	$17,120,758,337.94		**Output**		
8	4	$32,313,344,000.00	$16,113,654,906.29		P	$126,270,309,008.42	
9	5	$36,190,945,280.00	$15,165,792,852.98				
10	6	$40,533,858,713.60	$14,273,687,391.04				
11	7	$45,397,921,759.23	$13,434,058,720.98			=SUM(C5:C12)	
12	8	$50,845,672,370.34	$12,643,819,972.69	=B9/(1+$F$4)^(A9)			
13							
14	=$F$2+$F$3*(A10-1)						
15							

Figure 3.33 Present worth of geometric gradient series cash flows.

The mathematics for a negative gradient follows as before, with g taking on a negative value. As with the compound amount factor, it is rare that g and i will take on the same, negative value, as i is generally assumed to be nonnegative. The next section addresses our final equivalent-value analysis.

3.3 Equal-Payment Factors

Our final set of interest rate factors to be derived consists of the equal-payment factors. As opposed to discounting or compounding cash flows, we now look to spread them equally over time. Our goal is to transform any cash flow diagram into an equal-payment series of length N and size A. This new series is said to be equivalent to our original cash flow diagram. Specifically, we want to determine the value A over N periods, assuming periodic interest rate i, for the following four cash flow diagrams:

1. A single cash flow F at time N.
2. A single cash flow P at time zero.
3. N consecutive cash flows beginning with size 0 and increasing by G in each consecutive period.
4. N consecutive cash flows beginning with size A_1 and increasing at rate g in each consecutive period.

Mathematically, A is an unknown variable, while the other parameters, F, P, G, A_1, g, i, and N, are all known.

3.3.1 Single-Payment Analysis: Sinking-Fund Factor

The *sinking-fund factor* equates a future cash flow F at time N to an equal-payment series of size A over periods 1 through N, assuming a constant interest rate i. Figures 3.34(a) and (b) define our desired transformation, with part (c) placing both diagrams on the same axis. The question we are asking is

What is A, given F?

A related question is, What size of payment A must be invested into an account in order to remove all funds F at time N, assuming a constant periodic interest rate i. The term "sinking-fund factor" is used because one "sinks" money into an account to achieve level F.

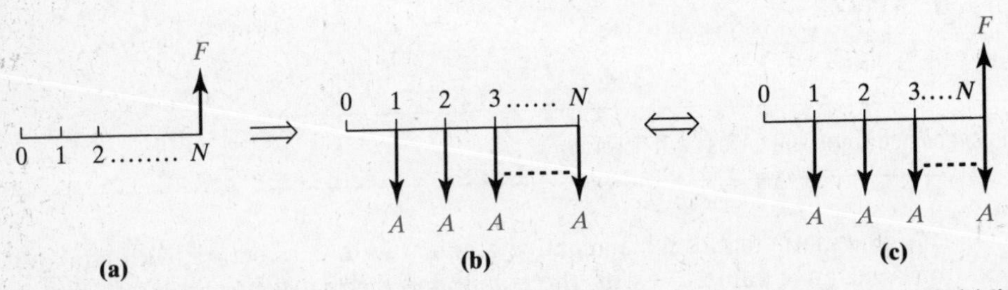

Figure 3.34 (a) Future value F at time N. (b) Equal-payment series of cash flows A from period 1 through period N. (c) Cash flow diagram for the equal-payment series sinking-fund factor.

We previously defined F in Equation (3.2) for the compound amount factor for an equal-payment series. Solving for A results in

$$A = F\left[\frac{i}{(1+i)^N - 1}\right]. \tag{3.9}$$

The bracketed term is our *sinking-fund factor* for a single payment. The shorthand notation yields

$$\boxed{A = F\left[\frac{i}{(1+i)^N - 1}\right] = F(^{A/F,i,N}).}$$

An example follows.

EXAMPLE 3.13

Equal-Payment Series Sinking-Fund Factor

Effiage Construction completed building the Millau Bridge, the tallest bridge in the world, at the end of 2004, in just 39 months. The bridge crosses a 1.5-mile gorge on the highway from Paris to Barcelona. A 780-foot cable-stayed pier is the focal point of the design, which features spans of up 1,150 feet between columns. Construction, which required the use of 64 hydraulic jacks to move prefabricated pieces into place, cost about \$525 million.[16] Assume that the \$525 million cost is a future value at the end of December 2004. If the monthly interest rate is 1%, what is the equivalent monthly payment series of the cost over the 39 months of construction of the bridge? (Assume that the final month is December 2004.)

Solution. The 39 equivalent payments for the construction cost are illustrated in Figure 3.35. Note that the payments are monthly.

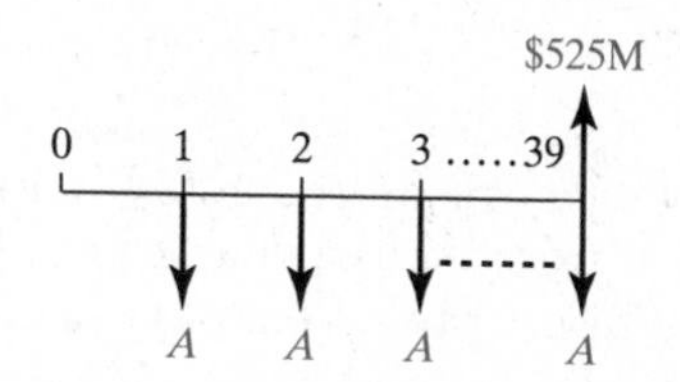

Figure 3.35 Equivalent monthly payments for cost of \$525 million bridge.

The shorthand notation defines A as

$$A = F(^{A/F,i,N}) = \$525\text{M}(^{A/F,1\%,39}).$$

Intuition: $A < \frac{F}{N}$ if $i > 0$

Using the sinking-fund factor in Equation (3.9), we calculate A as follows:

$$A = F(^{A/F,i,N}) = F\left[\frac{i}{(1+i)^N - 1}\right] = \$525\text{M}\left[\frac{0.01}{(1+0.01)^{39} - 1}\right] = \$11.07 \text{ million}.$$

[16] Stidger, R., "Private Financing Builds Millau Bridge," *Better Roads,* May 2005.

The sinking-fund factors are found in column seven of interest factor Table A.4 in the appendix. Substituting the factor for 1% and $N = 39$ gives

$$A = F(^{A/F,i,N}) = \$525\text{M}(\overset{A/F,1\%,39}{0.0211}) = \$11.08 \text{ million.}$$

Note the difference between using the formula and using the factor from the back of the textbook. Clearly, this is a minor difference.

The PMT function in Excel (or PAYMT in Lotus and Quattro Pro) can be used to find F, given A. The function call, similar to both PV and FV functions, is defined as

=PMT(rate,nper,pv,fv,type).

In our notation,

$$A = \text{PMT}(i, N, , -F) = \text{PMT}(0.01, 39, , -525) = 11.073,$$

in millions of U.S. dollars.

3.3.2 Single-Payment Analysis: Capital-Recovery Factor

The *capital-recovery factor* is similar to the sinking-fund factor in that it deals with the conversion of a single payment in time to an equal-payment series. Specifically, a single payment P at time zero (Figure 3.36(a)) is converted to an equal-payment series of size A (Figure 3.36(b)) over the ensuing N periods. Both diagrams are consolidated into one cash flow diagram in Figure 3.36(c).

The defining question is

What is A, given P?

This situation can be examined as an investment scenario in which P is deposited into an account paying interest at rate i, and in each ensuing period an amount A is removed from the account, leaving the account empty after the final period N. The term "capital-recovery factor" is used because N is

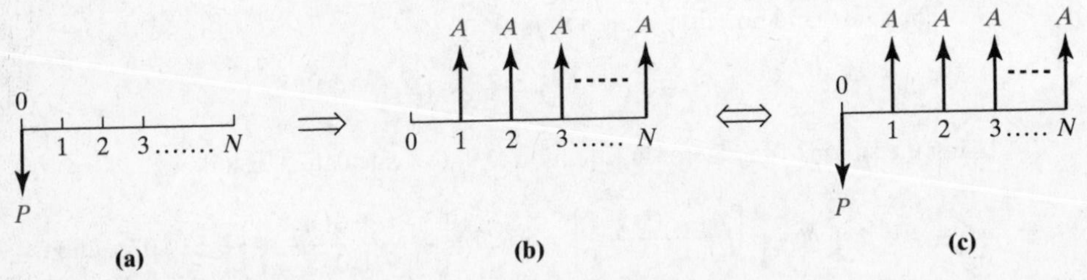

Figure 3.36 (a) Present value P at time zero. (b) Equal-payment series of cash flows A from period 1 through period N. (c) Cash flow diagram for the equal-payment series capital-recovery factor.

the number of periods that would be required in order to recover the initial investment P, often referred to as capital, with interest.

The definition of the *single-payment capital-recovery factor* is the reciprocal of the present-worth factor for a single payment. Solving Equation (3.6) for A results in

$$A = P\left[\frac{i(1+i)^N}{(1+i)^N - 1}\right]. \tag{3.10}$$

The value in the brackets is defined as the *capital-recovery factor* for a single payment. The shorthand notation follows as before:

$$\boxed{A = P\left[\frac{i(1+i)^N}{(1+i)^N - 1}\right] = P(A/P,i,N).}$$

The value of A is determined when the known value of P is multiplied by A/P. An example follows.

EXAMPLE 3.14

Equal-Payment Series Capital-Recovery Factor

Northrop Grumman and General Dynamics were awarded an \$8.4 billion contract to produce five submarines for the United States Navy. The contract, awarded at the beginning of 2004, will run through April 2014.[17] Assume that the \$8.4 billion is a time-zero (April 2004) amount and the annual interest rate is 6%. What is an equivalent equal annual payment series over the ensuing 10 years?

Solution. The cash flow diagram for our desired transformation is given in Figure 3.37.

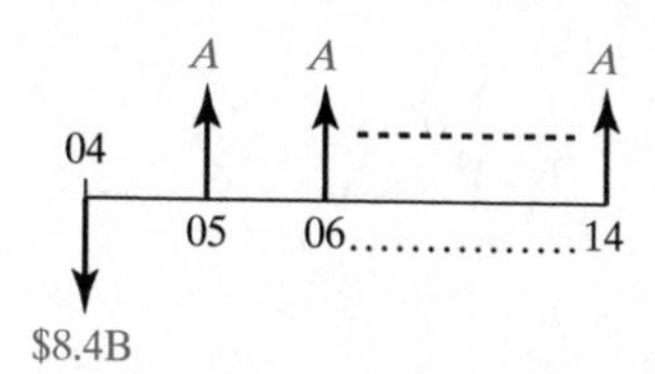

Figure 3.37 Equivalent equal-payment series over 10 years for \$8.4 billion contract.

Using our shorthand notation, we write

$$A = P(A/P,i,N) = \$8.4\text{B}(A/P,6\%,10).$$

Intuition: $A > \frac{P}{N}$ if $i > 0$

Using Equation (3.10), we compute the value of A as

$$A = P(A/P,i,N) = P\left[\frac{i(1+i)^N}{(1+i)^N - 1}\right] = \$8.4\text{B}\left[\frac{0.06(1+0.06)^{10}}{(1+0.06)^{10} - 1}\right] = \$1.141 \text{ billion}.$$

[17] Lee, K.M., "Northrop Grumman to Split \$8.4B Submarine Pact with General Dynamics," *Dow Jones Newswires*, January 20, 2004.

The appropriate factors are given in column eight in the appendix. For the 6% factors (Table A.12) and $N = 10$, the solution is

$$A = P\left(\overset{A/P,6\%,10}{\overset{}{A/P,i,N}}\right) = \$8.4\text{B}(0.1359) = \$1.142 \text{ billion.}$$

The spreadsheet function PMT in Excel can again be used in this case, to yield

$$A = \text{PMT}(i, N, -P) = \text{PMT}(0.06, 10, -8.4) = 1.1413,$$

in billions of U.S. dollars.

3.3.3 Arithmetic Gradient Series Analysis

Recall that the arithmetic gradient series begins with no cash flow in period 1 and grows by the amount G in each period to $(N - 1)G$ in period N, as shown in Figure 3.38(a). This series is equivalent to a series of N consecutive payments of size A, as described in Figure 3.38(b), using the *equivalent-payment series factor*.

Figure 3.38(c) describes our desired transformation in one diagram, defining an investment scenario (hypothetical) in which the arithmetic gradient payments are made into the account and the equal-payment series cash flows are removed. We are answering the question

What is A, given G?

We have derived expressions for both P and F for the arithmetic gradient series. Thus, we could easily convert either of these expressions to an equal-payment series. Using our arithmetic gradient series compound amount factor, Equation (3.3), we define A with our single-payment sinking-fund factor, Equation (3.9), as

$$A = G\underbrace{\left(F/G,i,N\right)}_{G\Rightarrow F}\underbrace{\left(A/F,i,N\right)}_{F\Rightarrow A}.$$

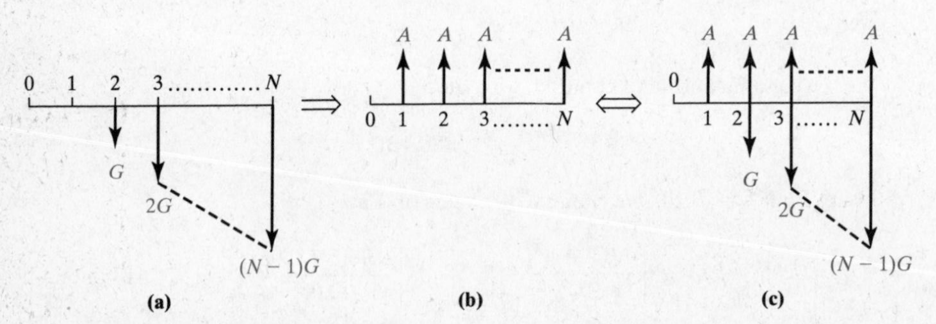

Figure 3.38 (a) Increasing series of cash flows G from period 1 through period N. (b) Equal-payment series of cash flows A from period 1 through period N. (c) Cash flow diagram for the equal-payment series factor for an arithmetic gradient series.

This equates to

$$A = G\left[\frac{(1+i)^N - Ni - 1}{i^2}\right]\left[\frac{i}{(1+i)^N - 1}\right].$$

Reducing this leaves

$$A = G\left[\frac{(1+i)^N - Ni - 1}{i\left((1+i)^N - 1\right)}\right]. \tag{3.11}$$

The *arithmetic gradient series equal-payment series factor* is defined in the brackets in Equation (3.11) and can be written in our usual shorthand notation as

$$\boxed{A = G\left[\frac{(1+i)^N - Ni - 1}{i\left((1+i)^N - 1\right)}\right] = G(^{A/G,i,N}).}$$

The term $(^{A/G,i,N})$ also represents the equal-payment series factor for an arithmetic gradient series. We illustrate with an example.

EXAMPLE 3.15

Arithmetic Gradient Series Equal-Payment Factor

Anglo American, a large Anglo–South African mining company, spent $454 million to open the Skorpion mine in Namibia in the summer of 2003. The mine and refinery complex is expected to produce 150,000 tons of refined zinc annually.[18] Assume that revenue is $750 per ton of refined zinc. Assume further that production builds with time, with zero production in the first half of 2003, growing by 50,000 tons each ensuing half-year until maximum capacity is reached. Given a semiannual interest rate of 7%, what is the equivalent (semiannual) equal-payment series of the expected revenues over the first two years?

Solution. The cash flow diagram for the arithmetic gradient series and desired equal-payment series is given in Figure 3.39. Note that the periods are semiannual.

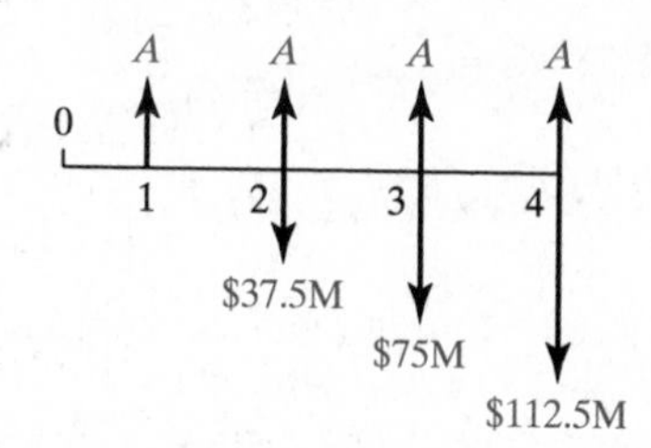

Figure 3.39 Equivalent equal semiannual cash flows for increasing revenues from zinc operations.

[18] Innocenti, N.D., "Anglo American Opens Low-Cost Mine," *Financial Times*, p. 29, September 15, 2003.

Examining Figure 3.39 and using our shorthand notation, we can quickly summarize our approach as

$$A = G(^{A/G,i,N}) = \frac{\$750}{\text{ton}} \times 50{,}000 \text{ tons } (^{A/G,7\%,4}).$$

Substituting our expression in Equation (3.11) yields

$$A = G(^{A/G,i,N}) = G\left[\frac{(1+i)^N - Ni - 1}{i\left((1+i)^N - 1\right)}\right] = \$37.5\text{M}\left[\frac{(1+0.07)^4 - 4(0.07) - 1}{0.07\left((1+0.07)^4 - 1\right)}\right]$$

$$= \$53.08 \text{ million.}$$

The relevant factors are found in the final column of the interest factor tables in the appendix. For the 7% interest rate (Table A.13) and $N = 4$,

$$A = G(^{A/G,i,N}) = \$37.5\text{M}(\overset{A/G,7\%,4}{1.4155}) = \$53.08 \text{ million.}$$

As noted earlier, there are no spreadsheet functions for arithmetic gradients. However, we can use the SUM function to aggregate the present worth of each individual cash flow and then convert the resulting sum to an annual equivalent with the use of the PMT function in Excel, as programmed in cell F7 in the spreadsheet in Figure 3.40. This is referred to as a "nested" function. The individual cash flows, as well as their present-worth values, are given in the spreadsheet.

	A	B	C	D	E	F	G
1	Example 3.15: Zinc Mine Production				**Input**		
2					G	$37,500,000.00	
3	**Period**	**Cash Flow**	**Present Worth**		Interest Rate	7%	per six months
4	0	--	--		Periods	4	(semi-annual)
5	1	$0.00	$0.00				
6	2	$37,500,000.00	$32,753,952.31		**Output**		
7	3	$75,000,000.00	$61,222,340.77		A	$53,082,607.14	
8	4	$112,500,000.00	$85,825,711.36				
9							
10	=$F$2*(A8-1)		=B7/(1+$F$3)^A7			=PMT(F3,F4,-SUM(C5:C8))	
11							

Figure 3.40 Equivalent periodic cash flows for increasing zinc operation revenues.

The next example couples the arithmetic gradient series with an equal-payment series. The cash flow diagram decomposes as it did in previous analyses.

EXAMPLE 3.16 *Increasing Arithmetic Gradient with Equal-Payment Series*

In the summer of 2003, Toshiba Corporation announced that it would spend YEN350 billion over four years to build a large, advanced facility in Kyushu, Japan, to produce chips used in game consoles and mobile phones.[19] Assume that YEN50 billion is

[19] Moffett, S., "Japanese Firms Spend More, Boosting Hopes for Rebound," *The Wall Street Journal Online*, August 11, 2003.

invested in 2004, with the amount invested increasing YEN25 billion per year over the next three years. What are the equivalent equal annual payments for this investment over the same four years, assuming an annual interest rate of 12%?

Solution. The cash flow diagram for the investment is given in Figure 3.41(a). As detailed in our earlier analyses for present-worth and compound amount factors, the cash flow diagram can be decomposed. Figure 3.41(b) illustrates the equal-payment series and (c) the arithmetic gradient series.

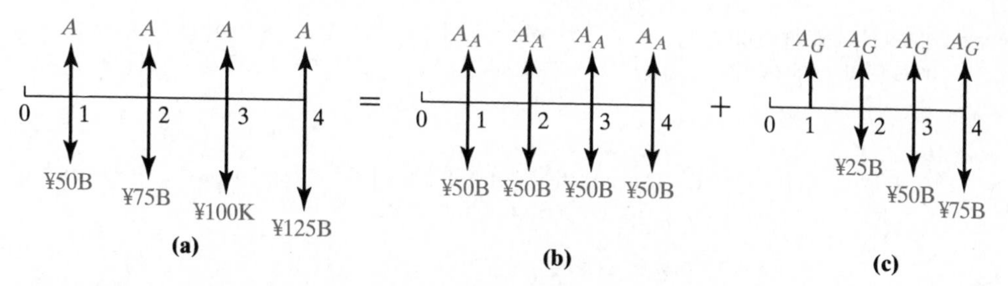

Figure 3.41 (a) Equivalent equal-payment series A of an equal-payment and an arithmetic gradient series decomposed into (b) equal-payment (YEN50B) and (c) arithmetic gradient (YEN25B) series.

We first focus our attention on the arithmetic gradient series, or part (c) of Figure 3.41. With G given and A_G unknown (we follow similar notation as before), we can write the necessary equation in our shorthand notation:

$$A_G = G(^{A/G,i,N}) = \text{YEN25B}(^{A/G,12\%,4}).$$

The remaining part of the cash flow diagram is already an equal-payment series of size YEN50 billion. Thus, we can write our final solution as

$$\begin{aligned} A &= A_G + A_A \\ &= \text{YEN25B}(^{A/G,12\%,4}) + \text{YEN50B} \\ &= \text{YEN25B}(\overset{A/G,12\%,4}{1.3589}) + \text{YEN50B} \\ &= \text{YEN33.97B} + \text{YEN50B} = \text{YEN83.97 billion.} \end{aligned}$$

As seen here, it is generally easy to convert an arithmetic gradient series to an equivalent equal-payment series. The reason is that any existing value at time 1, such as the YEN50 billion in this example, merely sums directly into the final answer. A spreadsheet analysis would resemble that of Example 3.15.

3.3.4 Geometric Gradient Series Factor

We now address our final conversion of a geometric gradient series into an equal-payment series. The geometric gradient series is depicted in

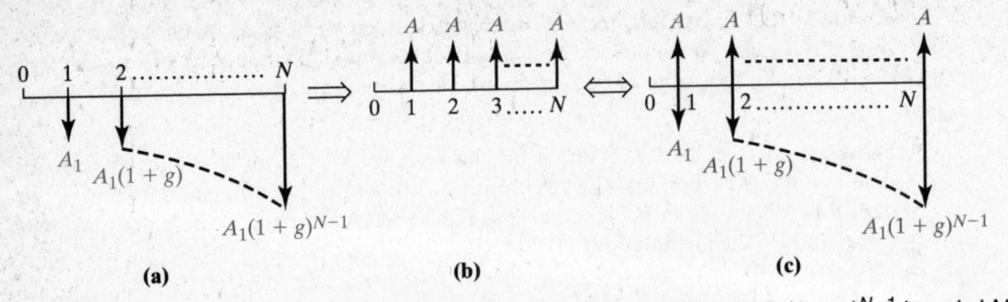

Figure 3.42 (a) Geometric series of cash flows growing from A_1 in period 1 to $A_1(1+g)^{N-1}$ in period N. (b) Equal-payment series of cash flows A from period 1 through period N. (c) Cash flow diagram for the equal-payment series factor for a geometric series.

Figure 3.42(a), with the equal-payment series depicted in part (b). Both are drawn in part (c). Again, we ask our question:

What is A, given A_1 and g?

Our desired transformation converts the geometric gradient series cash flows into an equal-payment series. We derived the compound amount factor for the geometric gradient series in Equation (3.4). We can convert this to an equal-payment series by multiplying by the equal-payment series sinking-fund factor, Equation (3.9). We get

$$A = A_1 \underbrace{(F/A_1,g,i,N)}_{A_1,\,g \Rightarrow F} \underbrace{(A/F,i,N)}_{F \Rightarrow A}.$$

This is defined mathematically as

$$\begin{aligned} A &= A_1 \left[\frac{(1+i)^N - (1+g)^N}{i-g} \right] \left[\frac{i}{(1+i)^N - 1} \right] \\ &= A_1 \left[\frac{i\left((1+i)^N - (1+g)^N\right)}{(i-g)\left((1+i)^N - 1\right)} \right]. \end{aligned}$$

For the case where $i = g$, we proceed in a similar manner, so that

$$A = \underbrace{(F/A_1,g,i,N)}_{A_1,\,g \Rightarrow F} \underbrace{(A/F,i,N)}_{F \Rightarrow A}.$$

The expression is written as

$$\begin{aligned} A &= A_1 N (1+i)^{N-1} \left[\frac{i}{(1+i)^N - 1} \right] \\ &= A_1 \left[\frac{Ni(1+i)^{N-1}}{(1+i)^N - 1} \right]. \end{aligned}$$

Thus, for the *geometric gradient series equal-payment series factor*, we define the following two cases:

$$A = \begin{cases} A_1\left[\frac{i\left((1+i)^N-(1+g)^N\right)}{(i-g)\left((1+i)^N-1\right)}\right] & \text{if } i \neq g \\ A_1\left[\frac{Ni(1+i)^{N-1}}{(1+i)^N-1}\right] & \text{if } i = g \end{cases} \tag{3.12}$$

Our shorthand notation follows:

$$A = \left\{ \begin{array}{ll} A_1\left[\frac{i\left((1+i)^N-(1+g)^N\right)}{(i-g)\left((1+i)^N-1\right)}\right] & i \neq g, \\ A_1\left[\frac{Ni(1+i)^{N-1}}{(1+i)^N-1}\right] & i = g. \end{array} \right\} = A_1(A/A_1,g,i,N).$$

We illustrate the use of the geometric gradient series equal-payment series factor in the next example.

EXAMPLE 3.17

Geometric Gradient Series Equal-Payment Factor

Hyundai Mobis, South Korea's largest auto-parts supplier, has been investing heavily in plants throughout the world in order to achieve its goal of becoming one of the 10 largest global suppliers. Hyundai Mobis supplies parts to a variety of companies, including Hyundai Motors, Kia Motors, General Motors, and DaimlerChrysler.[20] Assume that annual current (beginning 2004) capacity for Hyundai Mobis is 1.5 million modules. If capacity grows by 15% per year and sales generate \$100 per module, what is the equal-payment series equivalent of the revenue stream over the next six years? (Assume a 5% annual interest rate.)

Solution. The geometric growth in revenues is depicted in Figure 3.43.

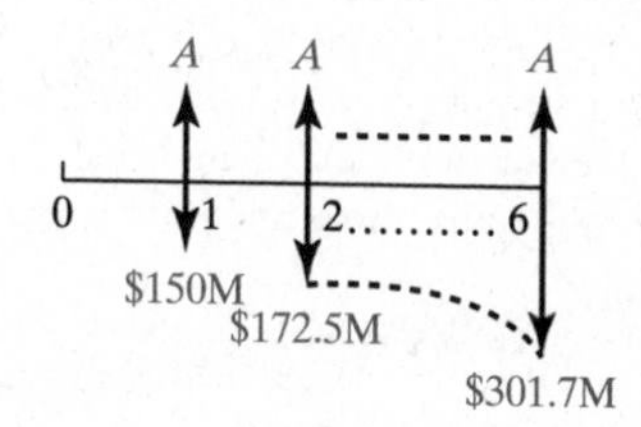

Figure 3.43
Equivalent equal-payment series A from revenues for auto-part modules.

The cash flow diagram is defined with $g = 15\%$, $i = 5\%$, $N = 6$, and $A_1 =$ \$100 × 1.5 million, or \$150 million. Define A with our shorthand notation as

$$A = A_1(A/A_1,g,i,N) = \$150\text{M}(A/A_1,15\%,5\%,6).$$

[20] Won Choi, H., "Hyundai Mobis's Narrow Focus Widens Its Reach, Aids Results," *The Wall Street Journal Online*, August 29, 2003.

Since i and g differ, we can solve for A:

$$A = A_1(^{A/A_1,g,i,N}) = A_1\left[\frac{i\left((1+i)^N - (1+g)^N\right)}{(i-g)\left((1+i)^N - 1\right)}\right]$$

$$= \$150\text{M}\left[\frac{0.05\left((1+0.05)^6 - (1+0.15)^6\right)}{(0.05-0.15)\left((1+0.05)^6 - 1\right)}\right] = \$214.56 \text{ million.}$$

The analysis is carried through on the spreadsheet in Figure 3.44. Instead of computing the present worth or future value of each individual cash flow, aggregating the results, and then computing A, we have programmed Equation (3.12) into the spreadsheet. Because there are two different equations, depending on the values of g and i, we use the IF function in Excel.

	A	B	C	D	E	F
1	Example 3.17: Automobile Part Module Sales			**Input**		
2				A1	$150,000,000.00	
3	**Period**	**Cash Flow**		g	15%	per year
4	0	--		Interest Rate	5%	per year
5	1	$150,000,000.00		Periods	6	years
6	2	$172,500,000.00				
7	3	$198,375,000.00		**Output**		
8	4	$228,131,250.00		A	$214,564,303.86	
9	5	$262,350,937.50				
10	6	$301,703,578.13	=IF(E3=E4,(E2*E5*E4*(1+E4)^(E5-1))/(((1+E4)^E5)-1),			
11	=$F$2*(1+$F$3)^(A9-1)		E2*(E4*((1+E4)^E5-(1+E3)^E5))/((E4-E3)*((1+E4)^E5-1)))			
12						

Figure 3.44 Spreadsheet solution for A, given A_1 and g and using the IF function.

The IF function is defined as

=IF(logical test, value if true, value if false)

and is programmed into cell E8 in Figure 3.44. Although the formula given in the figure is complicated, the logical test asks if i and g are equal (E3 = E4). If they are, then the case of $i = g$ is utilized from Equation (3.12), as it is programmed after the first comma. The second argument is the equal-payment series factor when $i \neq g$. Computing A in this manner eliminates the need to move individual cash flows to the present or future before calculating A, but it does require precise programming of Equation (3.12) into cell E8.

For the case where $i = g = 12\%$, we would solve for A as

$$A = A_1(^{A/A_1,g,i,N}) = A_1\left[\frac{Ni(1+i)^{N-1}}{(1+i)^N - 1}\right] = \$150\text{M}\left[\frac{6(0.12)(1+0.12)^5}{(1+0.12)^6 - 1}\right]$$

$$= \$195.45 \text{ million.}$$

As expected, this results in a much lower value for A, due to the lower growth rate and higher interest rate. This value could also be found with the spreadsheet by changing the values of cells E4 and E5 to 12%.

With the interest factors for discrete compounding and discrete cash flows that we have derived, we can convert any cash flow diagram into either an equivalent future value F at time N, an equivalent present value P at time zero, or an equal-payment series A over N periods. We summarize the factors in the next section and illustrate their use in more general scenarios in the section that follows.

3.4 Summary of Interest Factors

Table 3.1 summarizes the interest factors that we have just derived for discrete cash flows and discrete compounding. The table gives the cash flow diagrams, shorthand notation, and interest factors for all three cases we have examined.

3.5 Using Multiple Factors in Analysis

Earlier, we derived a number of expressions to help us move money through time. Our goal is to be able to convert a given cash flow diagram into another economically equivalent cash flow diagram. Presumably, the new diagram that we define will be easier to analyze than the original one.

As you may expect, it will be rare that you will draw a cash flow diagram in which the cash flows look exactly like those in the diagrams we have analyzed. Accordingly, you must carefully apply the factors defined previously, as a number of them may be required simultaneously over different periods. The most common error in analysis is to mistake N in the interest formulas as the period N, as opposed to the *number* of periods (N) for evaluation. That is, be strict with your reference to time zero. We illustrate the use of multiple factors in the next example and point out some potential pitfalls in analysis.

EXAMPLE 3.18

Multiple-Factor Analysis

In the fall of 2003, Standard Microsystems Corp. announced that Intel would pay it $75 million for intellectual property rights and other business. Specifically, the following payment plan was established: $20 million in 2003, $10 million in both 2004 and 2005, $11 million in 2006, and $12 million in both 2007 and 2008.[21] Find F for this payment plan, assuming a 15% annual rate of interest and end-of-year cash flows. The cash flow diagram for the agreed-upon payment series is given in Figure 3.45.

Solution. There are a number of ways in which to analyze the cash flow diagram. The interest rate factors were derived in order to make the analysis efficient. Here are *some* ways in which to find the desired result:

(1) Single-Payment Analysis

You may already have noticed that the interest rate factors we derived are not really necessary, especially if you are handy with a spreadsheet. The reason is because you

[21] DeLeon, C., "Intel/Standard Micro -2: To Be Paid Over Next Five Years," *Dow Jones Newswires*, September 8, 2003.

TABLE 3.1 Summary of interest factors for discrete payments and discrete compounding.

Question	Cash Flow Diagram	Notation	Factor
Compound Amount Factors			
What is F, given P?	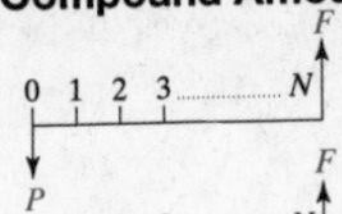	$(F/P,i,N)$	$(1+i)^N$
What is F, given A?	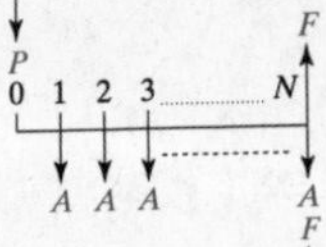	$(F/A,i,N)$	$\left[\frac{(1+i)^N-1}{i}\right]$
What is F, given G?	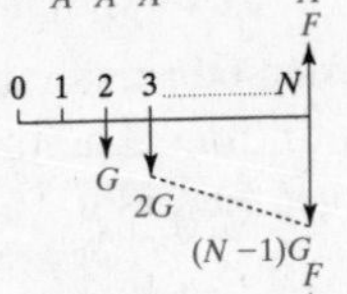	$(F/G,i,N)$	$\left[\frac{(1+i)^N-Ni-1}{i^2}\right]$
What is F, given A_1 and g?	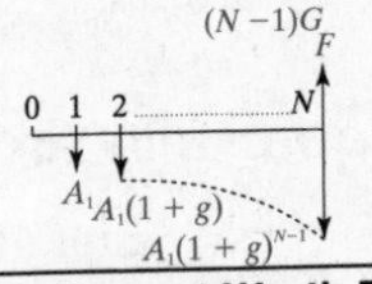	$(F/A_1,g,i,N)$	$\left[\frac{(1+i)^N-(1+g)^N}{i-g}\right]$, $i \neq g$ $N(1+i)^{N-1}$, $i = g$
Present Worth Factors			
What is P, given F?	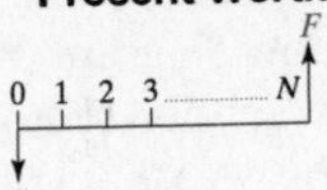	$(P/F,i,N)$	$\frac{1}{(1+i)^N}$
What is P, given A?	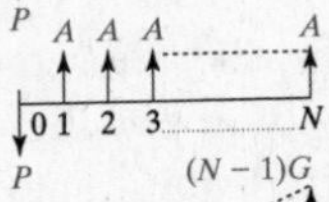	$(P/A,i,N)$	$\left[\frac{(1+i)^N-1}{i(1+i)^N}\right]$
What is P, given G?	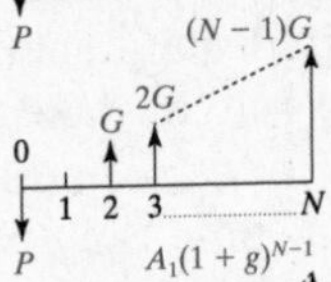	$(P/G,i,N)$	$\left[\frac{(1+i)^N-Ni-1}{i^2(1+i)^N}\right]$
What is P, given A_1 and g?	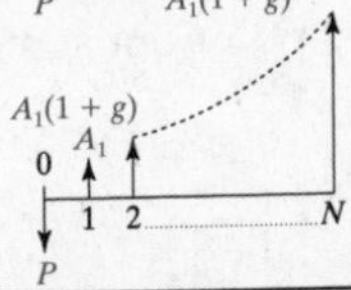	$(P/A_1,g,i,N)$	$\left[\frac{1-(1+g)^N(1+i)^{-N}}{i-g}\right]$, $i \neq g$ $\frac{N}{(1+i)}$, $i = g$
Annual Equivalent Factors			
What is A, given F?	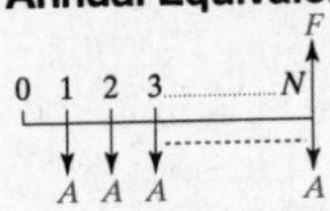	$(A/F,i,N)$	$\left[\frac{i}{(1+i)^N-1}\right]$
What is A, given P?	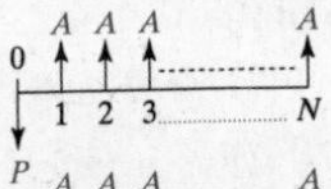	$(A/P,i,N)$	$\left[\frac{i(1+i)^N}{(1+i)^N-1}\right]$
What is A, given G?	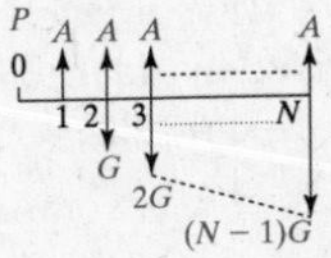	$(A/G,i,N)$	$\left[\frac{(1+i)^N-Ni-1}{i((1+i)^N-1)}\right]$
What is A, given A_1 and g?	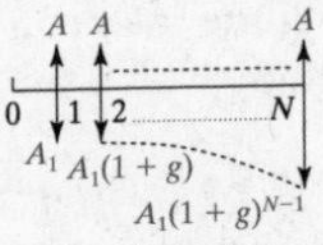	$(A/A_1,g,i,N)$	$\left[\frac{i((1+i)^N-(1+g)^N)}{(i-g)((1+i)^N-1)}\right]$, $i \neq g$ $\left[\frac{Ni(1+i)^{N-1}}{(1+i)^N-1}\right]$, $i = g$

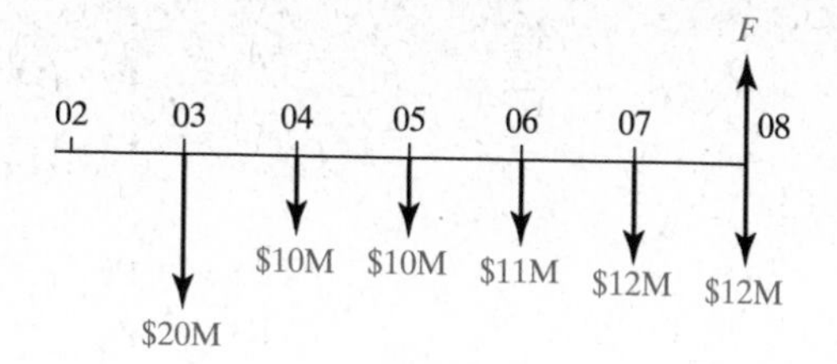

Figure 3.45 Payment plan in years 2003 through 2008 for intellectual property rights.

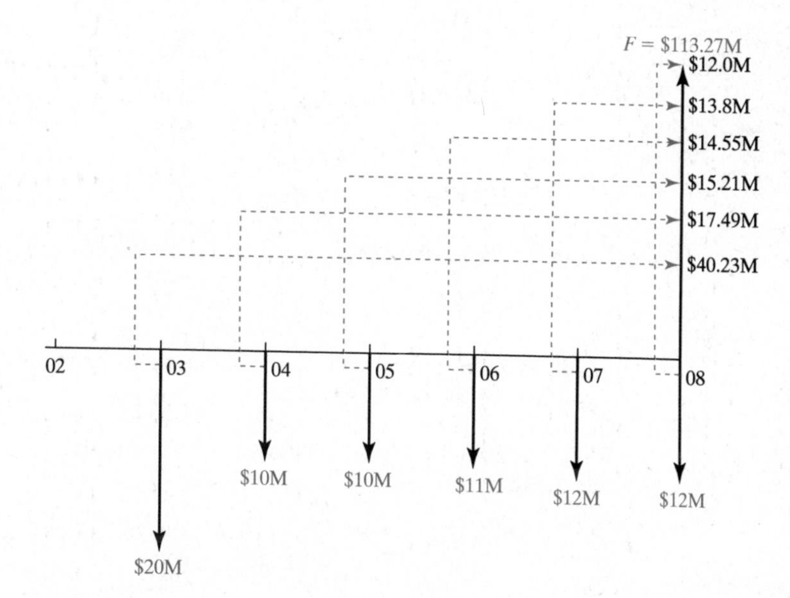

Figure 3.46 Equivalent future value of payments, using the compound amount factor for a single payment repeatedly.

can always treat each cash flow in a cash flow diagram individually. With this approach, either the compound amount factor (for future values) or the present-worth factor (for present values) can be used repeatedly and summed for the desired result. We emphasize that approach here and in Figure 3.46, which moves each cash flow to period N individually with the compound amount factor for a single payment.

Mathematically, we write the sum of the six compound amount values as

$$\begin{aligned}
F &= \$20\text{M}(^{F/P,15\%,5}) + \$10\text{M}(^{F/P,15\%,4}) + \$10\text{M}(^{F/P,15\%,3}) + \$11\text{M}(^{F/P,15\%,2}) \\
&\quad + \$12\text{M}(^{F/P,15\%,1}) + \$12\text{M}(^{F/P,15\%,0}) \\
&= \$20\text{M}(1+0.15)^5 + \$10\text{M}(1+0.15)^4 + \$10\text{M}(1+0.15)^3 + \$11\text{M}(1+0.15)^2 \\
&\quad + \$12\text{M}(1+0.15)^1 + \$12\text{M}(1+0.15)^0 \\
&= \$113.27 \text{ million.}
\end{aligned}$$

This is clearly the most tedious approach, but will always lead to the correct solution.

(2) Equal-Payment Series Analysis

We could shorten our work a bit by recognizing that the similar $10 million and $12 million payments are equal-payment series. We depict our approach graphically in Figure 3.47.

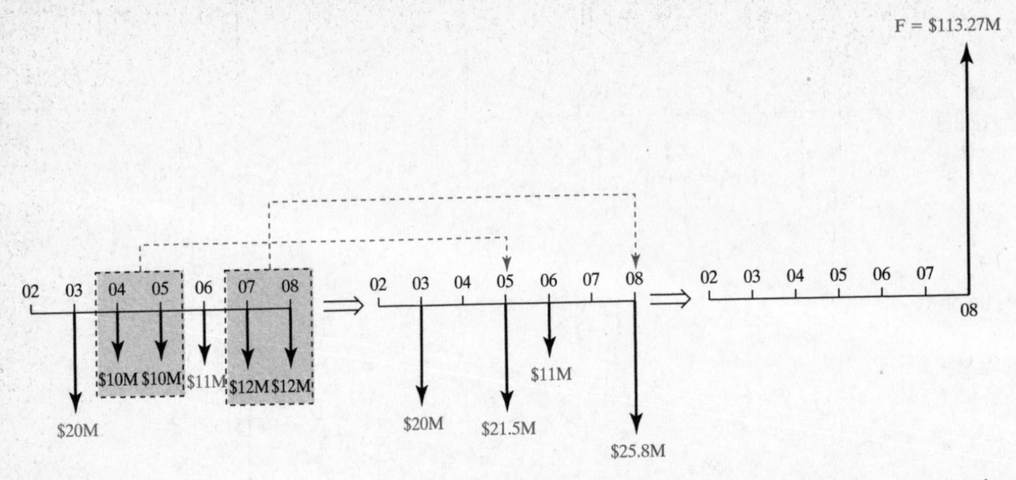

Figure 3.47 Equivalent future value of payments, using compound amount factors for equal-payment series and a single payment.

Note that, in analyzing an equal-payment series, N refers to the number of payments. In computing F with $(F/A,i,N)$, the value of F is defined at the period of the final cash flow in the equal-payment series. For the $10 million series, this is 2005, while it is 2008 for the $12 million series. The approach is summarized as follows in our shorthand notation:

$$F = \$20\text{M}(F/P,15\%,5) + \$10\text{M}(F/A,15\%,2)(F/P,15\%,3) + \$11\text{M}(F/P,15\%,2) + \$12\text{M}(F/A,15\%,2).$$

This will result in the same answer, $113.27 million, as before. The reader may verify the result.

(3) Equal-Payment Series Analysis Revisited

Equivalently, we could assume that the payments consist of one equal-payment series of $12 million and subtract (or add) the differences as

$$F = \$12\text{M}(F/A,15\%,6) + \$8\text{M}(F/P,15\%,5) - \$2\text{M}(F/P,15\%,4) - \$2\text{M}(F/P,15\%,3) - \$1\text{M}(F/P,15\%,2).$$

This approach is summarized in Figure 3.48, where the original cash flow diagram is rewritten as the sum of an equal-payment series and a number of single payments. A similar approach could assume a series of $20 million, $10 million, or $11 million payments. They all result in the same answer.

(4) Arithmetic Gradient Series Analysis

The flows in periods 3 through 5 (years 2005 through 2007) represent an arithmetic gradient series of $1 million. (This includes a flow of zero in the first year, commonly

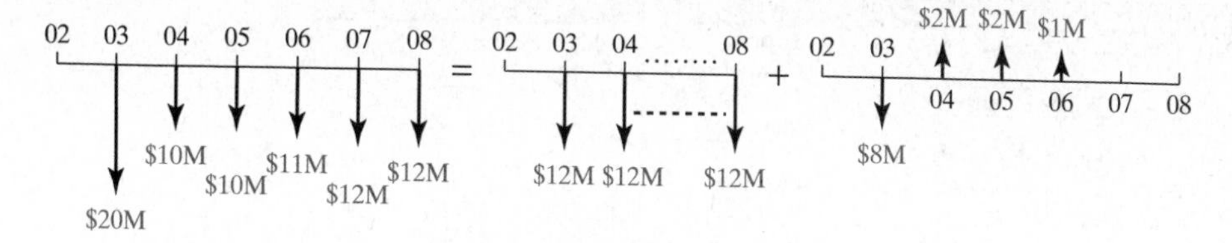

Figure 3.48 Redrawing of cash flows as an equal-payment series and four single payments.

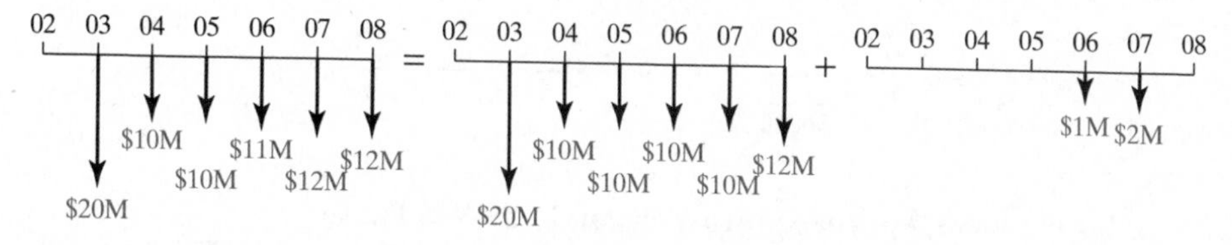

Figure 3.49 Redrawing of cash flows as the sum of an equal-payment series and an arithmetic gradient series.

forgotten in arithmetic gradient analyses.) That series is coupled with equal annual cash flows of $10 million in years 2 through 5. Thus, we could write

$$F = \$20\text{M}(^{F/P,15\%,5}) + \$10\text{M}(^{F/A,15\%,4})(^{F/P,15\%,1}) + \$1\text{M}(^{F/G,15\%,3})(^{F/P,15\%,1}) + \$12\text{M}(^{F/P,15\%,0})$$

$$= \$20\text{M}(^{F/P,15\%,5}) + \left(\$10\text{M}(^{F/A,15\%,4}) + \$1\text{M}(^{F/G,15\%,3})\right)(^{F/P,15\%,1}) + \$12\text{M}(^{F/P,15\%,0}).$$

This approach is depicted graphically in Figure 3.49.

(5) Spreadsheet Approach with Nonstandard Cash Flows and NestedFunctions

Neither Excel nor Lotus has a function that directly compute the equivalent future value for a general series of cash flows. (Quattro Pro allows for direct calculation with its FUTV function.) However, *all* spreadsheets allow for the present worth of a general series to be computed. For Excel, this is accomplished with the NPV function, defined as

=NPV(rate,value1,value2,...).

As before, the 'rate' is i. The arguments 'value1', 'value2', etc., are the cash flows, which are assumed to occur in consecutive periods. The function returns the equivalent present value *one period before the timing* of the cash flow defined by 'value1.'

To determine the future value of our sample cash flow, we *nest* the NPV function in our FV function, as follows:

$$F = \text{FV}(i, N, -\text{NPV}(i, A_1, A_2, \ldots, A_N))$$

$$= \text{FV}(0.15, 6, \text{-NPV}(0.15, 20, 10, 10, 11, 12, 12)) = 113.27.$$

The answer, of course, is in millions of U.S. dollars.

	A	B	C	D	E	F	G
1	Example 3.18: Intellectual Property Rights Payments				Input		
2					Cash Flow Diagram		
3	**Period**	**Cash Flow**	**Future Value**		Interest Rate	15%	per year
4	2002	--	--		Periods	6	years
5	2003	$20,000,000.00	$40,227,143.75				
6	2004	$10,000,000.00	$17,490,062.50		**Output**		
7	2005	$10,000,000.00	$15,208,750.00		F	$113,273,456.25	
8	2006	$11,000,000.00	$14,547,500.00		P	$48,971,241.02	
9	2007	$12,000,000.00	$13,800,000.00	=SUM(C5:C10)			
10	2008	$12,000,000.00	$12,000,000.00			=F7/(1+F3)^F4	
11							
12	=B5*(1+$F$3)^($F$4-A5+2002)						
13							

Figure 3.50 Evaluation of payment plan by means of a spreadsheet.

(6) Spreadsheet Approach with Nonstandard Cash Flows
In our final approach, we build a cash flow diagram as in Figure 3.50. This approach is similar to those seen previously, in which we compute the future value of each individual cash flow. Note that we define our periods as the actual year of occurrence in column A, so the future values in column C must be adjusted with the base year 2002 in the exponent of the compound amount factor programmed in cells C5 through C10.

We sum these individual future values with the SUM function in cell F7. We then convert the resulting future value to a present worth P at the end of period 2002 by programming cell F8 with the present-worth factor, Equation (3.5), referencing cell F7.

No single approach to analyzing cash flow diagrams can be deemed *correct*. We have illustrated a number of "correct" approaches to address this issue. There is absolutely nothing wrong with being creative in the process of converting a cash flow diagram into an equivalent diagram. The key is to get the correct answer and avoid the numerous *incorrect* approaches to the problem.

3.6 Discrete Payments and Continuous Compounding

In the case of discrete payments and continuous compounding, we rely on our conversion of the continuously compounded interest rate to an equivalent rate discounted over an appropriately defined discrete period. From Equation (2.5), the effective annual interest rate, given the nominal annual rate r and continuous compounding, is

$$i_a = e^r - 1. \tag{3.13}$$

Thus, to solve problems with an interest rate that is continuously compounded, we merely substitute this expression for our interest rate i, for each interest rate factor developed in the previous sections. We illustrate with a simple example.

EXAMPLE 3.19

Compound Amount Factor: Single Payment

The Russian airplane manufacturer Irkut, which is known for the production of the Sukhoi combat fighter, secured an order for eight Beriev Be-200 amphibious jets to be deployed in fire-fighting missions on the U.S. west coast by Hawkins and Powers Aviation. The plane, powered by Rolls-Royce BR175 engines, can scoop 12 tons of water in 20 seconds. The order, valued at \$200 million, was placed in early 2004, with delivery expected in 2007.[22,23] Assume that full payment was made at the time of the order (the beginning of 2004) and that delivery will occur at the end of 2007. What is the future value (at the end of 2007) of the payment if the interest rate is 7% per year, compounded continuously?

Solution. The time-zero cash flow of \$200 million is illustrated in the cash flow diagram in Figure 3.51. The equivalent future cash flow F is unknown.

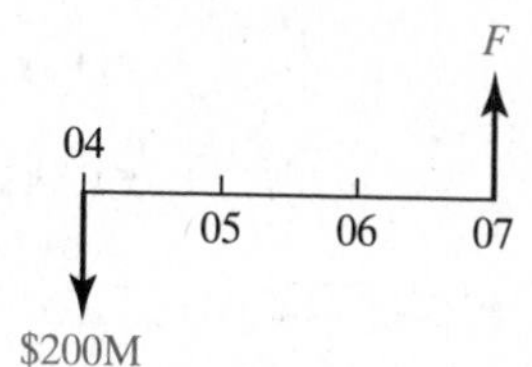

Figure 3.51 Equivalent future value F of time-zero sale of eight amphibious jets for \$200 million.

Previously, we defined the value of F in our shorthand notation for discrete compounding as

$$F = P(1+i)^N.$$

Substituting $i = e^r - 1$ results in

$$F = P(1+e^r-1)^N = P[e^{rN}] = P[^{F/P,r,N}].$$

For the current example,

$$F = P[^{F/P,r,N}] = \$200\text{M}[^{F/P,7\%,3}],$$

and we compute F as follows:

$$F = P[^{F/P,r,N}] = P\left[e^{rN}\right] = \$200\text{M}\left[e^{(0.07)(3)}\right] = \$246.74 \text{ million}.$$

Note the use of brackets to differentiate the use of r from that of i.

The spreadsheet functions defined for the discrete cash flow and discrete compounding problems can be utilized if one converts the continuously compounded interest rate into an equivalent effective interest rate over the necessary period. This can be achieved by nesting Equation (3.13) into the spreadsheet function. For Excel, this can be accomplished as follows:

$$F = \text{FV}(\text{EXP}(0.07) - 1, 3, , -200) = 246.74.$$

[22] Smith, G.T., "Irkut Wins First US Order for Firefighting Planes," *Dow Jones Newswires*, February 9, 2004;

[23] Orstrovsky, A., "Irkut Looks to Lighten its Military Load," *Financial Times*, p. 25, March 1, 2004.

TABLE 3.2 Summary of interest factors for discrete payments and continuous compounding.

Question	Cash Flow Diagram	Notation	Factor
Compound Amount Factors			
What is F, given P?	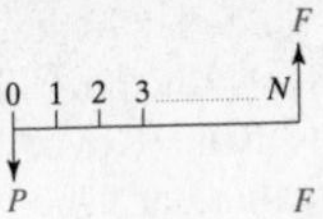	$[F/P,r,N]$	$[e^{rN}]$
What is F, given A?	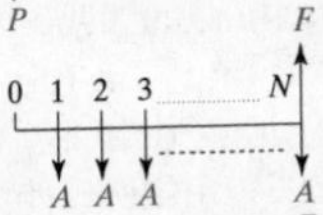	$[F/A,r,N]$	$\left[\frac{e^{rN}-1}{e^r-1}\right]$
What is F, given G?	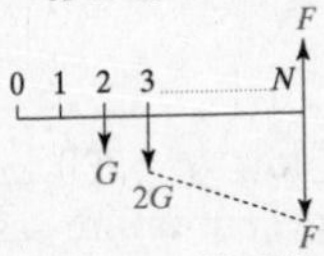	$[F/G,r,N]$	$\left[\frac{e^{rN}-N(e^r-1)-1}{(e^r-1)^2}\right]$
What is F, given A_1 and g?	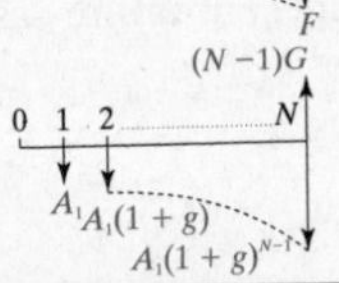	$[F/A_1,g,r,N]$	$\left[\frac{e^{rN}-e^{gN}}{e^r-e^g}\right]$ if $r \neq g$, $Ne^{r(N-1)}$ if $r = g$.
Present-Worth Factors			
What is P, given F?	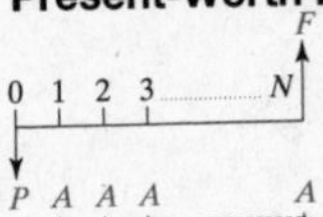	$[P/F,r,N]$	$\left[\frac{1}{e^{rN}}\right]$
What is P, given A?	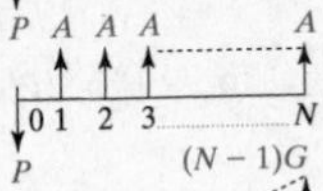	$[P/A,r,N]$	$\left[\frac{e^{rN}-1}{e^{rN}(e^r-1)}\right]$
What is P, given G?	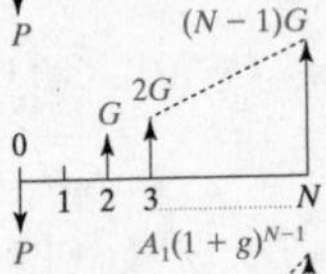	$[P/G,r,N]$	$\left[\frac{e^{rN}-N(e^r-1)-1}{e^{rN}(e^r-1)^2}\right]$
What is P, given A_1 and g?	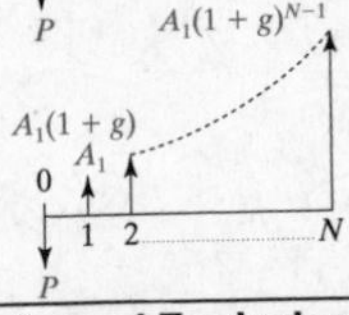	$[P/A_1,g,r,N]$	$\left[\frac{1-e^{(g-r)N}}{e^r-e^g}\right]$ if $r \neq g$, $\frac{N}{e^r}$ if $r = g$.
Annual Equivalent Factors			
What is A, given F?	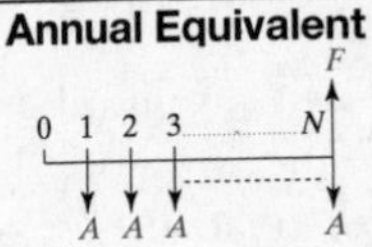	$[A/F,r,N]$	$\left[\frac{e^r-1}{e^{rN}-1}\right]$
What is A, given P?	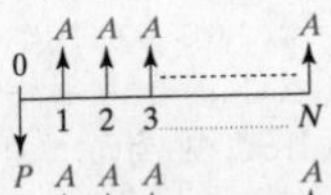	$[A/P,r,N]$	$\left[\frac{e^{rN}(e^r-1)}{e^{rN}-1}\right]$
What is A, given G?	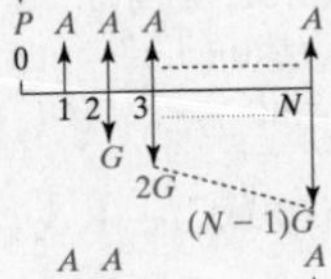	$[A/G,r,N]$	$\left[\frac{e^{rN}-N(e^r-1)-1}{(e^{rN}-1)(e^r-1)}\right]$
What is A, given A_1 and g?	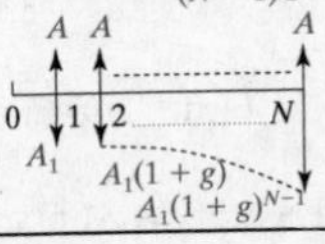	$[A/A_1,g,r,N]$	$\left[\frac{(e^{rN}-e^{gN})(e^r-1)}{(e^{rN}-1)(e^r-e^g)}\right]$ if $r \neq g$, $\left[\frac{Ne^{r(N-1)}(e^r-1)}{e^{rN}-1}\right]$ if $r = g$.

The EXP function returns e to the power r in Excel. Thus, nesting allows any of the functions previously illustrated to be utilized with continuous compounding. Note that the spreadsheet solution is in millions of U.S. dollars.

The other factors developed for discrete compounding can be converted similarly. These are summarized in Table 3.2. Note that it is assumed that the nominal interest rate r is defined according to the timing of the discrete cash flows. If we are given a nominal annual rate, we can convert it to a discrete rate over the appropriate period by raising e to the appropriate power, as illustrated in Section 2.5.2. The table gives the cash flow diagrams, shorthand notation, and interest factor for all three cases we have examined.

More detailed derivations of these interest factors with continuous compounding can be found on the website. Furthermore, derivations in which the interest rate is compounded continuously and the cash flows are assumed to be continuous, as defined in Section 2.1, are also provided on the website.

3.7 Examining the Real Decision Problems

We return to our introductory example concerning Apple's supply contracts with Hynix, Intel, Micron, Samsung, and Toshiba and the questions posed:

1. Given an 18% annual rate of interest, what annual payments at the end of years 2005 through 2010 would have been equivalent to the $1.25 billion prepayment (at the end of 2005)?

Because we want to spread a 2005 cash flow over the years 2005 through 2010, we must be careful not to rush to apply the single-payment capital-recovery factor. However, if we "move" the 2005 payment to 2004, the factor can be applied directly as

$$A = \$1.25\text{B}\overset{P/F,18\%,1}{(0.8475)}\overset{A/P,18\%,6}{(0.2859)} = \$302.9 \text{ million.}$$

This is the equivalent annual value in the years 2005 through 2010 (inclusive).

2. If Apple makes similar $1.25 billion payments at the end of each year from 2005 through 2010, what is the worth of the contract (payments) at time 2010?

The future value is found with the use of the compound amount factor for an equal payment series, or

$$F = \$1.25\text{B}\overset{F/A,18\%,6}{(9.4420)} = \$11.80 \text{ billion.}$$

3. What if the payments increase 12% per year due to increasing sales? Is this significant?

	A	B	C	D	E	F	G
1	Chapter 3: Flash Memory Contracts				**Input**		
2					A1	$1.25	billion
3	**Period**	**Cash Flow**	**Future Value**		g	12%	per year
4	2005	$1.25	$2.86		Interest Rate	18%	per year
5	2006	$1.40	$2.71		Periods	6	years
6	2007	$1.57	$2.58				
7	2008	$1.76	$2.45		**Output**		
8	2009	$1.97	$2.32		F	$15.12	billion
9	2010	$2.20	$2.20				
10						=SUM(C4:C9)	
11	=$F$2*(1+$F$3)^(A9-$A$4)						
12			=$F$2*(((1+$F$4)^$F$5-(1+A16)^$F$5)/($F$4-16))				
13							
14	g	F					
15	0%	$11.80					
16	2%	$12.29					
17	4%	$12.81					
18	6%	$13.34					
19	8%	$13.91					
20	10%	$14.50					
21	12%	$15.12					
22	14%	$15.77					
23	16%	$16.45					
24	18%	$17.16					
25	20%	$17.90	=$F$2*F5*(1+F4)^(F5-1)				
26							

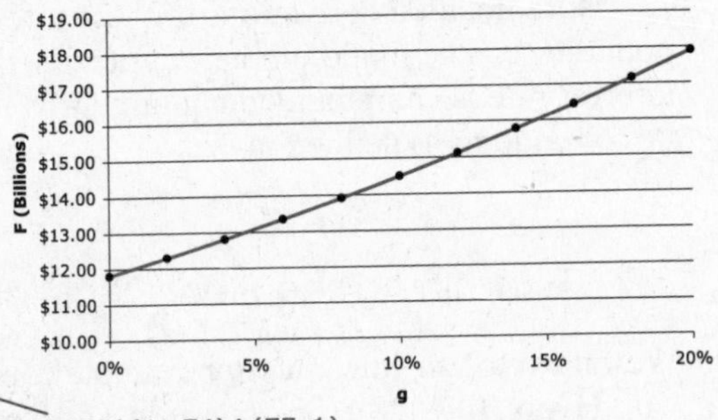

Figure 3.52 Analysis of flash-memory chip contract payments.

This represents a gradient series, such that F is now computed as

$$F = \$1.25\text{B}(\overset{F/A_1,12\%,18\%,6}{12.0955}) = \$15.12 \text{ billion.}$$

We also computed this value with the spreadsheet in Figure 3.52. The use of the spreadsheet allows us to answer the question of significance, as we have graphed the value of F with respect to the growth rate g over a range from 0% to 20%. The data in the table were generated by inputting the geometric series compound amount factor in cells B15 through B25, with the rate for g defined in cells A15 through A25. This is all shown on the spreadsheet. Note that cell B24 was programmed for the case of $i = g$.

The graph was created with the Chart Wizard, as noted in Section 2.9. However, an 'XY (Scatter)' graph was chosen in this case. Such a graph is similar to the bar chart shown previously, except that each data set can have its own x- and y-coordinates. The data and graph show a difference of $6.10 billion between growth rates of 0% and 20%. This graph presents the results of a sensitivity analysis—specifically, the sensitivity of F with respect to g. We will revisit the concept of sensitivity analysis numerous times in this textbook.

3.8 Key Points

- If two cash flow diagrams are economically equivalent, then their financial consequences are the same.
- If two cash flow diagrams are economically equivalent, then we are indifferent when asked to choose between them.
- Interest factors are used to convert a cash flow diagram into another, economically equivalent, cash flow diagram.
- Cash flow diagrams can be defined by a single payment, equal-payment series, arithmetic gradient series, geometric gradient series, or a combination of these cash flows.
- A cash flow diagram should always be drawn in an economic analysis. If a spreadsheet is to be used, input data should be contained in a data center such that it can be changed readily.
- Compound amount factors are used to convert cash flows into a single cash flow F at time N.
- Present-worth factors are used to convert cash flows into a single cash flow P at time zero.
- Equal-payment series factors are used to convert cash flows into a series of cash flows of size A over N consecutive periods.
- The interest rate factors for discrete payments and discrete compounding are summarized in Table 3.1.
- Cash flow diagrams can be converted into equivalent cash flow diagrams with use of the interest factor formulas, interest factor tables (beginning with Table A.1), or spreadsheet functions.
- The interest factors for discrete payments and continuous compounding are summarized in Table 3.2.
- There are a variety of ways to manipulate one cash flow diagram into another, economically equivalent, cash flow diagram.

3.9 Further Reading

Blank, L., and A. Tarquin, *Engineering Economy,* 5th ed. McGraw-Hill Co., Boston, 2002.

Eschenbach, T., "Using Spreadsheet Functions to Compute Arithmetic Gradients," *The Engineering Economist,* 39(3):275–280, 1994.

Fleischer, G.A., *Introduction to Engineering Economy*. PWS Publishing Co., Boston, 1994.

Grant, E.L., W.G. Ireson, and R.S. Leavenworth, *Principles of Engineering Economy*, 8th ed. John Wiley and Sons, New York, 1990.

Newnan, D.G., J.P. Lavelle, and T.G. Eschenbach, *Essentials of Engineering Economic Analysis,* 2d ed. Oxford University Press, New York, 2002.

Park, C.S., *Fundamentals of Engineering Economics.* Prentice Hall, Upper Saddle River, NJ, 2004.

Steiner, H.M., *Engineering Economic Principles,* 2d ed. McGraw-Hill, New York, 1996.

Sullivan, W.G., E.M. Wicks, and J.T. Luxhoj, *Engineering Economy,* 12th ed. Prentice Hall, Upper Saddle River, New Jersey, 2003.

Thuesen, G.J., and W.J. Fabrycky, *Engineering Economy,* 9th ed. Prentice Hall, Upper Saddle River, New Jersey, 2001.

White, J.A., K.E. Case, D.B. Pratt, and M.H. Agee, *Principles of Engineering Economic Analysis,* 4th ed. John Wiley and Sons, New York, 1998.

3.10 Questions and Problems

3.10.1 Concept Questions

1. Why might you get different answers when using the interest factor tables in the back of the textbook as opposed to using the formula? Will this present a problem in economic analysis? Explain.

2. Suppose you do not have the interest factor formula handy and there is neither an interest factor table for the desired interest rate (say, 6.5%) nor a spreadsheet available. How might you still determine the appropriate factor amount? Verify your method by using the compound amount factor for an equal-payment series, and then check against the result obtained from the formula.

3. Use a spreadsheet to derive your own interest factor table. Allow the user to input the interest rate into a cell such that the factors update automatically. What is the benefit of your spreadsheet table compared with those in the back of the textbook?

4. Derive another interest factor table in a spreadsheet for computing equivalent values from a geometric gradient series. In this case, the user must be able to enter an interest rate i and a growth rate g. In addition to requiring two inputs (i and g), what difficulty does this problem present? How can you overcome the difficulty?

5. Derive expressions for F, given an equal-payment series, an arithmetic gradient series, and a geometric gradient series, assuming that the interest rate is continuously compounded.

6. Derive an expression for F with uniform continuous cash flows (a continuous flow of funds) over the period and a continuously compounded interest rate.

3.10.2 Drill Problems

1. If, at time zero, you place $10,000 into an account that earns 12% per period, how much will be in the account after nine periods?

2. If you place $22,000 into an account at time zero, how much can you withdraw (in equal amounts) after each period for the next 12 periods if the interest rate is 17% per period and no money remains in the account after the last withdrawal?

3. If you place \$750 into an account at the end of each quarter for three years, how much is in the account after the third year if the nominal interest rate is 12% compounded quarterly?

4. Suppose you place \$22,500 into an account at time 3. The money earns 10% per period. How much will be in the account at period 7?

5. Suppose you place \$500 into an account at time zero. The money earns 14% every 6 months. How much will be in the account after 12 years?

6. What is \$55,000 in an account at 2010 worth at time 2005 if the annual rate of interest is 4.35% per year?

7. If you require \$150,000 at the end of seven periods, how much money must you place at time zero in an account that earns 5% per period?

8. If you require \$200,000 at the end of year 15, what minimum equal annual payments must be placed into the account beginning at time 1 if the interest rate is 8% per annum?

9. If you require \$50,000 at the end of 20 years, how much money must you place in the account at time zero if the account earns 3.78% per quarter?

10. Payments into an account start at \$10,000 and decrease 8% per year. What is the future value of the payments at time 12 if the account pays 8% per year?

11. If you require \$3650 in two periods, will \$2500 invested at 7% per period be enough?

12. If you make five equal annual payments into an account of size \$2000 and then leave the money for an additional five years, how much will be in the account if interest is 6% per year? What is the equivalent amount at time zero?

13. Payments into an account start at \$5000 and increase 4% per year. What is the present value of the account if the payments continue for a total of 15 years and the interest rate is 5%? Recompute for interest rates of 3% and 4%.

14. If you make a payment of \$2000 into an account at time 1 and increase the deposit by \$500 for the next five years, how much will be in the account at the end of all the payments? Assume an annual interest rate of 3%.

15. You invest annually into an account that pays 7.5% interest. You deposit \$5000 at time 1 and increase your deposit by 4.25% each year. What is the account worth at time 10? What are the equivalent equal annual payments over the same period?

16. Given the cash flow diagram

$$A_1 = \$200{,}000,$$

$$A_2 = \$300{,}000,$$

$$A_3 = \$400{,}000,$$

$$A_4 = \$500{,}000,$$

$$A_5 = \$100{,}000,$$

find

(a) F (at period 5), assuming a periodic interest rate of 13%.

(b) P (at time zero), assuming a periodic interest rate of 12%.

(c) An equivalent amount A over the same five periods, assuming a 10% periodic interest rate.

17. Given the cash flow diagram

$$A_1 = \$500,$$
$$A_2 = \$400,$$
$$A_3 = \$300,$$
$$A_4 = \$200,$$
$$A_5 = \$100,$$
$$A_6 = \$0,$$
$$A_7 = -\$100,$$

find

(a) F (at period 7), assuming a periodic interest rate of 22%.

(b) P (at time zero), assuming a periodic interest rate of 2.5%.

(c) An equivalent amount A over the same seven periods, assuming an 11% periodic interest rate.

18. Given the cash flow diagram

$$A_1 = \$0,$$
$$A_2 = \$250,$$
$$A_3 = \$400,$$
$$A_4 = \$550,$$
$$A_5 = \$700,$$
$$A_6 = \$550,$$
$$A_7 = \$400,$$
$$A_8 = \$250,$$
$$A_9 = \$150,$$

find

(a) F (at period 9), assuming a periodic interest rate of 6%.

(b) P (at time zero), assuming a periodic interest rate of 14%.

(c) An equivalent amount A over the same nine periods, assuming a 25% periodic interest rate.

19. Given the cash flow diagram

$$A_1 = \$200{,}000,$$
$$A_2 = \$200{,}000,$$

$$A_3 = \$200{,}000,$$
$$A_4 = \$400{,}000,$$
$$A_5 = \$600{,}000,$$
$$A_6 = \$800{,}000,$$
$$A_7 = \$1{,}000{,}000,$$

find

(a) P (at time zero), assuming a periodic interest rate of 4%.

(b) F (at time 7), assuming a periodic interest rate of 10%.

(c) An equivalent amount A over the same seven periods, assuming a 2% periodic interest rate.

20. Given the cash flow diagram

$$A_1 = \$200{,}000,$$
$$A_2 = \$0,$$
$$A_3 = \$200{,}000,$$
$$A_4 = \$0,$$
$$A_5 = \$200{,}000,$$
$$A_6 = \$0,$$
$$A_7 = \$200{,}000,$$
$$A_8 = \$0,$$
$$A_9 = \$200{,}000,$$
$$A_{10} = \$0,$$
$$A_{11} = \$200{,}000,$$

find

(a) P (at time zero), assuming a periodic interest rate of 7%.

(b) F (at time 11), assuming a periodic interest rate of 9%.

(c) An equivalent amount A over the same 11 periods, assuming a 21% periodic interest rate.

3.10.3 Application Problems

1. Siemens AG received a EUR230 million contract to build a 400-MW gas turbine power plant in Antwerp for chemical company BASF AG. The plant became operational in August of 2005.[24] What is the equivalent present value of the

[24] "Siemens Gets EUR230M Order to Build BASF Power Plant," *Dow Jones Newswires*, December 16, 2003.

payment at the time the contract was signed (at the end of December 2003), assuming that payment was made at the end of August and the interest rate was 0.75% per month?

2. IBM Corporation received a $60 million-per-year contract over 10 years with the U.S. Department of Defense to provide chip technology applications from its plant in Essex, Vermont. The contract runs through 2014.[25] Assuming an interest rate of 12% per annum and equal annual payments of $60 million beginning at the end of 2005, what is the future value of the contract at time 2014?

3. Vietnam Airlines paid between $50 million and $60 million each for three Airbus A321 planes delivered, respectively, in July 2004, October 2004, and early 2005.[26] Assuming that the final plane was delivered in January of 2005 (with all deliveries at the end of the month), what is the equivalent time-zero value (April 2004) of the payments? Assume that $55 million per plane is paid upon delivery and that the monthly interest rate is 0.5%. What is the equivalent future value (January 2005)? What are the equivalent monthly payments over the same time frame, assuming a monthly rate of 1%?

4. Hampson Industries was selected by Eclipse Aviation to build the tail section of the company's new corporate jet, a six seater that will cost roughly $950,000 to purchase. The company has 2100 orders booked through late 2008 and hopes to begin production in 2006. The agreement was valued at $380 million.[27] If production begins with 10 planes in January 2006 and then increases by 10 planes each month until reaching a capacity of 100 planes per month, what is the present value (December 2005) of revenues from the first three years of aircraft sales for Hampson? Assume that Hampson is paid $190,000 for each tail section and receives payment when an aircraft is sold. Assume further that cash flows are monthly, the monthly interest rate is 1.25%, and all planes produced are sold.

5. Chartered Semiconductor Manufacturing began installing equipment in its new $3 billion wafer fabrication plant in the spring of 2004.[28] Assume that production in May that year totaled 1000 12-inch wafers and that it increases at the rate of 20% per month for the first year. If revenues are $5000 per wafer, what is the future value of the first-year revenues, assuming a 2.5% monthly interest rate?

6. Volvo Aero has been awarded a contract to supply various components and expertise to General Electric Co.'s new LMS100 TM gas turbine. The contract is estimated to be worth SEK7 billion over 20 years. Volvo Aero is responsible for the design, development, and production of the power turbine case for the engine, which is expected to have a 10% increase in efficiency compared with today's most efficient model, the LM6000.[29] Assume that the contract begins in 2004, with the

---

25 "IBM Wins US Defense Dept Contract Worth up to $600 Mln," *Dow Jones Newswires*, November 11, 2003.

26 Hanoi Bureau, "Vietnam Airlines to Borrow $200M to Buy 5 Airbus-321," *Dow Jones Newswires*, December 3, 2003.

27 "UK's Hampson Signs $380M Mfg Pact with Eclipse Aviation," *Dow Jones Newswires*, December 10, 2003.

28 Lin, P.A., "Singapore's Chartered Starts Equipping New Fab Plant," *Dow Jones Newswires*, March 25, 2004.

29 "Volvo Aero Gets SEK7 Bln GE Turbine Deal," *Dow Jones Newswires*, December 10, 2003.

SEK7 billion spread evenly over 20 years. What is the present value (at the end of 2003) of the contract, assuming a 15% annual rate of interest?

7. CHC Helicopter Corporation, a supplier of helicopter services to the oil and gas industry, received a four-year, C$23 million contract to provide and maintain two new Sikorsky S76C+ helicopters to Transocean and Dolphin Drilling in support of programs off the shore of India. The agreement was reached in the fall of 2003.[30] If the value given is a present value (at the end of 2003), determine the equivalent equal annual payments over the four-year contract (2004–07), assuming a 17% annual interest rate.

8. Song Networks was awarded a four-year contract worth NOK100 million to provide fixed-line telephone services to all state-run universities and polytechnics in Norway.[31] If the contract is paid in equal annual payments (NOK25 million) over four years, what is the future value at the end of the contract, assuming a 3.25% annual interest rate?

9. The French engineering firm Technip SA is building a semisubmersible offshore oil production platform for Petrobras, a Brazilian oil company. The contract, signed in December of 2003, is valued at $775 million.[32] Assume that time zero is the end of 2003 and the rig will take two years to manufacture. If the interest rate is 12% compounded quarterly and the contract price is paid upon delivery, what is the time-zero equivalent of the $775 million contract? If the $775 million is paid at time zero, what is it worth at the time of delivery?

10. Canam Manac Group, Inc., received an order for 600 prepainted, 53-foot, aluminum vans with the tandem air-ride axle from trucking firm S.L.H. Transport of Kingston, Ontario. The order, placed in early 2004, is worth C$15 million.[33] Assume delivery of 50 vans per quarter over 12 quarters, with payment (C$25,000 per truck) upon delivery. What is the equivalent future and present value, given a nominal interest rate of 20% compounded quarterly?

11. In November of 2005, Sunoco, Inc., announced that it would invest $1.8 billion in its refineries over the next three years, to increase capacity to 1 million barrels of crude oil a day.[34] Assume that Sunoco invests $300 million in 2006 and that amount increases by $300 million in each of the next two years. What is the present value (at the end of 2005) of the investment plan, assuming an annual interest rate of 18%?

12. Bombardier Aerospace announced that it would invest $200 million over seven years, beginning in May of 2006, to establish a manufacturing facility in Queretaro, Mexico, for producing aircraft components.[35] Assume that investment in 2006

---

[30] Tsau, W., "CHC Helicopter Awarded New Contract in India," *Dow Jones Newswires*, October 28, 2003.

[31] "Song Networks Gets NOK100M Contract from Uninett," *Dow Jones Newswires*, November 18, 2003.

[32] Pearson, D., "Technip Wins $775M Contract for Brazil Offshore Platform," *Dow Jones Newswires*, December 19, 2003.

[33] King, C., "Manac Inc. Gets C$15M Contract from S.L.H. Transport Inc.," *Dow Jones Newswires*, March 15, 2004.

[34] Siegel, B., "Sunoco Inc. 3Q EPS $2.39 Vs 69c," *Dow Jones Newswires*, November 2, 2005.

[35] Zachariah, T., "Bombardier Aerospace Establishes Manufacturing Capability in Queretaro, Mexico," *Dow Jones Newswires*, October 26, 2005.

totals $40 million and declines by 10% per year over the next six years. Determine an equivalent equal-payment series for the investment stream over the same seven years, assuming an 8% annual interest rate.

3.10.4 Fundamentals of Engineering Exam Prep

1. If costs are expected to be $10,000 per year for the next five years, the equivalent present worth, assuming 12% per year, is most closely

(a) $30,370.

(b) $36,050.

(c) $41,110.

(d) $27,740.

2. A chip supplier signs a 10-year contract worth $250 million at time zero. The equivalent equal annual payments over the length of the contract, assuming an interest rate of 18% per year, is most closely

(a) $25.00 million.

(b) $58.08 million.

(c) $112.35 million.

(d) $55.63 million.

3. An LNG transport ship is ordered in 2005 at the cost of $150 million. The fee is to be paid upon delivery, in 2009. If the annual rate of interest is 12%, the present worth (2005) of the order is most closely

(a) $95.33 million.

(b) $106.76 million.

(c) $150 million.

(d) $133.94 million.

4. $12,500 is sunk quarterly into an account for equipment replacement. Assuming an interest rate of 16% per year compounded quarterly, the amount in the fund after 2.5 years is most closely

(a) $138,000.

(b) $150,000.

(c) $125,000.

(d) $262,500.

5. Revenues are expected to grow from $1.5 million in year 1 at the rate of 8% per year for the next 9 years. What are the equivalent equal annual revenues over the same 10 years, assuming an annual interest rate of 8%?

(a) $4.32 million.

(b) $13.88 million.

(c) $2.07 million.

(d) $1.50 million.

6. A cutting tool is to be replaced every five years at the cost of $75,000. The present worth of the next four replacements, assuming a 3.37% annual rate of interest and assuming that the first replacement occurs five years from now, is most closely

(a) $300,000.

(b) $226,400.

(c) $375,000.

(d) $201,750.

7. A bridge is expected to cost $23.5 million to construct, but last 20 years. The annual equivalent cost of the bridge over its lifetime, assuming a 2% annual interest rate, is most closely

(a) $1.44 million.

(b) $1.72 million.

(c) $1.18 million.

(d) $0.47 million.

8. A start-up firm has zero revenues in year 1, but they grow $15,000 per period over the next five years. The present worth of these revenues, assuming a 10% annual rate of interest, is most closely

(a) $75,000.

(b) $102,600.

(c) $33,300.

(d) $145,300.

9. How much should be put in a fund at time zero to cover the next four years of college expenses totalling $40,000 per year? The fund earns 8% per year.

(a) Less than $128,000.

(b) At least $132,500.

(c) Between $128,000 and $132,000.

(d) None of the above.

10. A contract is signed to deliver 1.5 million tons of natural gas a year at $1,800 a ton, increasing at 3.5% per year. The price is given in time-zero dollars, with delivery starting in year 1 and lasting 10 years. The present worth of the contract at 4% interest per year is most closely

(a) $27.5 million.

(b) $26.3 million.

(c) $22.1 million.

(d) $38.5 million.

11. The salvage value for an asset is expected to be $10,000 in seven years. Assuming a 4% annual rate of interest, the present value of the salvage value is most closely

(a) $8000.

(b) $1425.

(c) $2500.

(d) $7600.

12. An order is placed for custom machinery to be delivered in 18 months. The $2 million purchase price is due on delivery. Assuming a nominal rate of 12% per year, compounded monthly, for an account, the equal-sized monthly payments into the account between the time of the order and its delivery are, in millions,

(a) $\$2(A/F, 12\%, 1.5)$.

(b) $\$2(A/F, 12\%, 18)$.

(c) $\$2(A/F, 1\%, 18)$.

(d) $\$2(F/A, 1\%, 18)$.

13. After three years, money placed into an account paying 2% every six months has most closely

(a) Increased in value by 12.62%.

(b) Increased in value by 6.12%.

(c) Remained the same value.

(d) Increased in value by 2.00%.

14. A farmer purchases a new combine harvester for $200,000 with an expected lifetime of 15 years and a salvage value of $15,000 at that time. If the operating and maintenance costs are $3500 per year over the expected life, and the interest rate is 6% per year, the annual equivalent costs for ownership of the equipment are most closely

(a) $3500.

(b) $21,600.

(c) $23,400.

(d) $19,900.

15. An industrial robot is purchased and installed for $85,000. Annual operating and maintenance costs are expected to be $1500 a year, with software upgrades increasing by $500 per year from $5000 in year 1. If the robot has an eight-year life, what is the present value of the costs of owning it? Assume that the robot has no salvage value.

(a) $\$85{,}000 + \$6500(P/A, i\%, 8) + \$500(P/G, i\%, 7)$.

(b) $\$85{,}000 + \$6500(P/A, i\%, 8) + \$500(P/G, i\%, 8)$.

(c) $\$85{,}000(A/P, i\%, 8) + \$6500 + \$500(A/G, i\%, 8)$.

(d) $\$85{,}000 + \$6500(8) + \$500(7)$.

4 Economic Equivalence

(Courtesy of the Norfolk Southern Corporation.)

Real Decisions: Century Bonds

Transportation company Norfolk Southern Corporation sold $300 million worth of notes (bonds) on March 7, 2005. The notes carry an interest rate of 6% per year, with semiannual interest payments beginning September 15, 2005, and are due (i.e., they mature) March 15, 2105 (100 years after their issue!). Norfolk Southern said the proceeds would be used for general corporate purposes, as notes are often sold to fund capital investments.[1] Furthermore, the notes sold for $103.538 per $100 face value on the open market on December 28, 2005.[2] These facts lead to a number of interesting questions for an investor:

1. Assume that an investor bought $10,000 worth of bonds at par at the initial offering. What is the yield to maturity on the investment?

---

[1] "Norfolk Southern Prices $300 Million of 100-Year Notes," *News Releases,* www.nscorp.com, March 8, 2005.

[2] From www.nasdbondinfo.com on January 2, 2006.

2. If the investor sold the bonds on December 28, 2005, for the market price, what was the return on the investment? If the investor put all proceeds from the investment into an account earning 0.25% per month, would he or she have $10,800 on March 15, 2006?
3. What is the yield to maturity for the investor who purchased the bonds on December 28, 2005?

In addition to answering these questions, after studying this chapter you will be able to:

- Utilize various definitions of economic equivalence in order to establish the equivalence or nonequivalence of two sets of cash flows. (Section 4.1)
- Determine the interest rate or horizon time that equates two sets of cash flows. (Sections 4.2–4.3)
- Describe the behavior of the interest factors derived in the previous chapter, especially at their limits. (Section 4.4)
- Describe how loans, bonds, and stocks provide capital for funding investments. (Section 4.5)
- Compute the true cost of a loan, the yield to maturity for a bond, and the return on the investment in a stock. (Section 4.5)
- Compute payment schedules for standard loans, such as equal total payments and equal principal payments, and for nonstandard loans. (Section 4.5)
- Establish conditions stipulating when a bond will sell at a premium or discount. (Section 4.5)

In the previous chapters, we derived interest factors for a variety of cash flow diagrams with special structure. Our motivation was to change a given cash flow diagram into an economically equivalent diagram for decision-making purposes. In the current chapter, we define and examine the concept of economic equivalence more formally. In the second part of the chapter, we give examples of economic equivalence having to do with loans, bonds, and stocks. Through these examples, we gain an insight into how companies and government entities raise funds to support their investment projects.

4.1 Economic Equivalence Properties

When establishing economic equivalence between two sets of cash flows, the cash flows must be defined by a cash flow diagram. That is, the magnitudes and timing of each set of cash flows must be known. Further, establishing equivalence requires the following:

- *Interest rate.* In the previous chapters, we used the interest rate to move money through time. As we will see later in this chapter, equivalence can be established for a given set of interest rates over the study period N. However, if any of these rates change, equivalence must be reestablished.
- *Common time basis.* Equivalence requires that cash flows be defined according to similar units of time, generally defined by the compounding period of the interest rate.
- *Common unit of measure.* As seen in the numerous examples in previous chapters, we live in a world economy defined by numerous currencies. A single monetary unit must be chosen for analysis.

To say that two items are equal is to say that they are the same. We define economic equivalence similarly in a number of properties that follow, each illustrated with an example.

> Economically equivalent cash flows have the same monetary value at the same point in time.

If two sets of cash flows are redrawn as their single-payment cash flow equivalents at the same period in time, and these cash flow equivalents are the same, then the original sets of cash flows are equivalent.

EXAMPLE 4.1

Economic Equivalence

Heroux-Devtek, Inc., a manufacturer and repairer of aerospace and industrial products, received a contract from General Electric Aircraft Engines in 2003 to build components for CF34, CFM56, CF6, T700, LM2500, and CFE738 engines. The payments are scheduled as follows: C$2.5 million in 2004, C$7.3 million in 2005, C$6.8 million in

2006, and C$5.0 million in 2007.[3] Assuming end-of-year cash flows and an annual interest rate of 12%, show that this payment schedule is economically equivalent to a single cash flow of C$22.58 million at the end of 2006.

Solution. The cash flow diagram of the payment schedule between 2004 and 2007, inclusive, is given in Figure 4.1(a). To establish equivalence (or its absence), we need to transform this cash flow into a single flow at time 2006, as shown in Figure 4.1(b).

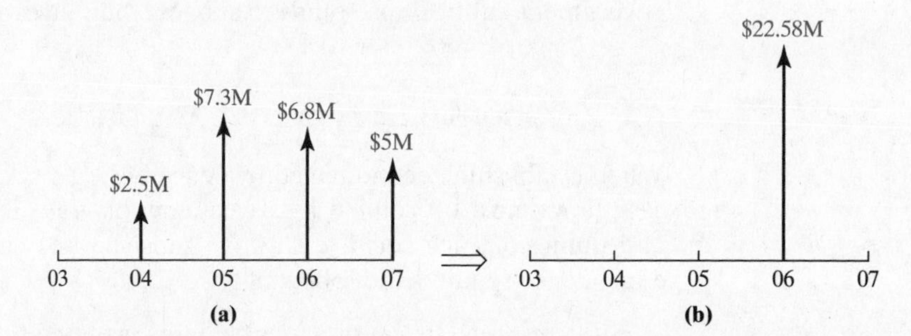

Figure 4.1 (a) Cash payments for deliveries of aircraft engine components and (b) year-2006 equivalent.

To obtain our single flow, we can merely move each cash flow individually to 2006 by using the compound amount factor for a single payment for the cash flows in 2004 and 2005 and the present-worth factor for a single payment for the cash flow in 2007. The cash flow in 2006 obviously does not need to be moved. We use the term F_{2006} to denote the future value of the cash flows at 2006, not 2007, as traditionally assumed. The calculation is

$$\begin{aligned} F_{2006} &= \text{C\$2.5M}(1+0.12)^2 + \text{C\$7.3M}(1+0.12)^1 + \text{C\$6.8M} + \text{C\$5.0M}\left[\frac{1}{(1+0.12)^1}\right] \\ &= \text{C\$22.58 million.} \end{aligned}$$

Thus, because the equivalent value of the payments is C$22.58 million in 2006, the payments are economically equivalent to the single payment of C$22.58 million in 2006.

Our definition of economic equivalence is really quite general. To illustrate this fact, we define a number of properties, each of which relates back to the previous example for further illustration.

> Economic equivalence can be established at any point in time.

There is no restriction as to what period we choose to establish equivalence. Once we have established equivalence, it is easy to restate it at any period, as it merely requires the application of a compound amount factor or present-worth factor for a single payment for any length of time.

[3] Moritsugu, J., "Heroux-Devtek Division Gets C$21.6M in Pacts," *Dow Jones Newswires,* September 22, 2003.

EXAMPLE 4.2

Economic Equivalence Revisited

In the previous example, we established equivalence by transforming the payment series into a single value at time 2006. Determine equivalence for the single C$22.58 million cash flow and the payments at time zero (the year 2003).

Solution. We must convert the payment series depicted in Figure 4.2(a) into an equivalent present value as in (b). For equivalence, the series must be equal to the time-zero equivalent of the 2006 cash flow as illustrated in (c).

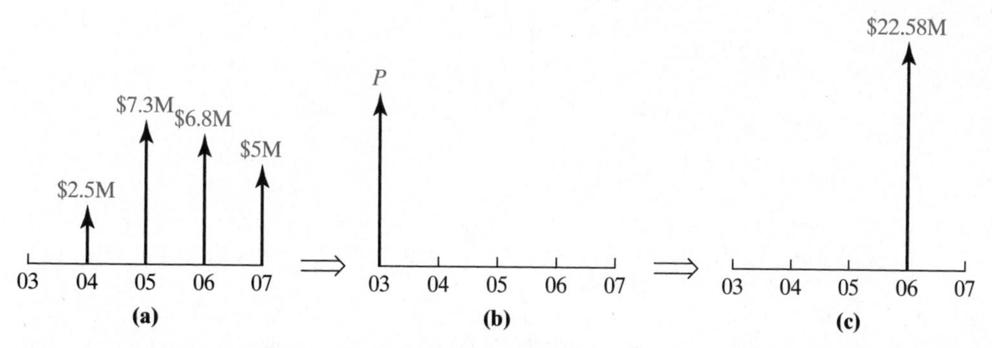

Figure 4.2 (a) Cash payments for deliveries of aircraft engine components, (b) time-zero (the year 2003) equivalent, and (c) the year-2006 cash flow.

The present value is defined with the use of the present-worth factor for the individual cash payments:

$$P = \frac{\text{C\$2.5M}}{(1+0.12)^1} + \frac{\text{C\$7.3M}}{(1+0.12)^2} + \frac{\text{C\$6.8M}}{(1+0.12)^3} + \frac{\text{C\$5.0M}}{(1+0.12)^4} = \text{C\$16.07 million.}$$

The single payment in 2006 is equivalent to

$$P = \text{C\$22.58M}\left[\frac{1}{(1+0.12)^3}\right] = \text{C\$16.07 million}$$

at time 2003, which establishes equivalence. In fact, we could choose any period in which to establish equivalence, even if it falls outside of our periods of analysis (2003 through 2007), assuming that we have defined the interest rate appropriately.

Economic equivalence can be established over any number of periods.

Equivalence need not be established with a single value at a single point in time. Rather, it can be established as a series of cash flows over time. The series does not have to be structured, but an equal-payment series would be the most common series used and follows most easily from analysis.

EXAMPLE 4.3

Economic Equivalence over Multiple Periods

Using the data from the previous examples, establish equivalence as an equal-payment series over the years 2004, 2005, and 2006.

Solution. Figure 4.3 is similar to our cash flow diagrams in the previous example, but the desired conversion is to an equal-payment series over the years 2004 through 2006.

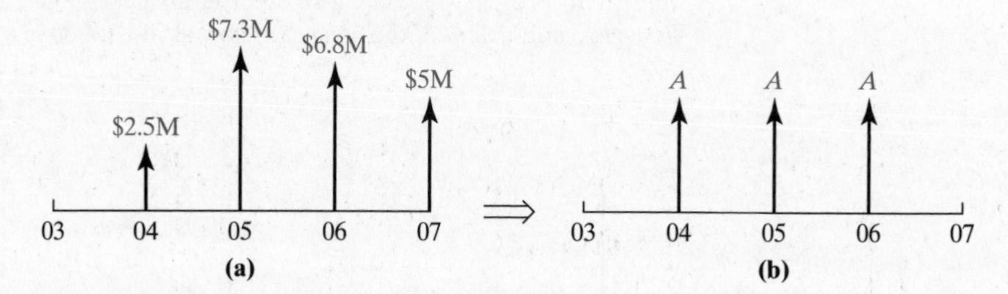

Figure 4.3 (a) Cash payments for deliveries of aircraft engine components and (b) equal-payment series over the years 2004 through 2006.

Previously, we found that the cash payments had an equivalent present value (in 2003) of C$16.07 million. This relationship can be rewritten as an equal annual series over the appropriate periods as

$$A = \text{C\$}16.07\text{M}(\overset{A/P,12\%,3}{0.4163}) = \text{C\$}6.69 \text{ million.}$$

Similarly, we could spread the single payment of C$22.58 million in 2006 over the same three periods:

$$A = \text{C\$}22.58\text{M}(\overset{A/F,12\%,3}{0.2963}) = \text{C\$}6.69 \text{ million.}$$

This again establishes equivalence of the respective sets of cash flows.

The difference between two economically equivalent cash flows is zero.

It should be obvious by now that the difference between two economically equivalent cash flows is zero. The reason is that they take on the same value at any point in time. This property is important in analysis, because what is important when comparing two sets of cash flows is their differences, not their similarities. It is these differences that define the worth of one project compared with another, as similarities merely cancel each other out.

EXAMPLE 4.4

Economic Equivalence Revisited

Continuing with Example 4.1, consider the original cash flow payments of C$2.5 million in 2004, C$7.3 million in 2005, C$6.8 million in 2006, and C$5.0 million in 2007. Consider further an equal series of payments of C$5.29 million over the same four years. Show that the two series are economically equivalent.

Solution. The difference between the cash flows defines the following new series of cash flows:

$$A_1 = \text{C\$}2.5 - \text{C\$}5.29 = -\text{C\$}2.79 \text{ million};$$

$$A_2 = \text{C\$}7.3 - \text{C\$}5.29 = \text{C\$}2.01 \text{ million};$$

$$A_3 = \text{C\$}6.8 - \text{C\$}5.29 = \text{C\$}1.51 \text{ million};$$

$$A_4 = \text{C\$}5.0 - \text{C\$}5.29 = -\text{C\$}0.29 \text{ million}.$$

This series is illustrated in Figure 4.4.

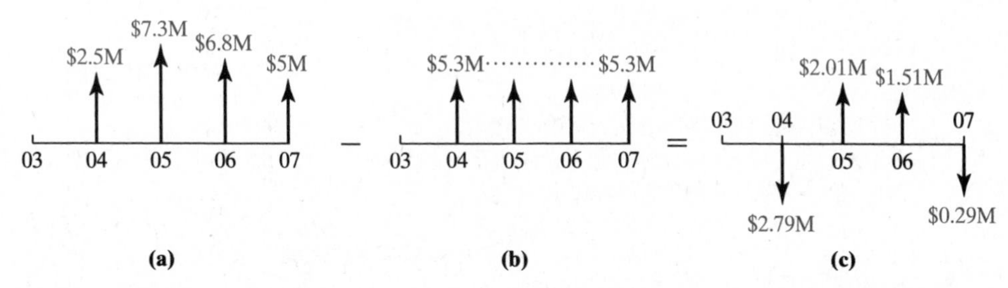

Figure 4.4 (a) Cash payments for deliveries of aircraft engine components, (b) equal-payment series of C$5.29 million, and (c) difference in cash flow series.

An equivalent present value of this new set of cash flows, defined in Figure 4.4(c) is

$$P = \frac{-\text{C\$}2.79\text{M}}{(1+0.12)^1} + \frac{\text{C\$}2.01\text{M}}{(1+0.12)^2} + \frac{\text{C\$}1.51\text{M}}{(1+0.12)^3} + \frac{-\text{C\$}0.29\text{M}}{(1+0.12)^4} = \text{C\$}0.$$

Thus, equivalence has been established by examining the differences of the cash flows.

Economic equivalence does not require a constant interest rate over the period examined.

Until now, we have analyzed cash flows under the assumption of a constant interest rate over N periods. In general, this restriction is not required, as the interest rate can fluctuate with time. Care must be taken, however, in analyzing cash flows with differing interest rates.

EXAMPLE 4.5

Economic Equivalence with Varying Interest Rates

Revisit our *original* payment plan in Example 4.1, but assume the following interest rates:

Year 2004:	12% compounded semiannually.
Years 2005–06:	1.25% per month.
Year 2007:	3% per quarter.

Note that we have now defined a completely different problem. We wish to show that the cash payments are equivalent to a single payment of C$15.25 million at time 2003.

Solution. With these interest rates, we can redraw the cash flow diagram to account for the different compounding periods, as in Figure 4.5. The figure also notes the differing interest rates over time.

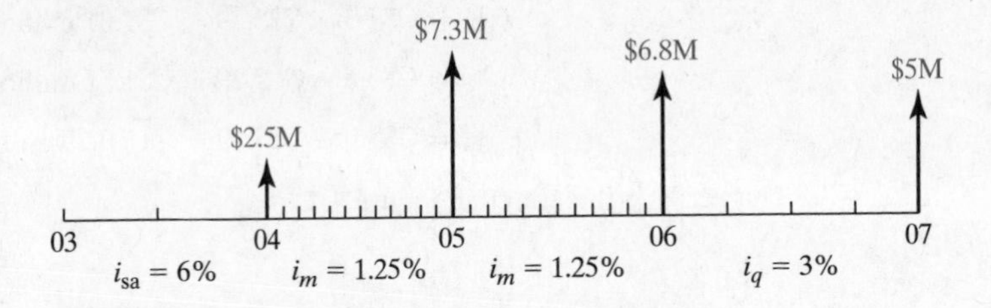

Figure 4.5 Cash payments for deliveries of aircraft parts with time redrawn to match the compounding period of the changing interest rates.

(1) Individual Compounding Period Approach

We will analyze this approach in stages. First, consider the C$5.0 million cash flow at time 2007. That cash flow is equivalent to

$$\text{C\$5.0M}\frac{1}{(1+0.03)^4} = \text{C\$4.44 million}$$

at time 2006. We can combine this value with the C$6.8 million flow at time 2006, totaling C$11.24 million. This total value could be moved to time 2004 (because the rate is constant over that period) as follows:

$$\text{C\$11.24M}\frac{1}{(1+0.0125)^{24}} = \text{C\$8.34 million.}$$

Note the use of $N = 24$, due to the definition of a monthly interest rate. This amount can be combined with the C$2.5 million cash flow at time 2004 and the C$7.3 million cash flow at time 2005 *after* it has been discounted one year (12 months) to yield

$$\text{C\$8.34M} + \text{C\$2.5M} + \text{C\$7.3M}\frac{1}{(1+0.0125)^{12}} = \text{C\$12.13 million}$$

at time 2004. Finally, that amount can be discounted one more year as

$$\text{C\$17.13M}\frac{1}{(1+0.06)^2} = \text{C\$15.25 million.}$$

Equivalently, we could have performed these operations in one step. The following equation illustrates bringing each cash flow back one year at a time:

$$P = \left(\text{C\$2.5M} + \left(\text{C\$7.3M} + \left(\text{C\$6.8M} + \frac{\text{C\$5.0M}}{(1+0.03)^4}\right)\frac{1}{(1+0.0125)^{12}}\right)\right.$$
$$\left.\times \frac{1}{(1+0.0125)^{12}}\right)\frac{1}{(1+0.06)^2}$$
$$= \text{C\$15.25 million.}$$

This establishes equivalence with a single cash flow at time 2003. As expected, the same answer is returned from both methods. Note that the value is slightly lower than our C$16.07 million when the interest rate was 12% per period. The average annual rate of interest in this example is higher, explaining the difference in the present values.

(2) Single Compounding Period Approach

Since the cash flows are defined annually, we could alternatively have defined each interest rate as an annual rate, calculated as follows:

Period	Original Rate	Annual Rate
2004	12% compounded semiannually.	$\left(1+\frac{0.12}{2}\right)^2 - 1 = 0.1236 = 12.36\%$
2005–06	1.25% per month.	$(1+0.0125)^{12} - 1 = 0.1608 = 16.08\%$
2007	3% per quarter.	$(1+0.03)^4 - 1 = 0.1255 = 12.55\%$

We redraw our cash flow diagram in Figure 4.6. Although the flows match those of our original example, the interest rate is defined for each period, as it is no longer constant.

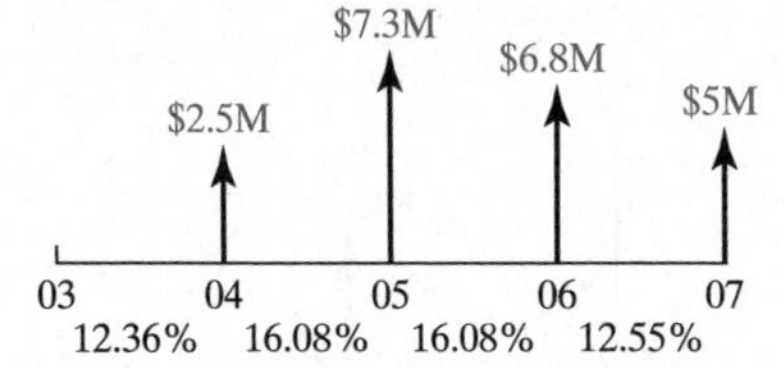

Figure 4.6 Cash payments to Heroux-Dektek, Inc., with varying interest rates over time.

This leads to the calculation of the equivalent present value as

$$P = \left(\text{C\$2.5M} + \left(\text{C\$7.3M} + \left(\text{C\$6.8M} + \frac{\text{C\$5.0M}}{(1+0.1255)}\right)\frac{1}{(1+0.1608)}\right)\times\frac{1}{(1+0.1608)}\right)\frac{1}{(1+0.1236)}$$

$$= \text{C\$15.25 million.}$$

While equivalence can be established with different interest rates, note that it must be reestablished if the rates change.

> Economic equivalence can be established with the use of current dollars and a market interest rate or real dollars and an inflation-free interest rate.

In Chapter 2, we noted that price changes are a result of inflation or deflation and are measured with our inflation rate f. Because the market and inflation-free interest rates are linked through the inflation rate, equivalence can be established with either current- or real-dollar cash flows. (See Section 2.7.3.) Note that the inflation rate is mathematically equivalent to our growth rate g for a geometric gradient. We illustrate this concept in the next example.

EXAMPLE 4.6

Economic Equivalence with Inflation

BHP Billiton, Ltd., together with its partners, agreed to supply four Chinese steel mills with 12 million metric tons of iron ore annually over the next 25 years. The mills will not pay more than the prevailing world ore price during the length of the contract. Revenues are estimated at roughly $9 billion over the life of the contract.[4] Assume a 2.75% annual rate of inflation for iron ore over the life of the contract. Determine an equivalent present value, assuming that the market interest rate is 12% per year and time zero is January 1, 2005. Further, assume end-of-year cash flows, and assume that the price for one metric ton of iron ore is $0.36 at time zero.

Solution. We have the choice of analyzing the revenue stream with either (1) real-dollar cash flows and the inflation-free rate or (2) current-dollar cash flows and the market interest rate. We first address the real-dollar analysis. Recall that this analysis ignores the effects of inflation. Thus, the revenue stream associated with analysis (1) is given in Figure 4.7. Because we ignore the effects of inflation, the revenue stream is constant over the 25 years. The annual cash flow is

$$12 \frac{\text{million tons}}{\text{year}} \times \frac{\$0.36}{\text{ton}} = \$4.32 \text{ million per year.}$$

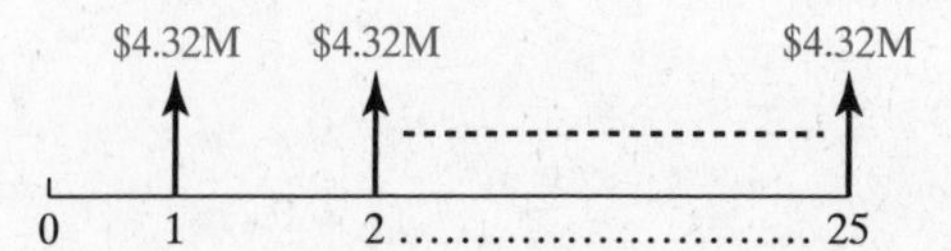

Figure 4.7 Real-dollar cash flow diagram for 25 years of iron-ore shipments.

To analyze the *real-dollar cash flows*, we require an inflation-free interest rate. Recall from Chapter 2 that the inflation-free rate is merely the market interest rate with the effects of inflation removed; thus,

$$i' = \frac{(1+i)}{(1+f)} - 1 = \frac{1+0.12}{1+0.0275} - 1 = .09002 = 9\%.$$

The equivalent present value for the real-dollar cash flow diagram can be found by using the present-worth factor for an equal-payment series. Recall that A'_n denotes real-dollar cash flows such that

$$P = A'(P/A, i', N) = \$4.32\text{M}(P/A, 9\%, 25) = \$4.32\text{M}(\overset{P/A,9\%,25}{9.8226}) = \$42.43 \text{ million.}$$

For the case of *current-dollar cash flows*, we need to convert the real dollars. As introduced in Chapter 2, this is accomplished by incorporating the effects of inflation:

$$A_n = A'_n(1+f)^n = \$4.32\text{M}(1+0.0275)^n.$$

For example, the current-dollar cash flows at the end of periods 1 and $N = 25$ are, respectively,

$$A_1 = \$4.32\text{M}(1+0.0275)^1 = \$4.439 \text{ million;}$$
$$A_{25} = \$4.32\text{M}(1+0.0275)^{25} = \$8.512 \text{ million.}$$

This conversion results in the current-dollar cash flow diagram given in Figure 4.8. The diagram defines a geometric gradient with $g = 2.75\%$.

[4] Brown, O., "BHP Sets China Iron-Ore Deal," *Dow Jones Newswires,* March 1, 2004.

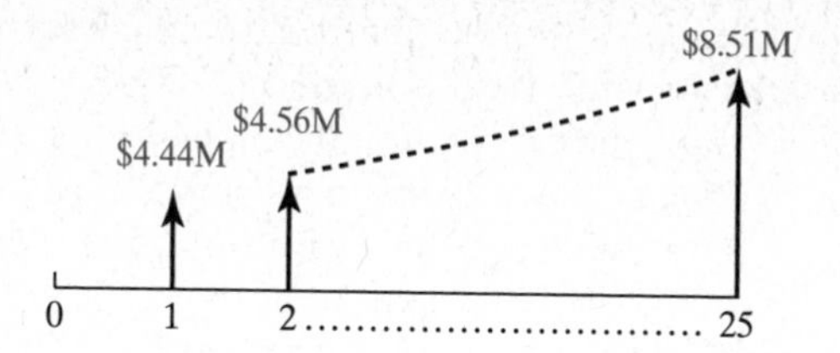

Figure 4.8 Current-dollar cash flow diagram for 25 years of iron-ore shipments.

Since $i = 12\%$ and $A_1 = \$4.439$ million, the equivalent present value of this series is

$$P = A_1(P/A_1,g,i,N) = \$4.439\text{M}(P/A_1,2.75\%,12\%,25) = \$4.439\text{M}(9.5578)$$
$$= \$42.43 \text{ million.}$$

As expected, using real or current dollars results in the same equivalent present value. However, this is assured only if the proper interest rates are used.

The reason we are confident that real and current cash flows will always generate the same solution is because they are tied together by the inflation rate. Let us look at the previous example in a more general way to convince you that the analyses will always be consistent. If we are dealing with current dollars over N periods of study, then the equivalent present value of those cash flows is

$$P = A_0 + \frac{A_1}{(1+i)} + \frac{A_2}{(1+i)^2} + \frac{A_3}{(1+i)^3} + \cdots + \frac{A_N}{(1+i)^N}. \tag{4.1}$$

As noted in the previous example, the inflation-free rate i' is related to the market interest rate i by the inflation rate f such that

$$(1+i) = (1+i')(1+f).$$

Substituting this expression into Equation (4.1) gives

$$P = A_0 + \frac{A_1}{(1+i')(1+f)} + \frac{A_2}{(1+i')^2(1+f)^2} + \frac{A_3}{(1+i')^3(1+f)^3} + \cdots + \frac{A_N}{(1+i')^N(1+f)^N}.$$

Noting that

$$A'_n = \frac{A_n}{(1+f)^n},$$

we can complete our derivation:

$$P = A_0 + \frac{A'_1}{(1+i')} + \frac{A'_2}{(1+i')^2} + \frac{A'_3}{(1+i')^3} + \cdots + \frac{A'_N}{(1+i')^N}. \tag{4.2}$$

Since the value of P has not changed, the analyses with current-dollar cash flows, Equation (4.1), and real-dollar cash flows, Equation (4.2), are equivalent. Note that one must be careful to utilize the correct interest rate for analysis.

These examples illustrate that economic equivalence between two sets of cash flows can be established in many ways. It is most common to establish equivalence with a single value at either period zero or period N or with an equal set of values over N consecutive periods. This is because our interest factors were derived for those specific situations.

4.2 Equivalence Involving the Interest Rate

Thus far, we have been interested in computing the values of F, P, or A, given a cash flow diagram with an interest rate i. It is possible that we will be faced with scenarios in which desired cash flow conversions are known, but we want to identify the interest rate that makes the conversion possible.

From a theoretical point of view, this should be easy, as we have only defined a different unknown variable that we could isolate algebraically. Consider a situation in which we have cash flows P at time zero and F at time N. What periodic interest rate i equates the values of P and F, given that they are N periods apart? Recall the compound amount factor for a single payment, Equation (3.1), which equates the cash flows as

$$F = P(1+i)^N.$$

Solving for i, we have

$$i = \left(\frac{F}{P}\right)^{\frac{1}{N}-1}.$$

This seems straightforward, but let us consider a different question: What periodic interest rate i equates the values of P and an equal-payment series of size A over N periods, beginning at time 1? Using the present-worth factor of an equal payment series, Equation (3.6), we can write

$$P = A\left[\frac{(1+i)^N - 1}{i(1+i)^N}\right].$$

Solving for i in this case is not a trivial matter. In situations like this, we can generally find i through trial-and-error methods and make use of a computer to ease the workload. We illustrate in the next example.

EXAMPLE 4.7 ***Finding i through Interpolation***

In the fall of 2003, French tire manufacturer Michelin SCA signed a EUR1.22 billion contract over eight years with IBM to maintain and manage its information technology infrastructure in Europe and North America.[5] Assume that payments of EUR152.5 million are to be made each year to IBM, with 2003 designated as time zero. What must the annual interest rate be in order to claim that the contract is worth EUR1 billion at time zero?

Solution. The cash flow diagram with the value of $P =$ EUR1 billion and $A =$ EUR152.5 million is given in Figure 4.9.

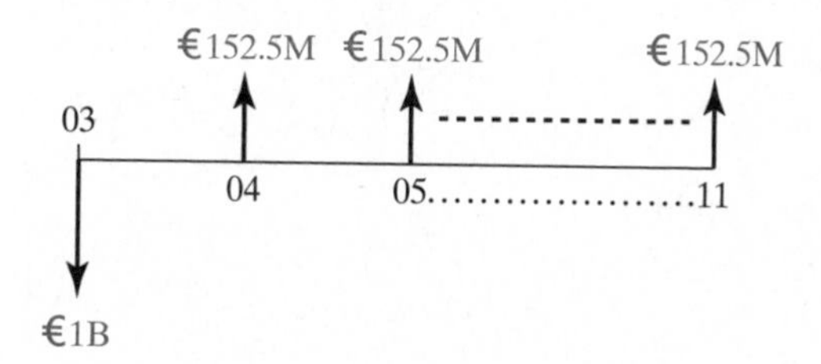

Figure 4.9 Time-zero equivalent of eight-year contract.

We can equate the present value to the annual series by using our equal-payment series present-worth factor. Since P and A are given and the interest rate is unknown, we can solve for the factor as the unknown variable:

$$P = A(^{P/A,i,N})$$

$$\text{EUR1B} = \text{EUR152.5M}(^{P/A,i,8})$$

$$\Rightarrow (^{P/A,i,8}) = 6.5574.$$

Examining the tables for different interest rates in the appendix, we find that

$$(^{P/A,4\%,8}) = 6.7327;$$

$$(^{P/A,5\%,8}) = 6.4632.$$

From these values, we know that the interest rate lies between 4% and 5%. We can get a good answer quickly by interpolating between these two values. Linear interpolation assumes that the function being estimated between two known points is a line. Thus, to determine the interest rate, we must merely construct a line with our two points (4% and 5%) and compute the interest rate that has a corresponding interest factor of 6.5574. This approach is illustrated in Figure 4.10.

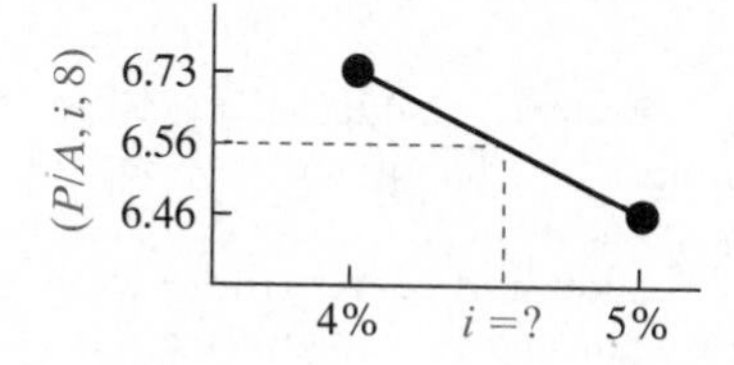

Figure 4.10 Linearly interpolating the equal-payment series present-worth factor between 4% and 5%.

[5] Delaney, K.J., "IBM Wins $1.22 Billion Contract from Tire Maker Michelin," *The Wall Street Journal Online*, December 12, 2003.

The figure is not necessary, but it does illustrate clearly what we are trying to accomplish. Since we assume that the function is linear between 4% and 5%, the slope should be constant. It is calculated as the ratio of the rise (movement in interest factor values) over the run (movement in interest rate):

$$\frac{6.7327 - 6.4632}{0.04 - 0.05} = -26.95.$$

We can equate the slope value to the rise over the run by using our unknown interest rate x:

$$\frac{6.5574 - 6.4632}{x - 0.05} = -26.95.$$

Solving for x yields

$$x = 0.05 + \frac{6.5574 - 6.4632}{-26.95} = 0.0465,$$

so that the interest rate which equates P and A is 4.65%.

Because we are using linear interpolation, we could have written, equivalently,

$$\frac{6.7327 - 6.5574}{0.04 - x} = -26.95$$

in solving for the interest rate.

Interpolation is a general tool. One can assume any function between two points to make an estimate. The use of linear interpolation here may not be justified, as the interest rate factors are not in fact linear. However, if the endpoints chosen (4% and 5%) have a small difference (say, less than 5%), then the error will be small and should suffice for decision-making purposes.

Although linear interpolation is a powerful tool, we can be more precise in our estimates by searching for the interest rate directly. In the previous example, our search would have been between the interest rates of 4% and 5%. The use of a computer makes this task quite simple, as illustrated in the next example.

EXAMPLE 4.8

Finding i using Goal Seek

Voisey's Bay Nickel Co. ordered multiple Caterpillar 3616, 4-megawatt generator sets for use at the company's mining site in Labrador, Canada, from Toromont Industries, Ltd.'s, Caterpillar dealership. The contract, valued at C$10 million, calls for the engineering, fabrication, and delivery of the diesel generator sets, controls, and accessories. The order is expected to be delivered in the fall of 2004, with a customer option for an additional unit in 2006.[6] Assume that C$10 million worth of equipment

[6] Li, J., "Toromont's Caterpillar Dealership Gets C$10M Generator Order," *Dow Jones Newswires,* October 10, 2003.

is delivered at the end of 2004 and the option is exercised such that C$5 million is delivered at the end of 2006. What is the interest rate that equates these revenues to a present value (at the end of 2003) of C$12 million?

Solution. The cash flow diagram (Figure 4.11) illustrates the present value P of C$12 million, together with two future values of C$10 and C$5 million.

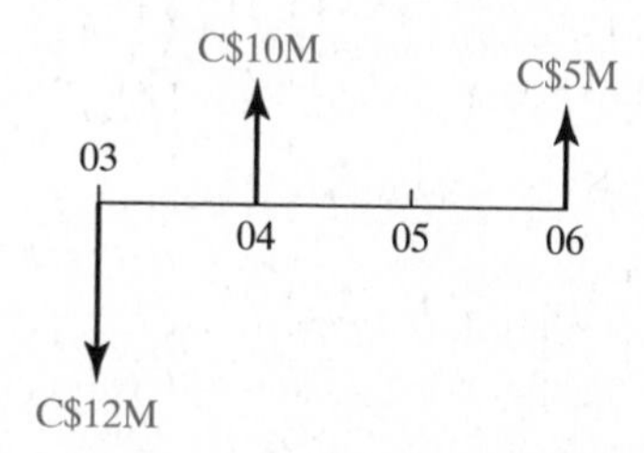

Figure 4.11 Revenues from the sale of diesel generators.

As there are two values in the future, we must use the present-worth factor for a single payment twice to equate the cash flows:

$$\begin{aligned} \text{C\$12M} &= \text{C\$10M}(^{P/F,i,1}) + \text{C\$5M}(^{P/F,i,3}) \\ &= \text{C\$10M}\frac{1}{(1+i)} + \text{C\$5M}\frac{1}{(1+i)^3}. \end{aligned}$$

We could solve for i "by hand," plugging in different values until we find one that makes the equation true, or we can turn to the computer. Using Goal Seek in Microsoft Excel, we can compute this value quite easily. Goal Seek is a search tool found under the Tools menu. It has three inputs in its dialogue box, as shown in Figure 4.12, which

	A	B	C	D	E	F	G
1	Example 4.8: Diesel Generator Purchases				Input		
2					Cash Flow Diagram		
3	**Period**	**Cash Flow**	**Present Worth**		Periods	3	years
4	0	-$12,000,000.00	-$12,000,000.00				
5	1	$10,000,000.00	$8,703,504.53				
6	2	$0.00	$0.00		**Output**		
7	3	$5,000,000.00	$3,296,495.47		P	$0.00	
8					Interest Rate	14.90%	per year
9			=B5/(1+$F$7)^A5			=SUM(C4:C7)	

Goal Seek
Set cell: $F$6
To value: 0
By changing cell: $F$7
OK Cancel

Sheet1 / Sheet2 / Sheet3

Figure 4.12 Goal Seek dialogue box for finding the interest rate that sets the equivalent present worth to zero.

also shows the cash flow diagram and data center. The function drives the value of a cell, defined by the user, to a given number, also specified by the user, by manipulating another cell, which is defined by the user as well. In economic analysis, this can be useful for determining break-even values.

Now, consider our current problem, in which we calculate the equivalent present worth by summing the individual cash flow present-worth values. If we enter the three inputs F6, 0, and F7 into the Goal Seek dialogue box, then it will drive cell F6 (P) to the value of zero by changing the value in cell F7 (the interest rate). As seen in Figure 4.12, the answer is 14.90%, found after hitting 'OK' in the Goal Seek dialogue box.

Note that Goal Seek is not infallible. It searches for a solution by trying a number of inputs until it identifies the parameter value that achieves the desired goal. (One does not have to drive the value of a cell to zero; any value is allowed.) However, it is possible that Goal Seek will not find a solution. If this occurs, you receive an error message. It is worth trying a different input into the cell that you are allowing to be changed in the Goal Seek function (cell F7 in our example). This often serves as a starting point for the Goal Seek algorithm and thus may lead the search to the correct answer.

Interpolation and Goal Seek are not the only two ways in which to determine an interest rate to establish equivalence. In Section 9.2.1, we outline a number of ways, in addition to the two mentioned here, in which to calculate the interest rate that forces the equivalent present worth to zero.

4.3 Equivalence Involving the Horizon

As with i, we may also be interested in determining the appropriate value of N to equate certain cash flows at a given interest rate. In the previous section, we learned that computing i is not trivial; the same argument holds for computing N.

There is an additional layer of complication in computing N that we have not experienced with any of the other variables. We have assumed in this chapter that the compounding period of the interest rate corresponds to the timing of the cash flows. This assumption defines N as a number of compounding periods and thus an integer value. However, when we solve for N, either algebraically or through trial and error, we are not guaranteed to find an integer. Consequently, we must be careful as to how the question concerning N is posed. Generally, we seek minimum or maximum values of N such that some cash flow is greater than or equal to another cash flow. This allows for an integer-valued solution. We illustrate in the next example.

EXAMPLE 4.9

Finding a minimum N

The U.S. Army Corps of Engineers awarded a \$5.5 million contract to Conrad Industries, Inc., to build a 255-foot crane barge for maintenance dredging and repair operations.[7] Assume that the Army placed \$5 million into an account paying 1.25% quarterly interest at the end of the third quarter in 2003 and that delivery is contingent upon payment. What is the earliest quarter in which the barge can be delivered?

Solution. The cash flow diagram (Figure 4.13) illustrates the \$5 million placed into an account at time zero with the future value F required to be at least \$5.5 million.

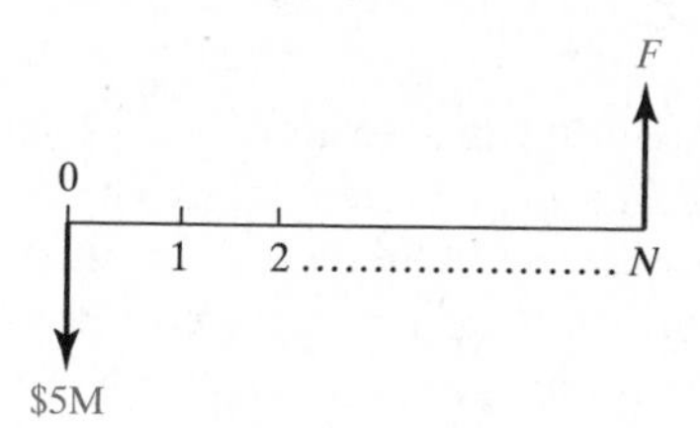

Figure 4.13 Equating \$5 million at time zero to a future amount that is at least \$5.5 million.

We can equate the values of P and F by using our compound amount factor or present-worth factor for a single payment. Using the compound amount factor yields

$$F = P(^{F/P,i,N});$$

$$\$5.5\text{M} = \$5\text{M}\,(1 + 0.0125)^N$$

$$\Rightarrow N = \frac{\ln(1.1)}{\ln(1.0125)} = 7.67.$$

We round the value of N to eight periods such that

$$F = P(^{F/P,i,N}) = \$5\text{M}(\overset{F/P,1.25\%,8}{1.1045}) = \$5.522 \text{ million}$$

is accumulated in the account. Only \$5.454 million is in the account after seven quarters; thus, delivery can occur in the eighth quarter or later, but not earlier.

	A	B	C	D	E	F	G	H
1	Example 4.9: Crane Barge Purchase				**Input**			
2					Cash Flow Diagram			
3	**Period**	**Cash Flow**	**Present Worth**		Interest Rate	1.25%	per quarter	
4	0	-\$5,000,000.00	-\$5,000,000.00					
5	1	- -	--		**Output**			
6	2	--	--		P	\$0.00		
7	3	--	--		Periods	8	quarters	
8	...	...	...		Quarters	7.672370808		
9	N	\$5,500,000.00	\$5,000,000.00					
10								
11								

=C4+C9

=ROUNDUP(F8,1)

Figure 4.14 Determining equivalent value of N by using Goal Seek and the ROUNDUP function.

[7] Huang, I., "Conrad Industries Gets \$5.5M Pact from US Army Corps of Engineers," *Dow Jones Newswires,* September 30, 2003.

Figure 4.14 illustrates how this solution can be found with the Goal Seek and ROUNDUP functions in Excel. The equivalent present worth in cell F6 is driven to zero by altering cell F8 with Goal Seek.

Because the Goal Seek search cannot be restricted to integer values, cell F7 is defined as

= ROUNDUP(F8,1),

which rounds the horizon value in F7 to the nearest integer.

4.4 Interest Factor Behavior

Before pressing on with additional examples of economic equivalence, it helps to build further intuition about interest factors by examining their behavior in the limits. That is, we want to examine our interest factors when the interest rate i and the time horizon N approach zero or infinity.

Consider the compound amount factor for a single payment, assuming discrete compounding:

$$(1+i)^N.$$

If we want to examine the behavior of the compound amount factor, we merely take its limit at the extremes:

$$\lim_{i \to 0}(1+i)^N = 1^N = 1.$$

This means that as the interest rate goes to zero, we can expect that $F \to P$, because

$$F = P(1+i)^N.$$

This should make sense, because if $i = 0$, then the value of P will not grow. Similarly, as i approaches infinity,

$$\lim_{i \to \infty}(1+i)^N = \infty.$$

Again, we should not be surprised by this result, as we would expect our future value to be infinite if we were paid interest based on an infinite interest rate.

We can perform similar analyses for the horizon N, so that

$$\lim_{N \to 0}(1+i)^N = 1.$$

Because there is no time for interest to accrue, we should expect that $F = P$ in this case. At the other extreme,

$$\lim_{N \to \infty}(1+i)^N = \infty.$$

Again, there is no surprise here, in that if there are an infinite number of compounding periods, then we would expect $F \to \infty$.

Table 4.1 summarizes the behavior of each of the factors in the limits, namely, $i \to 0$ and $N \to \infty$; the extremes where $i \to \infty$ and $N \to 0$ are not as interesting. The table lists the discrete compounding factors, but behavior in the limits is the same for continuous compounding. Note that the geometric gradients are not included, as they also require the definition of g.

TABLE 4.1

Behavior of interest rate factors for extreme *i* and *N* values.

Factor	$i \to 0$	$N \to \infty$
$(F/P,i,N)$	1	∞
$(F/A,i,N)$	N	∞
$(F/G,i,N)$	$\frac{N(N-1)}{2}$	∞
$(P/F,i,N)$	1	0
$(P/A,i,N)$	N	$1/i$
$(P/G,i,N)$	$\frac{N(N-1)}{2}$	$1/i^2$
$(A/F,i,N)$	$1/N$	0
$(A/P,i,N)$	$1/N$	i
$(A/G,i,N)$	$\frac{N-1}{2}$	$1/i$

Our reasons for studying interest rate factors in the limit are not necessarily practical. Rather, by studying these extreme values, we build or reaffirm our intuition that the factor behaves as we would expect if we were to increase or decrease the interest rate.

The case of the time horizon is also interesting. Although there are few instances where we might be faced with the case of $N = 0$, it is not surprising to see N approach very large—even infinite—values. The Millau Bridge, mentioned in Example 3.13, was built to last 120 years,[8] while the notes sold by Norfolk Southern in our chapter's introductory real decision problems have a 100-year life. Again, 100 or 120 are not infinity, but the values are quite large. Furthermore, there are examples in which N does equal infinity, such as the next example.

EXAMPLE 4.10

Perpetual Annuity

Industrialist Asa Packer, owner of the Lehigh Valley Railroad and controller of much coal mining in Eastern Pennsylvania, endowed Lehigh University with the gift of \$500,000 in 1865.[9] If the endowment had been placed into an account earning 7%

[8] Stidger, R., "Private Financing Builds Millau Bridge," *Better Roads,* May 2005.

[9] Yates, W.R., "The Beginning of a Lehigh Tradition," www3.lehigh.edu/about/luhistory.asp.

per year, what is the maximum amount that could be spent each year such that the principal of the investment does not decrease and the endowment would last forever?

Solution. Essentially, we are trying to determine the value A, given P, with a known interest rate and an infinite horizon, or

$$A = P(^{A/P,i,N}) = \$500{,}000(^{A/P,7\%,\infty}).$$

From Table 4.1, we know that the value of the $A/P \to i$ as $N \to \infty$; thus,

$$A = Pi = (\$500{,}000)(0.07) = \$35{,}000.$$

This is the maximum that can be spent each year to keep the principal intact.

The previous example is actually quite common. You may be familiar with the term "endowment." The idea behind an endowment is that an account (usually through a donation) is established to fund some activity, such as a student scholarship or chaired professorship at a college or university. The key is that these endowments are generally established to last forever. To achieve this, only the interest generated from the account can be spent each compounding period. If this steadfast rule is followed, the amount of money in the account will always be equal to the amount initially donated. If the interest rate is constant forever, then the amount that can be withdrawn from the account (the interest generated) will always be constant.

Mathematically,

$$A = \lim_{N\to\infty} P(^{A/P,i,N}) = \lim_{N\to\infty} P\left[\frac{i(1+i)^N}{(1+i)^N - 1}\right] = \lim_{N\to\infty} P\left[\frac{i\,\frac{(1+i)^N}{(1+i)^N}}{1 - \frac{1}{(1+i)^N}}\right]$$

$$= P\left[\frac{i}{1}\right] = Pi.$$

The reciprocal of this argument, or

$$P = \frac{A}{i},$$

is often referred to as the *capitalized equivalent amount*, or the amount that must be placed into an account in order to periodically pay A in perpetuity (forever), given the interest rate i per period.

4.5 Examples of Equivalence: Raising Capital

In this section, we look at three financial instruments that illustrate the concept of economic equivalence: loans, bonds, and stocks. Why should engineers study these topics? Most importantly, it is because firms that engage in engineering activities require funds—money to purchase equipment, expand production, and explore new technologies. Two sources of funds are lending institutions, such as banks, which provide loans, and investors, who may buy bonds or stocks. We briefly discuss stocks in the final subsection, but elaborate on loans and bonds in the next two subsections.

The reason we study loans and bonds closely, in addition to their usefulness, is that they are *deterministic*. That is, they are contracts such that when a bond is purchased or a loan is taken out, all of the information required for analysis is known with certainty (assuming that the owner keeps the bond until its date of maturity). This information includes all possible payments to and from the investor over time. Thus, loans and bonds are wonderful examples with which to perform analysis. Stocks are more difficult to analyze because information such as future stock values and dividend payments is not always known with certainty. However, stocks are a critical source of funding in addition to loans and bonds and are discussed so that our treatment is complete.

4.5.1 Loans

Loans are a traditional method for companies to acquire funds for financing projects. A lending institution, such as a bank, provides funds to a company. Over time, these funds are repaid along with interest payments. The interest is the fee, or price, the borrower pays for the use of the money.

Of course, loans are not restricted to companies, as people regularly take out loans to purchase big-ticket items such as homes, cars, and appliances. Loans can be taken from a variety of sources, including banks. However, people often get loans (or what is termed a payment plan) from dealers selling items, or they may even agree to a payment plan to repay funds borrowed from a relative.

A loan is essentially a contract in which funds are exchanged in an agreed-upon fashion. The party making the loan provides funds that are returned over time with interest payments. The key components of any loan are the **principal**, which is the amount being borrowed, the **interest rate**, which is used to calculate the interest payments, and the **payment plan**. Loan payments consist of both principal and interest over time.

As mentioned, payments on a loan can follow any agreed-upon pattern. However, payments generally follow either **equal total payment** or **equal principal payment** schemes. At a minimum, the interest payments must be made or the loan will go into default. If one defaults on a loan, the lending institution (e.g., a bank) may have the right to take some collateral, such as a piece of property, into its possession.

The total payment A_n made from the borrower to the lender at time n is composed of an interest payment IP_n and a principal payment PP_n. We define the total payment at time n formally as

$$A_n = IP_n + PP_n. \tag{4.3}$$

To calculate the payments to be made, regardless of the payment schedule, one must first calculate the **loan balance**, or the amount of principal that remains to be paid at any point in time. This is also referred to as the *outstanding balance* of the loan or the *principal outstanding*. We define B_n as the loan balance at time n. Note that B_0 is the principal amount P being borrowed, so that

$$B_0 = P; \tag{4.4}$$

$$B_n = B_{n-1} - PP_n. \tag{4.5}$$

With each principal payment PP_n, the amount owed declines, a fact that is reflected in the new loan balance. From the loan balance, the interest payment for any period is calculated as

$$IP_n = i B_{n-1}. \tag{4.6}$$

This requires that the periods n and the compounding period of the interest rate coincide. Payments are generally calculated in an iterative fashion, as illustrated in the next example.

EXAMPLE 4.11

Equal Principal Payments

In the summer of 2002, the U.S. federal government lent Amtrak, its national rail network, \$100 million in order to upgrade facilities and cover operating costs. It was hoped that the railroad could become self-sustaining after receiving the loan.[10]

While the terms of the loan were not disclosed, we make some necessary assumptions to illustrate the analysis. Assume that the loan is to be repaid with equal principal payments over a five-year period at a 6.5% annual interest rate. For simplification, we assume that the loan was made at the end of 2001, with the first payment due at the end of 2002.

Solution. The case of equal principal payments is straightforward, as the amount of principal to be paid with each payment is fixed. For this example, $PP_n = \$100/5 = \$20$ million is to be repaid each year. The interest to be paid, from Equation (4.6), is

$$IP_n = i B_{n-1} = (0.065) B_{n-1}, \quad \text{for } n = 1, 2, \ldots, 5.$$

The first interest payment is (0.065)(\$100M) = \$6.5 million, bringing the total payment in the first year to \$26.5 million. The complete five-year payment schedule is easily programmed into a spreadsheet as shown in Figure 4.15. The PP_n column is fixed and

[10] Machalaba, D., "Amtrak Is Seeking Heavier Funding for Rail Network," *The Wall Street Journal*, January 21, 2003.

	A	B	C	D	E	F	G	H	I	J
1	Example 4.11: Equal Principal Payments						Input			
2							Principal	$100,000,000.00		
3	Time n	Bn-1	PPn	IPn	An		PP	$20,000,000.00		
4	2002	$100,000,000.00	$20,000,000.00	$6,500,000.00	$26,500,000.00		Interest Rate	6.50%	per year	
5	2003	$80,000,000.00	$20,000,000.00	$5,200,000.00	$25,200,000.00		Periods	5	years	
6	2004	$60,000,000.00	$20,000,000.00	$3,900,000.00	$23,900,000.00					
7	2005	$40,000,000.00	$20,000,000.00	$2,600,000.00	$22,600,000.00		Output			
8	2006	$20,000,000.00	$20,000,000.00	$1,300,000.00	$21,300,000.00	=C7+D7	Payment Schedule			
9	Totals	--	$100,000,000.00	$19,500,000.00	$119,500,000.00					
10	=B5-C5		=B7*$H$4							
11										

Figure 4.15 Equal principal payment schedule for $100 million loan.

the IP_n column is merely a function of the B_{n-1} column. Finally, the A_n column is a sum of the PP_n and IP_n columns. The only other calculation that must be programmed is updating the loan balance B_{n-1}.

Tracking payments as in the spreadsheet in Figure 4.15 allows for straightforward analysis and keeps the account current. For example, if Amtrak wanted to pay off the loan early, it would merely have to pay the current loan balance and any outstanding interest.

The principal payments do not have to be equal in order to calculate the loan payments. Consider a graduated principal payment plan in the next example.

EXAMPLE 4.12 *Increasing Principal Payments*

Suppose the federal government wanted to give Amtrak more time to accumulate funds in order to pay back the $100 million loan. Analyze the loan, assuming principal payments of $0, $10, $20, $30, and $40 million over the five years.

Solution. The interest payment calculations follow as before for the 6.5% annual interest rate. All payments are given in the spreadsheet in Figure 4.16. Note that although the principal payments are not fixed per period, they have been defined by the contract. In this instance, the total interest paid increases because principal payments have been deferred until later.

	A	B	C	D	E	F	G	H	I
1	Example 4.12: Increasing Principal Payments						Input		
2							Principal	$100,000,000.00	
3	Time n	Bn-1	PPn	IPn	An		PP (G)	$10,000,000.00	
4	2002	$100,000,000.00	$0.00	$6,500,000.00	$6,500,000.00		Interest Rate	6.50%	per year
5	2003	$100,000,000.00	$10,000,000.00	$6,500,000.00	$16,500,000.00		Periods	5	years
6	2004	$90,000,000.00	$20,000,000.00	$5,850,000.00	$25,850,000.00				
7	2005	$70,000,000.00	$30,000,000.00	$4,550,000.00	$34,550,000.00		Output		
8	2006	$40,000,000.00	$40,000,000.00	$2,600,000.00	$42,600,000.00		Payment Schedule		
9	Totals	--	$100,000,000.00	$26,000,000.00	$126,000,000.00				

Figure 4.16 Increasing principal payment schedule for $100 million loan.

We contrast the increasing principal payment plan with that of a decreasing principal payment plan, as described in the next example.

EXAMPLE 4.13 *Decreasing Principal Payments*

We reverse our logic from the previous example and assume that Amtrak wants to pay the loan off faster. Accordingly, assume principal payments of $40, $30, $20, and $10 million over *four* years.

Solution. There is no change in the procedure for calculating the interest payments. All payments are summarized in the spreadsheet in Figure 4.17.

	A	B	C	D	E	F	G	H	I
1	Example 4.13: Decreasing Principal Payments						**Input**		
2							Principal	$100,000,000.00	
3	**Time n**	**Bn-1**	**PPn**	**IPn**	**An**		PP (G)	-$10,000,000.00	
4	1	$100,000,000.00	$40,000,000.00	$6,500,000.00	$46,500,000.00		Interest Rate	6.50%	per year
5	2	$60,000,000.00	$30,000,000.00	$3,900,000.00	$33,900,000.00		Periods	4	years
6	3	$30,000,000.00	$20,000,000.00	$1,950,000.00	$21,950,000.00				
7	4	$10,000,000.00	$10,000,000.00	$650,000.00	$10,650,000.00		**Output**		
8	Totals	--	$100,000,000.00	$13,000,000.00	$113,000,000.00		Payment Schedule		

Figure 4.17 Decreasing principal payment schedule for $100 million loan.

Compared with the increasing principal payments loan, this loan pays half of the amount of interest over time. This is not unexpected, due to the accelerated principal payment plan.

A more common payment plan has the same total payment made to the lender in each period. To determine the amount of this payment, an interest rate must be specified. Given the amount lent and the interest rate charged, we can calculate the equal total payments (principal and interest) by using our equal-payment series capital-recovery factor as defined in Chapter 3. Recall that this factor equates a single payment at time zero to a series of equal payments over N periods. The payments include the effects of interest, which represent the cost of the loan in our current examples. Using the equal-payment (capital-recovery) factor for a single payment, or Equation (3.10), we find that the total payment is

$$A_n = A = P(^{A/P,i,n}) = P\left[\frac{i(1+i)^N}{(1+i)^N - 1}\right].$$

The interest payment IP_n is calculated as before, with the interest rate and outstanding loan balance defined in Equation (4.5). The principal payment is merely the difference between the total payment A_n and the interest payment IP_n, from Equation (4.3). This is illustrated in the next example.

EXAMPLE 4.14

Equal Total Payments

Companhia Vale do Rio Doce SA (CVRD) of Brazil, the largest exporter of iron ore in the world, secured a seven-year, \$300 million loan from a group of nine international banks. The loan carries the six-month LIBOR rate plus 0.7%.[11] Although terms were not discussed, assume that the loan is to be paid over three years in equal semiannual total payments beginning in the summer of 2004. Assume further that the LIBOR rate is fixed at 1.6% compounded semiannually over the life of the loan. Define the payments (principal and interest) over the life of the loan.

Solution. With a 2.3% annual interest rate compounded semiannually, the six-month interest rate is (2.3%/2)=1.15%. The total payment to be made every six months over the three years is

$$A_n = P(^{A/P,i,n}) = \$300\text{M}(\overset{A/P,1.15\%,6}{0.1734}) = \$52.03 \text{ million}.$$

The interest paid in the first period is calculated as

$$IP_1 = i\,B_0 = (0.0115)\$300\text{M} = \$3.45 \text{ million}.$$

This leaves the first principal payment as

$$PP_1 = A_1 - IP_1 = \$52.03\text{M} - \$3.45\text{M} = \$48.58 \text{ million}.$$

As with our loans defined by principal payment schemes, we can easily construct payment tables in a spreadsheet, as in Figure 4.18. Here, the equal total payment is calculated with the PMT function in cell H7 and is then referenced in column E. The principal and interest calculations follow as before, but note that rounding intermediate calculations will lead to errors. These errors become obvious if the final principal payment does not equal the outstanding loan balance. In this example, we have calculated values out to the penny and computed the total annual payment with high accuracy in order to avoid the rounding issue.

For equal total payment loans, principal and interest payments can also be calculated directly by means of built-in spreadsheet functions. In Excel, the PPMT and

	A	B	C	D	E	F	G	H	I	J
1	Example 4.14: Equal Total Payments									
2							Input			
3	Time n	Bn-1	PPn	IPn	An		Principal	$300,000,000.00		
4	1	$300,000,000.00	$48,581,674.86	$3,450,000.00	$52,031,674.86		Interest Rate	1.15%	per six-months	
5	2	$251,418,325.14	$49,140,364.12	$2,891,310.74	$52,031,674.86		Periods	6	per six-months	
6	3	$202,277,961.02	$49,705,478.31	$2,326,196.55	$52,031,674.86					
7	4	$152,572,482.72	$50,277,091.31	$1,754,583.55	$52,031,674.86		Output			
8	5	$102,295,391.41	$50,855,277.86	$1,176,397.00	$52,031,674.86	=$H$7	An	$52,031,674.86		
9	6	$51,440,113.55	$51,440,113.55	$591,561.31	$52,031,674.86					
10	Totals	--	$300,000,000.00	$12,190,049.15	$312,190,049.15					
11	=B6-C6	=E9-D9	=B8*$H$3				=PMT(H3,H4,-H2)			
12										

Figure 4.18 Equal total payment schedule for \$300 million loan.

[11] Esterl, M., "Brazil's CVRD Obtains \$300 Mln, 7-Yr Syndicated Loan," *Dow Jones Newswires*, April 1, 2004.

IPMT functions calculate the principal and interest payment, respectively, for a given period, of an equal total payment loan. These functions are defined as follows:

$$= \text{PPMT(rate,per,nper,pv,fv,type)} = \text{PPMT}(i, n, N, -B_0);$$

$$= \text{IPMT(rate,per,nper,pv,fv,type)} = \text{IPMT}(i, n, N, -B_0).$$

We ignore the last two arguments (fv and type), as the present value is defined and end-of-period payments are assumed.

Thus, in our $300 million loan example, the first principal payment is calculated as

$$PP_1 = \text{PPMT}(0.0115, 1, 6, -300) = 48.582,$$

while the first interest payment is

$$IP_1 = \text{PPMT}(0.0115, 1, 6, -300) = 3.450,$$

in millions of U.S. dollars. In order to find values for other periods, the value of n must be changed accordingly. The functions in Quattro Pro and Lotus are defined as PPAYMT and IPAYMT, for principal and interest payments, respectively.

An interest rate is generally specified in a loan contract. However, it is possible that an advertisement for a loan will only define the payments. In order to calculate the principal and interest components of the payments, the interest rate, also defined as the **true cost of the loan**, must be calculated. This is merely the interest rate that equates the principal received to the payments at some common point in time. The next example illustrates this concept.

EXAMPLE 4.15 *Unspecified Interest Rate*

In order to expand its semiconductor production capacity from 40,000 8-inch wafers per month to 70,000, South Korea's Dongbu Anam Semiconductor, Inc., signed a syndicated loan worth KRW1.2 trillion won in the spring of 2004. The loan included a tranche (segment) denoted in U.S. dollars ($150 million).[12] Although no details were given, assume that the payment plan for the tranche was specified over four years, with total payments of $25M, $37.5M, $50M, and $62.5M in each succeeding year. What is the true cost of the loan?

Solution. The terms of the loan are described by the cash flow diagram in Figure 4.19. As the loan is for $150 million and there are $175 million in payments, we know that $25 million in total interest is paid over the life of the loan. However, without an interest rate, we do not know the timing of these payments.

We first must calculate the true cost of the loan by equating the loan principal to the payments at the same point in time. The payments consist of an equal-payment

---

[12] Kim, Y-H. and S-J. Cha, "S Korea's Dongbu Anam Signs KRW1.2T Syndicated Loan," *Dow Jones Newswires,* April 9, 2004.

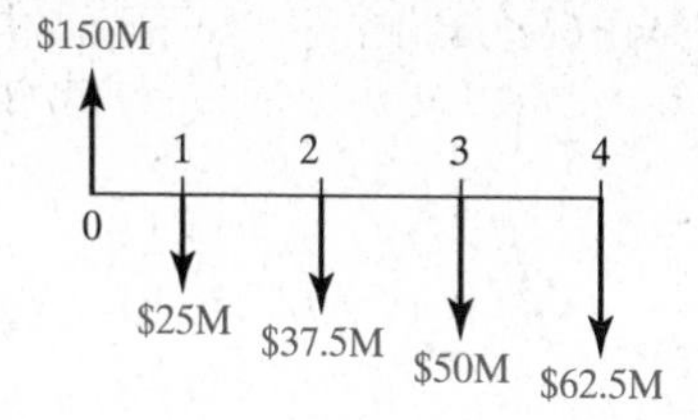

Figure 4.19 Initial loan and repayment over four years.

series with $A = \$25\text{M}$ and a uniform gradient series with $G = \$12.5\text{M}$. Equating the inflows to the outflows at time zero yields

$$P = A(^{P/A,i,n}) + G(^{P/G,i,n});$$

$$\$150\text{M} = \$25\text{M}(^{P/A,i,4}) + \$12.5\text{M}(^{P/G,i,4})$$

$$= \$25\text{M}\left[\frac{(1+i)^4 - 1}{i(1+i)^4}\right] + \$12.5\text{M}\left[\frac{(1+i)^4 - 4i - 1}{i^2(1+i)^4}\right].$$

Solving for i directly will prove difficult, because the interest rate is raised to the fourth power in this equation. Using the Goal Seek function in Microsoft Excel, as illustrated earlier in the chapter, results in $i = .0560566$ for this example.

Given the interest rate and the total payments, we can determine the periodic interest and principal payments, as in the spreadsheet in Figure 4.20. The interest rate was rounded to $i = 5.606\%$ for calculations.

	A	B	C	D	E	F	G	H	I
1	Example 4.15: Increasing Total Payments						**Input**		
2							Principal	$150,000,000.00	
3	**Time n**	**Bn-1**	**PPn**	**IPn**	**An**		A1	$25,000,000.00	
4	2004	$150,000,000.00	$16,591,510.00	$8,408,490.00	$25,000,000.00		G	$12,500,000.00	
5	2005	$133,408,490.00	$30,021,574.00	$7,478,426.00	$37,500,000.00		Interest Rate	5.61%	per year
6	2006	$103,386,916.00	$44,204,481.00	$5,795,519.00	$50,000,000.00		Periods	4	years
7	2007	$59,182,435.00	$59,182,435.00	$3,317,565.00	$62,500,000.00				
8	Totals	--	$150,000,000.00	$25,000,000.00	$175,000,000.00		**Output**		
9							Payment Schedule		

Figure 4.20 Payment schedule for $150 million loan.

Note that the spreadsheet functions called to determine the interest rate corresponding to the true cost of a loan generally assume equal total payment loans. When payments adhere to a different schedule, the analysis in Figure 4.20 must be performed and the use of functions cannot serve as a substitute.

The true cost of a loan is calculated by equating *all* outflows (payments) to inflows (principal) at some common point in time (generally, time zero). It is possible that when one calculates the true cost of a loan, it will not equal the interest rate quoted by the lender, even if it is an effective interest rate over the appropriate period. The reason is that there are often additional fees the borrower must pay. While this may not change how interest payments are calculated, these fees should be included in the calculation because they increase the cost of acquiring funds, which is reflected in the true cost of a loan interest rate. However, if there are no additional fees, then the true cost of the loan is defined by the payments. The possibility of additional fees can lead to various payment plans for the same true cost of a loan. We illustrate in the next example.

EXAMPLE 4.16

True Cost of a Loan

Compute the true cost of the loan for the four payment schemes analyzed for the $100 million Amtrak loan.

Solution. Three payment schemes were analyzed with defined principal payments, and the fourth plan was defined for equal total payments. For each of these loans, the total payments made by the borrower differed. What may not be clear is that the true costs of the loans for these payment schemes are the same. We can illustrate this by computing the time-zero equivalent of the four payment schemes.

For the equal principal payments plan,

$$P = \frac{\$26.5\text{M}}{(1+0.065)^1} + \frac{\$25.2\text{M}}{(1+0.065)^2} + \frac{\$23.9\text{M}}{(1+0.065)^3} + \frac{\$22.6\text{M}}{(1+0.065)^4} + \frac{\$21.3\text{M}}{(1+0.065)^5}$$
$$= \$100 \text{ million.}$$

For the increasing principal payments plan,

$$P = \frac{\$6.5\text{M}}{(1+0.065)^1} + \frac{\$16.5\text{M}}{(1+0.065)^2} + \frac{\$25.85\text{M}}{(1+0.065)^3} + \frac{\$34.55\text{M}}{(1+0.065)^4} + \frac{\$42.6\text{M}}{(1+0.065)^5}$$
$$= \$100 \text{ million.}$$

For the decreasing principal payments plan (the loan is actually paid off in four years under this plan),

$$P = \frac{\$46.5\text{M}}{(1+0.065)^1} + \frac{\$33.9\text{M}}{(1+0.065)^2} + \frac{\$21.95\text{M}}{(1+0.065)^3} + \frac{\$10.65\text{M}}{(1+0.065)^4} = \$100 \text{ million.}$$

Since the time-zero equivalents of all payments are $100 million, which is equal to the principal borrowed, the true cost of each loan is the same: 6.5% per year. Note that we need not analyze the equal total payments plan, because the annual payment A was computed with the 6.5% interest rate.

The preceding example highlights an interesting question: Are the loans the same? The answer is "Absolutely not," as they are defined by different payment plans. It is clear that there are other considerations that must be factored into how payment plans are established. A main consideration is cash flow. Companies (or individuals) may not have sufficient cash on hand to make payments; thus, alternative designs (that delay payments) are needed.

4.5.2 Bonds

Another manner in which companies or government entities can raise funds for capital projects is by issuing, or selling, what is called "debt." Specifically, a company sells bonds to investors whereby the investor essentially loans money to the seller of the bonds. The money is then available to the company to, for example, purchase equipment or expand production, while the investor is compensated in the form of interest according to a contract defined by the bond. The term "bond" means "contract," and it is a contract between the seller and buyer. Bonds are termed *marketable securities* if they can be sold on the market (to other investors) after they are purchased from the company or government entity.

Securities can be categorized according to their time to maturity or whether the investor receives coupon payments. The **time to maturity** is the length of time that the contract defined by the bond is active. A **coupon payment** is essentially an interest payment the investor receives as compensation for lending money (buying the bond).

In general, there are three types of securities, categorized according to length of life. **Bills** are short-term contracts that generally last 1 year or less. **Notes** have terms longer than 1 year, but no more than 10 years, **Bonds** have terms of 10 years or longer. Note that the term "bond" is used loosely and may often refer to any of these types of securities. We will also use the term "bond" in a general fashion.

As bills have such a short life, they generally do not pay interest in the form of coupon payments. Rather, the bill is purchased for less than **par** (face value). At the time of maturity, the par value is returned to the investor. Thus, the profit made by the investor is the difference between par and the price paid.

Notes and bonds are generally purchased at or near par, but unlike bills, they compensate the investor with interest payments. These payments are termed *coupon payments* and are defined by the **coupon rate** of the bond. For most bonds, the par value of the bond is returned to the investor at the time of maturity, regardless of what the investor paid for the bond.

Bonds are interesting to study because they are essentially a contract between an investor and an entity raising funds. Because they are a contract, most information such as the purchase price, coupon rate, coupon payment schedule, and time of maturity is known at time zero; thus, bonds can be easily analyzed. Note that many "complicated" bonds also exist on the market. For example, instead of receiving the par value of a bond at the maturity date, the investor may be given shares of stock in the company. Such a bond is termed a *convertible* bond. Bonds may also be recalled by the company before the maturity date. These are termed *callable* bonds. We will not examine these more complicated bond scenarios in great detail.

Zero-Coupon Bonds

Zero-coupon bonds tend to mature in a short time and are fairly easy to analyze, since there are only two cash flows in the transaction: The investor

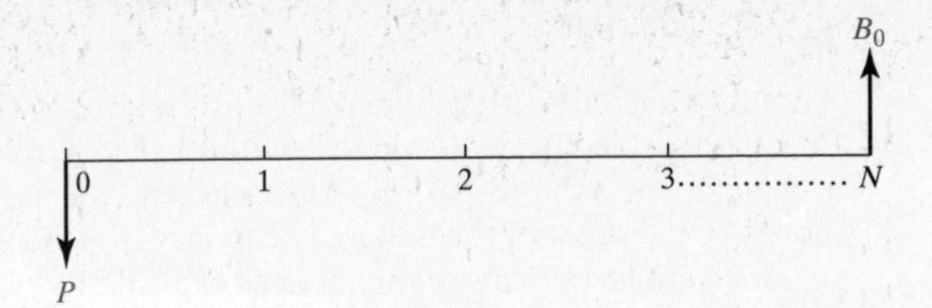

Figure 4.21 Typical cash flow diagram for a zero-coupon bond from the perspective of an investor.

pays the purchase price at time zero, and when the bond matures, the face value of the bond is returned to the investor. Clearly, the face value must be greater than the purchase price of the bond, so that the investor makes a return on the investment. The rate that equates the purchase price to the par value at an equal point in time is known as the **yield to maturity**.

From the perspective of the investor, the cash flow diagram that applies to purchasing a zero-coupon bond is given in Figure 4.21. The diagram is similar to that of the compound amount factor or present-worth factor for a single payment.

In this situation, the value P is the purchase price paid at time zero, and the value B_0 is the face value returned to the investor at the time of maturity, N. The values of B_0 and N are specified by the bond, as in a contract. The value of P is often defined by the market, but known to the investor at the time of purchase. Mathematically, the yield to maturity is the interest rate that equates P and B_0 at an equivalent point in time. Using time zero as our point of reference, we know from Chapter 2 that

$$P = B_0 \frac{1}{(1+i)^N}.$$

Solving for i, we find that the yield to maturity is

$$i = \left(\frac{B_0}{P}\right)^{1/N} - 1. \tag{4.7}$$

Note that the interest rate is compounded in each period over N.

EXAMPLE 4.17

Zero-Coupon Bond

The U.S. government sells securities every week, generally on Thursdays. The sale of securities provides funding for operations and "services debt" by replacing securities that are maturing with new securities. On December 22, 2005, the federal government sold 182-day Treasury bills (T-bills) at the price of \$97.866556 for every \$100 worth of bills. The maturity date was June 22, 2006.[13] What was the yield for the investor who held the bills until maturity?

[13] "Recent Treasury Bill Auction Results," *Bureau of the Public Debt Online,* U.S. Department of the Treasury, www.publicdebt.treas.gov, December 23, 2005.

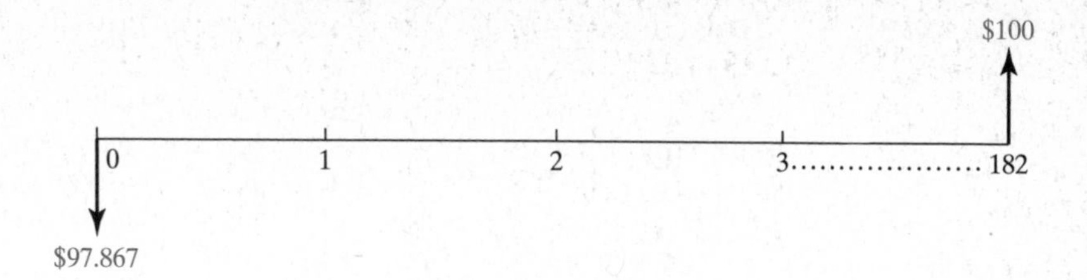

Figure 4.22
Cash flow diagram for a 182-day bill.

Solution. The cash flow diagram for the purchase of the T-bills is given in Figure 4.22. Regardless of the number of bills purchased, the yield to maturity is easily found from Equation (4.7). For a six-month yield,

$$i = \left(\frac{100.00}{97.866556}\right)^{1/1} - 1 = 0.02180 = 2.180\% \text{ per six months,}$$

or equivalently, for a daily yield,

$$i = \left(\frac{100.00}{97.866556}\right)^{1/182} - 1 = 0.0001185 = .01185\% \text{ per day.}$$

The annual yield is computed as

$$i = \left(\frac{100.00}{97.866556}\right)^{1/(1/2)} - 1 = 0.04407 = 4.407\% \text{ per year.}$$

The reader can verify that these interest rates are equivalent. (See Section 2.5.2 for a review.)

Bills are fairly straightforward to analyze, as there are only two cash flows defined in the transaction. Bonds and notes generally have coupon payments, complicating the analysis.

Coupon-Paying Bonds

Longer term securities generally compensate the investor with coupon, or interest, payments, defined by the coupon rate of the bond. If the coupon rate is i, then the amount of interest paid each year is iB_0, where B_0 is the face value of the bond. The amount of interest is divided equally according to the number of payments per year. For example, if semiannual payments are to be made, the payments are of size $iB_0/2$.

As before, the **yield to maturity** is the interest rate that equates the purchase price to the face value returned (at the time of maturity) *and* the coupon payments at a common point in time. This value is more difficult to calculate now due to the presence of additional cash flows. The next two examples illustrate coupon payments and yields, whether the bonds are purchased at par value or not.

EXAMPLE 4.18 ***Bond Example***

In September of 2003, Rayovac Corp. issued $350 million worth of 10-year bonds with a maturity date of October 1, 2013. The bonds sold at par and pay semiannual coupon payments of 8.50%.[14] Analyze the purchase of $10,000 worth of the bonds. Assume that they were purchased at par on October 1, 2003 and are kept to maturity. What is the yield to maturity?

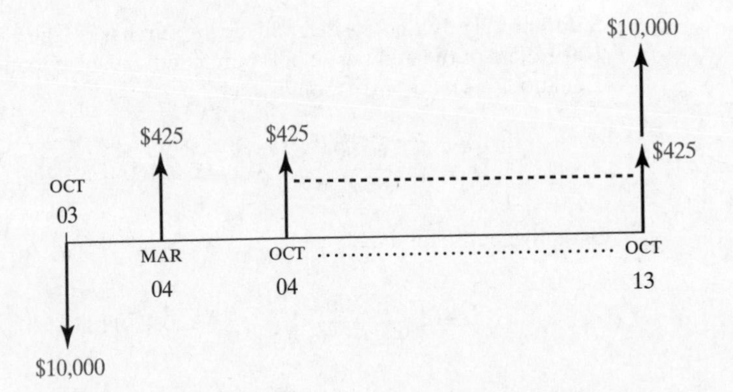

Figure 4.23 Cash flow diagram for 10-year bond investment with coupon rate of 8.50%.

Solution. The cash flow diagram for the investor is given in Figure 4.23, assuming that the bond is kept to maturity. The semiannual coupon payment is calculated as

$$IP_n = \frac{iB_0}{2} = \frac{(0.085)(\$10{,}000)}{2} = \$425.$$

The yield to maturity is calculated by equating the cash inflows to the cash outflows at time zero:

$$P = IP_n(^{P/A,i,N}) + B_0(^{P/F,i,N}).$$

$$\$10{,}000 = \$425(^{P/A,i,20}) + \$10{,}000(^{P/F,i,20})$$

$$= \$425\left[\frac{(1+i)^{20}-1}{i(1+i)^{20}}\right] + \$10{,}000\left[\frac{1}{(1+i)^{20}}\right].$$

Using Goal Seek in Microsoft Excel, we find that $i = 4.25\%$ per six months, or 8.68% per year. Note that the coupon rate (8.50%) and the yield to maturity (8.68%) do not match exactly, as we have assumed compounding every six months to calculate our yield to maturity, so that $(1.0425)^2 - 1 = 0.08681$.

It can be shown that the yield to maturity and the coupon rate are the same when a bond is purchased and sold at par and payments are made *annually*. Because the coupon rate defines the annual amount of interest paid, only annual payments ensure

[14] Rappaport, L., "Rayovac's $350M 10-Yr Debt Prices to Yield 8.50%," *Dow Jones Newswires*, September 26, 2003.

that the coupon rate and the yield to maturity are the same. Note that with semiannual payments, the nominal interest rate (compounded semiannually) will equal the coupon rate. In the previous example, the nominal (annual) interest rate for the yield to maturity is 2(4.25%) = 8.50% compounded semiannually, and the coupon rate was defined as 8.50%.

When a company decides that it is going to sell a bond at a given price and coupon rate, it may have to adjust one or more parameters according to market conditions at the time of sale, because markets are constantly changing. Often, the coupon rate is actually set as a variable based on a bank's rate. For example, in September of 2003, General Electric (GE) raised 85 billion yen through its Japanese funding arm in order to fund operations in Japan.[15] In the offering, ¥20 billion was raised by selling bonds with a September 24, 2013, maturity date. The coupon rate was quoted as LIBOR plus 40 basis points. (A basis point is one-tenth of a percentage point, or .01%. Again, LIBOR is the London Inter-Bank Offer Rate, or the interest rate that banks charge each other for loans; it is the rate most commonly used for pricing in Europe.) The LIBOR is officially fixed once per day by a small group of large London banks, but the rate does fluctuate throughout the day. In essence, the coupon rate of the bond was not set until the value of the LIBOR rate was known at the time of the sale. In the GE example, the rate was set at 2.03% when the LIBOR rate was known. This state of affairs is quite common when selling securities. U.S.-based companies will often tie their coupon rate to a government-issued Treasury note with the same time to maturity. This is why the "spread," or difference, between the yield to maturity for a bond and the yield to maturity for a similar Treasury note is often quoted.

A company may also decide to fix the coupon rate in order to make the coupon payments a "round" or "even" number. When this occurs, the price of the note or bond may actually fluctuate from par according to slight changes in the interest rates. In the next example, the coupon rate is fixed and the bond is consequently sold at a slight discount due to changes in current market rates.

EXAMPLE 4.19

Bond Example

On September 17, 2003, Cabot Finance, the financial arm of Cabot Corporation, a leading producer of specialty metals and chemicals (such as carbon black and tantalum), raised $175 million selling 10-year notes expiring on September 1, 2013. The notes were priced at $99.423 per $100, with a coupon rate of 5.25%.[16] What is the yield to maturity for a $10,000 investment?

---

[15] Seki, M., "WRAP: GE Raises Y85 Bln Through Novel Japan Bond Issue," *Dow Jones Newswires,* September 10, 2003.

[16] Geressy, K., "Cabot Finance $175M 10-Yr 144a Priced at Treasurys +1.15," *Dow Jones Newswires,* September 17, 2003.

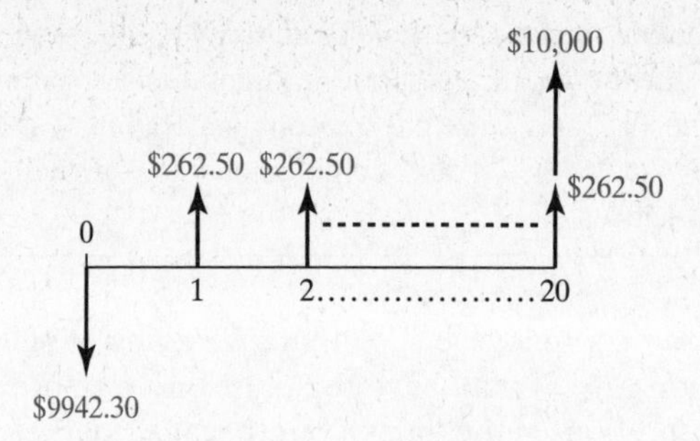

Figure 4.24 Cash flow diagram for a discounted $10,000 bond purchase with a coupon rate of 5.25%.

Solution. The cash flow diagram from the perspective of the investor who keeps the bond to maturity is given in Figure 4.24. The semiannual coupon payment IP_n is calculated as

$$IP_n = \frac{i B_0}{2} = \frac{(0.0525)(\$10{,}000)}{2} = \$262.50.$$

The yield to maturity is calculated by equating the discounted coupon payments and par-value payment to the purchase price:

$$\$9942.30 = \$262.50(P/A,i,20) + \$10{,}000(P/F,i,20)$$

$$= \$262.50\left[\frac{(1+i)^{20}-1}{i(1+i)^{20}}\right] + \$10{,}000\left[\frac{1}{(1+i)^{20}}\right].$$

Using Goal Seek in Microsoft Excel, we find that $i = 2.66\%$ per six months, or 5.40% per year. Note that the yield to maturity is greater than the coupon rate, as the discounted purchase price increases the return on the investment.

Marketable securities get their name because an investor is not obligated to hold the bond until the maturity date. Rather, he or she can sell the bond to another investor. When this occurs, the rights of the bond transfer to the new owner, such that the new owner of the bond receives the coupon payments and the par value on the maturity date. From the perspective of the company or government entity that sold the original bond, nothing changes (except the mailing address for the coupon payments!).

You can imagine the difficulty investors would have if they had to sell a bond on their own. After all, placing an ad for a bond in the local paper is not likely to get the same response as placing one for a used car. Accordingly, there is a market for selling bonds, and sales can easily be accomplished with most brokers (professionals who help investors purchase financial instruments such as stocks and bonds). The market is essentially a way to match buyers with sellers.

When an investor sells a bond, three things can happen: The investor can sell it (1) at par, which is the face value of the bond; (2) at a **discount**, in that the investor receives less than the face value; or (3) at a **premium**, in that the bonds fetch more than their par value. Before examining why this occurs, let us look at two examples of purchasing bonds on the market.

EXAMPLE 4.20

Bond Purchased at Premium

On September 19, 2003, Texas Instruments, Inc., bonds with a maturity of April 1, 2007, and a coupon rate of 8.75% paid in semiannual payments were sold on the market for \$114.50 per \$100.[17] What is the yield to maturity for the investment? For simplicity, assume that the bond is purchased on October 1, 2003, and the first coupon payment is April 1, 2004.

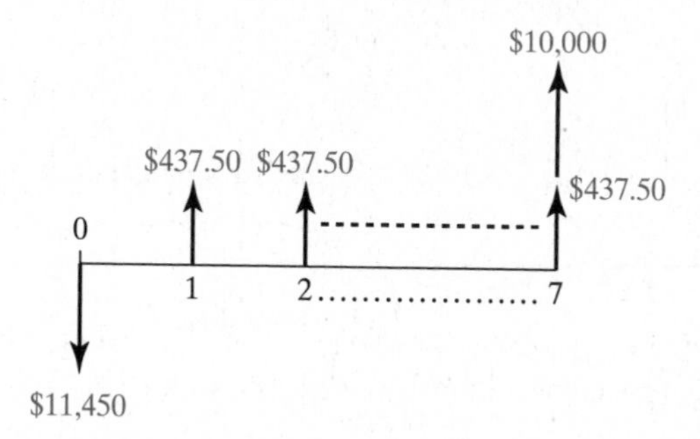

Figure 4.25 Cash flow diagram for an 8.75% bond purchased at a premium.

Solution. The cash flow diagram from the perspective of the investor who keeps the bond to maturity is given in Figure 4.25. The semiannual coupon payment is calculated as

$$IP_n = \frac{i\,B_0}{2} = \frac{(0.0875)(\$10{,}000)}{2} = \$437.50.$$

The yield to maturity is calculated as follows:

$$\begin{aligned}\$11{,}450 &= \$437.50(^{P/A,i,7}) + \$10{,}000(^{P/F,i,7}) \\ &= \$437.50\left[\frac{(1+i)^7-1}{i(1+i)^7}\right] + \$10{,}000\cdot\left[\frac{1}{(1+i)^7}\right].\end{aligned}$$

Using Goal Seek in Microsoft Excel, we find that $i = 2.12\%$ per six months, or 4.29% per year. We should expect a yield to maturity that is lower than the coupon rate, as the bond has been purchased at a premium.

Note that this purchase price for the buyer of the bond is also the selling price for the seller. Suppose the investor selling the bond purchased it at par value on April 1, 1997. The (historical) cash flow from the investor's perspective if he or she decides to sell the bond for the \$11,450 price is described in Figure 4.26.

---

[17] From www.nasdbondinfo.com on September 22, 2003.

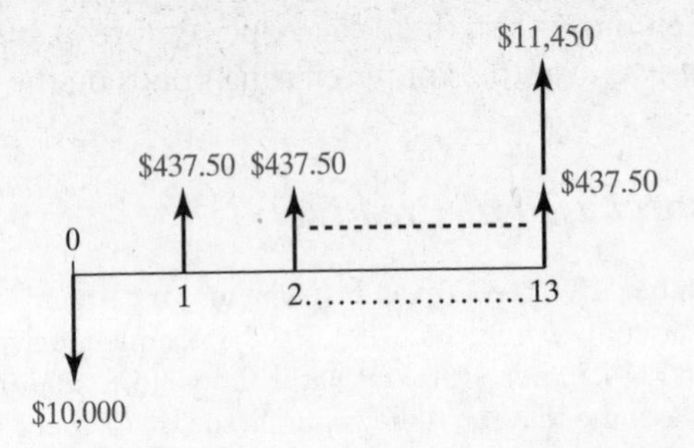

Figure 4.26 Cash flow diagram for a bond sold prematurely for a premium.

The yield for the original investor would be

$$\$10{,}000 = \$437.50(^{P/A,i,13}) + \$11{,}450(^{P/F,i,13})$$

$$= \$437.50\left[\frac{(1+i)^{13}-1}{i(1+i)^{13}}\right] + \$11{,}450\left[\frac{1}{(1+i)^{13}}\right].$$

The six-month rate is now 5.18% and the annual return is 10.64%, as the sale at a premium increases the yield above the coupon rate.

The bond described in the previous example was purchased at a premium. We now illustrate a case in which the bond is purchased on the market for a discount.

EXAMPLE 4.21 *Bond Purchased at Discount*

On November 22, 2005, bonds issued by Pitney Bowes, a producer of postage meters and business communications solutions, were sold for $96.83.[18] The bonds have a maturity date of January 15, 2016, and a 4.75% coupon rate with semiannual payments. Assume that an investor purchased a $10,000 bond for $9683 on July 15, 2005, received the first coupon payment on January 15, 2006, and retained the bond to maturity, with the final coupon payment received on January 15, 2016. What is the yield to maturity for the investment?

Solution. The cash flow diagram from the perspective of the investor is given in the spreadsheet in Figure 4.27, assuming that the bond is kept to maturity. The semiannual coupon payment is calculated as

$$IP_n = \frac{iB_0}{2} = \frac{(0.0475)(\$10{,}000)}{2} = \$237.50.$$

The yield to maturity is calculated by finding i from

$$\$9683 = \$237.50(^{P/A,i,21}) + \$10{,}000(^{P/F,i,21})$$

$$= \$237.50\left[\frac{(1+i)^{21}-1}{i(1+i)^{21}}\right] + \$10{,}000\left[\frac{1}{(1+i)^{21}}\right].$$

[18] From www.nasdbondinfo.com on November 27, 2005.

	A	B	C	D	E	F	G
1	Example 4.21: Bond Investment			Input			
2				P	$9,683.00		
3	Period	Cash Flow		B0	$10,000.00		
4	Jul-05	-$9,683.00		A	$237.50	per six months	
5	Jan-06	$237.50		Periods	21	(semi-annual)	
6	Jul-06	$237.50					
7	Jan-07	$237.50		Output			
8	Jul-07	$237.50		Yield to Maturity	5.14%	nominal	
9	Jan-08	$237.50		Yield to Maturity	2.57%	per six months	
10	Jul-08	$237.50		P	$0.00		
11	Jan-09	$237.50					
12	Jul-09	$237.50			=NPV(E9,B5:B25)+B4		
13	Jan-10	$237.50					
14	Jul-10	$237.50	=YIELD("7/15/2005","1/15/2016",0.0475,96.83,100,2,0)				
15	Jan-11	$237.50					
16	Jul-11	$237.50					
17	Jan-12	$237.50					
18	Jul-12	$237.50					
19	Jan-13	$237.50					
20	Jul-13	$237.50					
21	Jan-14	$237.50					
22	Jul-14	$237.50					
23	Jan-15	$237.50					
24	Jul-15	$237.50					
25	Jan-16	$10,237.50					

Figure 4.27 Cash flow diagram for $10,000 bond purchased for $9683 with a coupon rate of 4.75%.

This equation is programmed into cell E10, with the NPV function referencing the interest rate in cell E9. Using Goal Seek to drive cell E10 to zero, we find that $i = 2.57\%$ per six months, resulting in an annual yield to maturity of 5.21%. This value is higher than the coupon rate, due to the discounted purchase price.

We could also determine the yield to maturity by using the YIELD function in Excel, as programmed in cell E8 in Figure 4.27. The YIELD function is programmed as

= YIELD(settlement date,maturity,coupon rate,price,face value,coupon frequency, time basis),

resulting in a yield of 5.14%. (The inputs for our example are given in the spreadsheet in the figure.) Note that 5.14% is a nominal rate, equal to twice our six-month rate of 2.57%.

An interesting question is "Why are bonds sold at discounts or premiums?" The answer is actually quite complex, but most often the reason is explained by changes in interest rates. If an investor wants to sell an old bond that has a coupon rate of 5.5%, and new bonds are being introduced on the market at 6.0%, then the investor will have a hard time finding a buyer, as the coupon rate of the bond being sold is too low (compared with the rates of other available bonds). However, by lowering the purchase price (discounting the bond), the investor can sell the bond, as the yield will then be in line with other available bonds. The argument also goes in the other direction. If an investor has a bond with a coupon rate of 7.0%, then many investors would want to purchase the bond, because the current rates are near 6.0%. In this scenario, the investor can charge a premium for the bond and increase its return. In general, the price of a security rises as interest rates fall below the coupon rate of the security, and the price of the security drops if interest rates rise above the coupon rate of the security. (When we say "interest rates," we mean the average yields of new bonds being sold on the market.) The yields are equal to the appropriate Treasury note yields, plus some premium, which is the "spread" that was discussed earlier.

Of course, there are other factors. If two companies sell bonds with identical statistics (coupon rate and time to maturity), are they guaranteed to get the same purchase price? The answer is no, because the companies are not the same. Two firms, Moody's Investors Service[19] and Standard and Poor's,[20] rate companies according to their safety levels. A company with a better rating can usually sell bonds for lower coupon rates than other companies because they have a lower risk of default. A company that does not have the highest rating must provide incentives to an investor to take the risk in purchasing a bond, because if the company goes bankrupt, the bondholder could lose most or all of his or her investment. The incentive is provided when the company either offers a higher coupon rate or sells the bond at a lower price. Either of these moves results in a higher yield to maturity. Note that government-backed securities generally sell at the lowest rates because they are deemed to be the safest investment.

Other factors influence bond prices and how coupon rates are set. Among these factors are supply and demand, as companies may sell bonds that are "oversubscribed," in that demand is greater than supply (the amount sold). This may drive prices up or interest rates down. Note that interest rates and safety are generally the most important factors in determining the selling price and yield of a bond.

4.5.3 Stocks

A final method by which a company can raise funds is issuing, or selling, stock. Issuing stock essentially makes the company a public entity, in that each owner of stock is a part, albeit small, owner of the company. An investor can make money in two ways through owning stock: capital appreciation and dividends. If an investor purchases stock at a given price and sells the stock later at a higher price, he or she achieves a capital gain. This is known as *capital appreciation*. Also, the company issuing the stock may pay dividends (usually on a quarterly basis), which contribute to the return that an investor achieves.

---

[19] www.moodys.com.

[20] www.standardandpoors.com.

Most companies list their stocks on a given exchange where buyers and sellers can be paired. Stock exchanges are located throughout the world. Much like bonds, stocks can be purchased through a broker. It is not uncommon for a company to issue a number of different kinds of stock, such as common and preferred stock. *Preferred* shares generally pay a higher dividend than *common* stock.

Unlike bonds or loans, which have specified payments, stocks offer no guarantees. The price of a stock may fall, which can lead to capital losses. In addition, a company may decide to change or discontinue dividend payments.

Despite the great uncertainty regarding cash flows associated with stock transactions, we present a few examples to illustrate economic equivalence with stock transactions because issuing stock is a common method for companies to raise funds. In an initial public offering, or IPO, a company will sell a number of shares to the public for the first time at an initial price. As with bonds and loans, the money received can be used to fund activities and growth. The company may retain a number of these shares, called "treasury stock," which can be sold later to raise more cash.

From the perspective of an investor, the **return on investment** is the interest rate that equates the inflows (dividends and funds from the sale of stocks) and outflows (the purchase price of stocks) at the same point in time. This is illustrated in the next example.

EXAMPLE 4.22

Return on Stock Investment

Software giant Microsoft initially sold shares to the public on March 13, 1986, at the price of $21 per share.[21] Calculate the return on investment for an investor who purchased 100 shares of Microsoft at the IPO and sold all of the shares on March 13, 2004.

Solution. Determining the return on investment for the Microsoft stock purchase is not trivial, because, in the time span considered, Microsoft's stock has *split* nine times, and in 2003 the company started offering a dividend (which it did not do before). When a stock splits, the holder will receive additional shares in the company.

Consider a three-for-two stock split: For every two shares that an investor owns, the investor receives a third share. Hence, the investor now has three shares for every two that he or she formerly had. When this occurs, the market price of the stock changes such that the value (of all shares) of the stock owned by the investor does not change. For example, if the share price was $20 before the split, it would be $13.33 after the split, and the investor would have an additional share for every original two shares of stock. Note that an integer number of shares is not required, so an investor with only one share would have one-and-a-half shares after a three-for-two split. A company will often split its stock if its value increases substantially. The rationale is that more investors will want to purchase the stock if its price is lower. (One must purchase at least one share when buying stock.)

[21] All data are from www.microsoft.com.

A history of stock splits for Microsoft is given in Table 4.2. According to the table, 1 share of Microsoft would be worth 288 shares in March of 2004. Since our hypothetical investor purchased 100 shares at the IPO, he or she owned 28,800 shares in March of 2004.

TABLE 4.2 **Microsoft stock splits since its IPO in March of 1986.**

Date	Split	Stock Price (post split)
September 1987	2-for-1	$53.50
April 1990	2-for-1	$60.75
June 1991	3-for-2	$68.00
June 1992	3-for-2	$75.75
May 1994	2-for-1	$50.63
December 1996	2-for-1	$81.75
February 1998	2-for-1	$81.63
March 1999	2-for-1	$92.38
February 2003	2-for-1	$24.96

Source: www.microsoft.com/msft/stock.mspx

After its most recent stock split, Microsoft announced that it would pay a dividend for the first time in its history. The company declared that it would pay $0.08 per share on March 7, 2003, to shareholders of record on February 21 (people who owned Microsoft stock on that date). This means that our investor received ($0.08)(28,800) = $2304 on March 7, 2003. Microsoft followed this with a $0.16 dividend paid on November 7, 2003, for shareholders of record on October 17, 2003. Thus, a shareholder received an additional $4608 in November. Given these data and a closing price of $25.38 per share on March 13, 2004, and assuming quarterly cash flows, the cash flow diagram from the perspective of the investor is given in Figure 4.28.

The return on investment is determined by equating the investment cost with the returned cash flows:

$$\$2100 = \$2304(^{P/F,i,68}) + \$4608(^{P/F,i,71}) + (28{,}800 \text{ shares} \times \$25.38)(^{P/F,i,72}).$$

Note that the dividends were paid in the 68th and 71st quarters of the investment. Using Goal Seek in Excel, we find the interest rate to be 8.48% per quarter (38.51% per year).

It is often assumed that investors will reinvest their dividends. That is, instead of receiving cash at the time a dividend is paid, the company will issue more shares of

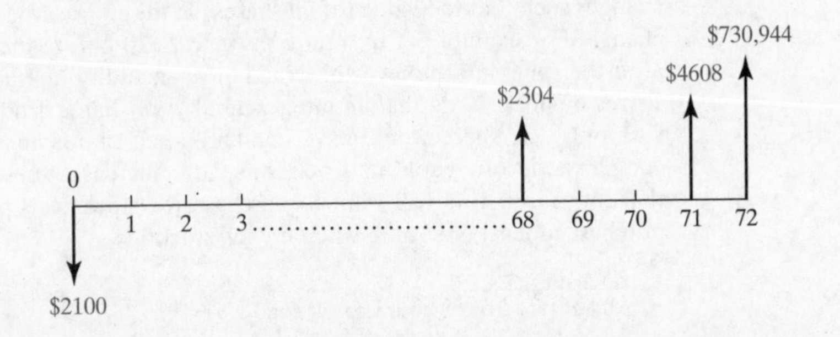

Figure 4.28 Cash flow diagram for stock investment.

stock to the stockholder. Microsoft's stock price closed at $23.56 on March 7, 2003, and at $26.10 on November 7, 2003. According to these values, 97.7929 ($2304/$23.56) shares would be given to the investor on March 7, and 176.5517 shares ($4608/$26.10) would be given to the investor on November 7. This leads to the following calculations for the return on the investment:

$$\$2100 = (29{,}074.3446 \text{ shares} \times \$25.38)(^{P/F,i,18}) = \$737{,}906.87(^{P/F,i,18}).$$

Note that the interest rate is defined as an annual rate in this instance because there is no dividend payment to incorporate into the calculation. Since there is only one uncertain term, we can solve for i directly, using the present-worth factor for a single payment. The solution is 38.49%.

A number of companies have much more difficult calculations because they have long dividend-paying histories and the dividend payments have changed with time. However, the approach to finding the return on the investment is the same.

The stock price of a company is a very important issue with investors and company managers. There are many reasons: (1) If the company retains treasury stock, it can sell the stock for the current market price to raise funds. Thus, a higher stock price means that the company can raise more funds. Treasury stock consists of shares retained by the company during some (initial) offering or shares reacquired (bought back) from the market. (2) A company can issue more (new) shares of stock in order to raise funds. Clearly, if the stock price is high, the company can raise funds more easily. Note that selling additional shares will generally lower the value of the stock, simply because of supply and demand (as supply increases without a change in demand, driving the price down). A common practice for a large company is to issue shares when buying another company, as opposed to using cash. This is often justified because the purchase of another company expands the business and in turn justifies an expansion in the number of shares available to the public. (3) Employees may have their compensation tied to the company's stock price such that they are rewarded if the price goes higher.

For the purposes of capital investment, reasons (1) and (2) are important, because they provide the means for a company to raise capital for making investments. Numerous examples of this exist, including some from companies that engage in engineering activities. Here are a few examples from different engineering sectors:

- IRobot, maker of military and consumer-use robots, sold 4.3 million shares in its initial public offering at $24 per share in November of 2005. The company was founded in 1990 by roboticists from MIT.[22]
- In the beginning of 2004, Rowan Cos. announced that it would spend $280 million over the next three years to construct four new offshore oil

[22] Cowan, L., "IRobot Opens Up 23% from IPO Price," *Dow Jones Newswires*, November 9, 2005.

rigs, upgrade equipment and facilities, and purchase new aircraft. To help fund these activities, Rowan Cos. planned on selling 10 million shares to the public. The company ultimately raised $230 million from the sale of 11.5 million shares. (It had allotted an extra 1.5 million shares to be sold if demand was strong.) Of the 11.5 million shares, 1.7 million came from its own treasury stock.[23,24]

- Rural market cell-phone company Western Wireless Corp. sold 12 million shares of Class A common stock to raise $228 million, less expenses, to fund capital expenditures, working-capital needs, and general corporate purposes. Before the sale in October of 2003, the company had 79.3 million shares outstanding.[25]
- Gibraltar Steel Corp., a metal processor and manufacturer of steel products, announced that it had sold nearly 4.5 million shares at $24.75 per share in a public offering. The sale resulted in net proceeds of approximately $75 million to reduce debt, fund acquisitions, and simply be available for general corporate uses.[26,27]
- In March of 2004, Uranium producer Cameco Corp. said that it would issue new stock in order to pay for its $333 million purchase of a 25.2% stake in the South Texas Project from American Electric Power. The project consists of two 1,250-megawatt nuclear units and has three other co-owners.[28]
- Spartech Corp., a producer of engineered thermoplastic materials and molded products, announced the sale of 7.3 million shares at $24 per share in January of 2004. The proceeds were used to pay down debt and for general purposes, including possible future acquisitions.[29]
- Telecommunications provider Level 3 Communications agreed to purchase WilTel Communications Group for 115 million Level 3 common shares (valued at $2.70 per share) and $370 million in cash.[30]

---

[23] Enrich, D., "Rowan Sees Spending $280M on Fleet Expansion in 2004-06," *Dow Jones Newswires,* January 28, 2004.

[24] Hadi, M., "Rowan Cos. Plans to Offer 10M Common Share," *Dow Jones Newswires,* January 27, 2004.

[25] Abejo, J., "Western Wireless to Raise $228M in Common Stk Offering," *Dow Jones Newswires,* October 29, 2003.

[26] Coyle, B., "Gibraltar Steel Files to Sell 4.13M Common Shares," *Dow Jones Newswires,* November 7, 2003.

[27] "Gibraltar Announces Sale of Over-Allotment Shares," *Gibraltar News Release,* www.gibraltar1.com, January 22, 2004.

[28] Olver, L., "Cameco Plans Equity Issue to Fund Most of Texas Deal," *Dow Jones Newswires,* March 1, 2004.

[29] "Spartech Announces Pricing of Public Common Stock Offering," *Spartech News Release,* www.spartech.com, January 29, 2004.

[30] "NEWS WRAP: Level 3 to Buy WilTel for about $680M," *Dow Jones Newswires,* October 31, 2005.

The analysis of stocks and bonds generally resides in finance departments. However, the calculations merely follow from cash flow analysis as described in the previous chapters. It is important that engineers understand bonds and stocks because they are a critical source of funds for many companies, including engineering firms.

4.6 Examining the Real Decision Problems

We return to our introductory example concerning Norfolk Southern's 100-year note offering and the questions posed:

1. Assume that an investor bought $10,000 worth of bonds at par at the initial offering. What is the yield to maturity on the investment?

From what we have learned, we know that the yield to maturity has a nominal interest rate of $r = 6\%$. This translates to an annual effective rate of

$$i = \left(1 + \frac{r}{2}\right)^2 - 1 = \left(1 + \frac{0.06}{2}\right)^2 - 1 = 6.09\%.$$

2. If the investor sold the bonds on December 28, 2005, for the market price, what was the return on the investment? If the investor put all proceeds from the investment into an account earning 0.25% per month, would he or she have $10,800 on March 15, 2006?

The $10,000 bond was purchased on March 15, a $300 coupon payment was received on September 15, and the bond was sold on December 28 (all in 2005) for $10,353.80. Let us assume that the sale of the bond occurred on January 15, 2006, so that we can work with a monthly interest rate. This is depicted in the cash flow diagram in Figure 4.29.

To find the return on the investment, we equate the outflows to the inflows as follows:

$$\begin{aligned}\$10{,}000 &= \$300(^{P/F,i,6}) + \$10{,}353.80(^{P/F,i,10}) \\ &= \$300\left[\frac{1}{(1+i)^6}\right] + \$10{,}353.80\left[\frac{1}{(1+i)^{10}}\right].\end{aligned}$$

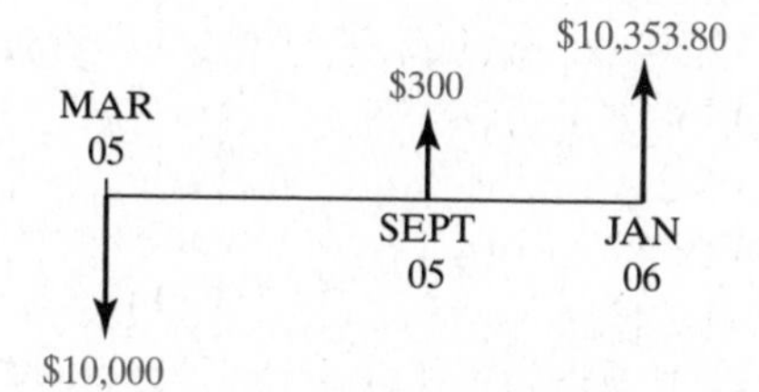

Figure 4.29 Cash flow diagram for the purchase and sale of a bond.

Solving for i with the use of Goal Seek in Excel leads to a monthly interest rate of 0.643%, or equivalently, a 7.99% annual interest rate.

If the proceeds were put into a bank account that earns 0.25% per month, then

$$F = \$300(\overset{F/P,0.25\%,6}{1.0151}) + \$10,353.80(\overset{F/P,0.25\%,2}{1.0050}) = \$10,710.16$$

would be in the account, an amount short of the target of $10,800.

3. What is the yield to maturity for the investor who purchased the bonds on December 28, 2005?

As with our previous question, we assume for simplicity that the purchase occurred on January 15, 2006. The first interest payment arrives on March 15, 2006, and is followed by 198 more payments and the principal repayment. The value of all these payments on March 15, 2006, is

$$P = \$300 + \$300(^{P/A,i_{sa},198}) + \$10,000(^{P/F,i_{sa},198}),$$

assuming a semiannual interest rate. We can equate this to the purchase price as

$$\$10,353.80 = \frac{P}{(1+i_m)^2},$$

using a monthly rate. Since $i_{sa} = (1+i_m)^6 - 1$, we can substitute i_m into our definition of F and solve for i_m. Using Goal Seek in Excel, we are left with a monthly interest rate of 0.48%, or 2.95% per six months, or 5.996% per year. Note that the extremely long horizon dampens the effect of buying the bond for a premium.

4.7 Key Points

- If two cash flows are economically equivalent, then we are indifferent when choosing between them.
- Economic equivalence can be established at any point or points in time.
- The difference between two economically equivalent cash flows is zero. When comparing cash flow diagrams, only the differences are relevant.
- Economic equivalence does not require a constant interest rate over the study horizon.
- Economic equivalence can be established with current dollars and a market interest rate or real dollars and an inflation-free interest rate.
- Determining the interest rate that equates two cash flow diagrams generally requires solving a polynomial equation.
- Solving for the study period N that equates two cash flows often results in a noninteger answer that must be subsequently rounded.

- Table 4.1 provides a summary of interest rate factors when the interest rate approaches zero or the study horizon approaches infinity.
- Loans, bonds, and stocks provide the means for a company to raise funds.
- A loan is a contract between a lending institution and the entity receiving funds. In return for a loan (an up-front cash payment), the entity receiving funds must pay back the principal with interest over time according to the contract.
- Loans are generally paid back through equal total payment or equal principal payment plans.
- The true cost of a loan defines the loan's cost as a rate of interest.
- Bonds are a contract between a company or a government entity and an investor. An investor provides money up front (used to fund capital expenditures) in return for payments over time, defined by the coupon rate, with the par (face) value returned at the bond's maturity.
- The owner of a zero-coupon bond does not receive interest (coupon) payments.
- Most bonds can be sold from one investor to another through an exchange.
- Bonds with higher coupon rates can be sold on the market at a premium (greater than face value), compared with similar bonds, while bonds with lower coupon rates are usually sold at a discount (less than face value).
- The yield to maturity for a bond describes the return earned by the investor from the time the bond is bought through its maturity.
- Companies sell stock in order to raise capital. In return, the investor becomes a part (albeit small) owner and hopes for a rise in the price of the stock or payments of dividends in order to make a profit on the investment.
- The return on investment for a stock is similar, mathematically, to the true cost of a loan and the yield to maturity for a bond.

4.8 Further Reading

Bodie, Z., A. Kane, and A.J. Marcus, *Investments,* 2d ed. Irwin, Inc., Homewood, Illinois, 1993.

Brealey, R.A., and S.C. Myers, *Principles of Corporate Finance,* 5th ed. McGraw-Hill Co., New York, 1996.

Fleischer, G.A., *Introduction to Engineering Economy*. PWS Publishing Co., Boston, 1994.

Fleischer, G.A., M.M. Dessouky, and S.M. Ng, "Loans for which the Acquired Asset Serves as the Collateral," *The Engineering Economist*, 40(2):127–144, 1995.

Park, C.S., *Contemporary Engineering Economics*, 3d ed. Prentice Hall, Upper Saddle River, NJ, 2002.

Park, C.S., and G.P. Sharp-Bette, *Advanced Engineering Economics*. John Wiley and Sons, New York, 1990.

Thuesen, G.J., and W.J. Fabrycky, *Engineering Economy*, 9th ed. Prentice Hall, Upper Saddle River, NJ, 2001.

4.9 Questions and Problems

4.9.1 Concept Questions

1. Summarize the concept of economic equivalence. Why is this concept important in economic analysis when we compare two different cash flow diagrams?
2. Is a constant interest rate required to establish economic equivalence? Explain.
3. If it is determined that two cash flow diagrams are economically equivalent, and the interest rate then changes, what can we say about their equivalence? Explain.
4. How must equivalence be established if the cash flows are in current dollars? What if they are in real dollars?
5. Why is establishing equivalence with an unknown interest rate difficult?
6. What difficulties arise when establishing equivalence with an unknown length of time, N?
7. How should one establish equivalence when comparing different currencies?
8. Why is it useful to understand the limits of the interest rate factors ($N \to \infty$ and $i \to 0$)?
9. Derive the value for $(^{A/G,i,N})$ as $N \to \infty$.
10. What is capital? Why do firms raise capital?
11. Explain the differences, in terms of cash flows, of acquiring loans, selling bonds, or floating (issuing) stock.
12. What are the components of a loan payment? How are they determined?
13. Is there one type of loan that banks give? Explain.
14. What is a zero-coupon bond? How does it differ from other typical bonds?
15. Explain why you may sell a bond for more than the face value.
16. Why may the initial sale of a bond from a company actually be for less than the par value?
17. What is the relationship among the yield to maturity, the true cost of a loan, and the rate of return on an investment?
18. What is treasury stock? Why is it important?

19. Why do some companies not pay dividends?

20. How does a stock split affect calculating the return on a stock investment?

4.9.2 Drill Problems

1. Are the following two cash flow diagrams economically equivalent if the interest rate is 10% per period?

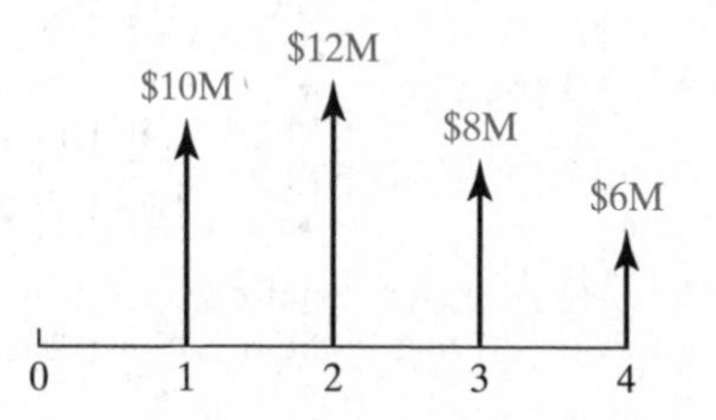

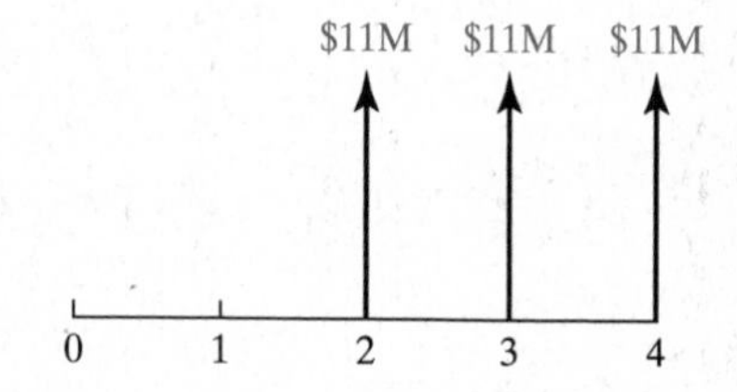

2. Are the following two cash flow diagrams economically equivalent if the interest rate is 14% per period?

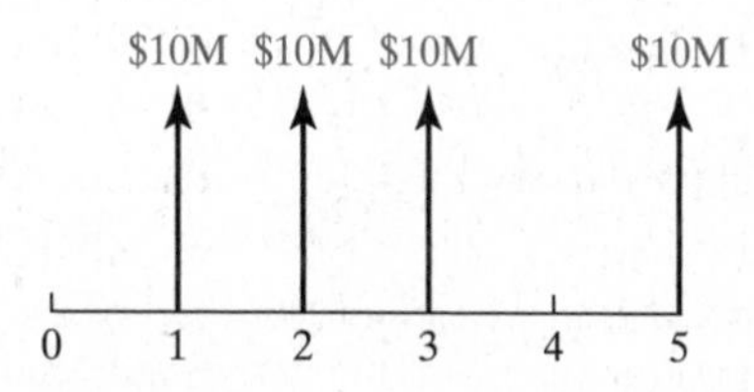

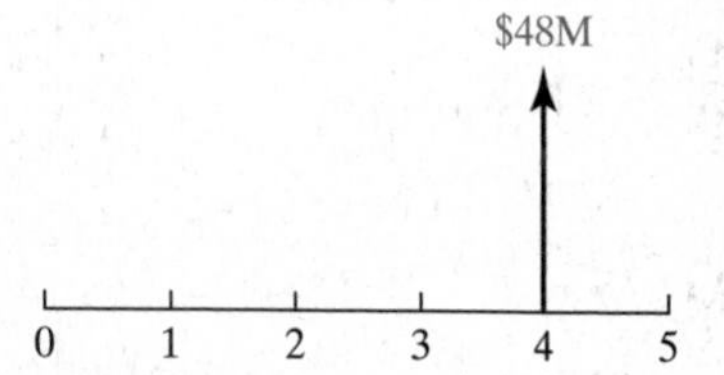

3. Show that an arithmetic gradient series with $G = \$500$ over a 10-year horizon is economically equivalent to a single cash flow of \$22,260 at time period seven if the interest rate is 8% per period.

4. Determine an equal annual payment series of three years' length that is economically equivalent to an equal annual payment series of \$1300 per year over seven years. Assume that the three-year series starts in the second year of the seven-year series and the interest rate is 4% per period.

5. Show that the cash flows

$$A_1 = \$0,$$
$$A_2 = \$1000,$$
$$A_3 = \$1500,$$
$$A_4 = \$2000,$$
$$A_5 = \$2500,$$
$$A_6 = \$3000,$$
$$A_7 = \$3000,$$
$$A_8 = \$3000,$$
$$A_9 = \$3000,$$
$$A_{10} = \$3000,$$

are economically equivalent to an equal-payment series of size \$4610 in periods 5 through 9 at an interest rate of 10% per period.

6. Compute the present value of the cash flow diagram in the previous question under the following assumptions:

(a) The interest rate is 18% per period.

(b) The interest rate is 10% per period from time zero through the end of period 6 and then 12% through the end of period 10.

(c) The interest rate is 1% at the end of time period 1 and grows 1% (arithmetically) each period thereafter.

7. What interest rate equates an equal annual series of cash flows of size \$500 over eight years to a single cash flow of size \$2319.43 at time zero?

8. What interest rate equates an arithmetic gradient series of cash flows with $G =$ \$1000 over six periods to a future value of \$15,718.63 at time 6? Use interpolation to find the answer.

9. What interest rate equates a geometric gradient series of cash flows with $A_1 =$ \$10,000 and $g = 6.5\%$ over 15 periods to a present value of \$65,688.42 (at time zero)? Use Goal Seek to find the answer.

10. If \$2000 is placed into an account that pays 12% per year, how many years must you wait to have at least \$12,500 in the account?

11. How many annual payments of size \$2000 into an account that pays 3% per year are required in order to accumulate at least \$16,000 in the account?

12. If payments are made into an account starting with zero in year 1 and increasing by \$2500 each year thereafter, how many payments must be made in order to have at least \$200,000 in the account? The interest rate is 25% per year.

13. A \$250,000 loan is taken out at time zero to be repaid in equal quarterly total payments. Develop a payment schedule, noting principal and interest payments, if the loan is to last five years and the interest rate is 3.5% per quarter.

14. If a \$100 handling fee is charged with each payment in the loan in Problem 13, what is the true cost of the loan?

15. Recompute the loan payments in Problem 13 under the assumption of equal principal payments. What is the present value of the two payment schemes, assuming a 5% quarterly interest rate?

16. Assume that the total payments made towards a \$250,000 loan over five years are \$100,000, \$25,000, \$50,000, \$75,000, and \$50,000. Determine the principal and interest payments from this schedule and compute the true cost of the loan.

17. Fill in the missing fields in the following loan payment schedule:

Time n	B_{n-1}	PP_n	IP_n	A_n
1		\$10,000,000.00	\$3,250,000.00	
2	\$40,000,000.00		\$2,600,000.00	
3	\$25,000,000.00	\$15,000,000.00		
4		\$5,000,000.00		\$5,650,000.00
5	\$5,000,000.00		\$325,000.00	

18. What is the true cost of the loan in Problem 17?

19. A zero-coupon bond with a face value of $100 and a maturity date of July 1, 2005, is purchased on January 1, 2005, for $98.50. What is the yield to maturity (as an annual rate of interest)?

20. A zero-coupon bond is purchased for $950 on July 1, 2005. If it matures in one year and has a face value of $1000, what is the yield to maturity?

21. A $10,000, 10-year bond with a 6.5% coupon rate is purchased at par. If payments are semiannual and the bond is held to maturity, what is the yield to maturity?

22. If the bond in Problem 21 can be sold for $10,500 after five years, what is the yield on the investment?

23. A bond is purchased on the market for $9800 with a coupon rate of 3.5% (semiannual payments) and five years until maturity. If the face value is $10,000 and the first coupon payment comes six months after purchase, what is the yield to maturity?

24. A $1000 bond is purchased on the market for $1200. If it carries a 7.5% coupon rate (semiannual payments) and matures in three years, what is the yield to maturity? Assume that the first coupon payment is received six months after the bond is purchased.

25. If, in Problem 24, the investor who sold the bond after receiving a coupon payment had purchased it at par two years earlier, what was the yield on his or her investment?

26. One hundred shares of stock are purchased at the end of 2003 for $20 each. They are sold after three years for $32 each. What is the return on this investment (as an annual interest rate)?

27. Fifty shares of stock are purchased at the end of 2004 for $25 dollars each. After each of the first three years, the stock splits two for one. After the seventh year, it splits three for one. If the stock is sold for $24 per share at the end of 2014, what is the annual return on the investment?

28. Ten shares of stock are purchased for $10 each. A dividend of $0.05 per share is paid each quarter. If the shares are sold for $12 each after the dividend is paid in the final quarter of the fifth year of ownership, what is the return on the investment? Calculate both the quarterly and equivalent annual return.

29. One thousand shares of stock are purchased at $20 each. The stock pays a $0.35 quarterly dividend that increases 5% per year. (All four dividend payments in the same year are always the same.) If the stock is sold for $22 per share after the second dividend payment in the sixth year of ownership, what is the quarterly return on the investment?

30. One share of stock is purchased for $50 at the end of 2005. The stock splits two for one at the end of 2007 and three for two at the end of 2010. Further, the stock pays a dividend of $0.15 each quarter for the first three years. It then increases by 15% per year. (Again, dividend payments in the same year are of the same size.) If the stock is sold for $62 at the end of 2012, what is the return on the investment as an annual rate of interest?

4.9.3 Application Problems

1. Construction firm Grupo Ferrovial SA was awarded a EUR319.1 million railway construction contract in northern Italy in early 2004. The work is expected to take 47 months.[31] Assume that the contract is paid with equal monthly payments of EUR6.8 million over the 47 months of construction.

(a) Find the equivalent present value (one month before the first payment), assuming a monthly interest rate of 0.25%.

(b) Find the equivalent future value (end of the 47th month), assuming a monthly interest rate of 0.25%.

(c) Find the equivalent single payment at the end of month 14, assuming a monthly interest rate of 0.50%.

(d) Find equivalent monthly payments from the end of month 20 through month 32, assuming a monthly interest rate of 0.50%.

(e) Find the equivalent present value, assuming a monthly interest rate of 0.25% for the first 10 months, 0.50% for the next 10 months, and 0.75% for the remaining months.

(f) The contract has been signed; thus, the payments are in current dollars. If the inflation-free interest rate is 0.10% per month and the inflation rate is 0.10% per month, determine the equivalent present value of the payments, using real dollars. Compare this with the present value, using current dollars and the market interest rate.

2. Boatmaker Brunswick Corp. purchased four aluminum boat brands (Crestliner, Lowe, Lund, and Lund Canada) from Genmar Holdings in the spring of 2004 for $191 million in cash and $30 million to be paid in three years if certain conditions are met. The boat brands had a combined annual revenue of $311 million before the deal, and Brunswick said that the deal included $35 million (time-zero dollars) in tax benefits.[32] Assume that the following investment payments are made: $156 million at time zero and $30 million at the end of year three. Answer the following:

(a) The purchase generates net revenues of $31 million per year for 10 years. What annual rate of interest equates these revenues to the investment payments of $191 and $30 million?

(b) Recompute the solution in part (a), assuming that net revenues are $31 million in year 1 and increase $2 million per year through year 10.

(c) Ignore the gradient in the previous problem. How long must the $31 million in net revenues be made in order for the cash flow diagram to be worth at least $170 million at time zero, assuming a 10% annual interest rate?

(d) Recompute part (c), assuming an interest rate of 8% per year.

3. A unit of Stanley Works, a provider of tools, hardware, and door systems, offered $45.3 million to purchase security systems company Frisco Bay Industries, Ltd.,

---

[31] "Ferrovial Wins EUR319.1M Italian Rail Contract," *Dow Jones Newswires*, December 17, 2003.

[32] Salisbury, I., "Brunswick Closes Deal for 4 Genmar Boat Brands," *Dow Jones Newswires*, April 2, 2004.

in early 2004 in order to diversify Stanley's business. At the time, Frisco Bay had annual sales of $40 million.[33] One way in which to value the purchase of another company is to assume that it will continue its normal operations indefinitely. If the annual rate of interest is 12%, determine whether the following cash flows are equivalent to the purchase price, which is assumed to be a present value:

(a) $5.4 million in annual profits that continue indefinitely.

(b) $4.0 million in annual profits that grow by $0.1 million per year thereafter indefinitely.

4. Drillers Technology Corp. arranged for an additional C$2.5 million credit facility through Socar Investments, Ltd., to fund working-capital needs. The facility bears an annual interest rate of 12%. Quarterly principal repayments of C$156,250 began on March 31, 2004, and continue through December 31, 2007.[34] Assume that the 12% annual interest rate is compounded quarterly, and answer the following:

(a) Determine the quarterly payments over the life of the loan, assuming that the equal principal payment plan of $156,250 per quarter is met. Construct a table such that interest and total payments are computed.

(b) If an equal total payment plan had been approved, what would the payment be each quarter?

(c) Compute the difference in the present value of the principal payments of each of the plans in parts (a) and (b), assuming a 15% interest rate.

5. A group of more than 30 banks granted Formosa Plastics Group an NT$88.9 billion syndicated loan to fund expansion of the petrochemical company's naphtha cracker complex. The complex already houses an oil refinery, two naphtha crackers, and dozens of petrochemical plants.[35]

(a) If the loan is to be repaid in equal total payments of NT$20 billion per year over a five-year period, what is the true cost of the loan? What are the principal payments for this loan schedule?

(b) Assume that the following payments are made over five years to repay the loan: NT$10 billion, NT$15 billion, NT$20 billion, NT$25 billion, and NT$30 billion. What is the true cost of the loan? How are the payments divided according to interest and principal?

6. Air Products and Chemicals sold $125 million of notes in November of 2003 with a December 1, 2010, maturity date. The bonds were sold at a discount of $99.721 per $100 with a coupon rate of 4.125%.[36] Answer the following questions:

---

[33] Jordan, J., "The Stanley Works to Offer to Buy Frisco Bay Industries Ltd. for U.S. $15.25 per Shr Cash; Transaction Valued at U.S. $45.3 M," *Dow Jones Newswires,* January 20, 2004.

[34] Moritsugu, J., "Drillers Technology Gets C$2.5M Credit Facility," *Dow Jones Newswires,* September 26, 2003.

[35] Sun, Y.H., "Taiwan's Formosa Plastics Group to Get NT$88.9B Loan," *Dow Jones Newswires,* April 1, 2004.

[36] Geressy, K., "Air Products $125M 7-Year Yields 4.171%; Treasurys +0.50," *Dow Jones Newswires,* November 14, 2003.

(a) Assume that bonds with a face value of $10,000 were purchased at the time of issue, semiannual coupon payments, and that the bonds are kept until maturity. What is the yield to maturity for the investment?

(b) Assume that it is known at time zero that the bonds could be sold for $10,350.50 after the second coupon payment in year 5. What is the yield until the time of the sale?

(c) Consider the purchaser of the bond at the end of year 5 (for $10,350.50). If he or she retains the bond through its maturity, what is the yield?

7. Level 3 Communications raised $500 million through the sale of eight-year bonds with a maturity date of October 15, 2011. The bonds sold at par with a coupon rate of 10.75%.[37] Answer the following questions:

(a) Assume $10,000 worth of bonds were purchased at the time of issue with semiannual coupon payments. If the bonds are kept until maturity, what is the yield?

(b) Consider again the previous question if the coupon payments were made annually.

(c) Assume that it is known at time zero that the bonds could be sold after the first semiannual coupon payment in year 4 for $9650.25. What is the yield until the time of the sale?

(d) Consider the purchaser of the bond for $9650.25 in part (c). If he or she retains the bond through its maturity, what is the yield?

(e) Why might the Level 3 Communications bonds described here have been sold with such a high coupon rate compared with that of the Air Products and Chemicals bonds in Question 6?

8. Cisco Systems, a provider of networking hardware and services, had an initial stock offering on February 16, 1990, at $18 per share. The stock has split nine times since then, as listed in Table 4.3. The stock closed at the price of $17.71 on February 16, 2005. Assume that you purchased 10 shares at the IPO.

TABLE 4.3 **Cisco Systems stock splits since its IPO in February of 1990.**

Date	Split	Stock Price (post split)
March 1, 1991	2-for-1	$28.50
March 6, 1992	2-for-1	$39.00
March 5, 1993	2-for-1	$46.00
March 4, 1994	2-for-1	$39.50
February 2, 1996	2-for-1	$44.50
November 18, 1997	3-for-2	$53.42
August 14, 1998	3-for-2	$64.42
May 24, 1999	2-for-1	$61.56
February 22, 2000	2-for-1	$72.19

Source: investor.cisco.com

[37] Geressy, K., "Level 3 $500M 8-Yr Yields 10.75%; Priced at Par," *Dow Jones Newswires*, September 26, 2003.

Answer the following:

(a) What was your original investment worth on February 16, 2005?

(b) If you sold all of the shares on February 16, 2005, what would have been your (annual) rate of return on the investment?

(c) Recompute your return, assuming that you sold your stock after the split on February 22, 2000. (Assume that this date is close enough to February 16 for computational purposes.)

(d) If you purchased an additional 10 shares after each split, what would be the return on your total investment strategy? Assume that you sell all shares on February 16, 2005, and assume end of month cash flows.

(e) If you had purchased only 100 shares after the split on February 22, 2000, what would be your return, assuming that you sold them on February 16, 2005?

(f) Cisco has never offered a dividend, but assume that it paid a quarterly dividend of $0.04 per share starting in 2002. (Assume that the first payment was on May 16, 2002.) Recompute your return on your original 10-share investment, assuming that cash was received and you sold all shares on February 16, 2005.

(g) If you had sold the original investment after a stock split, which date would have given you the highest annual return? Assume all transactions occur at the end of the month.

9. Toyota Motor Credit Corp. sold $750 million 4.35% coupon-rate seven-year notes in December of 2003 (maturing December 15, 2010) at a price of $99.885 per $100. The bonds are noncallable.[38] As of December 22, 2005, the bonds sold on the market for $98.181 per $100.[39] Assuming that $10,000 worth (face value) of bonds were purchased at the initial offering in December of 2003 with semiannual coupon payments, answer the following questions:

(a) What is the yield if the bonds are retained to maturity?

(b) What is the yield if an investor purchased the bonds in December of 2005 for the market price and retained them to maturity?

(c) What is the yield for the investor who purchased the bonds at the initial offering and then sold them after receiving the fourth coupon payment (in 2005)?

(d) What price was required in December of 2005 to achieve a yield of 6.0% per year?

10. Since Microsoft's initial dividend payments in 2003, noted in Example 4.22, the company has made several other dividend payments, listed in Table 4.4. The $3.00 dividend in December of 2004 was declared "special," as it could not be reinvested. Using the information here and from the original example, answer the following:

(a) Recalculate the return on an investment of 100 shares at the IPO if all dividends except for the special dividend were reinvested and all shares were sold on

---

[38] Richard, C., "Toyota Motor Credit $750M 7-Yr Notes Price at Tsys +0.50," *Dow Jones Newswires*, December 2, 2003.

[39] From www.nasdbondinfo.com on December 23, 2005.

TABLE 4.4 Microsoft dividends and closing stock prices.

Pay Date	Year	Dividend	Stock Price
March 7	2003	$0.08	$23.56
November 7	2003	$0.16	$26.10
September 14	2004	$0.08	$27.44
December 2	2004	$3.00	$27.09
December 2	2004	$0.08	$27.09
March 10	2005	$0.08	$25.43
June 9	2005	$0.08	$25.51
September 8	2005	$0.08	$26.61
December 8	2005	$0.08	$27.69
March 9	2006	$0.09	$27.00

Sources: www.microsoft.com/msft/stock.mspx, finance.yahoo.com

December 8, 2005, after the dividend was received. (Assume end of quarter flows for simplicity.)

(b) For the same assumptions, what stock price on March 9, 2006 (after the dividend payment) would increase the annual return computed in part (a) by 0.1%.

(c) Recompute the return on investment, assuming that all dividends were taken as cash.

11. Florida Power sold $300 million of 12-year notes due December 1, 2015. The notes were sold at $99.802 per $100 with a coupon rate of 5.10%.[40] As of December 21, 2005, they sold on the market for $98.271 per $100.[41] If $5000 worth (face value) of bonds were purchased on December 1, 2003, with semiannual coupon payments, answer the following questions:

(a) What is the yield to maturity on the investment?

(b) What is the yield to maturity on a similar investment made in December of 2005?

(c) Why did the bonds sell at a discount in December of 2005?

12. Apple Computer, Inc., was first sold on the market in December of 1980 at $22 per share. Table 4.5 gives a history of dividend payments and stock splits, most recently in February of 2005. Noting that Apple's stock closed at $72.18 on December 15, 2005, and assuming that all transactions shown in the table occurred at midmonth and dividends were paid before stock splits, answer the following:

(a) If you had purchased 100 shares of Apple at its IPO and sold the investment on December 15, 2005, what was the return on your investment? Compute both a

[40] Geressy, K., "Florida Power $300M 12-Yr Yields 5.122%; Treasurys +0.3," *Dow Jones Newswires*, November 18, 2003.

[41] From www.nasdbondinfo.com on December 23, 2005.

quarterly and an annual return, assuming that dividends were not reinvested. (Assume end of quarter flows.)

(b) Recalculate your return, assuming that you reinvested your dividends at the closing prices in Table 4.5.

TABLE 4.5

Apple dividends (per share) and stock splits since the IPO in December of 1980.

Date	Year	Action	Dividend	Stock Price (post split)
December 12	1980	IPO	–	$22.00
June 15	1987	Dividend	$0.03	$41.50
June 15	1987	2-for-1 Split		$41.50
September 15	1987	Dividend	$0.03	$51.75
December 15	1987	Dividend	$0.04	$37.50
March 15	1988	Dividend	$0.04	$45.00
June 15	1988	Dividend	$0.04	$45.75
September 15	1988	Dividend	$0.04	$41.63
December 15	1988	Dividend	$0.04	$39.50
March 15	1989	Dividend	$0.05	$35.00
June 15	1989	Dividend	$0.05	$47.50
September 15	1989	Dividend	$0.05	$45.00
December 15	1989	Dividend	$0.06	$33.75
March 15	1990	Dividend	$0.06	$36.75
June 15	1990	Dividend	$0.06	$39.50
September 15	1990	Dividend	$0.06	$33.75
December 15	1990	Dividend	$0.06	$40.13
March 15	1991	Dividend	$0.06	$66.25
June 15	1991	Dividend	$0.06	$42.00
September 15	1991	Dividend	$0.06	$47.25
December 15	1991	Dividend	$0.06	$50.50
March 15	1992	Dividend	$0.06	$63.38
June 15	1992	Dividend	$0.06	$52.63
September 15	1992	Dividend	$0.06	$48.25
December 15	1992	Dividend	$0.06	$56.38
March 15	1993	Dividend	$0.06	$57.00
June 15	1993	Dividend	$0.06	$42.00
September 15	1993	Dividend	$0.06	$24.40
December 15	1993	Dividend	$0.06	$29.75
March 15	1994	Dividend	$0.06	$37.63
June 15	1994	Dividend	$0.06	$27.81
September 15	1994	Dividend	$0.06	$36.00
December 15	1994	Dividend	$0.06	$37.13
March 15	1995	Dividend	$0.06	$35.00
June 15	1995	Dividend	$0.06	$43.63
September 15	1995	Dividend	$0.06	$35.88
December 15	1995	Dividend	$0.06	$35.25
June 20	2000	2-for-1 split	–	$55.63
February 28	2004	2-for-1 split	–	$44.86

Sources: www.apple.com/investor/; finance.yahoo.com.

(c) If you had sold your investment after the final dividend in 1995, what would have been the return? Assume dividends were not reinvested.

(d) If you had purchased 100 shares (for the first time) on December 15, 1995, and sold those shares December 15, 2005, what was the annual return on your investment?

4.9.4 Fundamentals of Engineering Exam Prep

1. An engineering firm had revenues of $250,000 last year, which are expected to grow $25,000 per year for the foreseeable future. Assuming an interest rate of 8%, the maximum that should be paid to purchase the firm is most closely

(a) $560,000.

(b) $7.03 million.

(c) $2.5 million.

(d) $1.3 million.

2. A scholarship is established at an engineering college to perpetually pay a student $15,000 per year. If a bond can be purchased that pays 6.5% annually, the endowment must be

(a) At least $230,800.

(b) Between $200,000 and $225,000.

(c) Between $180,000 and $195,000.

(d) None of the above.

3. A bridge is built with an expected lifetime of 120 years. If it cost $500 million to build, with no salvage value, the annualized capital costs over the life of the bridge are, at an interest rate of 6% per year, most closely

(a) $3.1 million.

(b) $4.2 million.

(c) $8300.

(d) $30 million.

4. A construction loan is secured for 4% in the first year, but the rate is expected to increase 0.25% per year due to increasing interest rates. If the $100,000 loan is to be repaid with equal principal payments of $20,000 per year, the total payment in year 3 is most closely

(a) $21,800.

(b) $20,000.

(c) $22,700.

(d) $22,400.

5. A $250,000 loan is secured, to be paid back with equal monthly payments over three years. The monthly payments for a rate of 24% per year, compounded monthly, are

(a) $250,000$(A/P, 24\%, 3)$.

(b) $250,000$(A/P, 2\%, 3)$.

(c) $250,000$(A/P, 2\%, 36)$.

(d) $250,000$(A/P, 24\%, 36)$.

6. Payments of $25,000 per year are made over four years to pay off an $85,000 loan. The loan's interest rate (the true cost of the loan) is most closely

(a) Between 6% and 8%.

(b) Greater than 8%.

(c) Between 4% and 6%.

(d) Less than 4%.

7. A five-year, $50,000 loan is secured at a nominal rate of 4% per year, compounded quarterly. If immediately after making your second quarterly payment you decide to pay off the loan, the amount paid is most closely

(a) $45,920.

(b) $45,440.

(c) $49,000.

(d) $43,560.

8. A fund is started with a $10,000 investment and $2500 is contributed each quarter thereafter. If the fund pays 2% per quarter, the number of payments required to amass $32,000 is at least

(a) 6.

(b) 7.

(c) 8.

(d) 9.

9. The sum of $2500 is contributed each quarter to a fund paying 2% per quarter. The number of payments required to amass $22,000 is at least

(a) 6.

(b) 7.

(c) 8.

(d) 9.

10. A $10,000 bond is purchased at par with a coupon rate of 5% and annual coupon payments. If the bond is held to maturity, what is the yield to maturity (in 10 years)?

(a) 5% per year.

(b) 5% per six months.

(c) 10% per year, compounded semiannually.

(d) None of the above.

11. A \$10,000 bond is purchased at par with a coupon rate of 5% and semiannual coupon payments. If the bond is held to maturity, what is the yield to maturity (in 10 years)?

(a) 5% per year.

(b) 5% per six months.

(c) 5% per year, compounded semiannually.

(d) None of the above.

12. A \$1000 bond is purchased at par at time zero. Coupon payments of size \$20 are received every six months. If the bond is sold for \$1050 after one year (immediately after the second coupon payment is received), the return earned by the investor is most closely

(a) 1.5% per year.

(b) 2.0% per year.

(c) Between 3.0% and 4.0% per year.

(d) Greater than 4.0% per year.

13. A \$1000 bond with a 5% coupon rate is purchased for \$990 at time zero. If the bond is sold for par after the first coupon payment is received, the yield on the investment is most closely

(a) 5% per year.

(b) 4.5% per year.

(c) Greater than 5% per year.

(d) None of the above.

14. A newspaper advertises loans with monthly payments of \$330 for every \$10,000 borrowed. What is the true cost of the loan if \$20,000 is borrowed and paid off over three years?

(a) $i: (P/A, i\%, 36) = .033$.

(b) $i: (A/P, i\%, 36) = .033$.

(c) $i: (A/P, i\%, 36) = 30{,}303$.

(d) $i: (A/P, i\%, 36) = .0165$.

15. The rate of interest that most closely equates \$10,000 at time zero and \$12,500 at time 1 is

(a) 25% per period.

(b) 12.5% per period.

(c) 14% per period.

(d) None of the above.

16. Given an annual interest rate of 4%,

(a) \$2000 per year for 10 years is preferred to \$21,000 at time zero.

(b) \$21,000 at time zero is preferred to \$2000 per year for 10 years.

(c) $21,000 at time zero is preferred to $2000 per year for 30 years.

(d) $2000 per year for 10 years is preferred to $31,000 at time zero.

17. A company offers $2.5 million for the rights to a patent. If the patent can bring in $500,000 in royalties over the next six years, should the offer be taken? Assume an interest rate of 6% per year.

(a) Not enough information is provided.

(b) Yes.

(c) No, the offer must be at least $3.0 million.

(d) No, the offer must be at least $4.0 million.

18. A company offers $2.5 million for the rights to a patent. If the patent can bring in $500,000 in royalties over the next four years, you would sell the rights if the interest rate was

(a) Greater than 6%.

(b) Below 4%.

(c) Between 4% and 5%.

(d) None of the above.

19. Ten shares of a software company's stock, which does not pay dividends, are purchased for $20 each. Six years later, the same 10 shares are sold for $47.50 each. The annual return on the investment is most closely

(a) 15%.

(b) 14.5%.

(c) 15.5%.

(d) 17%.

20. Ten shares of a software company's stock, which does not pay dividends, are purchased for $20 each. The stock splits three for two after two years. Six years later, all shares are sold for $47.50 each. The annual return on the investment is most closely

(a) 23.5%.

(b) 14.5%.

(c) 15.5%.

(d) 22.5%.

21. A company uses $100,000 annually of its $250,000 credit facility to fund working capital. If the facility charges a nominal rate of 16% per year, compounded quarterly, what is the minimum quarterly payment to be made?

(a) $40,000.

(b) $10,000.

(c) $16,000.

(d) $4,000.

22. A contract for parts delivery is worth $100,000 per year in time-zero dollars. If the inflation rate is 3% per year and the market interest rate is 7.12% per year, the worth of the contract at time zero, assuming a 10-year horizon, is most closely

(a) $853,000.

(b) $811,000.

(c) $1,000,000.

(d) $702,000.

23. A contract for parts delivery is worth $200,000 per year in time-zero dollars. If the inflation rate is 3% per year and the inflation-free interest rate is 4% per year, what is the worth of the contract at time zero, assuming a 10-year horizon?

(a) $853,000.

(b) $811,000.

(c) $1,000,000.

(d) $702,000.

24. Two four-year contracts are offered: payments increasing from $0 in period 1 by $100,000 per year or $150,000 per year. At a 10% annual rate of interest, which contract is preferred?

(a) The fixed-rate contract, by less than $10,000.

(b) The fixed-rate contract, by over $10,000.

(c) The increasing-rate contract, by over $5,000.

(d) Both contracts are equivalent.

25. Equipment can be purchased for $10,000 and last for *N* years (with no salvage value), or it can be leased for $2200 a year. If interest is 12% per year, purchasing is preferred if *N* is

(a) Less than 4 years.

(b) Between 5 and 6 years.

(c) At least 7 years.

(d) Leasing is always preferred.

26. Equipment can be purchased for $10,000 and last for five years (with no salvage value) or leased for $2200 a year. The lease is preferred if interest is nearest

(a) 0%.

(b) 1%.

(c) 2%

(d) 4%.

27. The lottery offers 20 years of $1 million payments or $10.5 million upon winning. You prefer the payments if interest is

(a) You would never prefer the payments.

(b) Between 8% and 10%.

(c) You always prefer the payments.

(d) Between 6% and 7%.

28. Payments of \$5000 are made annually into an account to save for college. The annual rate of interest is 10%. In order to remove \$40,000 per year for the final four years (at which time the account is empty or nearly empty), the minimum number of payments required is most closely

(a) 16.

(b) 17.

(c) 18.

(d) More than 20.

29. One maintenance contract is free for two years and is followed by three years of \$5000 payments while the other is \$2300 per year for five years. The contracts are equivalent if the interest rate i is

(a) $i: ((1+i)^3 - 1)/((1+i)^5 - 1) = 0.46$.

(b) $i: ((1+i)^5 - 1)/((1+i)^3 - 1) = 0.46$.

(c) $i: ((1+i)^3 - 1)/((1+i)^5 - 1) = 2.17$.

(d) None of the above.

30. A company donates \$100,000 to a university that invests in an account expected to earn 6% interest per year. How much can the university use annually in perpetuity?

(a) \$60.

(b) \$600.

(c) \$6000.

(d) \$60,000.

Part II

Decision-Making Preliminaries

5 Problem or Opportunity Definition

(*Courtesy of Chris Butler/Photo Researchers, Inc.*)

Real Decisions: Space Junk

An estimated 200,000 items larger than 1 centimeter in diameter are currently orbiting the earth, and most of it is garbage, including paint specks, discarded rocket stages, boosters, and fuel. Much of the debris was deposited when the United States and the Soviet Union were learning to explore space. Most of this "space junk" falls from orbit and burns off upon reentry to the atmosphere, but items in orbit can reach speeds of 30,000 kilometers per hour, which is fast enough to put holes in satellites, spacecraft, or space suits. Current systems can track objects greater than 10 centimeters, but an object one-tenth that size can destroy a satellite.[1] This leads to a number of interesting questions:

1. Is space junk a problem? If so, for whom and why?

2. Does space junk represent an opportunity? If so, for whom and why?

[1] Dasey, D., "Aussies help track space junk," *The Sun Herald,* p. 34, October 30, 2005.

3. What further information would be necessary to explore this as an opportunity?

In addition to answering these questions, after studying this chapter you will be able to:

- Define problems and opportunities. (Sections 5.1–5.2)
- Differentiate between problems and symptoms and between opportunities and prospects. (Sections 5.1–5.2)
- Delineate the relationship between problems and opportunities with respect to both the current status and a desired future status. (Section 5.3)
- Outline the characteristics and importance of being able to define problems and opportunities. (Section 5.4)

Before solving a problem or taking advantage of an opportunity, the situation must be clearly defined. This first step in the decision-making process is critical in that it initiates the process and limits the scope of all ensuing investigations and analyses. This may seem obvious, but it often requires extensive data analysis and investigation, usually because we do not recognize problems immediately. Rather, we witness symptoms, just as a doctor does with an ailing patient; and just like the doctor, who must make a diagnosis before any treatment or remedy can be suggested, we must define the problem before we can solve it.

5.1 Defining Problems

It can be argued that once a problem is defined, the ensuing steps follow a strict protocol and are generally analytical. This does not guarantee that these steps will be easy, but implementation of the decision process is clear. In this first step, there are no defined rules or strict protocols, because problems and their environments are unique. Thus, this is a challenging step for the engineer. It was Albert Einstein who said that "The formulation of a problem is far more often essential than its solution, which may only be a matter of mathematical or experimental skill."

Engineers are trained to use the tools of mathematics and science to solve problems. However, a problem cannot be solved unless it is defined and understood. The term "problem" has many connotations and elicits different definitions from different people. We employ the following definition, often referred to as *gap analysis*: A **problem** is the difference between the current situation and a desired situation. Note that defining a problem in this manner requires two measures: the current state and the desired, or sometimes ideal, state. Generally, an assessment is required to define the baseline, or current, situation. Defining the desired or optimal state often requires examining the objectives of the associated entity.

Interestingly, while problems are often "internal" to a company, they tend to surface due to "external" factors, including competition and regulation. Consider the following examples:

- In 2003, Airbus delivered more commercial jets to customers than Boeing.[2] This was the first time since Airbus' inception that this had occurred. This fact, brought about by competition, may be a symptom of a problem or problems at Boeing.
- In 2004 and 2005, the U.S. National Highway Traffic Safety Administration (NHTSA) was considering revamping the way it differentiates light

[2] Lunsford, J.L., "Bigger Planes, Smaller Planes, Parked Planes," *The Wall Street Journal Online,* February 9, 2004.

trucks and automobiles.[3,4] This development was significant, as rules which were in effect at that time required new-model cars sold in 2005 to average 27.5 miles per gallon and light trucks to average 21 miles per gallon. A vehicle is classified as a light truck if "a seat can be readily removed to allow for additional cargo space." This definition leads to the classification of numerous smaller vehicles as light trucks when they are arguably not. Since these smaller vehicles have better gas mileage rates, the overall average of the category is improved—allowing automakers to produce and sell a greater number of larger, more inefficient vehicles. It should be clear that this potential change in regulation could identify symptoms and, eventually, problems that automobile manufacturers need to rectify.

- In the first two months of 2004, General Motors recalled 2.4 million vehicles while three of its biggest competitors had not recalled a single vehicle.[5] This fact, highlighted by competitors with better quality ratings, is a symptom of a possible problem or problems at General Motors.
- On October 6, 2003, the European Commission banned the use of single-hull tankers for transporting heavy fuel oil in and out of European ports.[6] This regulation may lead to the identification of problems for both shipping companies and ship builders.

Note that no *specific* problems were identified in the preceding examples. Rather, "problem areas," such as low quality or bans on a product, were identified. These examples illustrate the difficulty of defining a problem. If GM says that its desired state is to be the highest-rated automobile manufacturer in terms of quality, then the company has defined a goal. However, no problem that can be solved has been identified or defined. Further investigation must determine *why* GM is not in its "ideal" state. This means that a problem or problems must be identified and solved.

Generally, there are two obstacles to defining a problem: Symptoms can be mistaken for problems, and the scope of a problem can be defined too broadly. We address both of these issues with simple examples.

---

[3] Power, S., "Rules Regulating Gasoline Mileage Face an Overhaul," *The Wall Street Journal Online,* December 22, 2003.

[4] "Oil Prices May Force Administration to Strengthen Fuel Economy Plan," *Energy Washington Week,* Vol. 2, No. 39, September 28, 2005.

[5] Carty, S.S., "General Motors Fights to Improve Quality Image," *Dow Jones Newswires,* March 1, 2004.

[6] Ang, E., "Europe Set to Ban Single-Hull Fuel Oil Tankers Oct 6," *Dow Jones Newswires,* September 15, 2003.

A symptom indicates that a problem (probably) exists, but it does not readily identify the problem. Rather, a symptom is an effect that results from the existence of a problem. Consider the following example:

> A machine is producing only 50 parts per hour.

Is this a problem? Alone, the statement is merely a fact that helps define the current situation. The word "only" hints that a problem may exist; however, it cannot be classified as a problem unless a desired situation is defined. Now consider the following rephrasing:

> The machine is producing only 50 parts per hour when we require 75 per hour.

This statement clearly identifies the existence of a problem, as it provides insight into both the current situation and the ideal situation. However, more specific information about the current situation is required in order to limit the scope of the problem. Obtaining such information may require an investigation. First, the maximum capacity of the machine must be determined; this will shed light on the type of problem that exists. Upon further investigation, it is found that

> The machine, with a maximum capacity of 80 parts per hour, is producing only 50 when we require 75.

This additional information has limited the scope of the problem, which should aid in the solution process. From this one statement, it can be determined that there is a problem with the machine or its operating environment. If the problem is fixed, the ideal situation can be achieved.

Note, however, that the true problem has still not been identified! Records must be checked to be certain that the machine has been properly maintained. The machine must be examined to see if any parts have failed. Manuals must be researched to see if capacity is limited by other factors, such as power supply or temperature. Workers must be quizzed to see if they are properly trained and if they are operating the machinery in its intended manner. The entire manufacturing process, including inventory levels, must be reviewed in order to determine whether the machine is at fault or the line is not producing at capacity (i.e., the machine is merely producing the parts that it is fed). It is through these and other investigations that the true source of the problem can be identified, perhaps as follows:

> The machine, which has a maximum capacity of 80 parts per hour, is producing only 50 when we require 75. The manufacturing process upstream from the machine seems to be functioning normally, with work-in-process inventory accumulating in front of the machine in question. According to maintenance records, the last two scheduled oil and filter changes were skipped due to overtime constraints. This may result in decreased efficiency and output, according to the operating manual.

This additional information indicates that the problem lies with the machine and its operation. Taking the machine down and performing a maintenance check may shed light on the true problem. A clogged filter may be causing the machine to overheat, preventing output, or a part may be wearing. Once the maintenance check is complete, the exact cause of the low output can be identified and the solution process may ensue. The key point to be gleaned from this small example is that symptoms are not problems. However, examining a trail of symptoms can lead to the identification of a problem.

Recall that defining a problem relies not just on the current situation, but also on the desired situation. Suppose further analysis had led to the following statement:

> The machine is producing parts at its maximum output rate (50 parts per hour) when we require 75.

Unlike the example we have thus far been examining, this statement sheds light on the fact that there is no problem with the machine, since it is operating at peak efficiency. Rather, there is a problem with the current situation, because the total capacity available does not meet the desired level. This fact has a direct bearing on possible solutions, which will undoubtedly require adding capacity.

The second obstacle to be avoided in defining a problem is allowing too wide a scope. Consider the following statement:

> Manufacturing costs are 20% higher than projected this period.

In this statement, the current situation and a desired state are defined (assuming that the projected costs are equivalent to the desired or ideal costs). However, the scope of this problem is far too large to be handled. In order to narrow its focus, we must perform an extensive investigation that may include charting the manufacturing process, determining where bottlenecks arise and work-in-process inventories grow, identifying where capacities are overloaded, etc. The data collected can be compared with data from other periods to aid in finding out where the anomalies are and to help pinpoint the problem or problems. A more manageable problem to tackle is expressed in the following statement:

> Newly installed machinery is proving to be more difficult to deal with than anticipated, with workers spending 10–15% more time setting up machines than previously. Also, the machines are requiring more maintenance than their predecessors required.

Again, more investigations must be performed. Long setup times and high maintenance costs are still symptoms. The long setup times may be caused by improperly trained workers, and the higher maintenance costs may be the result of improper preventive maintenance procedures or faulty parts. However, investigating these more specific symptoms will likely lead to the definition of a problem or, in this case, problems. It is the experience and training of the engineer that will outline the investigations needed for a problem to be accurately defined.

Problems define the difference between a current situation and a desired situation and thus are usually described by costs that are higher than desired or revenues that are below expectations. In the original example set forth in this section, the output of the machine was below the desired level. Presumably, revenues were also lower. Thus, the solution would be to remedy the situation by restoring revenues to expected levels. In the second example, manufacturing costs were seen as being higher than expected. Thus, the solution likely would consist of making an investment in order to reduce those costs.

Because problems are usually identified by an increase in costs or a decrease in revenues, solutions to problems tend to involve reducing costs or restoring revenues, respectively. Understanding these typical solution scenarios generally sheds light on how one should explicitly define the problem.

5.2 Recognizing Opportunities

Whereas problems generally present themselves, albeit unexpectedly, and require a solution that reduces costs or restores revenues, opportunities represent situations in which one can enhance revenues beyond current expectations. This often happens through the exploration of alternative applications of skill or the expansion of current outputs. As with defining problems, opportunities can be defined through gap analysis: An **opportunity** is the difference between current accomplishments and potential accomplishments. When an engineer discovers new ways in which to apply science or expand horizons, opportunities may arise whereby an investment can generate additional profits.

There is an old adage that says "opportunities are often created," in that they do not generally "fall out of the sky." Taking advantage of an opportunity requires creativity and an ability to think "outside the box." A creative engineer not only seeks to solve problems, but also initiates and proposes new ideas that may go beyond what is required. Such ideas may include examining possibilities beyond current capabilities or determining what other uses can be made from the resources and expertise that are currently available. These initiatives lead to the expansion and growth of companies and other entities.

Consider the following statement:

> A new residential area is being developed across town.

This may be an opportunity for numerous companies. Retailers may open new outlets in order to service the new market. Energy, cable, satellite, and telephone suppliers may invest in new infrastructure in hopes of expanding their customer base. Private schools may expand bus service in order to increase enrollments. Companies may locate in the area to take advantage of the influx in labor supply. Real-estate developers may draw plans for the surrounding area to take advantage of the new market.

It should be clear that this change in the environment can provide numerous opportunities for a wide range of companies. However, as with defining a problem, the need for an investigation is not precluded. A baseline is still required and information about the opportunity is needed in order to assess the situation.

Consider a telecommunications firm that may want to expand its operations to the new market. Several questions must be answered before specific investment opportunities can be identified: Is the new area currently served by other telecommunications (cable and telephone) firms? In other words, is this new development located near other developments and thus other established companies? Is the new residential area large enough to support an expansion? Have rights-of-way for laying cable already been established? Are the surrounding areas zoned similarly for further development? Although the opening of the new development seems like an opportunity, the investigation that answers the preceding questions will truly determine whether an opportunity exists. Much as symptoms can be mistaken for problems, prospects can often be mistaken for opportunities. The following description of the aforementioned opportunity is much more pointed and useful:

> A new residential area is being developed on the northwest side of town. The development, consisting of 120 single-family homes, is located 0.5 mile from the town border, 1.8 miles from the nearest residential development, and 7.8 miles from the nearest service area of the given company. The area immediately surrounding the development is currently farmland and is unzoned.

More information may be necessary in order to move to the next step in the decision-making process, but it should be clear that the preceding definition provides a clearer picture of an opportunity and not just a prospect. The following examples illustrate situations in which opportunities were presented to companies:

- In December of 2003, the United Kingdom's energy minister announced that, by 2015, 15.4% of all electricity in the United Kingdom is to come from renewable resources, such as wind and waves.[7]
- As of 2003 and 2004, it was expected that within three to five years China would overtake Germany and perhaps even Japan as the second-largest automobile market behind the United States.[8] This boom has led to an explosion of investment in China from automobile manufacturers, as well as from producers of raw materials.
- In early 2004, the U.S. Department of the Interior proposed a new rule to suspend royalties paid on the first 15 billion cubic feet of gas produced by wells between 15,000 and 18,000 feet deep. Furthermore, no royalties are to be paid on the first 30 billion cubic feet of natural gas from wells deeper than 18,000 feet.[9]

In order to take advantage of an opportunity, a company may have to expand production beyond its current means. Doing this could be as simple as acquiring more capacity or as complicated as shifting away from traditional production. We categorize how companies can take advantage of opportunities in the following three ways:

1. Expand current operations.
2. Develop new applications for current operations.
3. Develop new products or services that require new operations.

These methods in which to take advantage of opportunities clearly get more difficult as we move down the list. The case for expanding current operations is obviously the most common and natural form of expansion, requiring the addition of either facilities, equipment, or workers, but the work being performed is the same as (or similar to) that performed in previous operations. This kind of expansion is generally the result of robust demand.

The second step in the list is a slightly more drastic one. Some examples are as follows:

---

[7] Talley, I., and A. Chipman, "NEWS SNAP: UK Gives Lifeline to Renewable Energy Investors," *Dow Jones Newswires,* December 1, 2003.

[8] Bradsher, K., "In a Slow Start, Ford Opens an Auto Factory in China," *New York Times,* January 20, 2003.

[9] United Press International, "Royalty Relief Proposed for Gulf Gas Wells," *The Washington Times Online,* January 23, 2004.

- Honda, known for its production of engines for applications including automobiles, lawn mowers, and generators, developed an engine for small aircraft.[10] The engine achieves 40% higher fuel efficiency than conventional aircraft. Honda went on to produce its first jet (12.5 meters long with a 12.2-meter wingspan and a 2000-kilometer range) to further test the engine in late 2003. The jet made its first public demonstration flight in July of 2005.[11]
- Engineering group IMI redesigned one of its valves used in heavy-industry pumps to serve beer.[12] Specifically, the valve allows a pump to serve 10 pints of beer per minute, as opposed to 4, by addressing the problem of carbonated drinks foaming uncontrollably when depressurized. The valve was designed in 2003, with hopes of attracting sales at a variety of club and arena venues.
- French aerospace firm Dassault, known for its production of fighter jets, is also the creator of the computer-aided-design (CAD) software Catia, which is standard in the airplane-manufacturing industry.[13] The software was created for in-house projects, but has been released commercially to generate additional profits. The software is sold through an agreement with IBM such that Dassault focuses on producing new software and not on sales and marketing.

The final way in which a company can take advantage of opportunities is to expand operations beyond their traditional uses. This includes developing applications in areas that are novel, such as the following:

- While investigating new materials for its tires, Japanese tiremaker Bridgestone developed a high-quality silicon carbide.[14] Compared with standard silicon used in the manufacture of semiconductor wafers, the new product has a greater resistance to heat and electric charges, a property that is ideal for automobile applications. Because of the new development, Bridgestone announced that it would enter the chip-wafer business, hoping to book YEN10 billion in sales in 2010.
- General Motors has been investigating the use of fuel cells in automobiles. During its investigation, the company developed cells that are used by Dow Chemicals to produce energy at Dow's Freeport, Texas, plant, representing the largest commercial application of fuel cells to date

[10] Takahashi, Y., "Honda Says Small Business Jet Flight Tests Successful," *Dow Jones Newswires,* December 16, 2003.

[11] Shirouzu, N., "Honda to Showcase Jet in Development Next Week," *The Wall Street Journal Online,* July 19, 2005.

[12] Guthrie, J., "Beer Pump Points to IMI's Drive Towards Innovation," *Financial Times,* p. 24, September 9, 2003.

[13] Bulkeley, W.M., "Boeing to Use Software Package from Dassault," *The Wall Street Journal Online,* February 12, 2004.

[14] "Bridgestone to Make Semiconductor Wafers," *Dow Jones Newswires,* December 1, 2003.

(early 2004).[15] The cells are expected to generate 2% of the energy needed at the plant by using excess hydrogen produced by Dow. The agreement allows General Motors to expand its research and generate additional revenue, while Dow reduces its reliance on natural gas and expands its potential sources of energy.

- In 2003, Progress Energy of Raleigh, North Carolina, joined electric utilities in a number of U.S. cities, including Allentown, Pennsylvania, Washington, DC, and St. Louis, Missouri, in providing high-speed Internet services over electrical power distribution lines.[16] Signals from a home computer are relayed to the power line, where they are converted into a radio-frequency signal for transmission along the line. The application has excellent potential, as power lines reach virtually all homes in the United States.

In these last examples, the firms have examined their expertise and assets and seen an opportunity to expand their operations beyond current or traditional boundaries. Opportunities may have presented themselves—such as unexpected demand for a given product—but the company has set new goals in order to take advantage of those opportunities. Again, the type of opportunity we have just examined is more than an expansion of current practice; rather, it is the branching out of one's typical business.

In order to capitalize on an opportunity, a firm will undoubtedly have to make investments. For example, IMI may have to add production lines in order to produce its new valves. Dassault may have to hire additional engineers strictly to update its software products. Bridgestone will have to invest in a semiconductor wafer fabrication (wafer-fab) facility. These investments are inevitable if a company wants to take advantage of the opportunity presented to it. It is reasonable to expect that additional investment will increase revenues and hence profits.

5.3 Opportunities and Problems

We have defined problems and opportunities separately, but they are clearly related. Indeed, the gap analyses used in defining the two are similar. In both situations, a current baseline measure is required, in addition to some desired level that is to be attained. If the desired level has been reached previously, then the gap generally defines a problem. If the desired level has never been attained, then the gap defines an opportunity. In this manner, both problems and opportunities can be defined simultaneously, as illustrated in Figure 5.1.

[15] "NEWS WRAP: Dow Chemical Uses GM Fuel Cells to Power Plant," *Dow Jones Newswires,* February 10, 2004.

[16] "Progress Energy Tests Internet Service over Power Lines," *Dow Jones Newswires,* February 18, 2004.

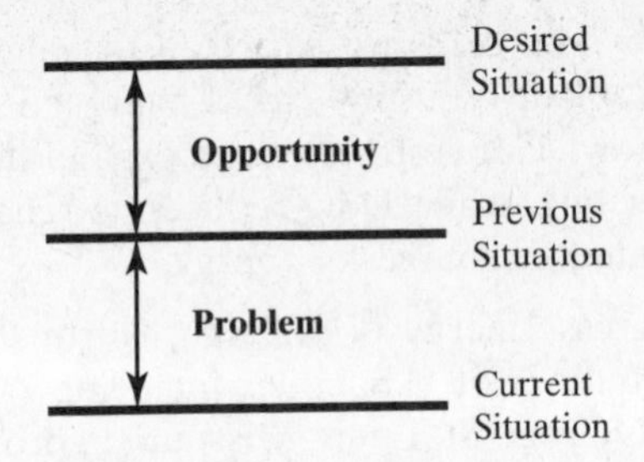

Figure 5.1
Differentiation between problems and opportunities according to a baseline.

According to Figure 5.1, solving a problem returns a situation to some defined baseline, whereas opportunities define a new baseline. The simple example of a machine's capacity in Section 5.1 can be viewed in this manner. Returning the machine to its capacity of 50 parts per hour can be seen as a problem, while increasing the capacity of the system to 75 parts per hour to deal with increased demand can be seen as an opportunity.

To further examine the relationships between problems and opportunities, recall the possible changes in regulations by the NHTSA whereby automobile manufacturers would be forced to improve the gas mileage of trucks and sport-utility vehicles (or risk cutting production of vehicles with poor mileage). Is this a problem or an opportunity? The answer depends on the party being asked and defines our first relationship between problems and opportunities:

> A problem for one entity can often be seen as an opportunity for another.

Recall that in our definition of a cash flow, an inflow for one party can be seen an outflow for another party. The situation is similar here, as whether one is presented with a problem or an opportunity is a matter of perspective.

Current producers of light trucks that do not meet the fuel efficiency rating criteria would probably view the new law as a problem because their trucks will become obsolete when the legislation goes into effect. However, this can also be viewed as an opportunity. Companies that develop technologies to improve gas mileage, such as using lighter materials or gas–electric hybrid motors, will be able to expand their operations. Alternative forms of transportation may also be able to take advantage of the new law. Automotive manufacturers can view it as an opportunity because the company that is first able to develop a truck that meets the new regulations will likely grab a large portion of the market share and see improved revenues. This will, of course, require additional investment. New designs, materials, fuels, or methods of production may be required to meet the new regulations. Under the assumption that meeting the regulations is not trivial, the number of competitors in this market is likely to decrease. Therefore, assuming that there is no drop-off in truck sales, the firm that is able to deliver the product stands to reap large financial rewards.

It is reasonable to assume that automakers which currently produce light trucks that do not meet the regulations must address the issue before they try to seek opportunities presented by the legislation. That is, before they can take advantage of the opportunity to be first to market, they must solve the problem of their vehicles not meeting the code. In other words,

> Problems, in general, must be solved before opportunities can be pursued.

The good news is that solving problems may lead to many benefits. Although the new government regulation may seem imposing, it can be a blessing in disguise. In an effort to meet the new regulation, car producers may expend numerous dollars in the research and development of new technologies. Not all of these investments will improve the gas mileage of trucks. However, some investments may define further means by which the company can profit. For example, the development of a new fuel cell may prove to be too inefficient for automobiles, but the technology may be well suited for developing the next small-scale power generator. This is clear through the relationship that General Motors and Dow Chemicals have established, as noted earlier. Corollary to our previous relationship is the idea that

> Problems, in general, must be solved, whereas opportunities can often go unpursued.

Earlier, we noted that Boeing had been surpassed in sales by Airbus in 2003. Again, this is not a problem, but the symptom of a problem. If the problem were to go unaddressed, however, Boeing could be pushed out of the commercial aviation market for large aircraft. In a related situation, while projections support growing markets for small regional jets, Airbus may not feel obligated to chase this opportunity in light of the company's focus on its current market for larger aircraft. That is, an opportunity does not have to be explored, although growth may be limited if most opportunities are bypassed. This is not the same for problems, as they must generally be addressed. Fortunately, the solution process may lead to more opportunities:

> Defining and solving problems often leads to new problems and opportunities.

The theory of constraints is a methodology for solving manufacturing problems. Its focus is on identifying the "bottleneck" operation, or the step in the process that is limiting current production. Once this bottleneck is removed (in other words, once the problem is solved), a new bottleneck emerges and becomes the focus of attention. In many ways, this is simply intrinsic to solving problems and finding opportunities. When one solves problems, new problems may surface. Solving these problems in turn allows for new applications of talent, and opportunities for growth may emerge.

5.4 The Definition

Whether you are faced with a problem or an opportunity, you need a clear definition to solve the problem or take advantage of the opportunity. Specifically, the definition of the problem or opportunity should:

- Be concise, narrow, and as clear as possible, even though the solution may have a broad impact.
- Define both the baseline of the current situation and the desired level of achievement. The desired level is dependent on whether a problem or an opportunity is being defined.
- Not be confused with symptoms or prospects. Rather, the investigation of symptoms and prospects will give rise to appropriate definitions.
- Align itself with the organization's objectives.

If problems or opportunities are defined in this manner, the solution process will progress more smoothly.

5.5 Examining the Real Decision Problems

We return to our introductory problems concerning space junk and the questions posed there:

1. Is space junk a problem? If so, for whom and why?

Because space junk can destroy equipment used in space, it is clearly a problem for those who explore space, including government entities such as NASA and private explorers such as Virgin Galactic.[17] It is also a problem for businesses (e.g., satellite TV providers or weather forecasters) and government agencies (e.g., the military) that use satellites. Thus, it is ultimately a problem for engineers who design spacecraft, satellites, and space suits.

2. Does space junk represent an opportunity? If so, for whom and why?

To understand possible opportunities, we may need some preliminary ideas about how to solve the problem. If doing so entails making stronger equipment that is able to withstand the impact of space junk, then it

---

[17] Baker, D., "N.M. gov calls state spaceport an investment in the future," *Associated Press Newswires,* December 14, 2005.

may be an opportunity for makers of space-flight equipment to distinguish themselves from other manufacturers of equipment. If the solution entails "avoiding" space junk, then there are opportunities for companies to provide mechanisms to track the junk and plot its trajectory. This is clearly an interesting engineering problem that would rely heavily on the laws of physics. Finally, if the solution entails destroying the space junk, that opens another realm of opportunities, although they would appear to be contingent on the opportunity to track the junk.

3. What further information would be necessary to explore this situation as an opportunity?

At the very least, what is required is a better understanding of the ramifications of space junk striking various pieces of equipment, such as rockets, satellites, spaceships, or space suits. For example, taking into account its size and type, does such junk cause minor damage, cause great damage, disable the equipment, etc.? This kind of understanding would allow for a better definition of a problem that needs to be solved and the opportunities available for enhancing equipment. It would also help to have an understanding of the current technology that is available to track the junk. Earlier, it was mentioned that junk 10 centimeters in diameter can currently be tracked. Questions must be answered as to how much this technology can be pushed or improved, and the answer will undoubtedly lead to more questions.

5.6 Key Points

- Defining a problem or opportunity is the first step in the decision-making process.
- A problem or opportunity is the difference between a current situation and one that is desired. Solving a problem restores a situation to a previous state, while taking advantage of an opportunity moves a situation to a new state.
- Symptoms identify the existence of problems, but further investigation is generally required to identify the cause of the symptoms.
- Limiting the scope of the definition of a problem greatly improves the decision-making process.
- Problems usually highlight situations characterized by increased costs or decreased revenues. Thus, solutions focus on reducing costs or restoring revenues.
- Opportunities are situations characterized by the potential for increased profits. Thus, solutions focus on increasing revenues.
- Prospects, like symptoms, must be investigated in order to define opportunities.

- Finding opportunities requires creativity and a willingness to go beyond "normal" operations.
- A problem for one entity may be viewed as an opportunity for another.

5.7 Further Reading

Couger, J.D., *Creative Problem Solving and Opportunity Finding*. International Thompson Publishing, New York, 1995.

Evans, J.R., *Creative Thinking in the Decision and Management Sciences*. South-Western Publishing Co., Cincinnati, Ohio, 1991.

Fogler, H.S., and S.E. LeBlanc, *Strategies for Creative Problem Solving*. Prentice Hall, Upper Saddle River, New Jersey, 1995.

Goldratt, E.M., and J. Cox, *The Goal: A Process of Ongoing Improvement*, 2d rev. ed. North River Press, Great Barrington, Massachusetts, 1992.

Golub, A.L., *Decision Analysis: An Integrated Approach*. John Wiley and Sons, New York, 1997.

Polya, G., *How to Solve It: A New Aspect of Mathematical Method*, 2d ed. Princeton Science Library, Princeton, New Jersey, 1985.

5.8 Questions and Problems

5.8.1 Concept Questions

1. Why is the first step in the decision-making process so critical?

2. Define a problem and contrast your definition with that of an opportunity.

3. What must be avoided when defining a problem?

4. How are problems and symptoms related?

5. John Dewey once said, "A well-stated problem is half solved." What did he mean and what is the relevance to decision making?

6. What must be avoided when defining an opportunity?

7. How are opportunities and prospects related?

8. How are problems and opportunities related?

5.8.2 Application Questions

For each of the statements that follow, identify a problem *and* an opportunity. If the problem or opportunity is ill defined, explain how you might define it more clearly.

1. Inspections reveal cracks in the brake equipment of 20 high-speed Acela trains used by Amtrak in its profitable northeast corridor.[18]

2. In May of 2003, Pennsylvania announces the Energy Harvest Grant Program, with funds and loans to "provide the last increment of funding for clean and renewable energy projects built in the commonwealth from sources such as biomass, wind, solar, small-scale hydroelectric, landfill methane, coal-bed methane, and waste coal."[19]

3. In December of 2003, the United Kingdom's energy minister announced that, by 2015, 15.4% of all electricity in the United Kingdom is to come from renewable resources, such as wind and waves.[20]

4. In 2004, it was expected that within three to five years China would overtake Germany and perhaps even Japan as the second-largest automobile market behind the United States.[21]

5. New York City's public-transportation system of subways and buses, which serves roughly seven million people daily, is idled for three days due to a strike.[22]

6. In December of 2005, the price of gold topped $510 an ounce, a 24-year high.[23]

7. In early 2004, the U.S. Department of the Interior proposed a new rule to suspend royalties paid on the first 15 billion cubic feet of natural gas produced by wells between 15,000 and 18,000 feet deep. Furthermore, no royalties are to be paid on the first 30 billion cubic feet of natural gas from wells deeper than 18,000 feet.[24]

8. The European Commission charged Microsoft with failing to comply with its ruling that the software giant "reveal to rivals information about the workings of its Windows operating system." The order was meant to allow other companies to design compatible software for Windows-driven computers more easily. Microsoft was warned that it could face financial penalties if it did not comply.[25]

9. In recent years, Bethlehem Steel and the holding corporation for Union Pacific Railroad vacated their headquarters from the 21-story Martin Tower in Bethlehem, Pennsylvania, leaving a large amount of available corporate office space. The tower was built in 1972.[26]

---

[18] "Amtrak returns full Acela fleet to service," *Trains,* Volume 65, Number 12, p. 26, December, 2005.

[19] "Pennsylvania Energy Harvest Program Investing $6 Million," *Industrial Environment,* Volume 16, Number 1, January, 2006.

[20] Talley, I. and A. Chipman, "NEWS SNAP: UK Gives Lifeline to Renewable Energy Investors," *Dow Jones Newswires,* December 1, 2003.

[21] Bradsher, K., "In a Slow Start, Ford Opens an Auto Factory in China," *New York Times,* January 20, 2003.

[22] Enrich, D., "On Wall St., NYC Transit Strike Prompts Early Vacations," *Dow Jones Newswires,* December, 21, 2005.

[23] Haselhurst, D., "Reliable resources; there's gold in them there drills as mining stocks finish the year looking as rosy as our portfolio," *The Bulletin,* Volume 123, Number 51, January 10, 2006.

[24] United Press International, "Royalty Relief Proposed for Gulf Gas Wells," *The Washington Times Online,* January 23, 2004.

[25] Buck, T., "Brussels charges Microsoft," *The Financial Times,* p. 1, December 23, 2005.

[26] Radzievich, N., "Builders seek zoning change," *The Morning Call,* December, 20, 2005.

10. Hurricanes Rita and Katrina took six oil refineries, used to convert oil to gasoline, out of operation in Beaumont and Port Arthur, Texas, and Lake Charles, Louisiana, for months. The six refineries converted 1.7 million barrels of oil per day.[27]

11. In 2000, the Australian carrier Qantas, purchaser of Boeing jets for 41 years, placed an order for 12 A380 superjumbo jets from Airbus.[28]

12. Total demand for solar cells used in solar energy panels was expected to grow by more than 30% in 2005, but the capacity for polysilicon, used to make the cells, has not grown.[29]

13. London's Heathrow Airport, Europe's busiest hub airport, handled 67.7 million passengers over 12 months in 2004 and 2005. The facility was designed to handle only 50 million.[30]

14. Guidant Corp. issued recalls of some of its implantable heart defibrillators after the devices failed, causing at least two deaths.[31]

15. The general manager for human resources at Hyundai Engineering of South Korea stated, "We are seeing a shortage of high-level engineers as we win a number of big plant orders from the Middle East. We need more supervisors who can take charge of overseas projects."[32]

16. Brown University announced that it would build a student fitness center adjacent to its athletic facility. Currently, the area has 200 parking spaces, and the ratio of people to parking spaces on campus is about 3 to 1 and growing.[33]

17. In September of 2005, DaimlerChrysler selected North Charleston, South Carolina, for its new Sprinter Van assembly plant. The plant is expected to cost $35 million.[34]

18. The MV-22 Osprey, an experimental military aircraft with engines that tilt 90 degrees to enable the craft to take off and land like a helicopter, but fly like a plane, crashed twice, killing 23 marines, during testing in 2000.[35]

19. Using commute-time estimates and current gas prices, Salary.com determined that commuters in Brownsville, Texas, were the hardest hit by rising gas prices in 2005.

---

27 "NEWS WRAP: Rita May Have Damaged More Oil Rigs than Katrina," *Dow Jones Newswires,* September 28, 2005.

28 Tucker, S., K. Dome, and D. Cameron, "Qantas opts for Airbus over Boeing," *The Financial Times,* pp. 19–21, December 15, 2005.

29 Hillé, K., "Motech seeks to produce its own polysilicon raw materials," *The Financial Times,* p. 26, November 28, 2005.

30 "Preparing for Take-off: Other Expansion Plans," *The Financial Times,* p. 2, December 10, 2005.

31 Loftus, P., "Update: J&J to Pay Lower Price for Guidant under New Deal," *Dow Jones Newswires,* November 15, 2005.

32 Song, J.-A., "South Korea looking for engineers in the pipeline," *The Financial Times,* p. 25, December 8, 2005.

33 Davis, K.A., "Residents turn out to hear Brown's 5-year plan," *The Providence Journal,* December 14, 2005.

34 Naseri, D., "DaimlerChrysler Selects North Charleston, S.C. for New Sprinter Van Assembly Plant," *Dow Jones Newswires,* November 28, 2005.

35 Wolf, J., "US Navy finds tiltrotor Osprey aircraft 'suitable,' " *Reuters News,* July 13, 2005.

Cities in California, Florida, Hawaii, Massachusetts, and New York made the top 10.[36]

20. Freight railroad CSX was blamed for 780 of the approximately 1650 commuter train delays on the line between Boston and Worcester, according to data provided by the Massachusetts Bay Transportation Authority. About 100 of those delays were described as "freight train conflicts."[37]

5.8.3 Fundamentals of Engineering Exam Prep

1. Solving an engineering problem will generally

(a) Restore revenues to previous levels.

(b) Reduce costs.

(c) Improve quality.

(d) All of the above.

2. Taking advantage of an opportunity will often

(a) Require investment.

(b) Reduce the capabilities of a firm.

(c) Increase revenues.

(d) Both (a) and (c).

3. A good definition of a problem

(a) Is limited in scope.

(b) Is clear and concise.

(c) Defines the desired result.

(d) All of the above.

4. Defective products sold by a company represent

(a) The symptom of a problem or problems for the company.

(b) An opportunity for a competitor.

(c) Both (a) and (b).

(d) None of the above.

5. Which of the following can be defined as pursuing new opportunities?

(a) A distribution-and-shipping company offers to provide logistics-planning services to its clients.

(b) A company abandons a production line.

(c) A company reduces its product lines.

(d) All of the above.

---

[36] "Salary.com Reports That American Workers Are Pumping Entire 2005 Pay Increases Down the Gas Tank," *Business Wire,* October 12, 2005.

[37] Shartin, E., "Freights Delay MBTA Trains," *The Boston Globe,* December 1, 2005.

6 Generating and Designing Feasible Solution Alternatives

(*Courtesy of Robert Holmes/CORBIS.*)

Real Decisions: Main Street Crawl

The small town of Bellevue, Ohio, examined 10 feasible routes to alleviate downtown traffic congestion on U.S. Highway 20. The downtown span runs for over 4 miles and takes, on average, 13 minutes to navigate, although the time lengthens considerably during rush hour.[1] This is a problem common to many cities and towns;[2] and thus, our analysis will be general.

As noted in the previous chapter, the phrase "traffic congestion" actually defines a symptom, not a problem that can be solved. Assume,

[1] Sanchez, Y. P., "Bellevue Mulls Options to Reduce Traffic Congestion," *Port Clinton News Herald*, www.portclintonnewsherald.com, December 26, 2003.

[2] Miller, L., "Traffic bottlenecks increasing, study says," *Associated Press Newswires*, February 18, 2004.

however, that investigating the symptom has led to the following definition:

> City engineers have determined that it takes an average of 13 minutes to navigate the city center along Main Street. This time increases to nearly 25 minutes during rush hour. It has been determined that the increase in rush-hour traffic is from commuters passing through town. As the number of cars increases during rush hour, the time to pass through invariably increases, since the speed limit through town is 30 miles per hour slower than the speed limit one sees upon approaching the town and there are a number of intersections with traffic lights. Thus, the rate of inflow is much greater than the throughput rate, causing backups and delays.

Given this definition, we can

1. Generate alternatives to solve the defined problem.
2. Comment on whether an alternative solves the problem at hand, on its feasibility of implementation, and on any positive or negative impacts that will result from its implementation.
3. Delimit the alternatives to a manageable set.

In addition to performing these functions, after studying this chapter you will be able to:

- Connect the engineering design process to that of generating solution alternatives. (Section 6.1)
- Promote creative thinking when working alone. (Section 6.2)
- Identify the pros and cons of generating ideas in a group. (Section 6.3)
- Perform the Delphi and nominal group process methods for generating alternatives in a group. (Section 6.3)
- Define the role of the do-nothing alternative as a viable option for investment. (Section 6.4)
- Reduce the possible alternatives generated to a feasible set for subsequent analysis. (Section 6.5)

Once a problem or opportunity has been clearly defined, the process aimed at solving it may begin. This step in the decision process—generating solution alternatives—cannot be overstated because, as noted in Chapter 1, it is the step at which engineers *design* solutions. That is why this step is referred to as the "creative step" in the decision-making process.

This step at which engineers generate solution alternatives is critical because it defines all possible engineering projects for subsequent analysis. In other words, this step defines the investment alternatives that are to be analyzed. If the step is ignored, the alternatives that are available to the firm may not be sufficient to solve problems or take advantage of opportunities and promote growth.

6.1 Generating Alternatives

The truth is, generating ideas to solve problems or take advantage of opportunities is hard, for a variety of reasons. One reason is that we are not asked to be creative very often. Rather, we spend most of our time examining previously defined options, and we use our brains to analyze the consequences of, and differences among, these specific options. It is rare that one is asked to *design* a new option and determine the best course of action. Thus, the creative process may need a "jump start."

Before we examine some methods for "jump-starting" the creative process of generating alternatives, we stress the fact that most solutions are not created from scratch. Rather, they are adaptations of previous ideas and designs. This is natural, as most developments progress incrementally.

The methods of adaptation we shall examine can generally be grouped into the following categories:

- Updating, renewing, or adapting a previous design.
- Applying an old design to a new area.
- Understanding and learning from the competition.
- Incorporating research and development.

This is not to say that adapted designs or solutions are not creative. On the contrary, it can take just as much creativity and skill to adapt a design from a different function to the current situation as developing a new design from scratch. The fact is, developing solutions is a creative process, regardless of the origins of one's inspiration.

6.2 Working Alone

Generating ideas is not an easy task. In the next section, we address the problem of generating solutions in a group. There, we define both the benefits and problems associated with working in a group. Clearly, the benefits become problems for those working alone, while the problems of working in a group can be avoided when one works alone.

Although most engineers work in groups, it is possible that one will have to develop ideas individually. There is no magic formula for doing so, and people with more experience generally create their own methods to encourage the generation of ideas.

A number of factors help promote a creative working environment. First, you should create a relaxed atmosphere that is free from distraction. This may entail getting away from the office (or the television, phone, refrigerator, or computer) in order to reduce distractions to your train of thought. Second, you should reserve time for creative thinking at an appropriate hour of the day. It is amazing how a day can be filled with meetings and e-mail. It is important to set aside time to "think" and give yourself the ability to be creative. You should take advantage of this time when you are best at thinking (e.g., probably not right after lunch!).

Again, there is no perfect method for generating ideas, but there are some general guidelines that should be followed in order to promote good thinking:

- *Record all your ideas.* Many ideas are lost or forgotten because they are never recorded. It is true that most ideas you generate to deal with a current problem are generally not used, but they may prove useful for future problems. Notebooks, either digital or paper, should be kept specifically for recording ideas. (Digital audio or video recording makes this quite easy!)
- *Generate, don't edit!* The greatest inhibitor to generating ideas is ourselves, in that we criticize and edit possible solutions before they are fully developed. This guideline ties in directly with the previous point about recording all your ideas. In other words, save your judgments for another time.
- *Seek input.* Seek information from trade books, journals, textbooks, patent submissions, competing company designs, etc., in order to stimulate the creative process.

Note that these principles help even when you work in a group, because you are still required to contribute in that situation.

6.3 Working in a Group

Engineers rarely work alone, especially when designing solutions. Team-based approaches are quite common in solving problems. Often, the teams consist of diverse individuals with different backgrounds, experience, fields of expertise, and personalities. The benefits of working in a group, especially in the context of generating solution alternatives, are numerous and include the following:

- *Greater sum of knowledge.* More people provide more knowledge, both in education and experience, leading to a greater number of ideas and approaches.
- *Greater diversity of approaches.* More people means greater diversity in education, perspectives, and goals, which should lead to a greater diversity in solution approaches. It can be quite stimulating to hear different approaches, providing more ways to look at problems.
- *Ability to build upon others' ideas.* People can generate new ideas by "piggybacking," or expanding, on the ideas of others. This includes the opportunity to ask questions in order to clarify ideas that may be misunderstood by members of the group.
- *Feeling of inclusion.* When people work together towards a common goal, there can be a feeling of inclusion and accomplishment for all in the group. People often like working in a group because they feel it can be fun to work with others. The result is a healthy working environment.

This is not to say that working in a group will inevitably lead to the best solution designs. Unfortunately, there are many negative aspects of working in a group. Often these negatives are tied to human nature. Among the potential problems are the following:

- *Dominance.* Persons with strong leadership qualities may dominate the proceedings and dictate the direction of the group. This can stifle creativity.
- *Intimidation.* Some members of the group may be reluctant to contribute because they are intimidated, for any number of reasons. For example, a member of the group may have previously criticized a person, leading to a situation in which the criticized individual does not want to contribute again. Or an engineer may be in a group with his or her supervisor and thus may feel reluctant to speak freely.
- *Social pressure to conform.* A member of a group may offer only ideas that conform to typical thinking, so as to not look out of place. This narrow focus, often termed "groupthink," can hinder creativity as bold new ideas are avoided, downplayed, or ignored altogether.
- "*Satisficing.*" Similar to the preceding idea, "satisficing" is the notion that contributions may reach only for solutions which satisfy a given need, as opposed to finding the optimal approach. This will result in lost opportunities or solutions that cost more than is necessary.
- *Social laziness.* If a group is too large or exclusive, some members may not contribute. Rather, they let the other members of the group perform the necessary duties. This obviously defeats the purpose of bringing people together.
- *Lack of consensus.* The larger the group, the harder it may be to reach a consensus if one is needed.

No doubt, there are many benefits to group decision making, due to the interactions among the individuals in the group. However, this comes at the cost of interacting as a group. Despite these problems, group decision making is in constant use in our society. Whether a group of four people must decide where they want to eat dinner or a board of directors must decide where to locate a new factory, groups face decisions on a daily basis.

To avoid some of the pitfalls of group interaction, many techniques have been developed to aid a group in developing, or arriving at, good solutions. Ideally, these methods would eliminate all of the pitfalls while capitalizing on the benefits. Obviously, this is not entirely possible. We present two techniques: the Delphi method and the nominal group process method. The two are similar, but have subtle differences with respect to how they allow a group to interact, leading to different outcomes.

6.3.1 The Delphi Method

The Delphi method was developed in 1949 by the Rand Corporation in an effort sponsored by the U.S. Air Force. Its first application was actually quite grim: it sought consensus from a group of experts to determine appropriate levels of nuclear weapons and viable targets. Our application of the method is not quite as dire, as we simply seek solution alternatives to a problem or opportunity that is presented to us.

Our ultimate goal in using the Delphi method is actually quite different from its original application, as the second step of our decision-making process is not consensus building. Rather, our goal is to generate as many solution alternatives as possible from our group, which probably consists of a number of professionals, including engineers with varied backgrounds.

The method, altered slightly from its original application, can be described in the following steps:

1. The problem or opportunity (defined in the previous chapter) is written down and given to each member of the group. Each member is asked to individually list specific solution alternatives.
2. Once completed, the lists are collected by the monitor, and the results are tabulated without editing, such that the final list is merely a compilation of all of the ideas that have been generated.
3. The list of ideas is distributed to the group.
4. Members of the group are once again asked to individually generate solution alternatives.
5. Once completed, the lists of ideas are collected by the monitor, and the results are again tabulated. The process continues until no new ideas are generated or a prespecified number of rounds has been completed.

In the original application of the Delphi method, the goal was to answer a specific question. Thus, each member of the group would provide an answer and possibly request further data, and in each round each member would see the answers of the other members. This iterative process would continue until the group came to a consensus.

In our application, the goal is to generate as many solution alternatives as possible. Thus, iterations can continue, but a point of diminishing returns will be reached wherein the number of new alternatives being generated is minimal. The process can usually be terminated after two or three iterations.

It may not be entirely clear from reading the steps of the process, but the Delphi method is designed to capitalize on group interaction *without* direct communication or physical interaction between the members. The negative aspects of group interaction, listed earlier, are generally the result of people communicating directly, such as people discussing issues around a table. However, the Delphi method is carried out without any direct communication between the members, thereby allowing for anonymity among members of the group, which helps eliminate many negative issues associated with group interaction.

Of course, the method is not flawless, as many researchers have pointed out. Some of the benefits of interacting in a group, such as its social aspects and a feeling of belonging, are lost because the group members interact only through a sheet of paper (or, more likely nowadays, a computer screen). In addition, there is a great burden on the monitor to tabulate and disperse results from each round. Finally, and most importantly, when ideas are not entirely clear, it may be difficult for other members to use the information they get, as they are not allowed to inquire about it or ask for more information. Despite these issues, the method does provide a process in which people's ideas are brought together without allowing the negative aspects of group interaction to interfere with the process.

6.3.2 The Nominal Group Process

The nominal group process has many variants, the first of which was devised by André L. Delbecq and Andrew H. van de Ven in 1968 when studying industrial engineering problems of program design at the National Aeronautics and Space Administration (NASA). The process can be described as a structured meeting of a group of people in which interaction is limited. We again tailor the process to our goal of generating solution alternatives:

1. The problem or opportunity is written down and given to each member of the group. The members silently generate their alternatives within a prespecified time frame.
2. Upon the completion of step 1, a member of the group serves as a monitor. Starting with any member of the group and going around the table, each member of the group offers one suggestion that the monitor tabulates and that all members can see. This process continues until all of the ideas of

every group member are listed. No discussion takes place between the participants during this phase.

3. Once all of the ideas have been compiled, the information listed is open to discussion by the group. The discussion may be limited by time.
4. In the second round, the group is asked to once again generate solution alternatives silently. The process repeats until no new ideas are generated or a prespecified number of rounds has been completed.

As might be expected, the nominal group process technique overcomes many of the shortcomings of the Delphi method by allowing some, albeit limited, interaction between group members. The advantages include a sense of accomplishment or involvement in the group and the socialization associated with working in a group. Also, the discussion period allows for the clarification of ideas, a feature that may not be possible without direct communication.

However, this interaction comes at a price: Once the group is allowed to interact, the problems associated with working in a group surface and may lead to subpar results with respect to generating solutions.

There have been a number of studies that have tried to compare the Delphi method with the nominal group process. Most notably, it has generally been concluded that the two methods are quite effective, especially compared with traditional groups making decisions (i.e., brainstorming a list of ideas in the absence of a process). The methods differ in their level of interaction, which should be the basis of deciding which one to employ. The time frame and the geographic dispersion of the group members are also important considerations. The nominal group process technique requires that the group members interact. While this can be accomplished through videoconferencing when the members are geographically dispersed, it still requires that the group members meet at a prespecified time, always a difficult task with a large group. The Delphi method does not require this level of interaction and can actually be accomplished through time-lagged interaction, such as e-mail. Unfortunately, this may lead to a process that is significantly longer than desired, as the monitor must wait for responses from all group members before proceeding to subsequent rounds.

The key point here, however, is that the processes are really quite similar, with both techniques requiring the individual generation of ideas and both pooling solutions for the final group of solutions. The most noticeable difference is that Delphi process members are (or can be) anonymous, whereas nominal group process members interact. Thus, convenience may drive the decision or the desire to include the social aspects of group decision making.

6.4 The Do-Nothing Alternative

It is entirely possible that the opportunities available are not worthy of investment. Thus, it is important to recognize that the **do-nothing alternative** is almost always a viable option. This alternative represents the status quo. In a situation in which only one investment opportunity is presented, the economic consequences of seizing that opportunity must be better than the status quo in order to be worthy of investment.

Note that adhering to the status quo does not mean that investment funds are put under a mattress until another day. Rather, it means that the funds are placed into investments that are always available. One can envision the do-nothing alternative as a risk-free investment alternative that is always available and that has a known and defined rate of return. Thus, investing in an opportunity must bring about returns that are better than the return that is always assumed to be available.

Note also that the definition of the do-nothing alternative is dependent on the situation. If the analysis is to determine whether a new product should be launched, then the do-nothing alternative represents the expected return that our invested funds would achieve elsewhere. Thus, the expected returns of the new project must be better than these other opportunities. However, if we are evaluating whether to keep a machine for another year or replace it with a new machine, then the do-nothing alternative is clearly different. In this situation, the do-nothing alternative is defined by keeping the current machine, as that represents the status quo. Clearly, the returns associated with this do-nothing alternative are different from those in the previous example of investing in a new product line. In the example of deciding whether to replace a machine, the economics of the do-nothing alternative are defined by the current asset. Finally, suppose we must choose between two processes or technologies for a given project. These processes are crucial to the project and the goal is to choose the least costly technology. In this situation, the do-nothing alternative is not valid, as *some* technology must be chosen.

In essence, the definition of the do-nothing alternative is related to whether a problem is being solved or an opportunity is being pursued. It is not an absolute rule, but the do-nothing alternative is generally feasible when new opportunities are being evaluated. In this case, the do-nothing alternative represents the expected return from other opportunities available to the company. In the case of solving problems, the do-nothing alternative may not be feasible, or it may be defined by the economics of the status quo, as the available alternatives will generally lead to lower cost solutions.

In later chapters, we will define the do-nothing alternative more explicitly. However, it is important to note that alternative here, as this is the step in the decision-making process where we are defining possible alternatives for future analysis.

6.5 Reduction to a Feasible Set of Alternatives

Once this creative step has been accomplished, a number of possible solutions lie before the decision maker(s). Ideally, each and every one of these alternatives would continue along the decision path, with cash flows being developed and information gathered for each such that they can be economically evaluated. However, cash flow estimation is an expensive and time-consuming process. Therefore, it is only natural that the number of alternatives to be evaluated be reduced to a reasonable set in order to conserve resources. This is not an easy process, as differing opinions on different methods may lead to disagreements over which alternatives to pursue. However, one should keep in mind that the goal here is the generation of *a number* of alternatives, not just one, for the latter would eliminate the need for economic evaluation and the need for a choice to be made. Unfortunately, it is this step in which many decision-making processes fail, because they reduce the number of alternatives to a group that is so small that what is left is a simplistic choice between two alternatives or, worse, one lone alternative that, by default, must be adopted.

The reduction of our set of choices to a reasonable one is truly a process that is defined by engineering design. Once alternative approaches have been defined, the engineering design process lays out steps that, when followed, determine feasibility. The key here is that alternatives should be eliminated due to engineering limitations, not due to economic factors. The analysis that ensues in further steps of the decision-making process will weed out projects that do not measure up economically. Having said that, we recognize that the world is not black and white. It is possible that, in reducing the number of alternatives, economic data may be roughly estimated, and it may become abundantly clear that some alternatives are so costly that they do not merit further analysis. Clearly, these decisions can be made with the use of sound judgment. The Delphi method and the nominal group process technique described earlier may also be used to pare down the number of alternatives to a reasonable set.

We defined the design process in Chapter 1. Determining the feasibility of an alternative is discipline and application specific. It may entail designing prototypes, programming an initial application, or simulating a new process. The following questions must be answered in order to determine whether the project should be researched any further:

1. Does the project meet its technical objective of solving the problem or taking advantage of the opportunity?
2. Does the project provide additional opportunities beyond those stated?
3. Will the project generate any negative impacts, either to the firm or others?

As opposed to stating specific steps for this process, which is generally not possible, we outline a hypothetical scenario to illustrate the idea with the example of an opportunity.

EXAMPLE 6.1

Opportunity: Increasing Demand for Natural Gas

The United States has long relied on its own supplies of natural gas to meet energy needs. The fields producing those supplies are beginning to age, however, and as a result, production is slowing. At the same time, demand from power plants is increasing, as there has been a shift to using clean-burning natural gas, as opposed to coal, to generate power. This has been accompanied by growth in other application areas, such as natural-gas-burning vehicles.[3]

Let us assume that we are an energy company with production assets in the Gulf of Mexico and off the coast of Australia. The supply of gas from current sources in the Gulf of Mexico is shrinking steadily, while the assets in Australia can still increase production. From this perspective, we summarize the opportunity as follows:

> Existing supplies of natural gas to the United States are dwindling, while demand is increasing and is expected to increase steadily for the foreseeable future. Because our company has assets in the United States, we have an advantage in that some infrastructure exists for bringing natural gas to this robust market.

Some possible investments, along with comments, to take advantage of this opportunity are as follows:

1. Increase the rate of production from current sources in the Gulf of Mexico.
   Comment: The application of new technologies has allowed for increased yields from older natural-gas fields. The benefits of these investments are that they are relatively low risk because the infrastructure is already in place. Unfortunately, there is a limit as to how much the rate of production can be increased (if at all), and there is also a limit to the length of time that supplies from current sources will be available.
2. Drill for new supplies in the Gulf of Mexico.
   Comment: Because most "easy" sources in the Gulf of Mexico have been explored, further drilling will require larger investments due to increased exploration costs and deeper drilling. However, the infrastructure for transporting gas to shore is essentially in place due to current operations. Possible investments include
   (a) Acquiring leases to explore other (deeper) waters in the Gulf of Mexico.
   (b) Purchasing assets of other companies in the Gulf of Mexico.
   (c) Drilling for new supply sources, assuming that the relevant leases are acquired.
3. Search for sources of natural gas in new regions.
   Comment: This is clearly the most risky investment, due to the fact that infrastructure and expertise are located in Australia and the Gulf of Mexico. However a number of options exist:

[3] Hoyos, C., "Demand Accelerates in the US, Europe and Asia," *The Financial Times,* FT Gas Industry, March 19, 2004.

(a) Acquire licenses to explore other regions of the world.

(b) Partner with another company (presumably with expertise in another region) or with a government entity to explore or produce natural gas.

(c) Purchase assets in another region from another company.

(d) Invest in new infrastructure in a new region (probably after extensive testing).

4. Transport gas from Australia to the United States to capture the growing market.
Comment: Gas is currently delivered to its markets via pipeline. To transport gas from Australia to America would require the use of liquefied-natural-gas (LNG) tankers, as well as the building of a liquefaction facility in Australia and a receiving facility in America to regasify the LNG. Possible alternatives include

(a) Selling natural gas to second parties for transport to the United States.

(b) Building facilities to liquefy natural gas and transport it to the United States. A receiving terminal must also be constructed, or the gas could be sold to a receiving facility operated by another energy concern.

5. Increase production of gas off the Australian coast.
Comment: The methods with which to accomplish this follow those of increasing production in the Gulf of Mexico. Obviously, however, there are many additional benefits here, as the supplies are not expected to shrink sharply in the near future.

6. Explore new technologies or new sources of energy.
Comment: Although the company has focused on natural-gas exploration and removal, it has acquired tremendous knowledge about the gas itself. With research-and-development funding, it may be possible to develop synthetic fuels. This has tremendous upside potential, but is clearly a risky proposition. Government aid in the endeavor may help alleviate some of the risk. Some infrastructure, such as research-and-development laboratories, must be established, and additional scientists and engineers will need to be hired.

7. Continue current operations (the do-nothing alternative).
Comment: Although this will not take advantage of increasing demand, higher per unit profits are possible due to expected rises in natural-gas prices. Investments must eventually be made, as current supplies will eventually dry up and the company will no longer function (i.e., the status quo output is decreasing with time).

We will not pare down these options, as they appear to be mostly feasible and each (with the exception of the do-nothing alternative) takes advantage of the increasing demand in the United States. It is reasonable to assume that the decisions will be made by applying a capital-budget limit, thereby reducing the number of alternatives pursued. This concept is explored in later chapters.

The type of analysis presented in Example 6.1 can also be utilized to solve a problem. We illustrate next by looking at the chapter's real decision problems.

6.6 Examining the Real Decision Problems

We return to our motivational decision problems and assume that the city engineers have used one of our idea-generating methods and proposed a number of solutions. No solutions have been eliminated from this step, and the options are presented in the order that they were developed.[4] A brief discussion is presented with each option in order to examine whether the technical aspects of the solution solve the problem at hand and whether the project has any additional positive or negative aspects. This information can be used to reduce the options to a manageable set. Here are the city engineers' proposed solutions:

1. Improve the configuration and timing of stoplights.
   Comment: This is a relatively inexpensive option that should help improve the flow of traffic through town, albeit not drastically.
2. Install an intelligent transportation system such that motorists receive real-time traffic information.
   Comment: This is another relatively inexpensive option that informs motorists about traffic conditions (such as high congestion). However, unless alternative routes are available, this option will not help alleviate congestion.
3. Build a city bypass to the north and east of town.
   Comment: This would clearly alleviate traffic through town, and assuming that the speed limit on the bypass is greater than the city speed limit, it should improve traffic flow in the region. The proposed route is generally farmland. Although it may be costly to build the road (because no infrastructure exists), the environmental impact of the solution appears to be minimal. This solution would increase the mileage of the typical commute, but should have safety benefits as the number of intersections is reduced and fewer cars are brought into contact with city-center pedestrian traffic.
4. Build a city bypass to the south and west of town.
   Comment: This solution is similar to the previous alternative, except that the area to the south is densely wooded forest. Thus, the costs would be higher to cut a road through the forest, and the environmental impact would be greater.
5. Eliminate or reduce the number of intersections.
   Comment: This would improve traffic flow through the city and also improve safety. Note that it could be accomplished in a number of ways (e.g., bridges, tunnels, or dead-end roads), each with its own benefits and costs.

---

[4] Ideas are from "Decade of Delay Offers Congestion Suggestions," *Texas Transportation Researcher,* Volume 34, Number 1, 1998.

6. Widen the road.
   Comment: A widened road would increase the flow of traffic through the city and possibly allow for a higher speed limit. It is questionable as to whether this is feasible, however, since the rights-of-way reside with the businesses located on the main street. A higher speed limit and wider road might not provide a safer solution.
7. Eliminate parking during rush hour.
   Comment: This would bring about benefits similar to those of the previous solution, but without the construction costs or land acquisition problems associated with widening the road. However, the removal of parking on main street may be detrimental to local businesses.
8. Add an HOV lane.
   Comment: This can bring about many improvements in both traffic flow and environmental benefits. (A smaller number of vehicles reduces pollution.) Analyses must be performed to see if the option requires an expansion of the road, which could be accomplished by eliminating parking on one side of the street.
9. Build a tram.
   Comment: This is an interesting option that would bring prestige to the city, but the lack of infrastructure in terms of a public city transportation system makes it technically infeasible. If the transit operations of the entire region could be examined, this option might be feasible and useful, as it would clearly address the traffic congestion problem. Note, however, that the problem is due to commuters passing through the city; therefore, it is unlikely that a tram system would reduce this type of traffic unless its endpoints were sufficiently far apart to allow commuters to abandon their cars altogether.
10. Charge a toll to pass through the city.
    Comment: Although tolls may help reduce traffic, commuters in this example do not appear to have alternative methods of transportation or routes in which to divert to avoid the toll. (In London, tolls have been quite successful in reducing traffic,[5] but there are many forms of public transportation for people to use instead of their own automobiles.) Tolls may also be detrimental to city-center business, although rebates may be possible for those not using the street as a throughway.
11. Increase the use of public transportation.
    Comment: The comments here are not unlike those pertaining to the tram. Again, the entire system must be examined in order to determine whether it can service the needs of the users. Only if it can will this option

[5] Giles, C., "Road charge greatly benefits London, research finds," *Financial Times,* p. 4, October 25, 2005.

help reduce congestion. The benefits of using public transportation are numerous.

12. Employ and equip emergency road traffic crews to quickly remove accidents from an affected area.
Comment: Accidents were not noted in the statement of the problem, but this option clearly would help improve traffic flow if accidents were deemed to be the cause of much congestion in the city.

13. Provide incentives to businesses to shift their working hours or to utilize telecommuting.
Comment: This is another idea with many possible benefits, such as decreased congestion, reduced pollution, and people spending more time at home or at work as opposed to being in transit. An examination of the types of businesses in the area (e.g., manufacturing establishments, at which people must be present, or call centers, for which it is possible to work remotely) must commence to determine whether the proposed incentives would help.

14. Do nothing.
Comment: This appears to be the option that is least costly to the government. However, public discontent may increase enough to make it an infeasible option.

Reducing the number of options to carry forward for analysis is not easy, but the list can be pared down by examining, for each option, our three questions of whether that option (1) solves the problem, (2) has additional benefits, and (3) has any negative impacts. It is interesting to note that one particular option may be dependent on another. For example, the idea of a toll may seem useless, as there are no alternative routes available to the commuters. However, building a bypass and charging a toll to go through the city center may be feasible.

In this situation, it would appear that the tram system could be eliminated due to infeasibility. The use of intelligent transportation systems and emergency removal crews can also be eliminated, because they do not address the problem defined. This leaves the bypass solutions, widening the road, the HOV option, reducing the number of intersections, and telecommuting as viable options to pursue, either separately or in combination.

The two examples of traffic congestion and energy exploration illustrate the number of diverse alternatives that can be developed, given a current problem or opportunity. While the processes that led to delineating the options in these examples were not explicitly detailed, the lists should illustrate how ideas complement each other and how "piggybacking" occurs. It is important to develop a rich list of alternatives to ensure cost-effective solutions to problems and good opportunities for growth.

6.7 Key Points

- The solution generation step is the creative step in the decision-making process, because that is the step at which engineers design solutions.
- The design of new solutions generally comes from the adaptation of old designs or alternative applications of previous technologies.
- The advantages of generating alternatives in a group include diversity and a greater abundance of ideas. The disadvantages of working in a group include dominance, social laziness, groupthink, and satisficing.
- The Delphi method and the nominal group technique both try to take advantage of working in a group while removing the negative aspects of group interaction. The Delphi method does not require any group interaction, while the nominal group technique limits the discussion time of groups.
- The do-nothing alternative generally represents a feasible option to the decision maker. In the case of opportunities, it often represents other available investments that produce an expected return. However, the do-nothing option may also represent the status quo, or it may not be feasible if a choice between two alternatives must be made.
- It is reasonable to expect that the number of alternatives available must be pared down due to the high cost of investigating and analyzing each alternative. To do this, it must be determined whether a proposal solves the problem or takes advantage of the stated opportunity. In addition, the (often noneconomic) costs and benefits of the project should be considered.

6.8 Further Reading

Dalkey, N.C., *The Delphi Method: An Experimental Study of Group Opinion.* Rand Corporation, Santa Monica, California, June 1969.

Dalkey, N.C., *Experiments in Group Prediction.* Rand Corporation, Santa Monica, California, 1968.

Delbecq, A., A.H. van de Ven, and D.H. Gustafson, *Group Techniques for Program Planning: A Guide to Nominal Group and Delphi Processes.* Scott Foresman & Co., Glenview, Illinois, 1975.

Fogler, H.S., and S.E. LeBlanc, *Strategies for Creative Problem Solving.* Prentice Hall, Upper Saddle River, New Jersey, 1995.

Hill, G.W., "Group versus Individual Performance: Are $N + 1$ Heads Better than One?" *Psychological Bulletin,* 91:517–539, 1982.

Holloman, C.R., and W. Hendrick, "Adequacy of Group Decisions as a Function of the Decision-Making Process," *Academy of Management Journal,* 15:175–184, 1972.

Linstone, H.A., and M. Turoff (eds.), *The Delphi Method: Techniques and Applications.* Addison-Wesley, London, 1975.

Maier, N.R.F., "Assets and Liabilities in Group Problem Solving: The Need for an Integrative Function," *Psychological Review,* 74:239–249, 1967.

Miner, F., Jr., "Group versus Individual Decision Making: An Investigation of Performance Measures, Decision Strategies and Process Losses/Gains," *Organizational Behavior and Human Performance,* 33:112–124, 1984.

van de Ven, A.H. *Group Decision Making and Effectiveness, an Experimental Study.* The Comparative Administration Research Institute of the Center for Business and Economic Research, Graduate School of Business Administration, Kent State University, Kent, Ohio, 1974.

van de Ven, A.H., and A.L. Delbecq, "Nominal versus Interacting Group Process for Committee Decision-Making Effectiveness," *Academy of Management Journal,* 14:203–212, 1971.

van de Ven, A.H., and A.L. Delbecq, "The Effectiveness of Nominal, Delphi and Interacting Group Decision Making Processes," *Academy of Management Journal,* 17:605–621, 1974.

6.9 Questions and Problems

6.9.1 Concept Questions

1. Why is the solution generation step in the decision-making process so critical?
2. How is the solution generation step related to the engineering design process?
3. Why is it hard to generate solutions? Where and how do engineers usually get their solutions?
4. What are the benefits of working in a group?
5. What are the drawbacks to working in a group? Order these according to severity and justify your answer.
6. What are the differences between the Delphi method and the nominal group process technique? Does one have advantages over the other? Explain.
7. What is the do-nothing alternative? What is its role in the capital-budgeting process? Is this an important role?
8. How should our list of feasible alternatives be trimmed to a reasonable set for evaluation?

6.9.2 Application Questions

For each of the application questions listed in Section 5.8.2 of Chapter 5, generate solution alternatives. If possible, do this in a group. When you are finished, reduce the feasible set for subsequent evaluation to a maximum of three alternatives.

6.9.3 Fundamentals of Engineering Exam Prep

1. When considering investment options, the do-nothing alternative should
   - **(a)** Capture the opportunity cost of making an investment.
   - **(b)** Not represent a viable option.
   - **(c)** Not be taken seriously.
   - **(d)** All of the above.
2. The do-nothing alternative
   - **(a)** Ensures a 100% return on investments.
   - **(b)** Represents the status quo.
   - **(c)** Forces budgets to be expended.
   - **(d)** All of the above.
3. When generating solution alternatives to an engineering problem,
   - **(a)** Never think "outside the box."
   - **(b)** Offer only suggestions that will be agreed upon by the majority.
   - **(c)** Never consult previous solutions or designs.
   - **(d)** None of the above.
4. Generating solutions to engineering problems often
   - **(a)** Relies on engineering designs.
   - **(b)** Requires teamwork.
   - **(c)** Relies on previous solutions.
   - **(d)** All of the above.
5. A possible solution to an engineering problem should be
   - **(a)** Eliminated before it is technically (seen if it solves the problem) or economically analyzed.
   - **(b)** Analyzed economically before being analyzed technically.
   - **(c)** Analyzed technically before being analyzed economically.
   - **(d)** It is irrelevant in which order a solution is analyzed.

7 Developing Cash Flows and Gathering Information

(*Courtesy of James L. Amos/CORBIS.*)

Real Decisions: Not Just Hot Air!

The Western Australian government approved the Gorgon gas project, a joint venture between Chevron, Exxon Mobil, and Royal Dutch Shell to extract some 40 trillion cubic feet of recoverable resources (especially natural gas) that was discovered 30 years previously. The approval allowed for initial engineering and design, marketing, and environmental impact studies to commence.[1] The project owners committed A$100 million to front-end engineering and design studies in July of 2005, having already spent A$1 billion in preliminary studies.[2] Chevron, the majority shareholder in the project, moved closer to final approval for the project in 2005 after securing A$28 billion in deals to supply 4.2 million

[1] Bell, S., "W Australia Govt Approves Chevron Texaco Gorgon Gas Plan," *Dow Jones Newswires,* September 8, 2003.

[2] Noonan, R., "UPDATE: Australia's Gorgon Gas Proj Moves a Step Closer," *Dow Jones Newswires,* July 1, 2005.

tons of liquefied natural gas (LNG) annually to Japan (1.5 million to Osaka Gas Co., 1.2 million to Tokyo Gas Go., and 1.5 million to Chubu Electric Co.) for 25 years, starting in 2010.

The deal replaced a tentative A$30 billion agreement with China National Offshore Oil Corporation (CNOOC) that was scrapped.[3,4] This was the second major contract that CNOOC had cancelled, citing the recent rise in LNG prices, despite needed supplies to feed its Fujian terminal. The contract was motivated by China's desire to diversify its energy usage beyond oil and coal. The price CNOOC contracted for with BP in 2002 was $0.98 per barrel of oil equivalent (boe) for the gas (with delivery starting in 2007), totaling $7.5 billion over the 25-year contract. However, in 2005 the market price was nearly $4 per boe for the gas, leading BP to pressure CNOOC into renegotiating its contract. In response, CNOOC scaled back its requests significantly.[5,6]

Initial estimates for the Gorgon project were A$11 billion in capital costs, but these were being revised higher due to surges in steel and labor costs. Chevron was investigating ways to reduce the use of steel in order to lower development costs and retain solid profit margins.[7] These developments lead to a number of interesting questions to be answered, assuming an interest rate of 15% per year:

1. In addition to the development (capital) costs and contracted sale price, what other estimates are required to evaluate the Gorgon project?
2. Which estimate is more critical, the development cost, which is spread over the years 2005 through 2009, or the 25-year contracted sale price?
3. How could the rise in steel prices (for Chevron) or LNG prices (for CNOOC) have been estimated more accurately? How crucial is the LNG price to CNOOC?

In addition to answering these questions, after studying this chapter you will be able to:

- Define the life cycle of a product or service and its associated costs and revenues, according to development, operational, and disposal phases. (Section 7.1)

[3] Noonan, R., "Chevron: Signs Gorgon LNG Deal with Japan's Chubu," *Dow Jones Newswires,* November 21, 2005.

[4] Chambers, M., "INTERVIEW: Chevron Targets US Not China for Gorgon LNG," *Dow Jones Newswires,* December 8, 2005.

[5] Donnan, S., and E. Tsui, "CNOOC cuts order from BP's Tangguh," *Financial Times,* p. 30, December 16, 2005.

[6] Chambers, M., "Shell Australia: China Caught Off Guard by LNG Prices," *Dow Jones Newswires,* December 6, 2005.

[7] Chambers, M., "Chevron to Use Less Steel on Gorgon to Curb Rising Costs," *Dow Jones Newswires,* December 1, 2005.

- Identify revenues and costs that must be estimated for typical engineering projects. (Section 7.1)
- Understand the trade-off between the accuracy of an estimation, the cost to achieve that accuracy, and the time at which an accurate estimate is needed. (Section 7.2)
- Perform sensitivity analysis in order to identify the critical parameters for a project that must be estimated accurately. (Section 7.2)
- Create spider plots and tornado diagrams in order to better visualize sensitivity analyses. (Section 7.2)
- Provide rough estimates of cash flows, using the unit method, index method, power law and sizing model, learning curve, curve fitting, and factor method. (Sections 7.3 and 7.5)
- Establish ranges of estimates and additional information to generate scenarios of possible outcomes for further study. (Sections 7.4 and 7.8)
- Estimate the interest rate, or minimum attractive rate of return (MARR), for discounting cash flows, using the weighted-average cost of capital, growth rate, and opportunity cost. (Section 7.6)
- Estimate the study horizon for a project, given the desired outcome for the project and the available data. (Section 7.7)
- Utilize a variety of resources to estimate the required data. (Section 7.9)

We have now finished with the first two steps in the decision-making process: Our problem or opportunity has been defined, and a number of solution alternatives have been generated and reduced to a reasonable set of feasible options. Each of these options must now be investigated in detail. Specifically, we need estimates for all components of the cash flow diagram, including the cash flows themselves, in terms of both magnitude and timing, the length of the project, and the interest rate to evaluate the cash flow diagram. In addition to these estimates, we need information regarding the possible outcomes of the project. This information may be in the form of ranges of cash flows that are tied to certain events, such as the good or poor performance of a technology or weak or strong demand. The more information that can be gathered concerning the possible outcomes of the project, including the probabilities of certain outcomes, the more informed the decision will be. Finally, we need to gather information about any noneconomic impacts of the project, such as the effect of an investment on worker morale. These play an important part of any decision.

7.1 Life-Cycle Costs and Revenues

In Chapter 2, we introduced the cash flow diagram as a way of visually depicting an investment scenario. This diagram is generally composed of "net" cash flows in that the costs from each period are subtracted from the revenues of each period, leaving a single cash flow for each period. Figure 7.1 illustrates the cash flow diagram with component cash flows, which are essentially all of the types of costs and revenues experienced with an investment project over time. The component cash flows shown are then reduced to net cash flows for subsequent analysis. Further, we categorize the cash flows according to their timing.

The initial cash flows, which are generally incurred in the first few periods of an engineering project, occur in the **development phase** and are followed by cash flows from the **operational phase**. The project ends at time N (time 12 in the figure), with the cash flows at that time referred to as the **disposal phase**. Taken in total, the cash flows from all three phases constitute the **life-cycle costs** of an investment.

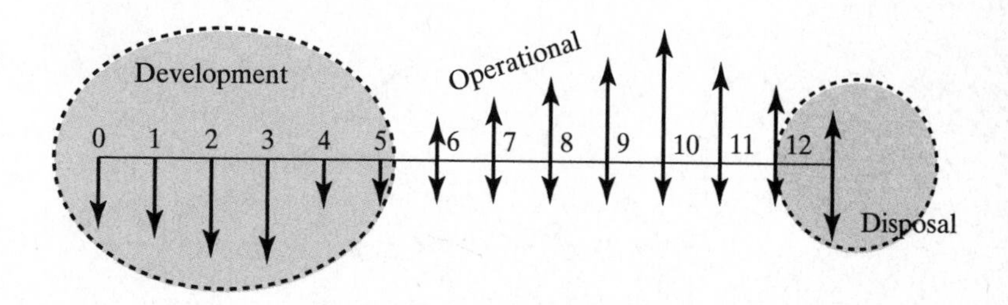

Figure 7.1
Cash flow representation of an investment opportunity, divided by phases.

The term "life-cycle" is a natural choice, since products generally proceed through a lifetime. In the development phase, the product is designed and tested. After a number of iterations and possible redesigns, which may involve market analysis and product-life testing, investments are made to begin production, which may entail building plants and facilities and acquiring production equipment. When these are ready, the operational phase begins and a product is produced or a service is provided. It takes time for production to reach full capacity, as the learning curve must be climbed, kinks in the manufacturing process must be smoothed, and the product must be introduced to various markets. After some time, sales of the product peak and then begin to wane as newer, more advanced products are introduced into the market, making the current model obsolete. This decline eventually leads to a point in time where it is determined that the product should be discontinued, in a disposal phase. The latter could entail selling a facility and equipment, converting it for another use, or actually scrapping the assets. Measures may also to be taken for the cleanup of environmentally sensitive materials used on the site.

Figure 7.2 depicts the life cycle of a product in two ways. (Note that we use the term *product* here, but it can also be a service that is provided.) Time is represented on the x-axis, while the y-axis has two pieces of data: money spent (invested) per unit time and the number of units sold per unit time. These two figures move in opposite directions. Money is quickly consumed by the project during the development phase, because the production process must be established. It is not until the operational phase of the project that revenues from sales are generated. The disposal phase can lead to either revenues or costs, depending on whether the used equipment and facilities can be salvaged.

This figure should illustrate the risks involved in a typical engineering project. There is *considerable* investment early in the life of the project. But the benefits—revenues from products sold or services provided—are not realized until a much later time. It is quite possible that investments will never be recovered, as there is a considerable lag between the time of the investment and the time that revenues arrive. The risks to the investor are defined by the magnitude of the investment and the delay in recouping it.

Figure 7.2
Life cycle of a product in terms of sales, together with associated investments and costs.

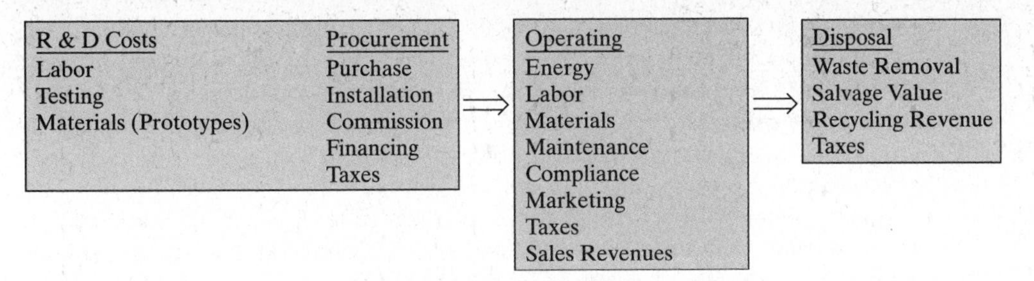

Figure 7.3 Relevant costs and revenues to be estimated over a product's life cycle.

Figure 7.2 also highlights the fact that in order to properly assess an engineering proposal, *all* life-cycle costs must be considered. Note that these costs are not easily identified. Before ground can break on a new production facility, costs may go into researching and developing new products. During operations, cost and revenues beyond those of just production and sales must be considered. These costs include the costs of maintenance of equipment, logistics, and warranties. Finally, costs incurred at disposal must also be represented. These costs may continue past the time a factory is closed for there may be environmental consequences, remedial actions, etc. In order to make an informed decision, all of these cash flows must be estimated and included in the analysis. Figure 7.3 provides a synopsis of costs and revenues to be estimated over the life cycle. Note that while research-and-development investment decisions are often made separately from decisions concerning investments in production, they are still an integral part of the development phase of a product's life cycle and are thus included here.

One special cash flow that appears at both the beginning and end of a project is termed **working capital**—the cash that is necessary to begin and sustain operations. For example, money is required to purchase materials and acquire inventory. Because this funding is necessary to begin operations, it is viewed as a capital investment. Generally, working capital is treated as an inflow to the project in initial periods and an outflow at the end of the project, as the money is returned to its source (usually the company). Working capital is termed "capital" and is not a cost because it is ignored for tax purposes, as noted in the next chapter.

The costs in Figure 7.3 can also be separated into direct and indirect costs. **Direct costs** are attributable solely to the project. For example, the cost of materials used to construct parts in a plant for subsequent sale is a direct cost to the project. Those costs which cannot be attributed to a single project are referred to as **indirect costs** or *overhead costs.* Costs that fall into this category are all those which are not direct costs, such as management and marketing costs. For example, the salaries of the executives of a company are clearly an indirect cost with regard to a single project. That is because the executives must oversee the entire company, and only a portion of their time is given to each individual project. Many other members of the workforce, including managers and engineers, also spend time on multiple projects. Marketing campaigns for companies (as opposed to campaigns for individual product lines) also represent overhead costs, since the costs cannot be directly applied to any one project. Essentially, any cost that cannot be directly attributed to a product can be considered an indirect or overhead cost.

There are many ways in which to assign overhead costs to a project. The most common form is through an overhead rate. For example, a company may define an overhead rate of 12%; in other words, 12% of the investment cost (or some other measure) is charged periodically as an overhead expense, which is then treated as a typical cost in the project.

The methods for determining overhead rates can be quite diverse. Total indirect (overhead) costs are generally determined for a period and are then divided by some measure of productivity, such as total direct labor hours or total direct labor and material costs. This defines a rate that can be applied to the appropriate measure of a project to determine its overhead burden (cost). For example, if the overhead rate is calculated to be $1.50 per direct labor hour and a project is estimated to have 5000 direct labor hours per period, then the overhead costs charged to the project would be $7500 per period. Unfortunately, the choice of method for allocating overhead can lead to different costs, which can greatly affect whether a project is acceptable or not. (This topic is addressed further in Chapter 15.)

One cost that we have not identified as a life-cycle cost is a **sunk cost**—any cost that has occurred in the past. We mention sunk costs because they should be ignored in any economic analysis, since they are generally unrecoverable. Unfortunately, people have trouble forgetting sunk costs, which explains why they often resurface in analysis.

7.2 Priorities in Estimation

We noted earlier that the reason for paring down the list of solution alternatives was that the investigative process of gathering information about an alternative, including estimating its cash flows, is time consuming and thus costly. The process is time consuming because it is difficult. A cash flow diagram is merely a single snapshot of an uncertain future. A typical engineering project can last for years. Unfortunately, the longer a project endures, the farther into the future the cash flows occur. As we move out farther into the future, forecasting and estimating become more difficult. The task becomes even more complicated when we require a great level of detail or accuracy for our estimates.

Figure 7.4 illustrates a typical curve representing the trade-off between estimation accuracy (in terms of error) and the cost to achieve such accuracy. It should be clear that estimates will never achieve perfect accuracy, especially estimates far in the future. Thus, the curve never achieves 100% accuracy, or 0% error. As the curve approaches this unattainable goal, the cost required to

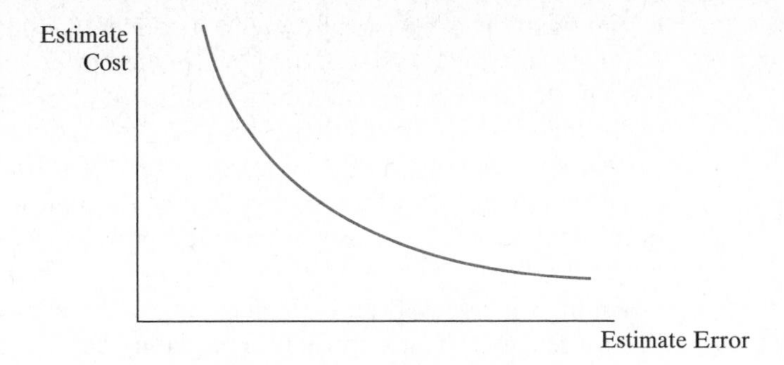

Figure 7.4 Trade-off between cost and accuracy in estimation.

reach that level of accuracy grows considerably—in direct contrast, of course, to the little cost required to achieve a low level of accuracy.

Examining the curve, the engineer must ask the following questions: (1) How accurate must our estimates be? and (2) Where should we focus our resources in the estimation process?

To answer (1), the desired level of accuracy requires knowing the probable outcome of a decision. That is, if the project to be evaluated is one in a group of proposals, then a high level of accuracy is required. However, if the decision is to determine whether further investigation is needed, then a lower level of accuracy is required. According to the American Association of Cost Engineers (AACE, now AACE International, for the Association for the Advancement of Cost Engineering), there are three levels of estimates differentiated by accuracy:

- *Order-of-magnitude estimates.* Before the scope of a project has been explicitly defined, these estimates provide a rough picture of the costs involved. It is expected that errors will be in the neighborhood of −30% to 50%.
- *Budgetary estimates.* These estimates represent the "second cut" at developing cash flow diagrams for a project. Detailed engineering analysis may not yet be performed at this stage, but more detailed estimates concerning equipment and facility costs are made. The accuracy in this situation is expected to be −15% to 30%.
- *Definitive estimates.* These estimates make up the final set of estimates used to construct the cash flow diagram. Definitive estimates represent the most stringent level of detail and accuracy. It is expected that errors will be limited to −5% to 15%.

There is direct correlation between the desired level of accuracy and the amount of effort required to achieve that accuracy, as noted in Figure 7.4. Order-of-magnitude estimates can be determined from historical data, simple estimation techniques (to be described soon), and intuition. More definitive estimates require more sophisticated techniques. While the ultimate goal of this text is to compare projects for which the estimates are definitive, it should be clear that the estimation process is incremental, so that order-of-magnitude estimates are made first and are then refined to be more accurate with time. Before elaborating on these techniques, we consider our second question regarding estimates.

To answer question (2) concerning the focus of the estimation process, consider the cash flow diagram in Figure 7.1 used to introduce the concept of life-cycle costs. The costs are divided into development, operational, and disposal phases. In general, none of these costs (or revenues) are easy to estimate, because they all happen in the future. However, it should be clear that those costs which lie further in the future are the most difficult to estimate because they are so distant. Clearly, at time zero we would have the least amount of information available about what occurs near period N. Consider, for example, the disposal of an asset. In order to forecast the asset's salvage value or disposal costs, we would have to assess its condition and use over time, in addition to any other relevant factors, such as the demand for that type of used asset. These are clearly all difficult to estimate at time zero. In contrast, the costs at the beginning of the horizon (in the development phase) are easier to estimate, because they are going to happen in the near future. For example, the purchase price of an asset that is to be bought in the near term should be a fairly accessible piece of data.

This small example of purchasing and eventually selling an asset highlights the differences in estimating difficulties with respect to time. The good news is that, from an estimation point of view, although costs in the future (during the disposal phase) are harder to estimate, they do not need to be overly accurate, because discounting minimizes their effect at time zero. In contrast, costs incurred in the near term (during the development phase) are the most influential, because discounting has little effect, but they are also the most easily estimated. This is fortunate, as the magnitudes of the cash flows in the development phase tend to be greater than those in the operational and disposal phases, as shown in Figure 7.1.

In general, discounting cash flows helps our estimation process because cash flows occurring near time zero are the most important (from a discounting point of view) and they are the easiest to estimate. But we can be much more formal about determining which project cash flows and parameters are most influential on the decision and thus should be more closely addressed in the estimation phase. We address this issue through a systematic process known as **sensitivity analysis**, a general tool in which the sensitivity, or impact, of changes in estimates on a decision can be gauged. The method is quite simple: Perturb a parameter from its expected value and measure the effect of the change. This effect can be captured by the change in some equivalent form of the cash flow diagram, such as the present value, before and after the parameter is perturbed. If a small change in the parameter leads to a significant change in the present value, then we say that the parameter is *highly sensitive*. The idea is to identify which parameters are most sensitive and thus should be addressed more closely in the estimation process.

The method is quite general, in that one can examine the effects of a change in a single parameter or in a multitude of parameters being changed simultaneously. Here, we consider the impact resulting from changing only one parameter at a time. For this discussion, we assume a parameter to be any cash flow, cash flow input, the interest rate, or the time horizon.

To perform a sensitivity analysis, we require an initial estimate and a plausible range of values for each parameter to be studied. The plausible range may be determined by the decision maker or may be defined by the expected range of accuracy for the estimate (i.e., −30% to 50% for an order-of-magnitude estimate). We also require some measure to gauge the effect of changes in our parameter estimates. In Chapter 3, we converted cash flow diagrams of various forms into either future values, present values, or equal-payment series values. As noted earlier, we use one of these measures to gauge the impact of our errors, since the present value, future value, or equal-payment series value is a single variable describing our cash flow diagram. Thus, changes can be tracked easily. For the discussion and example that follow, we will utilize the equivalent value at time zero, or present value, but the method is no different if one desires to examine future values or an equal-payment series.

Given this information, we can evaluate the impact of each individual parameter on the present value through the following steps:

1. Using initial estimates of all parameters, compute the baseline measure (i.e., the equivalent present value). This is defined as the "Original Measure," which uses the "Original Estimate" of each parameter.
2. Select a parameter to evaluate.
3. Perturb the parameter. This defines the "Perturbed Estimate." The percentage error of the estimate is calculated as

$$\text{Estimate Error} = \frac{\text{Perturbed Estimate} - \text{Original Estimate}}{\text{Original Estimate}} \times 100\%. \tag{7.1}$$

4. Using the perturbed estimate, recompute the Original Measure to gauge sensitivity. Define the result as our "Perturbed Measure" such that

$$\text{Measure Error} = \frac{\text{Perturbed Measure} - \text{Original Measure}}{\text{Original Measure}} \times 100\%. \tag{7.2}$$

5. If more perturbations are to be examined, go to step 3; otherwise stop.

Sensitivity is defined by the correlation between the Estimate Error and the Measure Error. It is a relative analysis, as one must evaluate all parameters to determine which are the most sensitive. Clearly, if Measure Error $\gg$ Estimate Error, the parameter is sensitive.

If this process is repeated for each parameter describing our engineering project, then we will establish which parameters are most sensitive to estimation errors. We will also determine how sensitive the errors can be. These two pieces of information answer our previously posed questions concerning the desired accuracy of estimates and where we should focus our resources in the estimation process. To illustrate the process and interpret the results, we perform a detailed sensitivity analysis in the next example.

EXAMPLE 7.1 *Sensitivity Analysis for Cost Estimation*

The motivation for this example comes from BHP Billiton, an Australian mining and energy company, analyzing a $550 million investment in a liquefied natural gas (LNG) import facility off the coast of Southern California. The company expects to supply 800 million cubic feet of natural gas per day beginning in 2008.[8]

Assume that the order-of-magnitude estimates for this example are given in Table 7.1. The (expected) project costs are depicted in the cash flow diagram in Figure 7.5(a), with the net cash flows given in (b).

TABLE 7.1 **Initial parameter estimates for LNG import facility.**

Parameter	Estimate
Investment Cost	$550 million
Construction Time	1 year
Annual Operating Costs	$15 million
Daily Production Level	800 million ft^3
LNG Material Costs	$2 per 1000 ft^3
LNG Price	$3.5 per 1000 ft^3
Disposal Cost	$100 million
Interest Rate	12%
Horizon	15 years

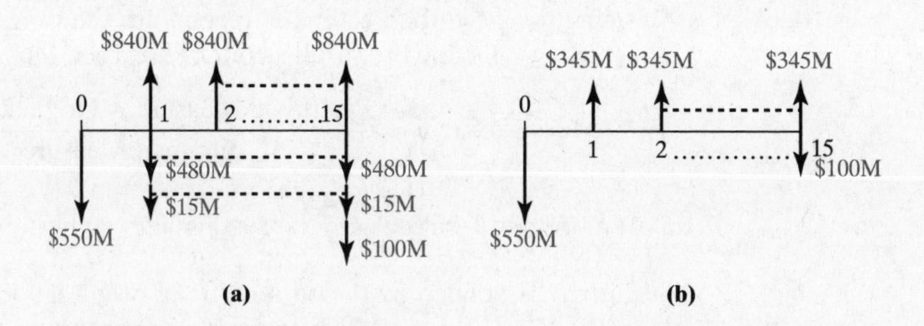

Figure 7.5 (a) Individual cash flow estimates for LNG import facility investment and (b) net cash flows for investment.

[8] Dowell, A., "BHP Proposes to Build LNG Import Facility in California," *Dow Jones Newswires,* August 14, 2003.

Solution. There are three components to the cash flow diagram: (a) a single cash flow at time zero; (b) 15 years of annual cash flows; and (c) a single cash flow at time 15. Note that the net revenues from the project are derived from the difference of the selling price and the material cost for production over the year, or

$$\left(\frac{\$3.50}{1000\text{ ft}^3} - \frac{\$2.00}{1000\text{ ft}^3}\right) \times \frac{800{,}000{,}000\text{ ft}^3}{\text{day}} \times \frac{300\text{ days}}{\text{year}} = \$360\text{ million per year.}$$

The operating costs must be subtracted from this total to calculate the net revenues over the life of the project. We can write the equivalent present value P of the entire investment as the sum of the present values of these individual components, or

$$P = \underbrace{-\$550\text{M}}_{\text{Initial Investment}} \underbrace{+(\$360\text{M} - \$15\text{M})(^{P/A,12\%,15})}_{\text{Annual Profits}} \underbrace{- \$100\text{M}(^{P/F,12\%,15})}_{\text{Disposal Cost}}$$

$$= -\$550\text{M} + \$345\text{M}\left[\frac{(1+0.12)^{15}-1}{0.12(1+0.12)^{15}}\right] - \$100\text{M}\left[\frac{1}{(1+0.12)^{15}}\right] = \$1.781\text{ billion.}$$

We now perform an analysis on each individual parameter to determine the sensitivity of its estimate with respect to the present value of the project.

Sensitivity of Investment Cost Estimate

We define x as the percentage error in the estimate of the investment cost. Then, using this variable, we can rewrite the equivalent present value, denoted as P' to differentiate it from P, of the investment with x as

$$\begin{aligned} P' &= -\$550(1+x)\text{M} + \$345\text{M}(\overset{P/A,12\%,15}{6.8109}) - \$100\text{M}(\overset{P/F,12\%,15}{0.1827}) \\ &= -\$550(x)\text{M} + \$1781\text{M}. \end{aligned}$$

This leaves a straightforward problem for analysis, as we merely check various error estimates (values of x) and compute the resulting present worth. To do that, we use a spreadsheet. Figure 7.6 illustrates a spreadsheet approach that allows for perturbations in each of the input parameters. The data center contains each perturbed estimate in column E according to the error defined in column F. For example, as given in the figure, the \$275 million investment cost (cell E3) is due to a −50% error (cell F3) in that estimate. The cash flows in column B are defined from the perturbed estimates.

The present worth of the cash flows is computed with the use of the NPV function in cell E16, with the error given in cell F16. For the perturbed investment cost, a present value of over \$2 billion is computed, which is 15.44% higher than the present worth without any perturbation in the investment cost (retained in cell E15). Estimate errors, as defined by Equation (7.1), between $x = -50\%$ and 50% are examined by changing cell F3. The results are summarized in Table 7.2. Examining this table, we see that a 30% error in the estimate of the investment cost results in an error of 9.27% in our measure of present value, as defined by Equation (7.2). Since this is our first parameter being analyzed, we do not have any other data with which to compare it. However, it is clear that the measure error is much lower than the estimate error. This is a bit surprising, as the investment occurs at time zero, but it is evident that the magnitude of the \$550 million investment cost is overshadowed by the 15 years of annual income of \$345 million. We examine the annual income next.

	A	B	C	D	E	F	G
1	Example 7.1: LNG Facility Investment			**Input**			
2				**Parameter**	**Perturbed Estimate**	**Error**	**Units**
3	**Period**	**Cash Flow**		Investment	$275,000,000.00	-50%	=15000000*(1+F4)
4	0	-$275,000,000.00		Operating Cost	$15,000,000.00	0%	per year
5	1	$345,000,000.00		Production Level	800,000,000	0%	cubic feet per day
6	2	$345,000,000.00		Production Days	300	0%	days per year
7	3	$345,000,000.00		LNG cost	$2.00	0%	per 1000 cubic foot
8	4	$345,000,000.00		LNG price	$3.50	0%	per 1000 cubic foot
9	5	$345,000,000.00		Disposal Cost	$100,000,000.00	0%	
10	6	$345,000,000.00		Interest Rate	12%	0%	per year
11	7	$345,000,000.00		Periods	15	0%	years
12	8	$345,000,000.00					
13	9	$345,000,000.00		**Output**			
14	10	$345,000,000.00		**Parameter**	**Perturbed Measure**	**Error**	
15	11	$345,000,000.00		P	$1,781,478,622.74		
16	12	$345,000,000.00		P'	$2,056,478,622.74	15.44%	
17	13	$345,000,000.00					
18	14	$345,000,000.00		=B4+NPV(E10,B5:B19)		=(E16-E15)/E15	
19	15	$245,000,000.00					
20		=((($E$8-$E$7)/1000)*$E$5*$E$6)-$E$4					

Figure 7.6 Spreadsheet for sensitivity analysis with perturbed investment cost.

TABLE 7.2 **Sensitivity of investment cost estimate on present value.**

Estimate Error (x)	Perturbed Estimate	Perturbed Measure (P')	Measure Error
−50%	−$275M	$2056M	15.44%
−30%	−$385M	$1946M	9.27%
−10%	−$495M	$1836M	3.09%
0%	−$550M	$1781M	0%
10%	−$605M	$1726M	−3.09%
30%	−$715M	$1616M	−9.27%
50%	−$825M	$1506M	−15.44%

Sensitivity of Production-Level Estimate

Next, we define x as the percentage error in the estimate of the annual output. We redefine our annual revenues less material costs as

$$\left(\frac{\$3.50}{1000\ \text{ft}^3} - \frac{\$2.00}{1000\ \text{ft}^3}\right) \times \frac{800{,}000{,}000(1+x)\ \text{ft}^3}{\text{day}} \times \frac{300\ \text{days}}{\text{year}}$$

$$= \$360(1+x)\ \text{million per year.}$$

Note that the investment cost returns to its original estimate when we perturb annual production. Using the annual production variable, we rewrite the equivalent present value as

$$P' = -\$550\text{M} + (\$360(1+x)\text{M} - \$15\text{M})\,(\overset{P/A,12\%,15}{6.8109}) - \$100\text{M}(\overset{P/F,12\%,15}{0.1827})$$

$$= \$1781\text{M} + \$2452(x)\text{M}.$$

This again leaves a straightforward problem for analysis, as we merely check various error estimates (values of x) and compute the resulting present worth. We can do so with the spreadsheet in Figure 7.6 by returning cell F3 to '0' and altering cell F5 accordingly. Estimate errors between $x = -50\%$ and 50% are summarized in Table 7.3.

TABLE 7.3 **Sensitivity of production-level estimate on present value.**

Estimate Error (x)	Perturbed Estimate	Perturbed Measure (P')	Measure Error
−50%	400M ft^3	\$556M	−68.82%
−30%	560M ft^3	\$1046M	−41.29%
−10%	720M ft^3	\$1536M	−13.76%
0%	800M ft^3	\$1781M	0%
10%	880M ft^3	\$2027M	13.76%
30%	1040M ft^3	\$2517M	41.29%
50%	1200M ft^3	\$3007M	68.82%

It is clear from Table 7.3 that the equivalent present value of the cash flows is much more sensitive to changes in the production levels than the initial investment. A 30% error leads to a 41.3% change in the final measure. At this point, we note that estimating the annual production levels must take precedence, in terms of required accuracy and allocated resources, over the initial investment estimate.

Sensitivity of Other Cash Flow Estimates

We can investigate the other parameters, including the operating costs, LNG cost, LNG price, and disposal costs, in a similar fashion. We follow a similar procedure of multiplying the initial estimate by $(1 + x)$, with x representing the percentage error. Table 7.4 gives a summary of the errors as a percentage deviation from the initial present value of \$1781 million. We will look at the interest rate and length of horizon separately, as the analysis for these cases is not straightforward.

Sensitivity of Interest Rate Estimate

Analyzing the interest rate is a bit more complicated. Consider the equation used to derive the equivalent present value at time zero:

$$P = -\$550\text{M} + \$345\text{M}\left[\frac{(1+i)^{15}-1}{i(1+i)^{15}}\right] - \$100\text{M}\left[\frac{1}{(1+i)^{15}}\right].$$

Following our approach of multiplying our parameter by $(1 + x)$ would clearly be difficult in this situation, as the interest rate appears four times in the present-value equation. To alleviate this concern, we can just substitute a range of values for i into the present-value equation and measure the result.

Table 7.5 lists different values of i and their impact. We list the values in this table in a different order than for previous analyses, because here we define the perturbed value of i before we calculate the percentage error of the estimate. In the investment cost case, we first defined the percentage error and then calculated the resulting perturbed estimate.

Note that, unlike our previous estimates for the cash flows, the impact of the interest rate is nonlinear. That is, the percentage of sensitivity cannot be assumed to be symmetrical such that a 10 percent error above or below the estimate will result in the same magnitude of impact (albeit with a different sign). The reason for this is that the interest factors $(P/A,i,15)$ and $(P/F,i,15)$ are nonlinear functions in the variable i.

TABLE 7.4 Sensitivity (Measure Error) of various parameter estimates on present value.

Parameter	−50%	−40%	−30%	−20%	−10%	10%	20%	30%	40%	50%
Investment Cost	15.44%	12.35%	9.27%	6.18%	3.09%	−3.09%	−6.18%	−9.27%	−12.35%	−15.44%
Operating Cost	2.87%	2.29%	1.72%	1.15%	0.57%	−0.57%	−1.15%	−1.72%	−2.29%	−2.87%
Production Level	−68.82%	−55.05%	−41.29%	−27.53%	−13.76%	13.76%	27.53%	41.29%	55.05%	68.82%
LNG Cost	91.76%	73.41%	55.05%	36.70%	18.35%	−18.35%	−36.70%	−55.05%	−73.41%	−91.76%
LNG Price	−160.57%	−128.46%	−96.34%	−64.23%	−32.11%	32.11%	64.23%	96.34%	128.46%	160.57%
Disposal Cost	0.51%	0.41%	0.31%	−0.21%	−0.10%	0.10%	0.21%	−0.31%	−0.41%	−0.51%

TABLE 7.5 **Sensitivity of interest rate estimate on present value.**

Perturbed Estimate	Estimate Error	Perturbed Measure (P')	Measure Error
5%	−58.33%	$2983M	67.44%
10%	−11%	$2050M	15.08%
12%	0%	$1781M	0%
15%	25%	$1455M	−18.32%
20%	66.67%	$1057M	−40.69%
25%	108.33%	$778M	−56.33%

TABLE 7.6 **Sensitivity of length-of-horizon estimate on present value.**

Perturbed Estimate	Estimate Error	Perturbed Measure (P')	Measure Error
5 years	−66.67%	$637M	−64.25%
10 years	−33.33%	$1367M	−23.26%
15 years	0%	$1781M	0%
20 years	33.33%	$1980M	11.12%
25 years	66.67%	$2150M	20.69%

Thus, it is possible, as seen in this example, that an error in one direction can have much more serious consequences compared with an error in the opposite direction. Table 7.5 shows that negative estimate errors have a more drastic effect on the present value than positive estimate errors.

Sensitivity of Length-of-Horizon Estimate

Much like the interest rate, the horizon poses a complication that is not easily analyzed with our $(1+x)$ error percentage approach, especially because we require the horizon to be an integer. We again use our present-value formula and allow for different values of N as:

$$P = -\$550\text{M} + \$345\text{M}\left[\frac{(1+0.12)^N - 1}{0.12(1+0.12)^N}\right] - \$100\text{M}\left[\frac{1}{(1+0.12)^N}\right].$$

Table 7.6 gives a summary of the sensitivity analysis for values of N ranging between 5 and 25 years. Like the interest rate, the length of the horizon provides nonlinear feedback when the measure errors are examined. Negative errors again are more influential on the present value.

In the true spirit of single-parameter sensitivity analysis, we have altered only the value of N in these equations. Clearly, the salvage value is closely tied to N; thus, it is reasonable that one would investigate simultaneous movements of these values. We will explore multiple-parameter sensitivity analysis later in the text.

Summary of Analysis

To truly determine which parameters are most sensitive with respect to present value, it sometimes helps to picture the errors simultaneously. Tornado diagrams and spider plots are two ways in which to do this.

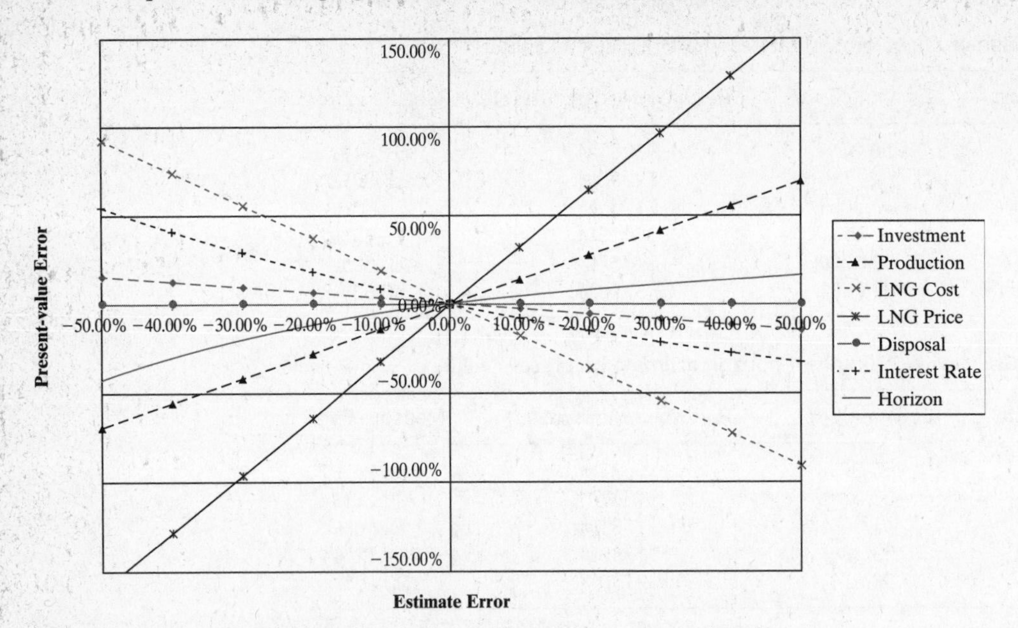

Figure 7.7 Plot of present-value error versus estimate error for a variety of inputs.

The spider plot illustrated in Figure 7.7 is easily generated as an 'XY (scatter)' graph in Excel, as illustrated in Sections 1.6 and 3.7. The only difference is that in a spider plot there are multiple sets of data. The individual x (estimate error) and y (measure error) coordinates can be input under the 'Series' tab when one uses the 'Chart Wizard.'

In the plot in Figure 7.7, the x-axis is the percentage error in the estimate, while the y-axis is the resulting percentage error in the present value. As expected, the functions are linear for our first five analyses, including the investment, disposal costs, production levels, materials price, and selling price. The plots of the length of the horizon and the interest rate are nonlinear.

Looking at the figure, we see more clearly our course of action in terms of estimation. The annual profit, which is directly affected by the price, the cost, and production levels of LNG are the most sensitive parameters in our analysis. The initial investment is not overly sensitive and does not need a great deal of accuracy in its estimate. Furthermore, estimates such as the disposal cost and the annual operating costs are essentially negligible. The interest rate is fairly interesting, but it is one of the few parameters that may not have great uncertainty, as discussed later in the chapter.

The spider plot provides a wealth of information for the decision maker, but can often be overwhelming. In addition, it may provide information beyond what is needed. To keep the graph symmetric, we plotted error ranges between −50% and 50%. However, it is unlikely that our estimate errors will be symmetric and the same for all parameters. If we can estimate true ranges of possible errors, those ranges may be used to construct the likely range of our measure (present value). Sorting these ranges according to the size of the spread, provides a quicker glimpse of the situation for the decision-maker, as given in the tornado diagram in Figure 7.8.

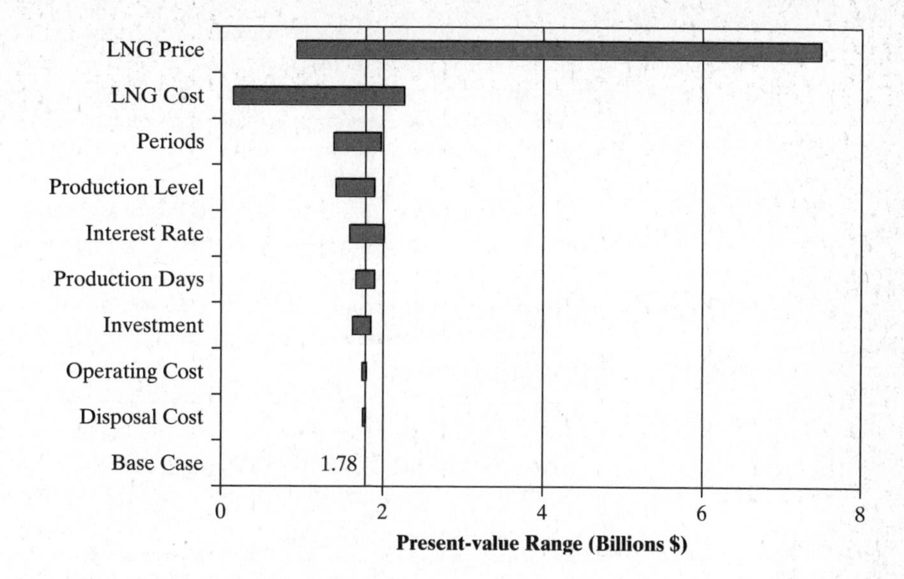

Figure 7.8 Tornado plot of present-value ranges for a variety of inputs.

TABLE 7.7 Errors used in constructing tornado diagram.

Parameter	Low Error	High Error	Low Measure ($B)	High Measure ($B)	Spread ($B)
Disposal Cost	−25%	100%	1.79	1.76	0.03
Operating Cost	−15%	30%	1.8	1.75	0.05
Investment	−15%	30%	1.86	1.62	0.24
Production Days	−5%	5%	1.66	1.9	0.24
Interest Rate	−15%	15%	2.02	1.58	0.44
Production Level	−15%	5%	1.41	1.9	0.49
Periods	−15%	30%	1.38	1.98	0.60
LNG cost	−15%	50%	2.27	0.147	2.123
LNG price	−15%	100%	0.923	7.5	6.577

The tornado diagram merely plots the present-value range for the given range of estimates. We used the data in Table 7.7 to construct Figure 7.8. The estimates in the table are assumed to be truer to the ranges of estimate errors experienced. (Note that the errors are not symmetric and differ for each parameter.) In essence, the tornado diagram is a spider plot with truncated data specific to each parameter, but displayed in sorted order (such that it resembles a tornado) for easy viewing. For example, we examined investment cost estimate errors of -15% and 30% which lead to present values of $1.86 billion and $1.62 billion for the investment. The value of $1.62 billion and the spread in these outcomes ($240 million) are used as the input to construct the diagram.

The tornado plot is constructed in Excel with use of the ‘Stacked Bar’ graph. For each input parameter, the minimum of the low and high measures (present values) and the spread are taken as inputs (highlighted as in the construction of a scatter graph for the spider plot) when creating the diagram. For the LNG Cost parameter in our data set, the graph generates a bar from 0 to 0.147 and then another bar from 0.147 to 2.27 by adding the spread of 2.123 and ‘stacking’ it onto the first bar. To get our final effect of a tornado, because we are interested only in seeing the spread, we double-click on the first bar generated (from 0 to 0.147). A dialogue box with various formatting options opens. Selecting the 'Border' and 'Area' options as 'None' under the 'Patterns' tab hides the first bar, leaving the diagram as shown in Figure 7.8. We have added a line to illustrate where the base-case present value of $1.78 billion resides such that relative changes in each parameter are more readily viewed.

It is clear in this example that the LNG price and cost—especially the price—are extremely critical parameters with respect to the decision. This was evident in the spider plot, but is clearer in the tornado diagram. (The two parameters are plotted only over viable ranges of errors.) From the information given in the tornado and spider plot diagrams, the decision maker understands where to allocate time and money in the estimation process.

Because rough estimates with plausible ranges of values can generally be determined for most parameters in a short period of time, it is suggested that the decision maker perform a sensitivity analysis early in the information-gathering phase. The information provided can be extremely useful in directing future efforts toward achieving accurate cost estimates. As seen in this example, the decision maker needs to focus this effort on the annual cash flows (operational phase) and not the development and disposal phases of the project. Of course, this conclusion may differ greatly depending on the application.

7.3 Estimation Techniques

Once the engineer has determined which data need to be estimated and the degree of accuracy required, he or she is ready to delve deeper into the issue of cost and revenue estimation. In this section, we introduce a number of simple estimation techniques and discuss some more complicated ones that go beyond the level of this textbook. Clearly, as the accuracy required grows, the difficulty in attaining the estimate grows.

A reassuring thought to have in considering the difficulty associated with cost estimation is that there is generally a lot of information available and a number of experts for any given problem. More importantly, because most work involves the use of outside contractors and vendors, both for construction and for the selection of equipment, the likelihood of getting accurate quotes is rather high. As it is quite likely that a company will outsource the actual construction of facilities, detailed estimates from potential contractors will always be possible. This does not excuse the engineer from any cost estimation duties—he or she will always be required to perform rough estimates without the help of a contractor—but it should provide some assurance when detailed estimates are required. For those engineering firms which engage in construction types of activities, the amount of knowledge accumulated over time has actually made the estimation problem fairly routine.

7.3.1 Unit Method

The unit method is a simple procedure that yields a rough estimate of the cost of a project. The method assumes that data are defined "per unit," examples of which are as follows:

- Construction cost per square foot of factory space.
- Construction cost per mile of new road surface.
- Cost per mile of pipeline.

With this information, which is often available from sources within a company or from trade journals, one can quickly get an estimate of a given project cost, given the size of various items in the project. In general, if we are given a unit cost C_r, or the cost of our reference unit, then the desired estimate is

$$C_x = C_r S_x, \tag{7.3}$$

where S_x is the size of the project (or piece of the project) under consideration. Note that use of the unit method is limited only by the definition of C_r. For example, in road construction, the reference value may describe the material cost per mile of road or the total construction cost (for materials, labor, and the use of equipment). Given the definition, the value of C_x will define the appropriate estimate. We illustrate with an example.

EXAMPLE 7.2

Unit Method

Stolt Offshore S.A. signed a contract to install 540 kilometers of 44-inch-diameter pipeline as part of the Omen Lange pipeline in the North Sea, expected to be the longest in the world. The contract, valued at $280 million, also includes an option to install 362 kilometers of 42-inch-diameter pipeline in 2006.[9]

We make the following assumptions: Guidelines and historical information suggest that it will cost $330,000 per kilometer (for materials and installation) for 44-inch-diameter pipe at the required depth and $280,000 per kilometer for the 42-inch-diameter pipe. Given these figures, estimate the cost of the pipeline project.

Solution. Using both estimates, we directly apply our cost formula (7.3):

$$C_x = \underbrace{\frac{\$330\text{K}}{\text{km}}(540 \text{ km})}_{\text{44-inch-diameter pipe}} + \underbrace{\frac{\$280\text{K}}{\text{km}}(362 \text{ km})}_{\text{42-inch-diameter pipe}} = \$280 \text{ million.}$$

Since the costs of materials and installation are included in the unit estimates, the $280 million represents the entire cost of the project.

---

[9] Rojas, T., "Stolt Offshore Signs $280M Pipeline Installation Pact," *Dow Jones Newswires*, November 14, 2003.

The unit method is most applicable in situations where units repeat. Miles of road, meters of pipe, and square feet of office space provide excellent examples of this application. Often, standards are published to conform to these measures. For example, guidelines may be published as to the cost of installing a pipeline of a unit length according to inputs such as the materials used and the location of the pipe (whether the pipe is submerged in water, installed underground, or suspended above the ground).

But the unit method is not limited to simple measures such as kilometers of pipe or miles of roadway. Rather, it can be used in any situation where items are repeated, such as stories on a building or pieces of equipment in a production line. All that is required is a per unit cost.

7.3.2 Index Method

You were already introduced to index methods in the discussion of inflation in Chapter 2. In the case of inflation, a price index provides a historical account of the movement in the cost of a certain commodity or item. Price indexes can be used to determine a rate of inflation that can subsequently be utilized to predict future costs.

Indexes exist for a number of commodities and products, with adjustments available to account for changes in size, time, population, and location. In general, we can write the cost of our desired value as

$$C_x = C_r \left(\frac{S_x}{S_r} \right). \tag{7.4}$$

The value of C_x is our desired estimate, which has an index value of S_x. The values of C_r and S_r are the reference cost and index values, respectively.

The Bureau of Labor Statistics (BLS) of the U.S. Department of Labor publishes a wealth of information and indexes—far too many to list here. Most are available through the BLS website (www.bls.gov). We have already introduced the CPI and PPI for estimating costs related to consumer and producer prices, respectively. Indexes of productivity, labor costs, and material costs are also available and are often broken out according to industry, occupation, and location. These indexes can be used separately or together, depending on the required estimate as illustrated in the next example.

EXAMPLE 7.3

Index Method

Boeing is currently developing the 7E7 Dreamliner, later dubbed the 787, an aircraft expected to go into production in 2008. The plane is expected to perform 20% more efficiently in terms of fuel costs, compared with the most efficient planes produced today.[10]

---

[10] Lunsford, J.L., "Boeing Gets Big Order for 7E7 Plane," *The Wall Street Journal Online,* April 26, 2004.

We make the following assumptions: Since the 787 is being designed to carry a similar number of passengers as previously designed planes, we can estimate the design cost by using an index. Assume that it cost \$6 billion to design the Boeing 777, which entered service in 1995. Assume further that this cost is in 1993 dollars and that the increase in cost can be approximated by the BLS Aircraft Manufacturing Producer Price Index, which has grown from 128.6 in 1993 to 164.2 in 2003. Use the index method to approximate the cost of designing the new aircraft in 2003.

Solution. A rough approximation of the design cost from (7.4) is

$$C_x = C_r\left(\frac{S_x}{S_r}\right) = \$6\text{B}\left(\frac{164.2}{128.6}\right) = \$7.7 \text{ billion.}$$

This rough estimate is an initial order-of-magnitude estimate.

7.3.3 Power Law and Sizing Model

The method of indexing can be expanded to the case of equipment when the asset being evaluated varies in size or capacity, but remains similar in technology and type of service. In general, the power law and sizing model is given as

$$C_x = C_r\left(\frac{Q_x}{Q_r}\right)^m. \tag{7.5}$$

The value of C_x represents our desired estimate, while the value of C_r is the cost of our reference data. Q_x/Q_r defines the ratio of the design specification capacity to the reference capacity. The exponent m is used to scale the factor accordingly. In nearly all applications, we would expect m to take on values less than 1 as there are generally **economies of scale** present when one purchases an item of greater scale. For example, if one were to purchase an engine with a horsepower rating of 110 for price C_r, we would not expect the price of a 220-horsepower engine to be $2C_r$. Naturally, we would expect the price to be higher than C_r, but less than twice the value.

The value of m is traditionally given in trade publications for a number of specific types of equipment and plants. The power law and sizing model is often referred to as the six-tenths rule because many of the exponents are about 0.6. But they are also known to vary between 0.3 and 1.0.

EXAMPLE 7.4

Power Law and Sizing Model

Scottish Power is constructing a 534-MW natural-gas-fired combined-cycle gas turbine power generation plant near Salt Lake City, Utah. The plant, with electricity grid connections, is expected to cost \$330 million.[11]

---

[11] "Scottish Power: PacifiCorp Approves Plant Development," *Dow Jones Newswires*, May 10, 2004.

We make the following assumption: A 200-MW natural-gas-fired combined-cycle gas turbine power generation plant was recently installed at the cost of $140 million (excluding grid connections). Assuming an exponent of 0.69 for these types of plants, estimate the cost of the 534-MW facility.

Solution. We substitute our data into the power law and sizing model (7.5) as follows:

$$C_x = \$140\text{M}\left(\frac{534\text{MW}}{200\text{MW}}\right)^{0.69} = \$275 \text{ million.}$$

This provides a rough estimate of the cost of the new plant.

The power law and sizing model can be utilized with an index method to account for inflation or changes in location.

EXAMPLE 7.5

Power Law and Sizing Model Revisited

Consider the previous example, but assume that the original data are from 1995 at a location in southern California. Assume further that the appropriate producer price index for power generators has risen from 122.5 in 1995 to 139.7 in 2004. Finally, assume that the location factor for southern California is 1.3 (according to the national average) and for Utah is 0.96. Revise the estimate accordingly.

Solution. While we have not analyzed location factors explicitly, their application is similar to the index method. Ignoring the location analysis, we find that our revised estimate, including the inflation index, is

$$C_x = \$275\text{M}\left(\frac{139.7}{122.5}\right) = \$314 \text{ million.}$$

This would be the appropriate cost if the plant were to be built on its original site in California. To adjust to a national average, we have

$$C_x = \frac{\$314\text{M}}{1.3} = \$242 \text{ million.}$$

Finally, to adjust from the national average to that for Utah, we obtain

$$C_x = \$242\text{M} \times 0.96 = \$232 \text{ million.}$$

This provides a much more accurate estimate, as time and location have been considered.

The power law and sizing model is especially useful for initial investment costs, such as the cost of plants or facilities that are defined by output or size. The model is also effective for equipment that is similarly defined.

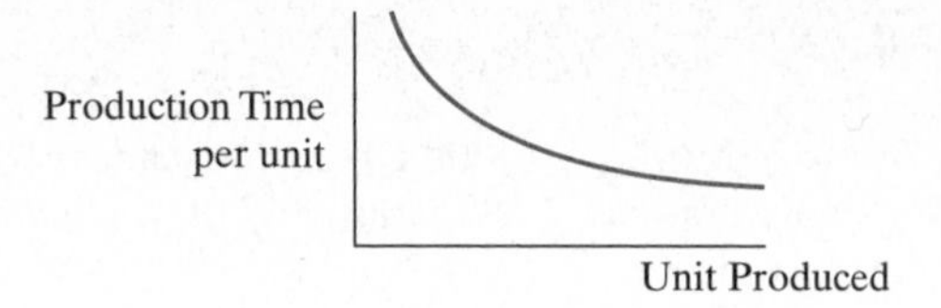

Figure 7.9
Typical learning curve: Per unit production time decreases with time.

7.3.4 The Learning Curve

Although companies often announce that they are building a facility with a given capacity (output level), they often do not achieve that level of production for quite some time. The reasons can be external, such as unrealized demand, or internal, including the need to train employees, to eliminate problems in the production process, or to test processes before operating at full capacity. We consider the internal reasons here. While these reasons could each be examined individually, the progression of output can generally be described by what is called a *learning curve*. A typical learning curve is illustrated in Figure 7.9, which shows the time required to produce each consecutive unit (or batch of units) over time. The idea is that as production progresses, the amount of time needed to produce the next batch decreases. The reason is that the problems highlighted earlier, including the need to train employees and the issues associated with manufacturing, are resolved with time.

The learning curve can be dramatic and is generally associated with the size of the object. For example, the time required to build a double-hulled tanker for shipping liquefied natural gas would presumably improve with time, since the tanker is an extremely large object. However, due to the long construction time, one would expect that substantial improvements in production could occur with every additional unit produced. Contrast this situation with the production of sparkplugs, and one would assume that that curve would not be very steep. In fact, most kinks in the process for producing sparkplugs could presumably be resolved immediately. Still, regardless of the application, the curves take on similar shapes.

Notice that the shape of the production-time curve in Figure 7.9 is exponential. Most often, learning curves are defined by the percentage of improvement (reduction in production time) every time output doubles. If we define K as the amount of time required to produce the first unit of production (again, note that one "unit" can refer to any size, whether it is one physical unit or one batch of 1 million units). Then the time required to produce the nth unit of production is

$$T = Kn^{\log\phi/\log 2}, \tag{7.6}$$

where $\log\phi/\log 2$ is defined by the exponential curve. If we expect a reduction of 20% in production time with every doubling of output, then $\phi = 0.8$. Note that T defines the production time of the nth unit, so cumulative totals must be summed accordingly for total production times.

EXAMPLE 7.6

Learning Curve

Samsung Heavy Industries Co. Limited of Korea entered into an agreement to supply three liquefied natural gas (LNG) transport ships with a capacity of 145,000 cubic meters to the BG Group. The order was placed in late 2003, with delivery expected three years later. The cost of the ships, together with options on an additional four is $430 million. Samsung is expected to build, equip, launch, and deliver the ships according to standard LNG ship specifications. The vessels will sail at an average speed of 20 knots on steam turbine propulsion.[12]

We make the following assumption: It takes one year to construct the first ship, and the learning curve is such that $\phi = 0.95$. How long will it take to produce the third ship, and how long will it take to produce the first seven ships combined (assuming that the options are taken)?

Solution. The data provided for the first ship plug directly into our learning model (7.6) to determine the time T_3 required to produce the third ship:

$$T_3 = Kn^{\log\phi/\log 2} = (1 \text{ yr.})(3)^{\log 0.95/\log 2} = 0.923 \text{ year.}$$

The time required to produce the first seven ships is

$$T = (1 \text{ yr.})\left[(1)^{\log 0.95/\log 2} + (2)^{\log 0.95/\log 2} + \cdots + (7)^{\log 0.95/\log 2}\right] = 6.4 \text{ years.}$$

Thus, the time expected to complete the first seven ships is almost six-and-a-half years. These estimates can be used to help in estimating the operating and maintenance costs of the shipyard. Note that these equations can be easily input into a spreadsheet to compute T.

Sometimes the meaning of the learning curve is unclear, as the formula is based on an exponential function. If we take the logarithm of both sides of Equation (7.6), we get

$$\log T = \log K + \frac{\log\phi}{\log 2}\log n.$$

This is a linear relationship within the logarithmic function such that the slope is defined by the ratio of $\log\phi$ over $\log 2$. The common definition of the slope is the rise over the run, which in this case is the percentage of time reduced for every doubling of output. That reduction comes from the initial value of K, which is the intercept term in the linear relationship. The next example illustrates one more use of the learning curve with respect to batch production.

[12] "BG Group Orders Three LNG Ships," *Dow Jones Newswires,* October 20, 2003.

EXAMPLE 7.7

Learning Curve Revisited

Sharp Corp. doubled the production capacity of the company's Takicho, Mie Prefecture, factory in Japan, which produces 2-inch LCD panels used in cellular telephones. The expansion took place in early 2004 at the cost of YEN40 billion and increased capacity by 4 million panels per month.[13]

We make the following assumption: Due to production glitches, just 2 million panels are produced in the first month of operation for the new line. If the appropriate learning curve in this situation is defined such that $\phi = 0.75$, how many months will it be until production reaches full capacity?

Solution. We can define a single production batch as 2 million units. Thus, the question we are trying to answer is, How long will it take until we can produce two consecutive batches in one month's time (or less)?

The time required to produce the second batch is

$$T_2 = Kn^{\log\phi/\log 2} = (1 \text{ mo.})(2)^{\log 0.75/\log 2} = 0.75 \text{ month},$$

while the time required for the third batch is

$$T_3 = (1 \text{ mo.})(3)^{\log 0.75/\log 2} = 0.63 \text{ month}.$$

If we continue in this manner, we note that the fourth batch takes 0.56 month, the fifth batch 0.51, and the sixth batch 0.48. Thus, production has reached its expected level, during the sixth batch of production.

The learning curve is useful for determining output and costs associated with production. Output must be compared with expected demand in order to determine revenues.

7.3.5 Curve Fitting

Historical data are often available for estimating numerous parameters, including costs and revenues. While many estimation techniques are available, it is often wise to plot the available data and see if simple curve-fitting techniques will suffice. Figure 7.10 illustrates a number of typical curves that are encountered in examining data. The associated analytic function is shown with each curve in the figure. The coefficients for a given data set and the appropriate curve can generally be produced with a regression analysis, available in the Analysis Toolpack in Excel. The application of this technique to curve fitting is beyond the scope of the text, but numerous references are available. The estimator should note that simpler models, which may not capture all of the complexity of a problem, are nonetheless generally more useful and robust

[13] "Sharp to Boost Output of Some LCD Panels—Nikkei," *Dow Jones Newswires,* October 19 2003.

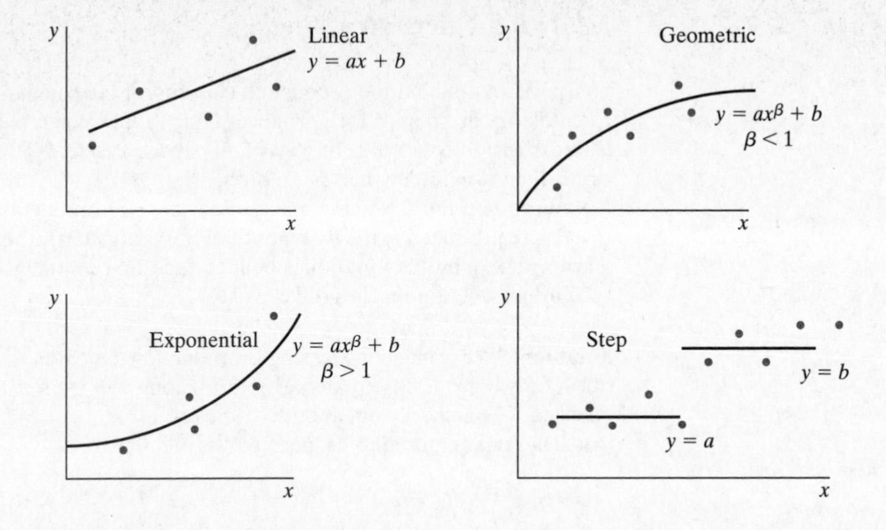

Figure 7.10 Typical curves for cost and revenue estimation, together with their function definitions.

for estimation purposes. Thus, simple models with only one or two input parameters are suggested.

7.4 Ranges of Estimates

It should be clear that data estimation is a critically important task, but also a terribly difficult one, as we are dealing with an uncertain future. It is often useful to give decision makers additional information, so that they have a "feel" for the problem they are studying. One manner in which to do this is by providing ranges of estimates for the data. This can be accomplished in a number of ways. For example, extreme values of the estimation parameters defined previously can be used to determine the ranges of the desired estimates. That is, extreme values of the exponential parameters in the power law and sizing model can be used to determine the possible ranges of estimates of costs for installing some level of capacity. This range of estimates can be further improved with an index method, which can also take on its extreme values.

Another approach in defining a range of estimates is to use reasoning and a bit of "Murphy's law," which says that whatever bad can happen, will happen. This reasoning can lead one to define the worst case for any possible parameter. Thinking in the opposite direction can lead to the definition of an optimistic scenario. While all this may sound a bit wild, examining these extremes allows for all possible ranges of values to be examined.

The reasons for defining a range of estimates should be obvious: We clearly cannot estimate the future with perfect accuracy, so we should not pretend that it is possible, and rather, we should address the issue and give a range of possible values. The reasons for changes in estimates are generally associated with changes in prices for items—some of which are highly unstable.

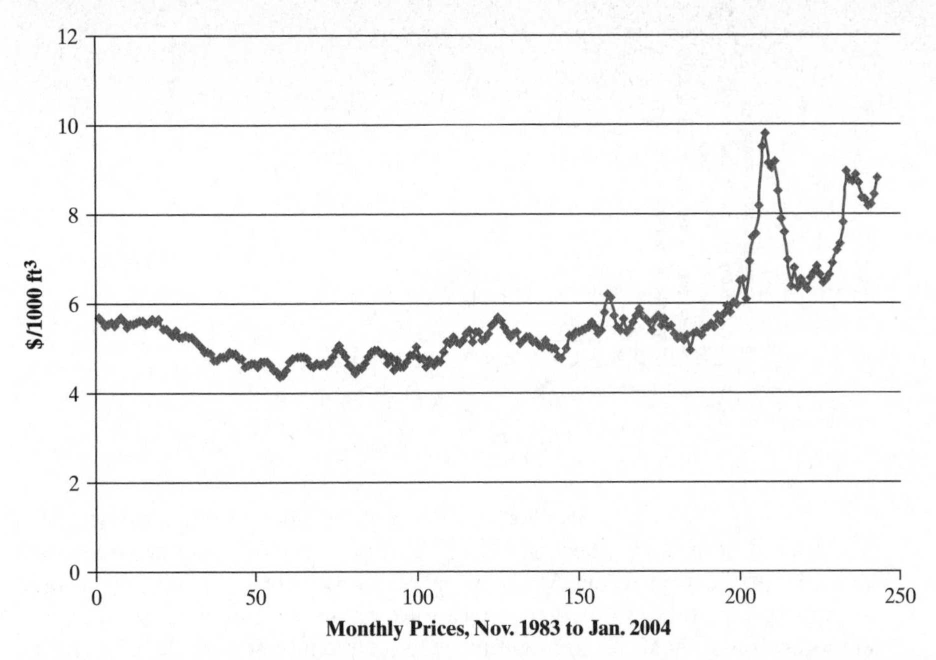

Figure 7.11 Price of natural gas paid by U.S. commercial consumers. Source: U.S. Energy Information Administration.

For example, in Example 7.1, the price of natural gas paid by commercial consumers over time, as charted in Figure 7.11, was shown to be the most sensitive parameter relating to an investment in an LNG handling facility.

Note that there is a variety of data available just for this subject. For example, the United States Energy Information Administration[14] publishes prices paid by residential consumers and electricity producers, as well as prices paid by importers and exporters of gas. Such data can be useful in developing a range of estimates, assuming that the right market is targeted. For our LNG example, if we assume that the gas is to be sold to commercial customers, then the historical data in Figure 7.11 provide a range of possible prices. From the end of 1983 to the present, the price has remained between $4.37 and $9.80 per thousand cubic feet. Further, curve-fitting techniques discussed earlier can be used to determine whether the historical range is sufficient for future estimates or whether a wider range may be necessary. Examining Figure 7.11, we see that the trajectory of the price appears to be positive. Running a simple linear regression on the last 50 data points might provide insight into a higher range.

Natural-gas prices are clearly an external factor in our analysis, because they are driven by the market. Another sensitive parameter in the same example of an LNG facility is the production level. The capacity of the system was slated to be 800 million ft^3 per day. Since this is the designed upper

[14] www.eia.doe.gov.

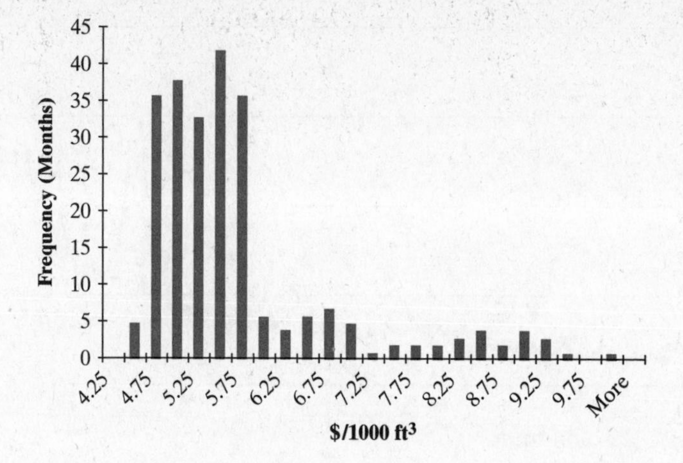

Figure 7.12 Histogram of price per thousand cubic feet of natural gas over time.

bound of the capacity of the system, the high end of the range of estimates has been determined. The question is how to estimate the low end of the range of production levels. This can be determined by examining downtime at similar facilities that are in operation (either at the same company or at other companies). The downtime should include scheduled downtimes, such as those for maintenance, and unexpected downtimes, such as those for equipment failures or production mishaps. This kind of study can give some insight into the level of production that can be expected over time.

Our final reason for defining ranges is to try and estimate the probability that each of the ranges, or scenarios, that have been defined will occur. There are advanced techniques for doing this, but we can take an initial stab at it by plotting the data on a histogram. Returning to our natural-gas example, we can plot the data shown in Figure 7.11 as a histogram in Figure 7.12.

The histogram is created by defining ranges of the parameter in question and then counting the number of times that the parameter is in that range for a given set of data. This count, or frequency, is then plotted against the ranges to produce the histogram. For the natural gas data, it is clear that the price has resided between $5 and $6 most of the time.

The histogram gives us some measure of the likelihood of an occurrence of certain values (natural gas prices), which can be used to define probabilities of given scenarios in our problem. We could crudely assign probabilities as follows: We note that there are 250 total months plotted and that about 43 have a price between $5.50 and $5.75 per thousand cubic feet. Thus, we could estimate the probability that the price of LNG will be in that price range as 43/250. This technique could then be used to develop a discrete distribution over a range of possible prices.

We could also try to match the data to a known continuous distribution. We will discuss distributions more extensively in Chapter 10, but Figure 7.13 illustrates a number of them. A probability distribution is essentially a function that describes the likelihood of a parameter lying within a certain range. The

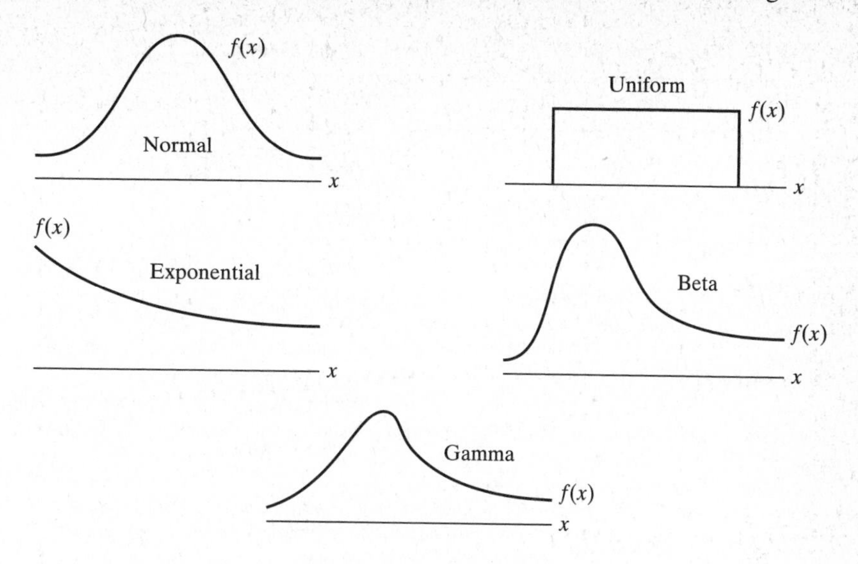

Figure 7.13
Typical probability density functions.

reason we study distributions is because the parameters being estimated are random variables that take on different values with different probabilities. A parameter defined by a probability distribution has an expected value (initial estimate) and a variance, which is a measure of the spread of the random variable.

The natural gas data from Figure 7.13 might be described by an exponential distribution with the proper parameters. Again, techniques for fitting distributions to data are beyond the scope of this text, but references on probability distributions and simulation will aid in this matter.

7.5 Generating the Cash Flow Diagram

The **factor method** is merely a name given to breaking work down into smaller pieces and applying the preceding concepts where necessary in order to get a more detailed estimate of the costs of a project. We illustrate the process in the next example.

EXAMPLE 7.8

Factor Method

Rolls-Royce won a $107 million contract from Dolphin Energy to supply six mechanical-drive industrial Trent Dry Low Emission compression packages. The packages will drive natural gas compressors to provide front-end boosting for gas through an under water pipeline (over 400 kilometers long) connecting Qatar and the United Arab Emirates. Each Trent package will drive a Dresser-Rand DATUM centrifugal gas

TABLE 7.8 **Factor method for estimating project cash flows.**

Parameter	Unit Cost or Price	Source	Units Required	Total	Error
Power Unit	\$12M	Sizing/Index	6	\$72M	8%
Compression Unit	\$6M	Catalog	6	\$36M	3%
Pipeline	\$300K per km	Manual/Index	400	\$120M	10%
Gas Cost	\$2.90 per 1000ft^3	Contract Price	Annual		3%
Operating Cost	\$100K $(1.05)^n$	Curve Fitting	Annual		10%
Maintenance	\$500K per year	Contract	Annual		3%
Overhead	10% of Operating Cost	Guidelines	Annual		3%
Output	—	Specifications	2 billion ft^3/day		10%
Gas Revenue	\$3.50 per 1000ft^3	Historical Average	Annual		30%
Remediation	\$100M	Index	1	\$100M	25%

Note: Output is expected to grow from 500 million ft^3 per day in year 1, with a learning curve defined by 0.85. The study horizon is estimated at 25 years after full production is reached, due to limitations on natural-gas reserves.

compressor. The pipeline will supply 2 billion cubic feet of natural gas per day for 25 years, starting in 2006.[15]

Assume that individual estimates have been procured, and use the factor method to aggregate the results. Note that this project is merely for the transfer of gas from the source (gas wells) to the customer. We are not analyzing the costs or revenues involved in exploring and extracting the gas.

Solution. In accordance with the factor method, we build our estimate of the construction (installation) cost incrementally and sum the results. This procedure is summarized in Table 7.8. Then, all of our information is compiled in one convenient place. The estimates of accuracy are based on experience and the level of detail available in the input data.

The factor method allows the engineering department to "put it all together" and estimate with accuracy, as individual components can be estimated with individual techniques as necessary. This information should form the basis of the cash flow estimates.

7.6 Choosing an Interest Rate

The choice of the interest rate can have a dramatic effect on our analysis, as evidenced by the sensitivity analysis presented in Example 7.1. This should come as no great surprise, as our discussion of the interest rate and discounting

[15] "Rolls-Royce Gets \$107M Pact from Dolphin Energy," *Dow Jones Newswires,* January 13, 2004.

has illustrated that as the interest rate increases, the greater is the discounting, and thus the impact of cash flows in the future is minimized.

We will discuss the meaning of the interest rate again in Chapter 9 when we discuss our measures of worth for a project. However, we cannot ignore this issue when we seek to determine what interest rate should be estimated for discounting purposes. There are two general concepts that we use in guiding our choice of the interest rate: (1) cost of capital and (2) growth rate. We address these two issues in the rest of this section.

7.6.1 Weighted-Average Cost of Capital

We introduced the idea of the cost of capital when we defined loans, bonds, and stocks—three means by which companies raise money for investment purposes. For example, if a company would like to grow by building new plants, it might sell bonds to investors in order to raise the capital necessary to construct the facilities. These bonds require the payment of interest to the investors. Interest payments represent the cost of capital. In general, a company may have acquired various forms of capital over time, including borrowing via loans, selling bonds, and floating stock. Each of these forms of capital comes at a cost, and together they define the cost of capital.

Why is the cost of capital important for a company? Interest is an expense that must be paid over time. Clearly, a company cannot grow unless it is able to pay off its interest and still retain funds (profits). Thus, determining the cost of capital for a company defines the minimum interest rate that the company must return on its investments in order to pay off its interest expenses. If the company earns money above the cost of capital, then it can grow.

A traditional method for calculating the cost of capital is to take a weighted average of all forms of acquired capital and their interest costs. For loans and bonds, this can be straightforward, but for stocks, it is not. The reason is that loans and bonds have a defined rate of interest that is being paid. Stocks (including preferred stock), by contrast, may pay a dividend, but investors also purchase stock for capital appreciation. The expected rate of growth can be viewed as the cost of capital for a stock (in addition to any dividends) as investors purchase the stock in hopes of achieving the necessary return (or better). Presumably, if the return is lower, then the stock will not be purchased. Since there is much debate about the calculation, we utilize both the dividend and capital appreciation (assuming that the latter is nonnegative), in our calculation.

To compute the weighted-average cost of capital (WACC), we calculate

$$\text{WACC} = k_e W_e + k_p W_p + k_d(1 - t)W_d, \tag{7.7}$$

where the coefficients k_e, k_p, and k_d represent the costs of equity (stock), preferred stock, and debt, respectively, as rates of interest and W represents the percentage of the given source of funds in the current capital structure at market values. The value of t represents the tax rate, as interest payments on debt are tax deductible.

To compute the WACC for a company, the number of shares and the current price of both common and preferred stock must be known. Further, all sources of debt, including bonds and loans, together with their amounts and relevant interest rates, must be known. We illustrate in the next example.

EXAMPLE 7.9

Cost of Capital

Unfortunately, the detailed data required for calculating the WACC can be difficult to compile. Let us assume that an engineering firm has the following capital structure:

- 250 million outstanding shares of common stock currently priced at $29.30. The stock pays a 2.3% dividend and has averaged a return (capital appreciation) of 18.25% over the past few years.
- 20 million preferred shares of stock paying a dividend of 5.9% and currently priced at $40.50.
- One bond issue worth $150 million with a 4.5% coupon rate, three years until maturity, and currently selling at $101.75 per $100 (premium) on the market.
- One syndicated loan for $500 million at 4.5% annual interest.

The effective tax rate is 40%.

Solution. There are many ways to proceed with the information given. One way is to compute the ratio of each input to the capital cost structure, as shown in Table 7.9.

From the information given, the cost of common equity is $18.25\% + 2.3\% = 20.55\%$ (capital appreciation plus dividend), and the cost of preferred equity is 5.9% (dividend). The cost of debt is found by computing the yield to maturity for the bonds. With three years until maturity, the cost of debt is 3.89%. The cost of the loan is given as 4.5%.

TABLE 7.9 **Capital structure of a company.**

Source	Amount	Price	Total	Weight
Common Stock	250,000,000 sh.	$29.30	$7,325,000,000	83.36%
Preferred Stock	20,000,000 sh.	$40.50	$810,000,000	9.22%
Bond	$150,000,000	$1.0175	$152,625,000	1.74%
Loan	$500,000,000	$1.00	$500,000,000	5.69%
Market Value			$8,787,625,000	100.00%

With all of this information, we find that the weighted-average cost of capital is

$$\begin{aligned}\text{WACC} &= k_e W_e + k_p W_p + k_d(1-t)W_d \\ &= \underbrace{.2055(.8336)}_{\text{common stock}} + \underbrace{.059(.0922)}_{\text{preferred stock}} + \underbrace{.0389(1-0.4)(.0174)}_{\text{bonds}} + \underbrace{.045(1-0.4)(.0569)}_{\text{loans}} \\ &= .1787,\end{aligned}$$

or 17.87%.

Note that using the cost of capital for the interest rate assumes that the company accepts this "historical" view of its cost structure as being representative of its future cost structure. This is important to note because the WACC is determined from funds raised in the past, whereas capital investments are generally being evaluated for the future. Ideally, we could compute the *marginal* cost of capital, or the change in capital costs with time, so as to be more accurate with the current cost of funds. This would capture the fact that interest rates change with time and it often costs more to acquire additional funds, due to the higher risks taken on by the lender.

7.6.2 Growth Rate and Opportunity Cost

We have not explicitly introduced the concept of **opportunity cost**. When one has to make a choice between two (or more) projects, the project that is not chosen represents a lost opportunity. This is true of simple decisions, such as choosing between a steak sandwich and a bowl of pasta for lunch, or, for example, the more complicated decision faced by the firm Handspring (now owned by Palm), which, due to limited resources, decided in 2003 to abandon the development of all of its lines of handheld computers and focus solely on Treo, its integrated computer and mobile telephone.[16] Choosing one path over the other entails benefits and risks. One risk is the potential loss from exercising that choice. We refer to these losses as *opportunity costs*. The reason we choose one path over the other is that we feel that the chosen path can bring about benefits greater than those defined by opportunity costs.

We mention this concept here because examining opportunity costs provides a lower bound on the expected growth rate of a company. When a company accepts a project and the project performs as expected, an expected return is generated. As we will see in succeeding chapters, a rule to follow when accepting projects is to accept only those which meet or exceed some threshold of growth. This ensures that the company will grow at its expected rate. The concept of opportunity cost provides some insight into the desired rate of growth. If a company can choose between two projects, the choice

[16] Tam, P., "No Room to Hedge: Hit by Downturn, Tech Firms Forced into Tough Choices," *The Wall Street Journal,* May 14, 2003.

provides some measure of the minimum desired growth rate: higher than that of the rejected project and less than or equal to that of the accepted project. This is why the term **minimum attractive rate of return,** or MARR, is often used to describe the interest rate for discounting purposes, because it is the minimum rate of growth that a company will accept from its invested projects.

Presumably, the growth rate or MARR is greater than the cost of capital computed in the previous section, as the MARR must allow for sufficient returns in order to pay back the cost of capital over time and still grow. This is similar to a bank charging more to lend money than it pays its depositors. How a company "decides" to grow—or defines its MARR—is not entirely straightforward. As with the calculation of the WACC, a company may look at the historical growth in the price of its stock, in addition to any returns from dividends, and prescribe the sum of the two as its desired growth rate. Of course, the rate may be adjusted for the size of the company, since smaller firms tend to grow more quickly than larger firms, and for the company's sector of business, as a technology company may be expected to grow faster than a provider of standard commodities to the market. Unfortunately, this perspective ignores future potential that can lead to increased growth. In other words, a company may change its strategic directions such that past data do not reflect the future.

In essence, determining a MARR is difficult because it requires that the company look at its potential returns. However, as with the WACC, if we accept historical returns as indicative of the future, then examining the traditional growth in the stock price in addition to paid dividends should provide a lower bound to the MARR. Note that this method ensures that the MARR is greater than the WACC under the assumption that the growth rate in the stock price is greater than any interest rates on debt. If the stock has not appreciated, then the company may choose to set a desired growth rate, which should be some margin of return above the weighted average or marginal cost of capital.

7.7 Choosing a Study Period

The study period N is a difficult parameter to estimate because it is highly dependent on the project itself. In Chapters 15 and 16, we examine the decisions faced after a project has been undertaken. Included is the decision to abandon, or end, a project. Thus, it should not be felt that the estimate of a project horizon is an end-all estimate, as it can clearly be altered once a project has begun.

Despite this reassurance, we require an estimate of N in order to perform an analysis of the project and to define a range over which cost and revenue estimates are required. For the case where we may be given a contract to perform engineering duties for a specified length of time, the problem is quite straightforward, because that length of time is defined in the contract. For other situations, there are rules of thumb to consider:

- *Estimation of product life cycle.* In discussing the relevant costs and revenues that occur over the life of a product, we discussed the product life cycle. In this cycle, revenues usually increase in the early periods, then level off, and finally decrease. For a number of products, this cycle can be estimated with some accuracy when historical data of previous similar products are available and the evolution of technology is somewhat predictable. For example, one can generally expect a doubling in the speed of microprocessors in personal computers every two years. This relationship has been referred to as "Moore's law," because the former Intel chief executive noted the trend. Remarkably, technology has kept up that pace for years.[17] This trend can be used to predict the life cycle of products that use microprocessors, such as personal computers. In general, trends can be developed from previous technologies to determine likely product cycles or at least viable ranges.

  In addition to technology, other relevant factors may limit the product life cycle. Clearly, the development of oil fields or mines is limited by the amount of the resource that can be extracted. Once an estimate is developed from seismic data or testing, an application of expected extraction methods generally yields the estimated life of the project.
- *Service lives.* Many assets have physical lives that are limited due to a variety of external factors, such as the inability to acquire spare parts after some period of time elapses or obsolescence. In these situations, the vendors are an excellent source of information regarding the maximum service lives of the assets, since they produce the products and, often, the spare parts. The maximum service life of an asset may coincide with the amount of time that warranties are made available (whether taken or not).
- *Infinite or indefinite horizons.* Many projects are expected to last for an indefinite or infinite period of time. This is certainly the case for engineering projects which provide companies with services that are required indefinitely. In these situations, information on the maximum service lives of the assets involved is generally required. Often, the maximum service life of an asset coincides with the maximum length of time that reasonable data estimates are available. Finally, further information, including how service will continue beyond the service life of the asset may be required.

Again, the estimates of the horizon are clearly project dependent.

---

[17] See www.intel.com.

7.8 Additional Information Gathering

We have discussed a number of topics in this chapter, including how to estimate costs and revenues, ranges of estimates, and associated probabilities. We have also discussed important parameters, such as the problem horizon and interest rate.

Unfortunately, one can never have enough information concerning an investment alternative. Many pieces of information besides the ones we have thus far mentioned can be useful in making a final decision, and usually it is easiest to gather this information when initial cost data are being acquired. Some other information that may be of use includes:

- *Engineering specifications.* For a given solution, multiple alternatives may be available that differ according to economics and engineering specifications, such as speed, strength, power, output, and capacity. These factors must be detailed clearly such that fair comparisons between alternatives can be made.
- *Environmental and social impacts.* Although estimates of end-of-life remediation costs may have been made, many zoning authorities require that studies be conducted on the impact of a process on the environment, especially when pollution must be controlled or noise abated.
- *Impacts on infrastructure.* In addition to environmental and social impacts, authorities may be interested in the impacts of operations on infrastructure. For example, will a proposed facility drain power from the public grid, or will local roads be swamped due to an influx of workers? These issues may have a direct bearing on the economics of a decision and are clearly important.
- *Noneconomic factors.* Obviously, information on noneconomic factors, such as those dealing with the workforce, should also be gathered and documented.

The preceding list presents an overview of information that is generally necessary in promoting a project. The exact details and requirements are clearly going to be project specific.

7.9 Sources of Data

The key to using any of the foregoing estimation methods is to have historical data with which to work. As with the design process, the estimation process is something that engineers have endured for years. Thus, a lot of data are available and there is no reason to work in the dark. The various sources of information can be divided into those which are within the firm and those from outside.

Sources inside the firm include the following:

- *People.* The experiences of others who have performed similar duties are invaluable.
- *Databases.* For industries in which turnover is high, databases of information are often utilized to supplant the availability of workers with experience.
- *Accounting records:* Aggregate-level accounting data are often tracked that can sometimes be used to provide historical information.
- *Research and development.* Preliminary testing of designs, materials, and products can provide insights into costs of production and operation.
- *Market surveys.* Much like research and development, survey data may provide information about future sales estimates.

Sources outside the firm include the following:

- *Reference books.* Aggregate-level data on a variety of products, processes, equipment, and facilities are often available. A number of indexes are maintained for engineering, including the *Engineering News-Record Indices*, the *Marshall and Swift Index*, and the *Chemical Engineering Plant Cost Index*.
- *Statistical surveys.* As noted earlier, the government collects and maintains large databases of cost information broken out by time, location, and industry.
- *Technical journals.* Journals in each field contain both design and operational data that can be useful for cost estimation. The journal *Cost Engineering*, published by AACE International, specifically discusses cost estimation.
- *Manufacturers, vendors, and contractors.* Because these entities are often competing for business, they produce a number of publications with efficiency and cost ratings. Prototypes are frequently made available at trade shows. Further, specialized quotes are often attainable in a reasonable amount of time.

The estimation process is not easy, but it is extremely important, since the analyses presented in the ensuing chapters assume that costs have been estimated. It is important that the engineer take advantage of all resources available, both internal and external, when developing estimates.

7.10 Examining the Real Decision Problems

We return to our introductory example concerning the Gorgon gas project and the various contracts to supply natural gas. The questions posed were the following:

1. In addition to the development (capital) costs and contracted sale price, what other estimates are required to evaluate the Gorgon project?

The development costs encompass all costs before production, so the remaining costs to be estimated are the annual production costs, which would most likely include fixed annual costs (for overhead and maintenance) and per unit (ton) production costs. These are critical estimates, since the difference between the contract prices and the annual production costs defines the profitability of the project over time and thus its worth.

2. Which estimate is more critical, the development cost, which is spread over the years 2005 through 2009, or the 25-year contracted sale price?

We could perform an extensive sensitivity analysis or note that the A$5.5 billion development cost (recall that Chevron owns 50% of the project), spread equally over the 25 years of contracts, is

$$\text{A\$5.5B}(\overset{A/P,15\%,25}{0.1547}) = \text{A\$851 million.}$$

Because the contracts, which may be fixed-price contracts, are for A$28 billion over 25 years, or A$1.12 billion per year, it is clear that annual costs must be less than A$269 million per year in order for the project to be feasible. Note that these figures have assumed a low-end estimate for development costs, but also have assumed that 4.2 million annual tons of gas have been sold. An extensive sensitivity analysis would determine the allowable error in both the development and production cost estimates.

3. How could the rise in steel prices (for Chevron) or LNG prices (for CNOOC) have been estimated more accurately? How crucial is the LNG price to CNOOC?

There are clearly historical price indexes of steel and LNG that may have proven helpful in defining acceptable ranges to examine. There have been times in the past when demand exceeded supply for these commodities; thus, the historical price ranges would add value. Furthermore, the price movements in the two commodities are related, as the higher demand for LNG has pushed prices higher, leading to further exploration and expansion of production by a number of companies, which in turn pushes raw materials prices higher. This is not to say that these prices could have been predicted with precision, but it appears obvious now that, historically, the original $0.98 boe price is cheap.

CNOOC is not extracting the gas and selling it to customers. Rather, it is purchasing the gas in liquid form from a company, converting it to the gaseous state, and then distributing it to customers. In theory, CNOOC can pass on any material costs to the customer. Thus, CNOOC's exposure is not as high as Chevron's. However, high LNG prices will generally hurt profit margins, since demand will likely decline as customers seek cheaper forms of energy.

7.11 Key Points

1. Life-cycle costs include all cash flows pertaining to a project over time. These costs may be classified according to phase: development, operational, or disposal.

2. Direct costs can be attributed to a single product, operation, or project, while indirect costs cannot. Indirect costs are also referred to as overhead.

3. A trade-off must generally be made between the accuracy of an estimate and the cost incurred to acquire the estimate.

4. The level of accuracy required for an estimate is dependent on its application: order-of-magnitude estimates are generally accepted for first-pass evaluations, while definitive estimates are generally required for detailed cash flow analysis.

5. Sensitivity analysis is a tool that identifies which parameters, such as cash flows, the interest rate, or the study period, are most influential on a project. The majority of resources should be spent on studying these critical parameters.

6. The unit method allows for estimates based on per unit costs.

7. The indexing method allows for previous estimates to be adjusted according to some index. Indexes exist for changes in price over time and for location.

8. The power law and sizing model estimates the cost of a new capacity on the basis of the cost of a previously owned asset with known capacity. Economies of scale are generally assumed in the estimate.

9. The learning curve method provides a way of estimating improvements in output over time.

10. Curve fitting is a general technique for estimating some function from a given set of data. This function can be used to estimate future costs or revenues.

11. The factor method allows a variety of estimates to be pulled together into one report.

12. Ranges of estimates can be useful for decision-making purposes, as they provide information about pessimistic and optimistic project outcomes or other viable scenarios.

13. The interest rate must also be estimated for discounting cash flows. The interest rate generally takes on the meaning of the cost of capital (the rate of interest paid on funds borrowed for investment) or the desired growth rate.

14. The weighted-average cost of capital is a method of estimating the historical cost of capital (the cost of borrowed funds) by taking a weighted average of all sources of funds for the company.

15. Identifying the desired growth rate is often an internal company decision that can include examining historical growth rates and future investment opportunities.

16. Study horizons can generally be estimated by examining typical product life cycles or equipment service lives.

17. Any relevant information about a potential project, including engineering specifications, noneconomic factors, environmental and societal impacts, and information concerning infrastructure, should be collected.

18. A variety of sources of information, from internal databases to technical journals, is available to aid in the estimation process.

7.12 Further Reading

Clark, F.D., and A.B. Lorenzoni, *Applied Cost Engineering,* 2d ed. Marcel Dekker, New York, 1985.

Eschenbach, T.G., "Technical Note: Constructing Tornado Diagrams with Spreadsheets," *The Engineering Economist*, Volume 51, Number 2, 2006.

Humphreys, K., and P. Wellman, *Basic Cost Engineering,* 2d ed. Marcel Dekker, New York, 1987.

Ostwald, P.F., and T.S. McLaren, *Cost Analysis and Estimating for Engineering and Management.* Pearson Education, Inc., Upper Saddle River, New Jersey, 2004.

Pratt, S.P., *Cost of Capital: Estimation and Applications.* John Wiley and Sons, New York, 1998.

Stewart, R.D., R.M. Wyskida, and J.D. Johannes, *Cost Estimator's Reference Manual.* John Wiley and Sons, New York, 1995.

Thuesen, G.J., and W.J. Fabrycky, *Engineering Economy,* 9th ed. Prentice-Hall, Upper Saddle River, NJ, 2001.

Westney, R.E. (ed.). *The Engineer's Cost Handbook: Tools for Managing Project Costs.* Marcel Dekker, New York, 1997.

7.13 Questions and Problems

7.13.1 Concept Questions

1. What is meant by the term "life cycle" as applied to a project or product? What costs are associated with a life cycle?
2. What are the three major phases of a capital investment? Illustrate these phases as they pertain to the construction of a semiconductor wafer fabrication facility.
3. Define indirect and direct costs.
4. What is working capital?
5. What are sunk costs? Are they considered life-cycle costs? Explain.
6. What is the relationship between the desired accuracy of an estimate and the work, money, or effort required to get that estimate? How does this relationship influence the cost estimation process?
7. Do all estimates need to be exact (all of the time) in economic analysis? Explain.
8. How can sensitivity analysis be used in the cost estimation process?
9. Explain when the unit method, index method, and power law and sizing model are appropriate techniques for estimation?
10. What is the purpose of the learning curve and how does it affect cost estimation?
11. What is the factor method?
12. Why might it be useful to develop a range of estimates? How else might this information be utilized?
13. How are the two methods discussed herein for establishing an interest rate mathematically related? Are these rates easy to determine? Explain.
14. Why is the weighted-average cost of capital often referred to as "historical" with respect to estimates of the cost of capital?
15. Describe inputs to consider when selecting a study period.
16. Besides the cash flow diagram and interest rate, what other information might be useful in making an investment decision?
17. Are data generally available for estimation? Explain.

7.13.2 Drill Problems

1. Assume that a project can be described by an investment $P = \$5$ million at time zero, five years of net cash flows $A = \$1.3$ million, and a salvage value $S = \$250{,}000$ at the end. Which is more sensitive, P or A? Use present worth and a MARR of 8% to make your decision.
2. Assume that a project has the following estimates: Investment $P = \$100$ million, spread over periods 0 and 1; revenues following a gradient (first positive flow in period 2) $G = \$10$ million; annual costs $A = \$2$ million; project life $N = 16$; and MARR = 25%. Rank the sensitivity of P, G, A, N, and the MARR. Assume that P is spread (as production lasts 15 years) evenly over periods 0 and 1 despite any changes in the estimate.

3. A project with a \$6.5 million investment cost at time zero is being considered. Operating and maintenance costs are expected to be \$1.5 million in the first year, growing by 1.5% per year, and overhead costs are assumed fixed at \$100,000 per year. The salvage value is expected to be \$500,000 at the end of the 15-year horizon. Finally, revenues are estimated to grow from \$500,000 in year 1 (A_1) according to an arithmetic gradient with $G = \$500{,}000$, until they reach $A = \$5$ million per year through the end of the horizon. The MARR is assumed to be 12%. Perform a sensitivity analysis on the following parameters:

(a) Investment cost.

(b) First-year O&M cost.

(c) O&M rate of increase.

(d) Initial revenue A_1.

(e) Revenue growth G.

(f) Maximum revenue level A.

(g) Salvage value.

(h) Horizon time.

(i) MARR.

Specifically, examine errors within 30% of the original estimates, and draw a spider plot showing the results. Then plot the result of errors between −5% and 15% for (a) through (c) and between −15% and 25% for (d) through (f), using a tornado diagram. Interpret your results.

4. A boiler is to be purchased with capacity 1.5 times that of its predecessor, which cost \$3.5 million. If the scaling factor is 0.75, estimate the cost of the new boiler.

5. Would an index method be relevant in answering the previous problem? Explain.

6. A new engine with 500 horsepower is going to be purchased. The previous engine was rated at 300 horsepower and cost \$5000. If the scaling factor is 0.85, estimate the cost of the new engine.

7. Would an index method be relevant in answering the previous problem? Explain.

8. A 500-MW power plant is to be built. An older 400-MW plant was constructed at the cost of \$400 million. If the scaling factor is 0.6, what is the estimate of the cost of the new plant?

9. Revise the estimate in Problem 8, taking into consideration the fact that the location is now different. The original location was in Iowa, with a cost factor of 0.8 of the national average, while the new plant is to be constructed near Denver, with a cost factor of 1.15 of the national average.

10. Now revise the estimate in Problem 9, taking into consideration the fact that the 400-MW plant was constructed 10 years ago. If the relevant inflation rate for power plants has been 4.2% per year, what is the new estimate? (Consider location factors, too.)

11. The plant described in Problems 8–10 is to be connected to the national grid through transmission lines. In total, 100 miles of transmission lines are to be laid at the cost of $42,500 per mile, including material and labor costs, given the terrain. What is the cost of the transmission component of this estimate?

12. The per unit transmission cost stated in the previous problem assumed that transmission towers would be spread a standard amount apart. However, due to the terrain, an additional tower is needed. Assume that a transmission tower costs $50,000 to construct. Assume further that the per unit cost for the transmission line is now $25,000 per mile. Finally, assume that two transmission substations are required at the cost of $350,000 each. Using the factor method, estimate the investment cost for the entire power plant investment. (Add the power plant estimates from Problem 10 to the analysis you perform here.)

13. A manufacturing plant produces specially crafted engines for high-performance automobiles. If it takes eight hours to produce the first engine and the learning curve is such that it takes an estimated 88% of the time to double output, determine the amount of time required to produce the first five engines.

14. A shipbuilding company can produce a large container ship in six months. If $m = 0.90$ for its learning curve, how much time should it take to produce the sixth ship?

15. A company produces large plasma screen TVs. Its line has a capacity of 1000 units per month. Assume that the firm produces 500 units in its first month as it irons out the production process, but the learning curve exponent is $m = 0.75$. How many months will it be until the company reaches full capacity?

16. Estimates are needed for probable O&M costs at a new facility being considered. According to company data, the past 10 years have resulted in the following costs at a similar facility: $1.57, $1.65, $1.35, $1.95, $2.05, $1.65, $1.50, $1.65, $1.75, and $1.35 million. Use these data to

(a) Determine an expected value and range of estimates.

(b) Define probabilities for the ranges of estimates.

(c) Plot the data and estimate whether there is a trend with simple regression.

17. Assume that the price (per ton) for a commodity has been $200, $300, $250, $250, $275, $300, $310, $400, $300, $325, $275, and $300 for the past number of periods. Use these data to form an expected value and range of estimates and define probabilities for the ranges.

18. Assume that a company has the following capital structure:

- 20 million outstanding shares of common stock currently priced at $20.50. The stock pays a dividend of 1.4% and has averaged a return of 12.8% over the past few years.
- Two bond issues, one consisting of $400 million notes with 2 years remaining until maturity, with a coupon rate of 3.75%, and selling on the market for a 99.56% discount, and the other consisting of newly issued 10-year notes (totalling $250 million) paying a 7.5% coupon rate and sold at par.
- A credit facility of $1 million. Currently, $500,000 is being used at 4.5% per year.

What is the WACC for a tax rate of 40%?

19. Assume that a company has the following capital structure:

- 200 million outstanding shares of common stock currently priced at \$33. The stock pays no dividend and has averaged a return of 17.5% over the past few years.
- One bond issue worth \$300 million, with a 3.5% coupon rate, three years until maturity, and currently selling at a 99.75% discount on the market.
- One credit facility for \$250 million, at an annual interest rate of 8.5%.

What is the WACC for a tax rate of 40%?

20. Assume that a company has the following capital structure:

- 100 million outstanding shares of common stock currently priced at \$42.30. The stock pays no dividend and has averaged a return of 8.25% over the past few years.
- 20 million shares of preferred stock that pays a dividend of 5.3% and a current price of \$52.33.
- One bond issue worth \$250 million, with a 6.5% coupon rate, five years until maturity, and currently selling at a premium of \$103.75 per \$100 face value on the market.
- One syndicated loan for \$500 million at 4.5% quarterly interest.

What is the WACC for a tax rate of 35%?

7.13.3 Application Questions

1. Trunkline LNG Co. announced that it would expand its LNG receiving and distribution terminal in Lake Charles, Louisiana, in 2003. A terminal has been in operation at Lake Charles since the 1950s.[18] Discuss the costs, revenues, and parameters that need to be estimated in order to analyze the investment. What noneconomic factors might be important?

2. Ford announced that it would expand its truck-part making capabilities (including fenders, sides, hoods, tailgates, etc.) in Louisville, Kentucky, in 2004.[19] Discuss the costs, revenues, and parameters that need to be estimated in order to analyze the investment. What noneconomic factors might be important?

3. The motivation for this problem is from Goldcorp, Inc., expanding its Red Lake Mine operations at the cost of \$92 million with a 7200-foot-deep shaft that will increase production by 200,000 ounces per year. The potential is for an additional 3.8 million ounces of gold in a new deposit. Production is expected to commence in 2007 at the cost of \$70 per ounce.[20] Perform an extensive sensitivity analysis of the project, analyzing the investment cost, production starting date (assume that expansion begins in 2003), production rate, price of gold (assume an estimated

---

[18] Reynolds, L., "CB&I Wins EPC Pact for Expansion of U.S. LNG Terminal," *Dow Jones Newswires,* October 8, 2003.

[19] Carty, S.S., "Ford to Invest \$73M in Kentucky Truck Plant Production," *Dow Jones Newswires,* March 22, 2004.

[20] Locke, T., "Goldcorp Sees Big Potential for Red Lake Mine," *Dow Jones Newswires,* September 23, 2003.

$390 per ounce), gold reserves, and interest rate (15% per year). How should this information guide the cost estimation process?

4. The motivation for this problem comes from Kia Automotive Group's decision to locate a new $870 million production facility in Zilina, Slovakia. Construction began in early 2004, with production set to begin in November of 2006 with a capacity of 200,000 vehicles per year.[21] Make the following assumptions: The investment cost is spread such that one-fourth of it is spent in 2004, one-fourth in 2005, and one-half in 2006. Demand is 25,000 in 2006, growing by 25,000 cars each half year until reaching the capacity of 200,000 cars produced. The profit per car is expected to average $2500 over six years, with fixed annual plant costs to total $10 million per year. Assume that the MARR is 12% per year and the plant is sold for $50 million after six years. Perform a sensitivity analysis on the data, and rank all parameters in order of their importance in estimation. Also, assume end-of-year cash flows in analysis.

5. Intel announced that it would expand its semiconductor chip manufacturing plant in Leixlip, Ireland, with the construction of 60,000 square feet of manufacturing clean-room space. Production is slated to begin in 2006. The clean room is to be equipped with 65-nm technology.[22] If a similar-sized facility for 90-nm technology costs $1.80 billion, estimate the cost of the new facility, using an exponent factor of 0.70 for the power law and sizing model. (Note that 65-nm technology is an improvement and jump in complexity over 90-nm technology.) Assume that the semiconductor producer price index was 91.1 in 2000 (the time of construction of the reference facility) and is 78.9 in 2004.

6. Alcatel SA of France and Fujitsu, Ltd., of Japan won a contract to build a 20,000-kilometer submarine cable network for a consortium of telecommunications operators. The cable will connect Southeast Asia, the Middle East, and Western Europe, carrying telephone, Internet, and broadband streams.[23] Assume the cable material costs $5000 per kilometer, while installation costs are expected to be $20,000 per kilometer. Estimate the cost of the investment.

7. Skanska, a Swedish construction company, received a contract to build 14,200 square meters of office space for the Norwegian Security Service. The customer is the Norwegian real-estate firm Avantor. The project began in the fall of 2003 and is scheduled for completion in 2005.[24] Assume that company data yield an estimated cost of NOK1000 per square meter of construction. Given this information, what is the estimated cost of the structure?

8. In October of 2003, Itron, Inc., agreed to install more than 120,000 automated water-reading meters in Houston at the cost of $82.5 per unit.[25] Assume that 5000 meters can be installed in the first month and installation follows a learning curve with $\phi = 0.85$.

---

[21] "Kia, Slovak Govt Sign Deal on New Auto Plant," *Dow Jones Newswires,* March 18, 2004.

[22] "Intel to Invest Up to $2 Billion for Future Manufacturing Capabilities," *Intel Pressroom,* www.intel.com, May 19, 2004.

[23] Man, S., "Alcatel Wins $500M SingTel Consortium Cable Network Deal," *Dow Jones Newswires,* March 29, 2004.

[24] "Skanska Gets SEK310M Norwegian Contracts,"*Dow Jones Newswires*, October 2, 2003.

[25] Beltran, E., "City of Houston Extends Pact with Itron; Continues Expansion of Automatic Water Meter Reading 'AMR' System," *Dow Jones Newswires,* October 2, 2003.

(a) How long will it take to install the next 5000 units?

(b) How long will it take to install all 120,000 units?

(c) If payments are made monthly according to the number of units installed for the month, draw the cash flow diagram for the first two months of installation.

(d) It turns out that Itron has installed over 300,000 units since December of 1998. If the company installed 60,000 in the first year and we assume the same ϕ for the learning curve, how long should the 120,000 units take with this new information?

9. Scottish and Southern Energy PLC announced that it would connect the Highlands and the Central Belt of Scotland with a 400,000-volt transmission line. The new 220-km line will replace an existing 132,000-volt line.[26] Make the following assumptions: The project will require two main transmission stations, 25 transmission substations (GBP2,000,000 each), 200 support towers (GBP50,000 each), and 220 km of transmission lines. The lines cost GBP25,000 per km for materials and installation. A previous main transmission station for the 132,000-volt line cost GBP20 million. The power law sizing model gives an exponent of 0.65, and the inflation rate is 1.5% per year over 20 years. Finally, removal of the old line will cost GBP10,000 per km. Perform a factor method analysis to determine the investment cost.

10. Given the spot price per ton for alumina in Table 7.10, develop a range of estimates for a potential producer or user of the material. Further, develop a probability estimate for each estimate in the range.

TABLE 7.10

Alumina spot prices in U.S. dollars per ton.

Year	Price
1992	$160
1993	$160
1994	$125
1995	$250
1996	$175
1997	$210
1998	$190
1999	$210
2000	$310
2001	$150
2002	$150
2003	$275

Source: *Metal Bulletin Research* through Morrison, K., "A Failure to Keep Pace," *The Financial Times*, Special Report: Aluminum, p. 2, November 5, 2003.

---

[26] "Scottish and Southern: Transmission Line to Cost GBP200M," *Dow Jones Newswires*, January 20, 2004.

7.13.4 Fundamentals of Engineering Exam Prep

1. Order-of-magnitude estimates generally

(a) Can be used for making investment decisions.

(b) Provide more accuracy than budgetary estimates.

(c) Provide preliminary information about a project before its scope has been entirely defined.

(d) Require extensive analysis and data gathering.

2. Working capital is used to

(a) Reduce taxes.

(b) Pay bank loans.

(c) Fund cash needs, such as procuring inventory.

(d) None of the above.

3. Given the time value of money, which of the following estimates, if assumed to be similar in magnitude, would generally require a more accurate estimate?

(a) Investments near time zero.

(b) Remediation costs at the end of the project.

(c) Salvage values at the end of the project.

(d) All of the above.

4. A pump rated at 50 horsepower was purchased in 1998 for $5000. If the cost index in 1998 was 153.5, the expected cost of a similarly rated pump in 2005, with the cost index at 172.3, is most closely

(a) $5000.

(b) $5600.

(c) $5800.

(d) $6000.

5. Revenues from a process are limited by capacity, which is rated at 10,000 parts per month. If 8000 parts are produced in the first month and the company expects a 10% reduction in production time for every doubling of output, the system is estimated to reach capacity in

(a) The second month.

(b) The third month.

(c) The fourth month.

(d) The fifth month.

6. For most engineering projects, the greatest costs are experienced

(a) At the end of a production run.

(b) After equipment is procured.

(c) During design and development.

(d) All of the above.

7. A new shipyard produces its first containership in nine months at a labor cost of \$15 million per month. If it is estimated that 10% of the production time will be reduced every time output is doubled, total labor costs for the third ship should be estimated at

(a) \$135 million.

(b) \$114 million.

(c) \$122 million.

(d) \$105 million.

8. Bids are requested on a 47-mile stretch of 18-inch-diameter submersible pipe (for depths of 50 feet) for transporting natural gas. Unit estimates are \$5 million per mile for the pipe and \$500,000 per mile of installation. An initial bid should

(a) Range between \$100 and \$150 million.

(b) Range between \$200 and \$250 million.

(c) Range between \$300 and \$350 million.

(d) Range between \$250 and \$300 million.

9. If a telephone pole costs \$500, wire is \$1000 per mile, installation costs \$25,000 per mile, and the terrain requires a pole every tenth of a mile, an estimate of the cost of installing telephone wires in a new community with 2.5 miles of roads (through all houses) should be

(a) Less than \$72,500.

(b) At least \$77,500.

(c) Between \$72,500 and \$77,500.

(d) None of the above.

10. A 250-MW power plant was constructed at the cost of \$200 million. If a 500-MW facility is being considered, a reasonable estimate of the cost, assuming a power law sizing exponent of 0.6, is

(a) \$325 million.

(b) \$350 million.

(c) \$250 million.

(d) \$300 million.

11. A 250-MW power plant was constructed in 2000 at the cost of \$200 million. The construction cost index was 6058 at that time and is 6500 in 2006. If a 500-MW facility is being considered in 2006, a reasonable estimate of the cost, assuming a power law sizing exponent of 0.6, is

(a) \$325 million.

(b) \$350 million.

(c) \$250 million.

(d) \$300 million.

12. The construction cost of a 10-story building was \$15 million in 1995. If the construction price index has risen from 150.8 in 1995 to 178.2 in 2005, an estimate for a 12-story building in 2005 should be

(a) Between \$15 and \$16.5 million.

(b) Between \$16.5 and \$17.7 million.

(c) No less than \$17.7 million.

(d) None of the above.

13. A small engineering firm operates with cash (\$2 million) from a credit facility that charges 2.5% interest per quarter and a mortgage (\$1.5 million) that charges 6% per year. The interest rate the firm should use when performing engineering economy studies should

(a) Range between 2% and 3%.

(b) Range between 3% and 4%.

(c) Be at least 4%.

(d) None of the above.

14. The life-cycle costs for an investment include

(a) Operating and maintenance costs.

(b) Salvage values or remediation costs.

(c) Investment costs.

(d) All of the above.

15. When analyzing an investment,

(a) All life-cycle costs should be considered.

(b) Only investment costs need to be estimated.

(c) Salvage values or remediation costs can be ignored.

(d) Overhead costs are irrelevant.

16. A 100,000-square-foot distribution facility cost \$45 per square foot to construct in 2003. Inflation has driven prices up 4.5% over time. An estimate of the cost of a 125,000-square-foot facility in a similar location in 2006, assuming an exponent of 0.8 in the power law sizing model, would be

(a) \$5.4 million.

(b) \$6.1 million.

(c) \$5.1 million.

(d) \$7.5 million.

17. An industrial motor with a 200-horsepower rating was purchased for \$2200 in 2000. If the cost index was 162.5 in 2000 and 172.3 in 2005, and cost trends are expected to continue, an estimate for the cost of a similar motor in 2007 would most closely be

(a) $2850.

(b) $3270.

(c) $2390.

(d) $2500.

18. Indirect costs

(a) Are generally allocated to projects according to some measure, such as direct labor hours.

(b) Are often called overhead costs.

(c) Cannot easily be attributed to any single product or process.

(d) All of the above.

19. A 100,000-square-foot distribution facility cost $45 per square foot to construct in 2003 in California. If the national construction cost index has risen from 180.5 in 2003 to 190.6 in 2006, adjustments for California are 1.3 times the national average and for Arkansas are 0.85 the national average, then an estimate for a similar facility in Arkansas in 2006, assuming that current trends continue, should be

(a) $3.1 million.

(b) $3.7 million.

(c) $4.8 million.

(d) Over $5 million.

20. Sunk costs are

(a) Irrelevant for analysis.

(b) Costs that have already been incurred.

(c) To be analyzed in time-zero dollars.

(d) Both (a) and (b).

8 Developing After-Tax Cash Flows

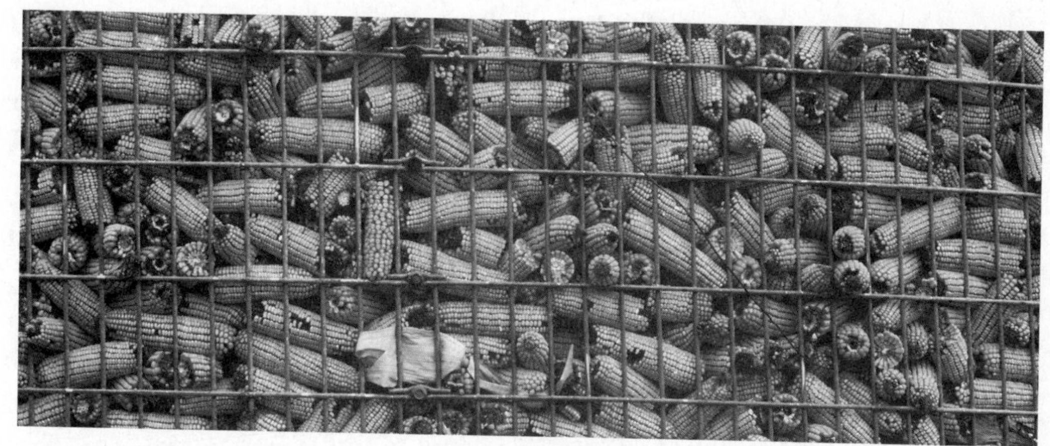

(*Courtesy of CORBIS.*)

Real Decisions: Cash Crop

Nordic Biofuels, a subsidiary of Nordic Energy Group, is building the first ethanol plant in Ohio. Ethanol, a gasoline additive made from corn, reduces air pollution. The facility is expected to cost $50 million and will produce 50 million gallons of ethanol annually, using 18.8 million bushels of corn from the state. The plant should be operational in 2006. In 2002, the governor of Ohio signed legislation to entice ethanol plant investments by offering tax breaks and making loans available.[1]

Make the following assumptions: Ethanol generates revenues of $1.90 per gallon and a bushel of corn costs $2.80. These are time-zero costs (at the end of 2005). The price of ethanol is expected to rise 2.5% annually over the project life, and corn prices are expected to remain steady. Time-zero energy costs are estimated at $15 million per year, with

[1] "Nordic Energy to Build Ohio's First Ethanol Plant," *Dow Jones Newswires,* February 3, 2004.

an annual inflation rate of 3.5%. Labor and overhead costs are estimated at $20 million, inflating at a rate of 2% per year. Finally, assume that working capital of $7.5 million (used for the procurement of inventory over time) is received at the beginning of the project and returned in the final period.

The facility is to be depreciated under the modified accelerated cost recovery system's (MACRS) maximum accelerated depreciation rules over a 10-year recovery period. The investment is to be analyzed over a 10-year horizon, at the end of which the facility will have a salvage value of $5 million. The federal income tax rate is 35%, but due to government incentives, there are no local or state taxes.

Develop after-tax cash flows for the following three scenarios:

1. The investment is made from retained earnings (a cash payment).
2. The investment is made with an investment credit such that $20 million is automatically depreciated in the first year and the remaining $30 million is depreciated normally under (MACRS).
3. The investment is made with a $30 million loan and cash. (There is no investment credit.) The loan is to be repaid over five years in equal total annual payments, assuming a 6.5% annual interest rate.

In addition to computing these after-tax cash flows, after studying this chapter you will be able to:

- Convey the importance of taxes in capital investment analysis. (Section 8.1)
- Use marginal federal and local tax rates or an effective tax rate to compute taxes on ordinary income. (Section 8.2)
- Compute periodic depreciation charges and the book value of an asset, assuming straight-line, declining-balance, or declining-balance switching to straight-line depreciation methods. (Section 8.3)
- Compute periodic depreciation charges in accordance with current U.S. tax law under MACRS. (Section 8.3)
- Select the depreciation method that maximizes worth. (Section 8.3)
- Compute periodic depreciation charges with historical methods, such as the sum-of-years'-digits and units-of-production methods. (Section 8.4)
- Understand adjustments to depreciation and tax calculations when an asset is retired or replaced. (Section 8.5)
- Compute taxes on capital gains. (Section 8.5)

- Compute depletion charges for natural-resource (mines and timber) investments. (Section 8.6)
- Adjust tax calculations for losses, amortization, and investment credits. (Section 8.6)
- Compute after-tax cash flows in current dollars for various investment scenarios. (Section 8.6)

The government serves the public in a number of ways. The funding required to perform these services comes from taxation. There are a number of taxes that affect people and companies alike. Purchases may be subject to sales taxes. Personal income is taxed. Profits from businesses, whether large corporations or individual contractors, are also subject to taxation. Property owners often pay taxes based on the worth of their property.

We focus on the taxes paid by corporations that affect investments in engineering projects. The estimation techniques shown in the previous chapter are for estimating before-tax cash flows. In this chapter, we convert these before-tax cash flows into after-tax cash flows.

While we try to maintain a global perspective in this textbook, taxes and tax policies are clearly a regional issue, as each government entity in each country operates differently throughout the world. In fact, a single country may have various rules at the provincial, state, or local levels. Thus, our focus will be on the U.S. federal government. Although the actual tax rates and rules differ greatly according to the country, topics such as depreciation and taxation are universal. Thus, while the work here focuses on the U.S. government taxing U.S. firms, this information can be used as a template to analyze other situations.

8.1 Taxation

In this chapter, we focus on taxes, and their implications, associated with capital investment decisions. Specifically, we address taxes paid on the following:

1. Ordinary income.
2. Gain or loss from the sale of depreciable property.
3. Capital gains.

To motivate our discussion of why we study taxes, consider United Parcel Service's (UPS) 2004 annual report.[2] The worldwide provider of logistics was chosen because it owns and operates numerous facilities and numerous pieces of equipment, including trucks, trailers, tractors, and airplanes. In 2004, UPS spent \$2.127 billion on capital expenditures. As of the end of 2004, the book value of the company's assets, including vehicles, aircraft, land, buildings, equipment, and technology, was almost \$14 billion, the result of nearly \$27 billion of investment and \$13 billion of depreciation over time. (The terms "depreciation" and "book value" will be defined subsequently in this chapter, but they relate to the accounting value of assets over time.)

One does not need to be an accountant to see the impact that taxes can have on an entity. In 2004, UPS reported an operating profit of \$4.989 billion.

[2] United Parcel Service of America, "Selected Financial Data," from *2004 Annual Report*, www.shareholder.com/ups/, 2004.

From this profit, the company paid $1.589 billion in taxes, resulting in net income of $3.333 billion. One expense that reduced the amount of taxes to be paid was $1.543 billion in depreciation and amortization. These numbers are significant and motivate the development of after-tax cash flows. Nearly all companies engage in engineering activities that require an investment in assets. The size, type, and timing of such investments greatly affect the amount of taxes paid. Thus, we must take these issues into account when considering investments at the project level.

8.2 Taxes on Ordinary Income

The first tax that we examine is charged when profits are made from "ordinary" operations. Ordinary operations are truly what they say: operations that you would expect a company to perform. When IBM sells mainframe computers, Dow sells chemicals, Lockheed Martin launches commercial satellites, or Pratt & Whitney sells aircraft engines, they are generating revenues from sales in their fields of expertise. Contrast this with ConocoPhillips selling its oil refinery assets in Colorado[3] or Corning selling its television picture-tube production equipment.[4] ConocoPhillips is an energy company that derives most of its income from recovering and distributing oil products. Corning makes a variety of materials and products for a variety of industries. Making money from the sale of assets is not "ordinary" for ConocoPhillips or Corning. Such profits are made outside of normal operations. We deal with these instances later in the chapter.

The concept of profit is actually quite straightforward. Operations generate revenues at some expense, or cost. **Profit** in a given period is the difference between revenues and the costs incurred in generating those revenues, assuming that the revenues are greater than the costs. If the costs exceed the revenues, then the firm suffers a **loss**. The government taxes profits (not losses), also referred to as income. We define income—or EBIT, for "earnings before income taxes," in period n as

$$\underbrace{\text{EBIT}_n}_{\text{Income}} = \underbrace{R_n}_{\text{Revenue}} - \underbrace{E_n}_{\text{Expenses}}.$$

For example, DaimlerChrysler receives revenue from every customer who purchases a car. Expenses are a bit more involved: The costs that go into making cars include materials (steel, rubber, plastic, etc.), labor (production workers, engineers, managers, etc.), logistics (the movement of goods), and support (marketing, etc.).

---

[3] Parker, L., "Smaller Oil Refiners Snap Up Assets Jettisoned by Bigger Firms," *The Wall Street Journal*, May 22, 2003.

[4] "Corning Sells Equipment to China Co for $45M," *Dow Jones Newswires*, February 26, 2004.

TABLE 8.1

U.S. federal tax rates (2006) for ordinary income for corporations.

Income Level	Tax Rate
$ 1 to $ 50,000	15%
$ 50,001 to $ 75,000	25%
$ 75,001 to $ 100,000	34%
$ 100,001 to $ 335,000	39%
$ 335,001 to $10,000,000	34%
$10,000,001 to $15,000,000	35%
$15,000,001 to $18,333,333	38%
Over $18,333,333	35%

Source: Internal Revenue Code, Title 26, Subtitle A, Chapter 1, Subchapter A, Part II, Section 11.

In addition to these operating costs, fixed costs must be taken into account. These costs include the purchase of equipment and the building of plants. Unfortunately, spending millions of dollars on a plant does not translate directly into an expense for the year in which the plant was built. Rather, strict rules define how such a purchase is expensed over time, termed **depreciation**. We discuss depreciation further, after we specify corporate tax rates.

8.2.1 Federal Tax Rates

The United States taxes ordinary income on an increasing scale, as depicted in Table 8.1. Note that this table was current for the year 2006 and is subject to change. The rate that a company pays is often referred to as the company's "tax bracket."

The tax table is not to be read as a single rate. That is, if a company earns $250,000 in one year, it does not pay 39% on those earnings. Rather, it pays 15% on the first $50,000, 25% on the next $25,000, 34% on the next $25,000, and 39% on the remaining $150,000. Thus, the rates published are *marginal* rates. We illustrate with an example.

EXAMPLE 8.1

Ordinary Income Tax

National Instruments Corporation, an engineering firm that specializes in virtual instrumentation with software products such as LabVIEW, reported income before taxes of $44.491 million in 2003.[5] Use Table 8.1 to determine the amount of taxes to be paid.

[5] National Instruments Corporation, "Form 10-K," United States Securities and Exchange Commission Filing, www.ni.com/nati/annual/, December 31, 2003.

Solution. Using Table 8.1, we compute the taxes owed as

$$\$50{,}000(0.15) + \$25{,}000(0.25) + \$25{,}000(0.34) + \$235{,}000(0.39) + \$9.665M(0.34) + \$5.0M(0.35) + \$3.33M(0.38) + (\$44.491M - \$18.333M)(0.35) = \$15.572 \text{ million.}$$

This is the total amount of taxes National Instruments owes to the federal government.

Examining the rates may lead one to think that they are strange, in that the marginal rates increase from 15% to 39% for the first $335,000 of income, only to drop to 34% and again repeat the process of increasing between $335,000 and $18.3 million to 38%, before dropping to the final rate of 35%. These rates were not created haphazardly and are not the product of Congress bickering over nickels and dimes. (Well, they actually may be, but there *is* a rationale to the movement in the rates.) Rather, the marginal rates are constructed so that most corporations do not need to use the table to compute their federal tax rates.

The effect of the increasing tax rate is this: If a company earns over $335,000 (but less than $10 million), it pays a flat rate of 34%. To illustrate, assume that a certain company earns $100,000, but instead of using the marginal rates for each level of income, it computes its tax liability (amount owed) with a flat rate of 34%. This totals $34,000 in payable taxes. The amount the company has overpaid is equal to the difference of the $34,000 and the amount computed from Table 8.1, or

$$\underbrace{0.34(\$100{,}000)}_{\substack{\text{Taxes Paid with} \\ \text{34\% Flat Rate}}} - \underbrace{(0.15(\$50{,}000) + 0.25(\$25{,}000) + 0.34(\$25{,}000))}_{\text{Taxes Paid from Table}}$$
$$= \$11{,}750.$$

It is no coincidence that the value of the 5% premium (the difference between 39% and 34%) paid on earnings between $100,000 and $335,000 is

$$(0.39 - 0.34)(\$335{,}000 - \$100{,}000) = \$11{,}750.$$

So, if a company earns over $335,000, but less than $10 million, it pays a flat tax rate of 34%. Now, recall that if the company pays a flat rate of 34% on the first $100,000, it would overpay by $11,750. If it continues to pay a flat rate of 34% on income earned between $100,001 and $335,000, it would underpay by $11,750. Thus, the overpayment and underpayment cancel, and the company would now be "on schedule." For income between $335,001 and $10 million, the rate is 34%. Hence, if a company earns between $335,001 and $10 million, it pays a flat tax rate of 34% and does not need to use the table to compute its tax liability.

But the flat rate applies only to income in excess of $335,000. If a company earns $335,000 or less, it will achieve savings by using the tax table as opposed to using a flat rate of 34%. This is because the $11,750 that is "lost" from using the flat rate for up to $100,000 in income is not gained back in its entirety until the threshold of $335,000 has been reached.

This entire process repeats itself at the $10 million-dollar level as

$$(0.38 - 0.35)(\$18{,}333{,}333 - \$15{,}000{,}000) = \$100{,}000.$$

This is exactly 1% of $10 million, which explains the difference in moving from the 34% bracket to the 35% bracket. Thus, if a company earns over $18.3 million, its tax rate is a flat 35%. In Example 8.1, we could have arrived directly at the value of $15.572 million by taking $44.491 million times 35%. Most companies fall into brackets whereby they pay either 34% or 35%.

8.2.2 Effective Tax Rates

In engineering economy studies, the analysis would be cleaner if we could deal with one tax rate. This is not a trivial issue, as a corporation may face taxes from a number of entities, including federal, state, and local governments. The first issue to deal with is the federal rate, since most taxes are paid to the federal government. We have already illustrated that if a company earns above $335,000, but less than $10 million, then it has a flat tax rate of 34%. Similarly, if the company earns in excess of $18,333,333, then it has a flat rate of 35%. Companies that earn money outside of these ranges can calculate a flat rate merely by using Table 8.1. To do so, they calculate their tax liability for the year from the table and then take the ratio of the taxes paid to their ordinary income. We illustrate this process in the next example.

EXAMPLE 8.2

Flat Corporate Income Rate

Software developer and service provider ILOG announced earnings before income taxes of $2.785 million in 2004.[6] Calculate ILOG's federal income tax rate from the schedule in Table 8.1.

Solution. We know from the previous section that the rate paid should be 34%, and we verify that here. From the table, we compute the following amount of taxes to be paid:

$$\$50{,}000(0.15) + \$25{,}000(0.25) + \$25{,}000(0.34) + \$235{,}000(0.39)$$
$$+ (\$2.785\text{M} - \$335{,}000)(0.34) = \$946{,}900.$$

[6] ILOG, S.A., "Fiscal 2004 20-F Annual Report," United States Securities and Exchange Commission Filing, www.ilog.com/corporate/profile/, October, 2004.

Thus, the percentage paid is

$$t_f = \frac{\$946{,}900}{\$2.785\text{M}} = 0.34 = 34\%.$$

The ratio of taxes paid to earnings before taxes is the flat corporate tax rate.

As we have noted, the federal tax rate is only one piece of the tax equation that companies face. In the United States, a company may also have to pay state taxes and/or local (county, district, borough, and/or city) taxes. States such as Nevada, Washington, and Wyoming did not charge taxes in 2005; Michigan, South Dakota, and Texas imposed small fees on corporations; and Iowa, Massachusetts, Minnesota, Pennsylvania, Vermont, and the District of Columbia had rates that reached over 9.5% for certain income levels.[7] Thus, it is rare that an analysis using only a flat federal income tax rate is adequate.

To be precise, one must compute revenues and expenses and calculate each layer of tax individually. Doing this, however, can be very tedious, and it provides information that is far more detailed than is required for most engineering economy studies. Generally, it is sufficient to utilize an **effective tax rate**—a single rate used to compute taxes on the basis of profits. The rate combines all federal, state, and local taxes into one value. To compute an effective rate, one must be aware that taxes paid at the local and state levels are considered expenses at the federal level. Since the taxes paid at all levels are interrelated, computing the effective tax rate is not as simple as adding the individual income tax rates together.

Assume that a company's gross income in period n is EBIT_n and the income taxes paid at the federal level are $t_f\text{EBIT}_n$, where t_f is the flat federal rate that we computed in Example 8.2. This value of $t_f\text{EBIT}_n$ is generally treated as an expense at the local and state levels. Thus, to compute the tax to be paid at the state and local levels, the gross income assumed is $\text{EBIT}_n - t_f\text{EBIT}_n$, or $(1 - t_f)\text{EBIT}_n$. Hence, the local tax paid is $t_l(1 - t_f)\text{EBIT}_n$, and the state tax paid is $t_s(1 - t_f)\text{EBIT}_n$. This leads to a total tax payment of

$$t_f\text{EBIT}_n + t_s(1 - t_f)\text{EBIT}_n + t_l(1 - t_f)\text{EBIT}_n,$$

or

$$\left(t_f + t_s(1 - t_f) + t_l(1 - t_f)\right)\text{EBIT}_n.$$

Given our income of EBIT_n and the tax rates at the federal (t_f), state (t_s), and local (t_l) levels, we can define our effective tax rate, as

$$t = \left(t_f + t_s(1 - t_f) + t_l(1 - t_f)\right).$$

[7] Federation of Tax Administrators, "Range of State Corporate Income Tax Rates," www.taxadmin.org/fta/rate/corp_inc.html, 2005.

Note that we are not adding the local and state rates to the federal rate. Rather, the local and state rates are *discounted*, because the taxes paid at the federal level are treated as expenses at the local and state levels. We illustrate calculating the effective tax rate in the next example.

EXAMPLE 8.3

Effective Tax Rate

Continuing our previous example, we note that ILOG is located in California. Assume that the company pays its taxes there at a rate of 8.84%, as it was in 2004. Assume further that ILOG pays 0.5% in local taxes. What is ILOG's effective tax rate?

Solution. As previously computed, ILOG's federal tax rate is 34%. We calculate the effective rate, then, as

$$t = t_f + t_s(1 - t_f) + t_l(1 - t_f) = 0.34 + (0.084)(1 - 0.34) + (0.005)(1 - 0.34)$$
$$= 0.3987 = 39.87\%.$$

Thus, the effective tax rate is about 6 percent higher than the federal tax rate, but considerably less than the sum of the federal, state, and local rates (43.34%).

In this textbook, unless specified otherwise, we assume that the tax rate given is an effective tax rate, so that it does not need to be adjusted for purposes of analysis. We now turn our attention to computing taxable income.

8.3 Depreciation and Book Value

Most expenses that a company incurs are straightforward. When building a product, materials are purchased, laborers and managers are paid, and energy is consumed. These costs are recorded as they are incurred and are subtracted from the revenues that are generated from the company's efforts, defining taxable income.

However, one expense is not as straightforward. Consider the purchase of an asset, such as a piece of machinery. While the "expense" of paying for the piece of equipment occurs at one point in time (when the supplier of the equipment is paid), it is not deemed an expense in the eyes of the government. Rather, the asset is expensed over time according to a predefined set of rules. Whenever a *capital asset* is purchased, it is expensed in this manner for the purposes of computing taxable income. This annual expense is termed **depreciation**. The idea is that capital assets are utilized over a number of years, and therefore the expense of purchasing a capital asset should be spread over the years it is utilized in generating ordinary income.

To reiterate, computing the taxable expense associated with the purchase of a capital asset is not simply a matter of adding the purchase price to a list of expenses and computing profit by subtracting these expenses from revenues. Rather, because it is assumed that capital assets such as machinery and plants will be used for a period greater than one year, the purchase price is "expensed" over time. That is, each year after the purchase of a capital asset, some amount of the purchase price can be written down as an expense for the purpose of computing taxable income. This amount is termed *depreciation*.

The amount of depreciation that a company can claim in a period is generally not subject to debate. Rather, given the definition of the asset and the depreciation rules, it is computed in accordance with strict guidelines. If D_n denotes the depreciation amount in period n, the **book value** of an asset is defined as its initial book value (purchase price and installation costs), less any depreciation charged thus far. With the book value after period n defined as B_n, we write

$$B_n = B_{n-1} - D_n.$$

The initial book value, B_0, is the purchase price of the asset, plus any expenses incurred in its procurement and installation. Thus, if a machine were purchased for \$450,000 and it cost \$2000 to deliver and install, then $B_0 = \$452,000$, and this amount is depreciated over time.

To compute the annual depreciation charges D_n, we must know (1) the **recovery period**, which is the number of years over which we can depreciate the asset, and (2) the **depreciation method**. In the next two subsections, we define the traditional straight-line and declining-balance methods of depreciation. These methods were used in deriving the current standards of practice that are operative in the United States. Other historical depreciation methods are discussed later.

8.3.1 Straight-Line Depreciation

The straight-line method of depreciation assumes that the asset is expensed an equal amount in each year over its recovery period. This results in a book value that decreases an equal amount each period.

Assume that the purchase cost of an asset is P. This value includes the purchase price and all procurement costs, such as delivery, taxes, and installation, as described earlier. We define the recovery period as N_D and the salvage value of the asset at the end of the recovery period as S. The amount $P - S$ is to be depreciated over N_D periods. With these definitions, the periodic depreciation is

$$D_n = D = \frac{P - S}{N_D}. \tag{8.1}$$

The initial book value is $B_0 = P$. The book value in subsequent periods is defined by a decrease in D in each period, such that

$$B_1 = B_0 - D = P - D$$

and

$$B_2 = B_1 - D = P - 2D.$$

Thus, in any period,

$$B_n = P - nD \tag{8.2}$$

and

$$B_{N_D} = P - N_D D = S. \tag{8.3}$$

We illustrate the preceding computations in the next example.

EXAMPLE 8.4

Straight-Line Depreciation

Brazilian oil company Petrobras ordered the construction and installation of a semisubmersible offshore oil production platform for use in the Roncador oil field. The rig is to be built by Technip SA of France for \$775 million.[8] Assume that the rig is delivered and installed on January 1, 2006. If it has a useful life of eight years with no salvage value, determine the depreciation schedule and book value over the recovery period, assuming straight-line depreciation.

Solution. Given $P = \$775$ million, $S = 0$, and $N_D = 8$, we compute the annual depreciation using the straight-line method with Equation (8.1) as

$$D = \frac{P - S}{N_D} = \frac{\$775M}{8} = \$96{,}875{,}000.$$

This amount can also be obtained from the Excel function SLN, as shown in the spreadsheet in Figure 8.1. The depreciation charges are calculated as

=SLN(cost,salvage,life)

	A	B	C	D	E	F	G
1	Example 8.4: Offshore Oil Platform Depreciation				**Input**		
2					P	$775,000,000.00	
3	**Period**	**Depreciation**	**Book Value**		Salvage Value	$0.00	
4	0	--	$775,000,000.00		Periods	8	years
5	1	$96,875,000.00	$678,125,000.00		Method	Straight-Line	
6	2	$96,875,000.00	$581,250,000.00				
7	3	$96,875,000.00	$484,375,000.00		**Output**		
8	4	$96,875,000.00	$387,500,000.00		Schedule		
9	5	$96,875,000.00	$290,625,000.00				
10	6	$96,875,000.00	$193,750,000.00	=C6-B7			
11	7	$96,875,000.00	$96,875,000.00				
12	8	$96,875,000.00	$0.00				
13							
14	=SLN($F$2,$F$3,$F$4)						
15							

Figure 8.1 Depreciation and book values over the recovery period for the offshore oil platform.

[8] Pearson, D., "Technip Wins \$775M Contract for Brazil Offshore Platform," *Dow Jones Newswires*, December 19, 2003.

in column B, while the book value is tracked in column C by merely subtracting the current depreciation charge from the previous book value. Note that the book value at the end of the recovery period is equal to the salvage value of zero.

8.3.2 Declining-Balance Depreciation

The declining-balance method of depreciation assumes that an equal percentage of the value of an asset is expensed in each period. Whereas straight-line depreciation defined a fixed-dollar amount of the book value to be expensed each year, declining-balance depreciation defines a fixed *percentage* of the book value to be expensed each year. Due to its definition, it is often called the "fixed percentage" method.

If we define the percentage of the book value to be depreciated each period as α, then the annual depreciation is

$$D_n = \alpha B_{n-1}.$$

Starting with our initial book value, we can compute

$$D_1 = \alpha B_0 = \alpha P.$$

For the second period,

$$D_2 = \alpha B_1 = \alpha (B_0 - D_1) = \alpha B_0 - \alpha D_1.$$

Note that D_1 is merely αB_0, so

$$D_2 = \alpha B_0 - \alpha (\alpha B_0) = \alpha(1-\alpha) B_0.$$

We can continue this progression to get

$$D_3 = \alpha B_2 = \alpha (B_0 - D_1 - D_2) = \alpha (B_0 - \alpha B_0 - \alpha(1-\alpha) B_0),$$
$$= \alpha B_0 \left(1 - \alpha - \alpha + \alpha^2\right) = \alpha(1-\alpha)^2 B_0.$$

This leads to defining the annual amount of depreciation in two ways:

$$D_n = \alpha B_{n-1} \tag{8.4}$$

or

$$D_n = \alpha(1-\alpha)^{n-1} B_0 = \alpha(1-\alpha)^{n-1} P. \tag{8.5}$$

So the amount of depreciation charged in each period decreases, but the percentage decrease in the book value remains the same. We can similarly derive an expression for the book value. In general,

$$B_n = B_{n-1} - D_n.$$

Substituting our expression for D_n defines

$$B_n = B_{n-1} - \alpha B_{n-1},$$

which leads to the following definitions of the book value:

$$B_n = (1 - \alpha) B_{n-1} \tag{8.6}$$

and

$$B_n = (1 - \alpha)^n B_0 = (1 - \alpha)^n P. \tag{8.7}$$

It is rare that a company would randomly choose an α to perform depreciation calculations. The most common form of declining-balance depreciation is **double-declining-balance depreciation**, where the value of α is defined as

$$\alpha = (2)\frac{100\%}{N_D},$$

with N_D defining the recovery period. The multiplier 2 is what leads to the definition of "double" declining balance. Multiplier values of 1.5 and 1.25 are also used. Note that a value of 1 defines a percentage equal to that of straight-line depreciation. Because values greater than 1 are generally used, this method is often called an **accelerated method of depreciation**, since the depreciation charges are higher in the first period compared with the straight-line method's first-period depreciation charges.

EXAMPLE 8.5

Declining-Balance Depreciation

Israel Electric Corp. ordered an open-recycling, 360-megawatt gas turbine from General Electric for delivery in the summer of 2005 at the cost of $42 million.[9] What are the depreciation charges and resulting book values, assuming a recovery period of 20 years and using double-declining-balance and 1.5-declining-balance depreciation?

Solution. For the case of double-declining-balance depreciation,

$$\alpha = (2)\frac{100\%}{20} = 10\%.$$

For the first year, the depreciation and book value are

$$D_1 = \alpha B_0 = (0.10)(\$42\text{M}) = \$4.2 \text{ million},$$

$$B_1 = (1 - \alpha) B_0 = (0.90)(\$42\text{M}) = \$37.8 \text{ million}.$$

[9] Pearson, D., "Israel Electric to Place $42M Order with GE," *Dow Jones Newswires,* January 4, 2004.

	A	B	C	D	E	F	G
1	Example 8.5: Gas Turbine Depreciation				**Input**		
2					P	$42,000,000.00	
3	**Period**	**Depreciation**	**Book Value**		Salvage Value	$0.00	
4	0	--	$42,000,000.00		Periods	20	years
5	1	$4,200,000.00	$37,800,000.00		Method	Declining-Balance	
6	2	$3,780,000.00	$34,020,000.00		Multiplier	2.0	
7	3	$3,402,000.00	$30,618,000.00				
8	4	$3,061,800.00	$27,556,200.00		**Output**		
9	5	$2,755,620.00	$24,800,580.00		Schedule		
10	6	$2,480,058.00	$22,320,522.00	=C6-B7			
11	7	$2,232,052.20	$20,088,469.80				
12	8	$2,008,846.98	$18,079,622.82				
13	9	$1,807,962.28	$16,271,660.54				
14	10	$1,627,166.05	$14,644,494.48				
15	11	$1,464,449.45	$13,180,045.04				
16	12	$1,318,004.50	$11,862,040.53				
17	13	$1,186,204.05	$10,675,836.48				
18	14	$1,067,583.65	$9,608,252.83				
19	15	$960,825.28	$8,647,427.55				
20	16	$864,742.75	$7,782,684.79				
21	17	$778,268.48	$7,004,416.31				
22	18	$700,441.63	$6,303,974.68				
23	19	$630,397.47	$5,673,577.21				
24	20	$567,357.72	$5,106,219.49				
25							
26	=DDB($F$2,$F$3,$F$4,A22,$F$6)						
27							

Figure 8.2 Depreciation and book values for gas turbine, using double declining balance.

A spreadsheet approach is given in Figure 8.2. The DDB function is used to calculate the declining-balance depreciation charges. Although the term DDB generally refers to double-declining balance, the function

=DDB(purchase,salvage,recovery,period,multiplier)

can be used for any multiplier value. The depreciation charges are calculated in column B with the DDB function, while the book values are calculated in column C. The figure illustrates calculations with the multiplier 2.0, as defined in cell F6.

By changing cell F6 from 2.0 to 1.5, the final argument in the DDB functions in column B change, resulting in new depreciation charges for 1.5-declining balance. For the case of 1.5-declining-balance depreciation,

$$\alpha = (1.5)\frac{100\%}{20} = 7.5\%.$$

For the first year, the depreciation and book value are

$$D_1 = \alpha B_0 = (0.075)(\$42\text{M}) = \$3.15 \text{ million},$$

$$B_1 = (1 - \alpha) B_0 = (0.925)(\$42\text{M}) = \$38.85 \text{ million}.$$

As expected, double-declining balance results in faster depreciation of the gas turbine.

Although not readily obvious from the previous example, there is a problem with declining-balance depreciation: One can never completely recover one's costs with that method. When we say "recover costs," we mean that the asset is never fully depreciated and thus never completely expensed. Examining the spreadsheet in Figure 8.2, we see that the book value of the turbine after 20 years is over $5.1 million for double-declining balance. Thus, the total amount of tax benefits the purchaser is entitled to are not received.

We can illustrate this shortcoming mathematically, but consider the calculation of the annual depreciation charge. We take a percentage of the book value. Then the book value is recomputed, and we take the same percentage of what remains. If we keep taking a percentage of what remains, some value will always remain. Thus, the book value is never driven to zero in N_D periods.

In fact, the book value cannot be driven to any specified salvage value at N_D unless it is altered. For example, if a specific book value is desired after N_D periods, then, with the declining-balance method, depreciation must stop if that value is reached before period N_D. If the value is not reached, then the depreciation calculated in the final period would have to be increased in order to achieve the desired book value. Clearly, neither of these "fixes" is clean. This problem leads us to introduce our next method.

8.3.3 Declining Balance Switching to Straight-Line Depreciation

We have identified a significant fault with declining-balance depreciation: The book value of the asset does not equal its salvage value upon completion of the depreciation charges through the recovery period. To overcome this problem, the two methods already described, declining balance and straight line, can be combined.

The idea is straightforward: Using both methods (straight line and declining balance), compute the depreciation charge at the start of each period and select the charge that is higher. Declining balance will always have a higher depreciation charge in the first year, due to its definition as an accelerated method of depreciation, while straight line will be greater in the later years. Thus, the problem is to determine when the switch occurs.

We do not need to define the new method mathematically, as it is implicit in the earlier definitions. Rather, we will just give an example to explain the switching method. The only difference between this method and using the depreciation charges defined in Equation (8.1), for straight line, and Equations (8.4) and (8.5), for declining balance, is that each period should be treated as "new," in that we begin fresh with a new book value and consider the question of starting from that point with another method.

EXAMPLE 8.6

Declining Balance Switching to Straight-Line Depreciation

Siliconware purchased 10 pieces of semiconductor production equipment from Agilent Technologies in the fall of 2003 for NT$5.452 million each.[10] Determine the depreciation schedule and book value of one asset over its five-year recovery life, using double-declining balance switching to straight-line depreciation. To illustrate the method, assume a salvage value of $200,000 at the end of the recovery period.

Solution. To compute double-declining-balance (henceforth, DDB) depreciation, the value of α is

$$\alpha = (2)\frac{100\%}{5} = 40\%.$$

For the first year, the DDB depreciation is

$$D_1 = \alpha B_0 = (0.40)(\text{NT\$5,452,000.00}) = \text{NT\$2,180,800.00}.$$

For straight-line (henceforth, SL) depreciation, the first year is calculated as

$$D_1 = \frac{P - S}{N_D} = \frac{\text{NT\$5,452,000.00} - \text{NT\$200,000.00}}{5} = \text{NT\$1,050,400.00}.$$

As expected, D_1 is larger for DDB than SL. Thus, we use the depreciation from DDB for the first year. This leads to the following book value at the end of the first year:

$$B_1 = (1 - \alpha)B_0 = (0.60)(\text{NT\$5,452,000.00}) = \text{NT\$3,271,200.00}.$$

For the second year, we continue our DDB calculations to obtain

$$D_2 = \alpha B_1 = (0.40)(\text{NT\$3,271,200.00}) = \text{NT\$1,308,480.00}.$$

For SL, we begin anew and assign the value of B_1 to P. Also, N_D is set to 4, as there are four years remaining in the recovery period. Thus, for SL,

$$D_2 = \frac{\text{NT\$3,271,200.00} - \text{NT\$200,000.00}}{4} = \text{NT\$767,800.00}.$$

Since D_2 is larger for DDB, in year 2 we again depreciate the asset with DDB such that

$$B_2 = (1 - \alpha)^2 B_0 = (0.60)^2(\text{NT\$5,452,000.00}) = \text{NT\$1,962,720.00}.$$

The spreadsheet in Figure 8.3 gives the complete depreciation schedule for the asset. We could have "constructed" a switching depreciation function in Excel with use of the MAX, SLN, and DDB functions, but this has already been accomplished with the function VDB, defined as

=VDB(cost,salvage,life,start_period,end_period,factor,no_switch).

[10] Nystedy, D., "Taiwan Chipmakers Siliconware, TSMC Buy up New Equipment," *Dow Jones Newswires*, November 13, 2003.

	A	B	C	D	E	F	G
1	Example 8.6: Semiconductor Equipment Depreciation				**Input**		
2					P	NTD 5,452,000.00	
3	**Period**	**Depreciation**	**Book Value**		Salvage Value	NTD 200,000.00	
4	0	--	NTD 5,452,000.00		Periods	5	years
5	1	NTD 2,180,800.00	NTD 3,271,200.00		Method	Switching	
6	2	NTD 1,308,480.00	NTD 1,962,720.00		Multiplier	2.0	
7	3	NTD 785,088.00	NTD 1,177,632.00				
8	4	NTD 488,816.00	NTD 688,816.00		**Output**		
9	5	NTD 488,816.00	NTD 200,000.00		Schedule		
10				=C6-B7			
11	=VDB($F$2,$F$3,$F$4,A7,A8,$F$6,FALSE)						
12							

Figure 8.3 DDB switching to SL for semiconductor production equipment purchase.

As highlighted in cell B8 in the spreadsheet, the function calls the purchase price (F2), salvage value (F3), and recovery period (F4). Relative references are used to define the starting and ending period, 3 (A7) and 4 (A8), respectively, for the period-4 calculation. The function call 'FALSE' requests a switch to straight-line depreciation at the optimal switching time. A call of 'TRUE' would result in typical DDB calculations.

The switch from DDB to SL occurs in period 4, and the asset is depreciated under SL for the remainder of the recovery period. The NT$488,816 depreciation charges in years 4 and 5 define a book value of NT$200,000 at the end of five years, as given in the spreadsheet. This was our desired goal, in that we could drive the book value to a specified value, a task that was not possible with the declining-balance method alone.

These three methods of straight-line depreciation, declining-balance depreciation, and declining balance switching to straight-line depreciation, form the basis of current U.S. tax law with respect to depreciation.

8.3.4 MACRS Depreciation

As you can imagine, the government would have a very difficult time processing tax returns if every company used different depreciation methods and recovery periods. Accordingly, the U.S. government has always provided guidelines as to how depreciation charges are to be calculated. Current law, although occasionally altered by various acts of Congress, is a result of the Tax Reform Act of 1986. Among a number of changes, the law reduced the number of allowable depreciation methods. The current system is known as the Modified Accelerated Cost Recovery System, or MACRS. Not surprisingly, the name comes from a modification of the Accelerated Cost Recovery System (ACRS), which was established in 1981.

In order to calculate periodic depreciation, a number of choices must be made and variables must be defined:

1. **Depreciation System.** There are two depreciation systems available: the General Depreciation System, or **GDS**, and the Alternative Depreciation System, or **ADS**. The choice of GDS or ADS defines the allowable

depreciation method and recovery period. GDS is generally preferred because it allows for the use of accelerated depreciation methods, but ADS may be chosen at the discretion of the owner. If the property is used predominantly outside of the United States, the ADS system must be used. Also, if the property listed is used less than 50% for business purposes, the ADS system must be used.

2. **Asset Classification.** Most *tangible* property, including buildings, machinery, vehicles, furniture, and equipment that is owned, is used in business, and has a determinable useful life greater than one year, can be depreciated. Certain *intangible* property, such as software, patents, and goodwill, may also be depreciated. These intangible assets are expensed according to the rules of amortization as discussed later.

   The category of depreciation into which an asset is classified is dependent on whether the asset is defined as personal or real. *Personal property* includes equipment, machinery, vehicles, and furniture. *Real property* consists of permanent property that cannot be (easily) moved, such as facilities, buildings, structures, roads, and bridges.

3. **Initial Basis.** This concept was previously defined as the initial book value, equal to the purchase price plus all delivery and installation costs. This value may be driven down by investment credits, discussed later in the chapter. Note that the *salvage value of an asset is assumed to be zero* for all basis, depreciation, and book-value calculations under current U.S. tax law.

4. **Placed-in-Service Date.** This is the date that defines when depreciation can commence. The placed-in-service date is the time at which an asset is available for service. Note that it is not necessarily the time an asset is first utilized, as the asset may sit idle. Also, the actual time depreciation can commence is governed by the timing convention discussed next.

5. **Timing Convention.** An asset can be depreciated according to midmonth, midquarter, or midyear conventions. The *midmonth convention* applies only to nonresidential real property and residential rental property. Depreciation is assumed to begin the first month the property is placed in service. Under this convention, one-half of the normal depreciation is allowed in the month in which the asset is placed in service or disposed.

   The *midquarter convention* is similar to the midmonth convention, but the asset is assumed to be placed in service in the middle of the quarter. This convention is to be used only if more than 40% of all depreciable assets (measured by their total initial bases) covered by MACRS are installed in the final three months of the tax year. The convention of applying half of the depreciation in the first and last quarters of service also applies here.

   For property that does not comply with the midmonth and midquarter conventions, as most will not, the *half-year convention* applies. Under this convention, it is assumed that all property is placed in service at the midpoint of the year. Thus, one-half year of depreciation is allotted in the first year the asset is placed in service and in the year of its disposal.

TABLE 8.2 **Sampling of class life and recovery periods for various assets.**

Equipment (or Equipment used in)	Recovery Period (N_D)		
	Class Life	GDS	ADS
Over-the-Road Tractors	4	3	4
Light Trucks	4	5	5
Semiconductor-Manufacturing Equipment	5	5	5
Information Systems (Computers)	6	5	5
Trucks (Heavy, Concrete-Mix, Ore), Trailers, Containers	6	5	6
Timber-Cutting, Construction, Onshore Drilling Equipment	6	5	6
Manufacture of Electronic Components, Products, and Systems	6	5	6
Offshore Drilling	7.5	5	7.5
Satellite Space Equipment	8	5	8
Buses, Manufacture of Apparel, Medical, Dental Supplies	9	5	9
Telephone-Switching Equipment, Manufacture of Chemicals	9.5	5	9.5
Office Furniture, Mining Equipment, Railroad Track	10	7	10
Manufacture of Wood Products, Furniture, or Aerospace Products	10	7	10
Waste Reduction and Resource Recovery Plants	10	7	10
Telephone Station, Cable Television (CATV) Distribution, and Satellite Ground Equipment	10	7	10
Telegraph, Ocean Cable, and Satellite Communications (TOCSC) Switching Equipment	10.5	7	10.5
CATV Head-End, Manufacture of Finished Plastic Products	11	7	11
Manufacture of Locomotives	11.5	7	11.5
Air Transport, Manufacture of Motor Vehicles, Ships, Boats, Railcars	12	7	12
Manufacture of Pulp and Paper	13	7	13
Manufacture of Glass Products, Railroad Equipment	14	7	14
Oil and Gas Exploration and Production, Natural-Gas Production	14	7	14
Manufacture of Primary Steel Mill Products	15	7	15
Petroleum Refining	16	10	16
TOCSC Control Equipment	16.5	10	16.5
Water Transportation, Telephone Central Office Equipment	18	10	18
Electric Utility Production (Turbine or Nuclear), Manufacture of Cement	20	15	20
Equipment Used in The Manufacture of . . .			
Steam and Electric Generation/Distribution, Pipeline Transport, LNG Plant	22	15	22
Wastewater Treatment Plant, Telephone Distribution Plant	24	15	24
TOCSC Cable and Long-line Systems	26.5	20	26.5
Electricity Transmission and Distribution	30	20	30
Gas Utility Distribution	35	20	35
Water Utilities	50	20	50

Source: Internal Revenue Service, "Table of Class Lives and Recovery Periods," Appendix B, Publication 946, pp. 92–102.

6. **Recovery Period.** The recovery period is the number of years over which an asset is depreciated. This length of time depends on the type of asset and the depreciation system chosen. Table 8.2 provides a sample of recovery periods from the Table of Class Lives and Recovery Periods produced by the Internal Revenue Service. Note that three periods of life are given in the table: (1) class life, (2) the GDS recovery period, and (3) the ADS recovery period. The class life recovery period is used for assets placed in service before 1981, as discussed later in the chapter. Note that the GDS and ADS recovery periods rarely coincide.
7. **Depreciation Method.** Once the system has been chosen, a depreciation method must be selected. For GDS, the choices include 200% declining balance switching to straight line, 150% declining balance switching to straight line, or straight-line depreciation. These choices are restricted according to the predefined recovery period. If ADS is chosen, only straight-line depreciation can be used over the predefined recovery period. Again, note that all depreciation methods assume that the asset has no salvage value at the end of the recovery period.

Once these choices have been made, the depreciation charge in each period can be calculated from equations presented earlier or from tables presented by the government (Tables 8.3 and 8.4). Reading the tables, one must merely determine the recovery period of the asset and its current year of use. This defines the percentage of the *initial* book value that is taken as the depreciation charge for that year.

We do not have tables for real property, because the choices are limited. Residential rental property is depreciated with the straight-line

TABLE 8.3

MACRS depreciation schedules, assuming GDS and 200% declining balance switching to straight line with the half-year convention.

Year	3-Year	5-Year	7-Year	10-Year
1	33.33%	20.00%	14.29%	10.00%
2	44.45%	32.00%	24.49%	18.00%
3	14.81%	19.20%	17.49%	14.40%
4	7.41%	11.52%	12.49%	11.52%
5	—	11.52%	8.93%	9.22%
6	—	5.76%	8.92%	7.37%
7	—	—	8.93%	6.55%
8	—	—	4.46%	6.55%
9	—	—	—	6.56%
10	—	—	—	6.55%
11	—	—	—	3.28%

Source: Internal Revenue Service, "Instructions for Form 4562," Table A, p. 10, 2003.

TABLE 8.4

MACRS depreciation schedules, assuming GDS and 150% declining balance switching to straight line with the half-year convention.

Year	5-Year	7-Year	10-Year	12-year	15-Year	20-Year
1	15.00%	10.71%	7.50%	6.25%	5.00%	3.750%
2	25.50%	19.13%	13.88%	11.72%	9.50%	7.219%
3	17.85%	15.03%	11.79%	10.25%	8.55%	6.677%
4	16.66%	12.25%	10.02%	8.97%	7.70%	6.177%
5	16.66%	12.25%	8.74%	7.85%	6.93%	5.713%
6	8.33%	12.25%	8.74%	7.33%	6.23%	5.285%
7	—	12.25%	8.74%	7.33%	5.90%	4.888%
8	—	6.13%	8.74%	7.33%	5.90%	4.522%
9	—	—	8.74%	7.33%	5.91%	4.462%
10	—	—	8.74%	7.33%	5.90%	4.461%
11	—	—	4.73%	7.32%	5.91%	4.462%
12	—	—	—	7.33%	5.90%	4.461%
13	—	—	—	3.66%	5.91%	4.462%
14	—	—	—	—	5.90%	4.461%
15	—	—	—	—	5.91%	4.462%
16	—	—	—	—	2.95%	4.461%
17	—	—	—	—	—	4.462%
18	—	—	—	—	—	4.461%
19	—	—	—	—	—	4.462%
20	—	—	—	—	—	4.461%
21	—	—	—	—	—	2.23%

Source: Internal Revenue Service, "Instructions for Form 4562," Table B, p. 10, 2003.

method, assuming the midmonth convention and a 27.5-year recovery period. Nonresidential real property, including elevators and escalators, is depreciated similarly over a 31.5-year recovery period.

We illustrate how to use the tables in the next example. Note that there are fluctuations in the tables after the switch from the declining-balance method to straight-line depreciation. This is due to round-off errors, because the tables are published only to two or three decimal places. The fluctuations ensure that the asset is 100% depreciated at the end of its recovery period.

EXAMPLE 8.7

Depreciation Charges with GDS

At the end of 2003, Boeing Co. purchased a Cray X1 supercomputer to run structural analyses and computational fluid dynamics codes at its Puget Sound data center.[11] Assume that the purchase price was $30 million. What is the depreciation schedule and resulting book value under MACRS, assuming the GDS system (double-declining-balance tables) and the half-year convention?

[11] Derpinghaus, T., "Cray Inc. Wins Boeing Order for Cray X1 Computer," *Dow Jones Newswires*, October 21, 2003.

	A	B	C	D	E	F	G	H
1	Example 8.7: Supercomputer Depreciation					**Input**		
2						P	$30,000,000.00	
3	**Period**	**MACRS %**	**Depreciation**	**Book Value**		Salvage Value	$0.00	
4	0	--	--	$30,000,000.00		Periods	5	years
5	1	20.00%	$6,000,000.00	$24,000,000.00		Method	MACRS GDS	
6	2	32.00%	$9,600,000.00	$14,400,000.00		Multiplier	2.0	
7	3	19.20%	$5,760,000.00	$8,640,000.00				
8	4	11.52%	$3,456,000.00	$5,184,000.00		**Output**		
9	5	11.52%	$3,456,000.00	$1,728,000.00		Schedule		
10	6	5.76%	$1,728,000.00	$0.00	=C6-B7			
11								
12		=$G$2*B8						

Figure 8.4 MACRS depreciation and resulting book values for computer system over six years.

Solution. From Table 8.2, we see that the recovery period for information systems according to the GDS system for MACRS is five years. The depreciation charge is calculated by multiplying the appropriate MACRS percentage by the initial book value ($30 million). We have a choice of using the 200%-declining balance (Table 8.3) or 150%-declining balance (Table 8.4) switching to straight-line depreciation. Assuming that we want to depreciate our asset as quickly as possible, we will use the double-declining-balance table (Table 8.3). Thus, D_1 is 20% of $30 million, or $6 million.

The spreadsheet in Figure 8.4 gives the depreciation charges and resulting book value for each period over the six years. The depreciation charges are calculated from the percentages in Table 8.3, which were input in column B. Note that the percentages, which add to 100%, are multiplied by B_0 in *each* period, not the current book value.

There are six years of depreciation charges, due to the half-year convention and a five-year recovery period. The book value is zero after the sixth year, as the MACRS percentages are derived from the double-declining-balance switching to straight-line depreciation method assuming zero salvage value.

EXAMPLE 8.8

Depreciation Charges with ADS

Intuitive Surgical, Inc., of California sells a robot, the da Vinci surgical system, to hospitals for $1.2 million to assist in surgery. The precise motions of the robot lead to reduced recovery times for patients.[12] Assume that the hospital is for profit and has elected to depreciate the robot with the ADS system. If the robot has a nine-year recovery period (from Table 8.2), what is the depreciation schedule, assuming the half-year convention and the ADS system?

Solution. The MACRS tables are defined for the GDS system. For the ADS system, we assume straight-line depreciation. The annual depreciation charge is

$$D_n = \frac{\$1.2\text{M}}{9} = \$133{,}333.33.$$

[12] Wysocki, B., Jr., "Robots Assist Heart Surgeons," *The Wall Street Journal Online*, February 26, 2004.

	A	B	C	D	E	F	G
1	Example 8.8: Surgical Robot Depreciation				**Input**		
2					P	$1,200,000.00	
3	**Period**	**Depreciation**	**Book Value**		Salvage Value	$0.00	
4	0	--	$1,200,000.00		Periods	9	years
5	1	$66,666.67	$1,133,333.33		Method	MACRS ADS	
6	2	$133,333.33	$1,000,000.00				
7	3	$133,333.33	$866,666.67		**Output**		
8	4	$133,333.33	$733,333.33		Schedule		
9	5	$133,333.33	$600,000.00				
10	6	$133,333.33	$466,666.67	=C6-B7			
11	7	$133,333.33	$333,333.33				
12	8	$133,333.33	$200,000.00				
13	9	$133,333.33	$66,666.67				
14	10	$66,666.67	$0.00				
15							
16	=SLN($F$2,$F$3,$F$4)						
17							

Figure 8.5 MACRS ADS depreciation and resulting book values for medical robot over 10 years.

This value is halved for the first year, resulting in the final half year of depreciation in year 10.

The complete depreciation schedule is given in the spreadsheet in Figure 8.5. The SLN value is halved in years 1 and 10 according to MACRS rules. Note that if GDS had been elected, then the recovery period would have been shorter.

There is no mystery surrounding Tables 8.3 and 8.4; they are provided by the government so that taxpayers do not have to compute the individual depreciation charges if they do not desire. To illustrate how these tables are derived, consider the MACRS method of depreciation for a five-year class asset. Assume that the GDS system is chosen together with 200%-declining-balance switching to straight-line depreciation with the half-year convention (since the tables assume the half-year convention for this class of asset).

We define the initial book value as 100% in order to derive the percentages in Table 8.3. For double-declining-balance (DDB) depreciation, the value of α is 2(100%)/5, or 40%. The resulting first year of depreciation would yield a deduction of

$$D_1 = 0.40(100\%) = 40\%.$$

However, we must halve this value to 20% due to the half-year convention.

For straight-line (SL) depreciation, the first-year deduction would be

$$D_1 = \frac{100\%}{5} = 20\%.$$

Again, we would halve the first year, resulting in a total of 10%. Thus, we would select the DDB value of 20%, which corresponds to the D_1 value for five-year class assets in Table 8.3.

Now consider year 2. Under DDB,

$$D_2 = 0.40(100\% - 20\%) = 32\%.$$

Due to the half-year convention, there are 4.5 years of depreciation remaining. Under SL,

$$D_2 = \frac{100\% - 20\%}{4.5} = 17.78\%.$$

DDB is chosen as the value of D_2 because 32% is greater than 17.78%. The value chosen corresponds to the value found in Table 8.3.

For year 3, the DDB depreciation is

$$D_3 = 0.40(100\% - 20\% - 32\%) = 19.2\%,$$

while SL gives

$$D_3 = \frac{100\% - 20\% - 32\%}{3.5} = 13.71\%.$$

Again, DDB is chosen and $D_3 = 19.2\%$. For D_4, using DDB results in

$$D_4 = 0.40(100\% - 20\% - 32\% - 19.2\%) = 11.52\%,$$

and using SL yields

$$D_4 = \frac{100\% - 20\% - 32\% - 19.2\%}{2.5} = 11.52\%.$$

Both methods result in the same value here, so we know that we have reached the switching point. Thus, we assign the value of 11.52% to D_4. To make certain that we have chosen correctly, we can compute the value of D_5 with both methods. Under DDB,

$$D_5 = 0.40(100\% - 20\% - 32\% - 19.2\% - 11.52\%) = 6.912\%,$$

and under SL, we have

$$D_5 = \frac{100\% - 20\% - 32\% - 19.2\% - 11.52\%}{1.5} = 11.52\%.$$

As if there were any doubt, the straight-line method is now clearly greater than the declining-balance method. Thus, we know that 11.52% will be the percentage for each remaining period. However, we must note that the final "year" of depreciation (year 6 in this example) is halved due to the half-year convention. Thus, the percentage for year 6 is half of 11.52%, or 5.76%. The values of 20%, 32%, 19.2%, 11.52%, 11.52%, and 5.76% for years 1 through 6, respectively, correspond to those in Table 8.3.

Note that the percentages in Table 8.3 are valid only if an asset is purchased and retained for all $N_D + 1$ years of its recovery period. If an asset is disposed of (sold) early, then the percentages may change. For an N_D-year class asset, disposal in years 2 through N_D results in the amount of depreciation being halved in the year of sale, as defined by the half-year convention. If the asset is sold after one year, then the depreciation charge is the same, as it is already halved according to the MACRS convention. Likewise, if the asset is sold in year $N_D + 1$, the depreciation is the same as in the table, since it is already halved by convention. We illustrate this calculation in the next example.

EXAMPLE 8.9

Depreciation Calculation with Early Asset Disposal

The West Basin Water Recycling Facility in El Segundo, California, placed an order for UV water treatment systems with Trojan Technologies of Canada for installation in the summer of 2004. The deal was valued at C$2.7 million. The system treats 47 million liters of water each day.[13] Assume that the asset (classified as a municipal wastewater treatment plant) is to be depreciated under the MACRS GDS system with a 15-year life (Table 8.4 percentages). Assume further that the cost of the system is $1,987,200 (U.S. dollars) and that it is sold after the fifth year of use. What is the depreciation schedule over the five-year lifetime of the asset, and what is its final book value?

Solution. The depreciation schedule for the water-treatment equipment is given in Table 8.5. Note that the percentage depreciation in the fifth year has been halved from the value defined by MACRS (Table 8.4).

TABLE 8.5 **MACRS depreciation with disposal of 15-year asset after 5 years.**

Year n	MACRS %	D_n	B_n
1	5.00%	$99,360.00	$1,887,840.00
2	9.50%	$188,784.00	$1,699,056.00
3	8.55%	$169,905.60	$1,529,150.40
4	7.70%	$153,014.40	$1,376,136.00
5	3.465%	$68,856.48	$1,307,279.52

The five years of depreciation, with the early asset sale, define a book value just over $1.3 million. Note that the depreciation would have been halved if the asset had been disposed of in any year between 2 and 15, inclusive.

8.3.5 Choosing a Depreciation Method

MACRS restricts the number of options available to a company for depreciating assets. However, there are still choices to be made. For one, either

[13] Moritsugu, J., "Trojan Technologies Gets C$2.7M California Pact," *Dow Jones Newswires,* October 14, 2003.

a declining balance switching to straight-line method or a pure straight-line method can be chosen. Further, it is possible to choose between 150% and 200% declining-balance methods for shorter lived assets.

Earlier, we noted that declining-balance methods are often referred to as "accelerated" methods of depreciation. This is because, according to these methods, the depreciation charges are higher early in the life of the asset compared with straight-line depreciation charges. Accelerated depreciation is significant because a higher depreciation charge means a higher expense. Although it sounds strange, a higher expense from depreciation is desirable because it lowers earnings before income taxes (EBIT). This means that the amount of tax to be paid is lowered. Because taxes are lower early in the life of an asset, the equivalent present value of savings is higher in terms of the time value of money. Thus, accelerated depreciation methods are generally preferred.

If it sounds confusing that it would be beneficial to have higher expenses (costs), there are two critical factors to understand. First, *depreciation is an expense, but it is not a cash flow*. Remember that depreciation is merely an accounting formality in order to "expense" capital equipment. The equipment has already been purchased; thus, from the taxpayer's perspective, the cash flow has already occurred. Consequently, it is beneficial for the taxpayer to get as much depreciation as quickly as possible in order to "recover" that expense, by lowering taxes paid on income. Second, regardless of what depreciation method is used, the same amount of tax is paid over time. However, the *timing* of the tax payments differs according to the depreciation method chosen. As we know, the principle of the time value of money states that money at time zero is worth more than money in the future for a positive interest rate. For this reason, higher depreciation expenses early in the life of an asset are preferred, because they generate greater tax savings earlier, which are worth more due to the time value of money under the assumption of a positive and fixed interest rate. We illustrate the differences with an example.

EXAMPLE 8.10

Comparing Depreciation Methods

In the winter of 2004, steelmaker Posco of South Korea announced that it would increase capacity with a new continuous galvanizing line to produce steel sheets (400,000 metric tons annual capacity) for the automotive industry. The expansion is expected to cost KRW189.9 billion and the line is to be operational by June of 2006.[14] Determine the annual book values of the investment over a 15-year recovery period, using straight-line, declining-balance (2.0 and 1.5), and switching (double-declining-balance to straight-line) depreciation methods. Assume that the salvage value of the line is zero at the end of 15 years and no half-year convention.

[14] Seon-Jin, C., "S Korea Posco to Invest KRW326.7B on Production Lines," *Dow Jones Newswires*, January 6, 2004.

Solution. The depreciation and book value calculations follow as before. Table 8.6 provides the depreciation schedules for straight line (column 2), double-declining balance (column 3), 1.5-declining balance (column 4), and switching (column 5).

TABLE 8.6

Depreciation charges (in billions of KRW) for steel plate production line.

Year n	SL D_n	DDB D_n	1.5DB D_n	Switch D_n
1	12.66	25.32	18.99	25.32
2	12.66	21.94	17.09	21.94
3	12.66	19.02	15.38	19.02
4	12.66	16.48	13.84	16.48
5	12.66	14.28	12.46	14.28
6	12.66	12.38	11.21	12.38
7	12.66	10.73	10.09	10.73
8	12.66	9.30	9.08	9.30
9	12.66	8.06	8.17	8.63
10	12.66	6.98	7.36	8.63
11	12.66	6.05	6.62	8.63
12	12.66	5.25	5.96	8.63
13	12.66	4.55	5.36	8.63
14	12.66	3.94	4.83	8.63
15	12.66	3.42	4.34	8.63
Sum	189.9	167.7	150.8	189.9
P	74.03	88.08	74.04	91.81

We already discussed the problem with using declining-balance methods in that they never completely depreciate the asset. This is clear from the summations at the bottom of Table 8.6, as only the straight-line and switching methods depreciate all 189.9 billion won.

To show that these methods are not economically equivalent, the equivalent present worth P of the depreciation charges is also given in Table 8.6, assuming an interest rate of 15%. The value for the straight-line method is calculated as follows:

$$P = \text{KRW}12.66\text{B}(\overset{P/A,15\%,15}{5.8476}) = \text{KRW}74.03 \text{ billion.}$$

The other present-worth values are found by bringing each individual depreciation charge back to time zero, using the $(^{P/F,i,N})$, and adding the results. This calculation is similar to our spreadsheet methods introduced in Chapter 3 and is easily accomplished with the NPV function.

The important conclusion to be drawn is that the values of P are not the same for any of the methods. Thus, the depreciation methods are not economically equivalent, because the timing of the depreciation charges is different. This difference is exacerbated by the methods that do not completely depreciate an asset. A fair comparison can be made between the straight-line and switching methods, as they both completely depreciate the asset: The equivalent present value with switching is nearly 20 billion won larger. Because the present value is larger, the switching method depreciates the asset faster than the straight-line method. As noted earlier, this leads to greater tax savings in terms of the time value of money and a positive interest rate.

The preceding discussion again explains why the declining-balance methods (including switching) are referred to as accelerated methods of depreciation. This is more easily seen by examining the book values in Table 8.7, which are also summarized in Figure 8.6. The higher depreciation charges lead to a rapid decline in the book value of the asset with the declining-balance methods. (The graph also illustrates the problem with declining-balance depreciation in that the book value approaches, but never reaches, zero.)

TABLE 8.7

Book values (in billions of KRW) for steel plate production line.

Year n	SL B_n	DDB B_n	1.5DB B_n	Switch B_n
1	177.24	164.58	170.91	164.58
2	164.58	142.64	153.82	142.64
3	151.92	123.62	138.44	123.62
4	139.26	107.14	124.59	107.14
5	126.60	92.85	112.13	92.85
6	113.94	80.47	100.92	80.47
7	101.28	69.74	90.83	69.74
8	88.62	60.44	81.75	60.44
9	75.96	52.38	73.57	51.81
10	63.30	45.40	66.21	43.17
11	50.64	39.35	59.59	34.54
12	37.98	34.10	53.63	25.90
13	25.32	29.55	48.27	17.27
14	12.66	25.61	43.44	8.63
15	0.00	22.20	39.10	0.00

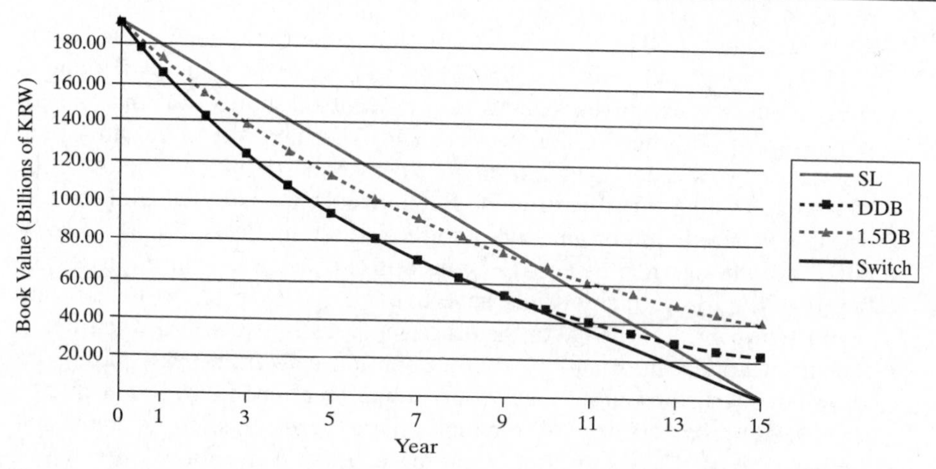

Figure 8.6 Decrease in book value with age of the asset.

Thus, if the goal is to select the depreciation method that provides the greatest equivalent present value of the depreciation charges (assuming a 15% interest rate for discounting), then the switching method with double-declining balance should be chosen.

This example is typical in that accelerated methods will always lead to greater tax savings compared with the straight-line method in terms of the time value of money. This is reflected in the fact that the equivalent present values of the depreciation charges are higher for switching methods compared with straight-line methods. Still, companies do have reasons for using straight-line methods, such as wanting to "even" out earnings, which clearly fluctuate with accelerated depreciation methods. There are also instances when straight-line depreciation must be used, such as with assets utilized outside of the United States. (In this case, ADS also defines longer recovery periods, which leads to even smaller depreciation charges.) Finally, the superiority of accelerated methods is dependent on assumptions of a constant tax rate, a constant discount rate, and the firm being profitable over the life of the asset. In general, however, the switching techniques are preferred.

8.4 Historical Methods of Depreciation

The MACRS system is designed for assets placed in service after 1986. For assets placed in service earlier, the depreciation rules are subject to the **Accelerated Cost Recovery System** (ACRS), which came into being in 1981. The differences between ACRS and MACRS are relatively minimal, with rules differing on the amount of depreciation received in the year an early asset is disposed of and in the construction of the depreciation tables. In principle, the two methods are quite similar.

For assets placed in service before 1981, the method of depreciation could be chosen from straight line, declining balance, declining balance switching to straight line, or sum of years' digits (to be discussed shortly). At that time, it was assumed that the decision maker could determine the useful life of the asset. If not, it was defined according to the asset depreciation range (ADR), which was a listing of recommended asset lives published by the U.S. Treasury.

It is generally important to be aware of these methods. Clearly, knowledge is required when assets are being utilized that were placed in service before 1981. (Since many assets have lives of 30 years, this is not uncommon.) However, knowledge of historical methods is also useful when one is seeking historical information from accounting data from years past. Such knowledge is also useful because many other government entities, including local, state, provincial, and foreign national governments, do not have systems such as MACRS, but rather require the use of conventional methods of depreciation.

One historically popular method, **sum of years' digits** (SOYD), is no longer a viable method under MACRS. According to SOYD, the depreciation in a given year is the ratio of the number of years of remaining life of the asset over the sum of the years in the asset's life (hence the name). To illustrate, note that if an asset can be depreciated over N_D years, then the sum of the individual years, or digits, is

$$\sum_{n=1}^{N_D} n = \frac{N_D(N_D+1)}{2}.$$

This value is the base of the ratio for determining depreciation in each year. For the first year of depreciation,

$$D_1 = \frac{N_D}{N_D(N_D+1)/2}(P-S),$$

while for year two,

$$D_2 = \frac{N_D-1}{N_D(N_D+1)/2}(P-S).$$

Generalizing, we obtain

$$D_n = \frac{N_D-n+1}{N_D(N_D+1)/2}(P-S), \quad n = 1,2,\ldots,N_D. \tag{8.8}$$

We will not show the derivation, but the book value can be calculated as

$$B_n = \left(\frac{N_D-n}{N_D}\right)\left(\frac{N_D-n+1}{N_D+1}\right)(P-S)+S. \tag{8.9}$$

The next example illustrates the method.

EXAMPLE 8.11

Sum of Years' Digits

Oceanex purchased a 150-meter ice-class containership for C$58 million in 2003 for delivery in the summer of 2005.[15] Assuming a 20-year recovery period and no salvage value, determine the depreciation schedule and book values for the ship.

Solution. For the 20-year class life, the sum of years' digits is

$$\frac{N_D(N_D+1)}{2} = \frac{(20)(21)}{2} = 210.$$

The first year's depreciation is then calculated as

$$D_1 = \frac{20}{210}\text{C\$58M} = \text{C\$5.523 million}.$$

The depreciation schedule is given in the spreadsheet in Figure 8.7. The depreciation charges are computed with the SYD function

=SYD(purchase,salvage,recovery period,period),

as shown in the figure. The solutions in the spreadsheet are in Canadian dollars.

[15] Moritsugu, J., "Oceanex to Purchase C$58M Container Ship," *Dow Jones Newswires*, December 17, 2003.

	A	B	C	D	E	F	G
1	Example 8.11: Containership Depreciation				**Input**		
2					P	$58,000,000.00	
3	**Period**	**Depreciation**	**Book Value**		Salvage Value	$0.00	
4	0	--	$58,000,000.00		Periods	20	years
5	1	$5,523,809.52	$52,476,190.48		Method	SOYD	
6	2	$5,247,619.05	$47,228,571.43				
7	3	$4,971,428.57	$42,257,142.86		**Output**		
8	4	$4,695,238.10	$37,561,904.76		Schedule		
9	5	$4,419,047.62	$33,142,857.14				
10	6	$4,142,857.14	$29,000,000.00	=C6-B7			
11	7	$3,866,666.67	$25,133,333.33				
12	8	$3,590,476.19	$21,542,857.14				
13	9	$3,314,285.71	$18,228,571.43				
14	10	$3,038,095.24	$15,190,476.19				
15	11	$2,761,904.76	$12,428,571.43				
16	12	$2,485,714.29	$9,942,857.14				
17	13	$2,209,523.81	$7,733,333.33				
18	14	$1,933,333.33	$5,800,000.00				
19	15	$1,657,142.86	$4,142,857.14				
20	16	$1,380,952.38	$2,761,904.76				
21	17	$1,104,761.90	$1,657,142.86				
22	18	$828,571.43	$828,571.43				
23	19	$552,380.95	$276,190.48				
24	20	$276,190.48	$0.00				
25							
26	=SYD($F$2,$F$3,$F$4,A22)						
27							

Figure 8.7 Depreciation schedule and book values using SOYD with a 20-year life.

Note that SOYD is another accelerated method of depreciation. Under straight-line depreciation, the first-period charge would have totaled only C$2.90 million.

Another historical method, called the **units-of-production** method, derives the allowable depreciation according to the asset's usage. The method is quite similar to straight-line depreciation, but instead of determining the amount of depreciation per year, the amount of depreciation per unit of output is determined. Thus, given an estimate of the life of an asset in terms of output, such as total tonnage, parts, or mileage, the depreciation per unit of output is

$$d = \frac{P - S}{N_D(\text{units})}. \tag{8.10}$$

The depreciation charge for a given year is the output for the year times d, as illustrated in the next example.

EXAMPLE 8.12

Units-of-Production Depreciation

In December 2004, Ford announced that it would invest 169 million pounds in its Dagenham engine plant in east London to meet a growing demand for diesel cars in Europe. The investment was expected to increase production by 400,000 units annually.[16] Assume that the investment included a stamping machine purchased for 10 million pounds with a life defined by a total output of 2 million parts and a salvage value of 500,000 pounds when that limit is reached. If the machine produces 150,000 parts in year 1 and 200,000 parts in year 2, what are the depreciation charges in the first two years, assuming units-of-production depreciation?

Solution. Given $P = \text{GBP10M}$, $S = \text{GBP500K}$, and $N_D(\text{units}) = 2$ million parts, we compute the depreciation per unit with Equation (8.10):

$$d = \frac{P - S}{N_D(\text{units})} = \frac{\text{GBP10M} - \text{GBP500K}}{\text{2M parts}} = \text{GBP4.75 per part.}$$

This leads to deprecation charges of GBP715,000 in the first year and GBP950,000 in the second year.

8.5 Taxes on Nonordinary Income

When a company profits by means other than sales, we refer to these profits as nonordinary income. The most common form of nonordinary income comes from selling assets, such as used machinery.

8.5.1 Retiring an Asset

We have already defined the book value of an asset as its initial book value, less any accumulated depreciation. This defines the worth, in an accounting sense, of the asset. Thus, if the asset is sold, the difference between the salvage value received, which represents the market value of the asset, and the book value of the asset at the time of the sale defines a profit or loss on the sale. If the sale results in a profit, taxes must be paid. No taxes are paid on a loss.

We define the gain from the sale of an asset as follows, where a loss is merely a negative gain:

$$S_n - B_n. \tag{8.11}$$

This gain is to be taxed. According to current U.S. tax law, the gain is actually treated as ordinary income. Thus, we can add the value to the profit from normal operations and tax it accordingly. However, we break out this calculation, as the law may change. For example, the government could change the rate at which gains are taxed, which would force the analysis to be computed separately from ordinary income. We illustrate calculating a loss or gain in the next example.

[16] Griffiths, J., "Ford to spend 169m on diesel engine plant," *The Financial Times*, p. 5, December 17, 2004.

EXAMPLE 8.13 *Selling an Asset for a Gain or Loss*

U.S. satellite operator PanAmSat Corp. and Japanese operator JSAT Corp. launched Galaxy 12/Horizons 1 in the fall of 2003. The satellite was manufactured by Boeing to distribute high-definition programming at the cost of about $250 million.[17] Assume that the asset is depreciated under MACRS GDS with a five-year recovery period. Assume further that the asset is sold after its fourth year of operation for either $50 or $60 million. Compute the tax on any gain, assuming a 34% effective tax rate.

Solution. The depreciation schedule and resulting book values for the first four years of operation of the satellite are given in Table 8.8. Note that the depreciation in the fourth year is halved due to the early disposal.

TABLE 8.8 **Depreciation of satellite (five-year asset) over four years.**

Year n	MACRS %	D_n	B_n
1	20.00%	$50.00	$200.00
2	32.00%	$80.00	$120.00
3	19.20%	$48.00	$72.00
4	5.76%	$14.40	$57.60

Since the book value of the asset at the time of sale is $57.6 million, a gain of $2.4 million is realized if the sale price is $60 million. The tax to be paid on this gain is

$$t(S_n - B_n) = (0.34)(\$60\text{M} - \$57.6\text{M}) = \$816{,}000.$$

Thus, the resulting after-tax cash flow of the transaction is the revenue minus the taxes paid, or

$$A_4 = \$60\text{M} - \$816{,}000 = \$59.184 \text{ million.}$$

In the case where the satellite is sold for $50 million, a loss of $7.6 million is realized. (Losses are discussed later in the chapter.)

8.5.2 Replacing an Asset

Replacing an asset is a two-stage process: A current asset is sold and another asset is acquired in its place. If the newly acquired asset performs the same functions as (or functions similar to those of) the sold asset, the government refers to the transaction as a "like-for-like" replacement. In this situation, no taxes are paid on the disposal of the old asset, as described in the previous section. Rather, the residual value of the old asset is transferred to the new asset, and depreciation charges continue.

17 Sheng, E., "New PanAmSat Satellite to Boost High-Definition Capacity," *Dow Jones Newswires*, September 30, 2003.

The gain or loss from the sale of the old asset is calculated as the difference between the salvage value and the book value, Equation (8.11). Because this value is not taxed, we adjust the initial book value of the newly acquired asset, which is usually the purchase price P, by subtracting the gain (or adding the loss) as

$$B_0 = P + B_n - S_n. \tag{8.12}$$

To see that this convention makes sense, consider the salvage value S_n. If $S_n = B_n$, then the asset is sold for its book value, and no taxes are paid when the asset is disposed. This has no impact on the initial book value of the newly acquired asset in the replacement case.

If $S_n > B_n$, then the asset is sold for a profit, which would be taxed in the disposal case. This positive value defines a gain on the sale and leads to a lower initial book value for the newly acquired asset. Thus, the replacement leads to lower depreciation charges for the newly acquired asset. Effectively, the tax that should have been paid on the gain from the sale is "paid" through lower depreciation charges over the recovery period of the replacement asset.

For the case where $S_n < B_n$, a loss is incurred. For the moment, assume that we would receive a tax credit equal to the tax rate times the loss (we discuss losses soon) if we had retired the asset. Instead of a tax credit, however, the initial book value of the newly acquired asset is increased, which in turn increases the depreciation charges for the new asset compared with purchasing the asset as a new investment (and not a replacement). Again, this increase in the book value effectively replaces the tax credits that would have been received from the disposal of the asset.

Generally, the asset that serves as a replacement must follow the same depreciation system and method as its predecessor. (If the original asset was purchased before MACRS rules were in effect, the replacement asset must follow MACRS rules.) The next example illustrates the situation.

EXAMPLE 8.14

Initial Book-Value Calculations for Replacement Assets

In the spring of 2004, Spirit Airlines of Miramar, Florida, announced that it would acquire 35 A321s and A319s from Airbus to replace its aging fleet of MD-80 aircraft. The deal was valued at $2 billion.[18] Assume that a new plane is purchased for $57 million and it replaces a plane that has a book value of $0, but is sold (traded in) for $2.5 million. What is the initial book value of the new plane?

Solution. The old aircraft is sold at a gain of $2.5 million, which is the difference between the salvage value and the book value. The initial book value of the new aircraft is

[18] Perez, E., "Spirit Airlines to Buy 35 Planes in $2 billion Deal," *The Wall Street Journal Online*, March 19, 2004.

$$B_0 = P + B_n - S_n = \$57M + \$0 - \$2.5M = \$54.5 \text{ million.}$$

This is the value to be used in all ensuing depreciation calculations.

8.5.3 Capital Gains

The gains from selling a capital asset are often confused with what are termed capital gains. The term *capital gains* refers to a situation in which one invests money (capital) and the value of the investment grows (gains). If the investment is sold at a price that is higher than the investment cost, a capital gain has occurred.

The majority of assets purchased by firms that engage in engineering activities are never sold for capital gains. It would be a rare occurrence if an oil rig, backhoe, computer, lithography machine, or delivery truck could be sold for more than its purchase price. Assets generally deteriorate or become obsolete with time. This is actually a requirement for an asset to be depreciated. Thus, one can expect that the market salvage values will always be less than the original purchase price.

This is not to say that engineering firms never register capital gains; it is just that these gains are generally the result of investments outside the firm, not investments in what we call capital assets. Here are some recent events that might lead to capital gains:

- In the fall of 2003, Intel invested $450 million in Micron Technology Inc., the second time that Intel had invested in the memory-chip maker. The investment gives Intel rights representing a 5.3% share in the company.[19] If Intel sells the purchased shares at a higher price, then they will be subject to a capital gain.
- Nokia paid $252.2 million to bring its stake in software maker Symbian to 63.3%. Symbian is a competitor of Microsoft in developing software for "smartphones," and the investment was seen as a move to guide future deliveries by Symbian.[20] It also was an investment that can lead to capital gains or losses if the investment shares are ever sold.

Capital gains are taxed as ordinary income, but it is possible that these types of investments will be taxed at different rates in the future. However, any losses incurred from such investments cannot be used to offset gains from ordinary operations. For individual investors, capital gains are taxed at different rates, depending on the length of time the shares are retained. This is not the case for corporations. We illustrate the calculation of a gain in the next example.

---

[19] Clark, D., "Intel's Investment in Micron Shapes Future of Chips," *The Wall Street Journal Online,* September 25, 2003.

[20] Nuttall, C., "Symbian buy is gamble for Nokia," *Financial Times,* p. 6, February 10, 2004.

EXAMPLE 8.15

Capital Gains

Energy concern Royal Dutch/Shell Group purchased 1.85 billion shares of Sinopec, or China Petroleum and Chemical Corp., in 2000 for HK$1.59 each. In March of 2004, Shell announced that it would sell its entire stake (all shares) for between HK$3.088 and HK$3.153 per share (a modest discount off the normal trading price).[21] Assuming that the selling price was HK$3.153 per share, calculate Shell's gain on the investment.

Solution. The (book) value of Shell's original investment was

$$B = (1.85 \text{ billion})(\text{HK\$}1.59) = \text{HK\$}2{,}941{,}500{,}000.00.$$

The (salvage) value of the investment at the time of sale was

$$S = (1.85 \text{ billion})(\text{HK\$}3.153) = \text{HK\$}5{,}833{,}050{,}000.00.$$

The value of $S - B$ is HK$2.89155 billion, defining the capital gain posted by Shell for the sale of the shares. Note that the length of time the shares are retained is immaterial for companies when a capital gain (or loss) is calculated under U.S. rules.

In this text, we are generally interested in the analysis of investments in capital assets that are used to generate ordinary income. We include this information to be thorough in our treatment of taxes and to note that companies often make investments of that sort (with engineering consequences, as noted in the brief example of the Intel investment in Micron). However, we will not deal with capital gains in this textbook beyond this chapter.

8.6 Notable Exceptions and Adjustments to Taxable Income

The problem with studying taxes is that they are subject to change with new laws. Governments often change tax rates or offer credits in order to promote growth (which generally entails enticing capital investments). The rest of this section presents a number of notable exceptions or changes that one should be aware of when computing an after-tax cash flow.

8.6.1 Units-of-Depletion Method of Depreciation

Mineral properties, such as oil and gas wells, mines, and geothermal deposits, or standing timber can be depreciated according to a convention known as **depletion**. The idea behind this approach is that the cost of the property is

[21] Chan, C., "Shell to Sell 1.85 Bln Sinopec Shares—Source," *Dow Jones Newswires*, March 17, 2004.

TABLE 8.9

Depletion percentages for a number of minerals for the percentage depletion method. Multiple listings of minerals generally means that the use of the mineral must also be established to define the percentage allowed.

Resource	Depletion Percentage
Gravel, peat, sand, stone, clay	5%
Clay, shale	7.5%
Coal, perlite, sodium chloride	10%
Clay, metal mines	14%
Gold, silver, copper, iron ore, oil shale, crude oil	15%
Sulphur, uranium, bauxite, natural gas	22%
Ores, including cobalt, lithium, manganese, nickel, platinum, titanium, zinc	22%

Source: Internal Revenue Code, Title 26, Subtitle A, Chapter 1, Subchapter 1, Part I, Sections 613 and 613A.

expensed according to the amount of resource depleted, or removed, in each year relative to the total amount of resource available, which is estimated. This is similar to the units-of-production method presented earlier. There are two general conventions used to compute annual depletion of the properties:

1. **Cost Depletion Method.** The adjusted basis, equal to the property's investment cost minus previous depletion amounts, is divided by the number of recoverable units of resources to determine a dollar rate per unit of depletion. This rate is then multiplied by the number of units of resource that are sold in the year in question to determine the amount of the cost depletion deduction.
2. **Percentage Depletion Method.** The deduction is taken as the minimum of the following two values: (1) 50% of *net taxable* income generated by the property before the depletion deduction is taken; and (2) a fixed percentage, defined for each mineral, of the *gross* income from sales for the year. Table 8.9 gives the percentages for a number of natural resources. Note that this is only a partial list and there are many exceptions, especially with the application of crude-oil and natural-gas deposits.

Generally, an analyst will compute the depletion deduction under both methods and then take the higher value. One notable exception is that timber can be depleted only with the cost method. We illustrate in the next example.

EXAMPLE 8.16

Depletion Deductions

Falconbridge, Ltd., invested $553 million in a nickel deposit at Nickel Rim South in Sudbury, Ontario, Canada. With the project beginning in 2004, production is expected to commence in 2008.[22] It is believed that there are 13.2 million metric tons grading 1.7% nickel, 3.5% copper, 0.04% cobalt, 0.8 gram gold per ton, 1.9 grams platinum per ton, and 2.2 grams palladium per ton. Operating costs are expected to be about \$0.66 per pound.[23]

Make the following assumptions: Annual production of 660,000 metric tons commences in 2008. Concentrate on the nickel and copper deposits, ignoring the other minerals. Assume that the price (revenue) of nickel is \$3.25 per pound and \$0.90 per pound for copper over the life of the mine. Determine the maximum amount of depletion allowed each year, assuming that U.S. tax law is followed.

Solution. There are 13.2 million tons of ore being mined at the rate of 660,000 tons per year. Because the investment cost is \$553 million, the per unit depletion rate is

$$\text{Unit depletion rate} = \frac{\$553 \text{ million}}{13.2 \text{ million metric tons}} = \$41.89 \text{ per ton.}$$

Since there are 660,000 tons mined each year, the annual depletion charge under the *cost depletion method* is

$$D_n = \left(\frac{\$41.89}{\text{ton}}\right)\left(\frac{660{,}000 \text{ tons}}{\text{year}}\right) = \$27.65 \text{ million.}$$

Note that, because this method considers only the investment cost and the total amount of mineral resource, we need not consider the value of the individual minerals involved.

Now consider the *percentage depletion method.* Assuming revenues of \$3.25 and \$0.90 per pound for nickel and copper, respectively, against costs of \$0.66 per pound (before depletion), we find that net taxable income for nickel is

$$\begin{aligned}\text{EBIT} &= \left(\frac{660{,}000 \text{ tons ore}}{\text{year}}\right)\left(\frac{(.017) \text{ tons nickel}}{\text{ton ore}}\right)\left(\frac{2204 \text{ lb}}{\text{metric ton}}\right)\left(\frac{\$3.25 - \$0.66}{\text{lb nickel}}\right)\\ &= \$64.048 \text{ million.}\end{aligned}$$

For copper,

$$\begin{aligned}\text{EBIT} &= \left(\frac{660{,}000 \text{ tons ore}}{\text{year}}\right)\left(\frac{(.035) \text{ tons copper}}{\text{ton ore}}\right)\left(\frac{2204 \text{ lb}}{\text{metric ton}}\right)\left(\frac{\$0.90 - \$0.66}{\text{lb copper}}\right)\\ &= \$12.219 \text{ million.}\end{aligned}$$

Thus, the maximum depletion amount is

$$D_n = (0.50)(\$64.048\text{M}) = \$32.024 \text{ million}$$

for nickel and

$$D_n = (0.50)(\$12.219\text{M}) = \$6.11 \text{ million}$$

for copper.

---

[22] McKinnon, J., "Falconbridge Announces 1st Phase of New Mine Devt in Sudbury," *Dow Jones Newswires*, March 11, 2004.

[23] "Falconbridge Announces First Phase of New Mine Development in Sudbury," *Media Release*, www.falconbridge.com, March 11, 2004.

For the copper deposit, the maximum depletion percentage is 15% (from Table 8.9), while it is 22% for nickel. For nickel, the gross income from sales is

$$\left(\frac{660{,}000 \text{ tons ore}}{\text{year}}\right)\left(\frac{(.017) \text{ tons nickel}}{\text{ton ore}}\right)\left(\frac{2204 \text{ lb}}{\text{metric ton}}\right)\left(\frac{\$3.25}{\text{lb nickel}}\right)$$

$$= \$80.368 \text{ million.}$$

Because 22% of $80.368 million is $17.681 million, the full $17.681 million can be depleted, as it is less than 50% of net taxable income before depletion charges.

For copper, the gross income from sales is

$$\left(\frac{660{,}000 \text{ tons ore}}{\text{year}}\right)\left(\frac{(.035) \text{ tons copper}}{\text{ton ore}}\right)\left(\frac{2204 \text{ lb}}{\text{metric ton}}\right)\left(\frac{\$0.90}{\text{lb copper}}\right)$$

$$= \$45.821 \text{ million.}$$

Since 15% of this value is $6.87 million, only $6.11 million can be depleted, as the percentage depletion charge exceeds 50% of net taxable income. Thus, the total depletion charge for the mine under the percentage depletion method is $17.681 + $6.11 = $23.79 million when only nickel and copper are considered. As this is less than the cost depletion method, the larger value of $27.65 million is used for the depletion charge.

8.6.2 Losses from Operations or from Sales of Assets

A company pays taxes on the basis that it is profitable as an overall entity. This is an important point, because the company's individual projects may not be profitable over their entire lifetime, yet they may not affect the overall profitability of the company in a given period.

When a company loses money, it is generally able to *carry forward*, and sometimes *carry backward*, the losses in order to reduce future (or previous) tax obligations. For example, if a company lost $10 million dollars in 2003, it would pay no taxes that year for their operations. Now, if the same company makes $15 million in 2004, it is obligated only to pay taxes on the difference between $15 and $10 million, as it can carry forward its losses. Similarly, companies may be able to carry losses backwards. Thus, according to U.S. tax law, a company is required to pay taxes only on profits over time. (Note that there are limits and exceptions to this rule, but, in general, it is true.)

This textbook is interested in evaluating investments at the project level. Thus, a convention must be adopted for each project as to how losses should be handled. We have two choices:

1. **Carry losses forward.** This convention assumes that a project mirrors a company. It is a good choice if the company is small, in which case the project's profitability reflects that of the company.

2. **Assume that a loss results in a tax credit.** Normally, we compute the amount of tax to be paid as some percentage of profits. Under this assumption, a tax credit is paid to the company as a percentage of the loss, with the percentage equal to the effective tax rate. The assumption is valid for larger organizations, in that overall profitability cannot generally be measured by the profitability of a single project.

The second convention may seem odd, as the government is not going to write rebate checks to each company taking a loss. However, recall that this is a project-level decision. If a company is profitable, but has a project that is unprofitable, the losses from the project actually reduce the taxes to be paid by the company by reducing its profits. For example, if a company has three projects such that one loses $150,000 while the others make a profit of $250,000 in a given period, then the company would have to pay taxes only on the net profits of $100,000 for that period. With a single tax rate, this is economically equivalent to earning a tax credit on the project with a loss and paying taxes on the two projects with gains. (We illustrate the after-tax cash flow effects of these two assumptions in the final section of the chapter.)

Losses from the sale of depreciable assets are treated in a similar manner, with the profits treated as ordinary income. However, capital losses from the sale of investments (such as stock) can be used to "eliminate" capital gains only in the same period. That is, if a company sold stock in another company and realized a $10 million gain from the sale, and if it also sold stock in another company and realized a $7 million loss, the selling company would pay taxes on only $3 million.

8.6.3 Investment Credits

The government often passes laws that stimulate growth in the economy by enticing companies to invest. A common approach is in the form of a tax credit. Essentially, tax credits speed the depreciation process in that costs are recovered more quickly than by MACRS. As with our discussion of accelerated depreciation and the time value of money, companies prefer credits because they generally lower taxes immediately and in turn can improve profits.

One convention is the **Section 179** clause. As an example of the way this deduction works, in 2003 through 2005 a business could *expense* up to $100,000 of equipment purchases in the year of purchase. That is, the equipment does not have to be depreciated over time. Previously, the deduction was limited to $24,000. However, the amount of deduction available decreases on a $1 to $1 basis if the total investments of a company exceed $400,000. That is, if a company invests $450,000 in a year, it can expense $450,000 − $400,000 = $50,000 in the year of purchase. The remaining $400,000 is depreciated as is any other property. This convention clearly does not affect large companies because, if investments exceed $500,000, no portion can be expensed.

The Job Creation and Workers Assistance Act of 2002 was passed in order to promote investment and job growth.[24,25] This act allowed for an "additional" depreciation charge of 30% in the first year for assets with 20-year lives or shorter. Thus, 30% of the asset could be depreciated in year 1, and the remaining 70% would be depreciated as if it were a normal asset. This act truly accelerates the depreciation process and provides tax savings to the company.

8.6.4 Amortization

Intangible property, such as goodwill,[26] trademarks, patents, and in some cases, software, can be depreciated through a process called *amortization*. Amortization follows the same rules as straight-line depreciation with no salvage value. The number of years is generally defined by the length or life of the asset (such as the length of a patent). In the case of goodwill, the recovery period is 15 years.

EXAMPLE 8.17

Amortization

In the fall of 2003, Sun Microsystems purchased CenterRun, Inc., which makes software for managing networks. In its filing with the Securities and Exchange Commission (SEC), the purchase price of $65 million included about $1 million in transaction costs. The price above the fair value of the acquired assets and liabilities would be recognized as goodwill according to the filing.[27] Assume that the fair value of CenterRun was $64 million and the remaining $1 million is defined as goodwill. Calculate the amortization over the 15-year allowance.

Solution. The $1 million of goodwill is spread equally over 15 years as

$$\frac{\$1M}{15} = \$66{,}666.66.$$

This amount is amortized each year over the next 15 years. Thus, it is an expense for Sun Microsystems over that time frame.

---

[24] Murray, S. and J.D. McKinnon, "Bush signs economic-stimulus law, offering big business tax breaks," *The Wall Street Journal,* A2, March 11, 2002.

[25] Whitman, J., "Some ailing companies may get a tax windfall," *The Wall Street Journal,* B5C, March 20, 2002.

[26] *Goodwill* is a term assigned to intangible property that has value, such as a brand name. For example, if a company pays $100 million for a company that is worth $90 million, $10 million is assigned to goodwill.

[27] Clinton, C., "Sun Microsystems Pays $65 Million to Acquire CenterRun," *Dow Jones Newswires*, September 29, 2003.

8.7 After-Tax Cash Flows

Having computed depreciation and tax liabilities, we are ready to put all of the resulting information together and compute after-tax cash flows. These are the cash flows that will ultimately be analyzed in the chapters that follow in order to make decisions with regard to capital investments. Developing after-tax cash flows can be complicated; there is a lot of information to process, so it is suggested that the cash flows be developed with the use of a spreadsheet.

Computing the after-tax cash flow can take many paths. The formulations that follow adopt a three-step approach:

1. *Compute taxable income.* Taxable income, also called earnings before income taxes (EBIT), is given by $\text{EBIT}_n = R_n - E_n$, where E_n includes all operating expenses, such as the costs of labor, materials, and energy, as well as depreciation and interest.
2. *Compute taxes payable.* Using the effective tax rate t, we compute the taxes to be paid, or $t\text{EBIT}_n$.
3. *Compute after-tax cash flow.* The after-tax cash flow is the earnings before taxes, less taxes, plus depreciation added back, since depreciation is not a cash flow, but rather, an expense. Mathematically, we have $A_n = \text{EBIT}_n - t\text{EBIT}_n + D_n$. The value of A_n is increased by any loan principal received, after-tax proceeds from sales of assets, or working capital, while it is lowered by loan principal payments and payments to working capital.

An after-tax cash flow A_n in any period n may also be computed directly as follows:

$$
\begin{array}{rclll}
A_n & = & -P_n & & \text{(capital investments)} \\
& + & S_n & & \text{(revenues from sales of assets)} \\
& - & t(S_n - B_n) & & \text{(taxes on gains from sales of assets)} \\
& - & W & & \text{(net working capital)} \\
& + & (1-t)R_n & & \text{(after-tax ordinary revenues)} \\
& - & (1-t)E_n & & \text{(after-tax operating expenses, including} \\
& & & & \quad \text{costs of labor, energy, and materials)} \\
& + & tD_n & & \text{(depreciation tax savings)} \\
& - & (1-t)\text{IP}_n & & \text{(after-tax interest payments)} \\
& - & \text{PP}_n & & \text{(principal payments)} \\
& + & B & & \text{(loans received)}
\end{array}
$$

Note that this approach assumes that tax credits are received in the case of a loss.

We now examine a number of detailed examples to illustrate the effects of depreciation methods, sales of assets, loans, working capital, and inflation on the computation of after-tax cash flows. It is these after-tax cash flows that are eventually used in analysis. Because each of these methods requires a significant amount of data, we will provide a real-life example before using our own estimates for the ensuing calculations.

EXAMPLE 8.18

Choice of Depreciation

In 2003, Sunoco, Inc., of Philadelphia, Pennsylvania, agreed to build a coke-making plant with an annual capacity of 550,000 tons per year in order to supply plants of International Steel Group, Inc. The facility was estimated to cost $150 million, and ISG agreed to purchase the coke (used to make steel) for the next 15 years.[28]

Make the following assumptions: The coke is sold for $350 per ton in time-zero dollars and increases 2% per year over the life of the 15-year contract. Annual expenses, in millions of time-zero dollars, are assumed to be $27.5 for materials (including coal, which is used to produce coke), $75 for labor, and $20 for energy. These are expected to inflate annually at rates of 1.5%, 4.5%, and 3.0%, respectively. Overhead costs are assumed to hold constant at $7 million per year over the life of the project. Assume that a $140 million investment is made at time zero and production begins in period 1. (Assume end-of-year cash flows.)

The investment is to be depreciated under the MACRS GDS system with a 7-year recovery period. Determine the after-tax cash flows over the 15-year contract assuming that the facility has no salvage value at the end of the horizon. The tax rate is 35%, and any losses are to be treated as tax credits. Compute the difference in the after-tax cash flows, using two depreciation methods available under MACRS rules: (a) alternative straight line; and (b) DDB switching to straight line (Table 8.3 percentages). Assume the half-year convention.

Solution. The calculation of the after-tax cash flow in period 1 includes the revenues from sales, less expenses, which include the costs of materials, labor, energy, and overhead, as well as depreciation and taxes. For straight-line depreciation, the depreciation in year 1, is one-half of one-seventh of the $140 million investment, or $10 million. The after-tax cash flow in year 1 is

$$\begin{aligned} A_1 = {} & \underbrace{(1-0.35)(\$192.5\text{M})(1+0.02)^1}_{\text{Revenues}} - \underbrace{(1-0.35)(\$27.5\text{M})(1+0.015)^1}_{\text{Materials}} \\ & - \underbrace{(1-0.35)(\$75\text{M})(1+0.045)^1}_{\text{Labor}} - \underbrace{(1-0.35)(\$20\text{M})(1+0.03)^1}_{\text{Energy}} \\ & - \underbrace{(1-0.35)(\$7\text{M})}_{\text{Overhead}} + \underbrace{(0.35)(\$10\text{M})}_{\text{Depreciation}} = \$44.10 \text{ million.} \end{aligned}$$

This dollar amount increases to $47.60 million with the MACRS percentages in Table 8.3. The calculations of A_2 through A_{15} follow similarly, although there are no savings from depreciation after period 8, as illustrated in the spreadsheet in Figure 8.8.

Cell C8 is highlighted to illustrate how the result of the after-tax cash flow computation can be input directly into the spreadsheet. We make use of the VDB

[28] Lowrey, D., "Sunoco Agrees to Build New Plant for Intl Steel Units," *Dow Jones Newswires*, November 6, 2003.

	A	B	C	D	E	F	G
1	Example 8.18: Coke Plant Investment				**Input**		
2					Investment	$140.00	million
3	**Period**	**Cash Flow (SL)**	**Cash Flow (DDB)**		Annual Prod.	0.55	million tons
4	0	-$140.00	-$140.00		Unit Price	$350.00	per ton
5	1	$44.10	$47.60		g (Price)	2%	
6	2	$47.19	$52.19		Materials	$27.50	million
7	3	$46.70	$48.28		g (Materials)	2%	
8	4	$46.15	$45.27		Labor	$75.00	million
9	5	$45.52	$42.89		g (Labor)	5%	
10	6	$44.81	$42.18		Energy	$20.00	million
11	7	$44.01	$41.38		g (Energy)	3%	
12	8	$39.62	$38.31		Overhead	$7.00	million
13	9	$35.14	$35.14		Salvage Value	$0.00	
14	10	$34.05	$34.05		Interest Rate	12%	
15	11	$32.86	$32.86		Dep Method	MACRS SL	col. B
16	12	$31.56	$31.56			MACRS DDB	col. C
17	13	$30.13	$30.13		Rec. Period	7	years
18	14	$28.59	$28.59		Tax Rate	35%	
19	15	$26.91	$26.91		Periods	15	years
20	Sum	$410.44	$410.44				
21	P	$142.86	$145.99		**Output**		
22	=(1-$F$18)*($F$3*$F$4)*(1+$F$5)^A8-				Cash flow diagrams		
23	(1-$F$18)*($F$6*(1+$F$7)^A8+$F$8*(1+$F$9)^A8+						
24	$F$10*(1+$F$11)^A8+$F$12)+$F$18*VDB($F$2,0,$F$17,A7-0.5,A8-0.5,2,FALSE)						
25							

Figure 8.8 After-tax cash flows under MACRS SL and switching methods for coke-making facility.

function in Excel, introduced in Example 8.6, as opposed to directly inputting the MACRS percentages (Table 8.3). One must be careful to call the function with proper timing, due to the half-year convention. Note that the after-tax cash flows in this spreadsheet were computed directly. In later examples, we build the after-tax cash flow values from individual cash flow components. Although the spreadsheets are larger, they are generally easier to follow.

The summations of the individual cash flow streams are given in row 20. Although the sums are the same, it should be clear that these methods are not economically equivalent. This is verified by finding an equivalent present value of these streams with the use of the NPV function in row 21 with an interest rate of 12%. The cash flows derived with accelerated deprecation have a higher present-worth value, as the higher depreciation charges early in the life of the asset provide greater tax savings when evaluated at time zero.

EXAMPLE 8.19

Losses in Cash Flows

Maxim Integrated Products, Inc., acquired a semiconductor wafer fabrication facility from Royal Philips Electronics in Texas for $40 million in the fall of 2003. The facility is capable of producing 8 million "moves" a year, or $500 million in annual revenues.[29]

[29] Park, J., "Maxim Acquires Submicron Wafer Fabrication Facility in San Antonio, Texas," *Dow Jones Newswires,* October 24, 2003.

Make the following assumptions: Revenues are $400 million in year 1, $450 million in year 2, and $500 million in each year thereafter through the horizon. All costs, excluding depreciation, total $450 million in each year through the horizon. The facility has no salvage value at the end of the six-year horizon and the effective tax rate is 40%.

The facility is to be depreciated under the MACRS GDS system (Table 8.3 percentages) with a five-year recovery period. Determine the after-tax cash flows over six years, assuming two cases for dealing with losses: (1) tax credits and (2) losses carried forward.

Solution. With a depreciation charge of (.20)($40M) = $8 million in year 1, a loss of $58 million is incurred in the first year. This leads to a credit of (.40)($58M) = $23.20 million, assuming our first method of dealing with losses, while no taxes are paid under the second method. In this case, the loss of $58 million is "carried forward."

As shown in the spreadsheet in Figure 8.9, the project does not realize a profit of $42.32 million until the third year. With our tax credit method, $16.93 million in taxes are paid. With the second method, we carry our losses from the previous periods forward (a total of $70.80 million). This is accomplished in Excel with the MAX and SUM functions, as illustrated for cell F18 in the spreadsheet. The SUM function carries losses forward, while the MAX function checks whether a profit has been made. Note that the SUM function also checks to see if previous taxes have been paid, as it can become tricky if a firm alternates between profitable and unprofitable periods.

	A	B	C	D	E	F	G	H	I	J	K
1	Example 8.19: Semiconductor Fab Purchase								**Input**		
2									Investment	$40.00	million
3	**Period**	**Pn**	**Rn-En**	**Dn**	**EBIT**	**Taxes**	**ATCF**		Revenues	$400.00	million
4	0	-$40.00	--	--	--	--	-$40.00		G (Revenues)	$50.00	million
5	1	--	-$50.00	$8.00	-$58.00	-$23.20	-$26.80		Max Revenues	$500.00	million
6	2	--	$0.00	$12.80	-$12.80	-$5.12	$5.12		Expenses	$450.00	million
7	3	--	$50.00	$7.68	$42.32	$16.93	$33.07		Salvage Value	$0.00	million
8	4	--	$50.00	$4.61	$45.39	$18.16	$31.84		Interest Rate	12%	per year
9	5	--	$50.00	$4.61	$45.39	$18.16	$31.84		Dep Method	MACRS GDS	2
10	6	--	$50.00	$2.30	$47.70	$19.08	$30.92		Rec. Period	5	years
11					=$J$11*E8		=C9-F9		Tax Rate	40%	
12									Periods	6	years
13	**Period**	**Pn**	**Rn-En**	**Dn**	**EBIT**	**Taxes**	**ATCF**				
14	0	-$40.00	--	--	--	--	-$40.00		**Output**		
15	1	--	-$50.00	$8.00	-$58.00	$0.00	-$50.00		Tax credits in A3		
16	2	--	$0.00	$12.80	-$12.80	$0.00	$0.00		Losses carried forward in A13		
17	3	--	$50.00	$7.68	$42.32	$0.00	$50.00				
18	4	--	$50.00	$4.61	$45.39	$6.76	$43.24				
19	5	--	$50.00	$4.61	$45.39	$18.16	$31.84				
20	6	--	$50.00	$2.30	$47.70	$19.08	$30.92				
21		=MAX(SUM($E$15:E18)*$J$11-SUM($F$15:F17),0)									
22											

Figure 8.9 Tax payments and after-tax cash flows under different assumptions about losses.

The $42.32 million profit is subtracted from the loss carried forward ($70.80 million) such that no taxes are paid in year 3 and the remaining $28.48 million in losses are carried to the next period. In the fourth year, earnings before taxes total $45.39 million. Since $28.48 million in losses are carried forward, taxes of $6.77 million are paid on the positive balance of $16.91 million.

Note that the choice of whether to take credits on losses or carry losses forward does not affect the total magnitude of the flows, as the sums of the after-tax cash flows and total taxes paid are the same. However, the differences in the timing of the tax payments reiterates the fact that these two methods (and the resulting cash flows) are not economically equivalent.

EXAMPLE 8.20

Gains and Losses from Sales of Assets

First Cellular of Southern Illinois placed a $5.5 million order with Telefon AB LM Ericsson of Sweden to deploy a complete EDGE-ready GSM/GPRS network over its existing 850-MHz CDMA network. The contract was agreed to in the fall of 2003 and was expected to take two years for deployment.[30]

Make the following assumptions: The $5.5 million investment is spread evenly over 2004 and 2005, with depreciation separated accordingly, as there are a number of assets being deployed. (Use the MACRS GDS system with percentages from Table 8.3 over a seven-year recovery period.) The investment leads to 115,000 customers in 2005, growing 15% annually. The average revenue per customer in 2005 is $40, which grows to $45 in 2006 for the remainder of the project (due to full deployment of the network). Maintenance costs are expected to be $500,000 in 2004 dollars, growing at 6% per year. Labor costs are estimated at $1 million per year, growing at the rate of 3% per year. Finally, overhead is assumed to be 7.5% of the total investment cost in each year of the project.

Determine the after-tax cash flow diagram under the assumption that the network will be retired at the end of four years (the end of 2008). Solve for two cases: (a) a salvage value of $2.5 million after four years and (b) $500,000 after four years. The effective tax rate is 38%.

Solution. The critical elements in this example are determining the depreciation schedules and the remaining book value at the time of the sale of the network assets. As seen in the spreadsheet in Figure 8.10, there are two depreciation schedules (cells C12 through F12 and D13 through F13), since half of the network is deployed at time zero and half is deployed at period 1. This leads to a total depreciation of $1.72 million from the first deployment (note that depreciation is halved in year 4 due to early disposal) and $1.31 million depreciation from the second deployment, defining a total book value of $2.47 million at the time of sale.

In the case where the network assets are salvaged for $2.5 million, a gain of $30,000 is realized. This gain is taxed as ordinary income, so we can merely add it to the revenues from sales in cell F6. The remaining $2.47 million (book value) that is received is not taxed and is merely added (in cell F22) to the after-tax profits when the after-tax cash flow is determined (cell F24).

In the second case, where the assets are salvaged for $500,000, a loss of $1.97 million is incurred. A credit of (0.38)($1.97) = $748,600 is received for the loss. The

[30] "Ericsson Gets $5.5M Deal with First Cellular of US," *Dow Jones Newswires,* December 2, 2003.

	A	B	C	D	E	F	G	H	I	J
1	Example 8.20: Cellular Network Deployment							**Input**		
2								Investment	$5.50	million
3	Year:	2004	2005	2006	2007	2008		Customers	0.115	million per yr.
4	Revenues							g (Customers)	15%	
5	Sales:		$4.60	$5.95	$6.84	$7.87		Revenue (Yr 1)	$40.00	per cust
6	Gain:		$0.00	$0.00	$0.00	$0.03		Revenue	$45.00	per cust
7	Total Inflows:		$4.60	$5.95	$6.84	$7.90		Maintenance	$0.50	million
8	Expenses							g (Maintenance)	6%	
9	Maintenance:		($0.53)	($0.56)	($0.60)	($0.63)		Labor	$1.00	million
10	Labor:		($1.03)	($1.06)	($1.09)	($1.13)		g (Labor)	3%	
11	Overhead:		($0.41)	($0.41)	($0.41)	($0.41)		Overhead	$0.41	million
12	Depreciation 1:		($0.39)	($0.67)	($0.48)	($0.17)		Salvage Value	$2.50	million
13	Depreciation 2:		--	($0.39)	($0.67)	($0.24)		Dep. Method	MACRS GDS	2
14	Total Expenses:		($2.37)	($3.10)	($3.26)	($2.58)		Recovery Period	7	years
15								Tax Rate	38%	
16	Taxable Income:		$2.23	$2.85	$3.59	$5.31	=I12-(I12+SUM(C12:F12)+SUM(D13:F13))			
17								**Output**		
18	Income Tax:		($0.85)	($1.08)	($1.36)	($2.02)		Cash flow diagram		
19							=F16+F18			
20	Profit A/T:		$1.39	$1.77	$2.22	$3.30	=I12-MAX(F6,0)			
21										
22	Purchase/Salvage:	($2.75)	($2.75)			$2.47	=F22+F20-F12-F13			
23										
24	**Cash Flow A/T:**	($2.75)	($0.97)	$2.83	$3.38	$6.18				

Figure 8.10 After-tax cash flow diagram for cellular network investment with early disposal.

total of this transaction ($1.2486 million) leads to an after-tax cash flow of $2.97 million in 2008.

In the next section, we look at the impact of investment credits, loans, and working capital on after-tax cash flows by closely examining the chapter's introductory real decisions problems.

8.8 Examining the Real Decision Problems

In this section, we revisit our introductory problems concerning the biofuels plant investment and develop after-tax cash flows for the following three scenarios:

1. *The investment is made from retained earnings (a cash payment).*
   This solution, given in the spreadsheet in Figure 8.11, follows as in our previous examples, with the exception of working capital. Working-capital payments do not affect taxes in any way and must merely be added to after-tax cash flows, much as are loan principal receipts and payments, to be discussed soon. The data center was not printed to conserve space due to the long horizon.
2. *The investment is made assuming a $20 million investment credit.*
   In the case with the tax credit, the depreciation charge in the first period is equal to the $20 million credit, plus the typical depreciation charge on the remaining $30 million (10%). This is clearly beneficial to the company, as

	A	B	C	D	E	F	G	H	I	J	K	L
1	Chapter 8: Ethanal Plant Investment (from Retained Earnings)											
2												
3	Year:	2005	2006	2007	2008	2009	2010	2011	2012	2013	2014	2015
4	Revenues											
5	Sales:		$97.38	$99.81	$102.30	$104.86	$107.48	$110.17	$112.93	$115.75	$118.64	$121.61
6	Gain:		$0.00	$0.00	$0.00	$0.00	$0.00	$0.00	$0.00	$0.00	$0.00	$1.72
7	Total Inflows:		$97.38	$99.81	$102.30	$104.86	$107.48	$110.17	$112.93	$115.75	$118.64	$123.33
8	Expenses											
9	Material:		($52.64)	($52.64)	($52.64)	($52.64)	($52.64)	($52.64)	($52.64)	($52.64)	($52.64)	($52.64)
10	Energy:		($15.53)	($16.07)	($16.63)	($17.21)	($17.82)	($18.44)	($19.08)	($19.75)	($20.44)	($21.16)
11	Labor:		($20.40)	($20.81)	($21.22)	($21.65)	($22.08)	($22.52)	($22.97)	($23.43)	($23.90)	($24.38)
12	Depreciation:		($5.00)	($9.00)	($7.20)	($5.76)	($4.61)	($3.69)	($3.28)	($3.28)	($3.28)	($1.64)
13	Total Expenses:		($93.57)	($98.52)	($97.69)	($97.26)	($97.14)	($97.29)	($97.97)	($99.10)	($100.26)	($99.82)
14												
15	Taxable Income:		$3.81	$1.29	$4.61	$7.60	$10.34	$12.88	$14.95	$16.65	$18.38	$23.51
16												
17	Income Tax:		($1.33)	($0.45)	($1.61)	($2.66)	($3.62)	($4.51)	($5.23)	($5.83)	($6.43)	($8.23)
18												
19	Profit A/T:		$2.48	$0.84	$3.00	$4.94	$6.72	$8.37	$9.72	$10.82	$11.95	$15.28
20												
21	Pur/Sal/Work Cap	($50.00)	$7.50									($4.22)
22												
23	**Cash Flow A/T:**	($50.00)	$14.98	$9.84	$10.20	$10.70	$11.33	$12.06	$12.99	$14.10	$15.22	$12.70

Figure 8.11 After-tax cash flow diagram for ethanol plant investment from retained earnings.

	A	B	C	D	E	F	G	H	I	J	K	L
1	Chapter 8: Ethanol Plant Investment (with Investment Credit)											
2												
3	Year:	2005	2006	2007	2008	2009	2010	2011	2012	2013	2014	2015
4	Revenues											
5	Sales:		$97.38	$99.81	$102.30	$104.86	$107.48	$110.17	$112.93	$115.75	$118.64	$121.61
6	Gain:		$0.00	$0.00	$0.00	$0.00	$0.00	$0.00	$0.00	$0.00	$0.00	$1.86
7	Total Inflows:		$97.38	$99.81	$102.30	$104.86	$107.48	$110.17	$112.93	$115.75	$118.64	$123.46
8	Expenses											
9	Material:		($52.64)	($52.64)	($52.64)	($52.64)	($52.64)	($52.64)	($52.64)	($52.64)	($52.64)	($52.64)
10	Energy:		($15.53)	($16.07)	($16.63)	($17.21)	($17.82)	($18.44)	($19.08)	($19.75)	($20.44)	($21.16)
11	Labor:		($20.40)	($20.81)	($21.22)	($21.65)	($22.08)	($22.52)	($22.97)	($23.43)	($23.90)	($24.38)
12	Depreciation:		($23.00)	($5.40)	($2.16)	($3.46)	($2.76)	($2.21)	($1.97)	($1.97)	($1.97)	($1.97)
13	Total Expenses:		($111.57)	($94.92)	($92.65)	($94.96)	($95.30)	($95.81)	($96.66)	($97.79)	($98.95)	($100.14)
14												
15	Taxable Income:		($14.19)	$4.89	$9.65	$9.90	$12.18	$14.36	$16.26	$17.96	$19.69	$23.32
16												
17	Income Tax:		$4.97	($1.71)	($3.38)	($3.47)	($4.26)	($5.02)	($5.69)	($6.28)	($6.89)	($8.16)
18												
19	Profit A/T:		($9.22)	$3.18	$6.27	$6.44	$7.92	$9.33	$10.57	$11.67	$12.80	$15.16
20												
21	Pur/Sal/Work Cap	($50.00)	$7.50									($4.36)
22												
23	**Cash Flow A/T:**	($50.00)	$21.28	$8.58	$8.43	$9.89	$10.68	$11.54	$12.54	$13.64	$14.76	$12.77

Figure 8.12 After-tax cash flow diagram for ethanol plant investment with investment tax credit.

they expense the investment more quickly. The other cash flows follow as in the previous problem. The cash flow diagram is given in the spreadsheet in Figure 8.12.

3. *The investment made with a \$30 million loan and cash. (There is no investment credit.)* The loan is to be repaid over five years in equal total annual payments, assuming a 6.5% annual interest rate. This case, depicted in the spreadsheet in Figure 8.13, requires the incorporation of interest and principal payments from the loan. The total loan payment is

$$A = \$30\text{M}(\overset{A/P,6.5\%,5}{0.2406}) = \$7.218 \text{ million},$$

	A	B	C	D	E	F	G	H	I	J	K	L
1	Chapter 8: Ethanol Plant Investment (with Loan)											
2		=-$O$20*SUM($B$22:C22)										
3	Year:	2005	2006	2007	2008	2009	2010	2011	2012	2013	2014	2015
4	Revenues											
5	Sales:		$97.38	$99.81	$102.30	$104.86	$107.48	$110.17	$112.93	$115.75	$118.64	$121.61
6	Gain:		$0.00	$0.00	$0.00	$0.00	$0.00	$0.00	$0.00	$0.00	$0.00	$1.72
7	Total Inflows:		$97.38	$99.81	$102.30	$104.86	$107.48	$110.17	$112.93	$115.75	$118.64	$123.33
8	Expenses											
9	Material:		($52.64)	($52.64)	($52.64)	($52.64)	($52.64)	($52.64)	($52.64)	($52.64)	($52.64)	($52.64)
10	Energy:		($15.53)	($16.07)	($16.63)	($17.21)	($17.82)	($18.44)	($19.08)	($19.75)	($20.44)	($21.16)
11	Labor:		($20.40)	($20.81)	($21.22)	($21.65)	($22.08)	($22.52)	($22.97)	($23.43)	($23.90)	($24.38)
12	Interest:		($1.95)	($1.61)	($1.24)	($0.85)	($0.44)					
13	Depreciation:		($5.00)	($9.00)	($7.20)	($5.76)	($4.61)	($3.69)	($3.28)	($3.28)	($3.28)	($1.64)
14	Total Expenses:		($95.52)	($100.12)	($98.94)	($98.12)	($97.59)	($97.29)	($97.97)	($99.10)	($100.26)	($99.82)
15												
16	Taxable Income:		$1.86	($0.31)	$3.37	$6.75	$9.90	$12.88	$14.95	$16.65	$18.38	$23.51
17												
18	Income Tax:		($0.65)	$0.11	($1.18)	($2.36)	($3.46)	($4.51)	($5.23)	($5.83)	($6.43)	($8.23)
19												
20	Profit A/T:		$1.21	($0.20)	$2.19	$4.39	$6.43	$8.37	$9.72	$10.82	$11.95	$15.28
21												
22	Loan Principal:	$30.00	($5.27)	($5.61)	($5.98)	($6.36)	($6.78)					
23												
24	Pur/Sal/Work Cap	($50.00)	$7.50	=PMT($O$20,$O$21,$O$18)-C12								($4.22)
25												
26	**Cash Flow A/T:**	($50.00)	$8.44	$3.18	$3.41	$3.78	$4.26	$12.06	$12.99	$14.10	$15.22	$12.70

Figure 8.13 After-tax cash flow diagram for ethanol plant investment with \$30 million loan.

which is readily calculated with the PMT function in Excel. The interest payment each period, shown in cells C12 through G12, is calculated by multiplying the interest rate by the outstanding loan balance. The principal payment (cell C22, for example) is calculated by subtracting the interest payment for the period from the \$7.218 million value returned from the PMT function. Note that cells O21 and O22 define the loan interest rate and length, respectively.

Note that the after-tax cash flow in each period takes the after-tax profits, adds in the depreciation charge, and subtracts the principal payment. Thus, principal received, principal payments, asset purchases, and working capital do not affect taxes.

8.9 After-Tax MARR

It should be clear that there is quite a difference between a before-tax cash flow diagram and the associated after-tax cash flow diagram. In general, the after-tax cash flows will be smaller, because the payment of income taxes reduces the cash generated in a given period. Earlier, we defined the MARR as the minimum rate of interest that we would want to see our investments earn. We must temper this desired return in light of taxes, so that we now define an after-tax MARR, noted as MARR′ and given by

$$\text{MARR}' = (1 - t)\text{MARR}. \tag{8.13}$$

This formula merely discounts, or reduces, the MARR and its associated expected return, as taxes reduce our before-tax cash flows. As opposed to using additional notation, we will use the terms "before-tax" and "after-tax" MARR to be clear.

8.10 Key Points

1. Taxes are paid on a variety of activities in order to fund government operations. Corporations pay taxes on ordinary income, such as sales of products or services, and nonordinary income, such as profits from investments or sales of assets.

2. Ordinary income tax is paid on profits, equal to the difference between revenues and expenses.

3. An effective tax rate incorporates federal, state, and local tax rates.

4. The purchase of an asset does not constitute an expense. Rather, an asset is expensed over a prespecified lifetime. The annual charge is defined as depreciation.

5. Depreciation is not a cash flow, but rather an expense. Thus, it reduces the amount of taxes to be paid in a period.

6. The initial book value of an asset is usually equal to the purchase and installation price of the asset. The book value at any point in the life of an asset is the difference between the initial book value and all accumulated depreciation at that point in time.

7. Straight-line depreciation provides an equal amount of depreciation each year. Declining-balance depreciation provides an equal percentage of depreciation (of the book value) each year.

8. Current U.S. tax law follows the Modified Accelerated Cost Recovery System, or MACRS. The system is based on straight-line, declining-balance, and declining balance switching to straight-line depreciation methods.

9. In order to depreciate an asset, the initial book value, depreciation method, and recovery period must be known. For MACRS, the timing convention and placed-in-service date must also be known.

10. Accelerated methods of depreciation provide faster depreciation in the early years of an asset's life, compared with the straight-line method. Accelerated methods are generally preferred because they reduce taxes paid early in the life of the project, a desirable outcome due to the time value of money.

11. If an asset is retired before the end of its recovery period, MACRS does not provide a full year of depreciation. The difference between the (market) salvage value and the book value defines either a gain or loss on the sale of the asset.

12. If an asset is replaced, its residual book value (equal to the book value minus the salvage value) is added to the initial book value of the new asset. No gains or losses are realized at the time of a replacement.

13. Capital-gains taxes are paid on investments that result in profits.

14. Mines and woods are depreciated according to their depletion. The cost or percentage depletion method can be used.

15. Losses from operations or sales of assets are not taxed. The losses may be carried forward in time to reduce future tax liabilities (reduce future profits). Alternatively, as a modeling approximation, losses may receive tax credits at the time of their occurrence.

16. Investment credits allow for more rapid depreciation of assets. The government uses such credits to induce investment.

17. Amortization is mathematically equivalent to straight-line depreciation and is generally applied to items such as software, goodwill, patents, and trademarks.

18. An after-tax cash flow is composed of revenues less expenses (including taxes). Loans received, principal payments, and working-capital payments do not affect taxes.

19. When analyzing after-tax cash flows, the MARR should be adjusted by the effective tax rate.

8.11 Further Reading

Fleischer, G.A., A.K. Mason, and L.C. Leung, "Optimal depreciation policy under the tax reform act of 1986," *IIE Transactions,* 22(4):330–339, 1990.

Hartman, J.C., "Technical Note: New Depreciation Rules from the Job Creation and Worker Assistance Act of 2002 and Their Impact on Capital Investment," *The Engineering Economist,* 47(3):354–367, 2002.

Kaufman, D.J., and L.J. Gitman, "The Tax Reform Act of 1986 and corporate investment decisions," *The Engineering Economist,* 33(2):95–108, 1988.

Lohmann, J.R., E.W. Foster, and D.J. Layman, "A comparative analysis of the effect of ACRS on replacement economy decisions," *The Engineering Economist,* 27(4):247–260, 1982.

Park, C.S., and G.P. Sharp-Bette, *Advanced Engineering Economics.* John Wiley and Sons, New York, 1990.

Thuesen, G.J., and W.J. Fabrycky, *Engineering Economy,* 9th ed. Prentice Hall, Upper Saddle River, New Jersey, 2001.

8.12 Questions and Problems

8.12.1 Concept Questions

1. How do taxes influence capital investment analysis?
2. What is profit and how is it related to taxes?
3. What is the difference between ordinary and nonordinary income?
4. What is an effective tax rate? Why do we want to compute such rates?
5. Define the initial book value of an asset and how it is determined.
6. What is the book value of an asset? Is it related to the market salvage value of the asset? Explain.
7. What is depreciation? Is it a cash flow? How does it affect taxes?
8. Why is the declining-balance method of depreciation referred to as an accelerated method?
9. Why are accelerated methods of depreciation generally preferred to straight-line methods?
10. Are there situations in which one would prefer not to use an accelerated method of depreciation? Explain.
11. Derive the three-year MACRS depreciation percentages.
12. Derive a general expression for the book value of an asset, assuming MACRS depreciation for a three-year class asset.
13. Derive the first three years of the MACRS depreciation percentages for real property with a 27.5-year class life. Assume that the property enters service in the first month of the year.
14. A recent act of Congress allowed for 20% bonus depreciation in the first year of qualifying assets. For a five-year class asset, determine the new six years of depreciation percentages, assuming that the 20% is taken first such that 80% of the initial book value needs to be depreciated over the *original* recovery period.
15. How are gains or losses from the sale of assets computed?
16. How are asset retirements different from asset replacements?
17. What constitutes a capital gain?
18. What is depletion and when is it computed?
19. When is it appropriate to carry losses forward in the economic analysis of a project? Explain. What is the other option?
20. How do investment credits motivate investment?
21. What is amortization and when is it computed?
22. Which cash flow components of an after-tax cash flow diagram are not affected by taxes?
23. How does repaying a loan influence an after-tax cash flow?
24. Are the MARR and the after-tax MARR the same?

8.12.2 Drill Problems

1. A company earns $225,000 in a given period. How much federal tax does it pay? Use the rates from Table 8.1. What is the company's flat federal tax rate?

2. A company earns $350,000 in a given period. How much federal tax does it pay? Use the rates from Table 8.1. If the company pays a total of 10% in local and state taxes, what is its effective tax rate?

3. A company earns $15 million in a given period. How much federal tax does it pay? Use the rates from Table 8.1. If the company pays a total of 14% in local and state taxes, what is its effective tax rate?

4. A company earns $25 million in a given period. How much federal tax does it pay? Use the rates from Table 8.1. If the company pays a total of 18% in local and state taxes, what is its effective tax rate?

5. An asset is purchased for $350,000. The asset has a five-year recovery period and $25,000 salvage value at age 5. Compute

   (a) The depreciation in year 1, using straight-line depreciation.

   (b) The depreciation in year 2, using double-declining-balance depreciation.

   (c) The depreciation in year 1, using sum-of-years'-digits depreciation.

   (d) The depreciation in year 5, using 150% declining-balance depreciation.

   (e) The book value after year 3 if straight-line depreciation is used.

   (f) The book value after year 5 if 200% declining balance is used?

   (g) The book value after year 2 if sum-of-years'-digits depreciation is used.

6. A seven-year asset is purchased for $1.5 million and has a salvage value of $300,000. Determine the depreciation schedule and resulting book values, assuming double-declining balance switching to straight-line depreciation.

7. A three-year asset is purchased for $2 million with no salvage value. Determine the depreciation schedule and resulting annual book values, using 150% declining-balance depreciation.

8. Repeat the analysis of Problem 7, but assuming straight-line depreciation. Show that the two methods are not equivalent in terms of depreciation.

9. A five-year asset is purchased for $750,000 with no salvage value. Depreciate the asset with straight-line and with double-declining-balance depreciation. What is the present value of each depreciation stream, assuming an interest rate of 12% per year?

10. A satellite operator in the United States launches a new satellite at the cost of $250 million (investment plus installation). Determine the depreciation schedule and periodic book values over a period of $N_D + 1$ years for the satellite, assuming that the method chosen is as accelerated and rapid (in terms of the recovery period) as possible under allowable MACRS rules. (Use Table 8.2 to determine the recovery period N_D.) Assume the half-year convention.

11. Apply the following instructions to the previous problem:

    (a) Assume that the satellite becomes obsolete after six years and is "switched off." Calculate the gain or loss on this "sale" of the asset.

(b) Recompute the gain or loss, assuming that the satellite is "switched off" after three years.

(c) Recompute the gain or loss if the satellite is sold to another satellite operator for $25 million after six years.

(d) Recompute the gain or loss if the satellite is sold to another satellite operator for $125 million after three years.

(e) Recompute the gain or loss if the satellite is sold to another satellite operator for $50 million after three years.

(f) If the satellite is replaced with a $300 million satellite after three years (the satellite is "switched off," with another asset launched in its place), determine the depreciation schedule and book values for the replacement satellite over the recovery period plus one year.

12. A steel manufacturer purchased $10 million in new equipment classified as a seven-year asset under MACRS GDS rules. Depreciate the asset according to (a) MACRS percentages in Tables 8.3 and 8.4 and (b) the alternative straight-line method. Assume the half-year convention with all methods. What is the difference in book value for each approach if the asset is sold after years 3 and 4?

13. A wastewater treatment facility is constructed for $30 million. Develop its depreciation schedule, using an accelerated method of depreciation under MACRS rules and assuming the half-year convention. (Use Table 8.2 to determine the recovery period.)

14. A trucking firm purchases new tractor trucks at a cost of $50,000 for transporting goods over the road. Depreciate the trucks with MACRS, using the ADS system without the half-year convention.

15. A trucking firm purchases 100 new tractor trucks at a cost of $45,000 per truck for use over the road. Depreciate the trucks according to MACRS using the GDS system, half-year convention, and 200% switching to straight-line depreciation. Determine the book value of a truck if it is sold after years 1, 2, 3, and 4, respectively.

16. A clay mine opened at the cost of $50 million is believed to contain 200 million tons of clay. The clay is to be extracted at the rate of 20 million tons per year, with revenues of $100 per ton at the cost of $75 per ton. What is the maximum amount of annual depletion allowed under U.S. law? Use a 7.5% depletion rate.

17. A platinum mine with 15 million tons of ore and producing 0.019% platinum per ton of ore is opened at the cost of $500 million. If 500,000 tons are mined each year at the cost of $250 per ton and platinum is valued at $400 per ounce, what is the maximum depletion amount per year?

18. A forest is procured for $150 million for logging. Of the forest's 250,000 million acres, 20 million are logged each year. What is the maximum amount of depletion that can be claimed each year?

19. A $600,000 five-year class asset is purchased to generate $1 million in annual revenues against $100,000 in labor and $200,000 in material costs. The effective tax rate is 34% and there is no salvage value. Develop after-tax cash flows over six years under the following assumptions:

(a) MACRS GDS depreciation with double-declining balance switching to straight-line method and half-year convention.

(b) MACRS GDS depreciation with straight-line method and half-year convention.

(c) Repeat Part (a), but assume that the asset is sold for $50,000 at the end of year 6.

(d) Repeat Part (a), but assume that the asset is sold for $250,000 at the end of year 3. (The project ends early.)

(e) Repeat Part (a), but assume that the asset is sold for $200,000 at the end of year 3. (The project ends early.)

(f) Repeat Part (a), but assume that a $500,000 loan is taken out at time zero and is repaid in equal total payments (at a 6% annual interest rate) over five years.

(g) Repeat Part (a), but assume that a $500,000 loan taken is out at time zero and is repaid in equal principal payments (at a 6% annual interest rate) over five years.

20. An asset is purchased for $100,000. It generates revenues of $85,000, growing at $10,000 per year, and expenses are $30,000 per year. The asset is a five-year asset with no salvage value after six years, the period of study for this problem. Assume MACRS GDS depreciation with double-declining balance switching to straight-line method and the half-year convention.

(a) Determine the after-tax cash flows, assuming that taxes are paid according to the tax bracket table (Table 8.1).

(b) Repeat Part (a), assuming that all costs and revenues inflate at an annual rate of 3%. (All costs and revenues are given in time-zero dollars.)

(c) Repeat Part (b), but assume that revenues are reduced to $40,000 in time-zero dollars in year 1 and then revenues return to $95,000 in time-zero dollars in year two. Assume that losses are carried forward.

(d) Repeat Part (c), but assume that a tax credit is paid on any losses. Use a flat 38% rate.

8.12.3 Application Problems

1. Swift Transportation ordered 4000 VN670 tractor trucks from Volvo.[31] Assume that Swift receives 800 of the trucks from Volvo in 2004. Assume further that the purchase price is $75,000 per tractor. Determine the depreciation schedule and any gains or losses on sales if the assets are depreciated in accordance with MACRS GDS (with a three-year recovery period and the half-year convention) under the following three scenarios:

(a) The trucks are sold for $25,000 each at the end of year 2.

(b) The trucks are sold for $25,000 each at the end of year 3.

[31] "PRESS RELEASE: Volvo secures major order in U.S.," *Dow Jones Newswires*, December 5, 2003.

(c) The trucks are sold for $10,000 each at the end of year 4.

Repeat the analysis, using the alternative straight-line method.

2. Delta Airlines ordered 32 fifty-seat CRJ200 jets from Bombardier, Inc., for its Delta Connection carriers. The order was valued at $780 million.[32] Consider a single asset at the cost of $24.4 million with a seven-year recovery period. Compute the depreciation schedules over the recovery period, assuming

(a) The straight-line method and no salvage value.

(b) The 150%-declining-balance method.

(c) The 200%-declining-balance method switching to straight line.

(d) The sum-of-years'-digits method and no salvage value.

For each of the preceding depreciation methods, compute the gain or loss from the sale if the aircraft is sold for

(a) $5 million after year 3.

(b) $0 after year 6.

(c) $15 million after year 4.

(d) $15 million after year 5.

3. An IBM z890 midsize mainframe computer can be purchased for $125,000 in some configurations.[33] Compute the depreciation and book value schedules after the $125,000 is paid, assuming a five-year recovery period. Use

(a) The straight-line method and a $10,000 salvage value.

(b) The straight-line method and no salvage value.

(c) The double-declining-balance method.

(d) The 150%-declining-balance method.

(e) The 150%-declining-balance method switching to straight line.

(f) The 200%-declining-balance method switching to straight line and a $10,000 salvage value.

(g) The sum-of-years'-digits method and a $10,000 salvage value.

(h) The sum-of-years'-digits method and no salvage value.

4. Duke Energy announced that it would sell eight power plants in the Southeastern United States for $475 million. Duke expected to receive $500 million in tax benefits from the sales.[34]

Make the following assumptions: Consider one power plant, and assume that it was placed in service at the end of 2000 at the cost of $350 million. Assume

---

[32] Lee, S., "Delta Connection Announces Aircraft Plans for 2004 and 2005," *Dow Jones Newswires,* March 2, 2004.

[33] Devine, N., "IBM Unveils World's Most Sophisticated Mid-Size Mainframe; IBM Celebrates 40 Years of Mainframe Technology with Launch of New Mainframe and Storage Systems," *Dow Jones Newswires,* April 7, 2004.

[34] "NEWS WRAP: Duke Energy to Sell 8 Power Plants for $475M," *Dow Jones Newswires,* May 4, 2004.

further that it was sold at the end of 2004 for $60 million. The plant is categorized as a 15-year asset under the GDS system, and the effective tax rate is 35%.

(a) Assume that the plant was depreciated with MACRS (GDS, declining balance switching to straight line, half-year convention). Compute the book value of the plant at the time of the sale and the associated after-tax cash flow.

(b) Assume that the plant was depreciated with MACRS (GDS, alternative straight-line method, half-year convention). Compute the book value of the plant at the time of the sale and the associated after-tax cash flow.

(c) Recompute the solutions from parts (a) and (b), assuming a selling price of $250 million.

5. Goldcorp, Inc., is expanding its Red Lake Mine operations at the cost of $92 million with another shaft that will increase production by 200,000 ounces per year. The potential is for an additional 3.8 million ounces of gold in a new deposit. Production is expected to commence in 2007.[35] Using U.S. law, determine the maximum annual depletion charges from the expansion. Assume a gold price of $390 an ounce and operating costs of $70 per ounce.

6. BHP Billiton is developing the Panda underground diamond mine in Canada at the cost of $182 million (split among the project's participants). The investment is expected to yield 4.7 million carats of diamonds over a six-year production life.[36] Assume equal production in all six years of the project and $100 revenue per carat against per carat costs of $80. What is the maximum depletion allowance each year, according to U.S. tax law? (Assume a 15% depletion rate.)

7. Chicago Bridge and Iron was awarded an $80 million contract to expand the Lake Charles LNG terminal for Trunkline LNG, a subsidiary of Southern Union Co. The work began at the end of 2003 and ran through 2005. The contract included engineering, procurement, construction, and the commissioning of a 140,000-cubic-meter storage tank, pumps and vaporizers, and civil, mechanical, and electrical works to increase output. The project increased output capacity by 570 million cubic feet per day.[37]

Make the following assumptions: The $80 million investment is spread equally over years 2004 and 2005, with operations commencing in 2006. Assume also average revenues from sales of 500 million cubic feet per day, 300 days per year, with a price of $4.5 per 1000 cubic feet in the first year of sales, growing 3% per year through the end of the project. Material costs total $2.5 per 1000 cubic feet in the first year of operations, increasing at a rate of 0.5% per year with operating and maintenance (O&M) and overhead costs totaling $5 million per year. Finally, assume that all losses receive tax credits, the effective tax rate is 40%, and the facility has a salvage value of $15 million at the end of the 15-year horizon.

(a) Compute the end-of-year after-tax cash flows of the expansion project over a 15-year horizon if the $80 million in assets are to be depreciated according to

---

[35] Locke, T., "Goldcorp Sees Big Potential for Red Lake Mine," *Dow Jones Newswires*, September 23, 2003.

[36] Johnston, E., "BHP Billiton: Approves US$182M Panda Project in Canada," *Dow Jones Newswires*, May 4, 2004.

[37] Reynolds, L., "CB&I Wins EPC Pact for Expansion of U.S. LNG Terminal," *Dow Jones Newswires*, October 8, 2003.

MACRS (declining balance switching to straight line) over a 15-year recovery period.

(b) What is the change in the final year's cash flow if the salvage value is $0?

(c) What is the change in the final year's cash flow if the salvage value is $30 million?

(d) Assume that the assets are depreciated with the alternative straight-line method over a (15-year recovery period). Recompute the end-of-year after-tax cash flows and contrast your answer with that of part (a).

8. Lockheed Martin Corp. announced that it would spend $24 million to update an old blimp factory, once used to construct airships for World War II, in Akron, Ohio.[38]

Make the following assumptions: The facility is to be funded by a $20 million loan from the U.S. government, repayable in equal annual payments over 5 years (at a 5.5% annual interest rate). Assume also that the investment is to be depreciated with the alternative straight-line method, with the half-year convention and a 7-year recovery period. Determine the after-tax cash flows over 10 years if the facility generates revenues of $15 million per year over that time and all other costs total $5 million per year. Finally, assume that the project receives $10 million in working capital in period 1 and returns that sum at the end of the project. The effective tax rate is 36%. Carry all losses forward, and assume that there is no salvage value at the end of the horizon.

9. Gillette Co. invested $20 million in its St. Petersburg plant in Russia to double the output of razors that are sent to the Middle East, Eastern Europe, and the United States.[39] Because the facility is being expanded outside the United States, depreciation must follow the ADS system in accordance with MACRS. Determine the depreciation schedule (straight line with the half-year convention), assuming that the assets are defined as equipment for "Manufacturing of Apparel, Medical, or Dental Supplies." (See Table 8.2.) Contrast this schedule with the most accelerated method allowable for this class of asset if it were located in the United States (the GDS system). For both depreciation schedules, determine the effect on after-tax cash flows in each period (often termed a depreciation shield), using a tax rate of 35%. What is the difference in the present worth of these depreciation shields, assuming an annual interest rate of 15%?

10. Ford expanded production capacity for truck parts in 2004 at its plant in Louisville, Kentucky, at the cost of $73 million. The expansion included a 95,000-square-foot facility with stamping presses to make fenders, roof panels, and hoods, among other parts, and opened in early 2006. Output is expected to total 100,000 parts per week.[40]

Make the following assumptions: The asset is classified as a seven-year property and that the salvage value declines $10 million with each year of use. Also,

---

[38] "Lockheed, Govt Announce Plan to Update Ohio Blimp Factory," *Dow Jones Newswires*, March 25, 2004.

[39] Moscow Bureau, "Gillette to Invest $20 Mln in Russian Plant - Vedomosti," *Dow Jones Newswires*, February 20, 2004.

[40] Carty, S.S., "Ford to Invest $73M in Kentucky Truck Plant Production," *Dow Jones Newswires*, March 22, 2004.

assuming MACRS, double-declining balance, and the *midquarter* convention, compute the after-tax cash flow strictly associated with the purchase, the depreciation schedule, and disposal of the asset. (There are no revenues or other expenses.) Assume further that the facility is installed in the first quarter of the first year. All losses are credited and the tax rate is 39%. Finally, assume study horizons of 2, 3, 4, 5, 6, 7, and 8 years. (The facility closes in the first quarter of the given year.)

11. In 2004, Coca-Cola finished constructing a bottling plant in northwest China (at Lanzhou, Gansu Province) at the cost of $12 million. The plant has a capacity of 24 million cases per year.[41]

Make the following assumptions: The plant is classified as a 12-year property according to ADS rules for MACRS and is depreciated according to the straight-line method with the half-year convention over that time. Construct an after-tax cash flow diagram over a 13-year period if the $12 million investment occurs at time zero and there is no salvage value at the end of the horizon. The following are time-zero values: Sales are expected to be $10 per case, growing at 3% per year, with costs of $9 per case, growing at 1.2% per year. Fixed costs are assumed to be $5 million per year (and will not inflate). The effective tax rate is 40% and all capacity is sold.

8.12.4 Fundamentals of Engineering Exam Prep

1. Under straight-line depreciation, the periodic depreciation charge for a $50,000 asset with no salvage value and a recovery period of seven years is

(a) $5000.

(b) $6250.

(c) $7140.

(d) $8000.

2. The following do not affect tax calculations in a given period:

(a) Depreciation charges.

(b) Loan principal payments.

(c) Working-capital payments.

(d) Both (b) and (c).

3. If the tax rate is a flat rate of 20% on profits and a company has $500,000 in revenues, $200,000 in expenses, and $50,000 in depreciation, the amount of taxes paid is

(a) $50,000.

(b) $100,000.

(c) $60,000.

(d) $80,000.

---

[41] Batson, A., "Coca-Cola to Open US$12 Mln Bottling Plant in NW China," *Dow Jones Newswires,* May 26, 2004.

4. The depreciation charge in the first year for a $400,000 asset defined as a three-year asset according to MACRS and with the use of the MACRS percentages is

   (a) $120,000.

   (b) $133,320.

   (c) $66,660.

   (d) $266,640.

5. Under straight-line depreciation, the book value after three years of a $25,000 machine with no salvage value and a recovery period of five years is

   (a) $0.

   (b) $5000.

   (c) $20,000.

   (d) $10,000.

6. If the tax rate is a flat 20% on profits and a company has $500,000 in revenues, $200,000 in expenses, and $50,000 in depreciation, the after-tax cash flow is

   (a) $300,000.

   (b) $250,000.

   (c) $450,000.

   (d) $200,000.

7. If 10% is paid on the first $100,000 of profits and 15% is paid on profits from $100,001 to $200,000, the amount of taxes paid on $175,000 of profits is most closely

   (a) $26,250.

   (b) $20,000.

   (c) $21,250.

   (d) $17,500.

8. If 10% is paid on the first $100,000 of profits and 15% is paid on profits from $100,001 to $200,000, the effective tax rate on $175,000 of profits is most closely

   (a) 10%.

   (b) 12.1%.

   (c) 15%.

   (d) 12.5%.

9. A $75,000 asset is depreciated over five years under straight-line depreciation. If it is sold for $33,000 after the third year,

   (a) A gain of $3000 is realized.

   (b) A loss of $2000 is realized.

   (c) A gain of $2000 is realized.

   (d) A loss of $3000 is realized.

10. An asset purchased for $100,000 and depreciated as a five-year asset with the accelerated MACRS percentages is sold after the third year. Its book value is most closely

(a) $48,000.

(b) $38,400.

(c) $80,000.

(d) $28,800.

11. Under double-declining balance deprecation, the first-year depreciation charge for a $2 million asset with no salvage value and a recovery period of 10 years is

(a) $200,000.

(b) $800,000.

(c) $400,000.

(d) None of the above.

12. Depreciation is

(a) An expense.

(b) A cash flow.

(c) A charge for leasing an asset.

(d) None of the above.

13. Under double-declining-balance depreciation, the book value after 2 years of a $2 million asset with no salvage value and a recovery period of 10 years is:

(a) $1.65 million.

(b) $1.72 million.

(c) $1.60 million.

(d) $1.28 million.

14. In accordance with the accelerated MACRS percentages, the depreciation charge in year 5 for a $250,000 asset with a recovery period of seven years and a salvage value of $10,000 is most closely

(a) $22,300.

(b) $21,400.

(c) $43,900.

(d) $33,900.

15. When computing depreciation with declining balance switching to straight line,

(a) The first year always uses declining-balance depreciation.

(b) The switch is from straight-line to declining-balance depreciation.

(c) The last year always uses straight-line depreciation.

(d) Both (a) and (c).

16. If the MARR (the interest rate used when evaluating projects) is 15% per year for before-tax calculations and the effective tax rate is 40%, then the after-tax MARR to be used in analysis should most closely be

(a) 21%.

(b) 6%.

(c) 9%.

(d) 15%.

17. In one period, a company experiences revenues of $5 million; material, labor, and energy expenses of $2 million; depreciation of $1 million; interest expenses of $500,000; and loan principal payments of $750,000. If the effective tax rate is 40%, the amount of taxes paid is

(a) $1.6 million.

(b) $600,000.

(c) $300,000.

(d) $450,000.

18. In one period, a company experiences revenues of $5 million; material, labor, and energy expenses of $2 million; depreciation of $1 million; interest expenses of $500,000; and loan principal payments of $750,000. If the effective tax rate is 40%, the after-tax cash flow in that period is most closely:

(a) $1.15 million.

(b) $2.15 million.

(c) $3.25 million.

(d) $1.85 million.

19. The book value at the end of three years of a $10,000 asset with no salvage value and a recovery period of three years is

(a) $0 under MACRS rules.

(b) $741 under MACRS rules.

(c) $0 under straight-line depreciation.

(d) Both (b) and (c).

20. The published depreciation percentages for MACRS (Modified ACRS) assume

(a) Declining balance switching to straight-line depreciation.

(b) That the asset has no salvage value.

(c) One-half depreciation in the first year.

(d) All of the above.

Part III

Making the Decision for a Single Project

9 Deterministic Evaluation

(*Courtesy of the BMW Group*.)

Real Decisions: Bigger Minis

In 2005, the BMW Group announced a EUR150 million expansion of its Mini production facility in Oxford, England, to increase capacity by 20% and ready production for the new Mini station wagon, likely to be called the Mini Traveller. The expansion will occur in 2006, with the station wagon to be released in 2008. Revamping of the facility will require cutbacks in production of the popular Mini in 2006 and 2007 (200,000 were built in 2005).[1]

Make the following assumptions: The investment is spread such that EUR100 million is paid in 2006 and EUR50 million in 2007. Upon completion of the project, the investment is to be depreciated over six years, assuming straight-line depreciation and no salvage value.

[1] Bauer, F.E., "Mini plant cuts output to prepare for station wagon," *Automotive News,* Volume 80, Number 6177, p. 28BB, November 21, 2005.

It is expected that 20,000 Travellers will be produced in 2008 and 40,000 per year for the next five years. Furthermore, there will be a decrease in Mini production of 30,000 vehicles in 2006 and 10,000 in 2007. The net revenues are EUR2000 per Mini and EUR2500 per Traveller. (Note that there is currently a waiting list for Minis, so that the BMW Group sells all production.) Fixed operating and maintenance costs, including overhead, for the facility are estimated at EUR10 million per year. The effective tax rate is 50% and the after-tax MARR is 12.5%.

The preceding scenario leads to a number of interesting questions:

1. What is the present worth of the investment before taxes?
2. What is the present worth of the investment after taxes?
3. Are there any concerns with using IRR analysis for this problem? If not, compute the IRR of the after-tax cash flow.

In addition to answering these questions, after studying this chapter you will be able to:

- Determine whether a project should be accepted or rejected when all relevant parameters are known with certainty. (Sections 9.1–9.2)
- Differentiate between absolute and relative measures of worth in evaluating an engineering project. (Sections 9.1–9.2)
- Compute the present worth, annual equivalent worth, future worth, and market value added as absolute measures of a project's worth. (Section 9.1)
- Compute the internal rate of return, external rate of return, and benefit–cost ratio as relative measures of a project's worth. (Section 9.2)

Now that we have completed the first three steps of the decision-making process, we are ready to analyze our solution alternatives. In this part of the textbook, we examine the case of a single project and begin our analysis under the assumption of complete certainty. For the stated project, we will analyze the following two choices:

- **Accept the project.** This decision implies that the project can generate an acceptable amount of wealth for the company or, for a public entity, that the benefits outweigh the costs. If the decision is acted upon, funds should be released to begin the project.
- **Reject the project.** This decision implies that there are better options available for investment than the one currently being analyzed. Rejecting a project is equivalent to accepting the do-nothing alternative, which assumes that the funds can be invested at the MARR.

In this chapter, we use estimated project data to compute a measure of a project's worth in order to determine whether the project is worthy of investment. If the measure of worth reflects well on the project, then we accept the project. If not, we defer the investment.

For this chapter, we make the following assumptions:

1. Cash flow magnitudes and timing (i.e. a cash flow diagram) are known with certainty.
2. The interest rate, generally referred to as the MARR, is known and constant for the length of the decision horizon.
3. The decision is to accept or reject the project. Rejecting the project implies doing nothing.

We divide our measures of worth as either absolute or relative. An absolute measure of worth specifically defines the amount of wealth that a project generates for a company. A relative measure of worth defines the worth of a project relative to its level of investment. As we will see, it is important to distinguish measures as such.

9.1 Absolute Measures of Worth

This section defines four absolute measures of worth. The term "absolute" refers to the fact that the value computed by the measure of worth is all that is required in order to make our decision. This is in contrast to a relative measure of worth, which requires additional data for decision making. Note that an absolute measure of worth provides a measure of the magnitude of wealth that the project will bring to (or take from) the company.

9.1.1 Present Worth

Consider an engineering project defined by a cash flow diagram with cash flows $A_0, A_1, \ldots, A_N$ in periods zero through N, respectively. The *present worth* of the project is mathematically defined as the sum of all of the project's cash flows, discounted to time zero with the MARR. Algebraically, the present worth is a function of the interest rate i:

$$\text{PW}(i) = A_0 + \frac{A_1}{(1+i)} + \frac{A_2}{(1+i)^2} + \cdots + \frac{A_N}{(1+i)^N}.$$

More compactly,

$$\text{PW}(i) = \sum_{n=0}^{N} \frac{A_n}{(1+i)^n}. \tag{9.1}$$

This value is also commonly referred to as *present value, net present value*, or NPV.

Once the present worth has been calculated, the decision is made according to the following relationships (where i is the MARR):

PW Value	Decision
$\text{PW}(i) > 0$	Accept the project.
$\text{PW}(i) = 0$	Indifferent about the project.
$\text{PW}(i) < 0$	Reject the project.

Before delving into the meaning of present worth and why these are our decision rules, let us illustrate the computation of present worth and how it determines our decision with an example.

EXAMPLE 9.1

Present-Worth Analysis

CSR, Ltd., an Australian producer of sugar, aluminum, and building products, spent A$100 million to build a new power plant that uses sugarcane trash as fuel. The 63-megawatt generating plant, composed of a new boiler, steam turbine generator, cooling towers, and filtering systems, was completed in 2005 at the company's sugar mill in Brandon, Australia. CSR supplies 80% of the power output from the plant to the national grid under a 10-year supply agreement, and the remaining 20% powers the mill.[2] Assume that the electricity sales and government subsidies for renewable energy together generate A$15 million annually and that savings for the mill for power generation total A$3.75 million per year at the cost of $1 million in annual operating and maintenance for the plant. Assuming that the plant is worth A$10 million at time 10, analyze the project over a 10-year life, as that is the length of the supply

[2] Brindal, R., "Australia's CSR: To Spend A$100M on Green Energy," *Dow Jones Newswires*, September 3, 2003.

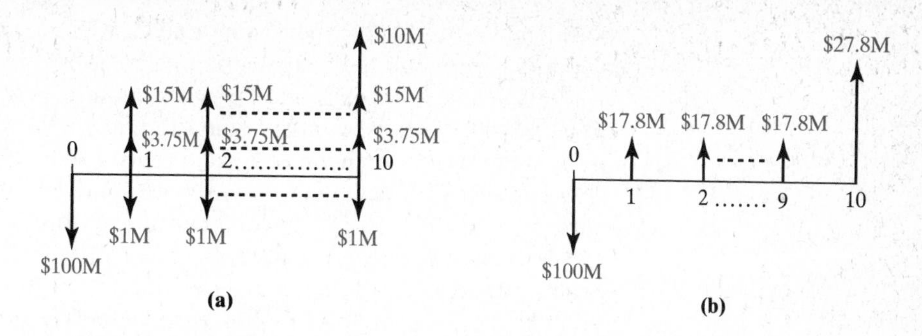

Figure 9.1 (a) Component cash flows and (b) net cash flows for power plant investment in Australian dollars.

contract. For this example, assume that the MARR is 12% per year and ignore any tax implications.

Solution. Assuming that the investment is paid for in 2004 and production begins in 2005, the investment is depicted according to its component cash flows in Figure 9.1(a) and its net cash flows in Figure 9.1(b).

Given the net cash flow diagram in Figure 9.1(b), we can determine whether the investment should be considered. There are three components to the cash flow diagram: (a) A single cash flow (−A\$100M) at time zero; (b) 10 years of annual cash flows (A\$17.75M = A\$15M + A\$3.75M − A\$1M); and (c) a single cash flow (A\$10M) at time 10. Using the present-worth factor for an equal-payment series, Equation (3.6), and the present-worth factor for a single payment, Equation (3.5), we can write the present worth of the entire investment as the sum of the present worth values of its individual components according to Equation (9.1) as

$$\text{PW}(12\%) = \underbrace{-\text{A\$100M}}_{\text{Initial Investment}} + \underbrace{\text{A\$17.75M}(\overset{P/A,12\%,10}{5.6502})}_{\text{Annual Profits}} + \underbrace{\text{A\$10M}(\overset{P/F,12\%,10}{0.3220})}_{\text{Salvage Value}}$$

$$= \text{A\$3.51 million.}$$

Since the PW(12%) is greater than zero, the decision should be to accept the project. Note that this decision is merely reflected by the present-worth criterion; other factors may influence the decision as well and are discussed later in this, and following, chapters.

As noted in Chapter 3, the present worth can be computed in Excel with the NPV function, which yields

$$= A_0 + \text{NPV}(i, A_1, A_2, \ldots, A_N) = -100 + \text{NPV}(0.12, 17.75, 17.75, \ldots, 27.75) = 3.51,$$

in millions of Australian dollars.

In the previous example, the present worth of the project came out to be A\$3.51 million, assuming a 12% MARR. We now want to explore the meaning of the A\$3.51 million and why this value leads us to a decision to accept the project.

It may aid in the discussion to clarify what present worth *is not*. Present worth is not profit. *Profit* is an accounting term defined by the difference between revenues and expenses in a given period, as noted in Chapter 8. If revenues are greater than expenses, then an entity is profitable for the period of study. Profit can be measured in one period or over the course of an entire project. Since there is no time value of money involved in calculating profit, profit and present worth are equivalent only if the interest rate is zero. This is why a project with negative present worth may be profitable, as discounting with a positive interest rate lowers the value of future profits at time zero.

Consider the previous example. The A$100 million investment led to a total of A$187.5 million in net revenues over 10 years. Clearly, this is a *profitable* project, as $187.5 million is considerably greater than $100 million. However, if we had raised the MARR to 25%, then

$$\begin{aligned} PW(25\%) &= -A\$100M + A\$17.75M(\overset{P/A,25\%,10}{4.1925}) + A\$10M(\overset{P/F,25\%,10}{0.1938}) \\ &= -A\$23.65 \text{ million.} \end{aligned}$$

Thus, in this case, while the project generates a profit, it is not acceptable according to our present-worth decision criterion.

We explore this concept further, considering what present worth *is*. Because periodic net cash flows represent the difference between revenues and expenses, present worth can be defined as "discounted" profit. The issue at hand is "What is the meaning of discounting?" The meaning depends on the interpretation of the interest rate (the MARR). We gave a number of reasons for choosing the MARR in Chapter 7, including (1) the cost of capital (funds) and (2) the minimum rate you expect your money to earn.

Consider (1) and recall how the periodic interest payment for a loan is calculated: multiplying the interest rate by the outstanding balance of the loan. Now, if we were to find the present worth of a cash flow of loan payments (interest plus principal) using the true cost of the loan as the interest rate, the remaining value would be the principal borrowed. This was illustrated when we defined the true cost of a loan in Chapter 4.

If we assume that the MARR represents the cost of funds, then it is analogous to representing the interest rate for a loan. Discounting the cash flows of a project with the interest rate leaves the principal amount, or remaining worth of the investment. Alternatively, discounting the cash flows at the interest rate "removes" the cost of funds from the flows, leaving the worth of the project after interest charges have been paid. This is why the interest rate used for discounting is often referred to as the "cost of capital."

If our calculations lead to a present worth of zero, then the project earns money exactly at that rate of interest. That is, the project earns enough money to pay off the cost of funds, but the project does not generate surplus money or wealth. If the present worth is negative, then the project does not generate enough money to cover the interest costs, such as interest on loans. If the present worth is positive, then we are able to pay off our loans *and* retain funds. The time-zero equivalent of these funds is the present worth.

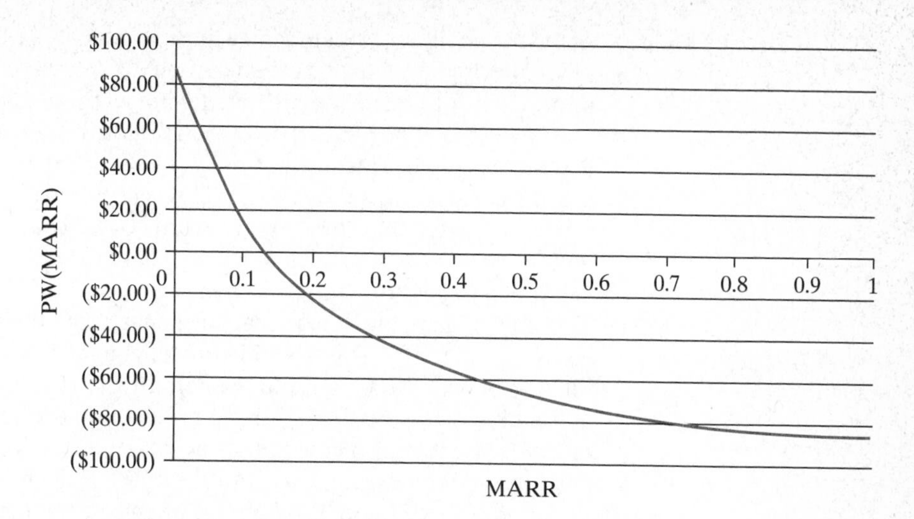

Figure 9.2 Graph of PW(MARR) for data from Example 9.1 for MARR > 0.

Consider the graph of the present worth for interest rate values between 0% and 100% in Figure 9.2. The data are from Example 9.1, but the curve follows that of a typical investment. Note that the curve is decreasing such that, as the interest rate increases, the present worth of the investment decreases. Examining Figure 9.2, we see that if we increase the interest rate (the cost of funds), then the worth of the project declines because we have to pay more interest back over time. This corresponds directly to our meaning of present worth as a measure of funds, or worth, after all expenses, including interest, are paid.

Defining the interest rate in this manner should make our decision rules clear. Assuming that the interest rate represents our cost of capital, we only accept projects that leave surplus funds after paying off our cost of capital. These surplus funds represent growth for the company. A present worth of zero means that the project earns just enough funds to pay off the cost of capital, but there is no growth and we are indifferent as to whether to accept or reject the project. Finally, a negative present worth leads to the decision to reject the project, because it cannot even pay the cost of capital, let alone provide an opportunity for growth.

Now consider (2), our second interpretation of the MARR, and recall that our investment options are to accept or reject the project. We can view the option to invest or not to invest as a choice between making the investment and doing nothing. Our goal is to create the most wealth.

Consider the do-nothing option. If it is chosen, we assume that the funds are placed in an investment that earns the MARR. We could write the present worth of this option as

$$\text{PW}(12\%) = -\text{A\$100M} + \text{A\$12M}(\overset{P/A,12\%,10}{4.1925}) + \text{A\$100M}(\overset{P/F,12\%,10}{0.1938}) = \text{A\$0}.$$

The A$12 million generated in each period is merely the value of the MARR times the investment of A$100 million. We assume that the investment is returned at the end of the project, defining a present worth of zero.

If we accept the project, then we are rejecting the do-nothing option. Thus, accepting a project means that the project will create greater wealth than an investment that earns the MARR. This is reflected in a positive present worth, which is greater than the present worth (wealth created) of the do-nothing option.

If we reject the project, then we accept the do-nothing alternative. That means that the project cannot generate as much wealth as one which earns the MARR. This should truly highlight the importance of, and the relation between, the do-nothing alternative and the MARR. The do-nothing alternative is essentially a mythical project that represents the average return of other investment opportunities available to the firm. Those opportunities are expected, on average, to earn the MARR. Thus, the MARR represents the opportunity provided by other possible investments. If the project being studied cannot generate more wealth than the do-nothing alternative, then we are better off investing in the do-nothing alternative (essentially, another project). This is reflected in the negative present worth of the project. A present worth of zero means that the project will generate exactly the same wealth as the do-nothing alternative. In this case, both projects are economically equivalent and we are indifferent about the choice.

We have taken the time to discuss present worth in detail because it is the basis of all our future analyses. Other measures of worth are introduced in this and later chapters, but their relevance is measured only according to their relationship to present worth. That is because accepting projects with positive present worth (in the case of single projects) will guarantee an increase in the worth of the firm, given our assumptions about knowing the interest rate and cash flow diagram with certainty. Under these assumptions, it is the optimal method for making an investment decision.

9.1.2 Future Worth

Present worth defines the monetary worth of a project at time zero. The *future worth* of a project carries the same meaning as present worth, but the measure is taken at a different point in time: the end of the project. Mathematically, for our investment cash flow diagram,

$$\begin{aligned}\mathrm{FW}(i) = {} & A_0(1+i)^N + A_1(1+i)^{N-1} \\ & + A_2(1+i)^{N-2} + \cdots + A_{N-1}(1+i) + A_N,\end{aligned}$$

which can be written as

$$\mathrm{FW}(i) = \sum_{n=0}^{N} A_n(1+i)^{N-n}. \tag{9.2}$$

We could also find the future worth from the present worth by using our compound amount factor for a single payment:

$$\text{FW}(i) = \text{PW}(i)(^{F/P,i,N}) = \text{PW}(i)(1+i)^N.$$

For any interest rate $i > -1$, the value of the compound amount factor, $(1+i)^N$, is positive. Thus, the future worth will always have the same sign as the present worth and therefore, for any project, the future worth will *always* lead to the same conclusion as present worth, so that we have the following relationships:

FW Value	Decision
$\text{FW}(i) > 0$	Accept the project.
$\text{FW}(i) = 0$	Indifferent about the project.
$\text{FW}(i) < 0$	Reject the project.

We illustrate the calculation of future worth with another example.

EXAMPLE 9.2

Future Worth

Carbo Ceramics, Inc., constructed a plant in Wilkinson County, Georgia, to produce ceramic proppants. Proppants are used in the energy industry to open larger pathways for extracting oil and gas in wells. Carbo's facility was built at the cost of $62 million[3] and is expected to be operational at the end of 2005, with an annual capacity of 250 million pounds.[4]

Make the following assumptions: The investment cost of $62 million is spent equally in years 2004 and 2005, with production running at full capacity in 2006. Revenues are estimated at $0.32 per pound of proppant at the end of 2005, at the cost of $0.20 per pound. The per pound revenue is expected to grow at 1.25% per year over the project horizon. Annual fixed production costs (including indirect charges) are estimated at 5% of the total investment cost over the life of the plant. Costs are not expected to inflate over time. The plant is to be depreciated under MACRS GDS (200%-declining balance switching to straight line) with a five-year recovery period. The effective tax rate is 40% and the after-tax MARR is 12% per year. Any losses result in tax credits. Find the future worth of the investment, assuming a seven-year horizon and assuming that the plant is sold for $5 million at that time.

Solution. The before-tax, current-dollar cash flow diagram for this example is given in Figure 9.3. The after-tax cash flow diagram is given in the spreadsheet in Figure 9.4. The depreciation of the facility cannot commence until 2006, as that is the first period in which the plant can be used. The after-tax cash flow in year 2006 is calculated as follows:

---

3 Wetzel, K., "CARBO Ceramics Inc. Announces New Plant Construction and Fourth Qtr Div," *Dow Jones Newswires,* January 14, 2004.

4 "Carbo Ceramics Inc. Announces New Plant Construction and Fourth Quarter Dividend," *Press Release #04-01,* www.carboceramics.com, January 14, 2004.

Figure 9.3 Before-tax, current-dollar cash flow diagram of investment in ceramic proppant facility.

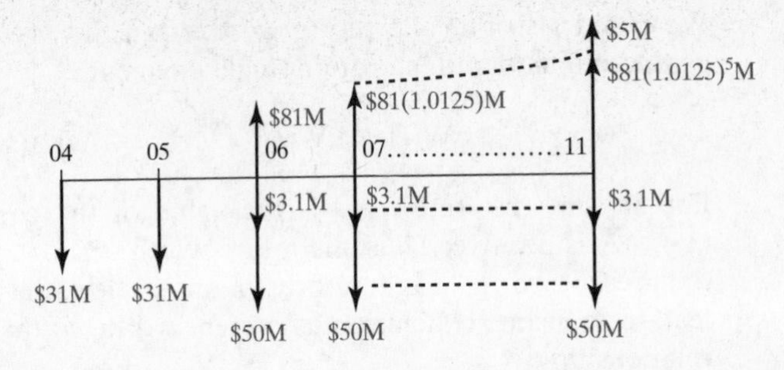

	A	B	C	D	E	F	G	H	I	J	K	L	M
1	Example 9.2: Ceramic Proppant Plant Investment										**Input**		
2											Investment	$62.00	million
3	Year:	2004	2005	2006	2007	2008	2009	2010	2011		Operating Cost	$3.10	million
4	Revenues										Production Level	250	million pounds
5	Sales:			$81.00	$82.01	$83.04	$84.08	$85.13	$86.19		Proppant Cost	$0.20	per pound
6	Salvage:			$0.00	$0.00	$0.00	$0.00	$0.00	$5.00		Proppant Price	$0.32	per pound
7	Total Inflows:			$81.00	$82.01	$83.04	$84.08	$85.13	$91.19		g (Price)	1.25%	per year
8	Expenses										Salvage Value	$5.00	million
9	Unit Costs:			($50.00)	($50.00)	($50.00)	($50.00)	($50.00)	($50.00)		MARR	12%	per year
10	Fixed Costs:			($3.10)	($3.10)	($3.10)	($3.10)	($3.10)	($3.10)		Periods	7	years
11	Depreciation:			($12.40)	($19.84)	($11.90)	($7.14)	($7.14)	($3.57)		Depreciation	MACRS	GDS
12	Total Expenses:			($65.50)	($72.94)	($65.00)	($60.24)	($60.24)	($56.67)		Recovery Period	5	years
13											Multiplier	2	
14	Taxable Income:			$15.50	$9.07	$18.03	$23.83	$24.88	$34.52		Tax Rate	40%	
15													
16	Income Tax:			($6.20)	($3.63)	($7.21)	($9.53)	($9.95)	($13.81)		**Output**		
17											FW(12%)	$56.13	million
18	Profit A/T:			$9.30	$5.44	$10.82	$14.30	$14.93	$20.71		=FV(L9,L10,,-B22-NPV(L9,C22:I22))		
19													
20	Purchase:	($31.00)	($31.00)										
21													
22	**Cash Flow A/T:**	($31.00)	($31.00)	$21.70	$25.28	$22.72	$21.44	$22.07	$24.28				

Figure 9.4 After-tax cash flow analysis of investment in ceramic proppant production facility.

$$\text{ATCF}_{2006} = \underbrace{(1-0.40)250\text{M}(\$0.32)(1+.0125)^1}_{\text{Revenues}} - \underbrace{(1-0.40)250\text{M}(\$0.20)}_{\text{Unit Expenses}}$$

$$- \underbrace{(1-0.40)(0.05)\$62\text{M}}_{\text{Fixed Expenses}} + \underbrace{(0.40)(0.20)(\$62\text{M})}_{\text{Depreciation}} = \$21.70 \text{ million.}$$

Calculating the after-tax cash flows in 2007 through 2010 is similar. For 2011, a gain is realized on the $5 million sale of the plant and is taxed accordingly. (The plant is completely depreciated at this point, so we treat the gain as ordinary income.)

The future worth can be calculated from Equation (9.2):

$$\text{FW}(12\%) = -\$31\text{M}(1+0.12)^7 - \$31\text{M}(1+0.12)^6 + \cdots + \$22.07\text{M}(1+0.12)^1$$

$$+ \$24.28\text{M} = \$56.13 \text{ million.}$$

Since the future worth of the investment is positive, we accept the project. Again, other factors may influence the decision, but, ignoring them and given these estimates, the company should pursue the investment.

There are no direct functions for computing the future worth of a general set of cash flows in Excel, as noted in Chapter 3. From the previous section, one can compute the present worth of the cash flows and then convert it to a future worth with the use of the compound amount factor. This is accomplished by nesting the NPV function in the FV function in Excel, as noted in Chapter 3, as follows:

$$= \mathrm{FV}(i, N, , -A_0 - \mathrm{NPV}(i, A_1, A_2, \ldots, A_N))$$
$$= \mathrm{FV}(0.12, 7, , 31.00 - \mathrm{NPV}(0.12, -31.00, 21.70, 25.28, 22.72, 21.44, 22.07, 24.28))$$
$$= 56.13.$$

Recall that the FV function returns the negative of the present-value input—hence our changing signs. Cell L16 in Figure 9.4 is programmed with this nested function, which uses the after-tax cash flows in row 22 as inputs. The spreadsheet values are in millions of U.S. dollars.

So why would one compute the future worth of a project if it always leads us to the same decision as the present worth? The answer is that is provides more information. The future worth tells us how much money our investment will accumulate over time and thus gives an indication of how much money the investment can generate over its lifetime. Calculating the future worth can also be useful in tracking an investment over its lifetime, as we shall see in Chapter 15.

9.1.3 Annual Equivalent Worth

One might want to define the worth of a project in some periodic fashion. Toward that end, we use the term *annual equivalent worth*, but it should be clear that we could determine the periodic worth of a project over any length of time. To determine the annual equivalent worth AE(i) for a given interest rate i, we merely compute the present or future worth and convert it to a periodic value by using the capital-recovery (Equation (3.10)) or sinking fund (Equation (3.9)) factor, respectively. Mathematically:

$$\begin{aligned} \mathrm{AE}(i) &= \mathrm{PW}(i)(^{A/P,i,N}) = \mathrm{PW}(i)\left[\frac{i(1+i)^N}{(1+i)^N - 1}\right] \\ &= \mathrm{FW}(i)(^{A/F,i,N}) = \mathrm{FW}(i)\left[\frac{i}{(1+i)^N - 1}\right]. \end{aligned} \tag{9.3}$$

As with the comparison of PW(i) and FW(i), the values of the capital-recovery and sinking-fund factors are always positive for $i > -1$. Thus, the decision with regard to AE(i) is consistent with those for present and future worth, so that we have the following relationships:

AE Value	Decision
$\mathrm{AE}(i) > 0$	Accept the project.
$\mathrm{AE}(i) = 0$	Indifferent about the project.
$\mathrm{AE}(i) < 0$	Reject the project.

The next example illustrates the use of annual equivalent worth.

EXAMPLE 9.3 *Annual Equivalent Worth*

China State Shipbuilding Corp. is expanding its Changxing Island Yard from an output of 3 million deadweight tonnes (dwt) to 12 million dwt by 2015 at the cost of $3.6 billion.[5] (Deadweight tons (dwt) are a common measure for shipping capacity output, as opposed to detailing the number of ships produced, because the ships may vary drastically in size.) Assume that China State's investment is spread equally over the years 2004 through 2014, with 1 million dwt of new capacity coming online each year from 2006 through 2015. Assume further that 1 million of dwt can generate $600 million of revenue at the cost of $500 million. Compute the annual equivalent worth of the project through 2020, assuming a 10% annual MARR. Finally, assume that time zero is the end of 2004 and that the salvage value of the yard is $300 million in 2020. Ignore any tax implications in the analysis.

Solution. The cash flow diagram for this investment is given in Figure 9.5. We assume that the yard operates at full capacity over its lifetime.

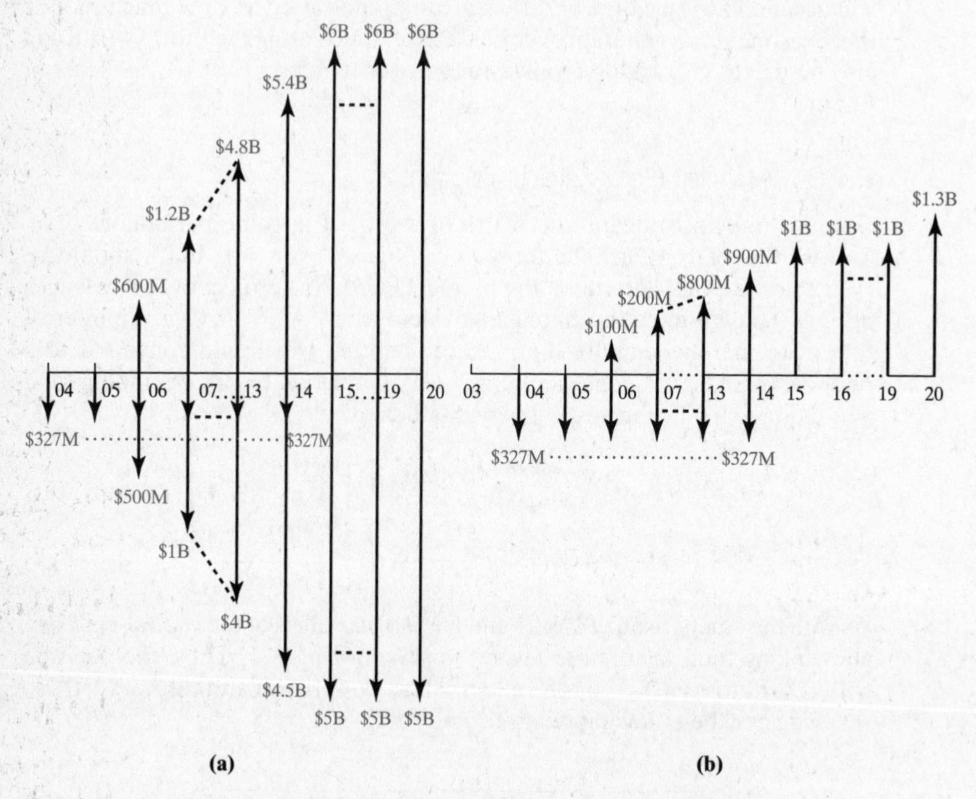

Figure 9.5 (a) Individual cash flows and (b) net cash flows for shipyard expansion.

[5] Lague, D., "China Seen Becoming World's Biggest Shipbuilder," *Dow Jones Newswires* via *The Far Eastern Economic Review,* September 10, 2003.

Examining Figure 9.5(b), we see that there are four distinct cash flow patterns to analyze. The $3.6 billion investment is an annual series over periods 2004 through 2014. The profits generated by the yard are defined by an arithmetic gradient series in periods 2005 through 2014, followed by an equivalent annual series in periods 2015 through 2020. Finally, the salvage value of the yard is a positive cash flow in 2015. We examine each of these groups individually and sum their result.

The investment of $3.6 billion is spread equally over 11 years, resulting in outflows of $327.27 million in years 2004 through 2014. The annual equivalent worth of this investment (over the 16-year study period) is

$$\text{AE}(10\%)_1 = \left(-\$327.27\text{M} - \$327.27\text{M}(\overset{P/A,10\%,10}{6.1446})\right)(\overset{A/P,10\%,16}{0.1278}) = -\$298.82 \text{ million.}$$

The $100 million gradient begins at zero in period 2005 and grows to $1 billion in period 2015. The annual equivalent worth of this portion is

$$\text{AE}(10\%)_2 = \left(\$100\text{M}(\overset{P/G,10\%,11}{26.3963})\right)(\overset{A/P,10\%,16}{0.1278}) = \$337.34 \text{ million.}$$

The equal payment series of the $1 billion profits between years 2016 and 2020 can be spread equally over the project's life as

$$\text{AE}(10\%)_3 = \left(\$1.0\text{B}(\overset{F/A,10\%,5}{6.1051})\right)(\overset{A/F,10\%,16}{0.02782}) = \$169.84 \text{ million.}$$

Finally, the salvage value of the facility can be spread back in time as

$$\text{AE}(10\%)_4 = \$300\text{M}(\overset{A/F,10\%,16}{0.02782}) = \$8.34 \text{ million.}$$

Summing the annual equivalent-worth components results in a total annual equivalent worth of $216.70 million per year. Since this number is positive, the project should be accepted.

As with future worth, there is no direct function for calculating the annual equivalent worth of a general set of cash flows. However, the present worth can be calculated and converted to an annual series with the use of the equal-payment series factor. To do so in Excel, we use the PMT function together with the nested NPV function as follows:

$$= -\text{PMT}(i, N, A_0 + \text{NPV}(i, A_1, A_2, \ldots, A_N))$$

$$= -\text{PMT}(0.10, 16, -327.27 + \text{NPV}(0.10, -327.27, -327.27, \ldots, 1300)) = 216.70.$$

In this approach, we adjust the sign of the number returned by the PMT function, as opposed to altering the input values. The solution is in millions of U.S. dollars per year over the 16 periods of study.

Determining annual equivalent worth leads us to the same conclusion that we would derive from determining present and future worth. Again, its usefulness compared with that of these other methods is that, with annual equivalent worth, we can have information displayed over different periods.

For example, it can be very useful to determine the annual equivalent worth of a project in order to learn the effect of the project over time. Although present and future worth also define the value of a project, the impact (in terms of scale) can be lost, as the project can span 2 to 20 years. Activity during this period is immediately captured in the annual equivalent worth.

9.1.4 Market Value Added

The metric economic value added, or EVA,[6] is a popular measure of a company's financial health. The metric, defined and copyrighted by Stern Stewart and Co., compares net after-tax operating profit against the allocated cost of capital for a given period. The net after-tax operating profit is merely the cash flow resulting from income less expenses, including depreciation and taxes. In a simple scenario where an asset or assets are purchased and used for a number of periods, the net after-tax operating profit is

$$(R_n - E_n - D_n)(1 - t).$$

Recall from Chapter 8 that R_n, E_n, and D_n represent revenues, expenses, and depreciation, respectively, in year n. The tax rate is defined as t. With this notation and noting that B_n is the book value of an asset at the end of period n, the EVA in any period $n < N$ is computed as

$$\text{EVA}_n = (R_n - E_n - D_n)(1 - t) - i(B_{n-1}). \tag{9.4}$$

For period N, which may include the sale of the asset, the EVA is computed as

$$\text{EVA}_N = (R_N - E_N - D_N)(1 - t) + (S_N - B_N)(1 - t) - i(B_{N-1}), \tag{9.5}$$

where $S - N$, the gain on the sale of the asset, is also assumed to be taxed at rate t. Note that the EVA in a given period is not equivalent to the after-tax cash flow in the same period. Rather, it is defined as the difference between the after-tax operating profit and a charge for the use of a capital asset. This is much different than present worth, either before or after tax, because with present worth, the charge for the capital asset is recognized at the time of purchase. We will discuss the implications of this difference shortly. First, we must define a decision criterion based on EVA.

Note that the number determined through EVA analysis is merely a single-period value. If this value is positive, then it is interpreted as earning profit above the cost of capital for the given period. A project typically has multiple periods and thus multiple EVA values. It would seem reasonable for a decision criterion that utilizes EVA to accept projects which have net EVA values over the life of the project that are positive, considering the time value of money. The term "market value added," or MVA, has been used to

[6] The term "EVA" is a registered trademark of Stern Stewart & Co., www.sternstewart.com.

describe the present value of the EVA values computed in each period. Thus, the MVA measure of worth is

$$\text{MVA}(i) = \sum_{n=1}^{N} \frac{\text{EVA}_n}{(1+i)^n}. \tag{9.6}$$

As before, the interest rate i is the MARR, although it may be better termed the "cost of capital," as the idea behind EVA is to "charge" for the use of capital in each period. The MVA decision criterion is as follows:

MVA Value	Decision
$\text{MVA}(i) > 0$	Accept the project.
$\text{MVA}(i) = 0$	Indifferent about the project.
$\text{MVA}(i) < 0$	Reject the project.

We illustrate with an example.

EXAMPLE 9.4

Market Value Added

Trans-Elect, an electricity transmission company based in Virginia, is expanding and upgrading Path 15, an 83-mile stretch of transmission lines in the Los Angeles area at the cost of \$300 million. The project, which will increase capacity by 1500 megawatts, was expected to be completed by the end of 2004. Trans-Elect owns 72% of transmission capacity rights. (Other entities also are involved in the investment.)[7] Assume that Trans-Elect receives annual revenues of \$35 million, with expenses totaling \$2 million in time-zero dollars. The revenues are expected to remain flat, since they are contracted, but expenses are expected to grow by 3 percent per year. Assume that construction takes only one year and the lines are to be depreciated over a 20-year period with MACRS GDS. Use a 20-period horizon for your analysis, and assume a salvage value of \$10 million, an after-tax MARR of 10%, and a 35% tax rate.

Solution. The cash flow diagram for the expansion of transmission capacity is given in Figure 9.6. Note that Trans-Elect purchased 72% of the lines; thus, the amount that they can depreciate is \$216 million.

The EVA_n values for years 1 through 19 can be computed from Equation (9.4). For $n = 1$,

$$\begin{aligned}
\text{EVA}_n &= (R_n - E_n - D_n)(1-t) - i(B_{n-1}); \\
\text{EVA}_1 &= \Big(\$35\text{M} - \$2\text{M}(1+0.03)^1 - 0.0375(\$216\text{M})\Big)(1-0.35) - (0.10)(\$216\text{M}) \\
&= -\$5.454\text{M}.
\end{aligned}$$

---

[7] "Construction to Start on Calif 1500MW Elec Line Expansion," *Dow Jones Newswires,* September 16, 2003.

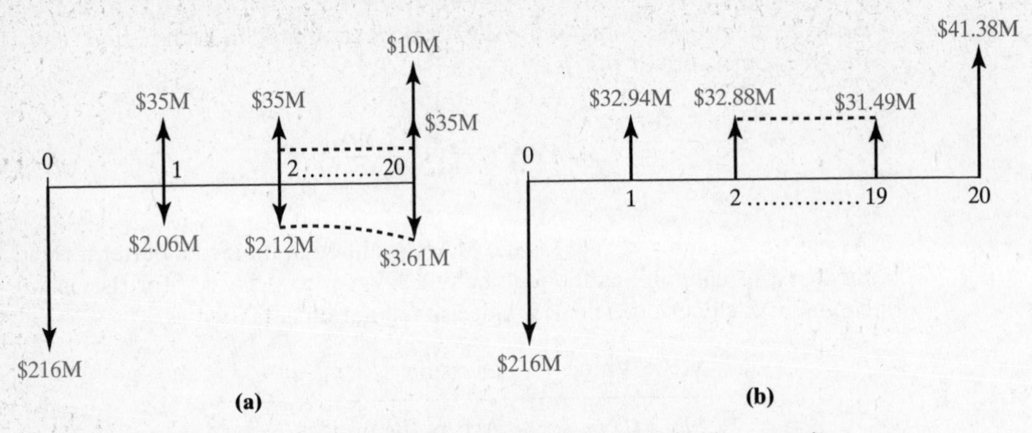

Figure 9.6 (a) Individual cash flows and (b) net cash flows for transmission line expansion.

	A	B	C	D	E	F	G	H
1	Example 9.4: Transmission Line Upgrade Investment				**Input**			
2					Investment	$216,000,000.00		
3	**Period**	**MACRS %**	**EVA**		Revenues	$35,000,000.00		
4	0	--	--		Operating Cost	2,000,000	years	
5	1	3.75%	-$5,454,000.00		g (Cost)	3%	per year	
6	2	7.22%	-$9,556,050.00		Salvage Value	$10,000,000.00	million	
7	3	6.68%	-$7,279,745.10		MARR	10%	per year	
8	4	6.18%	-$5,177,481.45		Periods	20	years	
9	5	5.71%	-$3,226,616.30		Method	MACRS GDS		
10	6	5.28%	-$1,434,747.99		Recovery Period	5	years	
11	7	4.89%	$206,723.97		Tax Rate	35%		
12	8	4.52%	$1,734,478.89					
13	9	4.46%	$2,745,634.86		**Output**			
14	10	4.46%	$3,658,108.71		MVA(10%)	-$1,874,661.65	million	
15	11	4.46%	$4,569,055.97					
16	12	4.46%	$5,478,430.85		=NPV(F7,C5:C24)			
17	13	4.46%	$6,386,186.17					
18	14	4.46%	$7,292,273.36					
19	15	4.46%	$8,196,642.36					
20	16	4.46%	$9,099,241.63					
21	17	4.46%	$10,000,018.08	=($F$3-$F$4*(1+$F$5)^A18-B18*$F$2)*(1-$F$11)				
22	18	4.46%	$10,898,917.02	-$F$7*(1-SUM($B$5:B17))*F2SUM($B$5:B17))*$F$2				
23	19	4.46%	$11,795,882.13					
24	20	2.23%	$16,031,855.39					

Figure 9.7 EVA values for all 20 periods of the electricity transmission investment.

For $N = 20$, we utilize Equation (9.5):

$$\text{EVA}_N = (R_N - E_N - D_N)(1-t) + (S_N - B_N)(1-t) - i(B_{N-1});$$

$$\text{EVA}_{20} = \left(\$35\text{M} - \$2\text{M}(1+0.03)^{20} - 0.0223(\$216\text{M})\right)(1-0.35)$$

$$+ (\$10\text{M} - \$9.677\text{M})(1-0.35) - (0.10)(\$14.494\text{M}) = \$16.03\text{M}.$$

All of the EVA values are programmed into column C in the spreadsheet in Figure 9.7. Recall that these are not cash flows. (Note that there are no direct methods for calculating EVA values in Excel. Alternatively, we could have developed columns for depreciation and before-tax cash flows for the calculations.)

We aggregate the EVA values into a single MVA value, as defined in Equation (9.6), in order to make a decision:

$$\text{MVA}(10\%) = \frac{-\$5.454\text{M}}{(1+0.10)^1} + \frac{-\$9.556\text{M}}{(1+0.10)^2} + \cdots + \frac{\$16.03\text{M}}{(1+0.10)^{20}} = -\$1.875 \text{ million.}$$

This value is also programmed into cell F14 in Figure 9.7 with the use of the NPV function. Since the MVA value is negative under these cash flow assumptions, the decision would be to reject this investment.

As we noted before we presented the example, the MVA measure of worth is different from previous measures of worth because we are not dealing directly with cash flows. That is, MVA discounts periodic EVA values, which are *not* equivalent to the project net cash flows. The most glaring difference is that instead of recognizing an investment, such as the purchase of equipment, as a cash flow early in the project, the MVA spreads the investment over time, using the interest rate times the book value (iB_n). Another difference is that the depreciation charge is not added back to the operating profit to determine the after-tax cash flow. The question must be raised, then, as to whether the MVA criterion will lead to different decisions compared with present worth. We explore the answer to that question now.

If we rewrite Equation (9.6) according to individual cash flow components, we obtain

$$\text{MVA}(i) = \sum_{n=1}^{N} \left(\frac{R_n - E_n - D_n - tR_n + tE_n + tD_n - iB_{n-1}}{(1+i)^n} \right) + \frac{(S_N - B_N)(1-t)}{(1+i)^N}. \tag{9.7}$$

Similarly, we can define the present worth of the same project (we assume that the amount of invested capital is P, which is being depreciated over N periods):

$$\text{PW}(i) = -P + \sum_{n=1}^{N} \left(\frac{R_n - E_n - tR_n + tE_n + tD_n}{(1+i)^n} \right) + \frac{S_N - (S_N - B_N)t}{(1+i)^N}. \tag{9.8}$$

Assuming that the values of MVA(i) and PW(i) are the same, taking the difference between Equations (9.7) and (9.8) results in

$$P = \sum_{n=1}^{N} \left(\frac{D_n + iB_{n-1}}{(1+i)^n} \right) + \frac{B_N}{(1+i)^N}.$$

We write out this expression and note that the purchase price P is equal to the initial book value B_0:

$$B_0 = \frac{D_1 + iB_0}{(1+i)} + \frac{D_2 + iB_1}{(1+i)^2} + \cdots + \frac{D_N + iB_{N-1} + B_N}{(1+i)^N}.$$

As with the definition of book value in Chapter 8, $B_n = B_{n-1} - D_n$. If we substitute this expression into the final term above,

$$\begin{aligned}\frac{D_N + iB_{N-1} + B_N}{(1+i)^N} &= \frac{D_N + iB_{N-1} + B_{N-1} - D_N}{(1+i)^N} \\ &= \frac{(1+i)B_{N-1}}{(1+i)^N} = \frac{B_{N-1}}{(1+i)^{N-1}}.\end{aligned}$$

We can now rewrite our complete expression for B_0 as

$$B_0 = \frac{D_1 + iB_0}{(1+i)} + \frac{D_2 + iB_1}{(1+i)^2} + \cdots + \frac{D_{N-1} + iB_{N-2} + B_{N-1}}{(1+i)^{N-1}}.$$

Utilizing the expression $B_n = B_{n-1} - D_n$, we can repeat the substitution we just performed for combining terms in period N into period $N-1$. Here, we can substitute those in $N-1$ to $N-2$, as follows:

$$\begin{aligned}\frac{D_{N-1} + iB_{N-2} + B_{N-1}}{(1+i)^{N-1}} &= \frac{D_{N-1} + iB_{N-2} + B_{N-2} - D_{N-1}}{(1+i)^{N-1}} \\ &= \frac{(1+i)B_{N-2}}{(1+i)^{N-1}} = \frac{B_{N-2}}{(1+i)^{N-2}}.\end{aligned}$$

This results in

$$B_0 = \frac{D_1 + iB_0}{(1+i)} + \frac{D_2 + iB_1}{(1+i)^2} + \cdots + \frac{D_{N-2} + iB_{N-3} + B_{N-2}}{(1+i)^{N-2}}.$$

If we continue substituting and aggregating through period $n = 1$, the result is

$$B_0 = \frac{D_1 + iB_0 + B_1}{(1+i)} = \frac{D_1 + iB_0 + B_0 - D_1}{(1+i)} = \frac{(1+i)B_0}{(1+i)} = B_0.$$

This shows that the difference between MVA and PW values is zero for a given interest rate i. In fact, it shows that the value produced by MVA is the same as that produced by PW. Thus, MVA is an alternative method of calculating the PW of an after-tax cash flow, and we are guaranteed to make the same investment decision if we use either MVA or PW.

It is interesting that the MVA and PW decision criteria achieve the same value for their respective measures of worth, because they go about computing that value in extremely different manners. In essence, using the interest rate, MVA spreads the initial investment cost over each period of the project. Thus, capital costs are addressed in a periodic fashion. In PW, the investment is recognized at the time of purchase.

One note of caution must be stated regarding the use of MVA to make an investment decision. Example 9.4 was straightforward in that the investment took one year to be installed and the first year of depreciation occurred in year 1. As we have seen with numerous examples in this textbook, investments may span many years. However, depreciation cannot commence until the asset is completely installed. The question is, What EVA values result in the periods between the time an investment commences and when the installation is complete? EVA can be used in this instance, but the book-value term in Equation (9.4) must have its meaning adjusted. If the definition of B_{n-1} is altered to mean "investment cost," then the analysis continues to produce values equivalent to present worth. Let us revisit the previous example and illustrate this alteration.

EXAMPLE 9.5

Market Value Added with Prolonged Investment

Consider the previous example, but assume that the $216 million investment is spread over two periods, with depreciation commencing in the period immediately after construction is complete.

Solution. Assuming a 21-period horizon, the first two periods of EVA calculations are now

$$\text{EVA}_1 = -(0.10)(\$108\text{M}) = -\$10.8\text{M},$$

$$\begin{aligned}\text{EVA}_2 &= \left(\$35\text{M} - \$2(1+0.03)^2\text{M} - 0.0375(\$216\text{M})\right)(1-0.35) - (0.10)(\$216\text{M}) \\ &= -\$5.494\text{M}.\end{aligned}$$

The value of EVA_{21} is now $15.96 million, resulting in an MVA value of −$11.904 million, leading us to the same conclusion that we should reject the investment. It is critical to note that the first-period EVA value is not calculated from the initial book value, which is not officially established until the end of year 2. Rather, the first-period EVA value is calculated from the investment.

Again, one must ask why we need to establish another measure of worth that is equivalent to present worth. The answer is that the usefulness of MVA is that it forces the decision maker to perform an after-tax analysis. No such analysis is required when one is using present worth, since before-tax cash flows can be analyzed. We have already stated that incorporating taxes into analysis may alter decisions, because taxes have a tremendous influence on cash flows. Thus, the MVA measure is worthy of note because the analysis must be performed in an after-tax environment.

9.2 Relative Measures of Worth

We now turn our attention to relative measures of worth. As noted earlier, relative measures of worth differ from absolute measures of worth in that they require additional information for decision making. This is a subtle difference, but it has strong implications for future chapters and analyses.

9.2.1 Internal Rate of Return

The absolute measures of worth that have been presented require an interest rate, the MARR, to transform a cash flow diagram into an equivalent measure of worth in order for us to make our investment decision. We now focus on a method that analyzes the investment according to its *return*, not the wealth it generates.

Define i^* as the interest rate, or set of rates, such that the present worth is equal to zero:

$$\text{PW}(i^*) = \sum_{n=0}^{N} \frac{A_n}{(1+i^*)^n} = 0. \tag{9.9}$$

If i^* is unique and real (not an imaginary number), we define it as the internal rate of return, or IRR, of the project. The IRR is the rate of return earned on an investment, or the rate at which money invested in a project grows. The term "internal" describes the fact that the rate, if one exists, is defined solely by the project's cash flows.

Given an IRR, our decision with regard to our investment project is as follows:

IRR Value	Decision
IRR > MARR	Accept the project.
IRR = MARR	Indifferent about the project.
IRR < MARR	Reject the project.

In Section 4.2, we showed that an interest rate could be found for equivalence calculations through a variety of methods, including trial and error (via hand or computer) or interpolation. It can also be found graphically or with the use of spreadsheet functions. We clarify these approaches in the next example, after which we will further discuss the meaning of the internal rate of return.

EXAMPLE 9.6

Internal Rate of Return

UPM-Kymmene Paper Industry Company, Ltd., added a new line of production capacity at the company's mill in Changshu, China, with the purchase of a new paper machine, including air systems, winders, an automation system, and mechanical drives, from Metso Corporation of Finland. The EUR100 million line increased the

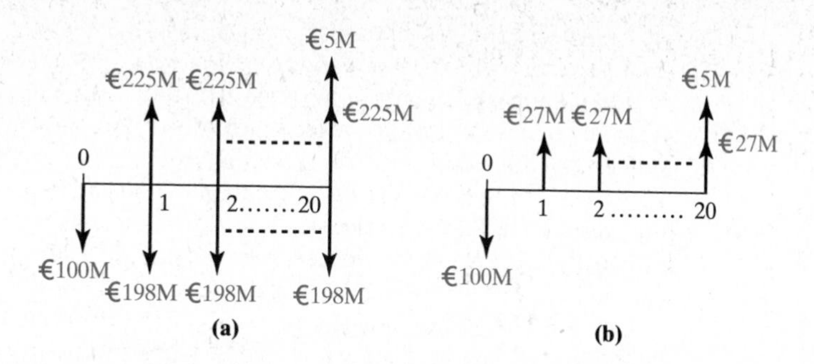

Figure 9.8 (a) Component cash flows and (b) net cash flows for paper plant expansion.

plant's annual capacity by 450,000 metric tons of uncoated copy and offset papers.[8] Assume that the production line was in place in the summer of 2005 and will produce at maximum capacity for 20 years, at which time it will be dismantled and sold for scrap for EUR5 million. Assume further that 1 metric ton of paper generates revenue of EUR500 and that profit margins are 12 percent; that is, profits (revenues minus expenses) total 12 percent of revenues. Find the internal rate of return of this investment, and compare it with the MARR of 20% per year for analysis, ignoring taxes.

Solution. The component and net cash flows of the investment are depicted in Figure 9.8(a) and (b), respectively. The present worth of the cash flows can be written as

$$\mathrm{PW}(i^*) = -\mathrm{EUR100M} + \mathrm{EUR27M}(^{P/A,i^*,20}) + \mathrm{EUR5M}(^{P/F,i^*,20}) = 0$$

$$= -\mathrm{EUR100M} + \mathrm{EUR27M}\left[\frac{(1+i^*)^N - 1}{i^*(1+i^*)^N}\right] + \mathrm{EUR5M}\left[\frac{1}{(1+i^*)^N}\right] = 0.$$

We highlight five ways in which to determine the value of i^*.

Trial and Error. Here, we emphasize a traditional method, meaning that it can be done by hand. There are many strategies for performing trial and error, which is essentially another phrase for "searching" for the answer. We illustrate the bisection method here, but there are many searching strategies that can be employed. The following table lists 10 iterations of the procedure:

Trial	i	PW(i)	Interpretation
1	10%	€130.6M	$i^* > .10$
2	20%	€31.6M	$i^* > .20$
3	30%	−€10.4M	$.20 < i^* < .30$
4	25%	€6.81M	$.25 < i^* < .30$
5	27.5%	−€2.54M	$.25 < i^* < .275$
6	26.25%	€1.93M	$.2625 < i^* < .275$
7	26.875%	−€352,000	$.2625 < i^* < .26875$
8	26.5625%	€778,000	$.265625 < i^* < .26875$
9	26.7188%	€210,000	$.267188 < i^* < .26875$
10	26.7969%	−€72,000	$.267188 < i^* < .267969$

[8] "Metso to Supply EUR100M Fine Paper Line to UPM-Kymmene," *Dow Jones Newswires*, September 8, 2003.

To begin the process, we choose an initial interest rate of 10%. Plugging this value into our present-worth function, we note that the resulting present worth of €130.6 million is positive. For a typical investment, we would expect the present-worth function to be decreasing as we increase the interest rate. Thus, we increase our interest rate to 20% for our second trial, resulting in a present worth of €31.6 million. It is not until the third trial of 30% that we compute a negative present worth. At this point, only three trials into the procedure, we know that i^* lies in the range (0.20, 0.30). This is actually enough information to make a decision, as the MARR is exactly 20%, but we would like to determine the actual rate of return for the project.

Now that we have a range in which to search, we begin the bisection method. This method merely reduces the search space in half (bisects it) in each iteration. Because we know that i^* is between 20% and 30%, we try 25%. This returns a positive present worth of €6.81 million, and our range for i^* shrinks to (0.25, 0.30), as we must increase the interest rate to move the present worth to zero. According to the bisection method, we now try 27.5% and so on, as summarized in the previous table.

After 10 iterations, we see that i^* is in the range (0.2672, .2680). This is more than enough accuracy for our purposes, as it is clear that the IRR is greater than the MARR and the project should be accepted.

Interpolation: Given any two interest rates and their corresponding present-worth values, we can interpolate to determine an approximate value for i^*. Linear interpolation assumes that the function between two endpoints is linear, and we only need two points to define a line. However, in order to ensure some level of accuracy, it is advisable to select two interest rates that produce one positive and one negative present-worth value. That way makes it easier to identify the relative range of i^*. In addition, the interest rates should not be too far apart (such as 5% or 10%). From our trial-and-error method, we obtain:

$$\text{PW}(20\%) = €31.6 \text{ million},$$

$$\text{PW}(30\%) = -€10.4 \text{ million}.$$

As noted in Section 4.2, linear interpolation merely draws a line between these two points, as shown in Figure 9.9, where the interest rate is the x-coordinate and the present worth is the y-coordinate.

The interest rate at which the line crosses the x-axis, or where the present worth is zero, is our estimate of i^*, computed as follows:

$$\frac{31.6 - (-10.4)}{0.20 - 0.30} = \frac{31.6 - 0}{0.20 - i^*}.$$

Solving for i^*, we get .2752, or 27.5%. Of course, we expect there to be error, as we are making an assumption that the present-worth function is linear between the interest rates of 20% and 30%. If we want to reduce the error, we must interpolate between

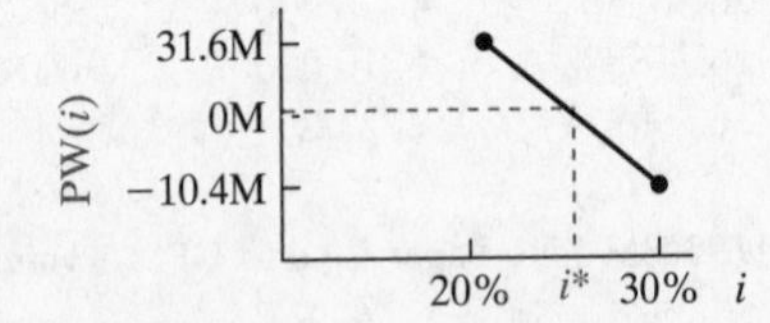

Figure 9.9 Endpoints for interpolating the IRR with our sample data.

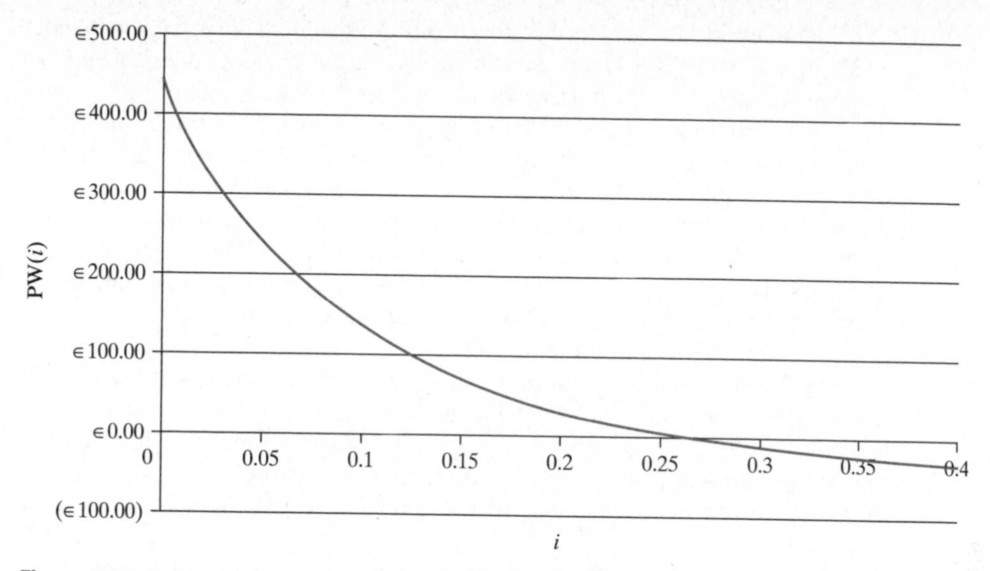

Figure 9.10 Graph of the present worth as a function of the interest rate.

closer interest rates, such as 25% and 27%. Regardless of our choice of endpoints, the method is the same.

Graphical Approach: Another approach is to graph the function PW(i) over an acceptable range of interest rates and observe where it crosses the x-axis. A plot like that in Figure 9.10 can easily be created as an 'XY (scatter)' graph in Excel, with the interest rate tabulated on the x-axis and present worth on the y-axis. Our accuracy in estimating the IRR is based on the accuracy of the x-axis (interest rate) scale. From the figure, it appears that the rate crosses the axis between 25% and 30%. Obviously, generating the data for the graph will provide a more accurate solution.

Computer Search: There are many ways in which to automate the trial-and-error (or search) process, including the bisection method illustrated earlier. Spreadsheet programs have simplified this procedure greatly.

We have already seen in Section 4.2 how the Goal Seek function in Excel can be very useful in determining the IRR. The solution with that function results in an IRR of 26.78%. In addition, the value is found in less than a second, far less time than it takes to input the cash flows into the spreadsheet.

Spreadsheet Function: All spreadsheet programs have an IRR function. In Excel, the function call is

$$= \text{IRR}(A_1, A_2, \ldots, A_N, \text{guess}) = \text{IRR}(-100, 27, 27, \ldots, 32,) = 26.78\%.$$

Note that the IRR function requires an initial guess. This is because the computer "searches" for the value of i^* in much the same way that we did with our bisection method. If a guess is omitted, then Excel uses 10% as an initial guess. A guess may, but is not guaranteed to, speed the search process.

Regardless of the method chosen to determine i^*, we have found that the IRR is 26.8%. Since this value exceeds the MARR of 20%, we would accept this project.

The internal rate of return is clearly different from the present worth in its evaluation of a project. For present worth, an interest rate is specified and the present-worth function returns a value equivalent to the worth of the project. For the internal rate of return, no interest rate is specified and no monetary value of the project is computed. Rather, the internal rate of return defines the interest rate at which all funds invested in a project grow. This feature illustrates why we define the internal rate of return as a relative measure of worth, rather than an absolute measure. The fact is, if you are told that an investment earns 26.8%, as in Example 9.6, you do not know how much money the project generates. Worse, we do not even know if the project is "good" without being told the MARR.

But do not let these comments detract from the information and utility the IRR provides. The IRR has always been very popular in industry because it offers a measure of efficiency for an investment. While it does not define the total worth of a project in terms of dollars, it defines the expected return for each dollar invested in the project. This is very useful information.

One truly has to understand the meaning of present worth in order to understand the meaning of the IRR, assuming i^* is unique and real. We stated that present worth represents the surplus of funds that a project generates after all expenses, including interest—represented by discounting at our interest rate—have been removed. The IRR defines the maximum rate at which the present worth is nonnegative. Thus, the IRR can be seen as a break-even value, which explains the resulting decisions. If the IRR is greater than the MARR, then a project returns more than a project that earns the MARR, and the project in question is therefore acceptable. Similarly, if the IRR is less than the MARR, then the IRR fails to generate a return that will cover all expenses, including interest.

The meaning of IRR and present worth are truly intertwined, and this is easy to see in a graph of the present worth as a function of the interest rate, such as Figure 9.10. In the figure, the present worth decreases with an increasing interest rate. Notice that for all interest rates less than the IRR, the present worth is positive. Thus, if the IRR is greater than the MARR, the project is acceptable, as the present worth is positive at the MARR.

In order to mathematically illustrate that PW and IRR will lead to the same decision, we must make some assumptions. Assume that the IRR is unique and real and that the present worth is a nonincreasing function of the interest rate, as in Figure 9.10. In this situation, we know that the present-worth function crosses the axis only once, at the IRR. Thus, if IRR $>$ MARR, then the present worth is positive for the MARR, and the IRR and present worth arrive at the same decision. The logic is similar if IRR $<$ MARR.

Unfortunately, as the next example shows, we cannot guarantee that an IRR exists (i.e., that i^* is unique and real).

EXAMPLE 9.7 *IRR Revisited*

It has been estimated that the cost of a new 1000-megawatt nuclear power plant can be \$1400 per kilowatt.[9] This results in a total investment of \$1.4 billion. Assume that the plant can operate for 20 years, generating \$100 of revenue per megawatt hour of power production, and that production runs for 16 hours per day, 300 days per year. Assume further that, of the \$100 in revenue generated, \$50 is retained as profit. (The other \$50 is used for operating and maintenance costs.) Finally, assume that costs to shut down the facility and dispose of the spent radioactive waste would total \$3.5 billion at the end of 20 years. Find the IRR of the investment proposal. Assume a MARR of 8%.

Solution. The net cash flow diagram for the investment is given in Figure 9.11. Thus, the annual cash flows represent profits. The present worth of the investment is

$$PW(i^*) = -\$1.4B + \$240M(^{P/A,i^*,20}) - \$3.5B(^{P/F,i^*,20}).$$

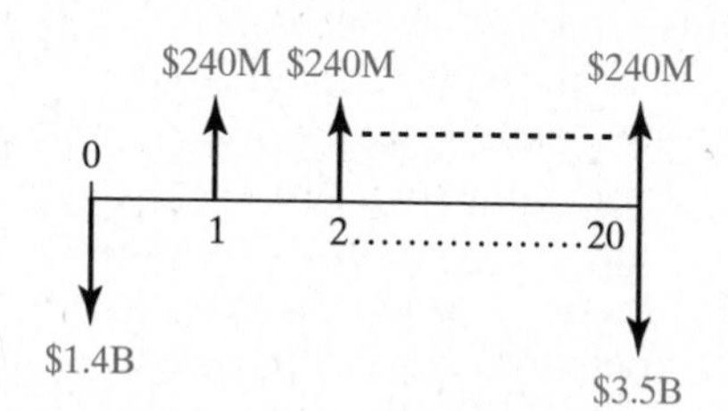

Figure 9.11 Net cash flow for investment in nuclear power plant.

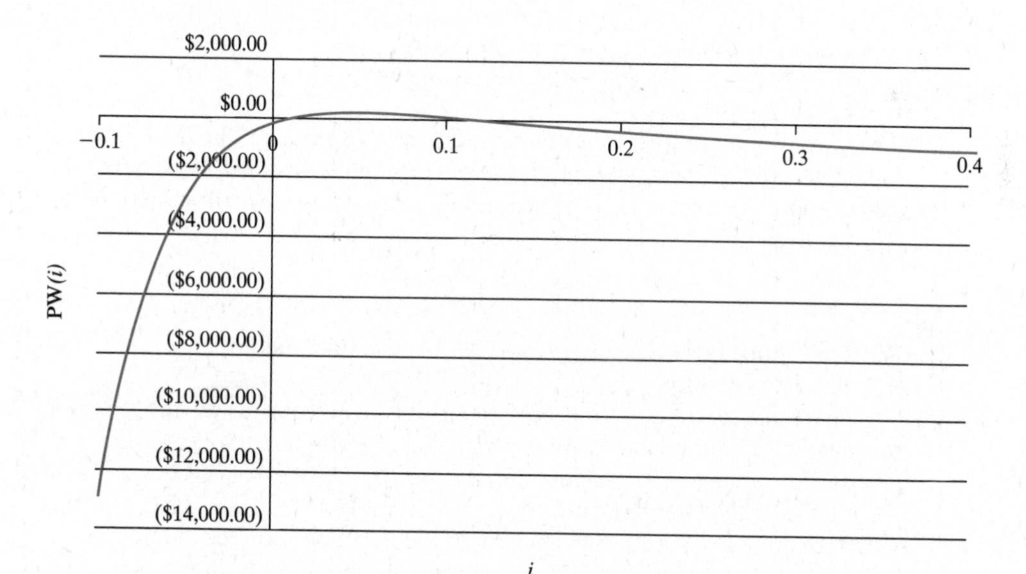

Figure 9.12 Present-worth graph for power-plant investment illustrating multiple rates of return.

[9] Nuclear Energy Institute, "CBO Report Draws Faulty Conclusions in Cost Analysis of New Nuclear Plants," www.nei.org, June 2003.

If we use the IRR function from Excel, an interesting dilemma arises. Using the default guess of 10%, we get

$$i^* = \text{IRR}(-1400, 240, 240, \ldots, -3260,) = 12.63\%,$$

but if we use a default of 0%, we obtain

$$i^* = \text{IRR}(-1400, 240, 240, \ldots, -3260, 0) = 0.57\%.$$

Solving for i^* results in two values: 0.5673% and 12.63%. This is verified in Figure 9.12, which shows the present-worth graph crossing the axis twice. Clearly, we have a problem, as we would like to make a decision with the IRR, but i^* is not unique.

The previous example has illustrated one dilemma that can arise when the IRR is used for investment analysis: The value of i^* is not unique. The intriguing questions then are, What do the values of 0.5673% and 12.63% mean? and Does either correspond to an IRR?

Before addressing meaning, let us look at the problem mathematically to understand why multiple values of i^* are not uncommon. The fact is, we cannot guarantee that an engineering project's cash flows will define a present-worth function as given in Figure 9.10. As you may recall from calculus, finding the root of an nth-order polynomial is not trivial. More importantly, there may be multiple roots. The question is, If we solve the equation

$$\text{PW}(i^*) = \sum_{n=0}^{N} \frac{A_n}{(1+i^*)^n} = 0,$$

for i^*, are we guaranteed to only have one rate of return? The answer is no. In fact, we are not guaranteed to have *any* real rates of return. If we make the substitution $x = 1/(1+i)$, then we see that our present-worth equation can be written as

$$\text{PW}(i) = A_0 + A_1 x + A_2 x^2 + \cdots + A_N x^N.$$

Written in this more typical polynomial form should make it clear as to why we might have multiple roots.

There are two rules that can be useful in alerting us to the possible presence of multiple roots. Descartes' rule of signs bounds the number of positive roots of a polynomial function by the number of sign changes in the coefficients of the function. Since we have substituted $x = 1/(1+i)$, this means that the number of sign changes in the A_n values bounds the number of interest rate values greater than -1 that set the present worth equal to zero. Thus, if there is only one sign change with respect to time in the cash flows, we are guaranteed to have at most one internal rate of return. However, with multiple sign changes, there is a chance of multiple roots and multiple internal rates of return.

Consider again Figure 9.19, the cash flow diagram for Example 9.7. The signs of the cash flows are negative for time zero, positive for periods 1 through 19, and negative for period 20. Thus, there are two sign changes in the cash flows, and multiple internal rates of return are possible.

Another useful rule is Norstrom's criterion, which states that if there is only one sign change in the *cumulative* cash flows through time, then a unique solution exists such that there is only one positive real root. For Example 9.7, the cumulative cash flows are (in million of dollars) −1400, −1160, −920, −680, −440, −200, 40, 280, . . . , 2920, 3160, and −100. Thus, there are two sign changes in the cumulative cash flow sequence (between periods 5 and 6 and between periods 19 and 20), and Norstrom's criterion cannot guarantee a unique rate.

Given these two rules, there are some typical engineering projects that occur frequently in practice and that *may* be subject to multiple rates of return because there are multiple sign changes in their cash flow diagrams. Among these projects are the following:

- **Investments with disposal/remediation costs.** As illustrated in Example 9.7, an investment that requires the removal of equipment or cleansing of a site may lead to a sign change in the final period(s) of a project. The nuclear power plant is one example, but in any situation in which a facility is to be dismantled or equipment disposed of, such a sign change is possible. Many manufacturing sites become future "brown fields," for which excessive costs can be incurred to return the area to its previous, usable state. A typical cash flow diagram for this situation is depicted in Figure 9.13.
- **Phased expansion.** At time zero, a company may decide that investments are to be made over a number of periods. For example, the company may decide to increase its output. However, instead of increasing capacity in one investment, it may increase capacity in increments over time. This can again lead to cash flow diagrams with multiple sign changes, such as that shown in Figure 9.14. The approach is often termed a phased expansion, because the investments occur in phases. We examine this type of problem in Example 9.9 in the next section.
- **Natural-resource extraction.** Mining for coal, ore, gold, silver, etc., cutting down trees, and pumping oil are problems closely associated with

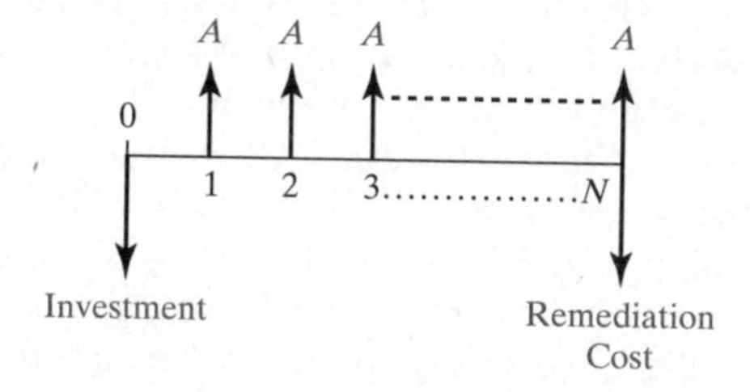

Figure 9.13 Typical net cash flow diagram for investments with extensive end-of-life costs.

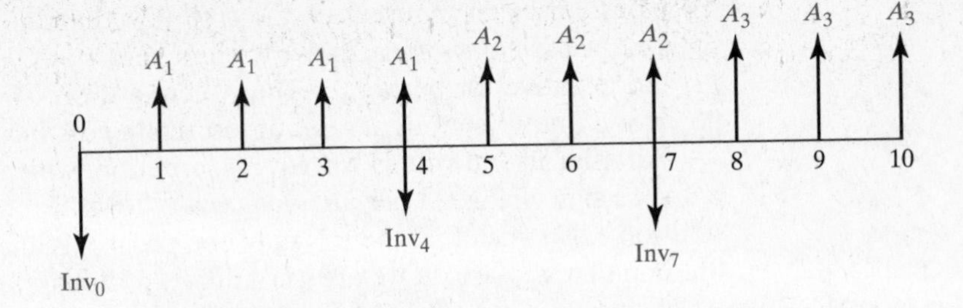

Figure 9.14 Typical net cash flow diagram for phased expansions of capacity.

engineering. Financially, they are all similar problems. In most cases, the extraction of these natural resources follows extensive surveying and testing of an area, generally leading to an estimate of the amount of the resource that is available for extraction. Given the investment, costs for removal, and probable revenue, an analysis would determine whether a mine shaft should be opened or an oil rig secured. If the company were to open one shaft or utilize one oil rig, then the cash flow diagram for the project would follow that of Figure 9.13, since remediation and cleanup costs associated with resource extraction are common.

However, most investments of this nature do not progress in such a fashion. Rather, over time, a company may expand by opening additional mine shafts, digging new oil wells, or acquiring more timber-cutting equipment. This would lead one to believe that the cash flow diagram would look more like Figure 9.14, since the expansions occur in phases. However, these investments are actually quite different.

With phased expansion, each addition of capacity generates additional revenues beyond current levels, assuming that the facility is operating at full capacity. In essence, the additions of capacity are somewhat independent of each other. (They are dependent in that they share costs, such as overhead.)

This is not the case with mineral extraction, in which the capacity—or potential revenue—of the system is defined by the amount of natural resource available. For example, when an oil field or coal deposit is discovered, the amount of resource that can be extracted is fixed by the amount of resource that is in the ground. When the first mine shaft or first oil well is installed and in full operation, one can estimate the length of time it will take to remove all of the resource, thereby defining a revenue stream, based on daily output, over that time horizon. Again, the cash flow diagram for this initial well would look like the one in Figure 9.13.

Now consider the installation of a second well in the same oil field. With the second well, the system (both wells together) produces more oil on a daily basis. However, the *total* amount of oil to be extracted over time *does not change*; its value is independent of the number of wells being utilized. The cash flow diagram associated with the second well is depicted in Figure 9.15. At first glance, the diagram may look a bit peculiar, as it illustrates an investment, followed by net revenues,

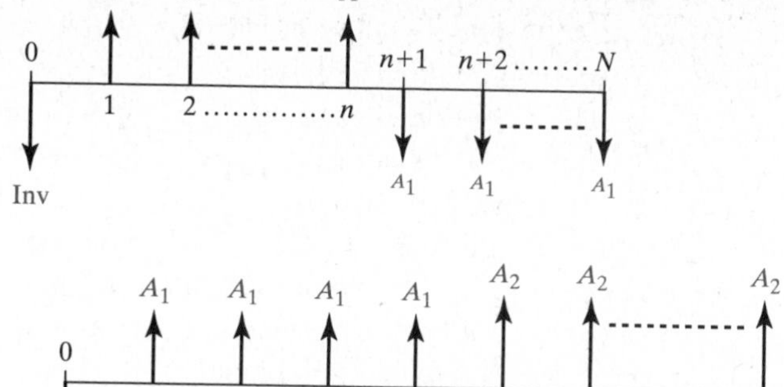

Figure 9.15
Typical net cash flow diagram for extraction of natural resources.

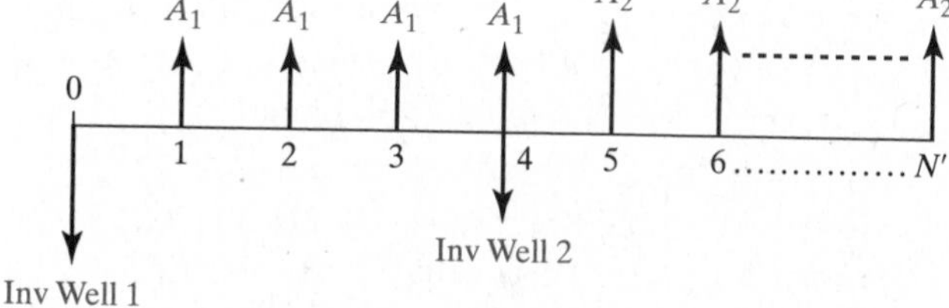

Figure 9.16
Net cash flows from the installation of two wells over time.

followed by what appear to be a number of expenses at the end of the horizon.

But the cash outflows at the end of the horizon are not expenses. Rather, they illustrate the difference between investing in the second well and doing nothing, which in this situation means operating only the first well. By adding a second well, the fixed amount of oil in the ground is being removed faster, such that all of the oil will be extracted sooner (presumably in half the time it would take to extract the oil with a single well). Figure 9.15 illustrates why we might have multiple rates of return: The cash flows change signs multiple times.

The net cash flows to the company over the history of the project, after the second well is installed, would be those in Figure 9.16, as they are sum of the of the two *individual* well investments. Note, however that the diagram is the result of two different investments and therefore is not a cash flow diagram that would in fact ever be analyzed.

In essence, bringing *additional* wells or mine shafts into operation for mineral extraction does not increase the *total* net revenues of a company over time. Rather, it brings the revenues to the company *sooner*. According to the time value of money, investing in this kind of project can be a very good decision in terms of generating wealth and thus is a common practice in mineral extraction.

Investing in additional means of extracting natural resources may seem like a strange idea, but many changes could lead to its being a good strategy: (1) A large resource, such as an oil field, may take 30 years to exploit fully, so it is only natural to expand production, as the time value of money generally makes the expansions financially beneficial. (2) Changes in commodity prices, such as a sharp rise in the price of oil, may make it worthwhile to remove resources more quickly. (3) Expanded production may lead to the discovery of additional resources. For example, a new mine shaft may help uncover an additional 30,000 ounces of diamond, further extending the life of the mine. (Indeed, finding additional resources for extraction may define a new investment project altogether.)

The aforementioned examples of adding oil rigs to producing fields and adding logging teams (with appropriate equipment) to forests being currently logged, lead to similar cash flow diagrams and again, perhaps to multiple rates of return. Descartes' rule of signs and Norstrom's criterion are merely sufficient conditions for the existence of multiple roots, not necessary conditions. This means that even if both rules indicate the existence of multiple roots, that is not guaranteed: there may still be only a single, real root.

Note that graphing the function over the interval $(-1, \infty)$ will identify all roots. We know that as $i \to \infty$, the present worth approaches A_0 asymptotically. This is because the present worth of all future cash flows go to zero as the interest rate gets larger. (Recall that the present-worth factor for a single payment, $1/(1+i)^N$, goes to zero as $i \to \infty$.) Examining the graph of the present-worth function in Figure 9.12 illustrates this concept. As the interest rate is increased beyond 40%, the present-worth function approaches the value $-\$1.4$ billion, the initial investment cost of the power facility. In general, there exists some interest rate at which we can guarantee that the present worth is negative and will remain negative for all higher interest rates. This also guarantees that there are no real roots greater than that interest rate. This is generally clear from the graph of the function, signaling the decision maker that all real roots have been identified.

Since we presented our example illustrating the case of multiple values of i^*, we have focused on the issue of when multiple rates can occur. Now we must address the issue of how to make a decision in this situation.

Recent research in the area suggests that decisions consistent with present worth can be made with one or all of the multiple rates. (These discussions are beyond the scope of this text.) Despite such research, the meaning of multiple rates is not easy to convey. One merely has to recall our cost-of-capital definition of the interest rate. With this interpretation of the interest rate i, it makes sense that, as the cost of capital (the interest rate) increases, the present worth of a project should decrease. This statement assumes that funds have been *invested*. However, the statement is not true in the case when funds are *borrowed*. Thus, the presence of multiple internal rates of return implies that a project is both investing *and* borrowing over its life. Clearly, this muddles the meaning of internal rates of return.

To avoid this complication, in the presence of multiple rates of return we turn to other methods of analysis. We strongly recommend the use of present worth, as it is the optimal decision method. However, for those who insist on generating an interest rate to describe an investment opportunity, we identify a modified rate of return in the next subsection.

9.2.2 External Rate of Return (Modified Internal Rate of Return)

While we recommend the use of present worth in situations when there are multiple rates of return, defining a rate of return for a project is very popular in industry. The reasons come from the desire to define a project according to its return, as a measure of efficiency, and not necessarily its monetary worth. Thus, we are compelled to be complete in our discussion and introduce the external rate of return, or ERR, which provides a rate of return that can be computed regardless of the nature of the series of cash flows defined by the project. The ERR goes by a number of names in the literature and has a number of variants, including the modified internal rate of return (MIRR) and the average rate of return (ARR). We use the term ERR in order to stress the fact that this rate is not internal to the project, as it is requires additional information, namely, an external interest rate, in its calculation.

In its most general form, the external rate of return requires *two* external interest rates: an investment rate r, which is the return we expect our investments to earn, and a borrowing rate k, often defined as the cost of capital. Further, we must classify our cash flows A_n as being either positive (defined as A_n^+) or negative (defined as A_n^-). To find the ERR, we compute the future worth (at time N) of all positive net cash flows, using the investment rate r. We compute the present worth of all negative net cash flows with the borrowing rate k. The ERR is the interest rate that equates these two values at time zero such that

$$\frac{\sum_{n=0}^{N} A_n^+(1+r)^{N-n}}{(1+\text{ERR})^N} + \sum_{n=0}^{N} \frac{A_n^-}{(1+k)^n} = 0,$$

or

$$\frac{\sum_{n=0}^{N} A_n^+(1+r)^{N-n}}{(1+\text{ERR})^N} = -\sum_{n=0}^{N} \frac{A_n^-}{(1+k)^n}.$$

As we have noted, the interest rate used for discounting (the MARR) can take on numerous meanings, including the cost of capital *and* the minimum acceptable level of return. If we assume that the MARR takes on both of these meanings, then we can assume that the preceding rates k and r are equivalent and equal to the MARR.

Accordingly, substituting i for both k and r, we can define the ERR in general as

$$\text{ERR} = \left(\frac{\sum_{n=0}^{N} A_n^+(1+i)^{N-n}}{-\sum_{n=0}^{N} \frac{A_n^-}{(1+i)^n}} \right)^{1/N} - 1. \tag{9.10}$$

This relationship is visually depicted in Figure 9.17, as the cash flows are moved to the future or present depending on their sign.

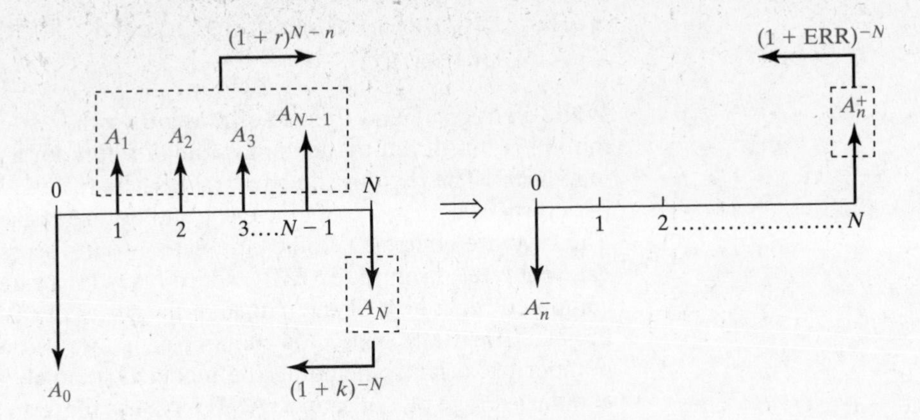

Figure 9.17 Visualization of the external rate of return.

With this method, we are guaranteed a single rate to evaluate. Our decision is then given by the following relationships:

ERR Value	Decision
ERR > MARR	Accept the project.
ERR = MARR	Indifferent about the project.
ERR < MARR	Reject the project.

Let us illustrate by revisiting our power-plant investment example with remediation costs.

EXAMPLE 9.8

External Rate of Return

For the data given in Example 9.7, determine the ERR and determine whether the investment should be undertaken. We assume that remediation costs total \$3.5 billion.

Solution. With a MARR of 8%, the ERR can be found from Equation (9.10):

$$\text{ERR} = \left(\frac{\sum_{n=0}^{N} A_n^+(1+i)^{N-n}}{-\sum_{n=0}^{N} \frac{A_n^-}{(1+i)^n}}\right)^{1/N} - 1 = \left(\frac{\$240\text{M}(\overset{F/A,8\%,20}{45.7620})}{\$1.4\text{B} + \$3.5\text{B}(\overset{P/F,8\%,20}{0.2145})}\right)^{1/20} - 1$$

$$= .085 = 8.5\%.$$

Since the ERR is greater than our MARR of 8%, we would accept the project. We illustrate this calculation with a spreadsheet function in the next example.

Each of the methods we have presented is consistent with present worth. We can make that same claim here with the ERR. We do so under the assumption that both k and r are equal to i. Using our notation for positive and negative cash flows, we know that a project is acceptable according to the present-worth criterion if

$$\mathrm{PW}(i) = \frac{\sum_{n=0}^{N} A_n^+(1+r)^{N-n}}{(1+i)^N} + \sum_{n=0}^{N} \frac{A_n^-}{(1+k)^n} > 0,$$

or, equivalently,

$$\frac{\sum_{n=0}^{N} A_n^+(1+r)^{N-n}}{(1+i)^N} > -\sum_{n=0}^{N} \frac{A_n^-}{(1+k)^n},$$

$$\sum_{n=0}^{N} A_n^+(1+r)^{N-n} > -\sum_{n=0}^{N} \frac{A_n^-}{(1+k)^n}(1+i)^N,$$

$$\frac{\sum_{n=0}^{N} A_n^+(1+r)^{N-n}}{-\sum_{n=0}^{N} \frac{A_n^-}{(1+k)^n}} > (1+i)^N,$$

$$\frac{\sum_{n=0}^{N} A_n^+(1+r)^{N-n}}{-\sum_{n=0}^{N} \frac{A_n^-}{(1+k)^n}} = (1+\mathrm{ERR})^N > (1+i)^N,$$

$$(1+\mathrm{ERR}) > (1+i).$$

If the PW of the project is positive, then the ERR is greater than i, or the MARR. Thus, the decisions are consistent with present worth.

The ERR allows us to compute a single rate for any set of cash flows, but we should not be misled by its meaning. It is not an internal rate of return, as we earlier defined the IRR. The ERR is an internal rate of return to a *modified* cash flow diagram (the original cash flow diagram modified with the MARR). Because of this requirement, the definitions and meaning of the IRR cannot be directly applied to the ERR. Thus, the ERR describes the return on an investment under explicit assumptions about how individual cash flows are costed. Let us examine one more example with use of the ERR and illustrate its computation with a spreadsheet function.

EXAMPLE 9.9

ERR Revisited

In late 2001, Comalco, Ltd., approved the construction of an Alumina Refinery in Gladstone, Queensland, Australia, with a capacity of 1.4 million tons of alumina per year at the cost of about $750 million.[10] (Alumina is produced from bauxite and is smelted to produce aluminum.) The plant was constructed such that it could be expanded to an annual capacity of 4.2 million tons, and in November of 2003 Comalco announced that it was considering expanding due to market conditions.[11]

[10] "Construction of Comalco Alumina Refinery to Commence," *News Article*, www.comalco.com.au, October 26, 2001.

[11] Sinclair, N., "Comalco Mulls More Capacity at Australia Alumina Refinery," *Dow Jones Newswires*, November 5, 2003.

Make the following assumptions: Consider this problem as a phased expansion problem with the cost to construct the original facility ($750 million) spread evenly over years 2002 through 2004. First-year (2005) revenues are $500 per ton against per unit costs of $150 per ton with a production of 1.4 million tons. The per ton revenues are expected to decrease 3% annually due to worldwide increases in capacity, while per unit costs are expected to increase 3% annually. Assume that expansion to 4.2 million tons of annual capacity is to commence in 2008 and 2009 at the cost of $700 million each year. This results in a capacity of 2.8 million tons in 2010 and 4.2 million tons in 2011 through the remaining plant life (ending in 2019). The plant carries an expected remediation cost of $500 million. Finally, annual fixed costs of $30 million are expected to begin in 2005 and extend through 2009, increasing to $70 million for the life of the project. Assuming a MARR of 16%, should the phased investment be considered?

Solution. The cash flows for this project are given in the spreadsheet in Figure 9.18, along with the cumulative cash flows. According to the cash flows and Descartes's rule, there is a possibility of multiple rates of return, as there is a change in the signs of the cash flows between periods 2004 and 2005, between 2007 and 2008, between 2009 and 2010, and between 2018 and 2019. Further, according to Norstrom's criterion, the change in the sign of the cumulative cash flows between periods 2005 and 2006, between 2008 and 2009, and between 2009 and 2010 also support the possibility of multiple roots.

Multiple roots can be confirmed by graphing the present-worth function for interest rates greater than −100%, as in Figure 9.19 for rates between −80% and 40%. Although it is difficult to establish where the present-worth function crosses the x-axis (the interest rate) from the figure, there are two roots: −74.4% and 34.1%.

Note that if the IRR function in Excel is called, it returns the value 34.1%. However, we cannot interpret this number as the IRR, because it is not unique. We can compute the external rate of return (ERR) from the MIRR function in Excel,

	A	B	C	D	E	F	G	H	I
1	Example 9.9: Phased Expansion of Alumina Refinery						Input		
2							Investment	Diagram	million
3	Period	Investment	Production	Cash Flow	Cumulative		Production Phase I	1.4	million tons
4	2002	-$250.00	$0.00	-$250.00	-$250.00		Production Phase IIa	2.8	million tons
5	2003	-$250.00	$0.00	-$250.00	-$500.00		Production Phase IIb	4.2	million tons
6	2004	-$250.00	$0.00	-$250.00	-$750.00		Production Cost	$150.00	per ton
7	2005		$460.00	$460.00	-$290.00		Production Revenue	$500.00	per ton
8	2006		$432.70	$432.70	$142.70		g (Revenues)	-3%	per year
9	2007		$405.84	$405.84	$548.54		g (Costs)	3%	per year
10	2008	-$700.00	$379.40	-$320.60	$227.94		Phase I Fixed Cost	$30.00	million
11	2009	-$700.00	$353.35	-$346.65	-$118.71		Phase II Fixed Cost	$70.00	million
12	2010		$645.33	$645.33	$526.62		Remediation Cost	$500.00	million
13	2011		$926.99	$926.99	$1,453.61		Interest Rate	16%	per year
14	2012		$851.94	$851.94	$2,305.55		Periods	17	years
15	2013		$777.80	$777.80	$3,083.35				
16	2014		$704.48	$704.48	$3,787.83				
17	2015		$631.92	$631.92	$4,419.75		Output		
18	2016		$560.07	$560.07	$4,979.81		ERR	21.02%	
19	2017		$488.84	$488.84	$5,468.65				
20	2018		$418.18	$418.18	$5,886.83			=MIRR(D4:D21,H13,H13)	
21	2019		-$151.98	-$151.98	$5,734.86	=D19+E18			
22									
23		=$H$5*($H$7*(1+$H$8)^(A15-2005)-$H$6*(1+$H$9)^(A15-2005))-$H$11							
24									

Figure 9.18 Cumulative cash flows, in millions of U.S. dollars, for expansion of alumina production.

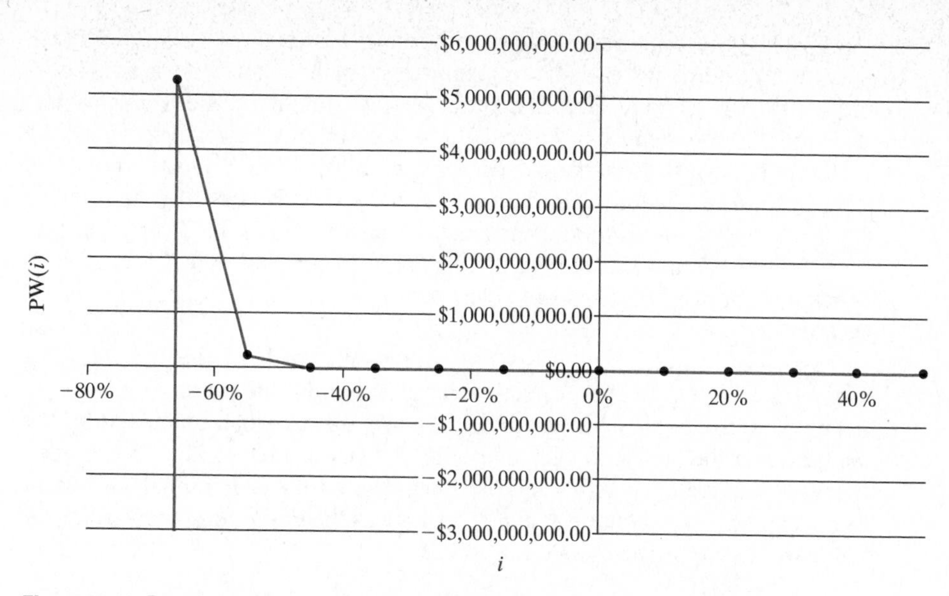

Figure 9.19 Present worth of investment in phased expansion of alumina capacity.

defined as

$$\text{ERR} = \text{MIRR(values, finance\_rate, reinvest\_rate)}.$$

Inserting the given data yields

$$\text{ERR} = \text{MIRR}(-250, -250, \ldots, 418, -152, 0.16, 0.16) = 21.0\%.$$

This value is programmed in cell H18 in the spreadsheet by using the cash flows in column D as inputs along with the MARR. Since ERR is greater than MARR, we accept the project and plan to expand production in phases.

9.2.3 Benefit–Cost Ratio

Thus far, we have focused our attention on the capital investment decisions of companies. Government entities also make investments, but with different goals, as a government entity does not ordinarily exist to generate profits. Investments made by government entities are usually designed to benefit the public in some way. Typical are investments in infrastructure, such as roads, highways, bridges, and dams; support for leisurely activities, such as parks, lakes, museums, and libraries; and defense, such as forts, tanks, ships, submarines, and fighter aircraft. Governments may also invest in companies in the form of tax subsidies in hopes of fostering growth and creating jobs.

Clearly, these investments are important. However, it is not clear how they should be analyzed, since maximizing worth is not their aim. A government exists to serve the public; thus, its investment decisions must reflect that goal. The question is, Should these investments be analyzed with the methods previously described? The answer is yes—the analysis should be rigorous and the decision should be financially sound—but from a different perspective. We differentiate this analysis from the ones we have presented thus far by defining the benefits and costs of a project, and if the analysis concludes that the benefits outweigh the costs, then the project is deemed acceptable.

Specifically, we define B as the equivalent worth of the benefits of a project to the user of the project, which is generally the public. Further, we define C as the equivalent worth of the costs of the project incurred by the facilitator of the project, which is usually the government entity. We employ the term "equivalent worth" because the present, future, or annual equivalent worth can be used to define B and C, as long as the definition is consistent for both. We define the benefit–cost ratio as

$$\mathrm{BC}(i) = \frac{B}{C}. \tag{9.11}$$

Note that the terms "benefits" and "costs" are a bit misleading, as the numerator comprises the benefits to the user (the public) and the denominator comprises the costs to the facilitator (the government). There may be negative benefits, or disbenefits, incurred by the user when the project is complete (such as an increase in pollution or noise from a new road). Although one may envision these as costs, they are placed in the numerator, because they are incurred by the user. Similarly, there may be situations in which the government entity generates revenue (such as an entrance fee) from a project. While this is not a traditional cost, we treat it as a negative cost in our denominator. Thus, although the ratio is defined as benefits over costs, it is more aptly defined as user over facilitator.

To be clear, the costs incurred by the facilitator are treated as positive values in the denominator, while any benefits for the facilitator are negative values in the denominator. This will generally lead to positive values for both B and C.

Once the BC ratio is computed, the decision as to whether to accept or reject the project depends on whether the benefits outweigh the costs, in accordance with the following table:

BC Value	Decision
$\mathrm{BC}(i) > 1$	Accept the project.
$\mathrm{BC}(i) = 1$	Indifferent about the project.
$\mathrm{BC}(i) < 1$	Reject the project.

The next example illustrates the use of benefit–cost analysis.

EXAMPLE 9.10

Benefit–Cost Analysis

Siemens Dematic received an $89.9 million contract from the U.S. Postal Service to install a ventilation and filtration system on the Postal Service's letter-sorting machines in 280 facilities in 2004. The system is designed to remove 99.7% of contaminants in the air by pulling exhaust through a set of filters.[12]

Assume that the system has a life of 25 years and that installation was completed in 2004 such that benefits of the system began in 2005. Assume further that the expected benefits of the system for the employees are as follows:

- Reduction of 100 sick days at each facility. A sick day costs each employee an estimated $80 in wages (8 hours at $10 per hour).
- Reduction, on average, of 8 major illnesses per year at a savings of $50,000 per illness.
- Reduction of 1 fatality per year at a savings of $500,000 per fatality.

The Postal Service itself receives benefits, too, including the following:

- Reduction of 100 sick days at each facility. A sick day costs, on average, 8 hours in loss of productivity, at an estimated $20 per hour at each facility.
- Annual savings of $550,000 in the cost of health care benefits.

Determine the benefit–cost ratio for this contract, assuming a 6% annual MARR. O&M costs are estimated at $2500 per year per system, and salvage values are assumed to be negligible.

Solution. The annual employee benefits can be tabulated as

$$280 \text{ facilities} \times \frac{100 \text{ days}}{\text{facility}} \times \frac{8 \text{ hours}}{\text{day}} \times \frac{\$10}{\text{hour}} = \$2.24 \text{ million},$$

leading to total annual benefits of

$$\underbrace{B}_{\text{Benefits:}} = \underbrace{\$2.24\text{M}}_{\text{Sick Days}} + \underbrace{(8)\$50,000}_{\text{Illness}} + \underbrace{(1)\$500,000}_{\text{Fatalities}} = \$3.14 \text{ million}.$$

The benefits to the post office (the facilitator) are:

$$\underbrace{\$4.48\text{M}}_{\text{Productivity}} + \underbrace{\$550,000}_{\text{Health Care}} = \$5.03 \text{ million}.$$

The annual costs total

$$C = \$89.9\text{M}(\overset{A/P,6\%,25}{0.07823}) + (280)(\$2500) = \$7.73 \text{ million}.$$

[12] Willetts, S., "United States Postal Service Awards $89.9M Pact to Siemens Dematic for New Ventilation and Filtration System," *Dow Jones Newswires,* November 17, 2003.

This leads to a benefit–cost ratio (recall that benefits to the facilitator are negatives in the denominator) of

$$BC(6\%) = \frac{B}{C} = \frac{\$3.14M}{\$7.73M - \$5.03M} = 1.16.$$

With the ratio greater than 1, the investment should be made.

Note that we defined the BC ratio according to the annual equivalent worth of benefits and costs. This was not necessary, as it is critical only that the benefits and costs be computed according to similar time frames and that the time value of money be considered. That is, we could have used the present or future worth of the benefits and costs, and the decision would not have changed. In fact, the value of the ratio would not have changed, since it is equivalent to multiplying both the numerator and denominator by the appropriate interest rate factor. Because these factors cancel each other, we are left with the same ratio.

The meaning of the benefit–cost ratio is quite clear: If the time-value-equivalent benefits exceed the time-value-equivalent costs, then the project should be pursued. This clear meaning does not, unfortunately, reconcile considerable debate about the use of the benefit–cost ratio. The debate is not about the analytical method, since it is sound and, as will be shown soon, is consistent with present worth.

Rather, the difficulty with using benefit–cost analysis is that the values utilized for both benefits and costs can be open to debate. Improvements in infrastructure often lead to savings in time and improvements in safety. These benefits are sometimes described as "noneconomic" benefits, but benefit–cost analysis requires economic quantities. For example, improvements in a road may speed traffic along more quickly. A benefit to users is that their travel time decreases. The question is, What is this time worth to the user? This could be computed by determining the types of users and possibly their wages, but that would not capture all of the worth associated with a reduction in time. The same road improvements may lead to a reduction in the number of accidents, thereby lessening the number of injuries and fatalities. What are these worth? Again, medical and insurance costs can be estimated, but estimating the worth of a life is a delicate issue.

There may also be questions as to whether a value belongs in the numerator or denominator. Road improvements may lead to increased tolls. Are these additional costs incurred by the user, or are they increased benefits (in the form of revenues) to the facilitator? This is another debatable issue.

Again, the debate is not with BC analysis, but rather with the *values used* in the analysis. Since these values are open to debate, it is easy to manipulate the ratio into producing desired results by skewing the values of different benefits or costs.

That the BC method is consistent with present worth is easy to show. Let us assume that B and C are defined as present-worth values. Then if

$$\text{PW}(i) = B - C > 0,$$

then we would accept the project. This implies that

$$B > C,$$

and since C is a positive value,

$$\frac{B}{C} > 1.$$

Thus, if the present worth of a project is positive, then the benefits of the project exceed the costs, and the decision generated by the benefit–cost ratio is consistent.

On a final note, although consistent with present worth, benefit–cost analysis provides much less information. The ratio merely conveys the factor by which benefits exceed costs, on a per dollar basis. As with the internal rate of return, the actual monetary impact is unknown unless the costs are specified. That is why we categorize the BC ratio as a relative measure of worth.

9.3 Examining the Real Decision Problems

We revisit our introductory example concerning the BMW Group's expansion of its Mini plant to accommodate the new station wagon model. As opposed to individually answering the questions posed, the spreadsheet in Figure 9.20 was designed to determine PW and IRR values for any tax rate or MARR values.

Recall that the after-tax MARR is equal to $(1 - t)$MARR, so that a 50% tax rate and a 12.5% after-tax MARR translates to a 25% MARR without taxes. This leads to a PW(25%) of –EUR21.83 million without taxes and a PW(12.5%) of EUR13.69 million with taxes. The IRR for both these cases, which is unique because we have only one sign change in the cash flows, is 21.60% and 14.5%, respectively. Thus, the project is acceptable when taxes are considered, but otherwise not.

A graph of both after-tax MARR and IRR values versus the tax rate is also given. It is clear from this graph that the tax rate is an important consideration, as it affects the MARR and the cash flows. The project appears acceptable for rates above 25%.

Note that we can generate the data in the range defined in cells A25:C35 "by hand." That is, we can enter tax-rate values in cell L18 and record the inputs (t and the after-tax MARR) and resulting IRR value in cells A25:C35. However, this process can be automated with another tool in Excel: VisualBasic for Applications. Entire books have been written about using VisualBasic and creating Macros in Excel. VisualBasic is an *extremely* powerful tool that combines the ease and advanced visual capabilities of a spreadsheet with the flexibility of general programming.

	A	B	C	D	E	F	G	H	I
1	Chapter 9: Mini Plant Expansion								
2									
3	Year:	2006	2007	2008	2009	2010	2011	2012	2013
4	Revenues								
5	Net Revs:	(€ 60.00)	(€ 20.00)	€ 60.00	€ 116.40	€ 112.91	€ 109.52	€ 106.24	€ 103.05
6	Salvage:	€ 0.00	€ 0.00	€ 0.00	€ 0.00	€ 0.00	€ 0.00	€ 0.00	€ 0.00
7	Total Inflows:	(€ 60.00)	(€ 20.00)	€ 60.00	€ 116.40	€ 112.91	€ 109.52	€ 106.24	€ 103.05
8	Expenses								
9	Fixed Costs:	€ 0.00	€ 0.00	(€ 15.00)	(€ 15.00)	(€ 15.00)	(€ 15.00)	(€ 15.00)	(€ 15.00)
10	Depreciation:	€ 0.00	€ 0.00	(€ 25.00)	(€ 25.00)	(€ 25.00)	(€ 25.00)	(€ 25.00)	(€ 25.00)
11	Total Expenses:	€ 0.00	€ 0.00	(€ 40.00)	(€ 40.00)	(€ 40.00)	(€ 40.00)	(€ 40.00)	(€ 40.00)
12									
13	Taxable Income:	(€ 60.00)	(€ 20.00)	€ 20.00	€ 76.40	€ 72.91	€ 69.52	€ 66.24	€ 63.05
14									
15	Income Tax:	€ 30.00	€ 10.00	(€ 10.00)	(€ 38.20)	(€ 36.45)	(€ 34.76)	(€ 33.12)	(€ 31.52)
16									
17	Profit A/T:	(€ 30.00)	(€ 10.00)	€ 10.00	€ 38.20	€ 36.45	€ 34.76	€ 33.12	€ 31.52
18									
19	Purchase:	(€ 100.00)	(€ 50.00)						
20									
21	**Cash Flow A/T:**	(€ 130.00)	(€ 60.00)	€ 35.00	€ 63.20	€ 61.45	€ 59.76	€ 58.12	€ 56.52

K	L	M
Input		
Investment	€ 150.00	million
Operating Cost	€ 15.00	million
Initial Production	0.02	million cars
G (Production)	0.02	million cars
Max Production	0.04	million cars
Net Revenues	€ 3,000.00	per car
g (Net Revs)	-3.0%	
Lost Production	-0.03	million cars
G (Lost Prod)	0.02	million cars
Net Revenues	€ 2,000.00	per car
Salvage Value	€ 0.00	million
MARR	25.0%	
A/T MARR	12.5%	
Periods	6	years
Depreciation	SL	
Recovery Period	6	years
Tax Rate	50%	
Output		
PW	€ 13.69	million
IRR	14.5%	

	A	B	C
24	**Tax Rate**	**A/T MARR**	**IRR**
25	0.00%	25.00%	21.63%
26	5.00%	23.75%	21.08%
27	10.00%	22.50%	20.50%
28	15.00%	21.25%	19.89%
29	20.00%	20.00%	19.25%
30	25.00%	18.75%	18.58%
31	30.00%	17.50%	17.86%
32	35.00%	16.25%	17.10%
33	40.00%	15.00%	16.30%
34	45.00%	13.75%	15.45%
35	50.00%	12.50%	14.53%
36			

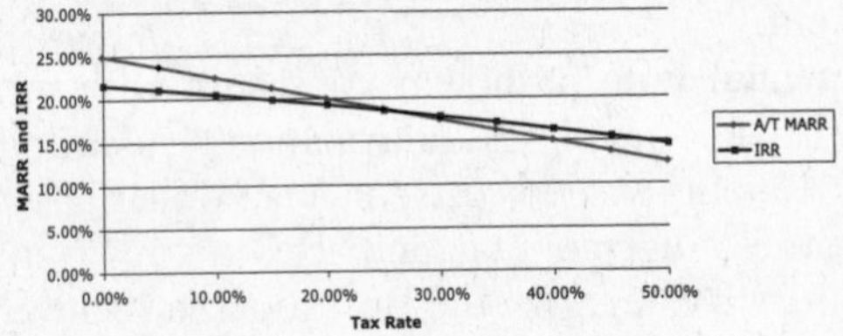

Figure 9.20 After-tax cash flow for expansion in production of station wagons.

Instead of discussing a number of techniques, we illustrate some basic functions in order to complete our task of generating the data for our graph. The best way to start using VisualBasic is to record a macro in Excel. First, go to the Tools menu and select 'Macro→Record New Macro.' After giving the macro a name, you will find that a box appears on the screen that, when pushed, ends the recording. However, a command is copied for every action taken between selecting the record option and stopping the recording. Going back to the Macro menu option will allow you to edit what has been recorded.

For our purposes, we require only that data be read from the spreadsheet into our Macro and that data be written from the Macro to the spreadsheet. In addition, we need to execute a loop to examine a number of possible values in a range.

To read data from the spreadsheet in Figure 9.20 into our VisualBasic Macro (see Figure 9.21), we use the command

$$\text{ATMarr} = \text{Cells}(14,12).\text{Value}.$$

This assigns the value in cell L14, designated as (Row,Column) by (14,12), to the variable ATMarr in the macro.

To write output to the spreadsheet, we reverse the command as follows:

$$\text{Cells}(25,2).\text{Value} = \text{ATMarr}.$$

This places the value of the variable ATMarr in the cell located at row 25, column 2 (cell B25).

There are many ways in which to loop. If the number of loops to complete is fixed, then the 'For Next' loop is the most straightforward. This type of loop looks like the following:

```
For Iteration=1 To 10 Step 1
        ⋮
Next Iteration
```

The code that resides between the For and Next lines is executed the requested number of times (10 in this case). Note that the default step is 1, but other steps may be useful, depending on the application. (The value of Iteration is 1 in the first loop, 1 + Step in the second loop, etc.)

After initializing the variables (with the Dim function) as Doubles (floating point), the VisualBasic code in Figure 9.21 executes a loop that increments the variable TaxRate from 0 to 0.50 by 0.05. (An apostrophe denotes a commented line, used to identify sections of code.) The TaxRate value is assigned to cell L18. This in turn changes the definition of the after-tax MARR in cell L14, which has been programmed as

$$= (1 - \text{L18}) * \text{L13}$$

in the spreadsheet. The after-tax cash flows in the spreadsheet are programmed with the tax rate in cell L18, and cell L22 computes its IRR. The macro then

Figure 9.21 Macro that generates graph data for expansion in production of station wagons.

```
Sub ATMarr()
'
' ATMARR Macro

  Dim TaxRate As Double
  Dim ATMarr As Double
  Dim IntRate As Double
  Dim MARR As Double

  MARR = Cells(13,12).Value

  For Iteration = 0 To 10 Step 1
    TaxRate = Iteration * 0.05
    'Get Input from Spreadsheet
    Cells(18, 12).Value = TaxRate
    ATMarr = Cells(14, 12).Value
    IntRate = Cells(22, 12).Value

    'Print Output to Spreadsheet
    Cells(25 + Iteration, 1).Value = TaxRate
    Cells(25 + Iteration, 2).Value = ATMarr
    Cells(25 + Iteration, 3).Value = IntRate

  Next Iteration

End Sub
```

assigns the after-tax MARR and IRR values computed in the spreadsheet to the variables ATMarr and IntRate, respectively. Finally, the variables TaxRate, ATMarr, and IntRate are written to consecutive cells in row 25. The 'Next' command increments the Iteration count by 1 and the process of calculating the TaxRate, ATMarr, and IntRate commences again with the results written in row '25 + Iteration,' The process repeats until Iteration = 11, thus terminating the loop and macro.

Once the macro is run, the data are available for graphing. The use of the Chart Wizard has already been discussed.

9.4 Key Points

- An absolute measure of worth is a measure of the amount of wealth a project generates, while a relative measure of worth is a measure of the expected return from the project, according to the amount invested.
- The present worth is the monetary value of a project at time zero. It can be viewed as discounted profit or the value of a project after the cost of capital has been paid.
- The future worth is equivalent in meaning to the present worth, but defined at time N.
- The annual equivalent worth of a project provides a periodic measure of the worth of the project.
- Market value added is a measure of worth that is equivalent to present worth and that is derived from income minus a charge for the use of capital.
- The internal rate of return is a relative measure of the return that an investment generates, expressed as an interest rate.
- The external rate of return, or modified internal rate of return, is the interest rate that equates the future worth of revenues (at the MARR) to the present worth of costs (at the MARR).
- Benefit–cost analysis is used to evaluate projects in the public sector. If a project's equivalent benefits exceed its equivalent costs, the project is deemed acceptable.
- Although each of the measures presented in this chapter provides slightly different information, they all provide consistent accept–reject decisions for a single project under certainty.

9.5 Further Reading

Bernhard, R.H., "'Modified' Rates of Return for Investment Project Evaluation—a Comparison and Critique," *The Engineering Economist*, 24(3):161–168, 1979.

Bussey, L.E., and T.G. Eschenbach, *The Economic Analysis of Industrial Projects,* 2d ed. Prentice Hall, Englewood Cliffs, New Jersey, 1992.

Fleischer, G.A., *Introduction to Engineering Economy.* PWS Publishing Co., Boston, 1994.

Hartman, J.C., "On the Equivalence of Net Present Value and Economic Value Added as Measures of a Project's Worth," *The Engineering Economist,* 45(2):158–165, 2000.

Hartman, J.C., and I.C. Schafrick, "The Relevant Internal Rate of Return," *The Engineering Economist,* 49(2), 2004.

Hazen, G.B., "A New Perspective on Multiple Internal Rates of Return," *The Engineering Economist*, 48:31–51, 2003.

Lin, S.A.Y., "The Modified Internal Rate of Return and Investment Criterion," *The Engineering Economist*, 21(4):237–247, 1976.

Lohmann, J.R., "The IRR, NPV and the Fallacy of the Reinvestment Rate Assumption," *The Engineering Economist,* 33(4):303–330, 1988.

Lorie, J., and L.J. Savage, "Three Problems in Capital Rationing," *The Journal of Business*, 28(4):228–239, 1955.

Park, C.S., *Contemporary Engineering Economics,* 3d ed. Prentice Hall, Upper Saddle River, New Jersey, 2002.

Shrieves, R.E., and J.M. Wachowicz, Jr., "Free Cash Flow, Economic Value Added, and Net Present Value: A Reconciliation of Variations of Discounted-Cash-Flow Valuation," *The Engineering Economist,* 46(1):33–52, 2001.

Teichroew, D., A.A. Robichek, and M. Montalbano, "An Analysis of Criteria for Investment and Financing Decisions under Certainty," *Management Science*, 13(3):150–179, 1965.

Thuesen, G.J., and W.J. Fabrycky, *Engineering Economy,* 9th ed. Prentice Hall, Upper Saddle River, New Jersey, 2001.

9.6 Questions and Problems

9.6.1 Concept Questions

1. What is the meaning of present worth and how is it dependent on the interest rate?
2. Why is present worth often referred to as discounted profit?
3. If a company uses the cost of capital to discount cash flows, what is the company's interpretation of present worth?
4. If a company uses a growth rate to discount cash flows, what is the company's interpretation of present worth?
5. What does the typical graph of present worth with respect to the interest rate look like? How does this answer relate to the one from the previous question?

6. Why do present worth, future worth, and annual equivalent worth produce the same decisions?
7. What is the difference between absolute and relative measures of worth?
8. When do present worth and market value added produce the same decision and value for the worth of a project?
9. Explain why one may want to calculate multiple measures of worth even if they produce the same decision. Use present worth, market value added, and the internal rate of return in your discussion.
10. What complications can arise when using the internal rate of return? If these problems are encountered, what should the decision maker do?
11. Can Descartes' rule or Norstrom's criterion guarantee the existence of multiple internal rates of return? Explain.
12. When should we be wary that a cash flow diagram will produce multiple internal rates of return?
13. What is the meaning of the internal rate of return? How is it related to present worth?
14. The benefit–cost ratio is often computed as the benefits minus the costs in the numerator and the equivalent costs of the investment in the denominator. Illustrate how this approach produces the same decisions as the ratio presented in the last section of this chapter.
15. Why is the benefit–cost ratio often debated in its application?

9.6.2 Drill Problems

1. An investment of $12 million is made at time zero, with net revenues of $3.3 million in year 1 growing at a rate of 8% per year. What is the present worth of the investment, assuming that the MARR is 18% and the project lasts seven years?
2. An investment of $400,000 is made at time zero. Revenues are expected to grow from $100,000 in period 1 by $50,000 each year for six years, while costs are expected to remain steady at $50,000 per year. If there is no salvage value after six years, what is the future worth of the project, assuming a MARR of 9%? What is the annual equivalent worth, assuming a MARR of 13%?
3. An investment has the following annual cash flows: –$100,000, –$50,000, $100,000, $100,000, $100,000, and –$50,000. Can the project have multiple internal rates of return? Find all internal rates of return. Also, calculate the external rate of return. Assume a MARR of 17%.
4. An investment has the following annual cash flows: –$100,000, –$50,000, $100,000, $100,000, $100,000, and –$175,000. Can the project have multiple internal rates of return? Find all internal rates of return. Also, calculate the external rate of return. Assume a MARR of 17%.
5. An investment of $1.25 million is made at time zero, with annual revenues of $1 million over 10 years and O&M costs of $350,000 in year 1, growing by $35,000 each year. The salvage value at the end of 10 years is $10,000. Graph the present

worth as a function of the interest rate. For what MARR values is the project accepted and for what MARR values is it rejected?

6. An investment of $1.25 million is made at time zero, with annual revenues of $1 million over 10 years and O&M costs of $350,000 in year 1, growing by $35,000 each year. The salvage value at the end of 10 years is $10,000. If the asset is classified as a 5-year asset and depreciated according to MACRS GDS (Table 8.3), the tax rate is 34% (credit all losses), and the after-tax MARR is 7%, should the project be accepted according to

   (a) Market value added?

   (b) Annual equivalent worth?

   (c) External rate of return?

7. A government entity makes an investment of $150,000 in a new park that is expected to receive 10,000 visitors per year. If the typical visitor stays for two hours and the stay is generally worth $15 per hour to the visitor, compute the benefit–cost ratio of the park. Assume a study period of 20 years, a salvage value of $50,000, $20,000 in annual maintenance costs, and a 3% annual rate of interest.

8. An investment of $100,000 is made at time zero, with net cash flow returns of $23,000 per year for the next eight years and no salvage value. Determine whether the project should be accepted or rejected, assuming

   (a) Present worth and a 15% MARR.

   (b) Future worth and a 10% MARR.

   (c) Annual equivalent worth and a 25% MARR.

   (d) Internal rate of return and a 12% MARR.

   (e) External rate of return and a 14% MARR.

9. An investment to improve the safety of a road at the cost of $2.5 million is being considered. The improvement will last 10 years with no salvage value, but is expected to reduce accidents by 50% each year. This amounts to 25 fewer accidents (valued at $10,000 per accident), 15 fewer cases of injury (valued at $25,000 per case), and 1 fewer fatality (valued at $500,000 per fatality) per year. Should the project be undertaken if the interest rate is 6% per year?

10. An investment of $4.5 million is made at time zero, with annual revenues of $500,000 in year 1, growing at the rate of 30% annually over a 7-year horizon. Annual O&M costs are estimated at $150,000 per year. The investment is to be depreciated over a 10-year period according to MACRS with the alternative straight-line method (and the half-year convention). Assume that the operation is sold for $750,000 after year 7. If the effective tax rate is 40% (credit all losses), should the project be accepted according to

    (a) Present worth with an after-tax MARR of 12%?

    (b) Market value added with an after-tax MARR of 8%?

    (c) Internal rate of return with an after-tax MARR of 14%?

9.6.3 Application Problems

1. Consider the data and after-tax cash flows developed in Application Problem 7(a) in Chapter 8, in which a contract was awarded to Chicago Bridge and Iron to expand an LNG terminal. Given the data from the original problem and assuming an after-tax MARR of 16%,

   (a) Determine the acceptability of the project, using present worth and after-tax analysis.

   (b) Determine the acceptability of the project, using annual equivalent worth and after-tax analysis.

   (c) Determine the acceptability of the project, using IRR or ERR analysis (after-tax analysis).

2. The New York City Transit Authority placed a $110 million order with Siemens AG to design and install computerized displays in 156 subway stations in the city.[13]

   Make the following assumptions: The displays are installed at time zero and have a life of three years with an annual maintenance cost of $2 million per year and no salvage value. Assume also that 2 million people use the displays each day for 250 days per year. Of those, 10% are able to change their plans, valued at $5 per transaction (for the value of the information), and 35% receive satisfaction from the information valued at $1 per transaction. Should the investment have been undertaken? Assume an annual interest rate of 4%.

3. Air Liquide SA of France agreed to supply oxygen and nitrogen to Union Carbide and Dow Chemical. The project calls for a $40 million pipeline investment to integrate Union Carbide's operations into Air Liquide's operations. Air Liquide will supply 5000 tons of oxygen and nitrogen a day under the contract.[14]

   Make the following assumptions: Although revenues from sales of oxygen and nitrogen can be approximated with a continuous-funds flow of 5000 tons per day and average revenues of $14 per ton, aggregate the revenues to end-of-year-cash flows. (Assume 250 production days per year.) The project is expected to last for 10 years and has no salvage value. Operating costs are $8.25 per ton, with fixed costs of $250,000 per year. Evaluate the project as follows with annual interest rates:

   (a) Using IRR and a MARR of 12%.

   (b) Using future worth and a MARR of 10%.

   (c) Using annual equivalent worth and a MARR of 15%.

4. Continental, a German tire manufacturer, announced that it would invest $270 million in a new factory in Brazil. The new factory would have a capacity between 5 and 8 million tires per year, of which 85% would be exported.[15]

   Make the following assumptions: The investment occurs in 2005, with production beginning in 2006. Demand grows from 4 million tires in 2006 at a rate of 15% per year until reaching the factory's capacity of 8 million tires. Analyze the

[13] "Siemens Gets New York Subway Contract," *Dow Jones Newswires,* October 2, 2003.

[14] Keller, G., "Air Liquide Secures Dow Chemical, Union Carbide Contracts," *Dow Jones Newswires,* December 9, 2003.

[15] Brasileiro, A., "Germany Continental to Invest $270M in Brazil Unit—Report," *Dow Jones Newswires,* February 2, 2004.

investment over a 10-year horizon assuming $20 million in remediation costs at the end of year 10. Assume also that the revenue per tire is $30, with costs of sales of $20 per tire in Brazil and $22 per tire for exports. Fixed plant costs are estimated at $500,000 per year. Should the project be accepted? Support your conclusion as follows:

(a) Using present worth and a MARR of 20%.

(b) Using annual equivalent worth and a MARR of 12%.

5. Consider the data from Application Problem 11 in Chapter 8, in which Coca-Cola announced opening a bottling facility in China. Given the after-tax cash flows and an after-tax MARR of 9%, evaluate the investment with (a) present worth, (b) internal rate of return, and (c) market value added analysis.

6. National Petroleum Corp. of China and Nelson Resources, Ltd., of Canada expanded their production in the North Buzachi oil field in Kazakhstan. The $82 million investment in 2004 included equipment to drill 15 new wells. The investment was designed to increase oil production from 7500 barrels to 10,000 barrels per day. Reserves in the field are estimated at 70 million tons (500 million barrels).[16]

Make the following assumptions: Oil revenues are $40 per barrel against $15 per barrel costs and $5 million in annual fixed costs. Assume also a remediation cost of $100 million at the end of the project life, which occurs when the oil field is dry. Draw the cash flow diagram resulting from the incremental recovery investment. Be sure to note that the do-nothing option is defined by a cash flow diagram representing 7500 barrels per day extraction. Might there be multiple internal rates of return for this problem? Explain. Aggregate oil production to end-of-year cash flows for analysis (350 production days per year), assuming a 22% annual interest rate. Justify the investment, using internal (if possible) and external rate-of-return analysis.

7. Consider the depreciation data for Ford's expansion of its truck parts production from Application Problem 10 in Chapter 8. Make the following additional assumptions: Analyze the investment over an eight-year horizon, assuming a tax rate of 40% and a MARR of 13% (annual rate). Assume further that output is 400,000 parts per month, with each part generating a profit of $10. Fixed operating costs are $2 million per month. Use future worth to determine the project's acceptability.

8. The South Korean government invested $16 million (a sum matched by IBM) in a new laboratory in South Korea for the development of software for mobile communications devices. The investment is being spread over four years to fund researchers and engineers.[17] If the interest rate is 4% per year, what annual benefits must the South Korean government receive to accept this project? What might these benefits entail?

9. Norwegian firm Yara International ASA undertook a $550 million expansion of Qafco, the world's largest fertilizer factory. The project includes building an ammonia plant and a urea plant with daily capacities of 2000 tons and 3200 tons, respectively.[18]

[16] Kistauova, Z., "CNPC, Nelson Resources to Invest $82M in Kazakh Oil Field," *Dow Jones Newswires,* February 26, 2004.

[17] Kim, Y.-H., "IMB, S Korea Govt to Invest $16M Each to Open Software Lab," *Dow Jones Newswires,* October 24, 2003.

[18] "Yara Expansion of Qatar Fertilizer Site Completed," *Dow Jones Newswires,* April 26, 2004.

Make the following assumptions: The expansion took place in 2003, with production at full capacity in 2004. Revenues are $160 per ton of ammonia and $135 per ton of urea. Costs are $110 per ton of ammonia and $85 per ton of urea, in addition to plant costs of $5 million per year, increasing 5% per year. The study horizon is 10 years, (350 production days per year) and the project has a salvage value of $50 million at that time. Analyze the investment with both present worth and annual equivalent worth, assuming a 10% annual interest rate and end-of-year cash flows.

10. Canadian oil and gas producer EnCana Corporation is expanding its coal-bed methane development on 700,000 acres of land in Alberta. The property is estimated to have more than 2 trillion cubic feet of recoverable natural gas from coal. The company already extracts 3 million cubic feet of gas per day from 35 wells on the northern part of the property. In the second half of 2003, Encana opened 100 new wells, bringing production to 10 million cubic feet per day. The company expects to drill another 300 wells in 2004 to bring production to 30 million cubic feet per day. The 2004 expansion is expected to cost C$90 million.[19]

Analyze the 2004 expansion, making the following assumptions: Natural gas revenues are C$7.50 per 1000 cubic feet against per unit costs of C$3.50 per 1000 cubic feet and annual fixed production costs are C$5 million per year. Assume also that remediation costs are C$50 million when the field is expended. Note that the do-nothing option is not trivial. Does this problem pose difficulties for IRR analysis? Explain. Analyze the expansion, using IRR, ERR, and PW analysis and assuming a MARR of 25% per year. (Aggregate daily cash flows to annual flows assuming production of 365 days per year.)

9.6.4 Fundamentals of Engineering Exam Prep

1. New buses for a mass-transit agency cost $10 million and are expected to last for 10 years with a $1 million salvage value. The buses cost $500,000 to operate and maintain each year. If the system is used for 750,000 trips per year, with the average cost to the rider $1.50, but providing $4.00 in benefits, and the interest rate is 4% per year, the benefit–cost ratio for the investment is most closely

(a) 1.82.

(b) 1.08.

(c) 1.14.

(d) 0.55.

2. A new city park is developed at the cost of $500,000. It is expected to last for 20 years, having a salvage value of $50,000 at that time, with annual maintenance costs of $25,000. If there are 5000 visitors a month to the park, with each visitor receiving estimated benefits of $5 per visit, the benefit–cost ratio, assuming a 6% annual interest rate, is

(a) 4.46.

(b) 0.37.

[19] Moritsugu, J., "Encana Expands Coal-Bed Methane Development," *Dow Jones Newswires*, November 18, 2003.

(c) 1.00.

(d) 2.78.

3. Benefit–cost analysis should

   (a) Take the ratio of present-worth benefits to the user over equivalent annual costs to the facilitator.

   (b) Take the ratio of equivalent benefits to the user over equivalent costs to the facilitator.

   (c) Take the ratio of equivalent costs to the facilitator over equivalent benefits to the user.

   (d) None of the above.

4. A producer of plastic parts is looking at adding another extruder at the cost of $250,000. The machine is expected to last seven years and produce revenues of $75,000 a year against costs of $25,000 per year. At 10% per year, the present worth of the investment is most closely

   (a) $86,400.

   (b) $243,000.

   (c) –$6600.

   (d) $493,000.

5. If an analysis produces multiple internal rates of return, an analyst should

   (a) Use any of the rates to make an investment decision.

   (b) Compute the present worth.

   (c) Compute the modified internal rate of return (sometimes call the external rate of return).

   (d) Either (b) or (c).

6. Consider an investment with net cash flows of –$110,000, $20,000, $25,000, $30,000, $35,000, and $40,000. The internal rate of return for the investment is most closely

   (a) 8%.

   (b) 10%.

   (c) 12%.

   (d) 14%.

7. Consider an investment with net cash flows of –$110,000, $20,000, $25,000, $30,000, $35,000, and $40,000. The present worth suggests that the project should be accepted for interest rates

   (a) Less than 10%.

   (b) Between 8% and 12%.

   (c) Less than 12%.

   (d) All of the above.

8. A steel producer procures a new stamping machine for \$100,000. The machine is used for five years, increasing net annual revenues by \$25,000 per year. To achieve an internal rate of return of 12% per year, the salvage value at the end of five years must be nearly

(a) \$17,000.

(b) \$12,000.

(c) \$15,000.

(d) \$22,000.

9. A steel producer procures a new stamping machine for \$100,000. The machine is used for five years, increasing net annual revenues by \$25,000 per year. A salvage value of \$10,000 at the end of five years produces a present worth (8% per year) closest to

(a) \$1000.

(b) \$35,000.

(c) \$0.

(d) \$6600.

10. A machine is purchased for \$50,000 and retained for five years before it is salvaged for \$0. It produces revenues of \$20,000 at annual costs of \$5000. If the machine is depreciated in accordance with straight-line depreciation and the effective tax rate is 30%, what is the present value of the after-tax cash flow? Assume a 10% annual interest rate.

(a) \$6800.

(b) –\$31,000.

(c) \$1200.

(d) \$0.

10 Considering Risk

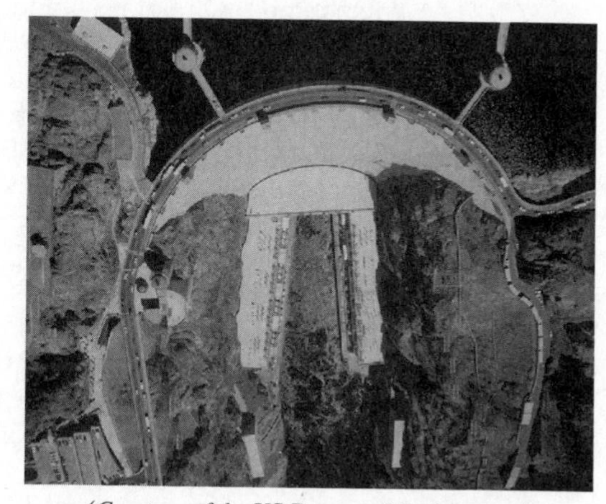

(*Courtesy of the US Bureau of Reclamation.*)

Real Decisions: Let It Rain, Let It Rain, Let It Rain

In 2003, Aecon Group, Inc., signed a C$108 million contract with James Bay Energy Corp. to build a 480-megawatt hydroelectric plant on the Eastmain River in northern Quebec. The contract calls for the construction of a power intake structure and three horizontal penstocks that lead the flow of water to the generators. Construction began in the spring of 2004 and was completed at the end of 2005.[1]

Make the following assumptions: Assume that the investment cost is split evenly over 2004 and 2005, with operations beginning in 2006. Assume further that the C$108 million cost is merely for the construction of infrastructure and that the generators will cost another C$200 million in investment.

[1] Li, J., "Aecon Unit Gets C$108M Pact to Build Hydroelec Power Plant," *Dow Jones Newswires*, December 18, 2003.

Although the plant has the capacity to generate 480 megawatts of power, it is constrained by the flow of water from the river such that it can generate more electricity in rainy, wet seasons and less during dry seasons. Assume that the amount of rainfall in a given year is between 3000 and 12,000 cubic meters. If more than 10,000 cubic meters falls, then the plant can generate an average of 480 megawatts of power for the year. However, for every 1000 cubic meters below 10,000, the plant generates 50 megawatts less power, such that it generates only 130 megawatts if rain totals 3000 cubic meters in a year. (Thus, megawatts generated $= 0.05 \times$ rainfall $- 20$ megawatts or zero, whichever is larger.)

We assume that rainfall is highly correlated with the previous season in this area and that 2006 has a 50% chance of being a wet or a dry season. From then on, if a wet season occurs, there is a 65% chance of the next season being wet and a 35% chance of it being dry. Furthermore, if a dry season occurs, there is a 65% chance of the next season being dry. The amount of rainfall in a wet season follows a Normal distribution, with a mean of 8500 cubic meters and a standard deviation of 1500 cubic meters. For a dry season, the mean is 5000 cubic meters with a standard deviation of 500 cubic meters.

Due to the high potential for erosion, maintenance costs for the plant are high and estimated at C$5 million in the first year. Assume that maintenance costs increase 6% in wet seasons and only 3% in dry seasons. The plant and structure have an expected service life of 20 years with no salvage value, assuming that no more than 15 rainy seasons occur during that time. Thus, the expected life of the plant is 20 years or a total of 15 rainy seasons, whichever occurs first.

Finally, assume that power is generated for 300 days per year, 24 hours per day, and that *net* revenues total C$25 per megawatt-hour. (All other operating, transmission, and fixed charges are included in this figure.) The interest rate is 12% per year. Given that the costs and revenues are dependent on rainfall, should this investment be made?

In addition to answering this question, after studying this chapter you will be able to:

- Define risk in the context of investment analysis.
- Compute the payback period, the payback period with interest, and the project balance of an investment in order to gauge the risk of a project, assuming that all estimates are known with certainty. (Sections 10.1–10.3)
- Perform sensitivity analysis and break-even analysis in order to gauge which input parameters present the greatest risk to a project and specifically what values of these parameters will "make or break" the investment. (Sections 10.4–10.5)

- Perform scenario analysis to measure the impact of simultaneous changes in multiple parameters in the form of pessimistic, optimistic, and average outcomes. (Section 10.6)
- Compute the mean and variance for a given investment project defined by various scenarios and probabilities of occurrence. (Section 10.7)
- Compute the expected return, the variance in return, and the probability of loss in simulation analysis, both with cash flows and with scenario outcomes, for projects in order to further gauge their risk. (Sections 10.6–10.7)

As we continue to analyze projects, we may feel weary of our progress. Until now, we have assumed that we know everything with certainty, including the timing and magnitude of all cash flows. But the future is unclear and it is quite likely that our predicted future will not come true. We begin to face that reality in this chapter. Specifically, we want to devise methods with which we can address the risk associated with investment projects.

In discussing risk thus far, we have learned that the risk with investing is that money can be lost. This risk is heightened when investments are large, because losing the investment may put the financial security of the firm in jeopardy. Most investments that we have discussed up to now have been described by an initial outlay of funds, such as the purchase of equipment or the construction of a facility, followed by returns from the investment, such as revenue from the sales of a product developed from the investment. In this scenario, the delay in time between the investment and its returns represents a risky situation. If the project terminates early, the investment will be lost. Even if the investment is recovered, the returns may not reach their expected level. Because the money could have been invested elsewhere to achieve higher returns, the opportunity cost also represents a risk. In total, the time lag between an investment and its return, as well as the possibility of not achieving a great enough return, contributes to the risk of losing all or part of an investment.

Mathematically, **risk** is defined by a situation with a number of possible outcomes, each with a known probability. With respect to an investment, the possible outcomes are defined by different cash flows. A cash flow can, of course, deviate positively or negatively from a prediction. If we know the possible deviations and their probabilities, then we are considering a decision under risk. This definition differs from the mathematical definition of **uncertainty**, in which the probability of none of the outcomes is known. In general, however, the definitions of risk and uncertainty get lumped together. In this chapter, we provide means by which to analyze a project under risk.

Risk can be measured in a variety of ways. Decision makers have generally accepted the following three measures:

1. *Variance in return.* This is generally regarded as the most common measure of risk. The idea is that the more your expected return varies, the greater is the risk. Note that this is a conservative measure, as it penalizes variance in either direction and most people do not mind variance in a positive direction, which increases the return.
2. *Semivariance in return.* This measure overcomes the conservative nature of variance in that it measures variance from the mean only in the negative direction. It recognizes that variance in the positive direction is good. Unfortunately, semivariance is difficult to incorporate into analyses.
3. *Probability of loss.* This measure clearly addresses the question of whether the project will be successful. In general, we define the probability of a loss as the probability of the present worth being negative. We have

already discussed the relationship between present worth and profit, so it is not necessarily the case that a project with a negative present worth will bankrupt a company. However, it should be clear that investing in a project with a negative present worth is not a sound strategy, as the costs associated with raising funds cannot be paid.

Risk is an extremely important issue with respect to investment analysis. A cash flow diagram describes *one* possible future for an engineering project. We know that it is highly unlikely that we will be able to predict this future accurately, so we must take that fact into account in our analysis in order to make an informed decision.

Before continuing, we must realize that while we may develop methods to help evaluate decisions under risk, we cannot see the future, and there is no certainty that our decisions will be correct for the evolving situation. This is not to say that we should throw up our hands in despair and just ignore the future. Rather, we should strive to reduce the risk by gathering and processing as much information as possible so that we can make the best decision possible under these difficult circumstances.

10.1 Payback Period

A simple measure of the risk involved in a project, and one that is quite popular in industry, is the payback period. Mathematically, it is the first period in which the cumulative inflows from a project exceed the cumulative outflows, or

$$n^* = \min_t \sum_{n=0}^{t} A_n \geq 0. \tag{10.1}$$

The term "payback period" is used because it identifies the first period in which you recover the outflows that generally occur at the beginning of an investment. Managers may often ask, "How long will it take for us to get our money back?" The manager is referring to the payback period, since most investments are characterized by cash outflows near time zero, followed by positive net cash flows for the duration of the project. Let us illustrate with an example.

EXAMPLE 10.1

Payback Period

In December of 2003, Arcelor SA of France, Nippon Steel of Japan, and Bao Steel of China agreed to build a plant in Shanghai to produce flat carbon steel for the automotive industry. The cold rolling and galvanizing operations have an annual capacity of 1.7 million metric tons per year, and production began in the second quarter of 2005.[2] Assume that operations commenced in April of 2005 with monthly production of 141,667 metric tons of output, generating *net* cash flows of \$31.2 million per month, based on income of \$220 per metric ton, over a 20-year life with no salvage value for the plant. Assume further that the \$800 million investment is spread evenly, in monthly payments, from January 2004 through March of 2005. What is the payback period for the investment. The interest rate is 15% per year, compounded monthly.

Solution. The cash flow diagram for the steel plant investment is given in Figure 10.1. Note the cash flows are given in months.

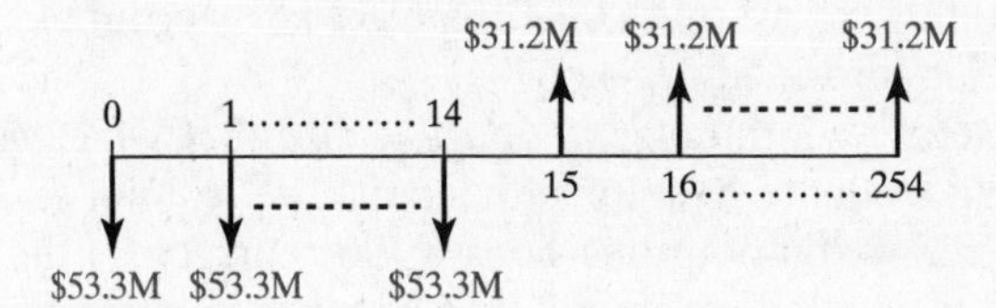

Figure 10.1 Expected monthly cash flows for flat carbon steel plant investment.

To compute the payback period, we sum the net cash flows in each period:

$$A_0 = -\$53.3 \text{ million},$$

$$A_0 + A_1 = -\$53.3\text{M} - \$53.3\text{M} = -\$106.6 \text{ million},$$

$$A_0 + A_1 + A_2 = -\$53.3\text{M} - \$53.3\text{M} - \$53.3\text{M} = -\$160 \text{ million}.$$

The progression continues until the investment in the plant is complete, such that the sum of the cash flows through the first 15 months is \$800 million. The pattern continues as follows when production begins:

$$\sum_{n=0}^{15} A_n = -\$800\text{M} + \$31.2\text{M} = -\$768.8 \text{ million},$$

$$\sum_{n=0}^{16} A_n = -\$768.8\text{M} + \$31.2\text{M} = -\$737.6 \text{ million}.$$

Progressing to period 39:

$$\sum_{n=0}^{39} A_n = -\$800\text{M} + (25)\$31.2\text{M} = -\$20 \text{ million},$$

$$\sum_{n=0}^{40} A_n = -\$20\text{M} + \$31.2\text{M} = \$11.2 \text{ million}.$$

Thus, the payback period $n^* = 40$ months, as the 40th month is the first period in which the total cash inflows of the project exceed the cash outflows. Note that finding the payback is easily accomplished with a spreadsheet, as one can merely compute the cumulative sum of the cash flows until it is positive.

---

[2] Pearson, D., "Arcelor, Nippon Steel in \$800M China Steel Plant Venture," *Dow Jones Newswires,* December 22, 2003.

We mentioned the risks involved in investing in a project, one of which is that the project may fail to proceed, which can result in the loss of the investment. The payback period gives us some gauge of this risk, as it measures the amount of time required to recoup an investment. We would prefer to invest in a project with a short payback period. For certain projects, such as buying a piece of equipment, a company may mandate that the payback period be less than a certain number of periods (usually measured in months, quarters, or years), or the investment will not be approved. The idea is that even if a project does not proceed as expected, a shorter payback period implies a lower risk of losing an investment because there is a better chance of getting returns sooner.

In Example 10.1, we found that the payback period for a certain investment was 40 periods. As often happens with new information, more questions arise: What does $n^* = 40$ mean? Is this enough information to make a decision? Is it enough information to ease our concerns about a risky future?

To delve into these issues, consider the definition of the payback period. It is merely a counting of income that identifies when the cash inflows from a project first exceed the cash outflows. There are many pitfalls in this type of analysis:

1. *It does not take into account the time value of money.* The procedure merely sums cash flows through time, regardless of when they occur. This is contrary to our principle of the time value of money.
2. *It does not consider cash flows past period n^*.* The definition of the payback period stops when we reach the first period in which the cumulative inflows exceed the cumulative outflows. If all net cash flows beyond period n^* are positive, then the payback period captures our desired information. But if the cash flows beyond n^* are both positive and negative, cash outflows may still exceed inflows over the life of the project, signifying that the investment is not recovered.
3. *It does not provide a measure of worth or return.* In solving Example 10.1, we computed the payback period, but it offered no information, in either an absolute or a relative sense, as to how much money we would earn by investing in the project.

This does not mean that the payback period does not have merit. The positive attributes of the payback period are as follows:

1. *It provides some measure of the risk associated with a project.* The measures of worth that we introduced earlier, such as present worth, do not provide any information concerning risk. The payback period tells us the minimum number of periods that the project must continue in order to recoup investment costs, ignoring interest. This, "break-even point" clearly provides more information about the risk associated with a project than we have had previously. Recall that we identified two risks associated with an investment, defined by the time lag between the cash outflows associated

with the initial investment and the level of cash inflows associated with project revenues. The payback period provides some measure of both of these risks.

2. *It can be used to enhance the evaluation of a project with a traditional measure of worth.* Since the payback period does not provide a measure of worth and our measures of worth do not provide a measure of risk, they clearly can be used together to help make a more informed decision.

But does the payback period give us enough additional information to ease our concerns about a risky future? No, but it is a start in the right direction and, as mentioned, can augment present worth analysis. For example, the present worth of the steel plant investment in Example 10.1 is \$1.39 billion. This datum can be coupled with the payback period of 40 months to make a more informed decision.

10.2 Payback Period with Interest

We can alleviate one of our concerns about the payback period by including interest in our calculation such that the time value of money is considered. We define the payback period with interest as

$$n^* = \min_t \sum_{n=0}^{t} \frac{A_n}{(1+i)^n} \geq 0. \qquad (10.2)$$

Note that our new definition includes the time value of money such that future cash flows are discounted to the present. As long as our interest rate is positive and we make an investment at time zero, the payback period with interest will always be greater than or equal to the traditional payback period without interest. Let us illustrate by revisiting our previous example.

EXAMPLE 10.2

Payback Period with Interest

For the plant in Example 10.1, we compute the payback period with interest.

Solution. The annual interest rate was given as 15% compounded monthly. This translates to a monthly rate of 1.17%. Computing the payback period with interest follows computing the payback period, but the cash flows must be discounted. Thus,

$$A_0 = -\$53.3 \text{ million},$$

$$A_0 + \frac{A_1}{(1+i)} = -\$53.3\text{M} - \frac{\$53.3\text{M}}{(1+0.0117)} = -\$105.98 \text{ million},$$

$$A_0 + \frac{A_1}{(1+i)} + \frac{A_2}{(1+i)^2} = -\$105.98\text{M} - \frac{\$53.3\text{M}}{(1+0.0117)^2} = -\$158.05 \text{ million}.$$

The computation continues in this manner such that

$$\sum_{n=0}^{47} \frac{A_n}{(1+i)^n} = -\$16.06 \text{ million},$$

$$\sum_{n=0}^{48} \frac{A_n}{(1+i)^n} = -\$16.06\text{M} + \frac{\$31.2\text{M}}{(1+0.0117)^{48}} = \$1.79 \text{ million}.$$

Hence, the payback period with interest, n^*, is 48 months, or four years. As expected, this value is greater than the payback period without interest.

Thus, we are armed with our first measure of risk. Note again that the payback period (with or without interest) is not to be confused as a measure of worth: It gives us no information concerning the worth of a project, but merely tells us when we expect to get our money back. Since investments are risky, we prefer short payback periods because we will get our money back sooner. In other words, the risk associated with a project with a shorter payback period is lower than that associated with one with a higher payback period.

10.3 Project Balance

We can stretch the idea of the payback period a bit further. The payback period told us when we would recover our investment. To get a better idea of the risk associated with a project, we may want to "track" the investment over time. Toward that end, we define the project balance $\text{PB}_t(i)$ as the future worth of a project at any given point in time t, defined by the cash flows up to and including those at time t, or

$$\text{PB}_t(i) = \sum_{n=0}^{t} A_n(1+i)^{t-n}. \tag{10.3}$$

Note that the project balance at time t analyzes the cash flows up to and including time t. For a project that is N periods long,

$$\begin{aligned}
\text{PB}_0(i) &= A_0, \\
\text{PB}_1(i) &= A_0(1+i) + A_1, \\
\text{PB}_2(i) &= A_0(1+i)^2 + A_1(1+i) + A_2, \\
&\vdots \\
\text{PB}_N(i) &= A_0(1+i)^N + A_1(1+i)^{N-1} + \ldots + A_{N-1}(1+i) + A_N.
\end{aligned}$$

Writing this last equation out, we see that

$$\mathrm{PB}_n(i) = \mathrm{PB}_{n-1}(i)\,(1+i) + A_n,$$
$$\mathrm{PB}_N(i) = \mathrm{FW}(i).$$

Since the project balance at period N is defined as the future worth of the project, we can use the project balance to make an accept or reject decision in accordance with the following relationships:

PB Value	Decision
$\mathrm{PB}_N(i) > 0$	Accept the project.
$\mathrm{PB}_N(i) = 0$	Indifferent about the project.
$\mathrm{PB}_N(i) < 0$	Reject the project.

The project balance can be seen as redundant, as it is merely the future worth of the project at time N. But this is not why we have introduced the measure. Rather, the reason for using the project balance lies in its ability to examine a project over time. We illustrate this property after showing how to compute the project balance in the next example.

EXAMPLE 10.3 *Project Balance*

LG Philips LCD, Inc., a joint venture between electronics manufacturers LG Electronics of South Korea and Philips Electronics of the Netherlands, spent KRW3.3 trillion in 2003 and 2004 to construct and equip a facility in South Korea to produce large (30 inches or more diagonally) LCD panels for use in computer monitors and television sets. The factory produces 90,000 panels per month.[3]

Make the following assumptions: The investment takes one year to complete and production runs at maximum capacity for six years. Assume initially that KRW1.2 million is net (revenues minus expenses) for each panel, but that this amount declines 12% each year due to competition and changing technologies. Analyze the project balance for this investment with an annual interest rate of 18%.

Solution. The cash flow diagram for the investment is given in the spreadsheet in Figure 10.2. As we are given an annual interest rate, we aggregate the monthly cash flows into annual flows.

The project balances over the life of the project follow directly from the cash flows, as shown in the spreadsheet. The calculations of project balance for the first two

[3] Kim, Y.H. and E. Ramstad, "LG Philips Slates $2.46 Billion for LCD Flat-Panel Factory," *The Wall Street Journal Online,* November 21, 2003.

	A	B	C	D	E	F	G
1	Example 10.3: LCD Production Investment						
2					Input		
3	Period	Cash Flow	Project Balance		Investment	KRW 3,300,000.00	million
4	0	-KRW 3,300,000.00	-KRW 3,300,000.00		Net Revenues	KRW 1.20	million/panel
5	1	KRW 1,296,000.00	-KRW 2,598,000.00		g (Revenues)	-12%	per year
6	2	KRW 1,140,480.00	-KRW 1,925,160.00		Sales	1,080,000	per year
7	3	KRW 1,003,622.40	-KRW 1,268,066.40		Interest Rate	18%	per year
8	4	KRW 883,187.71	-KRW 613,130.64		Periods	6	years
9	5	KRW 777,205.19	KRW 53,711.03				
10	6	KRW 683,940.56	KRW 747,319.58		Output		
11	=$F$3*(1+$F$4)^(A9-1)*$F$5				Project Balance		
12				=C5*(1+$F$6)+B6			

Figure 10.2 Cash flows and project balances from LCD panel factory investment.

periods are as follows:

$$PB_0(18\%) = -\text{KRW}3.3 \text{ trillion},$$

$$PB_1(18\%) = -\text{KRW}3.3\text{T}(1+0.18) + \text{KRW}1.296\text{T} = -\text{KRW}2.6 \text{ trillion},$$

$$PB_2(18\%) = -\text{KRW}2.6\text{T}(1+0.18) + \text{KRW}1.140\text{T} = -\text{KRW}1.9 \text{ trillion}.$$

The final value of KRW747 billion represents the future worth of the project at the end of the sixth year with an annual interest rate of 18%.

Computing the project balances, as in Example 10.3, provides a fairly clear picture of how the investment is expected to progress over time, but graphing the project balance gives the decision maker additional information. Consider, for example, the solution to Example 10.3, which is graphed with respect to time in Figure 10.3. The graph illustrates numerous concepts we have already discussed, including the payback period with interest, n^*, and the future worth FW(i).

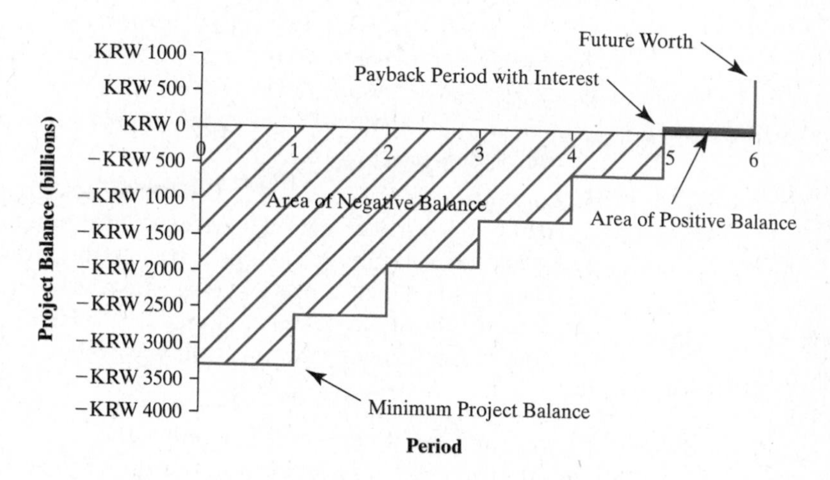

Figure 10.3 Graph of project balance (in billions of won) over time for Example 10.3.

More importantly, the graph tells us *how* these values are derived. We can partition the areas under the project-balance curve according to positive and negative values. When an investment is made, the project balance becomes negative and stays negative until the investment with interest, is recovered. Thus, the point at which the graph moves from an area of negative balance to one of positive balance is the payback period with interest (year 5). The graph grows until terminating at its future worth (KRW747 billion).

Given this information, we can derive some qualities of a project that we would find ideal:

1. *Large future worth.* The final point on the curve represents the future worth of the project. We know that if this value is positive, then the investment is worthy of consideration. The larger this value, the greater wealth the project will generate for the company.
2. *Small payback period with interest.* The period in which the axis is crossed represents the payback period with interest.
3. *Small area of negative balance.* This is generally the area under the curve preceding the payback period with interest. (It is possible to have negative areas later, depending on the cash flows.) This area represents the period of time that a project is in the "red" such that if the project is terminated during this period, money will be lost. While the project balance identifies the amount of money that can be lost at a given point in time, the area under the project-balance curve gives a "feel" to these numbers. A large negative area represents greater risk, as it considers both the time and magnitude of the possible losses. Note that the total area of negative balance can be calculated merely by summing the negative project balances. An ideal investment would have a small area of negative balance.
4. *Large minimum project balance.* Because the project balance identifies the amount of money that can be lost at any time over the course of the project, we wish the minimum project balance to be as large as possible, as it represents the most that can be lost in a project. If there are successive periods of investment, the minimum project balance will be greater (in absolute value) than the total investment, since the interest rate compounds the investments.
5. *Large area of positive balance.* The area of positive balance, generally to the right of the payback period, represents the amount of time that the project is generating wealth for the company as the investment (with interest) is being recovered. If a project terminates early during a period defined by an area of positive balance, money is not lost, but rather gained, by the project. Again, an ideal project will have a large area of positive balance.

The project balance gives us a relatively clear picture of the risk of a project, and it encompasses our only other measure of risk: the payback period with interest. Graphing the project balance provides additional information through areas of negative and positive balance, which provide the "feel" of a project with respect to risk. In order to try to capture this numerically, we can take a ratio of the areas of positive (KRW801 billion) and negative (KRW9.70 trillion) balance. Note that this is merely the sum of the positive project balances over time, divided by the sum of the negative project balances over time (using absolute values in both numerator and denominator), or

$$\frac{\text{APB}}{\text{ANB}} = \frac{\text{KRW0.801T}}{\text{KRW9.70T}} = 0.0825,$$

where APB is the area of positive balance and ANB the area of negative balance, respectively. A ratio greater than zero ensures that the project returns its investment with interest for at least one period, as this is captured by at least one period of positive project balance. A ratio greater than one implies that the project is "more positive than negative" and provides some assurance that the project will be successful. The ratio of 0.0825 for the LCD panel factory summarizes the fact that there is significant risk in the project, highlighted by its minimum project balance value of −KRW3.3 trillion from the initial investment. It should be clear that while this ratio cannot capture all of the information that is present in the graph of the project balance, it provides a quick synopsis of what the project balance captures over time.

10.4 Break-Even Analysis

The payback period also leads us to the definition of break-even analysis, which determines the value of a given parameter for which a project breaks even. In the context of project evaluation, that value is the value which leads to a present worth of zero. The parameters in question may be from the cash flow diagram, such as a specific cash flow, a set of cash flows, or the time horizon, or the parameters may be relevant inputs, such as demand or output, which help define the cash flow diagram.

The IRR represents the interest rate at which a project breaks even, since it is the interest rate at which the present worth is equal to zero. The payback period also represents a break-even value, but with a different goal. Instead of determining a value at which the present worth is equal to zero, the payback period determines when cash inflows equal (or first exceed) cash outflows.

To perform a break-even analysis, the parameter in question is altered, while all other parameters are held constant, in order to determine when the project breaks even (i.e., when the present worth is zero). Determining the break-even value provides more information to decision makers, as they know the critical value for the parameter in question and ranges of possible values according to which the project is acceptable. In the next example, we examine a critical cash flow.

EXAMPLE 10.4

Break-Even Analysis

Helicopter manufacturer AgustaWestland, owned by GKN PLC of Great Britain and Finmeccanica SpA of Italy, opened a \$6.8 million production line for its A119 Koala helicopter in Philadelphia, Pennsylvania, in the summer of 2004. The 40,000-square-foot facility includes flight hangars and manufacturing and assembly areas.[4,5] Assume that the company expects to generate \$30 million in sales from the facility. What is the break-even point for annual operating and maintenance costs (with respect to present worth) if the MARR is 13% per year over the 10-year life of the facility (with a salvage value of \$500,000)?

Solution. The partial cash flow diagram for this project is given in Figure 10.4. Note that the annual outflows (A) are unknown, but assumed to be constant over the life of the facility.

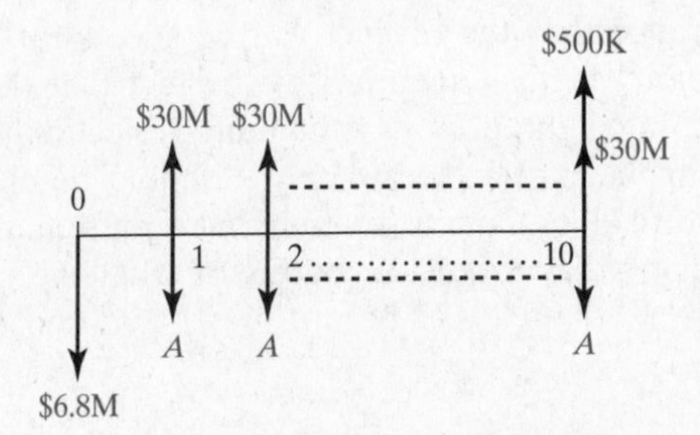

Figure 10.4 Partial cash flow diagram for helicopter production facility.

The present worth of the cash flows is

$$PW(13\%) = -\$6.8M + (\$30M - A)(\overset{P/A,13\%,10}{5.4262}) + \$500K(\overset{P/F,13\%,10}{0.2946})$$

To determine the break-even point, we set the present worth to zero and solve for A:

$$0 = -\$6.8M + (\$30M - A)(\overset{P/A,13\%,10}{5.4262}) + \$500K(\overset{P/F,13\%,10}{0.2946})$$

$$\Rightarrow A(\overset{P/A,13\%,10}{5.4262}) = \$156.13 \text{ million},$$

or

$$A = \$28.77 \text{ million}.$$

Thus, in order to break even with a 13% rate of interest, annual expenses must be \$28.8 million or less. This figure translates to over \$575,000 per employee (of the 50 to be hired). Considering the high skill of the workforce in aviation manufacturing and potentially high overhead costs, this is a number that should be examined closely.

---

[4] Lunsford, J.L., "Helicopter Seeks Made-in-U.S. Label," *The Wall Street Journal Online*, March 2, 2004.

[5] "AgustaWestland Expands Philadelphia Manufacturing Facility," *Press Release*, www.agustawestland.com, March 3, 2004.

In determining the critical value of the cash outflows in Example 10.4, we have provided more information to the decision maker, who can now readily compare expected costs with the critical value and see whether the cash flow contributes to the risk of the project. That is, if the critical value and its estimate are close, then there is a strong possibility of not breaking even. Similarly, if the break-even value deviates greatly from the expected value, then the decision maker should not be overly concerned. (We examine this issue a bit more deeply in the next section.)

Earlier, we noted that we did not have to limit our analysis to cash flows. Rather, we can look at other parameters, including inputs to the cash flows. This is actually quite common in break-even analysis and provides a different look at a similar question. Let us now reexamine a parameter that influences the cash flow diagram.

EXAMPLE 10.5

Break-Even Analysis Revisited

Consider again the previous example, but make some more detailed estimates in order to determine a break-even point based on the sales of helicopters. Make the following additional assumptions. The list price of the A119 Koala described in the previous example is \$2 million. Assume further that fixed costs for the facility are \$7 million per year and that it costs \$1.25 million to produce one helicopter. What is the break-even point, in terms of annual helicopter sales, assuming all other relevant data from the previous example?

Solution. With these new estimates, the cash flow diagram for the facility is given in Figure 10.5. The uncertain variable here is annual helicopter sales (x).

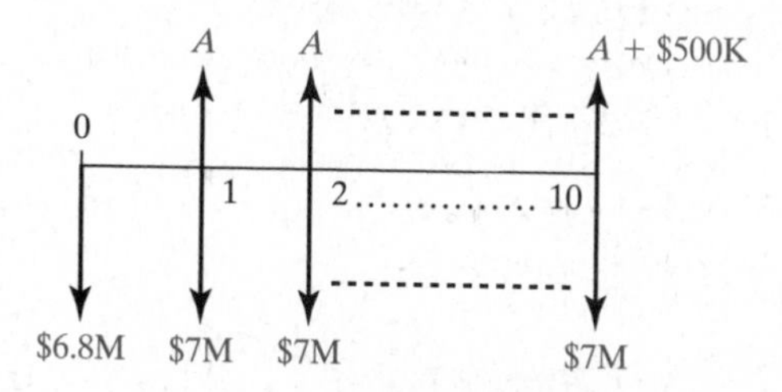

Figure 10.5 Partial cash flow diagram for helicopter production facility.

With a per helicopter net revenue of \$750,000, the present worth is now calculated with x as

$$\text{PW}(13\%) = -\$6.8\text{M} + (\$0.75\text{M}(x) - \$7\text{M})\overset{P/A,13\%,10}{(5.4262)} + \$500\text{K}\overset{P/F,13\%,10}{(0.2946)}.$$

Solving for x with PW $= 0$ defines the break-even point for required sales. We could also use Goal Seek in Excel to determine this value, noting that we require an integer solution. Doing so yields $x = 11$, so that 11 helicopters must be sold at \$2 million apiece in order to break even.

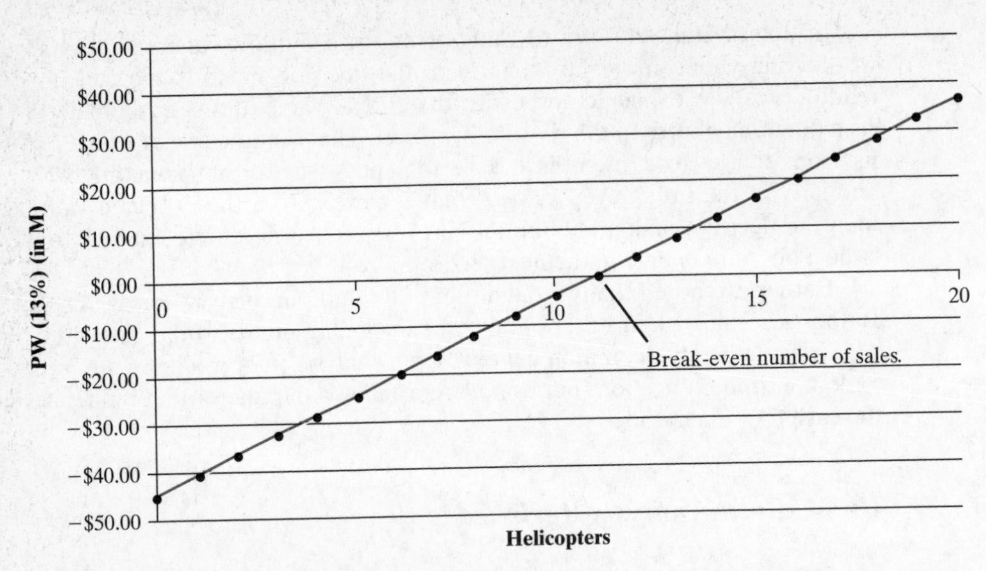

Figure 10.6 Present worth versus helicopter sales.

Are the analyses in Examples 10.4 and 10.5 different? Mathematically, they are the same, as a function is set to zero and solved for a given parameter. However, by allowing ourselves to look beyond the cash flow diagram, we gain greater insight into the risk at hand. This is a lot more information than we had before. We can look at demand estimates and expected output and get a better sense of our anxieties about the investment. The tendency in break-even analysis is to ask the following type of question: How much do I have to produce in order to break even?

If desired, we could plot the function with respect to the parameter in question to get even more information with regard to its break-even point. For Example 10.5, the curve representing the change in the present worth with respect to the number of helicopters sold annually is given in Figure 10.6. The curve is merely a plot of

$$\text{PW}(13\%) = -\$44.63\text{M} + \$4.07\text{M}(x),$$

where x is the number of helicopters sold annually. This equation follows directly from the present-worth equation in Example 10.5. The slope is fairly steep, in that each helicopter sale contributes $4.07 million to the present worth. The point at which the axis is crossed, $x = 10.97$, establishes 11 as the minimum number of helicopters that must be produced in order to break even. A steep curve would be worrisome to a decision maker, since it would signal that the parameter under study is quite critical.

Note that the graph produced from the data in Example 10.5 is linear. This is because we assumed a fixed price per unit sold. Break-even analysis does not require that relationships be linear, and it is often the case that they are not.

In Example 10.5, the list price was \$2 million. However, it may be reasonable to assume that a large order would receive some sort of quantity discount, a common practice in fleet management. We could model this condition in a number of ways, but will illustrate one.

Assume that if only one unit is sold, it generates \$2 million in revenue. However, if two units are sold, they each generate \$1.985 million in revenue. If three units are sold, they generate \$1.97 million each. Continuing this progression, the per unit sale price can be captured by the following equation, based on the total number x of units sold:

$$\text{Per Unit Revenue} = \$2.015\text{M}(.9925^x).$$

This equation produces our desired effect, since if one helicopter is sold, $x = 1$ and \$2 million in revenue is received. If $x = 2$, then \$1.985 million per helicopter is generated. The assumption here is that although total revenue increases per sale, the per unit revenue decreases. This relationship is common to most supply and demand curves in economics, in that as the price drops, demand generally rises. Similarly, as supply increases, prices fall.

Assuming that all other factors and costs remain as before, we now define the present-worth equation according to the number of helicopter sales $x (x \geq 0)$ as

$$\begin{aligned} \text{PW}(13\%) &= -\$6.8\text{M} + \big(\$2.015\text{M}(0.9925^x)(x) - \$1.25\text{M}(x) - \$7\text{M}\big) \\ &\quad \times \overset{P/A,13\%,10}{(5.4262)} + \$500\text{K}\overset{P/F,13\%,10}{(0.2946)} \\ &= -\$44.63\text{M} + \big(\$10.93\text{M}(0.9925^x) - \$6.78\text{M}\big)(x), \quad x \geq 0. \end{aligned}$$

The definition of our purchase price as a nonlinear function of the number of units sold leads to a nonlinear break-even curve, as depicted in Figure 10.7.

The assumption of a change in the selling price forces the break-even point to annual sales of 15 helicopters. The present worth, assuming annual sales of 14, is $-\$1.79$ million; for sales of 15, it is \$0.16 million. As with the linear relationship, the graph provides more information. Even though the relationships can get complicated, especially when factors such as economies of scale in sales and production levels (which are often nonlinear) are considered, the break-even curves are easy to generate with a spreadsheet.

10.5 Sensitivity Analysis

An analysis that is quite similar to break-even analysis, and one that we introduced in Chapter 7, is sensitivity analysis. In that chapter, revenues and costs were estimated for an engineering project in order to develop net cash flows for evaluation. We utilized sensitivity analysis in order to identify which cash flows or other relevant parameters, such as the interest rate or the project

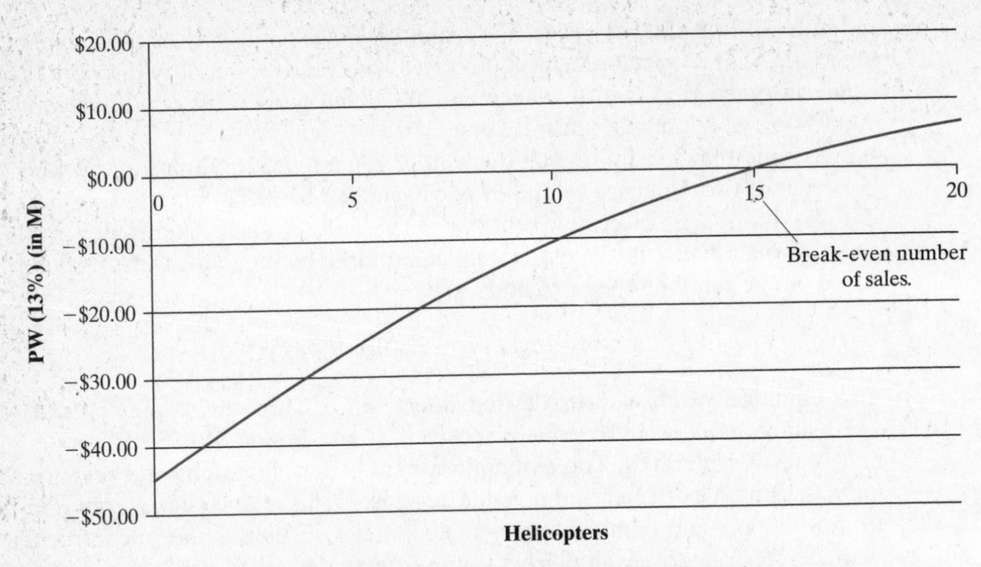

Figure 10.7 Present worth versus helicopter sales, assuming quantity discounts.

length were most important, so that we could concentrate our efforts and resources to make good estimates. Naturally, sensitivity analysis also defined which estimates were not critical and could be ignored.

In the context of project evaluation, sensitivity analysis does not change, but we expand upon our goals for its use. As before, we are interested in identifying the critical parameters, but we take the analysis a step further to determine the range over which parameters can change without affecting our decision. This is why sensitivity analysis is similar to break-even analysis. In breakeven analysis, we determined the critical *value* of a parameter such that the project breaks even. In sensitivity analysis, we are interested in examining a *range* of values that a parameter can take on with respect to our decision. In addition, we are interested in analyzing the sensitivity of *all* input parameters in order to assess the level of risk each brings to the project. This assessment gives us more information with which to make an informed decision.

Sensitivity analysis and break-even analysis are similar because, in both methods, we are examining a range of data in order to garner more information. A given parameter, or possibly a set of parameters, is altered while the remaining inputs are held constant. The effect of these changes in the parameters is measured by evaluating the corresponding change in some measure of worth. The difference between break-even analysis and sensitivity analysis is really only in our desired result. Break-even analysis is seeking the answer to a specific question: What parameter value will move a project from a negative measure of worth to a positive measure of worth? Sensitivity analysis is not as focused, telling us which parameters are most sensitive to our decision and identifying ranges of a variable that are acceptable.

10.5.1 Single-Parameter Analysis

The steps involved in single-parameter sensitivity analysis were outlined in Section 7.2. In essence, the analysis entails varying one parameter while holding the other parameters constant and examining the consequences. In this context, for a given parameter, we are interested in the range of values that would change our decision. We will not repeat the general steps here, but merely illustrate the procedure in the next example.

EXAMPLE 10.6

Single-Parameter Sensitivity Analysis

Glencairn Gold Corp. of Canada began constructing its Bellavista gold mine in Costa Rica in December of 2003. The mine was expected to be completed in one year, with a recovery rate of 78.6% of 555,000 ounces of mineable gold. Production capacity is 60,000 ounces of gold per year at the cost of $179 per recoverable ounce, including operating and royalty costs. The cost to open the mine was $25.9 million. The mine will create a footprint of 19 hectares, with total operations covering 85 hectares. Plans are to reforest the area upon depletion of the mine and donate the area to a local community association for parkland.[6,7] Assume that gold is worth $350 per ounce, the annual interest rate is 10%, and reforestation will cost $1 million. Analyze the risk of this investment through sensitivity analysis.

Solution. Given mineable gold reserves of (0.786)555,000 = 436,230 ounces, production is expected to net annual cash flows of

$$\left(\frac{\$350 - \$179}{\text{ounce}}\right)\left(\frac{60{,}000 \text{ ounces}}{\text{year}}\right) = \$10.26 \text{ million}$$

for the first seven years. This defines a final (eighth) year of production of 16,230 ounces, since 16,230 = 436,230 – (7)(60,000). Together with the remediation costs, this leads to a final cash flow of

$$\left(\frac{\$350 - \$179}{\text{ounce}}\right)\left(\frac{16{,}230 \text{ ounces}}{\text{year}}\right) - \$1\text{M} = \$1.78 \text{ million}$$

in year 8. The net cash flow diagram for the gold mine investment is shown in Figure 10.8.

The present worth of the investment for the assumed data is

$$\text{PW}(10\%) = -\$25.9\text{M} + \$10.26\text{M}(\overset{P/A,10\%,7}{4.8684}) + \$1.78\text{M}(\overset{P/F,10\%,8}{0.4665}) = \$24.88 \text{ million}.$$

There are a number of parameters to examine, including (1) the initial investment, (2) the price of gold, (3) the production levels (which could be affected by production, the total amount of mineable gold, or the recovery rate), (4) the cost of production, (5)

[6] Tsau, W., "Glencairn Gold Begins Construction of Bellavista Gold Mine," *Dow Jones Newswires,* December 11, 2003.

[7] www.glencairngold.com/bellavista.html.

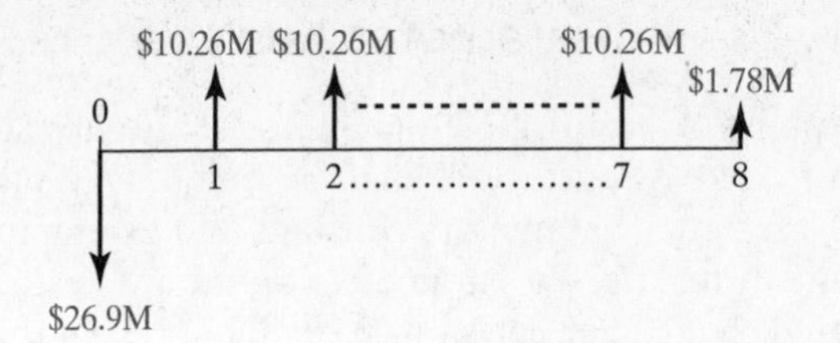

Figure 10.8
Cash flow diagram for gold mine investment.

the cost of remediation, (6) the interest rate, and (7) the horizon. We look at each of these parameters individually. As in Section 7.2, we assign a percentage error, where appropriate, and examine its impact on the present worth of the investment.

Investment Cost. We define x as a percentage error of the estimate, such that the present worth can be written as

$$PW(10\%) = \$24.88M - \$25.9M(x).$$

This leads us to the results in Table 10.1 for the sensitivity of the investment cost on the present worth of the investment. As noted in the table, an error in the initial investment estimate leads to roughly an equivalent error in the present worth of the project, in terms of percentages. This is because the worth of the project is very close to that of the initial investment and the relationship is linear.

TABLE 10.1

Sensitivity of investment cost on present worth of mine investment.

Error (x)	Investment Cost	PW(10%)	PW(10%) Error
−30%	$18.13M	$32.65M	31.23%
−10%	$23.31M	$27.41M	10.17%
0%	$25.90M	$24.88M	0%
10%	$28.49M	$22.29M	−10.17%
30%	$33.67M	$17.11M	−31.23%

Gold Price. The price of gold has traded in a wide band over the past decade, with prices ranging from $250 to over $500 per ounce. We analyze this range with the given data. As opposed to defining a percentage error x, we substitute the price to be analyzed directly (and compute the percentage error afterwards). The results are given in Table 10.2.

Clearly, the price of gold (as would be expected) has a dramatic impact on the present worth of the project. Changes in the price of gold of nearly 30% lead to present-worth errors greater than 120%. Further, a price of $250 per ounce defines a project with a negative present worth.

Annual Production. Annual production is an interesting parameter in this example, because it is dependent on a number of factors, including the capacity of the mineral extraction process and the amount of gold in the mine. We look at a number of these factors.

TABLE 10.2

Sensitivity of the price of gold (per ounce) on present worth of mine investment.

Gold Price	Error (x)	PW(10%)	PW(10%) Error
$250	−28.57%	−$5.09M	−120.46%
$300	−14.29%	$9.89M	−60.23%
$350	0%	$24.88M	0%
$400	14.29%	$39.86M	60.23%
$450	28.57%	$54.85M	120.46%
$500	42.86%	$69.83M	180.69%

TABLE 10.3

Sensitivity of annual production on present worth of mine investment.

Annual Production	Error (x)	PW(10%)	PW(10%) Error
50,000	−16.67%	$21.92M	−11.90%
55,000	−8.33%	$23.51M	−5.51%
60,000	0%	$24.88M	0%
65,000	8.33%	$26.05M	4.72%
70,000	16.67%	$27.14M	9.11%

Annual Production Capacity. Let us assume that the 60,000 ounces of production represents the capacity of the system. Since the amount of gold to be mined (436,230 ounces) is fixed, a change in the production rate actually defines a new problem, because the horizon will change. For example, if the production rate drops, then the project will last longer, as it will take longer to remove the gold. Conversely, if the production rate increases, the gold will be extracted sooner. We would not expect wild deviations in this estimate; thus, we examine a band between 50,000 and 70,000 ounces of annual production.

Table 10.3 provides the results of the analysis. To illustrate how the present-worth values were derived, assume that annual production is 50,000 ounces per year. Then production will last for 8.7 years, where the annual revenues in the first 8 years are $8.55 million per year and year 9 produces 36,230 ounces of gold, for a total year-9 revenue of $5.20 million, including the cost of remediation. The present worth is now

$$\text{PW}(10\%) = -\$25.9\text{M} + \$8.55\text{M}(\overset{P/A,10\%,8}{5.3349}) + \$5.20\text{M}(\overset{P/F,10\%,9}{0.4241}) = \$21.92 \text{ million.}$$

Note that the cash inflows over the 9 years of production total $73.6 million. This is the same amount for the presumed annual level of production of 60,000 ounces. Thus, the critical difference here is in the timing of the cash flows.

For an annual production of 70,000 ounces per year, production spans 7 years, and

$$\text{PW}(10\%) = -\$25.9\text{M} + \$11.97\text{M}(\overset{P/A,10\%,6}{4.3553}) + \$1.78\text{M}(\overset{P/F,10\%,7}{0.5132}) = \$27.14 \text{ million.}$$

Mineable Gold. If we assume that production is stable at 60,000 ounces per year, the length of the project can vary with the number of (actual) mineable ounces of gold in the mine. The value of 436,230 ounces was determined from the amount of estimated gold (555,000 ounces) and the recovery rate of 78.6%. If either of these numbers decreases, then the length of the project decreases. To capture the impact, we vary the amount of recoverable gold, as in Table 10.4.

TABLE 10.4

Sensitivity of total recoverable gold (ounces) on present worth of mine investment.

Error (x)	Mineable Gold	PW(10%)	PW(10%) Error
−30%	305,361	\$12.95M	−47.96%
−10%	392,607	\$21.13M	−15.05%
0%	436,230	\$24.88M	0%
10%	479.853	\$28.36M	13.99%
30%	567,099	\$34.59M	39.03%

The results in Table 10.4 require an explicit analysis due to the production level being fixed at 60,000 ounces per year. If there are only 305,361 ounces in the mine (30 percent less than estimated), then production will last for just over 5 years. The present worth in this case is

$$\text{PW}(10\%) = -\$25.9\text{M} + \$10.26\text{M}(\overset{P/A,10\%,5}{3.7908}) - \$83.4\text{K}(\overset{P/F,10\%,6}{0.5645}) = \$12.95 \text{ million.}$$

The analysis continues similarly to explore the impact of changes in per unit production costs, remediation costs, the interest rate, and the horizon. Changes in production costs and remediation costs can be examined by looking at percentage errors, as in the earlier investment cost analysis. Changes in the interest rate and horizon require explicit analysis with estimated values, as in the foregoing annual-production analysis.

Rather than review all of these parameters, we summarize the data with the use of a spider plot, as shown in Figure 10.9. In this example, the price of gold and the cost of

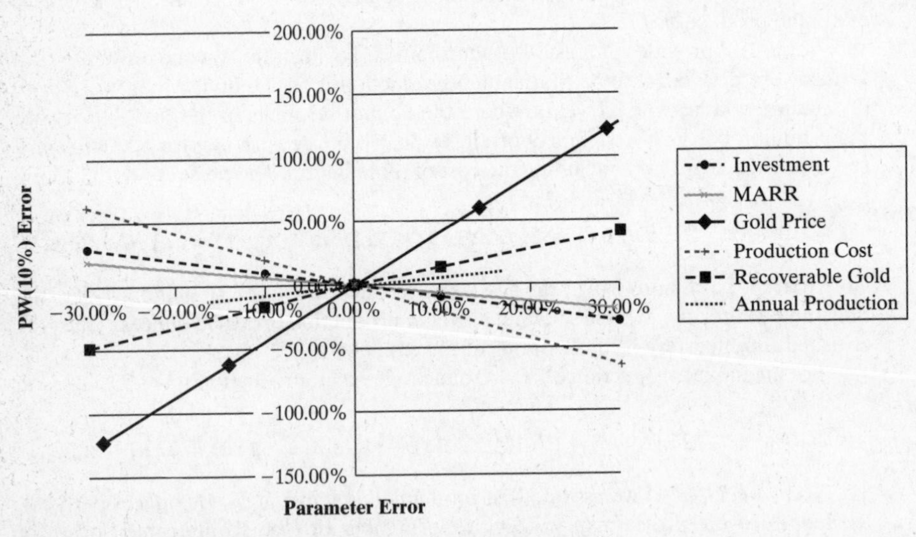

Figure 10.9 Sensitivity analysis spider plot for various parameters.

production are most sensitive, illustrated by their steep slopes. The parameters defined by flat lines on the spider plot, such as the MARR and remediation costs, have little impact on the problem. We could have arrived at similar conclusions with a tornado diagram (see Chapter 7), which orders the parameters according to sensitivity.

We already noted that by examining the slopes of the lines in the spider plot, it is easy to determine which parameters are more critical in the analysis. It is also important to note which parameter values can lead to negative measures of worth such that the project is not worthy of investment. In the previous example, only the price of gold had this type of impact, as an error of less than 30 percent leads to a negative present worth. This finding should signal to the decision maker that the price of gold (and its history) needs to be thoroughly examined before proceeding with the project.

Finally, note that the spider plot may contain both nonlinear and linear relationships. The graph in Figure 10.9 seems to have only linear data, but that is because the range of errors graphed is small. In general, the MARR and the change in horizon will lead to nonlinear relationships, which may require a wider range of error analysis, as a linear extrapolation may not be valid. In this example, changes in the production rates influence the length of the project, which may also define nonlinear relationships. This information can readily be captured on the spider plot.

10.5.2 Multiple-Parameter Analysis

Single-parameter sensitivity analysis, in our investment analysis context, alters one parameter and examines its effect on some measure of worth, while holding all of the other parameters constant. This is not a complete characterization of sensitivity analysis, which is not relegated just to single-parameter perturbations. Sensitivity analysis is a general tool that allows one to look at the change in a variety of parameters, whether simultaneously or not.

One could argue that in most cash flow diagrams, parameters do not change independently. For example, if we change the time horizon for a project, it is probable that the salvage value or disposal cost will change. If there are parameters that are expected to "move" together, then it may be advisable to analyze simultaneous perturbations.

In general, the approach used with one parameter holds for multiple parameters. Unfortunately, the workload increases dramatically. For the case of two parameters, we must perform a complete sensitivity analysis (check all perturbations) for the first parameter, for *every* feasible perturbation of the second parameter. This is clearly much more tedious a task than single-parameter analysis. For m different perturbations of a parameter, sensitivity analysis requires only m calculations for single-parameter analysis. In the two-parameter situation, we will have to make m^2 calculations. We illustrate in the next example.

EXAMPLE 10.7

Multiple-Parameter Sensitivity Analysis

We return to the previous example of the gold mine, as we now have a good understanding of the impacts of perturbations of individual parameters. We illustrate multiple-parameter sensitivity analysis for just one set of parameters, as doing so for more than one set can be quite tedious.

Solution. In theory, we could examine the impact between any two parameters. We will examine the impact of changing gold prices and the interest rate simultaneously. Defining x_1 as the MARR and x_2 as the price of gold, we find that the present-worth function is

$$\text{PW}(x_1\%) = -\$25.9\text{M} + (x_2 - \$179)(60{,}000)\left[\frac{(1+x_1)^7 - 1}{x_1(1+x_1)^7}\right] + \frac{(x_2 - \$179)(16{,}230) - \$1\text{M}}{(1+x_1)^8}.$$

We examine three levels of the interest rate (5%, 10%, and 15%) versus the range of gold prices that we examined in the single-parameter analysis, namely, $250 to $500 per ounce. The results are given in Table 10.5.

The information is also captured in the plot in Figure 10.10, which shows more clearly than the table does, that the impact of the changing gold price is minimized by

TABLE 10.5

Sensitivity of the price of gold and the MARR on the present worth of the mine investment.

MARR	Error (x_1)	Gold Price	Error (x_2)	PW(10%)	PW(10%) Error
5%	−50%	$250	−28.57%	−$1.15M	−104.61%
	−50%	$300	−14.29%	$16.76M	−32.63%
	−50%	$350	0%	$34.67M	39.36%
	−50%	$400	14.29%	$52.58M	111.34%
	−50%	$450	28.57%	$70.49M	183.33%
	−50%	$500	42.86%	$88.39M	255.31%
10%	0%	$250	−28.57%	−$5.09M	−120.46%
	0%	$300	−14.29%	$9.89M	−60.23%
	0%	$350	0%	$24.88M	0%
	0%	$400	14.29%	$39.86M	60.23%
	0%	$450	28.57%	$54.85M	120.46%
	0%	$500	42.86%	$69.83M	180.69%
15%	50%	$250	−28.57%	−$8.13M	−132.67%
	50%	$300	−14.29%	$4.62M	−81.43%
	50%	$350	0%	$17.37M	−30.19%
	50%	$400	14.29%	$30.11M	21.04%
	50%	$450	28.57%	$42.86M	72.28%
	50%	$500	42.86%	$55.61M	123.51%

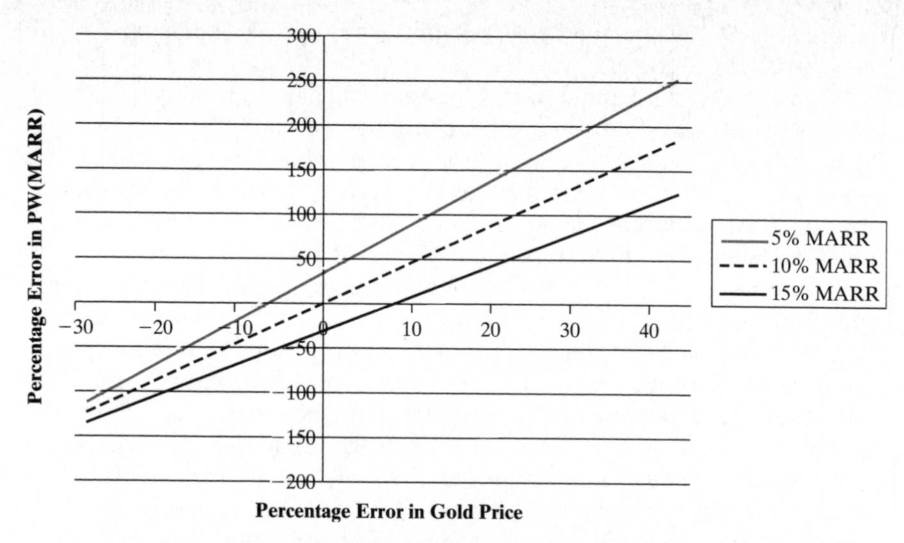

Figure 10.10 Impact of changing MARR and gold price on present worth of mine investment.

decreasing the MARR. This does not overshadow the fact that the project's success is highly dependent on the price of gold.

The power of sensitivity analysis with multiple parameters is that it allows the decision maker to examine the impact of simultaneous movements of parameters. If you know that some data are correlated, in that their parameters tend to move simultaneously, then it may be beneficial to examine their effects. For example, a decrease in the time horizon may lead to an increase in the salvage value. This relationship cannot be examined with single-parameter analysis.

We could extend this general procedure beyond that of two parameters, but looking at multiple parameters simultaneously comes at a cost. If we have n parameters and m perturbations for each parameter, then we will have m^n calculations to perform. The decision maker must truly have good reasons for taking on this additional workload, as it is generally unnecessary to go beyond examining two parameters at a time, and indeed, single-parameter analysis is often sufficient for decision-making purposes.

10.6 Scenario Analysis

As we noted earlier, we may disagree with the assumption in single-parameter sensitivity analysis that all other parameters remain constant when we examine perturbations. However, we also indicated that analyzing more than one parameter at a time can be very costly, as the number of comparisons required—and thus the number of computations—grows exponentially with the number of parameters.

To overcome this computational issue and still have the ability to examine multiple parameters simultaneously, we turn to scenario analysis. As with our other analyses, "scenario analysis" means exactly what the name implies: The decision maker devises scenarios that are likely to occur if the investment is made. These different scenarios allow one to examine the effects of different parameters in different settings. That is, the decision maker is free to alter any of the parameters in each scenario simultaneously. Often, we might try to determine three scenarios: likely, pessimistic, and optimistic. (We discussed this approach in Section 7.4.) Unlike the situation in single-parameter sensitivity analysis, we are free to alter any (and even all) of the parameters simultaneously in order to produce a wide variance of possible results to examine. For the pessimistic case, we might envision the worst situation for each and every parameter, whereas the optimistic case would allow all "hopeful" values to be achieved. The amount of work involved is tied to the number of scenarios to evaluate, since each requires the estimation of a cash flow diagram.

This is not a complicated analysis, as we merely repeat our deterministic analysis for each of the scenarios that we have devised. Again, the goal is to put more information in the hands of the decision maker. We illustrate in the next example.

EXAMPLE 10.8

Scenario Analysis

Toyo Tire and Rubber Co., Japan's fourth-largest tire maker, announced that it would build a plant in the United States to produce up to 2 million tires per year for passenger cars and light trucks. The plant would be built at the cost of $150 million, would employ 350 workers, and is to open in early 2006.[8] Provide a scenario analysis of this investment, assuming an 18% interest rate.

Solution. We examine optimistic, pessimistic, and average cases. Once we define each of the cases, it is merely a matter of computing the annual equivalent worth for the cash flow estimates.

Average Case. We assume that an investment of $150 million is made at the beginning of 2005, with production starting at the beginning of 2006. Sales of 1.75 million tires per year, at an average of $30 per tire, are expected over a 10-year life, with the plant having a $15 million salvage value. Total operating and maintenance costs are expected to be $10 per tire. With this information and a MARR of 18%, the annual equivalent worth of the project is

$$\begin{aligned} AE(18\%) &= -\$150M(\overset{A/P,18\%,10}{0.2225}) + 1.75M(\$30 - \$10) + \$15M(\overset{A/F,18\%,10}{0.0425}) \\ &= \$2.26 \text{ million.} \end{aligned}$$

[8] "Japan's Toyo Tire to Build New $150M Plant in US," *Dow Jones Newswires,* February 17, 2004.

Thus, in the average case, the project should be accepted.

Optimistic Case. In the optimistic case, sales reach a capacity of 2 million tires per year at a cost of $35 per tire, and operating costs are driven down to $9 per tire. In addition, the plant investment totals $140 million, and the plant remains in operation for 12 years. The salvage value is $14 million at this time. All other data remain the same. The annual equivalent worth of the project is then

$$\begin{aligned} AE(18\%) &= -\$140M(\overset{A/P,18\%,12}{0.2086}) + 2.0M(\$35 - \$9) + \$14M(\overset{A/F,18\%,12}{0.0286}) \\ &= \$23.20 \text{ million.} \end{aligned}$$

Clearly, the investment should be made with these optimistic estimates.

Pessimistic Case. In the pessimistic case, the investment costs $150 million, but production rises slowly due to manufacturing difficulties and lax demand. Production in the first year reaches 1 million tires, growing to 1.25, 1.5, and 1.75 million in the ensuing three years, at which level it remains throughout the rest of the 10-year life of the plant. As opposed to the plant having a salvage value, remediation costs of $5 million are to be paid at the end of the life of the plant. Further, the average price for a tire drops to $27 per unit, while costs rise to $11.50 per tire, for net revenues of $15.50 per tire. The annual equivalent worth in this case is

$$\begin{aligned} AE(18\%) &= \left[-\$150M + 1.0M(\$15.50)(\overset{P/A,18\%,4}{2.6901}) + 0.25M(\$15.50)(\overset{P/G,18\%,4}{3.4828})\right] \\ &\quad \times (\overset{A/P,18\%,10}{0.2225}) + \left[1.75M(\$15.50)(\overset{F/A,18\%,6}{9.4420}) - \$5M\right](\overset{A/F,18\%,10}{0.0425}) \\ &= -\$10.42 \text{ million.} \end{aligned}$$

The decision maker must now determine how to use this additional information. If the pessimistic case is likely to occur, then the investment may have to be reconsidered.

You can imagine how complicated these scenarios can become. Consider a "truly" optimistic plan in which the purchase price is as low as possible and annual returns are as high as possible, along with a high salvage value after a longer life of sales than was previously determined. Clearly, the analysis of this problem would give the decision maker the best outcome possible. On the flip side, a truly pessimistic plan would lay out the worst of all possible outcomes. From these outcomes, we would have a measure of worth that would be the lowest possible. With these two values, the decision maker would have a feel for the maximum range of the expected worth of the project. It should be clear how a scenario analysis will lead to further questions, such as How likely is the pessimistic (or optimistic) scenario to happen? Questions like these were discussed in Section 7.4 and will be addressed shortly.

10.7 Probabilistic Analysis

The previous sections provided a number of methods with which to address the risk involved in an investment decision. These methods are classified as deterministic because they do not take advantage of any probabilistic information that may be available. Probabilistic information is required to define a parameter as a *random variable*—a variable that takes on certain values, or a range of values, with some probability.

For example, if we assume that the investment cost in a project will be either \$1.5 million or \$1.75 million and we can assign a probability to each of these two possible values, then we say that the investment cost is a discrete random variable. If we believe that an investment cost will fall in the range of values *between* \$1.5 and \$1.75 million, then we say that the investment cost is a continuous random variable.

In general, a random variable is defined by a probability density function $f(x)$ and its cumulative distribution function $F(x)$. Generally, we are interested in knowing the expected value and the variance of a random variable, as they can be very useful in decision making. We usually assume that the random variable will take on a value near its expected value, while the variance provides a measure of the spread of likely values. (We introduced the concept of variance when we defined risk.)

For a discrete random variable, the expected value is merely the sum of each possible value times its probability. Formally, this is written as

$$E(x) = \sum_x x f(x).$$

For our simple investment cost example, if the probability is 0.35 that the investment will cost \$1.5 million, and thus 0.65 that it will cost \$1.75 million, then the expected value is

$$E(\text{Investment Cost}) = 0.35(\$1.50\text{M}) + 0.65(\$1.75\text{M}) = \$1.66 \text{ million.}$$

The variance is defined as the weighted average of the squared deviation between each possible value and the expected value. Formally,

$$\text{Var}(x) = \sum_x (x - E(x))^2 f(x).$$

For our simple example, the variance is

$$\begin{aligned}\text{Var(Investment Cost)} &= 0.35(\$1.5\text{M} - \$1.66\text{M})^2 + 0.65(\$1.75\text{M} - \$1.66\text{M})^2 \\ &= \$0.0142 \text{ million}^2.\end{aligned}$$

Thus, the variance is a measure of the spread of the distribution of the random variable, with a larger variance occurring when values are far from the expected value, or mean. Often, the standard deviation, which is the square root of the variance is reported. The reason for using the standard deviation rather than the variance is that large numbers frequently result from squaring the deviations. We will use the following notation:

$$\sigma_x = \sqrt{\mathrm{Var}(x)}.$$

The standard deviation for our small example is \$0.119 million, or \$119,000. As a rule of thumb, we would prefer if the standard deviation were less than half of the mean, signifying a very safe investment.

For a discrete distribution, we define the probability that x can take on a given value as

$$\Pr(x = a) = f(a),$$

while the probability that it lies in a range is

$$\Pr(a < x \leq b) = F(b) - F(a) = \sum_{x>a}^{b} f(x).$$

This formula can be useful in determining the probability of a loss.

Random variables can also be described by continuous distributions, as in our earlier example, where the investment cost is defined by the range [1.5, 1.75]. In general, the expected value of a continuous random variable is defined as

$$E(x) = \int_{-\infty}^{\infty} x f(x)\, dx,$$

and the variance is

$$\mathrm{Var}(x) = \int_{-\infty}^{\infty} (x - E(x))^2 f(x)\, dx.$$

These parameters have been defined in general for a variety of distributions. For example, if any value in the range is equally likely, then the variable is defined by a Uniform distribution, for which the expected value is merely the average of the endpoints, or

$$E(\text{Investment Cost}) = \frac{\$1.75\text{M} + \$1.50\text{M}}{2} = \$1.625 \text{ million},$$

while the variance is defined by

$$\mathrm{Var}(\text{Investment Cost}) = \frac{(\$1.75\text{M} + \$1.50\text{M})^2}{12} = \$.0052 \text{ million}^2,$$

or a standard deviation of $72,168. The probability of a continuous variable taking on a particular value is undefined, but its probability over a range (a, b) is defined as

$$\Pr(a \leq x \leq b) = F(b) - F(a) = \int_a^b f(x)\,dx.$$

Random variables can be defined in numerous ways. Discrete distributions, such as that in our first example, can be constructed over any range of possible values. Some common discrete distributions are the Binomial and Poisson distributions. Some common continuous distributions are the Normal, Uniform, and Exponential distributions. Some of these distributions were pictured in Chapter 7.

During our information-gathering phase in the cash flow estimation step of our decision-making process, we may acquire probabilistic information about the outcomes of a variety of parameters, including the cash flows themselves or the scenarios of possible outcomes. We will examine how this additional information can be useful in answering a number of further questions. Specifically, we examine methods for analyzing problems in which we have probabilistic information about the cash flows for a given project or probabilistic information about possible outcomes (scenarios) of a given project.

10.7.1 Probabilistic Scenario Analysis

Previously, we defined scenario analysis as a tool in which we could vary a number of parameters simultaneously, thereby defining a scenario. If we defined a number of scenarios, such as likely, pessimistic, and optimistic, then we could analyze each and have a possible range of expected outcomes. If we are able to define all possible scenarios, or at least a reasonable subset thereof, then we could compute an expectation and variance of our measure of worth if we have probabilistic data concerning the various scenarios. That is, if we know the probability of each possible scenario occurring, we can combine the information into an expected present worth and also compute the variance.

EXAMPLE 10.9 ***Mean and Variance of Scenarios***

Nokia Corp. was contemplating a EUR50 million expansion of its mobile-telephone handset-manufacturing facility in Komarom, Hungary. Together, the 20,000-square-meter plant and 10,000-square-meter logistics center were expected to double output to roughly 80 million sets per year by September of 2004.[9]

---

[9] Koranyi, B., "Nokia to Invest EUR50M to Expand Hungarian Unit—Report," *Dow Jones Newswires,* January 22, 2004.

Make the following assumptions: Define average, optimistic, and pessimistic cases for the expansion. Assume that the initial investment cost is 50, 40, and 60 million euros, respectively. Sales are expected to be between 30 million and 40 million euros per year, with per unit revenues between 40 and 45 euros per handset at costs between 30 and 35 euros per unit. Handset sales will last for either three or four years, with the salvage value of the facility either EUR500,000 or zero, respectively, for the two horizons. If the average, optimistic, and pessimistic scenarios have probabilities of 0.50, 0.20, and 0.30, respectively, calculate the expected present worth and its standard deviation. Assume a MARR of 22%.

Solution. We first analyze each scenario as follows.

Average Case. Assume an investment of EUR50 million with annual sales of 35 million handsets at 42.5 euros per unit at the cost of 32.5 euros per unit. Sales will last for three years with a EUR500,000 salvage value. With this information, and a MARR of 22%, the present worth of the project is defined as

$$\begin{aligned} \text{PW}(22\%) &= -\text{EUR50M} + 35\text{M}(\text{EUR10})\overset{P/A,22\%,3}{(2.0422)} + \text{EUR0.5M}\overset{P/F,22\%,3}{(0.5507)} \\ &= \text{EUR665.05 million.} \end{aligned}$$

Optimistic Case. Assume an investment cost of EUR40 million, with maximum sales lasting for four years (no salvage value on the facility) and with per unit revenues of EUR45 against per unit costs of EUR30. The present worth in this case is

$$\text{PW}(22\%) = -\text{EUR40M} + 40\text{M}(\text{EUR15})\overset{P/A,22\%,4}{(2.4936)} = \text{EUR1.456 billion.}$$

Pessimistic Case. The investment cost rises to EUR60 million, while per unit earnings are EUR40 – EUR35 on annual sales of 30 million handsets. The salvage value and horizon mimic those of the average case. The present worth in this case is

$$\begin{aligned} \text{PW}(22\%) &= -\text{EUR60M} + 30\text{M}(\text{EUR5})\overset{P/A,22\%,3}{(2.0422)} + \text{EUR0.5M}\overset{P/F,22\%,3}{(0.5507)} \\ &= \text{EUR246.61 million.} \end{aligned}$$

The expected present worth is then

$$\begin{aligned} E(\text{PW}) &= (0.30)(\text{EUR246.61M}) + (0.50)(\text{EUR665.05M}) + (0.20)(\text{EUR1.456B}) \\ &= \text{EUR697.71 million.} \end{aligned}$$

The variance is

$$\begin{aligned} \text{Var(PW)} &= (0.30)(\text{EUR246.61M} - \text{EUR697.71M})^2 + (0.50)(\text{EUR665.05M} \\ &\quad - \text{EUR697.71M})^2 + (0.20)(\text{EUR1.456B} - \text{EUR697.71M})^2 \\ &= \text{EUR174,968 million}^2. \end{aligned}$$

The variance corresponds to a standard deviation of EUR418.29 million. With this expected present worth and standard deviation, the decision maker should feel fairly confident that the project will be successful.

10.7.2 Simulation Analysis

It is possible that each individual cash flow and parameter for an investment opportunity is defined by a random variable. This can greatly complicate analysis. In this situation, Monte Carlo simulation is well suited to perform the analysis. Essentially, a simulation allows us to take *one* glimpse of the future—by generating possible outcomes of probabilistic parameters, such as future cash flows. (That is, it *simulates* the future.) We can analyze this glimpse and determine our best course of action. As this is only *one* glimpse of the uncertain future, we repeat the process. In fact, we repeat the process numerous times—in order to determine what we expect to happen in the future, and thus, plan accordingly. We next illustrate how simulation can be used to analyze the case where the cash flows follow distributions. This is followed by the case where scenarios follow distributions.

Cash Flow Analysis

In our application that is to follow, each cash flow or parameter is defined by a probability distribution. A simulation produces one possible realization of these parameters, which can then be analyzed with our traditional deterministic tools of computing a measure of worth. This process is repeated a number of times, resulting in a number of possible outcomes. From these outcomes, we can gather statistics, such as the expected value and standard deviation of the present worth of the project, in order to make a more informed decision as the expected value gives us an idea of what to expect from our investment in an uncertain future.

Formally, the steps involved in the process are as follows:

1. Generate a random value for each cash flow from their individually defined distributions. This results in the definition of $A_0, A_1, \ldots, A_N$.
2. Compute a measure of worth, such as PW(i), for the generated cash flow stream.
3. If the specified number of iterations have been completed, go to the next step; else, go to step 1.
4. Compute the sample mean and variance of PW(i).

Note that for each simulation (step 1), we generate a cash flow diagram and find its present worth (step 2). Computing the sample mean and sample standard deviation in the final step of the process allows us to aggregate the information from all of the simulation runs.

Define $\widetilde{\text{PW}}(i)_s$ as the present worth resulting from simulation s, or the sth simulation. If we perform m simulations, then the sample mean is merely the average of all of the present-worth values for each simulation run, or

$$\overline{\text{PW}} = \frac{\sum_{s=1}^{m} \widetilde{\text{PW}}(i)_s}{m}. \tag{10.4}$$

The sample variance can be calculated by taking the difference between the present worth of each simulation run and the sample mean. These values are squared and summed. Then, dividing by $m - 1$, because there is one fewer degree of freedom for use of the sample mean, leaves us with our sample variance, or

$$\text{Var}(\overline{\text{PW}}) = \frac{\sum_{s=1}^{m}\left(\tilde{\text{PW}}(i)_s - \overline{\text{PW}}\right)^2}{m-1}. \tag{10.5}$$

This information summarizes what we have learned from the simulations. For those who may be unfamiliar with the concept of simulation, the following simple example illustrates a simulation that can be performed by hand.

EXAMPLE 10.10

Simulating Cash Flow Diagrams

China's Galanz Electrical Apparatus Co, Ltd., built a new complex at the expected cost of 2 billion yuan in order to produce 12 million air-conditioning units annually. The site, located outside of Zhongshan, was completed in 2004.[10]

Make the following assumptions: The actual investment cost is either 1.9, 2.0, or 2.1 billion yuan, with respective probabilities of 0.25, 0.50, and 0.25. The plant operates for 15 years, with the salvage value being either 50 million, 0, or −100 million (remediation costs) yuan at that time, with probabilities of 0.20, 0.50, and 0.30, respectively. Finally, the net cash flow resulting from operations and sales is 60 yuan per unit. The number of units sold in each year is either 9 (0.1), 10 (0.2), 11 (0.3), or 12 (0.4) million. The figures in parentheses represent the probabilities of the given level of production. Assume that these are the only relevant cash flows and the interest rate is 18% per year. Perform five simulations and use the output of each to determine the expected present worth and variance of the project.

Solution. We will perform this simulation "by hand" in order to illustrate how it is done. A single simulation produces one possible outcome which defines a cash flow diagram that can be evaluated. We need a method by which to generate the random events that are possible.

First consider the investment cost. There are three possible values, with respective probabilities of 1/4, 2/4, and 1/4. These are written in fractional form with a common denominator so that it is clear how to generate the possible outcomes. Assume that we have a four-sided coin (pretend they exist!) that is perfectly balanced such that if we flip it in the air, any one of the four sides is equally likely to end facing up. We assign one side of the coin the cost of YUAN1.9 billion, two other sides of the coin the cost of YUAN2.0 billion, and the final side the cost of YUAN2.1 billion. If we flip the coin, there is a 1/4 chance that the YUAN1.9 billion cost will be facing up, a 2/4 chance that a YUAN2.0 billion cost will be facing up (since there are two such sides), and a 1/4 chance that the YUAN2.1 billion cost will be facing up. To simulate a possible outcome for the investment cost, we merely flip the coin and note which side is facing up.

[10] "China's Galanz to Build Giant Air Conditioner Factory," *Dow Jones Newswires,* October 22, 2003.

For both the annual returns and the salvage values, we would require a 10-sided coin (keep pretending), since a common denominator of all probabilities is 10. If we had a 4-sided coin (with sides numbered 1 through 4) and a 10-sided coin (with sides numbered 1 through 10), the simulation would be performed as follows:

1. *Generation of Investment Cost.* Flip the four-sided coin and take note of the resulting cash flow. The coin toss may be summarized as follows:

Probability	Coin Toss	Resulting Cash Flow
1/4	1	−YUAN1.9 billion
2/4	2 or 3	−YUAN2.0 billion
1/4	4	−YUAN2.1 billion

2. *Generation of Annual Income.* Flip the 10-sided coin 15 consecutive times and note the resulting cash flow each time. The coin tosses may be summarized as follows:

Probability	Coin Toss	Resulting Cash Flow
1/10	1	YUAN540 million
2/10	2 or 3	YUAN600 million
3/10	4, 5, or 6	YUAN660 million
4/10	7, 8, 9, or 10	YUAN720 million

3. *Generation of Salvage Value.* Flip the 10-sided coin again and note the resulting cash flow. The coin toss may be summarized as follows:

Probability	Coin Toss	Resulting Cash Flow
2/10	1 or 2	YUAN50 million
5/10	3, 4, 5, 6, or 7	0
3/10	8, 9, or 10	−YUAN100 million

As you will undoubtedly have trouble finding a perfectly balanced 10-sided coin, the RANDBETWEEN function in Excel can be used to generate a random integer. For the 10-sided coin, the call "=RANDBETWEEN(1,10)" generates an integer number between 1 and 10 inclusive.

The results of our first simulation are given in Table 10.6. (The sides of the coin are numbered from 1 to 10.) A coin flip of 2 for the investment cost results in a cash flow of −YUAN2.0 billion. A coin flip of 9 on the 10-sided coin results in net income of YUAN720 million in the first period. The simulation concludes with a flip of 6 on the 10-sided coin, defining a cash flow of zero for remediation costs.

We repeat the procedure in Table 10.6 an additional four times, with the results given in Table 10.7. The table shows the 17 coin tosses and the resulting cash flow diagram. Each diagram is analyzed with the 25% interest rate to define the present worth in the final column of the table.

As seen in the data in Table 10.7, the present worth of the project ranges between 411.75 million and 698.03 million yuan. Using Equation (10.4) to take the average of

TABLE 10.6 **Simulation of cash flow diagram for air-conditioner-manufacturing plant.**

Cash Flow	Coin Flip Result	Cash Flow Outcome (millions of YUAN)
Investment	2	−2000
Time-1 Income	9	720
Time-2 Income	6	660
Time-3 Income	4	660
Time-4 Income	7	720
Time-5 Income	8	720
Time-6 Income	4	660
Time-7 Income	2	600
Time-8 Income	5	660
Time-9 Income	9	720
Time-10 Income	4	660
Time-11 Income	6	660
Time-12 Income	4	660
Time-13 Income	4	660
Time-14 Income	1	540
Time-15 Income	3	600
Salvage Value	6	0

TABLE 10.7 **Results of five simulations of outcomes for air-conditioner-manufacturing plant.**

Simulation	Coin Tosses	Cash Flow Diagram	PW(25%) (yuan)
1	2 9 6 4 7 8 4 2 5 9 4 6 4 4 1 3 6	0; 720 660 660 720 ... 600; 1 2 3 4 ... 15; 2000	527.68M
2	4 4 3 4 9 3 7 5 8 9 9 9 5 4 5 4 3	0; 660 600 660 720 ... 660; 1 2 3 4 ... 15; 2100	459.32M
3	2 5 5 9 6 4 3 9 4 10 1 1 4 6 3 8 1	0; 660 660 720 660 ... 770; 1 2 3 4 ... 15; 2000	461.03M
4	1 5 4 5 2 10 10 10 5 9 9 8 10 4 7 6 1	0; 660 660 660 600 ... 710; 1 2 3 4 ... 15; 1900	698.03M
5	2 6 1 2 9 6 8 9 10 3 7 8 6 9 6 7 7	0; 660 540 600 720 ... 720; 1 2 3 4 ... 15; 2000	411.75M

the five present-worth values results in a sample present worth of

$$\overline{\text{PW}} = \frac{527.68\text{M} + 459.32\text{M} + 461.03\text{M} + 698.03\text{M} + 411.75\text{M}}{5}$$

$$= \text{YUAN}511.56 \text{ million.}$$

Using the preceding sample mean, we calculate the sample variance:

$$\text{Var}(\overline{\text{PW}}) = \frac{(527.68\text{M} - 511.56\text{M})^2 + \cdots + (411.75\text{M} - 511.56\text{M})^2}{4}$$

$$= \text{YUAN}287{,}627 \text{ million}^2.$$

The sample standard deviation is 536.31 million yuan. We discuss the appropriate number of simulations to perform and how the summary data can be utilized further for decision making shortly.

We compute the sample mean and variance in order to summarize our simulations, but this information can be put to practical use in decision making as noted in the next chapter. We could also determine the probability of the project *not* being successful (having a *negative* present worth) directly from our simulation by calculating the ratio of the number of simulation runs that generate a negative present worth over the total number of simulation runs. In our previous example with limited runs, this was zero. The next example illustrates a more comprehensive simulation analysis with the use of Excel.

EXAMPLE 10.11

Monte Carlo Simulation Continued

Toyota expanded its production facilities in Adapazari, Turkey, at the cost of EUR180 million, from an annual capacity of 100,000 cars to 150,000. The expansion was for the production of the new Corolla Verso, a compact minivan. Toyota expected the expansion to be operational by August of 2004 and hoped for sales of 62,000 Versos that year.[11]

Make the following assumptions: The investment to produce the new minivan will cost either EUR185 million or EUR195 million (each with equal probability of occurrence). Assume that sales in each year following the release of the new minivan are Normally distributed. In year 1, the mean is 60,000 cars, with a standard deviation of 10,000. The mean is expected to increase by 5000 each subsequent year, with the standard deviation holding steady. It is anticipated that the net return will be 900 euros per sale. Assume a five-year study horizon, with the salvage value of the facility equal to a percentage of the EUR20 million. The percentage is defined by an Exponential distribution with a mean of 1. Perform 300 simulations of the investment, and compute the sample mean and variance, as well as the probability that the investment will have a positive present worth. Assume an 18% annual rate of interest.

---

[11] Reiter, C., "Toyota Relaunches Corolla Verso, Expands Turkish Plant," *Dow Jones Newswires,* January 26, 2004.

Solution. We take advantage of various Excel functions in order to generate our random variables. To generate sales from a normal distribution, we use the function

NORMINV(probability,mean,standard_dev).

This function returns the inverse of the cumulative Normal distribution defined by the mean and standard deviation. That is, with the given probability, the random variable takes on a value less than or equal to the returned value. For example, if we call the function with our sales data in the first year (mean 60,000 and standard deviation 10,000) and a probability of 0.5, the NORMINV function returns 60,000, since the probability that sales in the first year will be less than or equal to 60,000 is 1/2. A probability of 0.25 returns a value of 58,255, meaning that sales will be less than or equal to 58,255 with a probability of 1/4.

Because we would like to simulate sales, we call the NORMINV function with the RAND function, which generates a Uniformly distributed random number between 0 and 1. Thus, sales in the first year are generated with

=NORMINV(RAND(),60000,10000).

Note that the RAND function has no arguments and it *recalculates* a new value every time new data are entered anywhere on the spreadsheet. This is important to remember in generating simulations.

Samples from an exponential distribution are generated with the EXPONDIST function, defined as

EXPONDIST(x,lambda,cumulative),

where lambda is the mean (1.0 in this example) and the 'cumulative' function call is 'TRUE', since we desire the return from the cumulative distribution function. The value 'x' is called with RAND in order to simulate our percentage. The salvage value (in millions) is then generated with

=20 * EXPONDIST(RAND(),1,TRUE).

The investment cost follows a discrete distribution. We could use the RANDBETWEEN function as we did in the previous example, or we can use the RAND function, as it generates a Uniformly distributed value between 0 and 1. Nesting the RAND function in a logical IF function, namely,

=IF(RAND()<=0.5,185,195),

produces our desired result. Recall that the IF function returns the second argument if the first argument is true. Thus, if RAND returns a value of 0.35, then the investment cost generated is 185 million euros.

The spreadsheet in Figure 10.11 illustrates the approach. The top of the spreadsheet follows our typical layout, although the data center is expanded greatly due to the wealth of information available for this example. The cash flows depicted in cells B4 through B9 follow our traditional form, but are also generated (separately) horizontally starting in row 16 in order to facilitate performing a number of simulations. (These rows were designed to be copied repeatedly.)

Specifically, the results of 10 simulation runs are presented, with references (cells B17, H17, and G24) illustrating how the individual cash flows and resulting

	A	B	C	D	E	F	G	H
1	Example 10.11: Production Expansion			**Input**			Distribution	Parameters
2				Investment	€ 185	million	Discrete	0.5
3	**Period**	**Cash Flow**			€ 195	million		0.5
4	0	-€ 185.00		Sales (mean)	60,000		Normal	60000+(n-1)*G
5	1	€ 56.91		G (mean)	5,000			10000
6	2	€ 60.91		Unit Net Revs	€ 0.0009	per million		
7	3	€ 59.59		Salvage Value	€ 20	million	Exponential	1
8	4	€ 56.35		Interest	18%			
9	5	€ 56.71		Periods	5	years		
10								
11				**Output**				
12				PW(18%)	-€ 2.91	million		
13								
14	IF(RAND()<=$H$2,-$E$2,-$E$3)					=B17+NPV($E$8,C17:G17)		
15	Run	0	1	2	3	4	5	PW(18%)
16	1	-€ 185.00	€ 46.25	€ 78.10	€ 64.47	€ 79.81	€ 88.64	€ 29.43
17	2	-€ 195.00	€ 37.80	€ 52.82	€ 62.86	€ 82.34	€ 75.60	-€ 11.26
18	3	-€ 185.00	€ 51.06	€ 55.21	€ 77.51	€ 60.51	€ 78.20	€ 10.49
19	4	-€ 195.00	€ 69.27	€ 46.85	€ 84.43	€ 64.88	€ 61.38	€ 9.03
20	5	-€ 185.00	€ 51.36	€ 64.88	€ 75.79	€ 63.74	€ 85.18	€ 21.36
21	6	-€ 185.00	€ 37.87	€ 48.19	€ 63.41	€ 70.44	€ 85.92	-€ 5.81
22	7	-€ 195.00	€ 48.34	€ 53.46	€ 55.23	€ 69.14	€ 68.26	-€ 16.53
23	8	-€ 195.00	€ 49.53	€ 75.46	€ 58.99	€ 62.46	€ 91.45	€ 9.26
24	9	-€ 195.00	€ 55.35	€ 68.92	€ 63.16	€ 87.27	€ 75.26	€ 17.76
25	10	-€ 195.00	€ 36.93	€ 60.65	€ 65.48	€ 57.69	€ 86.88	-€ 12.56
26	=NORMINV(RAND(),$E$4+$E$5*(G$15-1),$H$5)*$E$6+EXPONDIST(RAND(),$H$7,TRUE)*$E$7							
27								

Figure 10.11 Resulting cash flows and present worth (millions of euros) of first 10 simulation runs.

present-worth values were calculated. Note that the cash flow in period 5 includes the salvage-value term, which is not present in periods 0 through 4.

Note that we can copy row 16 the necessary 300 times to complete our simulation trials. It is suggested that, once that row is copied, the present-worth values (column H) be copied and pasted to another worksheet or location. (Use the 'Paste Special' command under the 'Edit' menu option. When the dialogue box appears, paste 'Values' such that the results of the simulation are retained.) Once complete, the runs can be analyzed and graphed if desired. If you do not perform this copy-and-paste operation, the values in column H will change every time an input changes on the spreadsheet due to the use of the RAND function. Furthermore, note that a simple 'Paste' operation will not suffice, as it will merely copy the functions in column H, which are derived from the RAND function values, to a new location. One must use the 'Paste Special' menu option.

The 300 simulations produced an expected present worth (sample mean) of EUR4.55 million and a sample variance of EUR187.67 million^2, or a sample standard deviation of EUR13.69 million. Examining Figure 10.11, we note that, of the 10 simulation runs, 4 resulted in a negative present worth and 6 resulted in a positive present worth. We can use this information to determine the probability that a project will be unsuccessful (have a negative present worth). Of the 300 simulation runs, 187 produced a positive present worth. This translates to a 0.623 probability of success and a .377 probability of the project not achieving a positive present worth.

The histogram in Figure 10.12 provides a further breakdown of the 300 simulation runs. The graph shows that, of the 300 trials, 2 produced present-worth values below

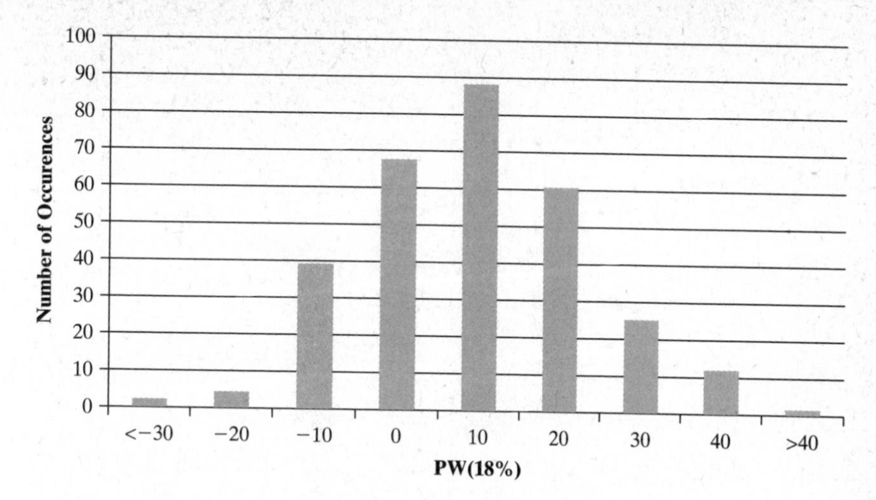

Figure 10.12
Distribution of PW(18%) from 300 simulations.

-EUR30 million, 4 between -EUR30 and -EUR20 million, 39 between -EUR20 and -EUR10 million, etc. This information can be used to compute the probability that the present worth lies in a particular range (such as greater than EUR10 million). Note that it is typical for the output histogram to take on the shape of a Normal distribution.

There remains a lingering question as to how many simulation runs must be completed in order to perform the preceding analyses. In the ideal case, we would run an infinite number, because the law of large numbers says that we will calculate the correct value of the present worth if we do so. But we don't have that kind of time! So how many is enough?

This question could be answered formally or informally. We will take the informal approach. As we continue to generate scenarios, we can calculate the sample mean for the total number of trials, up to and including the current trial. If we plot this value, we can see that the sample mean stabilizes after a number of runs. The sample mean, computed after each simulation run, for the previous example is plotted in Figure 10.13 for the first 100 trials. The graph illustrates that the sample mean stabilizes after time, signaling that the simulation can be halted.

A final question that needs to be addressed concerns the input distributions: How did we determine that annual sales would be Normally distributed? Or how did we know that the salvage value of the facility would be Exponentially distributed? These are not easy questions to answer, and techniques to address them are beyond this text. However, histograms are a good starting point for determining possible distributions of data, as noted in Section 7.4.

Scenario Analysis

We now expand the use of simulation to the case when we have probabilistic information about any of our parameters, including dependencies between them. Note the use of the term "parameters," and not just cash flows, for

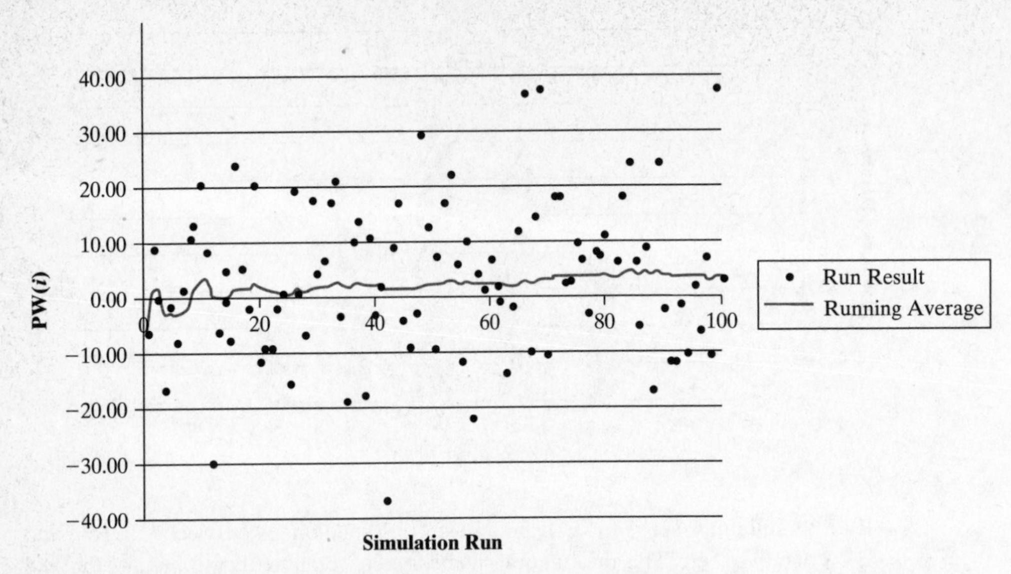

Figure 10.13 Sample means over first 100 simulation runs from example.

we can be very general in analyzing our projects with scenarios that allow for changes in any relevant parameter, including the interest rate and the time horizon. The decision maker, however, must be careful in the design of the simulation here, as it may require a predefined order in which to generate estimates. For example, if the horizon can take on the value of 8 or 9 periods and the salvage value is age dependent, then, clearly, the value of the horizon must be generated before the salvage value is generated for a given scenario. We highlight this change in our outline of the simulation procedure:

1. Generate random values for each parameter (cash flow, interest rate, time horizon, etc.) from their individually defined distributions. Note that the order in which parameters are estimated is determined by the dependencies of parameter estimates on previous estimates.
2. The values generated define one deterministic scenario that can be evaluated by computing a measure of worth.
3. If the specified number of iterations have been completed, go to the next step; else go to step 1.
4. Compute the sample mean and variance of the measure of worth.

We illustrate this procedure in the next section, which presents the solution to our introductory real decision problems. It should be clear from that example and the one we just completed that simulation is an extremely powerful tool for evaluating risky investment proposals.

10.8 Examining the Real Decision Problems

We revisit our chapter's introductory real decision problems concerning the construction of a hydroelectric plant in Quebec. As the amount of rainfall is a key driver of project costs and revenues, we simulate the rainfall each season and estimate costs on the basis of the amount of rainfall generated.

The spreadsheet in Figure 10.14 presents one simulation run from the given data. We first determine whether it is a rainy or dry season. The number 1 identifies a wet season, while the number 2 identifies a dry season in column B. In the first period of operations, the season is generated with the IF and RAND() functions in Excel as

$$\text{SEASON}_1 = \text{IF}(\text{RAND}() \leq 0.5,1,2).$$

This formula is programmed into cell B6. Recall that the IF function returns the second argument in its call if the first argument is true. If the first argument is false, the third argument is returned. The RAND() function returns a value between 0 and 1.

Each ensuing period of rain is dependent on the previous amount of rainfall and must therefore be simulated in order. SEASON_t is generated for cells B7 through B26 as follows:

$$= \text{IF}(\text{OR}(\text{AND}(\text{RAND}() \leq 0.65, \text{SEASON}_{t-1} = 1), \text{AND}(\text{RAND}() \leq 0.35, \text{SEASON}_{t-1} = 2)),1,2).$$

Note that the variable SEASON_{t-1} is called by the appropriate cell in column B. This formula invokes our assumption that there is a 65% chance that the season will not change and a 35% chance it will, regardless of which season just occurred. The AND function returns the value "TRUE" if all arguments in its call are true, while the OR function returns the value "TRUE" if at least one argument in its call is true. (The reader can verify that this works by assuming that it was just a rainy season and the RAND() function generates a 0.50 or a 0.85.)

Once the season has been generated, the amount of rainfall (column D) is simulated with

$$\text{RAIN}_t = \text{IF}(\text{SEASON}_t = 1,\text{NORMINV}(\text{RAND}(),8500,1500), \text{NORMINV}(\text{RAND}(),5000,500)).$$

Thus, a sample is taken from the correct Normal distribution, depending on the season that has been generated. The amount of electricity generated (column E) is calculated as

$$\text{ELEC}_t = \text{MAX}((0.05)(\text{RAIN}_t) - 20\text{MW},0).$$

This value is multiplied by 300 days $\times$ 24 hours $\times$ C\$25 to calculate the revenue (column F). Finally, the operating and maintenance costs (column G)

	A	B	C	D	E	F	G	H	I	J	K	L
1	Chapter 10: Hydo-Electric Power Generation									**Input**		
2										Investment	$308.00	million
3	**Period**	**Season**	**Cumulative**	**Rainfall**	**Power**	**Net Revenues**	**O&M Cost**	**Cash Flow**		O&M	$5.00	million
4		**Wet=1**	**Wet**	**Cubic Meters**	**MW**	**Revenues**				Wet Rise	6%	
5	0	--	--	--	--	--	--	-$154.00		Dry Rise	3%	
6	1	1	--	--	--	--	--	-$154.00		MW Net Rev	$25.00	per MWhr
7	2	2	0	5665	263	$47.39	$5.15	$44.24		Annual Prod	0.0072	million hrs
8	3	1	1	9551	458	$82.36	$5.46	$77.90		Normal Wet	8500	1500
9	4	2	0	5177	239	$42.99	$5.62	$39.37		Normal Dry	5000	500
10	5	1	1	7325	346	$62.32	$5.96	$57.36		Same Season	65%	
11	6	2	0	5651	263	$47.26	$6.14	$43.12		Interest Rate	12%	per year
12	7	2	0	4990	229	$41.31	$6.32	$36.98		Operating Pds	20	
13	8	2	0	4553	208	$37.37	$6.51	$32.86				
14	9	2	0	4636	212	$38.12	$6.71	$33.41		**Output**		
15	10	1	1	7095	335	$60.25	$7.11	$54.14		IRR	14.62%	
16	11	1	1	11407	480	$86.40	$7.54	$79.86		PW(12%)	$53.72	
17	12	1	1	7777	369	$66.39	$7.99	$59.40				
18	13	1	1	8877	424	$76.29	$8.47	$68.82				
19	14	1	1	9320	446	$80.28	$8.98	$72.30				
20	15	1	1	7322	346	$62.30	$9.52	$53.78				
21	16	1	1	9764	468	$84.27	$10.09	$75.19				
22	17	1	1	4929	226	$40.76	$10.69	$31.07				
23	18	1	1	8666	413	$74.40	$11.33	$64.06				
24	19	1	1	7670	364	$65.43	$12.01	$54.42				
25	20	2	0	5296	245	$44.06	$12.37	$33.69				
26	21	2	0	4085	184	$33.16	$12.74	$22.42				
27			12									

Figure 10.14 One simulation of hydroelectric power plant investment based on probabilistic rainfall.

are generated by multiplying the previous period's cost by 1.06 if a wet season occurs or 1.03 if a dry season occurs. For the first period of operation, the rate is multiplied by C$5 million.

The only other detail that must be attended to is to sum the number of wet seasons. If the number reaches 15 before the 20-year expected life of the plant is reached, then the project is terminated early and cash flows are analyzed only through the 15th wet season.

The average IRR for the 50 simulation runs was 14.66%, with a standard deviation of 1.75%. Note that 6 simulations produced 15 wet seasons over a 20-year period. In these runs, the life of the project was cut short accordingly. Because the MARR is only 12%, we would accept the project according to the average IRR. Furthermore, the average PW(12%) for the 50 simulation runs was $52.08 million, with a standard deviation of $33.06 million. Only 3 of the 50 simulations produced a negative present worth. Thus, the project appears to carry little risk, according to our assumptions about the data. This fact may help explain why there have been a number of power plants built on that river over time.

The simulation from this example was actually quite simplified. We could clearly have dealt with issues such as revenues (price per megawatt-hour) in a probabilistic fashion. Revenues may even be correlated with rainfall, which is dependent on the weather. The point is that simulation is a flexible and powerful tool for decision making. In addition, it is fairly easy to execute with the use of a spreadsheet.

10.9 Key Points

- The risk associated with an investment is that money can be lost. This can occur when costs are higher than expected, revenues are lower than expected, or some combination of these two outcomes occurs.
- Risk is often measured by the variance or semivariance in the return on the investment or the probability of loss (negative present worth).
- The payback period is the first period in which accumulated revenues exceed accumulated costs. A long payback period is generally associated with a risky investment.
- The payback period with interest includes the time value of money in its calculation.
- The project balance is the periodic future worth of a project's cash flows over time. The graph of the project balance presents the decision maker with a measure of risk, as it incorporates the payback period and the future worth, as well as identifying the amount by which a project is "in the red" during its lifetime.
- Break-even analysis determines the value for which a given parameter (cash flow or cash flow input) moves a project from being accepted to being rejected.

- Sensitivity analysis determines which parameters are most critical to a project's worth. These are the parameters that must be examined closely in any ensuing economic analysis.
- Scenario analysis provides a method by which to examine multiple parameter changes simultaneously. Pessimistic, optimistic, and average cases are generally examined. These may also be examined probabilistically.
- A random variable takes on different values (or ranges of different values) with different probabilities. The variance provides some measure of the spread of a random variable.
- Monte Carlo simulation is a technique used to evaluate cash flows or cash flow inputs defined by a variety of distributions. The sample mean and variance from a simulation provide more information for decision-making purposes. Finally, the probability of success of a project can be computed directly from the output of the simulation.

10.10 Further Reading

Bussey, L.E., and T.G. Eschenbach, *The Economic Analysis of Industrial Projects,* 2d ed. Prentice Hall, Englewood Cliffs, New Jersey, 1992.

Eschenbach, T.G., "Spiderplots vs. Tornado Diagrams for Sensitivity Analysis," *Interfaces,* 22(6):40–46, 1992.

Eschenbach, T.G., and L.S. McKeague, "Exposition on Using Graphs for Sensitivity Analysis," *The Engineering Economist,* 34(4):315–333, 1989.

Hajdasinski, M.M., "The Payback Period as a Measure of Profitability and Liquidity," *The Engineering Economist,* 38(3):177–192, 1993.

Law, A.M., and W.D. Kelton, *Simulation Modeling and Analysis,* 2d ed. McGraw-Hill, New York, 1991.

Lohmann, J.R., and S.N. Baksh, "The IRR, NPV and Payback Period and Their Relative Performance in Common Capital Budgeting Decision Procedures for Dealing with Risk," *The Engineering Economist,* 39(1):17–48, 1993.

Marshall, K.T., and R.M. Oliver, *Decision Making and Forecasting.* McGraw-Hill, New York, 1995.

Park, C.S., and G.J. Thuesen, "Combining Concepts of Uncertainty Resolution and Project Balance for Capital Allocation Decisions," *The Engineering Economist,* 24(2):109–127, 1979.

Sartori, D.E., and A.E. Smith, "A Metamodel Approach to Sensitivity Analysis of Capital Project Valuation," *The Engineering Economist,* 43(1):1–24, 1997.

Thuesen, G.J., and W.J. Fabrycky, *Engineering Economy,* 9th ed. Prentice Hall, Upper Saddle River, New Jersey, 2001.

Walpole, R.E., R.H. Myers, and S.L. Myers, *Probability and Statistics for Engineers and Scientists,* 6th ed. Prentice Hall, Upper Saddle River, New Jersey, 1998.

10.11 Questions and Problems

10.11.1 Concept Questions

1. What are the differences among decision making under (a) certainty, (b) uncertainty, and (c) risk?

2. How can one measure the risk associated with a project? Are these measures equivalent?

3. How does the payback period provide some measure of the risk associated with a project?

4. What are the criticisms of the payback period? Are they justified?

5. What information does computing the project balance over time provide for the decision maker?

6. Can the project balance be used to make a decision about whether a project should be accepted or rejected? If so, is the project balance consistent with present worth? Explain.

7. How are break-even and sensitivity analysis related? How are they different?

8. Is sensitivity analysis relegated to examining one parameter at a time? Why is this often the case in analysis?

9. Contrast scenario analysis with sensitivity and break-even analyses.

10. What is a random variable? Why is it appropriate to model a cash flow as a random variable?

11. What does the mean of a distribution tell us? What about the variance?

12. How can knowing the variance of the present worth of a project aid in decision making?

13. Why is Monte Carlo simulation such a powerful tool? What can be simulated in a Monte Carlo simulation?

14. How many simulations should we run for a project?

15. What should we calculate to summarize a simulation?

10.11.2 Drill Problems

1. An investment of $12 million is made at time zero with net revenues of $2.3 million in year 1, growing at a rate of 8% per year. What is the payback period of the investment? What is the payback period with interest, assuming a MARR of 14%? Is this enough information to make a decision about the investment?

2. An investment of $2.5 million is spread evenly over two years and is followed by revenues of $750,000 in year 1, increasing 10% each year thereafter, against constant costs of $150,000 per year. What are the project balances over the life of the project, which runs for seven years after the investment is completed, assuming a 15% interest rate? What information does all this provide?

3. A $250,000 machine is purchased that can produce up to 10,000 parts per quarter over a life of six years. If each part generates a *profit* of $2.50 and annual fixed costs are $15,000, what is the break-even production level per quarter for the machine? Assume a 12% annual interest rate.

4. Consider again Problem 3, and graph the project balance over time, assuming production runs at full capacity. On the same graph, plot the project balance over time, assuming that production starts with 10,000 parts in the first year and increases by 10,000 parts each year until reaching capacity. Use the graphs to define the differences in risk between the two scenarios.

5. An investment of $400,000 is made at time zero. Revenues are expected to grow from $100,000 at time 1 by $50,000 each year for the next six years while costs are expected to remain steady at $50,000 per year. The seven-year project has no salvage value. Is the present worth (use a MARR of 18%) more sensitive to the investment cost or to the value of the gradient? Construct a spider plot to defend your answer.

6. Consider again Problem 5, and construct a tornado diagram to illustrate sensitivity if it is believed that the investment cost will be within −10% to 10% of its estimate, the gradient value will be within −15% to 30% of its estimate, the first year's revenue will be within −15% to 20% of its estimate, and the annual cost will be within −15% to 20% of its estimate.

7. An investment of $1.25 million is made at time zero, with annual revenues of $1 million over 10 years and O&M costs of $350,000 in year 1, growing by a rate of 6% each year. The salvage value at the end of 10 years is $50,000. The MARR is 12%. Perform a sensitivity analysis, assuming an error range of −30% to +30%, on the following variables:

   (a) Investment cost.

   (b) Annual revenues.

   (c) Initial O&M cost.

   (d) Increasing rate of O&M cost.

   (e) Salvage value.

   (f) Study horizon.

   (g) MARR.

8. Consider again Problem 7, and draw the project balance over time. Does the project appear to be risky? Defend your answer.

9. Consider again Problem 7, and determine the level of annual revenues for which the project breaks even.

10. Consider again Problem 7, and determine the investment cost for which the project breaks even.

11. Repeat Problem 7, but with an after-tax analysis assuming a 40% effective tax rate and straight-line depreciation with a five-year recovery period and the half-year convention. In addition to testing the previous parameters, examine the sensitivity of the tax rate.

12. An asset is needed for a short amount of time (three years). Its cost is $250,000, but its salvage value is dependent on how much it is used over time. If the asset produces 10,000 parts per year, its salvage value is $100,000 at the end of year 3, but if it produces 20,000 parts per year, its salvage value drops to $50,000, if it produces 30,000 parts per year, its salvage value falls to $10,000. If net revenues are $8 per part, what is the present worth of the three scenarios presented (based on usage), assuming that the MARR is 15%?

13. A government entity makes an investment of $150,000 in a new park that is expected to receive 10,000 visitors per year. The typical visitor is expected to stay for 2 hours, and each stay is generally worth $15 per hour to the visitor. Assume a study period of 20 years, a salvage value of $50,000, and a 3% annual rate of interest. Is the BC ratio more sensitive to the number of visitors or the worth of their stay? What number of visitors results in the investment breaking even (in terms of the BC ratio)?

14. An investment to improve the safety of a road at the cost of $2.5 million is being considered. The improvement will last 10 years with no salvage value, but is expected to reduce accidents by 50% each year. This amounts to 25 fewer accidents (valued at $10,000 per accident), 15 fewer injuries (valued at $25,000 per injury), and 1 fewer fatality (valued at $500,000 per fatality) per year. If the interest rate is 5% per year, what is the break-even value for this project that concerns the value of a life? How sensitive is the accept–reject decision to this estimate?

15. An investment of $1.5 million is made at time zero with annual revenues of $600,000 in year 1, growing at the rate of 30% annually over a seven-year horizon. Annual O&M costs are estimated at $150,000 per year. The salvage value of the investment is $1 million at the end of year 7. How sensitive is the answer (PW(15%)) to the salvage value? Perform the following two-parameter sensitivity analyses:

(a) Sensitivity of salvage value and horizon.

(b) Sensitivity of salvage value and interest rate.

(c) Sensitivity of annual revenues and horizon.

Given the information obtained from your dual-parameter sensitivity analyses, construct pessimistic, average, and optimistic scenarios for the investment. What is the annual equivalent worth of each scenario?

16. An investment of $12 million is made at time zero, with net revenues of $2.3 million in year 1, growing at a rate of 8% per year. These are the expected values. The standard deviation of the net revenues is $500,000 in year 1, increasing by $25,000 each year thereafter. Assume $N = 10$ and MARR = 12%. If these flows are defined as independent and Normally distributed, perform a simulation (100 runs) to compute the expected present worth and variance. Are these values expected? Explain.

17. For the simulation in Problem 16, what is the probability that the present worth will be greater than $5 million? $7.5 million?

18. An investment of $400,000 is made at time zero. Revenues are expected to be either $80,000, $90,000, or $100,000 in year 1, with equal probability. In addition, revenues will grow either $10,000 per year, with probability 0.6, or $5000 per year, with probability 0.4. Costs will either be $50,000 (probability 0.6) or $60,000 (0.4) per year. There is no salvage value. Perform five hand simulations (use Excel to "roll the dice"), assuming a 12 year horizon with no salvage value. Compute the mean and standard deviation of the present worth, and compute the probability of success. The interest rate is 18%. Plot the mean after each simulation run. Is this enough information to make a rational decision? Explain.

19. Repeat the previous problem, but generate 250 simulation runs with Excel.

20. An investment of $750,000 is made at time zero, with annual revenues of $600,000 over 10 years and O&M costs of $200,000 in year 1, growing by a rate of 6% each year. The salvage value at the end of 10 years is $50,000, and the MARR is 12%. Perform a simulation, assuming the following:

- Revenues are Normally distributed, with mean $600,000 and standard deviation $400,000.
- Operating costs start at $200,000 in year 1 and grow at a rate that compounds on previous periods. Assume that the operating costs grow uniformly according to U[0.5%, 1.5%] in the second period, and that the rate increases such that it grows at the realized rate plus U[0.5%, 1.5%] in the third period, following this pattern in each period thereafter.
- The salvage value has a 0.8 probability of being $50,000, 0.15 of being $60,000, and 0.05 of being $70,000 at the end of the horizon.
- The horizon is either 10 or 11 periods, with probability 0.5 for each.

Determine the expected mean, its standard deviation, and the probability of success. How does this analysis compare with the deterministic one?

10.11.3 Application Problems

1. In May of 2004, Airbus SAS opened a $435 million assembly plant in Toulouse, France, for manufacturing its A380 jumbo jet. The first aircraft, with a capacity of 555 seats, is to enter service in 2006. The facility, built on 220 hectares, will accept parts produced throughout Europe, including France, Germany, Spain, and the United Kingdom, according to specialized transportation methods. The list price for the jumbo jet is $280 million, and Airbus had 129 firm orders as of the opening of the factory. The plant's capacity is four aircraft per month.[12]

Make the following assumptions: Annual fixed costs of operating the facility are expected to be $20 million per year. The plant has a 20-year life (the life cycle of the airplane), with a $20 million salvage value, and the interest rate is 15%.

(a) What must the per unit profit be in order for the investment to break even, assuming that sales remain at capacity over 20 years? Is this profit feasible,

[12] "Airbus Launches Production of A380 'Superjumbo' Airliner," *Dow Jones Newswires*, May 7, 2004.

given the list price of the airplane? Would it be feasible with quantity discounts given for large orders?

(b) Assume a per unit cost of $278 million per plane (price is $280 million). What is the sensitivity of the present worth of the project to this estimate?

(c) Consider the purchase of the plane by an airline for $280 million. Assume that the average revenue per seat is $250 per flight and that the plane is capable of three flights per day, 300 days a year. If fixed costs of operation are $2 million per year, what must the profit per flight be to break even, assuming a 14% annual interest rate, a 15-year horizon, and a salvage value of $15 million?

(d) Assume that per flight costs are $20,000 and average selling 65% capacity at $250/seat. What would be the increase in present worth for flying an additional route per day?

2. LANTA, the Lehigh and Northampton Transportation Authority, which serves Allentown, Bethlehem, Easton, and surrounding areas in Pennsylvania, has an operating budget for 2005–06 that is built off estimated revenues of $16.5 million. Of this amount, $3 million is projected from passenger fares and the remainder from grants and other government subsidies. Costs to operate the system include $1.2 million for fuel, $2.2 million for outsourcing some transportation functions, and $13.1 million in other costs (such as salaries, benefits, etc.). The system provided 4.6 million trips in the last year, with ridership up 44% since 1997.[13]

Examine this project simply as a single-year problem, since no explicit capital investments are being analyzed, and assume that 4.6 million trips are completed, paying total fares of $3 million for the year. Using our strict definition of the BC ratio (user benefits over facilitator costs, where the only users in this problem are the riders and all government entities are facilitators), determine the per trip break-even benefit for a BC ratio of 1. Does your answer seem reasonable? If not, what other benefits may help reduce this number? If the per trip benefit is $5 at a cost of $2, how much must ridership grow? Does this seem reasonable, given historical trends.

3. Kyocera Corp. expanded its solar cell production facility in Shiga Prefacture in Japan in June of 2004 at a cost of around ¥2 billion. The expansion increased annual capacity by 72,000 kilowatts.[14]

Make the following assumptions: The plant has a 10-year life with no salvage value and the interest rate is 8% per year.

(a) If sales match the increase in capacity, what must the per kilowatt unit profits be in order to achieve an IRR of 12% on the investment?

(b) Solve Part (a) again, but this time assuming annual fixed costs of ¥20 million per year.

4. Sasol Synfuels International of South Africa and Qatar Petroleum entered into a joint venture to construct a $900 million ORYX gas-to-liquids facility in Qatar. The facility will consume 330 million cubic feet of gas per day and convert it into

---

[13] Darragh, T., "Transit agencies face rough road ahead," *The Morning Call,* p. A1, January 1, 2006.

[14] "Kyocera Eyes Solar Panel Output in Europe, Mexico—Kyodo," *Dow Jones Newswires,* October 1, 2003.

34,000 barrels of liquids, including 24,000 of diesel, 9000 of naphtha, and 1000 of liquefied petroleum gas per day.[15]

Make the following assumptions: Production can continue for 20 years (with no salvage value at the end of that time), due to the facility's location near natural-gas reserves. The cost of natural gas is $2 per 1000 cubic feet. Revenues of $35 per barrel diesel, $30 per barrel naphtha, and $28 per barrel LPG are estimated. Other annual costs are estimated at $10 million in the first year, growing at 2% per year. The interest rate is 17% per year and production is 360 days per year.

(a) What is the payback period for the investment?

(b) What is the payback period with interest?

(c) Chart the project balance over the life of the project. Interpret the results.

(d) How sensitive are the price of natural gas, liquid revenue prices, the output level of each liquid, the investment cost of $900 million, and the initial O&M costs? Use annual equivalent worth for the analysis.

(e) Perform a two-parameter analysis examining the simultaneous movement in diesel and naphtha per unit revenues.

(f) Determine an average price for all production in order for the project to break even (in terms of annual equivalent worth).

(g) Repeat the foregoing analyses, assuming that the project lasts for only 15 years.

5. In 2004, Rio Tinto approved an expansion of its QIT Fer et Titane slag plant in Quebec, Canada. The process produces a high-grade titanium dioxide for use in the manufacture of white pigments for paint, paper, plastics, and titanium metal. The $76 million expansion increased capacity by 125,000 tons per year in 2005.[16]

Make the following assumptions: The expansion has a life of 10 years, with no salvage value. Production of 125,000 tons a year began in 2005 ($76 million was invested at the end of 2004), generating revenues of $1900 per ton while costing $1750 to produce the pigments, with an additional $500,000 in fixed annual operating costs. The interest rate is 12% per year.

(a) Determine the present worth and the payback period.

(b) Historical revenues for titanium have averaged between $1870 and $2200 per ton, while production margins (the difference between per unit revenues and costs) have fluctuated between 5% and 15%. Generate pessimistic, optimistic, and average scenarios with this information. Compute the present worth of each scenario.

(c) Perform a sensitivity analysis on the difference between per unit costs and revenues. What is the break-even difference?

(d) Perform a simulation on the per ton margins and historical revenues, assuming uniform distributions. What is the expected present worth, assuming that the other data are deterministic?

---

15 Abdallah, A., "Qatar Crown Prince Launches ORYX Gas-to-Liquids Project," *Dow Jones Newswires,* December 7, 2003.

16 "Rio Tinto Approves Expansion of QIT UGS Plant," *Dow Jones Newswires,* January 19, 2004.

6. Consider again the example of the Coca-Cola bottling plant in Application Problem 11 of Chapter 8. Using the original data, perform a sensitivity analysis on sales, the increase in sales, the per case profits, and the tax rate. Is the tax rate significant? Assume an after-tax MARR of 12%.

7. In late 2003, the French oil group Total SA invested EUR15 million in a windfarm in northern France. The farm has five windmills with a total capacity of 12 megawatts.[17]

Make the following assumptions: Assume that Total has agreed to sell all of the power generated to a local power distributor. However, the amount received is dependent on the amount of wind. Annual production follows a Normal distribution (mean 8.0 MW and standard deviation 4.0 MW), and per megawatt profit follows that of a Uniform distribution with U[EUR250K, EUR350K]. Simulate the project, assuming a 20-year horizon and remediation costs of EUR5 million to dismantle the windmills. The quarterly interest rate is 3%. Note that production is capped at 12MW, which can be controlled in the simulation. Compute the mean and standard deviation of the present worth and the probability of success.

8. Consider again the data from Application Problem 11 in Chapter 8. Assume that these data are correct, but assume further that the price of a case of cola follows a time series. Specifically, assume that the price in year 1 is Normally distributed with mean $10 and standard deviation $0.25. The price for the second year is also Normally distributed with the same standard deviation, but the mean is *equal to the price of the first period.* Generate prices over the horizon in this fashion, in that they are dependent on the previous period. Assume that the unit cost is Uniformly distributed as a percentage of the price (U[0.75, 0.95]) in each period. All other costs are assumed as estimated. Simulate sales as U[16M, 24M] over time and compute the average present worth of the investment, assuming an after-tax MARR of 9% per year. Is this a risky investment?

9. Consider again the ORXY gas-to-liquids facility investment from Problem 4. Make the following assumptions:

- Gas procurement prices start at $1.10 per 1000 cubic feet at time 1, but have a 0.60 probability of increasing $0.10, a 0.30 probability of staying the same, and a 0.10 probability of declining $0.05 from the previous period.
- Sales of diesel, naphtha, and LPG are expected to move in a correlated fashion. Generate the per barrel revenue for diesel as U[$25, $35]. Assume that the naphtha revenue is U[$4, $6] less per barrel and the LPG is U[$6, $8] less per barrel.
- Fixed annual costs are assumed to be Normally distributed with a mean of $10 million and standard deviation of $500,000 in each period. The mean grows by 2% each period.

All of the remaining data are as before. Produce a simulation model and perform a minimum of 100 iterations.

(a) What is the expected present worth and standard deviation of the project?

(b) Compute the probability of success for the project.

---

[17] "Total Opens Its First Ever Windfarm in Northern France," *Dow Jones Newswires,* November 14, 2003.

(c) Compute the average payback period and the payback period with interest for the project.

(d) On the basis of your results, comment on the riskiness of the project.

10. In 2004, OAO RusAl, Russia's largest aluminum producer, began a feasibility study examining the possible expansion and modernization of its Friguia alumina refinery in Guinea. The project is estimated to cost $350 million over three years (beginning in early 2005), increasing production by 700,000 tons per year.[18]

Construct a simulation according to the following assumptions: The investment cost is Uniformly distributed between $330 and $370 million and spread evenly over three years. Sales of tons of alumina in each period are also expected to follow a Uniform distribution, U[600K, 700K], with the lower limit increasing by 10,000 tons each year. The plant will remain in operation for 15 years after the expansion is complete (no revenues will be generated until then), with the salvage value following a Normal distribution with mean $20 million and standard deviation $5 million. Fixed operating costs are expected to be $10 million in year 1 increasing annually at a rate that is Normally distributed with mean 0.02 and standard deviation 0.005. Per ton costs are expected to be Uniformly distributed as a percentage of revenues (U[0.75, 0.85]). Generate data concerning revenue prices from U[$250, $650] per ton.

Given the preceding information, perform a Monte Carlo simulation and generate 500 runs. For each run, compute the present worth, assuming an interest rate of 10% per year.

(a) What is the expected present worth and standard deviation of the project? Does the project seem risky?

(b) Compute the probability of success for the project. Is it still a risky project?

(c) From your simulation data, plot the present worth of a given simulation outcome as a function of the average margin per ton of alumina over the life of the project. Recall that it was assumed that per unit costs would be between 75% and 85% of revenues. Does this plot provide more information regarding sensitivity?

(d) Repeat the previous analysis, using the average revenue (price) per ton over the life of the project.

10.11.4 Fundamentals of Engineering Exam Prep

1. A customized piece of equipment is purchased for $100,000 and has no salvage value, regardless of its salvage date. For what horizon does the investment break even if it nets $15,000 per year and the interest rate is 10%?

(a) 14 years.

(b) 12 years.

(c) 10 years.

(d) 8 years.

---

[18] Galitsina, A., "RusAl Begins Feasibility Study of Guinea Plant Upgrading," *Dow Jones Newswires,* February 26, 2004.

2. A customized piece of equipment is purchased for $100,000 and has no salvage value, regardless of its salvage date. What is the payback period if it nets $15,000 per year and the interest rate is 10%?

(a) 14 years.

(b) 12 years.

(c) 10 years.

(d) 7 years.

3. A steel producer procures a new stamping machine for $100,000. The machine is used for five years, at which time it has negligible salvage value. If operating and maintenance costs are $10,000 per year, the interest rate is 8% per year, revenues are $15,000 in the first year, and revenues grow by some amount G for the next four years, then the value of G at which the investment breaks even is most nearly

(a) Between $10,000 and $12,000.

(b) Between $12,000 and $14,000.

(c) Between $14,000 and $16,000.

(d) Between $16,000 and $18,000.

4. Consider an investment with net cash flows of –$100,000, $20,000, $25,000, $30,000, $35,000, and $40,000. What is the payback period for this investment?

(a) Three years.

(b) Four years.

(c) Five years.

(d) It does not pay back the amount invested.

5. A producer of plastic parts is looking at adding another extruder at the cost of $250,000. The machine is expected to last seven years and can produce up to 100,000 parts per year at a net revenue of $2 per part and fixed costs of $25,000 per year. At 10% per year, what is the minimum number of parts required to be produced annually to make the investment viable?

(a) Less than 35,000.

(b) Between 35,000 and 40,000.

(c) Between 40,000 and 45,000.

(d) Greater than 45,000.

6. A new city park is developed at the cost of $500,000. It is expected to last for 20 years, having a salvage value of $50,000 at that time, with annual maintenance costs of $25,000. If each visitor receives estimated benefits of $5 per visit and interest is 6% annually, the minimum number of monthly visitors at which the investment breaks even is most nearly

(a) 1130.

(b) 1000.

(c) 13,500.

(d) 14,700.

7. An investment of $150,000 produces net revenues of $35,000 per year for eight years and has no salvage value. For a 12% annual rate of interest, a 10% error (i.e., the investment is 10% higher than predicted) in the estimate of the investment cost produces an error in the present worth most nearly

(a) 20%.

(b) 40%.

(c) 60%.

(d) 80%.

8. An investment of $150,000 produces net revenues of $25,000 per year for eight years and has no salvage value. For a 12% annual rate of interest, the present worth is

(a) More sensitive to errors in net revenues than errors in the investment cost.

(b) More sensitive to errors in the investment cost than errors in net revenues.

(c) Both (a) and (b).

(d) None of the above.

9. A machine is purchased for $50,000 because it is expected to save $12,000 per year. The payback period is most nearly

(a) Five years.

(b) Four years.

(c) Six years.

(d) None of the above.

10. A machine is purchased for $50,000 because it is expected to save $12,000 per year. The payback period with interest (12% per year), is most nearly

(a) Four years.

(b) At least five years.

(c) Three years.

(d) None of the above.

11 Considering Noneconomic Factors and Multiattributes

(*Courtesy of Alan Spencer/Alamy.*)

Real Decisions: A Working Waterwheel

At the cost of GBP17 million and two years of planning and construction (expected, however, to take five years), her majesty the Queen opened the Falkirk Wheel on May 24, 2002. The Wheel connects the Union Canal to the Forth and Clyde Canal, 115 feet below, utilizing 10 hydraulic motors to turn two gondolas that can each carry up to four 66-foot-long boats. Since the gondolas are perfectly balanced, the amount of electricity required to move the 660 tons of steel and water is essentially negligible (estimated at GBP10 per day). The Wheel was part of a GBP84.5 million investment to restore 68 miles of canals between Edinburgh and Glasgow, Scotland, called the Millennium Link. The Wheel replaced a staircase of 11 locks. A total of 200,000 visitors was expected in the first year, paying GPB6.50 to take the 45-minute Wheel ride.[1,2]

Make the following assumptions: Consider the entire GBP84.5 million investment. For the Wheel, 200,000 visitors are expected per year over its life, assumed to be 50 years, at which time it has no salvage value. A visitor's utility is estimated to be GBP20 for the experience, at the cost of GBP6.50. Operation and maintenance of the Wheel and canals costs GBP1.5 million per year. It is expected that another 50,000 people derive GBP10 of utility from the canal each year. The scenario just presented leads to a number of interesting questions:

1. What is the BC ratio for the investment in the Wheel? Does this ratio support the decision to invest?
2. What is the break-even value for the number of visitors to the Wheel?
3. What other benefits or disbenefits, economic or otherwise, come from the investment? Do these factors influence the decision?

In addition to answering these questions, after studying this chapter you will be able to:

- Identify noneconomic factors, also called intangibles or irreducibles, that can heavily influence the project selection decision. (Section 11.1)
- Incorporate noneconomic factors into the decision-making process separately from the economic evaluation process. (Section 11.2)
- Utilize minimum-threshold analysis and scoring methods for evaluating projects with multiple attributes. (Section 11.3)
- Understand that judgment plays a critical role in concluding whether a project should be accepted or not. (Section 11.4)
- Properly write a business case that justifies a capital investment decision. (Section 11.5)

---

[1] Fettes, M., "Falkirk Wheel to give country a lift," *Evening News—Scotland*, p. 14, May 24, 2002.

[2] Butterley, Bachy Soletanche, Bennett Associates, ARUP, British Waterways Scotland, TGP, Morrison, and RMJM, "Falkirk Wheel Project Study," *New Civil Engineer*, www.nceplus.co.uk, 2002.

We have scrupulously analyzed projects from a number of angles and under uncertainty and risk. While the topic of "engineering economy" would seem to imply that we are concerned only with the economic and financial considerations of capital investment decisions, the decision-making focus of this textbook compels us to consider the noneconomic factors affecting our decisions. That is, investments can be influenced by, or lead to, a number of consequences that may not necessarily be captured through economic analysis. In this chapter, we examine some of these aspects that affect our decision, and we present methods by which to incorporate their effects into our decision. We conclude the chapter by bringing all of our analyses together for a final decision on our single project.

11.1 Noneconomic Factors

By now, it should be obvious that we have been considering very important, yet very difficult, decisions. Their importance comes from the fact that the decisions are being made to either solve problems or take advantage of the opportunities defined by the entity. Their difficulty comes from the fact that these problems are real and consider uncertain futures. Thus, there is a considerable risk of losing money in the investments being examined.

We have introduced a number of analyses to deal with the complexity of the problems and the risk at hand. But the fact remains that the decisions are real and therefore difficult. Our analyses have been geared toward examining the economics of the problem. However, many factors that can influence a capital investment decision are not *easily quantified* for economic analysis. These factors are often referred to as irreducibles or intangibles. Unfortunately, they only further complicate the problem, because they should be considered, but remain difficult to deal with in any analysis.

Let us consider some real examples:

- **Political stability.** In December of 2003, Magadi Soda Company, the largest producer of soda ash in Africa, announced that it would invest $97 million in building a new plant in Kenya. Soda ash is used to make glass and is also an ingredient in a variety of detergents and chemicals. According to the company's managing director, the timing of the investment was directly influenced by the political climate in Kenya, which has improved due to the installation of a new government with new policies.[3]
- **Worker morale and quality of life.** Boeing relocated its headquarters from Seattle, Washington, to Chicago, Illinois, after examining "a big matrix" of each city's attributes. (Denver, Colorado, and Dallas, Texas,

[3] "African Soda Ash Producer to Spend $97M on New Kenya Plant," *Dow Jones Newswires*, December 1, 2003.

were also considered.)[4] The quality of life and the cost of living for employees are often cited as important considerations when locating a facility or headquarters.[5]

- **Worker stability.** Ford announced that it would make a significant investment in its Genk plant in Belgium in order to end weeks of labor unrest that had closed the factory down in October of 2003 after Ford had originally announced that it would cut the workforce significantly at the facility.[6,7]
- **Prestige.** Central Japan Railway Co. and the Railway Technology Research Institute of Japan set the world speed record with their magnetically levitated train that clocked 361 miles per hour (581 kph) in tests west of Tokyo in late 2003.[8] Clearly, there is a level of prestige associated with building the fastest train in the world, even if the speeds approached are not utilized in everyday operations.
- **Exceeding specifications.** Boeing set a record for single-engine certification testing by flying the latest Boeing 777-300ER on one engine for $5\frac{1}{2}$ hours while crossing the Pacific Ocean in October of 2003. The 242-foot jet flew on one General Electric GE90-115B engine, the most powerful engine ever used on a commercial airplane. While regulations on transatlantic flights restrict twin-engine aircraft operations to keep within three hours of an emergency landing strip, Boeing conducted the test in hopes that the regulated range would be extended in the future.[9]
- **Infrastructure.** Canadian National Railway Co. invested C$24.2 million in its Halifax-to-Montreal line to improve transit times and on-time train performance. The investment was divided such that C$15 million was for tracks, bridges, and signals, while the remaining portion was for four new sidings, which allow trains to pass each other in single-track areas.[10]
- **Future considerations.** Helicopter manufacturer AgustaWestland, owned by GKN PLC of Great Britain and Finmeccanica SpA of Italy, opened a production facility for its A119 Koala helicopter in Philadelphia, Pennsylvania, in the summer of 2004 in order to compete for replacing

---

[4] Lyne, J., "$63 Million in Incentives, Last-Second Space Deal Help Chicago Land Boeing," *Site Selection*, www.siteselection.com, June, 2001.

[5] Sweeney, M., "The Corporate Headquarters Challenge," *Business Facilities: The Location Advisor*, www.facilitycity.com, April, 2003.

[6] Mackintosh, J., and R. Minder, "Ford Axes 3,000 Jobs in Belgium," *The Financial Times*, p. 23, October 2, 2003.

[7] "Ford Promises Major New Investment in Belgian Plant," *Dow Jones Newswires*, October 21, 2003.

[8] Woo, R. (ed.), "Japan's Maglev Train Sets Another World Speed Record," *Dow Jones Newswires*, December 2, 2003.

[9] "New Boeing Model Crosses Pacific on One Engine in Test," *Dow Jones Newswires*, October 16, 2003.

[10] Thomas, S., "Cdn Natl Rail to Invest C$24M in Halifax, Montreal," *Dow Jones Newswires*, October 23, 2003.

the helicopter fleet that transports the president of the United States of America. The company teamed with Lockheed Martin Corp. and Textron Inc.'s Bell Helicopter on the bid.[11,12]

- **Safety and environmental considerations.** The U.S. Army awarded Arotech Corp. a $5.2 million contract for zinc-air batteries in December of 2003. Reasons for the choice included the facts that the battery had a long life, was safe in combat situations, and was more environmentally friendly.[13]

In each of the foregoing examples, there are benefits and costs associated with a given decision. For example, an investment may bring prestige to a firm. This is clearly worth something, as a higher level of prestige may lead to improved sales. However, such prestige would be extremely hard to quantify, if not estimate, at best.

Thus, the question remains as to how to incorporate these factors into analysis. There is no magic answer to this question, but we outline how they should be considered in the next section.

11.2 Including Noneconomic Factors in Analysis

Without an analysis of *intangible* benefits, some investments might not be funded. Similarly, *intangible* costs may lead to the rejection of projects that have positive economic aspects. However, one must be careful when incorporating these intangible justifications, as they may overshadow the necessity of firms to examine the financial implications of the investments in question.

Thus, for any engineering project, a rigorous economic analysis *must* take place first. Such an analysis includes using the methods presented in the previous two chapters. We become interested in the analysis of noneconomic factors if the economic analysis leads to one of two conclusions:

1. The economic analysis is positive, but the noneconomic impacts are negative.
2. The economic analysis is negative, but the noneconomic impacts are positive.

In other words, the noneconomic impacts are irrelevant if they are redundant to the decision to accept or reject a project. However, they can be critical and should be investigated further if they can *change* the outcome of an economic analysis.

[11] Lunsford, J.L., "Helicopter Seeks Made-in-U.S. Label," *The Wall Street Journal Online*, March 2, 2004.

[12] "AgustaWestland Expands Philadelphia Manufacturing Facility," *Press Release*, March 3, 2004.

[13] Wallmeyer, A., "Arotech Awarded New $5.2 M US Army Battery Contract," *Dow Jones Newswires,* December 19, 2003.

Suppose the economic analysis is positive; that is, it yields a positive present worth. This defines a threshold value for the noneconomic impacts if they are to change the decision. The question to be answered is whether the (negative) noneconomic impacts from a project are worth more than the threshold defined by the positive present-worth value.

Conversely, if the economic analysis leads to a present worth that is negative, a threshold has been defined for the (positive) noneconomic impacts of the project. Again, the question to be answered is whether those positive noneconomic impacts are worth more than the absolute value of the negative present worth.

Answering the questions in these two cases may seem obvious (and therefore not overly useful), but note that we are identifying a threshold value for analysis that is defined by the project itself, *not the intangible consideration.* Thus, we ask whether the intangible has a worth that covers the gap. This avoids the problematic issue of assigning a value to the intangible and incorporating it into the economic analysis.

EXAMPLE 11.1 *Including Noneconomic Factors in Analysis*

The first coal-to-clean-fuel power facility to be built in the United States, in Frackville, Pennsylvania, was started in 2003 at a cost of $612.48 million (including a $100 million grant from the U.S. Department of Energy) and is due to be completed in 2007. The engineering expertise is being provided by Sasol Ltd., a South African synthetic fuels and chemicals company. The plant will gasify coal waste to produce power and steam, with a portion of the waste converted into synthetic hydrocarbon liquids (oil) that can be sold. It is expected that 4700 tons of coal waste will produce 5000 barrels of oil equivalents daily and 41 megawatts of power.[14]

Assume that annual revenues will total $18 million for electricity and $35 million for oil, with $15 million per year in operating and maintenance costs. What is the future worth of this investment, assuming an after-tax MARR of 8%, a 20-year study period (January 2003 is time zero), with a salvage value of $100 million at the end of that time. The plant (a $512.48 million investment due to the grant) is to be depreciated over 10 years under MACRS alternative straight-line depreciation. Note that construction of the plant takes 5 years, with annual cash flows (and depreciation charges) beginning after its completion. Assume a 3% annual rate of inflation for revenues and 2% for expenses. All revenues and costs are given in year-2003 dollars. A 34% tax rate is to be used, with losses carried forward.

Solution. The before-tax, constant-dollar cash flow diagram for this example is given in Figure 11.1. The after-tax cash flow diagram is given in Figure 11.2. The depreciation of the facility cannot commence until 2008, as that is the first period in which the plant

[14] Aljewica, A., "Sasol Provides Tech for $612M US Coal-to-Liquid Plant," *Dow Jones Newswires,* August 15, 2003.

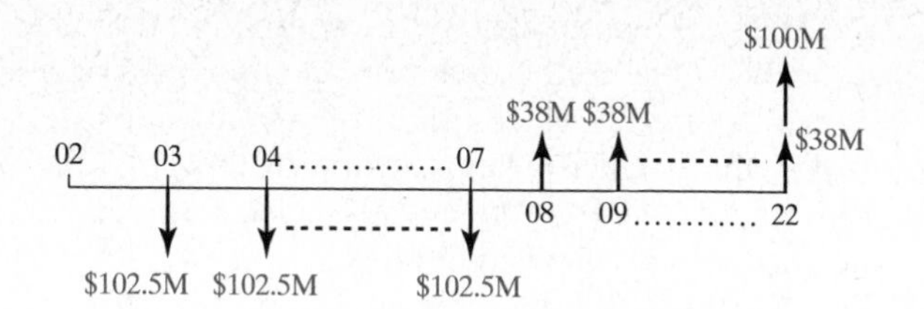

Figure 11.1 Before-tax cash flow diagram of investment in coal-to-clean-fuel power facility.

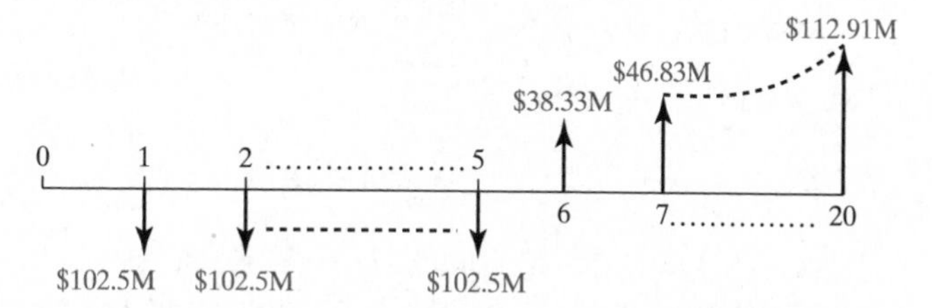

Figure 11.2 After-tax cash flow diagram of investment in coal-to-clean fuel power facility.

can be used. The after-tax cash flow in 2008 is calculated as follows:

$$\underbrace{(1 - 0.34)\$53M(1 + .03)^5}_{\text{Revenues}} - \underbrace{(1 - 0.34)\$15M(1 + .02)^5}_{\text{Expenses}} + \underbrace{(0.34)(\$25.62M)}_{\text{Depreciation}}$$
$$= \$38.33 \text{ million.}$$

Calculating the after-tax cash flows from 2009 through 2018 is similar. The $100 million salvage value at the end of the horizon is treated as ordinary income for tax purposes as the facility will be fully depreciated at that time.

The present worth of the facility can be found in Excel as follows:

$$\text{PW}(8\%) = -102.5 + \text{NPV}(0.08, -102.5, -102.5, -102.5, -102.5, 38.33, \ldots, 112.91)$$
$$= -118.3.$$

The solution is in millions of dollars. Since the value is negative, the economic analysis leads us to conclude that the facility should not be built, as it does not generate wealth for the company.

However, there are clearly a lot of noneconomic and intangible factors associated with this investment, including the following:

- Environmental benefits of little or no pollution compared with burning coal for power generation. These environmental benefits may include economic benefits, in that pollution permits do not need to be acquired.
- The prestige associated with building the first coal-to-clean-fuel power facility in the United States.
- The opening of future considerations by testing a new technology.

These are some very powerful factors that must be included in our analysis. If we can show that they are worth at least $118.3 million in present-worth terms, then the project should be pursued. Note that the $118.3 million figure amounts to only $12.05 million per year (annual equivalent worth) over the life of the project.

The analysis can also tilt the decision in the other direction. Consider the earlier example of the soda ash company building a new plant in Kenya. The managing director noted that the company had explored the project a decade earlier, but decided against it due to the political climate. Thus, it is quite possible that the economic analysis led to a decision to invest in the project (i.e., it produced a positive present worth). However, it was judged that the (negative) value of the political instability of the region was greater than the economic value. Thus, the decision at that time was to reject the project.

11.3 Multiattribute Analysis

As we conclude our analysis of a single project, it helps to take note of where we stand. We have been presented with a problem or opportunity, and solutions with estimates have been posed for subsequent analysis. In the previous chapters, we analyzed the project under conditions of certainty and risk. Now we have included noneconomic factors.

In our analysis under certainty, we could have computed the following measures of worth:

- Present worth.
- Future worth.
- Annual equivalent worth.
- Market value added.
- Internal rate of return.
- External rate of return.
- Benefit–cost ratio.

Taking risk into account, we could have computed the following measures:

- Payback period (with and without interest).
- Project balance over the project horizon.

Or we could have performed the following analyses:

- Break-even analysis of each parameter.
- Sensitivity analysis of each parameter.
- Scenario analysis of possible investment outcomes.

Continuing our risk analysis, we could have computed the following measures:

- Mean and variance of project scenarios.
- Mean, variance, and probability of loss, given a simulation of project outcomes.

We may want to further our analysis by considering alternative decisions for the investment over time, such as the decision to delay. This defines multiple options, to be analyzed in Chapters 12 and 13.

In the previous section, we added noneconomic factors into our analysis. This is clearly a lot of information and leads us to the most important question: How do we make a final decision—accept or reject—that takes *all* of this information into account. It is likely that we have not gathered all that information, but it is conceivable, and even suggested, that we have gathered at a least some of it. A reasonable subset would be some measure of worth (present worth or internal rate of return), the project balance (which provides information on the payback period and on the risk associated with the project), a sensitivity analysis of key parameters, and possible (probabilistic) scenario analyses of a range of possible outcomes.

Even with all of this information, we are still faced with a difficult decision, as not all of it is similar in terms of scale or scope. There are essentially two approaches that we could take in order to incorporate all of our analyses into our calculations and render a final decision:

1. **Minimum-threshold analysis.** Project criteria must meet minimum thresholds, or else the project is rejected. This kind of analysis requires that a standard be set for each of the criteria. Certainly, we have standards for the present, future, and annual equivalent worth in that they must be greater than zero when calculated at the MARR. However, other standards may have to be defined, for the payback period and the variance in the return, for example.
2. **Consolidate information into a single number.** This can be performed in numerous ways. For instance, one can assign weights to each piece of data (scaled appropriately) and sum the result. In addition, these weights can be assigned subjectively or through a formal rating method. Alternatively, instead of summing the result, the data can be aggregated according to some function that gauges risk, generally called a *utility function*.

There are numerous ways in which to apply these approaches.

EXAMPLE 11.2 *Multiattribute Analysis*

Recently, NCR Corp. opened its fourth plant that builds automated teller machines (ATMs) in São Paulo, Brazil. The plant, which will initially employ 75–100 people, has a capacity of 7000 units per year. NCR said that it is investing several millions of dollars in the facility.[15]

Make the following assumptions: An investment of $3 million is invested at time zero, with production increasing by 1500 ATMs per year after the production of 1000

[15] "NCR Building ATM Plant in Brazil," *Dayton Business Journal,* dayton.bizjournal.com, March 4, 2004.

units in year 1, until reaching the maximum capacity of 7000 units per year. Operating and maintenance costs are estimated at \$2000 per unit in the first year, increasing between 5% and 6% each year thereafter, with annual fixed costs of \$300,000 per year. Per unit revenues are expected to follow a Uniform distribution in each year as follows: U[\$2500,\$3000] in year 1 and U[\$2400,\$2900] in year 2, continuing this pattern with U[\$2000,\$2500] in year six.

Gather information concerning the expected present worth, its variance, and the payback period in order to determine whether the investment should be pursued. Assume a six-year horizon and a MARR of 20%.

Solution. We performed 50 simulations, generating data for the per unit revenues and expenses, and we computed the expected present worth (\$1.01 million), its standard deviation (\$674,000), and the expected payback period (3.8 years).

Minimum-Threshold Analysis. If it is declared that the payback period must be less than or equal to three periods or the standard deviation of the present worth must be less than \$500,000, then the project would be rejected according to minimum-threshold analysis, as all criteria being evaluated must meet their minimum thresholds. If these thresholds increase to four periods and \$1 million, respectively, then the project is to be accepted, since it generates a positive expected present worth.

Scoring Method: To provide a single number, or score, we must assign weights to each attribute. This is generally a difficult task that can seem quite arbitrary unless some formal method is followed. If we assign a rank order to the attributes, such as (1) present worth, (2) payback period, and (3) standard deviation of present worth, then we can provide weights of $\frac{3}{6}$, $\frac{2}{6}$, and $\frac{1}{6}$, respectively, where the numerator is the inverse order of the ranking and the denominator is merely the sum of the number of attributes ($1 + 2 + 3 = 6$). This ensures the weights correspond to the ranking.

Given this weighting, we need to scale our data appropriately, or extremely large numbers will dominate the decision and small values will be ignored. In this example, we have a present worth in the millions and a payback period of 3.8 years. Clearly, these are incompatible.

We can scale the data by dividing by an appropriate factor. If the "desired" (and realistic) present-worth value for this project is \$1.3 million, then we could divide the project's present worth of \$1.01 million by \$1.3 million to achieve a scaled score of 0.777. This scaling idea, however, is not as straightforward for data that we want to minimize, such as the payback period and variance. In this case, we might scale in the opposite direction. The best (but realistic) value for the payback period might be 2.5 years, such that a score of $2.5/3.8 = 0.658$ is achieved. Finally, a goal for the standard deviation would be \$500,000, so that a realistic score would be $\$500/\$674 = 0.742$. Now our weighted evaluation is

$$\text{SCORE} = \frac{3}{6}(0.777) + \frac{2}{6}(0.658) + \frac{1}{6}(0.742) = 0.732.$$

This has achieved our goal of reducing our three input data points to a single number. We have weighted the attributes according to our perceived importance of each one, and we have scaled the data according to a reference that rewards the appropriate behavior. (That is, large present-worth values, small payback periods, and small standard deviations are rewarded with higher scaled values).

The final consideration is whether a project with a score of 0.732 should be accepted. Just as the decision maker must determine the threshold values in minimum-threshold analysis and ideal values for scoring, he or she must determine a threshold score. Clearly, a number greater than 0.5 would be desired. Since this number we have calculated is above 0.7, the project appears to be worth pursuing.

It should be clear that this method can come under scrutiny, as there are many ways in which projects can be assigned weights and projects can be scaled. The methods described here represent just one way to accomplish that goal, and it is understood that other procedures may be valid as well.

Utility Theory: Utility functions are merely a different type of scoring method, with inputs weighted to generate a single value for evaluation. We term this single value the *utility* of the investment. In general, the amount of utility that we receive increases with each additional unit of wealth, but the rate of increase declines. This is known as the *law of diminishing returns* and is usually no different for investments. The idea is tied to risk: In order to generate higher returns, we frequently have to take on more risk. Thus, the utility we receive from increasing our returns generally increases more slowly with each additional unit of return (money). This relationship is often described by a logarithmic or exponential function, which is concave, as illustrated in Figure 11.3.

The concave utility function describes the typical risk-averse investor. In our capital-investment context, we are generally averse to projects with a high variance, as the potential for high rewards is joined by the potential for loss. A common utility function that captures risk aversion is an exponential function. One such function that incorporates both expected return and variance was given by Freund.[16] Assuming Normally distributed returns:

$$U(x) = 1 - e^{-Bx} \Rightarrow E(U(x)) = 1 - e^{-0.5\left(2B\mu - B^2\sigma^2\right)}.$$

If we ignore the payback period information and concentrate on the mean and variance of the return, we compute our utility with this function as follows (assuming that $B = 0.05$):

$$E(U(x)) = 1 - e^{-0.5\left(2(0.05)(1.01) - 0.0025(0.674)^2\right)} = 0.0488.$$

Since this value is greater than zero (albeit barely), we would accept the project. Many other utility functions exist and they can be tailored, through their different parameters, to match the level of risk aversion of the decision maker. The key is that

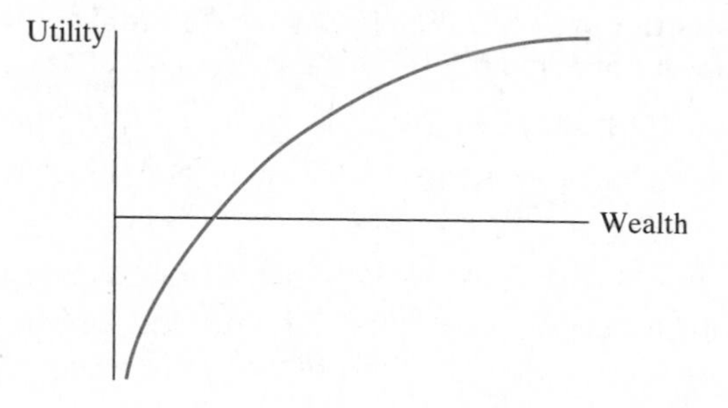

Figure 11.3
Concave function describing utility as a function of the return on an investment.

[16] From Bussey, L.E. and T.G. Eschenbach, "The Economic Analysis of Industrial Projects," 2nd Edition, Prentice-Hall, Englewood Cliffs, NJ, 1992.

we have provided a method which incorporates multiple attributes, and their attitudes towards risk, into the decision.

For the noneconomic factors discussed earlier in this chapter, a value can be subjectively assigned such that their evaluation can be included in the analytical methods (at least, in consolidation and threshold analysis). Unfortunately, this circumvents our desire to perform economic analysis independently of noneconomic factor analysis. Alternatively, we could perform our economic analysis (or analyses) and *then* examine the noneconomic factors to see if they might change the decision. As noted earlier in the chapter, this approach is preferred because the economic analysis comes first.

11.4 Judgment in Decision Making

Although we have tried to formalize methods by which to deal with the tremendous amount of data available to us, the fact remains that there is *no infallible approach* to solving an investment problem. "Boiling down" all of our analyses into a single number or eliminating options by means of minimum thresholds does not provide the perspective needed to deal with a project and its intricacies. Simply put, there is no method or system that can grasp the meaning of all of the available information at once.

Thus, it is imperative that you, the decision maker, include your own judgment when coming to a final decision to accept or reject a project. This is something that comes with experience and is something that generally cannot be taught. The hope is that, with all of these analytical tools, the decision will be somewhat obvious. Unfortunately, the hardest, and often most important, decisions are the ones that are generally not obvious. If the analysis points to accepting the project, you should review some fundamental questions before submitting a proposal to invest in the project:

1. Does the project adequately solve the problem or take advantage of the opportunity at hand?
2. Does the project fall within the scope of the company's goals and objectives?
3. Does the project have additional benefits that go beyond solving the problem at hand?
4. Does the project avoid costs that are beyond the norm?
5. Can the decision to invest be defended with rigorous analysis that also considers subjective judgment?

In others words, if the decision is a "go," can you "make a case" for it? Answering each of the preceding questions in the affirmative will help build that case.

11.5 Writing the Business Case

Assuming that we have performed all of the required analyses and answered the questions from the previous section in the affirmative such that we believe we can "make a case" for a project to be funded, the "business case" must be prepared. The term *business case* is used extensively in industry to describe the reports generated in support of an investment. As engineers, we often lose sight of the importance of this step. It should be somewhat clear that no matter how detailed or accurate the analysis, if the business case is not presented succinctly and clearly, the project will not be funded.

A business case may often be presented orally, but is almost always presented in report form, as that is easily circulated to the appropriate people. Because there may be hundreds of projects being evaluated at any given time, the business case for a single project must be to the point and readable. (Note that this format is also suitable for results from analyses in Part IV of this textbook.) A good business case has the following sections:

1. **Executive summary.** This is the most important section of the business case, because the reader will see it first and it may be the only section that he or she reads. The executive summary boils the entire report down to a few short paragraphs. These paragraphs must state the problem clearly and present the solution approach, any assumptions made, the financial aspects of the solution approach, and the likelihood of success of the investment. Any questionable characteristics, such as critical parameters that may cause concern or noneconomic factors that do not support the recommendation, should be noted here. Noneconomic benefits that support the recommendation can also be briefly mentioned. (A summary of other options examined may be included for the case of multiple projects, as analyzed in the next part of the text.)
2. **Introduction.** This section states the problem to be solved or opportunity to be pursued. A clear and succinct definition, followed by an overview of the solution approach, is given. Note that no cash flows are presented here. Rather, the discussion concerns the actual decision (i.e., purchasing some type of equipment or constructing a facility).
3. **Assumptions.** It has been stressed that an economic analysis is only as solid as the data in it. Thus, all estimated data and assumptions made in the analysis must be clearly stated. Furthermore, sources of the data should be referenced.
4. **Synopsis of analysis or analyses.** This section details the analysis (or analyses) performed and presents a justification for the approach(es) taken. As there are a number of analyses available to deal with risk, the approach should show coherent thought and not just be a potpourri of analyses. This section should also note which parties were involved in performing the analyses and coming to a conclusion. If the work was divided among

several people, then that should be made clear. Once this groundwork is laid, the results of the analyses are presented.

Note that the reader does not need to see equations on how one derives the present worth or payback period; rather, the reader must be presented with the result. Further, charts and graphs should be used in lieu of tables, as the former are often easier to read and provide more information simultaneously. If there is an abundance of information, put only what is critical to the decision into this section, and place any supportive information in an appendix.

5. **Additional considerations.** We have discussed the influence of noneconomic factors on investment decisions. These factors can be critical and must be included in the business case. As with our method for dealing with noneconomic factors, any additional factors should be given more credence if they have a strong influence on the decision. For example, if the economic analysis is positive and all noneconomic impacts are positive, then a great deal of discussion of these issues is not necessary, as they further support the proposal. In this case, they should just be mentioned in their supportive role. However, as we noted earlier, if the noneconomic factors are not supportive of the proposal, then they should be given a greater role in the document. In addition, a plan, if necessary, to mitigate potential conflicts should be discussed.
6. **Discussion and conclusions.** The final section of the business case should bring all of the information together with a recommendation. The discussion should be tailored to the situation. If the conclusion is not clear cut, as many decisions are not, a discussion of how each individual piece of information entered into the final decision may be warranted. It is appropriate to state concerns with the decision taken at this time. For example, "The committee recommends pursuing the investment under the assumption that the price for..." lets the reader know the risks involved and, more importantly, directly identifies the critical variable.
7. **Appendix.** If deemed necessary, data acquired and sample calculations may be contained in an appendix. Material in this section merely supports any analyses and conclusions described in the report.

There is one final issue to discuss that is related to writing the business case and to our analyses in general: *significant digits*. We have been lax in this textbook in dealing with significant digits, often to ensure that the reader understands how an answer is derived, as it may be unclear due to rounding.

Despite this fact, an answer should not have any more significant digits than the least significant input. For example, if we estimate investment costs to be $325 million per year and expected revenues to be $250 million against costs of $130 million, then the number of significant digits in the final solution should be two. The reason is that the investment cost is quoted in three significant digits, but the remaining data use two significant digits, the number of nonzero digits before a "trailing zero." If the $250 were meant to be read as having three significant digits, then would require a decimal after the "0" or should be quoted as $.250 billion. Any zeros that trail nonzero numbers after the decimal are significant. Many spreadsheets automatically compute monetary units to the penny. Clearly, this accuracy is neither warranted nor justified.

11.6 Examining the Real Decision Problems

We return to the introductory real decision problems of the Falkirk Wheel and canal restoration and the questions posed:

1. **What is the BC ratio for the investment in the Wheel? Does this ratio support the decision to invest?**

   The benefits to users are straightforward, as 200,000 people a year derive GBP13.50 of utility and 50,000 derive GBP10 of utility, for an annual total of GBP3.2 million. Spreading the GBP82.5 million cost over 50 years at a 3.5% annual rate of interest is GBP3.6 million per year, defining annual costs of GBP5.1 million to the facilitator. This defines a BC ratio of $3.2/3.6 = 0.63$, which would not support investment in the wheel.

2. **What is the break-even value for the number of visitors to the Wheel?**

   The break-even value is determined by increasing the number of visitors to the Wheel such that the BC ratio is driven above 1 as

$$\text{BC}(3.5\%) = \frac{x(\text{GBP13.50}) + 50{,}000(\text{GBP10})}{\text{GBP82.5M}(\overset{A/P,3.57\%,50}{0.0436}) + \text{GBP1.5M}} \geq 1$$

   The answer is $x = 341{,}000$ visitors per year.

3. **What other benefits or disbenefits, economic or otherwise, come from the investment? Do these factors influence the decision?**

   There are a number of noneconomic benefits to the project, including the following:

   - The Falkirk Wheel is a prestigious engineering achievement (the world's first revolving boat lift) that is equally attractive (to the eyes of most beholders) as it is efficient.
   - According to sources for this example, the canal had been in great disrepair since its closure in the 1960s. Thus, its reconstruction is a considerable upgrade to a significant stretch of land.
   - The project preserves an important piece of history, as the shipping canal was the world's first that traversed from coast to coast. It was completed in 1790.

   There are also a number of potential economic benefits that are difficult to quantify, but clearly are important. The canal's enhancement of the beauty of many urban areas may spur economic growth through investment in housing and retail outlets, perhaps to the tune of GBP400 million. If this number was (or is being) reached, then the canal was clearly an excellent investment. Finally, there have been discussions about using the canal system to transport refuse from Edinburgh to Glasgow that is now handled by truck. Since this is not a time-sensitive transport, it may be economically viable and also help reduce pollution by conserving fuel. It is clear that these "other" factors could change the decision resulting from the BC analysis.

11.7 Key Points

1. Noneconomic factors, also termed intangibles and irreducibles, can have a great influence on the project selection decision. The term "noneconomic" is used because it is often difficult to assign an economic value to the factor. (This is not, of course, to say that such factors do not have any economic influence.)

2. Common noneconomic factors include political stability, worker morale and stability, prestige, exceeding specifications, infrastructure, future considerations, and safety and environmental considerations.

3. Noneconomic factors should be considered only after a rigorous economic analysis is complete.

4. Noneconomic factors are important if they are deemed positive and the economic analysis is negative (i.e., has a negative present worth), or if they are negative and the economic analysis is positive. In either of the two cases, the noneconomic attributes should be considered in the analysis. In each case, the negative or positive present worth provides the gap that the noneconomic factors must bridge in order to overturn the decision.

5. The project selection decision is difficult when multiple criteria are considered.

6. Minimum-threshold analysis rejects a project if it does not meet minimum values for a given set of decision criteria.

7. Project scoring, or consolidation methods, such as utility theory, produce a single value from a given set of multiple attributes. If the single value is acceptable, then the project should be accepted.

8. Before a project is finally accepted, the decision maker must make sure that the project meets the goals and objectives of the company and must consider other opportunities the project may or may not bring.

9. Judgment comes with experience and is essential in coming to a final decision about a project.

10. Writing up conclusions in a business case can be as important as the actual analysis of a project, as it is generally the only means by which others learn about the decision.

11.8 Further Reading

Canada, J.R., and W.G. Sullivan, *Economic and Multiattribute Evaluation of Advanced Manufacturing Systems*. Prentice Hall, Englewood Cliffs, NJ, 1989.

Fabrycky, W.J., G.J. Thuesen, and D. Verma, *Economic Decision Analysis,* 3d ed. Prentice Hall, Upper Saddle River, NJ, 1998.

Lavelle, J.P., H.R. Liggett, and H.R. Parsaei (ed.), *Economic Evaluation of Advanced Technologies: Techniques and Case Studies*. Taylor and Francis, New York, 2002.

Nahmias, S., *Production and Operations Analysis,* 4th ed. McGraw-Hill Irwin, Boston, 2001.

11.9 Questions and Problems

11.9.1 Concept Questions

1. When should noneconomic factors be considered in the decision-making process? Why are they important?
2. Do noneconomic factors always need to be considered explicitly? Explain.
3. Of all the common noneconomic factors, which is most influential in decisions involving the expansion of capacity? Explain.
4. Why is the decision to accept or reject a project clouded with multiple criteria? Why then do we calculate numerous measures of performance?
5. What is the difference between threshold analysis and consolidating data into a single number?
6. Why is utility theory generally applied to expected-value and variance analysis?
7. What is the relationship between a scoring method and utility theory?
8. Why is judgment so important in making a final accept-or-reject decision?
9. Why it is important that the business case be written well?
10. What is the most important section in the business case? Why?

11.9.2 Drill Problems

1. An investment of $12 million is made at time zero with net revenues of $2.3 million in year 1, growing at a rate of 8% per year. Compute the present worth, payback period, and project balance over time (the ratio of positive area to negative area under the balance curve) for an interest rate of 18% per year and a 10-year horizon. Should the investment be made?

(a) Assume that the ideal values of the preceding three measures are $40 million, 4 years, and 4.0, respectively. Provide a score, using a weighted average of scaled measures. Does the project still appear acceptable?

(b) If the threshold values for the aforementioned three measures are $0, 6 years, and 1.0, determine whether the project is acceptable according to minimum-threshold analysis.

(c) If the standard deviation of the present worth is $0.15 million, use the present worth, the standard deviation, and the function suggested by Freund ($B = 0.05$) to calculate the utility. Is the project still acceptable?

2. A project has the following attributes (in order of importance): (1) present worth of $150 million; (2) internal rate of return of 16.5%; (3) payback period of 6 years; (4) variance in return of $60 million.

(a) Assume that the ideal values of each measure are $200 million, 25%, 4 years, and $25 million, respectively. Provide a score, using a weighted average of scaled measures. Does this project appear acceptable?

(b) Suggest another way of weighting each measure. How can that change influence the decision? Illustrate.

(c) If the threshold values for the foregoing measures are $0, 12%, 5 years, and $50 million, respectively, determine whether the project is acceptable according to minimum-threshold analysis.

(d) Explain how judgment might influence this decision.

3. Assume that the present worth of a project is $250 million, but there are negative noneconomic issues. How should they be addressed?

4. An investment of $4.5 million is made at time zero with annual revenues of $500,000 in year 1, growing at the rate of 30% annually over a seven-year horizon. Annual O&M costs are estimated at $150,000 per year. The salvage value is $1 million at the end of year 7. Assume that some negative noneconomic issues are related to this project. Should they be ignored or addressed? How? Assume that the MARR is 14% per year.

5. Consider again the previous problem, assuming that the noneconomic issues are positive.

11.9.3 Application Problems

1. This problem is motivated by the reemergence of the shale oil industry. An estimated 1 trillion barrels of oil is buried deep in shale rock formations in the western United States. The problem is that it is difficult to extract: Either the rock has to be heated to extremes in aboveground facilities called retorts (using a lot of water for processing), or it is heated below the surface with less water, but may put groundwater at risk. Shell Exploration and Production Co. said that the process would be economical if the price of a barrel of oil remained above $30, but the firm is at least five years from determining whether it will build a commercial facility to process the shale.[17]

What appears to be the critical factor in determining whether a shale oil extraction facility is to be built? What other factors may be important? If it

[17] Shore, S., "Shale oil industry heats up again," *Associated Press Newswires,* October 2, 2005.

were determined that the underground facility was economical, what noneconomic factors should be considered? Are these factors positive or negative?

2. The motivation for this example comes from Dell opening a $100 million, 527,000-square-foot assembly plant for its OptiPlex and Dimension desktop computers near Winston-Salem in October of 2005. At full capacity, the plant is expected to produce 700 computers an hour for distribution to the East Coast. The plant originally opened with 350 employees, but may grow to 1500 over time.[18,19]

Make the following assumptions: The $100 million investment occurs in year 2005 and the production of 300 computers per hour begins in 2006 with the plant operating one 8-hour-a-day shift, five days a week, 50 weeks per year. Output will increase by 100 computers per hour each subsequent year until reaching a capacity of 700 per hour. Each computer generates net revenues of $35, and annual fixed costs are $2.0 million. The interest rate is 20%.

(a) What is the payback period for the plant if it operates for 10 years? Plot the project balance over time and gauge its risk.

(b) If Dell expects a positive present worth (at 20%), an IRR greater than 25%, and a payback period with interest less than or equal to four years, would the company invest in the plant? Use minimum-threshold analysis.

(c) It was reported that the city and state offered large financial incentives (more than $37 million) to attract Dell to North Carolina. If these incentives reduce the investment cost by $37 million, how does the risk profile of the investment change? Revisit the minimum-threshold analysis.

3. In 2004, Calpine Corp. withdrew plans for a $1 billion liquefied natural gas terminal in Eureka, California, after encountering local opposition to the project. Calpine said its decision was based on community feedback and that it sought a "clear majority of support for proposed projects." The facility was to supply 1 billion cubic feet of gas each day to the state of California.[20]

Make the following assumptions: The facility was to be constructed over a 2-year period and operate for 19 years with no salvage value. Assume that the spread (margin) between the cost of purchasing the LNG and selling it to customers is $2.10 per 1000 cubic feet, with demand for 1 billion cubic feet per day. Assume 300 days of production a year and $200 million in annual fixed costs.

(a) Assume a MARR of 30% and determine the annual equivalent worth of the facility over the 20-period horizon.

(b) Discuss how the value you obtained for the annual equivalent worth in part (a) relates to the opposition of the public. Can the value of the opposition be discerned? Explain.

4. General Motors Corp. spent $10 million on a new plant outside of Detroit, Michigan, for low-volume production of high-performance engines for assorted vehicles.

---

[18] Nowell, P., "Ground broken on Dell's new $100 million plant near Winston-Salem," *Associated Press Newswires,* February 23, 2005.

[19] "Formal opening for Dell plant set for today," *Associated Press Newswires,* October 5, 2005.

[20] Kellaher, C.M., "Calpine Withdraws Plans for LNG Project in Eureka, Calif.," *Dow Jones Newswires,* March 18, 2004.

Production is expected to start in the summer of 2005 after construction of the plant terminates in 2004.[21]

Make the following assumptions: The investment is a bit risky because the plant is being constructed to build a low-volume component. Assume further that yearly demand over the next six years is Normally distributed with mean 1000 and standard deviation 300, with the average profit per engine $4000. The costs to run the plant are $1.25 million per year.

(a) Perform a simulation of the plant's performance over six years. Calculate the expected present worth, the standard deviation, and the expected payback period. From this information and a MARR = 11%, is this a risky project?

(b) Assume that the ideal values of each of the three measures mentioned in part (a) are $15 million, $2 million, and 3 years, respectively. Provide a score, using a weighted average of scaled measures. Does the project appear acceptable?

(c) If the threshold values for the three measures are $0, $10 million, and 6 years, respectively, determine whether the project is acceptable according to minimum-threshold analysis.

(d) Use the present worth, the standard deviation, using the function suggested by Freund to calculate the utility. Is the project acceptable for $B = 0.05$?

5. Consider again Application Problem 9 in Chapter 10 concerning the gas-to-liquids plant. The success of the plant is highly dependent on material costs and per unit revenues. Consider the expected present worth, standard deviation, probability of success, and average payback period from the simulation solution to that problem.

(a) Assume that the ideal values of the preceding four measures are $200 million, $50 million, 0.90, and 5 years, respectively. Provide a score, using a weighted average of scaled measures. Does the project appear acceptable?

(b) If the threshold values for the aforesaid measures are $0, $100 million, 0.75, and 8 years, respectively, determine whether the project is acceptable according to minimum-threshold analysis.

(c) Use the present worth, the standard deviation, using the function suggested by Freund to calculate the utility. Is the project acceptable for $B = 0.01$?

11.9.4 Fundamentals of Engineering Exam Prep

1. An intangible that may affect an investment decision is

(a) Prestige.

(b) Worker morale.

(c) Political stability.

(d) All of the above.

2. The decision to invest in an engineering project requires judgment because

(a) Intangibles may be hard to evaluate economically.

---

[21] "NEWS WRAP: GM to Invest $10M in New Center for Engines," *Dow Jones Newswires*, December 17, 2003.

(b) All estimates will be accurate.

(c) The decision is generally not strategically important.

(d) None of the above.

3. Intangibles should be studied closely when

(a) They are considered beneficial and the economic analysis suggests rejecting the project.

(b) They are considered costly and the economic analysis suggests accepting the project.

(c) They are considered beneficial and the economic analysis suggests accepting the project.

(d) Both (a) and (b).

4. If a project has a high internal rate of return, but a very high payback period,

(a) The project should be accepted outright.

(b) Further analysis and judgment must determine whether the project is deemed too risky.

(c) The project should be rejected outright.

(d) None of the above.

5. Analysis shows that, for expected production levels of 50,000 parts per year, a project has a present worth of $150,000 and an internal rate of return of 14.5%. Break-even analysis shows that the project would be accepted for annual production levels greater than 35,000 parts. A minimum-threshold analysis requiring a positive present worth, an IRR greater than 12%, and a break-even production rate of 40,000 parts would

(a) Reject the project for having too low of a rate of return.

(b) Accept the project.

(c) Reject the project for having too low of a break-even production rate.

(d) None of the above.

6. Analysis shows that, for expected production levels of 50,000 parts per year, a project has a present worth of $150,000 and an internal rate of return of 14.5%. Break-even analysis shows that the project would be acceptable for annual production levels greater than 35,000 parts. An ensuing analysis might

(a) Accept the project outright.

(b) Reject the project outright.

(c) Further investigate the likelihood of producing fewer than 35,000 parts.

(d) None of the above.

Part IV

Making the Decision for Multiple Projects

12 Deterministic Evaluation

(*Courtesy of BlueScope Steel Direct*.)

Real Decisions: Men of Steel

In late 2003 and early 2004, Bluescope Steel, Ltd., of Australia announced four separate investments: (1) a A$280 million investment in a new flat steel metallic coating and painting facility in China. The facility was to be completed in mid-2006, providing 250,000 tons of metallic coating and 150,000 tons of painting capacity annually;[1] (2) a similar facility with 125,000 and 50,000 tons of metallic coating and painting capacity, respectively, to be installed in Vietnam at the cost of A$160 million, with operations commencing in early 2006;[2] (3) an A$80 million investment in a new metallic coating line in Thailand with a capacity of 200,000 annual

[1] Johnston, E., "Bluescope: Plans A$280M China Investment," *Dow Jones Newswires,* February 17, 2004.

[2] Miller, B., "Australia's Bluescope: To Build Plant in Vietnam," *Dow Jones Newswires,* December 16, 2003.

TABLE 12.1 **Costs and present-worth values for steel company investment alternatives.**

Option	Investment	Investment Cost	PW(17%)
1	China Facility	A$280M	A$12.76M
2	Vietnam Facility	A$180M	A$6.13M
3	Thailand Facility	A$80M	A$9.24M
4	Purchase of Manufacturing Company	A$260M	A$2.07M

tons, expected to be in production in mid-2005;[3] and (4) the acquisition of Butler Manufacturing Company, a leading manufacturer and provider of preengineered steel building systems, for A$260 million in early 2004.[4]

Make the following assumptions: Construction of all of the facilities can occur in one year such that all of the investments begin at the end of 2004. Assume further that all costs and revenues have been estimated over the respective lives of the investments and the present worth of each individual alternative has been computed with a 17% annual rate of interest. The results are summarized in Table 12.1, leading to a number of interesting questions:

1. If there are no budgeting constraints, what is the optimal set of projects in which to invest?
2. What if the following constraints were imposed?
   (a) Variable budget limits of A$600, A$500, or A$300 million.
   (b) The Vietnam and China facilities are mutually exclusive, serving the same customers.
   (c) The Vietnam facility is dependent on the China facility opening and a A$500 million budget.
3. Do the preceding investment decisions concur when the projects are ranked on the basis of the PW, IRR, or PI?

In addition to answering these questions, after studying this chapter you will be able to:

- Identify situations in which projects must be analyzed simultaneously.
- Classify an investment proposal as either a service or revenue project. (Section 12.1)
- Define mutually exclusive investment portfolios consisting of individual projects. (Section 12.2)

3 "Bluescope Steel Continues Downstream Growth by Announcing New Thailand Investment," *News Release,* www.bluescopesteel.com, January 16, 2004.

4 "Bluescope Steel Announces Intention to Acquire World's Leading Manufacturer of Pre-Engineered Steel Buildings," *News Release,* www.bluescopesteel.com, February 16, 2004.

- Evaluate a set of investment portfolios with revenue projects, using (1) total investment analysis with any absolute measure of worth or (2) incremental investment analysis with any absolute or relative measure of worth. (Sections 12.3–12.4)
- Determine the minimum-cost project from a set of service projects with equal or unequal lives. (Sections 12.5–12.6)
- Evaluate a mix of projects, both revenue and service, under capital budgeting constraints. (Section 12.7)
- Solve large-scale problems with the use of integer programming or heuristic ranking methods. (Section 12.8)

In the previous chapters, we focused our attention on determining whether a single project was acceptable for investment. We still consider that question, but have the added complication of evaluating multiple projects. If the projects are *independent*, in that the decision to accept or reject one project does not affect any other projects, then we can analyze each project individually as in Chapters 9 through 11. If the decision to accept or reject one project affects the same decision for another project, then the projects are *dependent* and must be considered simultaneously. This occurs in the following situations:

1. **Multiple projects satisfy the same goal.** We encouraged the generation of many solution alternatives in the second step of the decision-making process. If multiple options have been developed to solve a problem, generally only one needs to be chosen and implemented. Thus, analysis must determine which of the multiple options is best. These options are called *mutually exclusive alternatives*, because the selection of one project precludes the selection of another.
2. **Projects have interdependencies.** The acceptance of one project may be contingent on the acceptance of another project. This dependence requires that the two projects be considered simultaneously.
3. **Resources are limited.** Companies may have many options for investment, but resources are often limited. The most common form of a limited resource is a cap on the amount of money that can be invested in a given period. However, other resources, such as research and development, engineering expertise, capacity, and labor, may also be limited, forcing the decision maker to choose among acceptable investments. Making this choice (or choices) is often referred to as the **capital budgeting process**, or simply, capital budgeting.

If we find ourselves in one of these situations, then we need to analyze the projects simultaneously to ensure that we invest in the best possible *set* of projects. In this chapter, we focus on deterministic analysis and assume that all cash flows for all projects are known with certainty. We also focus on the decision whether to invest. We address the issues of delaying a project, risk, and dealing with multicriteria in the ensuing chapters.

12.1 Classifying Engineering Projects

Before embarking on our analyses, we must define the types of projects that are to be analyzed. We first assume that we are analyzing investment projects, not sources of funds for projects. Identifying sources of funds was discussed in Chapter 4, and the cost of capital was discussed in Chapter 7. Funding decisions should be made separately from decisions to invest in a project, because the source of funding helps determine the interest rate (MARR) used in subsequent analysis. Given that we are analyzing investment options, the types of projects to be analyzed can be categorized as follows:

1. **Revenue Projects**. These are projects in which there are known benefits, or revenues, other than a salvage value, that are directly attributable to the investment. Revenue projects are "typical" investments, in that money is spent in hopes of generating revenues.
2. **Service Projects**. These are projects in which there are no revenues or benefits that are directly attributable to the investment, with the exception of a possible salvage value. A common example is the choice between two pieces of equipment that are needed in a manufacturing setting, but the choice does not affect expected revenues.

We separate our analyses of these two types of projects because they are fundamentally different. This should seem reasonable. Service projects do not generate revenues, and our decision criterion of present worth would always lead to a "reject" decision for service projects, which have no (or, at best, few) positive cash flows. However, these projects are generally not evaluated in the same context as an investment that generates revenues. Rather, a *lower cost* solution is being sought. This can lead to an alternative definition of the *do-nothing alternative,* as noted in Section 6.4. For example, if a piece of equipment must be installed for operations to commence and two vendors are being evaluated, then the do-nothing alternative is generally not a feasible option. Further, the do-nothing alternative may represent the status quo, which cannot always be summarized by a present worth of zero. This is why we break out service projects from revenue projects for purposes of analysis.

12.2 Forming Mutually Exclusive Alternatives

We defined mutually exclusive alternatives earlier. In this case, investing in one alternative out of a set of options eliminates the need to invest in any of the other options—because doing so would be redundant (and we need to be efficient with our money). This is natural when considering competing technologies or equipment required to solve a problem or take advantage of an opportunity. In this case, technology has defined the projects as being mutually exclusive. Consider, for example, the Sunrise gas project in waters separating Australia and East Timor, being developed by a number of partners, including Royal Dutch/Shell Group (United Kingdom), Woodside Petroleum (Australia), ConocoPhillips (United States), and Osaka Gas (Japan). In 2004, three proposals were being evaluated in order to develop the gas field, which is estimated to hold 7.7 trillion cubic feet of gas:[5]

[5] Bell, S., "Woodside Studies East Timor Pipeline for Sunrise Gas," *Dow Jones Newswires,* March 10, 2004.

1. Liquefy the natural gas at a floating facility on the site.
2. Pipe gas to a proposed LNG facility in East Timor, 150 kilometers from the site.
3. Pipe gas to an LNG facility under development in Darwin, Australia, 500 kilometers from the site.

Since all of these options achieve the same goal, Woodside (the field operator) must either choose *one* of them or do nothing.

The choice between competing technologies is not the only way in which to define mutually exclusive options. In fact, anytime we have multiple projects, we can define mutually exclusive options. Consider Caltex Australia, Ltd., which owns two fuel refineries in Australia. Due to new emissions standards, the refineries are both being upgraded at a combined cost of A$295 million in order to reduce sulfur and benzene levels beginning in 2006.[6] The refineries are

1. Kurnell, Australia, with a daily capacity of 124,500 barrels of crude oil.
2. Lytton, Australia, with a daily capacity of 105,500 barrels of crude oil.

Before the decision was made to upgrade both refineries, the investment options available were to (a) do nothing, which would have led to the closure of both plants in 2006; (b) upgrade the Kurnell refinery; (c) upgrade the Lytton refinery; or (d) upgrade both refineries. These four options are mutually exclusive, as investment in any one precludes investment in the other three. Note that a capital budgeting constraint might reduce the total number of options available for investment. For example, if a budget of A$200 million had been imposed, then option (d) would have been ruled infeasible due to its A$295 million cost, leaving only options (a), (b), and (c) for evaluation, assuming, of course, that none of them costs more than A$200 million individually. Other constraints may also limit the total number of options available.

We term each of these investment options a *portfolio*, as it gathers multiple projects into one option (just as the original option (d) invested in two projects). To be sure that we do not miss a portfolio option for investment, the first step in our evaluation process is to identify all possible investment portfolios. Given k projects, there is a total of 2^k possible portfolios for investment, since it is possible to accept or reject each project individually. This is "Step 0" in our evaluation process. Once the feasible investment portfolios have been defined, analysis can commence to select the best portfolio from the set of mutually exclusive portfolios.

[6] Pemberton, I., "Caltex Australia: Plans A$295M Refinery Upgrade," *Dow Jones Newswires,* February 24, 2004.

12.3 Evaluation of Revenue Projects with Equal Lives

We first turn our attention to projects that generate revenues, such as constructing a plant to manufacture products to be sold or opening a facility to provide services. We start with the easiest analysis: We assume that all of our options generate revenues and all have equal lives—equal project horizons. We deal with more complicated situations later in the chapter.

12.3.1 Total Investment Analysis

Total investment analysis is our first technique for selecting the best portfolio from our set of mutually exclusive portfolios. The steps of this approach are as follows:

1. Select an absolute measure of worth, such as present worth, as defined in Chapter 9.
2. Compute the measure of worth for each portfolio.
3. Select the portfolio with the greatest measure of worth for implementation.

This approach essentially evaluates all feasible combinations of projects. After evaluating each portfolio with one of our *absolute* measures of worth, we accept the portfolio that generates the greatest wealth.

EXAMPLE 12.1

Total Investment Analysis

In 2004, Sempra Energy proposed investing in two liquefied natural gas (LNG) receiving terminals with operations beginning in 2007:[7]

- $700 million investment in the Cameron LNG plant in Hackberry, Louisiana, with the ability to process up to 1.5 billion cubic feet of gas per day beginning in early 2007.[8]
- $600 million LNG receiving terminal in Baja California, Mexico, to supply 1 billion cubic feet of natural gas per day, also beginning in early 2007.[9]

Make the following assumptions: The Baja California terminal can be pursued alone or in a 50/50 joint venture with Shell International Gas Limited. In this case, all revenues and costs are split evenly between the two entities. However, an additional $10 million in annual expenses is incurred due to increased coordination and communications costs. Construction and development costs for all options are spread evenly over 3

---

[7] Bogoslaw, D., "TALES OF THE TAPE: Sempra Bets on LNG Future in U.S.," *Dow Jones Newswires,* November 7, 2003.

[8] "Sempra Energy LNG Corp. Completes Acquisition of Louisiana LNG Project," *News Release,* www.sempra.com, April 23, 2003.

[9] "Sempra Energy LNG Corp. and Shell Propose to Develop Mexican LNG Receiving Terminal," *News Release,* www.sempra.com, December 22, 2003.

TABLE 12.2 Feasible investment portfolios for LNG investments.

Label	Portfolio	Feasible?	PW(14%)
A	Do nothing.	Yes	$0
B	Hackberry LNG investment.	Yes	$771.2M
C	Baja LNG investment.	Yes	$741.4M
D	Baja LNG (split) investment.	Yes	$319.7M
	Both Hackberry and Baja LNG investments.	No, Budget Limit	
E	Both Hackberry and Baja LNG (split) investments.	Yes	$1.091B
	Both Baja LNG and Baja LNG (split) investments.	No, Mutually Exclusive	
	All three LNG investments.	No, Mutually Exclusive	

years, with operations commencing at the beginning of 2007 and lasting for 20 years. There are 300 production days per year.

For the Hackberry facility, revenues of $5 per 1000 cubic feet are expected over the life of the project against per unit costs of $4.35 per 1000 cubic feet and $20 million in annual operating, maintenance, and overhead costs. The terminal has no salvage value.

For the Baja facility, revenues of $5.25 per 1000 cubic feet are expected against $4.25 per 1000 cubic feet in costs and $50 million in annual expenses. A remediation cost of $60 million is expected at the end of the project life due to the delicate ecosystem of the area.

Finally, assume that Sempra has imposed a $1 billion capital budget for its LNG investments and that only one LNG investment can be pursued in Baja California, due to environmental constraints. Using present worth and a MARR of 14%, determine which investments, if any, should be pursued.

Solution. From our assumptions, the feasible investment portfolios of the eight possible combinations are given in Table 12.2. The investments in the Baja California location are mutually exclusive, while the budget limit constrains a number of combined investments. This leaves five viable portfolios, labeled A through E.

The present worth of the do-nothing option is zero. We can compute the present worth at the end of 2005 of each portfolio as follows:
For portfolio B,

$$\begin{aligned}\text{PW}(14\%) &= -\$233.33\text{M}(\overset{P/A,14\%,3}{2.3216})(\overset{F/P,14\%,1}{1.1400}) \\ &\quad + \left(1500\text{M ft}^3\left(\frac{300\text{ days}}{\text{yr}}\right)\left(\frac{\$5.00-\$4.35}{1000\text{ ft}^3}\right) - \$20\text{M}\right)(\overset{P/A,14\%,20}{6.6213})(\overset{P/F,14\%,2}{0.7695}) \\ &= \$771.2\text{ million.}\end{aligned}$$

For portfolio C,

$$\begin{aligned}\text{PW}(14\%) &= -\$200\text{M}(\overset{P/A,14\%,3}{2.3216})(\overset{F/P,14\%,1}{1.1400}) - \$60\text{M}(\overset{P/F,14\%,22}{0.0560}) \\ &\quad + \left(1000\text{M ft}^3\left(\frac{300\text{ days}}{\text{yr}}\right)\left(\frac{\$5.25-\$4.25}{1000\text{ ft}^3}\right) - \$50\text{M}\right)(\overset{P/A,14\%,20}{6.6213})(\overset{P/F,14\%,2}{0.7695}) \\ &= \$741.4\text{ million.}\end{aligned}$$

Portfolio D is defined by halving the present worth of portfolio C and subtracting \$10 million in additional annual costs, or \$741.4M/2−\$10M($\overset{P/A,14\%,20}{6.6231}$)($\overset{P/F,14\%,2}{0.7695}$) = \$319.7 million. Portfolio E is the combination of portfolios B and D with a resulting present worth of \$1.091 billion, as we merely sum the individual present-worth values because the projects are independent. Because portfolio E results in the largest present worth, it contains the set of investments to be pursued.

Total investment analysis guarantees that the greatest amount of wealth is generated for the firm. This is true because all possible combinations of projects have been enumerated and each has been analyzed, with the one with the greatest value having been chosen. Note that implementing total investment analysis does not permit the use of *relative* measures of worth, such as the internal rate of return or benefit–cost analysis.

EXAMPLE 12.2

Total Investment Analysis Revisited

Let us consider the last example again, but perform our analysis with the internal rate of return (which we previously stated we could not do, as it is a *relative* measure of worth). The internal rate of return for the do-nothing option is 14%, which is equal to the MARR. The internal rates of return for the other portfolios are given in Table 12.3.

TABLE 12.3

IRRs for each LNG investment portfolio.

Portfolio	IRR
A	14.00%
B	29.29%
C	30.92%
D	28.91%
E	29.17%

If you were to select the best portfolio on the basis of the highest internal rate of return, you would select portfolio C. Recall, however, that our actual choice with present-worth analysis was portfolio E.

If you were misguided, you might be tempted to choose portfolio C, because it has a higher internal rate of return compared with that of portfolio E. This is known as the *ranking problem* for internal rate of return. That is, it is possible to rank portfolios in a different order with the internal rate of return (or any relative measure of worth) as opposed to present worth through total investment analysis.

Why does this happen? Present worth is an absolute measure of worth. Given a cash flow diagram, the present-worth function defines the "worth" of the portfolio (or project) at time zero for a given MARR. By contrast, the internal rate of return is a relative measure of worth. It will not tell you how much wealth will be generated from the investment. Rather, the internal rate of return tells you the rate at which money invested in the project grows.

Consider Example 12.2 again. The fact that portfolio C earns 30.92% does not tell you anything about how much money it makes. It merely says that you will make a 30.92% return on money invested in the project. Similarly, portfolio E earns 29.17% on its investment. What does this mean? It means that portfolio C is more efficient than portfolio E. That is, for every dollar placed into each investment, portfolio C returns more money than portfolio E. However, you do not get to "pick" your level of investment in these projects. Each project is defined by a set of cash flows; thus, the level of investment is not a choice. (The choice is whether to invest a *given* amount of funds; the decision is not to determine the amount of money to invest in a project). For these individual portfolios, the amount of investment in portfolio E is greater than that in portfolio C. That is, given the level of investment, portfolio E achieves a greater present worth at the given MARR. Thus, although portfolio E may not be as efficient as portfolio C, it provides a greater increase in wealth for the firm. If you are going to use the internal rate of return, you need to use the incremental approach, as discussed in the next section.

12.3.2 Incremental Investment Analysis

Incremental investment analysis is our second technique for selecting the best portfolio from a set of mutually exclusive portfolios. The steps of the approach are as follows:

1. Select any relative or absolute measure of worth.
2. Order the projects according to increasing initial investments. Label the portfolios $1, 2, \ldots, p$. Define portfolio 1 as the *incumbent* k.
3. For $j = 2$ to p,
   - **(a)** Define portfolio j as the *challenger*. Compute the net cash flow diagram of portfolio $j - k$.
   - **(b)** Compute the designated measure of worth for $j - k$.
   - **(c)** If the measure of worth of $j - k$ is acceptable (i.e., if PW > 0, IRR $>$ MARR, or BC > 1), then the challenger j wins; then reject portfolio k, and define portfolio j as the new incumbent k ($j \to k$). Else the incumbent k wins; then reject portfolio j and keep k as the incumbent ($k \to k$).
   - **(d)** If $j < p$, then $j = j + 1$ and go to (a). Else ($j = p$), portfolio k is accepted as the best portfolio.

The analysis thus looks at each pair of portfolios in succession, with the best portfolio residing in the "incumbent" spot after all portfolios have been examined. The idea behind incremental investment analysis is to examine the *difference* between successive portfolios. We can view this difference as an entirely new portfolio. If that new portfolio is worthy of investment, then you will receive the benefits of the incumbent portfolio *and* this additional investment (the difference), which together define the challenger portfolio. Note that we only require knowing whether the difference is worthy of investment, because the incumbent has already been positively evaluated. (We usually begin with the do-nothing option.) Thus, the difference between the challenger and incumbent portfolios must only be compared with the do-nothing option in order to make a decision. This condition allows for the use of relative measures of worth. (Note that the procedure requires slight modifications when the BC ratio is used, as explained later.)

EXAMPLE 12.3 *Incremental Investment Analysis*

Let us reexamine the investment alternatives from the previous example. The options are listed in Table 12.4. Note that the portfolios are ordered according to increasing investment, as specified by the procedure just outlined.

TABLE 12.4 **Portfolios ordered for incremental analysis.**

Portfolio	Investment Project(s)	Investment	PW(14%)
1	Do nothing.	$0	$0
2	Baja LNG (split) investment.	$300M	$319.7M
3	Baja LNG investment.	$600M	$741.4M
4	Hackberry LNG investment.	$700M	$771.2M
5	Both Hackberry and Baja LNG (split) investments.	$1B	$1.091B

Solution. To illustrate the method, we strictly follow the procedure outlined. First, define portfolio 1 as the incumbent k.

Iteration 1:

1. Define portfolio 2 as the challenger j. The cash flow diagram of $2 - 1$ is merely the cash flow diagram for portfolio 2, as portfolio 1 is the do-nothing option.
2. The present worth of $2 - 1$ is $319.7 million.
3. Because $319.7 million > $0, the incumbent 1 is rejected and the challenger 2 becomes the new incumbent k ($2 \rightarrow k$).

From the first iteration, it is clear that investing in the Baja facility (split investment) is better than the do-nothing option.

Iteration 2:

1. Define portfolio 3 as the new challenger j. The cash flow diagram of 3 – 2 is given in Figure 12.1. It resembles portfolio 2, as portfolio 3 is nearly twice that of 2, with additional costs.

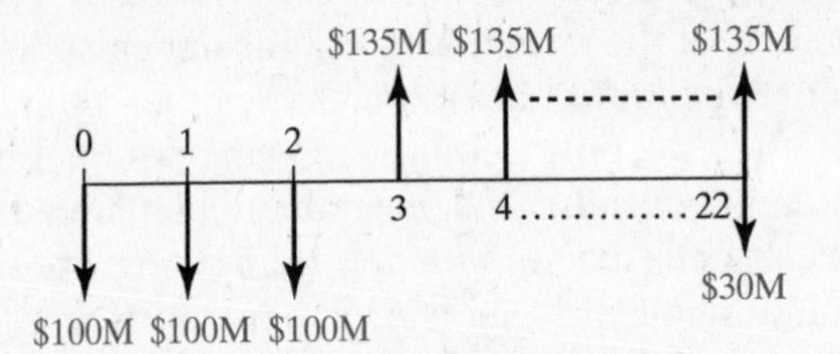

Figure 12.1 Net cash flow diagram of portfolio 3 – 2 for iteration 2.

2. The present worth of 3 – 2 is

$$\text{PW}(14\%) = -\$100\text{M}(\overset{P/A,14\%,3}{2.3216})(\overset{F/P,14\%,1}{1.1400}) + \$135\text{M}(\overset{P/A,14\%,20}{6.6213})(\overset{P/F,14\%,2}{0.7695})$$
$$- \$30\text{M}(\overset{P/F,14\%,22}{0.0560}) = \$421.7 \text{ million.}$$

3. Because \$421.7 million > \$0, the incumbent 2 is rejected and the challenger 3 becomes the new incumbent k ($3 \rightarrow k$).

The preceding equation says that the sole investment in the Baja facility is worth an additional \$421.7 million compared with sharing the investment. Thus, investing in the challenger would be worth the incumbent's return (\$319.7 million), plus the return defined by the difference of the challenger and incumbent (\$421.7).

Iteration 3:

1. Define portfolio 4 as the new challenger j. The cash flow diagram of 4 – 3 is defined in Figure 12.2.

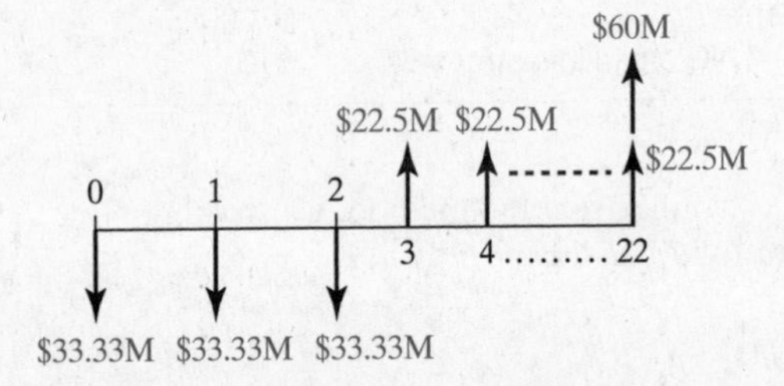

Figure 12.2 Net cash flow diagram of portfolio 4 – 3 for iteration 3.

2. The present worth of 4 – 3 is

$$\text{PW}(14\%) = -\$33.33\text{M}(\overset{P/A,14\%,3}{2.3216})(\overset{F/P,14\%,1}{1.1400}) + \$22.5\text{M}(\overset{P/A,14\%,20}{6.6213})(\overset{P/F,14\%,2}{0.7695})$$
$$+ \$60\text{M}(\overset{P/F,14\%,22}{0.0560}) = \$29.81 \text{ million.}$$

3. Because \$29.81 million > \$0, the incumbent 3 is rejected and the challenger 4 becomes the new incumbent k ($4 \rightarrow k$).

The previous equation says that the additional investment, defined by the difference between the Hackberry and Baja facilities, is worth \$30 million in present-worth dollars. This is expected, as it is merely the difference between the present worth of the original portfolios 4 and 3.

Iteration 4:

1. Define portfolio 5 as the new challenger j. The cash flow diagram of $5 - 4$ is defined by the cash flow diagram of portfolio 2 (Baja LNG split investment), since the Hackberry investments in portfolios 4 and 5 cancel each other out.
2. The present worth of $5 - 4$ is \$319.7 million.
3. Because \$319.7 million > \$0, the incumbent 4 is rejected and the challenger 5 becomes the new incumbent k $(5 \rightarrow k)$.

There are no other portfolios ($p = 5$), so the incumbent k, portfolio 5, is designated as the best investment. The value of the investment in both facilities is worth (in present-worth terms) \$319.7 million more than just investing in the Hackberry facility. This also represents a \$349.5 million difference in value over just investing in the Baja facility (\$29.8 + \$319.7 million) and \$771.2 million (\$421.7 + \$27.8 + \$319.7 million) over the smaller Baja investment. Finally, the chosen option represents a \$1.091 billion difference compared with the do-nothing option (\$319.7 + \$421.7 + \$29.8 + \$319.7 million). The \$1.091 billion figure is the same figure we derived with total investment analysis.

Hopefully, this example has illustrated the concept behind incremental investment analysis. As we have already tentatively accepted the incumbent portfolio, and the portfolios are ordered according to increasing investment, the cash flow diagram defined by the difference of the challenger and incumbent portfolios represents an additional investment and its associated returns. If this additional investment is better than the do-nothing alternative, then it is accepted. Accepting this difference in essence accepts the challenger, since its cash flows are defined by the incumbent's cash flows (already accepted), together with the cash flows defined by the difference between the challenger and incumbent. Mathematically, the incumbent cash flow diagram plus the difference (the challenger cash flow diagram minus the incumbent cash flow diagram) results in the challenger cash flow diagram. This is a critical point to understand, because comparing the difference in cash flows with the do-nothing option allows for the use of relative measures of worth, as illustrated in the next example.

EXAMPLE 12.4

Incremental Investment Analysis Revisited

We revisit Example 12.3 to illustrate the application of the internal rate of return in incremental analysis. Recall that the MARR is 14%.

Solution. The internal rates of return for the individual portfolios are given in Table 12.3. We again strictly follow the procedure outlined for incremental analysis. First, define portfolio 1 as k.

Iteration 1:

1. Define portfolio 2 as the challenger j. The cash flow diagram of 2 – 1 is merely the cash flow diagram for portfolio 2, as portfolio 1 is the do-nothing option.
2. The internal rate of return of 2 – 1 is 28.91%.
3. Because 28.91% > 14%, the incumbent 1 is rejected and the challenger 2 becomes the new incumbent k ($2 \rightarrow k$).

As with our present-worth analysis, it is clear that investing in the Baja (split) project is better than doing nothing.

Iteration 2:

1. Define portfolio 3 as j. The cash flow diagram of 3 – 2 is given in Figure 12.1.
2. The internal rate of return of 3 – 2 is found with Excel as

$$= \text{IRR}(-100,\ -100,\ -100,\ 135, \ldots,\ 135,\ 105,) = .3286.$$

3. Because 32.86% > 14%, the incumbent 2 is rejected and the challenger 3 becomes the new incumbent k ($3 \rightarrow k$).

The analysis up to now says that investing in a project defined by the cash flows of portfolio 3 – 2 is better than the do-nothing option. Thus, 3 – 2 is accepted, which is equivalent to accepting the challenger portfolio 3, because portfolio 2 has already been accepted and the cash flows of portfolio 3 – 2 plus the cash flows of portfolio 2 define portfolio 3. In other words, the additional investment required to execute portfolio 3 produces sufficient return (better than the do-nothing option) to warrant the acceptance of portfolio 3 over portfolio 2.

Iteration 3:

1. Define portfolio 4 as j. The cash flow diagram of 4 – 3 is defined in Figure 12.2.
2. The internal rate of return of 4 – 3 is found to be 18.49% with Excel.
3. Because 18.49% > 14%, the incumbent 3 is rejected and the challenger 4 becomes the new incumbent k ($4 \rightarrow k$).

With the same logic as before, the incremental investment in 4 – 3 is better than the do-nothing option. We accept 4 – 3 and thus accept portfolio 4.

Iteration 4:

1. Define portfolio 5 as j. Again, the cash flow diagram of 5 – 4 is defined by the cash flow diagram of portfolio 2 (Baja LNG split investment).
2. The internal rate of return of 5 – 4 is 28.91%.

3. Because 28.91% > 14%, the incumbent 4 is rejected and the challenger 5 becomes the new incumbent k ($5 \rightarrow k$).

We arrive at the same conclusion as that obtained from present worth, in that the best investment is portfolio 5, which invests in both the Hackberry and Baja (split) facilities.

We use the incremental investment approach because it leaves us with a single project to compare against the do-nothing option, thus allowing for the use of a relative measure of worth. The approach can also be used with benefit–cost analysis, with two slight modifications: First, we order the projects according to increasing denominators; second, for a given incumbent k and challenger j, we compute the BC ratio incrementally as

$$\text{BC} = \frac{B_j - B_k}{C_j - C_k}.$$

If BC > 1, the challenger is accepted and the incumbent is rejected. We illustrate in the next example.

EXAMPLE 12.5

Incremental Analysis with BC Analysis

In 2003, Impco Technologies, Inc., of California signed a contract with Hino Motors, Ltd., of Japan to develop compressed natural gas (CNG) fuel systems for a Hino JO8C natural-gas bus engine that will allow buses to meet Euro II emission standards.[10]

Make the following assumptions: A local government is considering the purchase of either traditional diesel buses or CNG buses for its mass-transit system. The diesel bus costs $280,000 to purchase and has a life of 10 years, with a salvage value of $50,000 at that time. It costs $1.90 per mile in operating and maintenance costs, which increase 2.5% per year of use. The diesel bus emits high counts of particulate matter, carbon monoxide, and hydrocarbons into the air, contributing to pollution. This pollution of the environment is viewed as incurring an annual cost of $50,000.

A CNG bus costs $500,000, with a salvage value of $100,000 after 10 years of use. Its operating costs are $2.50 per mile and increase at a rate of 4% per year of use. The CNG bus emits a negligible amount of particles into the air.

Both buses provide a benefit to society by providing transportation service, especially for those who do not have their own means of transportation. This benefit is valued at $1 million per year. (It is assumed that the buses will be used in a similar fashion.) Using benefit–cost analysis and a 4% annual rate of interest, which bus should be selected if they are operated for 30,000 miles per year?

Solution. The annual equivalent cost (AEC)—a concept similar to annual equivalent worth (AE), but one which assumes that costs are positive values—to operate the

[10] Jordon, J., "IMPCO Signs Pact with Hino Motors of Japan to Develop CNG Buses for Southeast Asian Market," *Dow Jones Newswires,* October 7, 2003.

diesel bus over 10 years is

$$\text{AEC}(4\%) = \$280\text{K}(\overset{A/P,4\%,10}{0.1233}) + \left(\frac{\$1.90}{\text{mile}}\right) 30\text{K miles}(\overset{A/A_1,2.5\%,4\%,10}{1.1114}) - \$50\text{K}(\overset{A/F,4\%,10}{0.0833})$$

$$= \$93{,}700.$$

The annual equivalent cost for the CNG bus is

$$\text{AEC}(4\%) = \$500\text{K}(\overset{A/P,4\%,10}{0.1233}) + \left(\frac{\$2.50}{\text{mile}}\right) 30\text{K miles}(\overset{A/A_1,4\%,4\%,10}{1.1855}) - \$100\text{K}(\overset{A/F,4\%,10}{0.0833})$$

$$= \$142{,}200.$$

We now must consider the emissions, which can be viewed as a cost, or disbenefit, *to society*, putting them in the numerator of our benefit—cost ratio. Note that the benefits of providing bus transportation can be ignored in this analysis, since both buses provide the same service. (In incremental analysis, these values negate each other.)

The do-nothing option is not a valid option in this case, as a bus must be selected for operations. The CNG bus has the larger denominator for the BC ratio and is thus denoted as option j (the challenger), while the diesel bus is denoted as option k (the incumbent). We define our incremental BC ratio as the difference in benefits over the difference in costs:

$$\text{BC}(4\%) = \frac{B_j - B_k}{C_j - C_k} = \frac{\$0 - (-\$50{,}000)}{\$142{,}200 - \$93{,}700} = 1.031.$$

Since the ratio is greater than one, challenger j, or the CNG bus, is chosen for operations.

12.4 Evaluation of Revenue Projects with Unequal Lives

We continue to focus on projects that generate revenues. However, in defining our mutually exclusive alternatives, we may define a number of portfolios or projects with different products, and it is quite possible that the study horizons for these projects will be different. That is, one facility may be estimated to produce a product for 12 years, while another facility may be estimated to produce a product for 5 years, due to rapid changes in technology.

This is a level of complication that must be addressed, as unequal lives require explicit assumptions about the investment of funds over time. Consider our simple example of facilities with expected lives of 5 and 12 years. If, due to budget limitations, only one of these projects can be chosen, how can we evaluate each project fairly, since they span different time horizons? Clearly, a project with 12 years of potential cash flows seems to have an advantage in such an analysis.

To answer this question, we must consider what will happen with the funds generated from the short-lived (5-year) project during the next 7 years in which the long-lived project continues. Toward that end, consider the cash

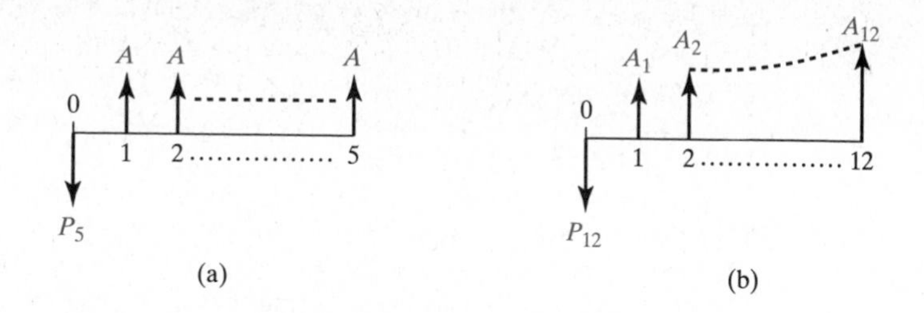

Figure 12.3 Cash flow diagrams for revenue projects with unequal lives.

flow diagrams in Figures 12.3(a) and (b). Part (a) refers to our short-lived project, while (b) refers to the 12-year project. Now assume a 12-year period for analysis. This method suggests that if we compute the future worth of each of these projects at the end of their respective horizons, we further compound the future worth of the 5-year project to time 12, using the MARR. If this is an acceptable assumption, then comparing the future worth for each project at time 12 is equivalent for decision-making purposes to comparing the present worth for each project at time zero. The reason is that the present worth is merely a transformation of the future worth, dividing each number by a constant.

If we assume that the funds generated from the short-term project are invested at the MARR for those seven periods, then we can merely compare the present worth of each project, since that is equivalent, in terms of decision making, to comparing their future worth values at time 12. Such a comparison is possible because we have explicitly stated that the funds in year 5 for the short-lived project will be invested at the MARR until period 12. This is called the *explicit reinvestment assumption* for unequal lives in an analysis of *revenue projects*.

EXAMPLE 12.6

Revenue Projects with Unequal Lives

In 2004, Argentine food producer Molinos Rio de la Plata SA announced that it was investing $80 million to expand a soy-processing plant from a daily grinding capacity of 5500 tons to 17,500 tons by the end of 2005. In addition, a new port was to be constructed to expedite exports with an annual capacity of 9.6 million tons of grain.[11]

Make the following assumptions: The cost of the expansion is $70 million, spread equally over 2004 and 2005, with production commencing in 2006. Assume also that the processed soy can be sold for $95 per ton at the cost of $85 per ton for raw materials and $5 per ton for processing, with $5 million per year for all other expenses (including transportation by current methods without ports). The expanded facility is expected to produce 12,500 tons of processed soy per day, with 340 days of production each year. The life of the plant is eight years once it is completed, and the plant has no salvage value.

Construction of the port will also take two years, at the cost of $10 million, and requires $775,000 in annual maintenance. The port will reduce shipping costs by

[11] Wong, W., "Argentina Molinos to Invest $80 Million for Plant, Port," *Dow Jones Newswires,* January 7, 2004.

$500,000 per year if the plant is not expanded and by $1.5 million per year if the plant is expanded. Furthermore, the port will bring in additional revenues of $1.5 million per year if the facility is not expanded and $500,000 per year if the plant is expanded. Note that this assumes that the port provides $2 million in annual benefits, regardless of whether the plant is expanded or not.

Assume further that a smaller port could be constructed with an annual capacity of 2 million tons of grain at the cost of $2.75 million. This facility will take one year to complete and requires $200,000 in annual maintenance. The port will not provide additional revenues, but will produce annual savings of $500,000 per year, regardless of whether the facility is expanded or not.

Assume the service lives of ports are 15 (small) and 20 (large) years and the savings produced for the plant can be translated into other revenues if the plant ceases to operate during their service lives.

Note that the investments in the ports are mutually exclusive. (Only one option can be pursued.) Also, the expansion of the plant cannot be undertaken unless some expansion of the port is undertaken in order to support the additional exports. Identify the feasible investment options if there is no constraint on spending. Assume a 10% annual rate of interest, ignore taxes, and take January 1, 2004, or equivalently, the end of 2003, as time zero.

Solution. All of the investment possibilities in this example are pared down to feasible options and enumerated in Table 12.5. The do-nothing alternative represents the status quo of processing and distributing 5500 tons of soy per day. This carries a present worth of zero. As each project generates revenue, we choose the portfolio with the largest present worth. This assumes that all cash flows are invested at the MARR through the horizon (22 years for the large port).

TABLE 12.5 **Feasible portfolios for soy processing and port expansion.**

Portfolio	Feasible	Option
Do nothing.	Yes	A
Small port.	Yes	B
Large port.	Yes	C
Expansion.	No, port required.	
Expansion and small port.	Yes	D
Expansion and large port.	Yes	E
Small and large port.	No, mutually exclusive.	
Expansion and both ports.	No, mutually exclusive.	

The present worth for constructing the small port is

$$\begin{aligned}\text{PW}(10\%) &= -\$2.75\text{M}(\overset{P/F,10\%,1}{0.9091}) + (\$500\text{K} - \$200\text{K})(\overset{P/A,10\%,15}{7.6061})(\overset{P/F,10\%,1}{0.9091}) \\ &= -\$425{,}610.\end{aligned}$$

Note that the $500,000 in benefits is realized regardless of whether the expansion is or is not undertaken with this option.

The present worth for constructing the large port is

$$PW(10\%) = -\$5M(\overset{P/A,10\%,2}{1.7355}) + (\$2M - \$775K)(\overset{P/A,10\%,20}{8.5136})(\overset{P/F,10\%,2}{0.8264}) = -\$58{,}580.$$

For the large port, the annual benefits of $2 million are assumed to be constant regardless of whether the expansion is or is not undertaken, as the benefits are made up of a combination of savings on shipping costs and additional revenues.

Expanding the processing facility has the following present worth:

$$\begin{aligned} PW(10\%) &= -\$35M(\overset{P/A,10\%,2}{1.7355}) \\ &\quad + \left(\left(\frac{12{,}500\text{ tons}}{\text{day}}\right)\left(\frac{340\text{ days}}{\text{year}}\right)\left(\frac{\$95 - \$90}{\text{ton}}\right) - \$5M\right)(\overset{P/A,10\%,8}{5.3349})(\overset{P/F,10\%,2}{0.8264}) \\ &= \$10.9\text{ million.} \end{aligned}$$

It is assumed that the expansion of the plant cannot take place without some expansion of the port. Thus, the $10.9 million present worth must be added to either of the present-worth values resulting from construction of the ports to determine the value of expanding production. Again, the port benefits have been estimated such that they are independent of the expansion of production; thus, the present-worth values need only be added as in total investment analysis. (If the savings and revenues for the ports had been different with and without expansion of the plant, then we would have had to examine two expansion projects: one with a small port and one with a large port, with the respective cost and revenue estimates.)

We make the assumption that funds are reinvested at the MARR. Thus, we can compare the present worths of each of our feasible investment options. These present worths are summarized in Table 12.6.

TABLE 12.6

Present-worth values for feasible investment portfolios.

Portfolio	PW(10%)
A	$0
B	−$425K
C	−$58.6K
D	$10.47M
E	$10.84M

Given the data in Table 12.6, the portfolio with the greatest present worth is the one that seeks to expand the production facility and construct the large port. This example illustrates that it may be necessary to make investments that are not economically viable (e.g., constructing just a port, which yields a negative present worth) in order to facilitate profitable investments. This is further motivation for examining multiple projects simultaneously.

12.5 Evaluation of Service Projects with Equal Lives

We now turn our attention to service projects. Recall that service projects have cash flow diagrams that do not explicitly define revenues or benefits other than possible salvage values. For the case of equal lives, in which each investment has a similar study horizon, we can turn to any of the analyses defined previously for revenue projects with equal lives. Again, the solutions will generally compute negative values of (present, annual equivalent, or future) worth.

In these situations, the best (minimum-cost) alternative is chosen. (The do-nothing alternative either is not feasible or is not defined by a present worth of zero.) It is common to define the *annual equivalent costs*, or AEC. We originally introduced this concept in Example 12.5. The AEC are also commonly referred to as equivalent uniform annual costs (EUAC), equivalent annual costs (EAC), or annual costs (AC). Since the only difference between AEC and AE is a sign change, it is only a matter of convenience to discuss the AEC in these terms. We illustrate with an example.

EXAMPLE 12.7

Service Projects with Equal Lives

The motivation for this example is All Nippon Airways and its order of 50 new 7E7 Dreamliner (now 787) planes in Boeing's launch of its new aircraft. The order is valued at about $6 billion according to list prices, with delivery of the first aircraft in 2008.[12] All Nippon will also have to purchase engines for the aircraft, as airlines purchase engines separately from airplanes. Both General Electric and Rolls-Royce were selected by Boeing to supply engines for the 787.[13] An engine generally carries a list price of $10 million, but this is rarely charged in full, in return for lucrative maintenance and spare-parts contracts.[14]

Make the following assumptions: The engine provided by manufacturer A costs $9.5 million and has a life of 20 years. Maintenance costs are expected to be $2.25 million in year 1, increasing 3 percent each year over the life of the engine. Operating costs (fuel and oil) are expected to be $40 million per year (time-zero cost) for a given level of demand, growing at a rate of 0.005% per year. The engine has a salvage value of $500,000 at the end of 20 years.

The engine supplied by manufacturer B costs $11 million with no maintenance costs over the first 5 years. Maintenance costs are expected to be $3 million per year over the remaining 15 years of engine life. Operating costs are expected to be $38 million per year (again, time-zero cost), growing at a rate of 0.0075% per year. The engine has a salvage value of $1 million at the end of its service life. Assuming an

---

[12] Lunsford, J.L., "Boeing Gets Big Order for 7E7 Plane," *The Wall Street Journal Online,* April 26, 2004.

[13] Souder, E., "Boeing Based 7E7 Engine Decision on Technical Merits," *Dow Jones Newswires,* April 7, 2004.

[14] Lunsford, J.L., and K. Kranhold, "GE, Rolls-Royce Gain Boeing Deal," *The Wall Street Journal Online,* April 7, 2004.

18% annual rate of interest, which engine should be chosen? It should be clear that revenues can be ignored, because the engines will provide the same service.

Solution. Since the service lives of the respective engines are the same, we can compute a number of measures of worth. Because we seek the minimum-cost solution, we compute the annual equivalent cost for each engine over its service life. The cash flow diagrams for utilizing each engine over 20 years are shown in Figure 12.4.

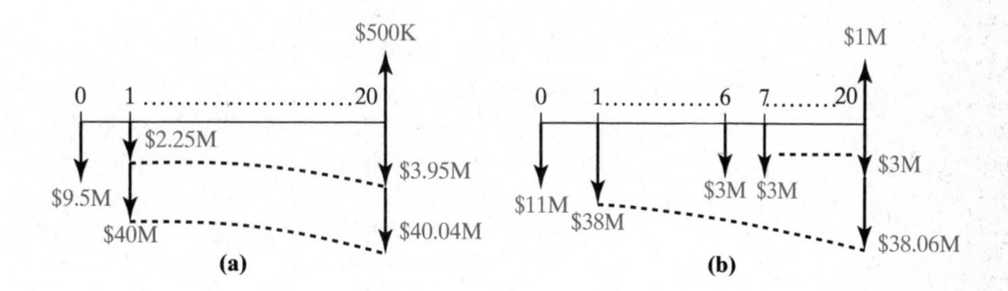

Figure 12.4 Net cash flow diagrams for choice between aircraft engines (a) *A* and (b) *B*.

For engine *A*,

$$\begin{aligned}\text{AEC}(18\%) &= \$9.5\text{M}(\overset{A/P,18\%,20}{0.1868}) + \$2.25\text{M}(\overset{A/A_1,3\%,18\%,20}{1.1634})\\ &+ \$40\text{M}(1+0.00005)(\overset{A/A_1,0.005\%,18\%,20}{1.0002}) - \$500\text{K}(\overset{A/F,18\%,20}{0.0068})\\ &= \$44.40 \text{ million.}\end{aligned}$$

For engine *B*,

$$\begin{aligned}\text{AEC}(18\%) &= \$11\text{M}(\overset{A/P,18\%,20}{0.1868}) + \$3\text{M}(\overset{F/A,18\%,15}{60.9653})(\overset{A/F,18\%,20}{0.0068})\\ &+ \$38\text{M}(1+0.000075)(\overset{A/A_1,0.0075\%,18\%,20}{1.00036}) - \$1\text{M}(\overset{A/F,18\%,20}{0.0068})\\ &= \$41.31 \text{ million.}\end{aligned}$$

Since the engine from manufacturer *B* has a lower annual equivalent cost, that is the engine of choice.

12.6 Evaluation of Service Projects with Unequal Lives

The analysis of service projects with unequal lives is the most complicated analysis we have discussed thus far. Consider the previous example, in which the decision was to choose between two engines. In engineering, we are repeatedly faced with the decision of choosing between multiple options in which the differences are not in the associated benefits, but rather in the costs.

The question we address now is "What if the two engines did not have the same service lives?"

The critical issues here are the service lives of the assets and the desired study horizon. Since we are not generating revenues, we cannot make the same argument we made with revenue projects—that funds generated from the project are invested at the MARR—as there *are no* funds being generated. Rather, we must carefully examine the needs of each project and clearly understand our assumptions about horizons in order to construct the correct analysis for a fair comparison.

In general, there are two approaches to the analysis:

1. **Explicit Study Period Approach:** In this method, we explicitly state how long the project will last. For example, if the choice is between competing technologies, the determination of a study period can generally be made independently of the choice. That is, a determination must be made as to how long the assets will be required to provide service. The ensuing analysis depends on the relationship between the service life of an alternative, N_A, and the study period N. The three possibilities are as follows:

   (a) If $N_A = N$, then no changes or further assumptions are needed.

   (b) If $N_A < N$, then we need to make some assumptions as to what will occur between periods N_A and N. Generally, we either assume repeatability or estimate cash flows from N_A to N explicitly.

   (c) If $N_A > N$, then we must explicitly estimate the salvage value for terminating the project or salvaging the asset early, thereby assigning the project or asset a life of N.

   Applying this logic, we see that all of the options will have the same horizon, and the problem is reduced to that from the previous section: service projects with equal lives.

2. **Infinite Horizon Study Period Approach:** In this situation, we cannot state a certain number of period's for a project to last. This is often true in applications where an investment is made and is expected to last for some time, but a terminal date is uncertain. We assume this means that the study horizon N is greater than the lives of all possible options ($N > N_A$). Because the only information we have is that N is large, we cannot explicitly estimate cash flows as in our previous approach.

   In this situation, it is common to assume that the options available at time zero are also available over time such that they can be *repeated.* Thus, to compare the two alternatives fairly, we must determine the lowest common multiple of the service lives of the respective alternatives. Doing so establishes two alternatives with equal horizons, which can be compared fairly.

   To illustrate the concept, assume that we have two options with respective service lives of four and seven years. The lowest common multiple of these options is 28 years. Clearly, it is cumbersome to analyze a project over 28 years (even with a spreadsheet!). The good news is that

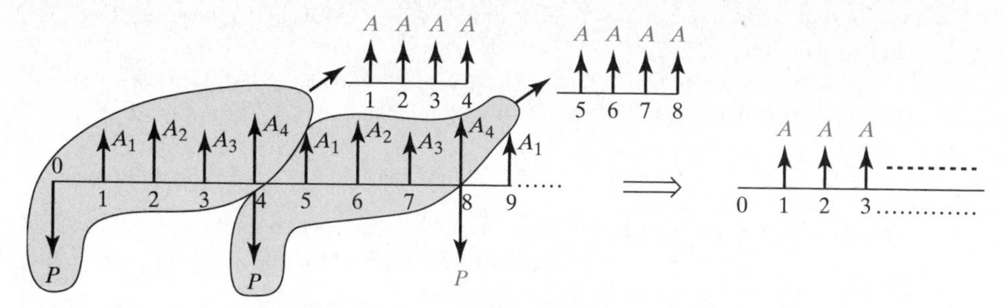

Figure 12.5 Repeatability and infinite horizon assumptions for unequal lives analysis.

this is not necessary. Consider Figure 12.5. If we were to determine the annual equivalent costs (AEC) for utilizing the 4-year asset over a 28-year horizon, it would be the same value as if it were evaluated over 4 years assuming repeatability. This is because the cash flows repeat themselves every 4 years. Similarly, the 7-year asset repeats four times over the 28-year horizon, and the AEC also repeats four times over those 28 years. Thus, we need to compare the AEC of the assets over their respective service lives only if we can assume a long (i.e., infinite) horizon and repeatability.

The best way in which to illustrate these different rules is by way of example. In doing so, we will also note the assumptions made with each approach.

EXAMPLE 12.8

Service Projects with Unequal Lives

In 2004, the District of Columbia chose to install Cisco Systems, Inc., equipment in its $93 million deployment of a voice-and-data broadband network to connect various sites in the city. The city had been paying Verizon Communications $30 million a year to use its network. Among the services provided by the network is emergency 911 phone calls. The city expects to save $10 million per year in communications costs by running its own network.[15]

Make the following assumptions: the network costs $93 million to install, with $15 million in annual maintenance and upgrades over a lifetime of seven years. The network has a residual value of $25 million after five years and $0 after seven years. Leasing the network costs $30 million per year for a five-year contract. Determine the lowest-cost option for the network, assuming an interest rate of 5% per year.

Solution. We solve this problem for a variety of horizons to illustrate the possible methods. Examining the data, we have a number of choices for analysis. We have explicit estimates of five- and seven-year horizons for purchasing the network. If we

[15] Baker, N., "UPDATE: Cisco Equipment Picked for DC Government Network," *Dow Jones Newswires,* February 9, 2004.

assume that the leasing option is available on an annual basis, then we have explicit costs for a number of periods.

First, assume that $N = 5$. Under this assumption, leasing the network costs \$30 million per year, while purchasing the equipment costs

$$\text{AEC}(5\%) = \$93\text{M}(\overset{A/P,5\%,5}{0.2310}) + \$15\text{M} - \$25\text{M}(\overset{A/F,5\%,5}{0.1810}) = \$31.96 \text{ million.}$$

In this situation, we choose the leasing option, as it saves \$1.96 million per year.

For the case of $N = 7$, the annual equivalent costs of purchasing and running the network are

$$\text{AEC}(5\%) = \$93\text{M}(\overset{A/P,5\%,7}{0.1728}) + \$15\text{M} = \$31.07 \text{ million.}$$

If we assume that the annual cost of the leasing contract does not change, we choose to lease the network, as leasing leads to savings of over \$1 million per year.

If we assume a longer horizon, such as $N = 10$, then we must make some explicit assumptions about the future. It is reasonable to assume that leasing could be extended to 10 years at the continued annual cost of \$30 million per year. The purchase decision is much more difficult. The question is, What do we do for the final 3 years of required service after the purchased equipment has reached its maximum life? The leasing option provides some measure of annual costs that could be used to fill this gap. Thus, we could assume that the purchased equipment is used for 7 years, and then the network could be leased for \$31 million per year. We estimate a higher cost here, as we cannot negotiate a longer contract. (Earlier, we assumed that the 7-year lease was the same price as the 5-year lease. It is reasonable to assume that a 7-year lease would cost the same as or less than a 5-year lease on an annual basis. Similarly, it is reasonable to assume that a shorter lease would cost more on an annual basis.)

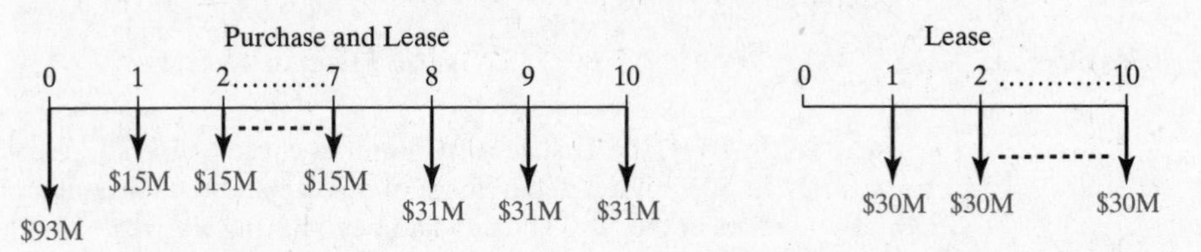

Figure 12.6 Explicit estimates for 10-year horizon analysis.

Figure 12.6 illustrates the cash flows for our purchase–lease option over 10 years. The cost of this option is

$$\text{AEC}(5\%) = \$93\text{M}(\overset{A/P,5\%,10}{0.1295}) + \$15\text{M} + \$16\text{M}(\overset{F/A,5\%,3}{3.1525})(\overset{A/F,5\%,10}{0.0795}) = \$31.05 \text{ million.}$$

Again, the scenario with ownership costs more than the lease option.

Our final analysis would assume that the horizon is infinite. If we assume repeatability of each option, then we can compare the annual equivalent costs of the options over their respective service lives. This was assumed in the seven-year horizon analysis, which selected the lease option.

The preceding example highlights the assumptions that must be made in considering service projects over unequal lives. The projects must be converted such that an evaluation of equal lives is undertaken in order to ensure fair comparisons. This requires either explicit estimates that shorten or lengthen assets' service lives, depending on the study horizon, or an assumption concerning repeatability over long horizons. As we saw in this last example, assumptions about repeatability can be difficult to estimate. Note that we did not answer the question of how long to retain the asset (or project). This problem, common when $N > N_A$, is known as replacement analysis and is examined in Chapter 16.

12.7 Mixing Revenue and Service Projects

In defining revenue and service projects, we proceeded with their analyses separately, because the returns from service projects are quite different from those of revenue projects. If a company places a single budgeting constraint on all investment projects, it is clear that we must be able to analyze these projects simultaneously.

Recall that service projects are defined by cash flows with few, if any, revenues. An example is the previous one, dealing with the choice of voice-and-data network the District of Columbia government should employ. The only revenues involved were the salvage values of the procured equipment. In these situations, the choices are generally between technologies that solve a problem or take advantage of an opportunity; thus, the options are mutually exclusive. Furthermore, the "do-nothing" alternative is often not viable. For the network example just mentioned, a choice of technology had to be made. For the choice of engine on the aircraft in Example 12.7, again, a choice *had* to be made. Thus, it is clear that service projects are different from revenue projects in that an investment choice *must* be made because the do-nothing option is usually not feasible. Since an investment decision must be made in these situations, one can "remove" the service projects from the usual capital-budgeting decisions and perform the analyses separately. This is because these decisions do not affect other decisions in terms of investment choices. Because such projects may also fall under capital-budget constraints, decisions can be made on the appropriate choices of technology, and the corresponding investment amounts can be removed from the budget for further consideration (separately) of revenue projects.

12.8 Solutions to Large-Scale Applications

Even when service projects are removed from the pool for analysis, the number of revenue projects to be studied can be daunting. This is because total investment analysis and incremental investment analysis require enumerating all possible mutually exclusive investment portfolios. If there are k projects available for investment, that could lead to defining 2^k investment portfolios to be analyzed. For three projects, we are left with a relatively small number of portfolios (8) to analyze in the worst-case scenario. However, this number grows exponentially. For 10 projects, a maximum of 1024 portfolios must be evaluated. While the number of feasible alternatives may be much lower due to budget constraints, many large companies routinely analyze hundreds of projects in a quarter or year. In this case, enumerating all possible portfolios may not be efficient. (For example, Royal Dutch Shell announced that it would spend $19 billion on capital investments in 2006 alone.[16] This clearly translates to a large number of projects for evaluation.)

We present two procedures that avoid enumerating all possible combinations of investments. The first approach is based on the concepts of mathematical (integer) programming, a field of study in applied mathematics and operations research. Used properly, this approach produces the same solution as total or incremental investment analysis.

The second approach, commonly referred to as a project-ranking method, ranks *individual* projects according to some criteria and selects the top projects, ensuring that no constraint is violated. Ranking methods are heuristic in that they are not guaranteed to give us the best answer. However, the analysis is relatively easy to perform, and while it may not give the best solution in terms of generating wealth, it usually provides good answers.

12.8.1 Integer Programming Approach

We will not dwell on the theory behind integer programming; rather, we introduce the concept of developing an integer programming model in our capital-budgeting context. These can be solved with Excel's Solver, as illustrated shortly.

An integer program requires the following:

1. **Decision Variables.** When looking at each individual project, the decision is to either accept or reject the project. Toward that end, we define the following variable:

$$x_j = \begin{cases} 1 & \text{if project } j \text{ is accepted} \\ 0 & \text{if project } j \text{ is rejected} \end{cases}.$$

   This variable definition should illustrate why the method is referred to as *integer programming*: The solutions are required to be integers.

2. **Objective Function.** The objective function drives our decision in that we want to maximize or minimize some measure. When a project is accepted, the expected cash flows are experienced and the company receives the

[16] Barker, A., "Shell boosts capital investment to $19bn for 2006," *The Financial Times*, www.FT.com, December 13, 2005.

present worth of the project as value. For capital budgeting, our objective is to select projects in order to maximize present worth.

Assuming that the present worth of project j at interest rate i is $\text{PW}(i)_j$, the objective is to maximize the present worth, as

$$\max \sum_j \text{PW}(i)_j x_j.$$

This is the amount of wealth generated by our decisions, as the x_j variables take on values of zero or one, such that the present worth of project j is either included, or excluded, from the summation, depending on whether it is accepted or rejected.

3. **Constraints.** The constraints are what tie these projects together. Capital budgets constrain the amount of money we are allowed to spend. Other constraints capture our mutually exclusive constraints and project interdependencies.

   (a) *Capital-budgeting constraints.* Because we are only making decisions for a single period, we are generally interested in limiting spending at time zero. With A_{0j} representing the cash flow at time zero for project j and B representing the budget limit, the budget constraint is written as

   $$\sum_j A_{0j} x_j \leq B.$$

   (b) *Mutually exclusive constraints.* This is where the use of integer programming becomes important. If we have two projects that are mutually exclusive—say, projects 1 and 2—then *at most* one of the projects can be accepted. Since the projects are represented as decision variables x_1 and x_2, we can write the mutual exclusivity condition mathematically as

   $$x_1 + x_2 \leq 1.$$

   Because x_1 and x_2 take on binary values of 0 or 1, this constraint will allow only one variable, x_1 or x_2, to take on a value of one. This type of constraint can be repeated for any set of mutually exclusive projects.

   (c) *Project interdependency constraints.* Interdependent projects suggest that an investment must be made in one project before another, dependent project can be attempted. For example, if project 2 is dependent on project 1 being selected, then, in terms of our binary variables, x_1 can be 1 or 0, but project 2 can be 1 only if project 1 is 1. Mathematically, this constraint can be written as

   $$x_2 - x_1 \leq 0.$$

   Since x_1 and x_2 are binary, the only feasible solutions are $x_1 = 0$ and $x_2 = 0$, $x_1 = 1$ and $x_2 = 0$, and $x_1 = 1$ and $x_2 = 1$. In each of these cases, project 2 can be accepted ($x_2 = 1$) only if project 1 has already been accepted.

A number of mathematical programming approaches have been developed similarly to the integer programming approach just presented, which provides the best portfolio (as with total and incremental investment analysis) if it is assumed that funds are loaned and borrowed at the MARR. More complicated models have been developed in the literature that consider individual cash flows over time, allowing for budget constraints over multiple periods. Models have also been developed to allow for different rates of interest for borrowing and investing funds. These models are beyond the scope of this text; however, they are mentioned because the capital-budgeting process that we have studied generally looks only at a single period in time which may not be adequate for many corporate-planning purposes.

We illustrate solving the integer program using Solver in Excel in the chapter's introductory real decision problems. However, we first introduce an alternative to integer programming.

12.8.2 Heuristic Methods: Project-Ranking Approaches

As noted earlier, project-ranking methods provide good answers to the capital-budgeting decision with relative ease, as we only have to compute one value for each project, not a value for each possible combination of projects (portfolio), as in our enumeration procedures. Although we do not advocate these approaches, because they do not guarantee an optimal solution, we understand that they may be useful when we need to consider an extremely large number of projects and the concepts of integer programming may be unfamiliar.

You have already been exposed to a ranking approach in Example 12.2. There, we observed that ranking multiple projects with the IRR would not produce the optimal solution. Noting this concern, we can still utilize the IRR to rank projects and select among them on the basis of the ranking of the IRRs. While this does not guarantee an optimal solution in terms of generating wealth, it does provide for an extremely efficient solution in that the funds invested will generate the highest percentage return.

Projects can be ranked by any number of criteria, such as present worth or IRR. Another popular measure for ranking projects is the *profitability index*, defined as the ratio of the present worth to the project's investment cost. The idea behind the profitability index is to provide some measure of efficiency in calculating the "bang for the buck." We define the profitability index as

$$\mathrm{PI}(i) = 1 + \frac{\mathrm{PW}(i)}{\mathrm{PW}(i)_{\mathrm{investment}}}. \tag{12.1}$$

Note that the denominator is defined as a cost term and takes on a positive value, much as we defined costs in the BC ratio from Chapter 9. The present worth of the project is placed in the numerator, while the denominator is the present worth of the investment cost(s). We are explicit about the definition of the denominator, as we have illustrated numerous examples in this text in which the investment cost is spread over multiple periods.

If the present worth of the project is negative, then the ratio is negative. Thus, with the profitability index, the decision criterion for a single project is defined in accordance with the following relationships:

PI Value	Decision
$\text{PI}(i) > 1$	Accept the project.
$\text{PI}(i) = 1$	Indifferent about the project.
$\text{PI}(i) < 1$	Reject the project.

It should be clear that the profitability index produces the same decision as present worth for a single project. This can be verified by substituting $\text{PW} > 0$ into the definition of PI.

Once a measure has been selected, the procedure for ranking multiple projects is quite straightforward. The measure is calculated for each project, and the projects are ranked from best to worst (i.e., highest to lowest PI; this is easily accomplished with 'Sort' under the 'Data' menu in Excel). Then, each project is accepted in rank order until the budget limit is exhausted. The only "catch" is that all other constraints must be accounted for in maintaining the list of ranked projects. For example, if two projects are mutually exclusive and one is selected, the other (which comes later in rank order) is deleted from the list. We illustrate in the next section by solving the real decision problems presented at the beginning of the chapter.

12.9 Examining the Real Decision Problems

We revisit our introductory real decision problems concerning Bluesteel Corporation's expansion plans and the questions posed. Before delving into the solution, however, note that we have four individual projects listed. This means that we would have a maximum of $2^4 = 16$ possible portfolio combinations to evaluate through total or incremental investment analysis. This is a tedious number of portfolios to evaluate, so we turn to integer programming and ranking procedures.

While Goal Seek allows one to change the value of a cell while manipulating another cell, Solver allows multiple cells to be manipulated while one is trying to achieve some goal: either driving a certain cell value to a given number or maximizing or minimizing the value of a cell. In addition, constraints can be added to limit the manipulation in certain cells. Thus, we will utilize Solver in Excel when determining the optimal portfolio under a variety of constraints. (We will look at the ranking procedure later.) In what follows, we develop the integer programming formulation and then we illustrate its solution with Solver:

1. *No budgeting constraints.* This problem is trivial, as all of the projects have positive present-worth values. Thus, each investment should be pursued. However, we will set up the problem anyway, as the cases that follow merely add constraints to this situation. Our decision variables are x_1 through x_4, with the following objective function:

$$\max \text{A\$12.76M}x_1 + \text{A\$6.13M}x_2 + \text{A\$9.24M}x_3 + \text{A\$2.07M}x_4.$$

Since there are no explicit constraints, we only impose integrality such that

$$x_j = \{0,1\} \text{ for } j = 1,2,3,4.$$

Solving this problem results in $x_1 = x_2 = x_3 = x_4 = 1$, with a present worth of A$30.2 million.

2. *Budgeting constraint of A$600M.* For this case, we add the following constraint:

$$\text{A\$280M}x_1 + \text{A\$180M}x_2 + \text{A\$80M}x_3 + \text{A\$260M}x_4 \leq \text{A\$600M}.$$

Solving this problem results in $x_1 = x_2 = x_3 = 1$ and $x_4 = 0$, with a resulting present worth of A$28.1 million for this portfolio. The investment for this solution totals $520 million, which is clearly under our constraint.

3. *Budgeting constraint of A$500M.* Lowering the budget to this level requires us to alter our previous constraint so that

$$\text{A\$280M}x_1 + \text{A\$180M}x_2 + \text{A\$80M}x_3 + \text{A\$260M}x_4 \leq \text{A\$500M}.$$

This results in only the first and third projects being pursued, with $x_1 = x_3 = 1$ and $x_2 = x_4 = 0$. The resulting present worth is A$22 million. The cost of this investment is $360 million.

4. *Budgeting constraint of A$300M.* Lowering the budget further requires us to alter our previous constraint such that

$$\text{A\$280M}x_1 + \text{A\$180M}x_2 + \text{A\$80M}x_3 + \text{A\$260M}x_4 \leq \text{A\$300M}.$$

This results in only the second and third projects being pursued, so that $x_2 = x_3 = 1$ and $x_1 = x_4 = 0$, with a resulting present worth of A$15.4 million. The cost of this investment is $240 million.

5. *The Vietnam and China facilities are mutually exclusive, servicing the same customers.* This condition requires that we add the following constraint:

$$x_1 + x_2 \leq 1.$$

Solving results in $x_1 = x_3 = x_4 = 1$ and $x_2 = 0$, with a present worth of A$23.1 million.

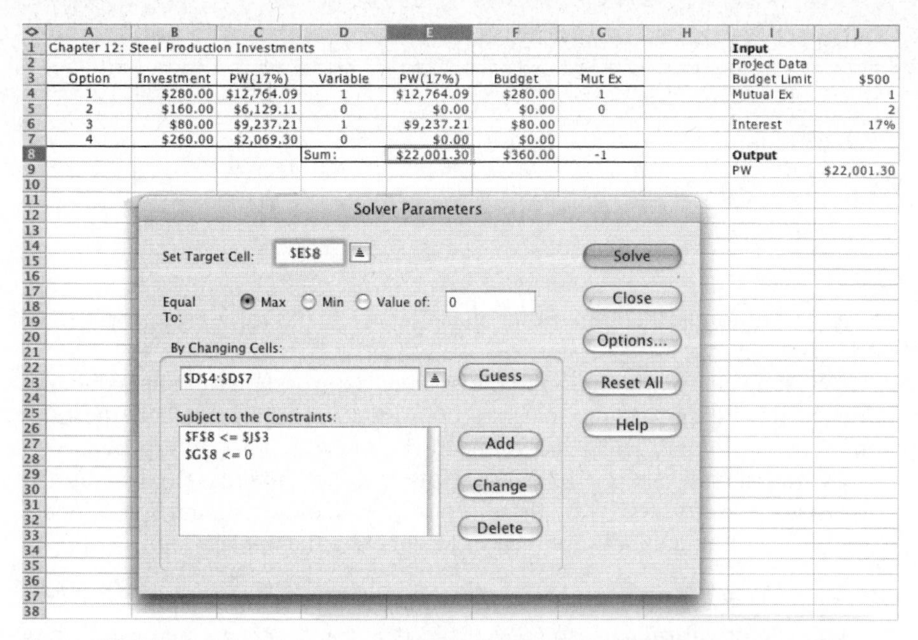

Figure 12.7 Spreadsheet setup for using Solver to solve capital-budgeting integer program.

6. *The Vietnam facility is dependent on the China facility opening and a A$500 million budget.* This condition requires the constraint

$$x_2 - x_1 \le 0,$$

in addition to the A$500 million budgeting constraint added previously. With these constraints, the solution is $x_1 = x_3 = 1$ and $x_2 = x_4 = 0$, with a worth of $22 million.

Figure 12.7 illustrates one approach to using Solver in Excel to solve the integer program. Cells D4 through D7 represent the decision variables, with column C indicating the present worth for each decision. Column E is merely the resulting objective function input, or the product of the decision variable (column D) and its present worth (column C). The sum of these values (cell E8) is the objective function value.

The dialogue box asks for a cell to optimize (you can select maximum, minimum, or a target value) and the cells to manipulate. These are given in Figure 12.7. Constraints are added by pushing the 'Add' button leading to a smaller dialogue box. This box requests cell references for both the left and right side of an equation, which can be selected as $\le$, $\ge$, or $=$. For part 6 of our example, the constraints are enforced with the summations in cells F8 (budget expended) and G8 ($x_2 - x_1$). Hitting 'Solve' leads to the solution

TABLE 12.7 **Rankings of projects in accordance with PW, PI, and IRR.**

Option	Investment Cost	PW(17%)	Rank	PI(17%)	Rank	IRR	Rank
1	A$280M	A$12.76M	1	1.046	2	18.24%	2
2	A$180M	A$6.13M	3	1.034	3	18.05%	3
3	A$80M	A$9.24M	2	1.12	1	20.11%	1
4	A$260M	A$2.07M	4	1.01	4	17.22%	4

given in the spreadsheet: investing in projects 1 and 3 for a present worth of A$22 million.

We now turn our attention to the heuristic approach of ranking individual projects according to some measure of worth. Table 12.7 presents the rank order of the four projects according to the present worth, the profitability index, and the internal rate of return. Note that the ranking is the same for PI and IRR, but differs for PW in that the positions of projects 1 and 3 are flipped.

We reexamine the different cases to illustrate the application:

1. *No budgeting constraints.* All projects are accepted.
2. *Budgeting constraint of A$600M.* Projects 1, 2, and 3 are accepted according to all measures.
3. *Budgeting constraint of A$500M.* Projects 1 and 3 are accepted by all measures.
4. *Budgeting constraint of A$300M.* For PI and IRR, projects 2 and 3 are accepted. Note that when project 3 is accepted, there is not enough funding available for project 1. Moving down the rank ordering, we see that project 2 is accepted to complete the selection process. With PW, projects 1 and 3 are accepted, leaving insufficient funds for any other investments. The PI and IRR agree with the integer programming solution.
5. *The Vietnam and China facilities are mutually exclusive, servicing the same customers.* Since the China facility (Option 1) is ranked higher than the Vietnam facility (Option 2), all measures select projects 1, 3, and 4.
6. *The Vietnam facility is dependent on the China facility opening and a A$500 million budget.* This does not change the result of the A$500 million constraint problem, as the China option is ranked higher than the Vietnam option and funding is exhausted before the Vietnam option is considered.

The ranking procedures in this example accept the same projects as our integer programming approach, with the exception of ranking with PW and the small budget. This agreement, however, is not guaranteed to always occur. As the number of projects increases, differences are more likely to appear, although the ranking procedure will generally provide good answers, especially with measures such as PI and IRR, as they account for the investment level in their calculation (relative measures of worth) and absolute measures (such as PW) do not.

12.10 Key Points

- The process in which a set of projects is accepted from a set of potential projects is often termed the capital budgeting process.
- Multiple projects that are interdependent (not independent) must generally be analyzed simultaneously in the capital budgeting process. Among such projects are those which provide the same service, compete for resources (such as budget dollars), and depend on the selection of another project.
- If two projects are mutually exclusive, then the acceptance of one precludes the acceptance of the other.
- Mutually exclusive portfolios of projects can be determined by enumerating all feasible combinations of investment options. Only one portfolio must be selected from this set.
- Revenue projects generate revenues for the company and thus are evaluated by maximizing present worth. Service projects do not generate revenues, other than a possible salvage value, and thus are analyzed by minimizing costs.
- The analysis aimed at selecting a project is dependent on the type of project (revenue or service) and the study horizon (project life).
- For the case of revenue projects with equal lives, total investment analysis selects the portfolio with the maximum (present) worth from a set of mutually exclusive portfolios. This approach can be used only with absolute measures of worth.
- Alternatively, for the case of revenue projects with equal lives, incremental investment analysis performs a systematic, pairwise comparison of all mutually exclusive portfolios. The portfolio that bests all other portfolios in this procedure is accepted. This approach can be used with either relative or absolute measures of worth.
- For the case of revenue projects with unequal lives, the portfolio with the largest present worth is selected, under the assumption that returns are invested at the MARR through the longest project horizon.
- For the case of service projects with equal lives, incremental or total investment analysis may be performed, noting that equivalent costs are minimized.
- For the case of service projects with unequal lives, the analysis is dependent on the study period of the projects in question. Either cash flows must be estimated such that the resulting (new) cash flow diagrams have equal lives, or assumptions of repeatability allow for the horizon to be a common multiple of the projects in question.

- When revenue and service projects are under simultaneous consideration with budgeting constraints, the service projects should be removed and evaluated separately.
- Enumerating all possible investment portfolios from k projects can define as many as 2^k possible investments. Since this number can be quite large, integer programming or heuristic approaches to the project selection decision may be necessary.
- Heuristic methods are not guaranteed to provide the optimal solution. However, a good heuristic should provide a good answer with relatively little work. Project-ranking procedures are often termed heuristic approaches.
- Integer programming is a mathematical technique that allows a function (the present worth of the projects selected) to be maximized, while abiding by system constraints, such as budgeting or dependency constraints. The approach gives the same solutions as that given by total or incremental investment analysis.
- Previously derived measures of worth, present worth, and internal (or external) rate of return, can be used to rank projects. These are heuristic approaches. Relative measures of worth tend to rank more consistently with integer programming approaches.
- The profitability index is a measure of the efficiency of a project, because it is defined by the ratio of the project's present worth to the present worth of the project's investment cost (plus 1). If the value is greater than one, then the project is acceptable. The value can also be used to rank projects for subsequent selection.

12.11 Further Reading

Blank, L., and A. Tarquin, *Engineering Economy,* 5th ed. McGraw-Hill, Boston, 2002.

Bernhard, R.H., "A Comprehensive Comparison and Critique of Discounting Indices Proposed for Capital Investment Evaluation," *The Engineering Economist,* 16(3)157–186, 1971.

Bussey, L.E., and T.G. Eschenbach, *The Economic Analysis of Industrial Projects,* 2d ed. Prentice Hall, Englewood Cliffs, New Jersey, 1992.

Charnes, A., W.W. Cooper, and M. Miller, "An Application of Linear Programming to Financial Budgeting and the Costing of Funds," *Journal of Business,* January 1959.

Fleischer, G.A., *Introduction to Engineering Economy.* PWS Publishing Co., Boston, 1994.

Hajdasinski, M.M, "NPV-Compatibility, Project Ranking, and Related Issues," *The Engineering Economist,* 42(4):325–340, 1997.

Lorie, J., and L.J. Savage, "Three Problems in Capital Rationing," *Journal of Business,* October 1955.

Park, C.S., *Contemporary Engineering Economics,* 3d ed. Prentice Hall, Upper Saddle River, New Jersey, 2002.

Park, C.S., and G.P. Sharp-Bette, *Advanced Engineering Economics.* John Wiley and Sons, New York, 1990.

Thuesen, G.J., and W.J. Fabrycky, *Engineering Economy,* 9th ed. Prentice Hall, Upper Saddle River, New Jersey, 2001.

Weingartner, H.M., *Mathematical Programming and the Analysis of Capital Budgeting Problems.* Prentice-Hall, Englewood Cliffs, New Jersey, 1963.

Weingartner, H.M., "Criteria for Programming Investment Project Selection," *Journal of Industrial Economics,* November 1966.

White, B.E., and G.W. Smith, "Comparing the Effectiveness of Ten Capital Investment Ranking Criteria," *The Engineering Economist,* 31(2):151–163, 1986.

12.12 Questions and Problems

12.12.1 Concept Questions

1. What is meant by the statement that "two projects are mutually exclusive."

2. What can cause two projects to be mutually exclusive?

3. When should multiple projects be examined simultaneously?

4. What is the difference between a revenue project and a service project? Do they have different objectives?

5. Should the analyses of revenue and service projects always be separated when budgets are limited?

6. Why is the assumption of the horizon time so critical when analyzing multiple alternatives? Explain.

7. Given k investment proposals, how are possible investment portfolios formed? Why is the term "portfolio" used?

8. What is the difference between total and incremental investment analysis? Do they come to the same conclusion?

9. Why must relative measures of worth use incremental analysis?

10. Explain the meaning of IRR analysis in an incremental analysis framework.

11. When is it appropriate to assume repeatability for study horizon considerations?

12. What is assumed when present worth is used to analyze multiple revenue projects with different horizons? Is this feasible with service projects? Why or why not?

13. In general, what are the approaches for projects with different service lives?

14. What is the difficulty with using total and incremental investment analysis techniques?

15. If there are *k* projects available for investment, what is the maximum number of portfolios to evaluate? How large must *k* become for this to become unmanageable?

16. What is integer programming? What does it do?

17. What is a heuristic? Why might a decision maker want to use one?

18. Why are project ranking and integer programming better suited than other methods for analyzing problems with large numbers of projects?

19. The net present-value index, or NPV index, is often denoted as the ratio of the net present value of a project over the project's investment cost. How is the NPV index related to the profitability index? Do they give the same result?

20. Can the integer programming model be extended to consider risk attributes such as variance in return? Explain and describe the data necessary to solve such a problem.

12.12.2 Drill Problems

1. Consider the following list of projects:

Option	Investment (M)	PW(15%) (M)
1	$297	$241
2	$118	$221
3	$297	$377

Determine the number of feasible portfolios if

(a) All projects are independent.

(b) All projects are independent, but the budget is $500 million.

(c) Projects 1 and 3 are mutually exclusive.

(d) Project 1 is contingent on project 2 being accepted.

2. If the projects in the previous problem are all revenue projects (which they must be, since they all generate positive present-worth values), do their lengths of study matter? Explain.

3. Solve Problem 1 (all four parts) using total investment analysis.

4. Consider the following list of projects:

Option	Investment (M)	PW(15%) (M)
1	$471	$716
2	$465	$757
3	$405	$179

If all three projects are mutually exclusive, which, if any, should be selected, using incremental investment analysis.

5. Consider the following list of projects:

Option	Investment (M)	PW(15%) (M)
1	$100	$573
2	$303	$283
3	$317	$647
4	$235	$665

Find the best portfolio with total investment analysis, assuming that

(a) All projects are independent.

(b) All projects are independent, but the budget is $800 million.

(c) Projects 2 and 3 are mutually exclusive and the budget is $500 million.

6. Consider the 10 projects given in Problems 1, 4, and 5. Write the integer programming formulation, assuming that (listed in order) project 1 is contingent on project 2 being accepted, projects 4, 5, and 6 are mutually exclusive, and projects 8 and 9 are mutually exclusive. Solve with Excel under the assumption of no budget constraint, a $1.5 billion budget, a $750 million budget, and a $300 million budget.

7. Consider the following projects and their cash flows (in millions):

Project	Period					
	0	1	2	3	4	5
1	−$322	$178	$228	$278	$328	$378
2	−$427	$122	$122	$122	$122	$122
3	−$314	$157	$131	$109	$91	$76
4	−$398	$118	$184	$183	$117	$138

Determine the number of feasible portfolios if

(a) All projects are independent, but there is a $1.1 billion budget?

(b) All projects are independent, but the budget is $750 million?

(c) The budget is $1.1 billion, but projects 3 and 4 are mutually exclusive?

8. Find the optimal portfolio in Problem 7 (c), using incremental investment analysis with the internal rate of return, assuming a MARR of 18%.

9. Find the optimal portfolio in Problems 7(a), (b), and (c) using total investment analysis, present worth, and a MARR of 22%.

10. Two mutually exclusive assets are being considered for installation to perform a job. Asset A costs $500,000 to purchase and $300,000 per year to maintain, lasts for eight years, and has a salvage value of $50,000. Asset B costs $350,000 to purchase and $400,000 in the first year to maintain, increasing by 15% each year thereafter. Asset B lasts for six years with no discernible salvage value. Assume a 9% annual rate of interest. Which asset should be purchased?

(a) Assume a study period of six years, and assume that Asset *A* has an estimated salvage value of $125,000 at that time.

(b) Assume a study period of eight years, and assume that the life of Asset B can be extended by increasing the maintenance costs in year six by 30% in each of the final two period.

(c) Assume a study period of 24 years.

(d) Assume an infinite horizon. What assumptions must be made in parts (c) and (d)?

11. A new asset can be purchased to perform a job, or the current machine can be upgraded. The new asset costs $1.2 million, generates revenues of $500,000 per year over its seven-year life against costs of $100,000 in year 1, increasing by $10,000 each year. It has a salvage value of $50,000 at that time. The current asset can be upgraded for $800,000 and will generate just $400,000 in revenues, as it does not have all of the capabilities of the new machine. O&M costs are expected to be $150,000 per year over the asset's remaining five-year life. If both assets are to be depreciated as five-year assets under MACRS alternative straight-line depreciation (with the half-year convention), which should be chosen? Assume a tax rate of 35% and an after-tax MARR of 12%. What assumptions, if any, did you make about the horizon time?

12. Consider the following projects and their cash flows (in millions):

Project	Period							
	0	1	2	3	4	5	PW(14%)	IRR
1	−$418.00	$541.31	$556.31	$571.31	$586.31	$601.31	$1,529.96	129.92%
2	−$229.00	$65.43	$65.43	$65.43	$65.43	$65.43	−$4.38	13.20%
3	−$362.00	$181.00	$150.83	$125.69	$104.75	$87.29	$105.03	27.33%
4	−$311.00	$53.00	$183.00	$72.00	$159.00	$200.00	$122.92	27.14%

Assume that project 1 is dependent on project 2 being accepted and that projects 3 and 4 are mutually exclusive. With a MARR of 14%,

(a) Identify the feasible investment portfolios.

(b) Determine the best portfolio for investment, using present worth and total investment analysis.

(c) Determine the best portfolio for investment, using IRR analysis.

12.12.3 Application Problems

1. In 2005, XM Satellite Radio Holdings, Inc., selected Loral Space & Communications Co. over longtime customer Boeing Co. to build and launch XM's next satellite. One of the most powerful commercial satellites ever built, the new craft is designed to have more than 20 kilowatts of power and will serve as a spare to back up XM's fleet of three orbiting spacecraft. It is expected to remain in operation for 15 years and cost between $200 and $300 million.[17]

Make the following assumptions: Two suppliers are offering satellites with the specifications listed in Table 12.8. One operator is known for its satellite hardware

[17] Pasztor, A., and S. McBride, "XM Radio Picks Loral Space to Build New Satellite," *Dow Jones News Service,* May 27, 2005.

TABLE 12.8

Cost data for different satellite design, launch, and operator contractors.

Parameter	Supplier A	Supplier B
Initial Cost	$250M	$225M
Annual Cost	$1.25M	$2.5M
Service Life (Years)	15	12

prowess, while the other has lower overhead and thus lower operating costs. The salvage values decline according to a straight line over the first five years of the satellite's operation, after which they vanish. The interest rate is 12%.

Given the preceding information, which operator should be chosen? In answering assume horizons of 3 years, 5 years, 12 years, 15 years, and an infinite horizon. State any assumptions you make.

2. Mitsubishi Electric Corp. of Japan was considering spending ¥10 billion to boost its global production of automotive parts. The company said it would build five new production facilities—two in Japan and one each in the United States, the Philippines, and Thailand—in hopes of growing sales from ¥350 billion in 2004 to ¥500 billion in 2008.[18]

Make the following assumptions: The five investments and their attributes are listed in Table 12.9.

(a) Assuming no constraints, how many possible portfolios are there?

(b) Assuming that Mitsubishi has a ¥14 billion budget limit, how many feasible portfolios are there?

(c) How many feasible portfolios are there with ¥12 and ¥10 billion budgets.

(d) Do any of the preceding answers change if the second Japan option can be undertaken only if the first is?

(e) Consider the original set of options with a ¥7 billion budget. Use total investment analysis and present worth to determine the best investment portfolio.

(f) Does the answer change if selecting the second Japan option depends on selecting the Thailand investment? If so, recompute the optimal portfolio.

(g) Redo the analysis, using incremental analysis and present worth to illustrate the method.

(h) Use integer programming to formulate and verify your solution.

3. In 2004 and 2005, Stora Enso Corp., a Finnish–Swedish forestry products company modernized a paper machine at its publication-paper mill in Summa, Finland, at the cost of EUR53 million.[19]

---

[18] Nishio, N., "Mitsubishi Elec to Spend Y10B to Boost Auto Parts," *Dow Jones Newswires,* January 27, 2004.

[19] "Stora Enso to Modernize Paper Machine at Summa Plant," *Dow Jones Newswires,* October 27, 2003.

TABLE 12.9 **Auto parts production expansion options, costs and present-worth values.**

Option	Investment (M)	PW(20%) (M)
Japan 1	¥3031.00	¥8433.26
Japan 2	¥3154.00	¥11788.77
United States	¥2692.00	¥3214.75
Philippines	¥3152.00	¥6219.45
Thailand	¥3128.00	¥3649.81

Make the following assumptions: The current paper machine (valued at EUR10 million) can be retained for another five years. Operating and maintenance costs over the five years are expected to be EUR6 million in the first year, increasing by 20% each successive year. The salvage value is expected to be zero after the five-year life of the machine. The paper machine can also be modernized at the cost of EUR53 million. It will have O&M costs starting at EUR50,000 due to advanced technologies and will increase 2.5% per year. The salvage value is expected to be EUR5 million after a maximum service life of eight years. The interest rate is 7% per year.

(a) Assume that revenues are the same regardless of which machine is chosen. How would we classify these projects? What are the decisions to be made?

(b) Assume a study horizon of five years. If the salvage value of the modernized machine after year 5 is EUR15 million, what is the best choice of technology?

(c) Assume a study horizon of three years, a salvage value of EUR1 million for the used machine, and EUR25 million for the modernized machine. What is the best choice of technology now?

(d) Assume a study horizon of eight years. Since it is unlikely that the used machine will be repeatable, assume that the modernized machine can be acquired at the end of its service life and determine the best option.

(e) Assume an infinite horizon. Again, assume that a modernized machine is purchased after the old one is sold at the end of its service life. What is the best choice of technology this time?

4. The City Council members of Visalia, California, decided to purchase CNG-fueled garbage trucks, as opposed to traditional diesel trucks, at the cost of $260,000 each ($50,000 more than equivalent diesel trucks). The trucks are expected to cost the same in terms of operation, and the city has operated CNG buses before, thus alleviating concerns about maintenance costs.[20]

Make the following assumptions: The costs of owning and operating the buses are given in Table 12.10, and the interest rate is 3.5% per year.

(a) If cost were the only consideration, which technology should be chosen?

[20] Sheehan, T., "Visalia trucks and buses go green; purchase of 13 compressed natural gas vehicles approved by City Council last week," *The Fresno Bee,* p. B1, December 17, 2005.

TABLE 12.10 **Ownership and annual operational costs for CNG and diesel garbage trucks.**

Parameter	Diesel	CNG
Initial Cost	$210,000	$260,000
O&M Cost	$15,000	$15,000
Annual Increase	$1000	$3000
Salvage Value	$10,000	$30,000
Service Life	8	8

TABLE 12.11 **Media company expansion options and annual cash flows (in millions).**

Option	Year								
	0	1	2	3	4	5	6	7	8
1	–$21	$5	$10	$15	$15	$15	–	–	–
2	–$26	$6	$7	$8	$9	$10	$11	$12	$13
3	–$17.5	$5	$7	$9	$11	$13	–	–	–
4	–$16	$12	$12	$12	$12	$12	$12	$12	$12

(b) Perform a benefit–cost analysis. Assume that the cleaner air is valued at $15,000 per year. Which technology should be chosen?

(c) To help pay for the CNG garbage trucks, the city received a $500,000 grant (to be divided among 13 trucks). Does this change the choice from (a)? How should it be incorporated into (b)?

5. Univision Communications, a Spanish-language media company, announced that it planned capital expenditures of $122.4M in 2004. Among these expenditures were (1) a $21 million buildup of Univision's Houston, Puerto Rico, and Austin, Texas, stations; (2) $26 million for expanding Univision Network upgrades and facilities; (3) $17.5 million for radio station facility upgrades; (4) $16 million for expanding TeleFutura Network upgrades and facilities; (5) $8 million for towers, transmitters, antennas, and digital technology; and (6) $34 million for capital improvements and information systems.[21] Assume that projects 1 through 4 are defined by the cash flows in Table 12.11.

(a) Assume that projects 5 and 6 are meant to improve efficiency in the office and do not affect revenues. If the other projects enhance revenues, should 5 and 6 be evaluated together with them, assuming that there are budget constraints? Explain.

(b) If there is a capital budget of $40 million, what are the feasible investment portfolios? Consider the projects in Table 12.11.

(c) Identify the best portfolio, using PW analysis and assuming a MARR of 16%.

[21] Siegel, B., "Univision Communications Sees 2004 Capex of $122.4M," *Dow Jones Newswires,* March 17, 2004.

(d) If project 4 requires project 2, what are the new feasible investment options.

(e) Identify the best portfolio from (d), using IRR and incremental investment analysis. (Assume a MARR of 16%).

12.12.4 Fundamentals of Engineering Exam Prep

1. Two projects are mutually exclusive if

(a) Both can be chosen.

(b) Choosing one means that the other cannot be chosen.

(c) Neither can be chosen.

(d) One must be chosen before the other can be chosen.

2. A service project

(a) Is the same as a revenue project.

(b) Is defined by costs and possibly a salvage value.

(c) Cannot be defined economically.

(d) None of the above.

3. Consider the following set of revenue projects for the four questions that follow:

Option	Investment (millions)	PW(17%) (millions)
A	$290	$240
B	$120	$220
C	$140	$190
D	$200	$150

With no budget constraints, and assuming that the projects are independent, the total number of feasible investment options is

(a) 18.

(b) 15.

(c) 16.

(d) 8.

4. If project C is contingent on project D being accepted, the total number of feasible investment options is

(a) 3.

(b) 7.

(c) 16.

(d) 12.

5. If the capital-budget limit is $300 million, the total number of feasible investment options is

(a) 6.

(b) 10.

(c) 16.

(d) 18.

6. If the capital-budget limit is $300 million, the investment that maximizes present worth (17%) is

(a) A and B.

(b) B and C.

(c) A, B, and D.

(d) Do nothing.

7. Consider the following set of revenue projects for the two questions that follow:

Option	Investment (millions)	PW(10%) (millions)
A	$10	$14
B	$12	$22
C	$8	–$5

With no budget constraints, and assuming that the projects are independent, the optimal set of projects to select is

(a) A, B, and C.

(b) B and C.

(c) A and B.

(d) A and C.

8. If B is contingent on C being chosen, the optimal set of projects to select is

(a) A, B, and C.

(b) A and B.

(c) B and C.

(d) A and C.

9. Two mutually exclusive investment options with the following cash flows are being evaluated:

Alternative	Period							
	0	1	2	3	4	5	PW(8%)	IRR
A	–$200	$75	$75	$75	$75	$75	$99.45	25%
B	–$100	$35	$40	$45	$50	$65	$76.61	32%

The optimal investment choice is

(a) A, according to present-worth analysis.

(b) B, according to internal-rate-of-return analysis.

(c) Both (a) and (b).

(d) Do nothing.

10. A mass-transit system is evaluating two buses. One operates on diesel gas and the other on compressed natural gas. Vital statistics for the two projects are as follows:

Parameter	Diesel	CNG
Initial Cost	$350,000	$450,000
Annual Cost	$30,000	$30,000
Annual Increase	$3000	$5000
Salvage Value	$10,000	$50,000
Annual Equivalent Costs	$82,900	$99,700
Benefit–Cost Ratio	3.01	2.71
Service Life	10	10

Each bus provides $5 million (utility minus fare) in net annual benefits to their users for transportation. The CNG bus provides an additional $2 million in benefits due to its lower emissions, promoting cleaner air. Assuming that an interest rate of 4% was used to derive the benefit–cost ratio shown in the table, which bus should be selected?

(a) Diesel, due to its benefit–cost ratio of 3.01.

(b) CNG, due to its benefit–cost ratio of 1.19.

(c) Diesel, due to its benefit–cost ratio of 1.19.

(d) None of the above.

11. Two machines are stated to be implemented in a production process whose costs are given in the following table:

Parameter	Machine A	Machine B
Initial Cost	$150,000	$250,000
Annual Cost	$10,000	$2000
Salvage Value	$0	$5000
Service Life	5	10

Use these data and a 10% annual rate of interest to answer the next two questions. The lower cost alternative is

(a) Machine A, with $72,500 lower costs (present worth).

(b) Machine B, with $44,200 lower costs (present worth).

(c) Machine A, with $100,000 lower investment cost.

(d) None of the above.

12. The lower cost alternative is

(a) Machine B, with $44,100 lower annual equivalent costs.

(b) Machine A, with $7200 lower annual equivalent costs.

(c) Machine B, with $7200 lower annual equivalent costs.

(d) None of the above.

13. Two mutually exclusive investment options with the following cash flows are being evaluated:

Alternative	Period					
	0	1	2	3	PW(8%)	IRR
1	−$250	$80	$75	$70	$99.45	25%
2	−$100	$45	$45	$45	$76.61	32%

The optimal choice according to IRR analysis requires finding the rate such that

(a) $-\$100 + \$45(P/A, i\%, 3) = 0$.

(b) $-\$250 + \$80(P/A, i\%, 3) - \$5(P/G, i\%, 3) = 0$.

(c) $-\$150 + \$35(P/A, i\%, 3) - \$5(P/G, i\%, 3) = 0$.

(d) $-\$350 + \$120(P/A, i\%, 3) - \$5(P/G, i\%, 3) = 0$.

14. A company is considering outsourcing a process. To purchase a machine would cost $200,000, $15,000 in annual fixed costs and $3.00 per part. Outsourced, the parts could be purchased for $4.00 each. If the horizon is five years with no salvage value, the interest rate is 10%, and 50,000 parts are required each year,

(a) Outsourcing would be cheaper by more than $30,000 per year.

(b) Producing in-house would be cheaper by $17,800 per year.

(c) Both options would cost the same.

(d) Outsourcing would be cheaper by less than $20,000 per year.

15. A machine can be purchased for $20,000 and used for three years with a salvage value of $0. It would cost $500 per year to maintain it during that time. If the same asset could be leased for $8500 per year and the interest rate is 6%, then

(a) Leasing would be cheaper by $500 per year.

(b) Leasing would be cheaper by more than $1000 per year.

(c) Buying would be cheaper by less than $600 per year.

(d) Buying would be cheaper by more than $1000 per year.

13 Considering Options in Time

(*Courtesy of Bayer Material Science AG.*)

Real Decisions: Paper or Plastic?

Bayer intends to slowly build up capacity over time for its product Makrolon®, a high-tech plastic used in the production of CDs, DVDs, headlight lenses, and roofing systems. Bayer announced that it would build capacity incrementally according to market demand, until reaching an annual capacity of 200,000 tons per year, at the cost of $450 million. The company began construction of the facility in 2003, with production expected to start in 2006.[1,2]

Make the following assumptions: Demand for the high-tech plastic is growing linearly at the rate of 50,000 tons per year. Thus, average demand in year 1 is 25,000 tons and growing 50,000 tons per year until a

[1] "Bayer Expands its Activities in China," *News Release*, www.news.bayer.com, November 26, 2003.

[2] "Bayer: Investments to Strengthen Growth in China," *News Release*, www.news.bayer.com, May 4, 2004.

maximum of 200,000 tons. The cost to add capacity follows a power law and sizing model with $m = 0.8$, and it costs $150 million to build 50,000 tons of annual capacity. Capacity of any size can be added in one year. O&M costs are estimated to be $15 million per year per 50,000 tons of capacity, and net revenues (revenues less expenses) are $4500 per ton, decreasing 10% for every additional 50,000 tons of capacity. Note that net revenues are received only on the minimum of capacity and demand. If the interest rate is 16% per year, consider these interesting questions:

1. What expansion options should be considered at time zero? (Assume capacity increments of 50,000 tons.)
2. As demand grows to 200,000 tons per year, what expansion decisions should be considered over the next four years? Draw the cash flow diagram for one of these paths, assuming a 10-year horizon with no salvage value.
3. What probabilistic information should be incorporated to improve decision making? How does this information alter the decision network?

In addition to answering these questions, after studying this chapter you will be able to:

- Illustrate how many engineering projects can be analyzed as sequential decisions with the ability to delay a viable option that may help mitigate risk.
- Analyze whether an investment should be delayed due to changes in external factors, such as market prices or demand. (Section 13.1)
- Perform probabilistic scenario analysis with a single-stage decision tree. (Section 13.2)
- Compute the value of perfect information to bound the cost on acquiring additional information. (Section 13.2)
- Analyze sequential investments with the use of multistage decision trees. (Section 13.2)
- Describe real options and their relationship to decision trees. (Section 13.3)

We have discussed the importance of the time value of money at length and ensured that we make sound decisions by incorporating that principle into all of our analyses. However, we have not been as adamant about incorporating time into the actual decisions themselves. That is, when a company or government entity is going to make an investment decision, that company or entity may give itself various options over time. In this chapter, we expand our decisions to include the following options: (1) invest; (2) wait; and (3) do not invest.

The decision to wait in an uncertain environment is obviously a reasonable one. It is also an interesting proposition. The time value of money says that it is better to invest sooner in an investment that will generate positive returns, as the money will be received sooner, leading to a higher present worth. However, waiting gives the decision maker time to gather information and learn about the future, such as market conditions, demand, and prices. Thus, waiting may help reduce the risk associated with an investment. This chapter considers time in our decisions and the options that it provides.

We restrict our decisions here to "sequential" ones: decisions where one decision can lead to another. For example, the decision to wait *requires* another decision to be made at a later time, namely, the decision to invest, not to invest, or delay (wait) further. This requirement of a subsequent decision defines a sequential decision or decisions.

Here are some examples of sequential decisions that are associated with engineering economy:

- Engineers drill exploratory wells when searching for oil deposits before progressing with full-scale oil-field developments.
- Bids are placed to acquire licenses for a variety of rights, including providing communications services to an area or searching for natural resources.
- Seismic data are used to identify mineral deposits before full-scale mining begins.
- Market analysts survey potential customers in order to estimate demand before production starts.
- A small increment of capacity is built to determine potential demand before expanding to larger production.

What all of these examples have in common is that there is a sequential set of decisions. An investment is made to learn about the future. This investment could be as simple as sending out a marketing survey to customers or as complicated as building a prototype for testing. The information received from this investment is then analyzed to determine the next best course of action.

For some of these decisions, such as considering whether to invest in capacity today or delay for a period and whether to drill an exploratory well before investing in an oil field, it may be argued that really only *one* decision is being analyzed. Thus, the analysis of that decision should be included in

Part III of this textbook. However, decisions to invest now or delay for a period are really two *mutually exclusive* decisions as defined in the previous chapter. Also, the sequential decisions to explore an oil field (perform testing) and then develop it are dependent and, furthermore, mutually exclusive from investing in the field without testing. Thus, problems that consider time clearly fall under the category of multiple project analysis. Therefore, if time provides mutually exclusive options, the analyses presented should be performed.

13.1 Decision Networks

It is often helpful to visualize possible decisions over time. The terms *graph*, *network*, and *tree* have been used synonymously to define figures that represent sequential decisions. We utilize the phrase **decision network** to describe possible investment decisions over time under the assumption of **certainty** with respect to information. Let us define the following terms pertaining to describing a decision network:

- *Decision Node.* A point in time at which the decision maker must choose from a given set of alternatives, represented by arcs emanating from the decision-node.
- *Option Arc.* A possible decision available at a given decision node. Each arc has an associated (expected) monetary worth or cost or leads to the definition of worth.

The arcs and nodes define a network that represents possible sequential decisions over time. That is, a decision at one node leads down an arc to another node. Presumably, the choices available to the decision maker change over time, depending on the path chosen.

Figure 13.1 illustrates a possible decision network for the three general options to invest, to delay, or not to invest. Note that the decisions to invest or not to invest terminate the path (with different values), but the decision to delay brings about these same options in the future.

We first examine this scenario under the assumption of certain information. Later, we will use a similar structure to incorporate risk, so that there

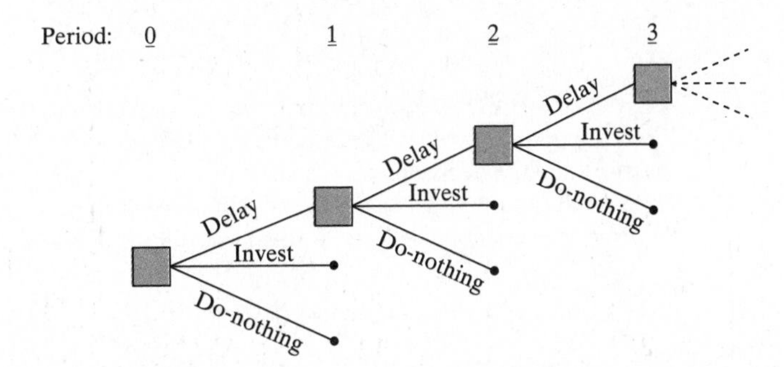

Figure 13.1 Delay, invest, or do-not-invest decisions over time in a decision network.

are probabilistic outcomes in the network. The next example illustrates the decision network for an expected future.

EXAMPLE 13.1

Decision Network

In early 2004, Peabody Energy announced that it would delay construction of a $2 billion power plant and coal mine in Washington County, Illinois, due to a recession. The plan was first announced in 2001 with the plant slated to open in 2007, but as of 2004, no construction had begun and officials said that the plant might not become operational until as late as 2011. The 1500-MW plant is expected to provide electricity for up to 1.5 million households by burning 6 million tons of coal a year.[3]

Make the following assumptions: With the end of 2004 as time zero, consider three options for the power plant (invest, do not invest, and delay) over the next four years. Assume that construction takes three years, with the investment costs spread evenly over that period. It is assumed that the 1.5 million homes will each generate revenue of $900 in 2007, with this rate expected to rise 1% per year for the foreseeable future as demand for energy increases with an improving economy. Assume annual costs of $780 million per year when the plant and mine are operational, and assume that the plant will operate for 20 years, with a $50 million remediation cost incurred after the final year of operation. Finally, the MARR is 25% per year.

Solution. There are numerous possible decisions that can be examined with respect to the power plant. We consider those depicted in the decision network in Figure 13.2 over four periods. The options available over time include investing or not investing in the project. In addition, the decision to delay allows one to reexamine these same decisions in the next period.

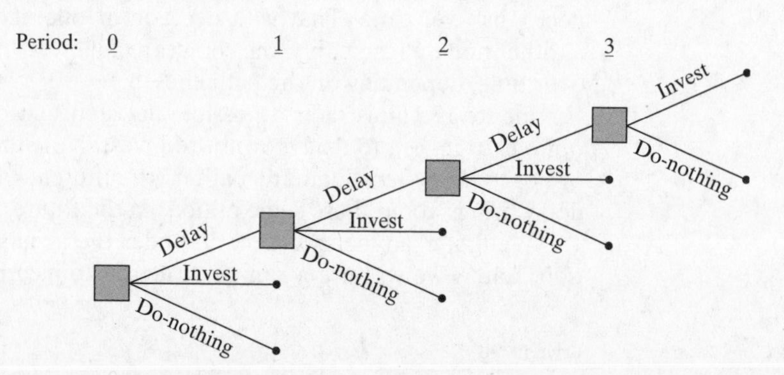

Figure 13.2 Decision network for power plant investment.

Each path in the network that terminates represents a possible investment decision for the company. Note that a path only terminates with either an invest or do-not-invest decision, and the decisions are mutually exclusive, as only one investment can be made. Each of these decisions is represented by a cash flow diagram that is dependent on the time the decision is executed. First consider the time-zero (end of 2004) option to

[3] "Plans for $2B Ill. Power Plant and Coal Mine Delayed," *Dow Jones Newswires,* February 5, 2004.

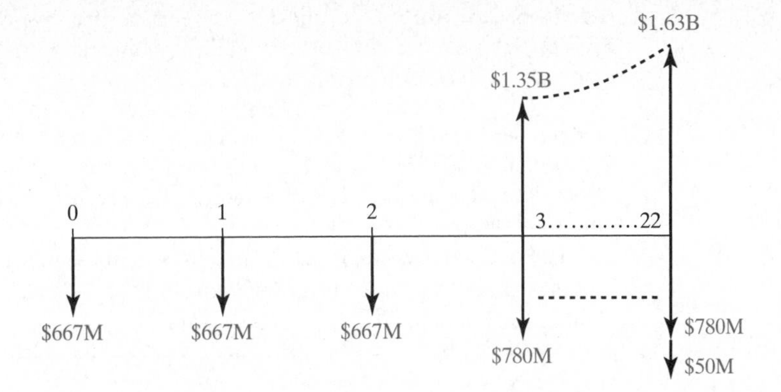

Figure 13.3 Cash flow diagram for investing in power plant at time zero.

invest in the project that initiates construction. The cash flow diagram for this option is given in Figure 13.3.

The present worth of this option is defined as

$$\text{PW}(25\%) = \underbrace{-\$666.67\text{M}(\overset{F/A,25\%,3}{3.8125})(\overset{P/F,25\%,2}{0.6400})}_{\text{Initial Investment}} + \underbrace{\$900(1.5\text{M})(\overset{P/A_1,1\%,25\%,20}{4.1081})(\overset{P/F,25\%,2}{0.6400})}_{\text{Annual Revenues}}$$

$$\underbrace{-\ \$780\text{M}(\overset{P/A,25\%,20}{3.9539})(\overset{P/F,25\%,2}{0.6400})}_{\text{Annual Costs}} - \underbrace{\$50\text{M}(\overset{P/F,25\%,22}{0.0074})}_{\text{Remediation Costs}} = -\$51.43 \text{ million.}$$

If this were the only option being considered, then the decision would be to reject.

If we consider the option to delay one period (until the end of 2005), then the cash flow diagram shown in Figure 13.3 would shift forward one period *and* the expected revenue in the first year of operation would be $900(1.01) = $909 per household. This

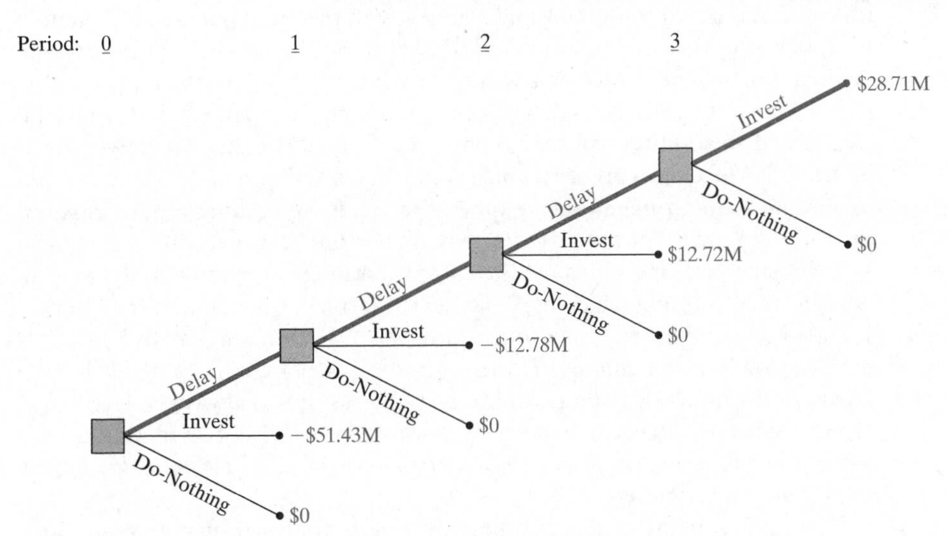

Figure 13.4 Decision network with present-worth values and optimal path.

is a critical assumption that validates the need for time-based decisions. Evaluating this decision defines a present worth of –$15.97 million in period 2005, which translates to a time-zero (2004) present worth of –$12.78 million.

The option to delay yet another period (until the end of 2006) defines the starting revenue per household at $918.09, which in turn defines a present worth (in 2004) of $12.72 million. Figure 13.4 updates our initial decision network with the present-worth values derived from the four cash flow analyses necessary to evaluate the four possible options in time.

The optimal decision is highlighted in bold in the figure, as it leads to the highest present worth. This decision, to delay for three periods and then construct the power plant in 2007, is defined by a present worth of $28.71 million.

The previous example illustrates why it might be prudent to delay an investment. By waiting a few periods, the market price of electricity improved, leading to decisions with positive present-worth values. Note that this movement in price was assumed to occur regardless of whether the plant was or was not built. Theoretically, we could have extended our analysis by delaying the investment indefinitely. The additional options that arose thereby would be modeled with additional nodes and arcs in the network, leading to additional computational burdens. These burdens must be examined in the context of how many periods the data are dynamic and require exploration.

Let us take note of what has been illustrated and, hopefully, learned. Clearly, if there is no change in the decision setting—that is, no change in any inputs required to make a decision—then we do not need to consider decision options that change with time. This is because the decision at time zero is the best decision over time. Why? If the present worth at time zero of an investment is positive and nothing changes with time, the only result obtained from delaying the investment an additional period would be a decline in the present worth. The reason is that the present worth is the same in the next period (since the cash flows do not begin until one period later), but it must be discounted an additional period to time zero. With an assumed positive rate of interest, this leads to a present-worth value that is lower than the value derived for the investment beginning at time zero. Thus, something must change in the future in order for the positive present worth to improve with a delay.

Similar reasoning holds for the case when the present worth of a project is negative. If nothing changes, delaying the project for an additional period will lead to a higher present worth, *but* the sign will remain negative and thus the decision will not change. This is because if we discount a negative value one period with a positive rate of interest, the result will always be larger, but remain negative. Hence, the decision not to invest will not change. *It is only when inputs to the cash flow diagram change over time that the delay option should be considered.*

There are many cash flow diagram inputs that can change over time. In the previous example, the market price of electricity changed with time, leading to a decision to delay investment in a power plant. These changes are often external to the company making the decision, and that company may not be able to influence a change. Some examples include demand, market prices, legislative rules, commodity prices, and input costs. Changes in these parameters point to the use of the delay option.

Note that our modeling of the delay option as a network of decisions is actually quite general. As opposed to just delaying decisions, it can be used to model phased expansions, which can be viewed as an initial investment with an additional, delayed investment. Thus, the methodology can be used to model capacity expansion decisions. We illustrate when solving the motivational example at the end of the chapter.

While the foregoing analysis is interesting, we should be wary because we are now assuming that we know *even more* about the future. That is, in the analyses before this chapter, we made assumptions to define a cash flow diagram at time zero. We are now assuming that we can define cash flow diagrams for events that will happen further in the future—and with certainty. In Example 13.1, we assumed that we could predict the change in the future prices of electricity. Since these events happen in the future, they are clearly quite uncertain. We now turn our attention to this issue in which there is uncertainty in the future. Often in these situations, delaying an investment may improve our knowledge of those uncertainties and thus lower the risk of undertaking the investment. That is, another benefit of waiting to invest is that there is time to learn about the future, either by testing or by observing.

13.2 Decision Trees

We used the term "decision network" in the previous section to represent possible decisions over time under the assumption of certainty. We will use the term **decision tree** to define the case where information is probabilistic. This is in fact the general term used in the literature. In addition to our previous definitions of decision nodes and option arcs, we define the following terms used in describing a decision tree:

- *Chance Node.* A point in time at which nature must take its course, resulting in an outcome from a set of possible outcomes represented by arcs emanating from the chance node.
- *Outcome Arc.* A result from a set of possible outcomes at a given chance node. Each arc has an associated probability.

We begin our discussion in the next section with a simple single-stage decision tree in order to introduce sequential decisions under risk over multiple periods.

13.2.1 Single-Stage Decision Tree

A single-stage decision tree can be used to visually depict the probabilistic scenario analysis defined in Section 10.7. In a single-stage tree, there is only

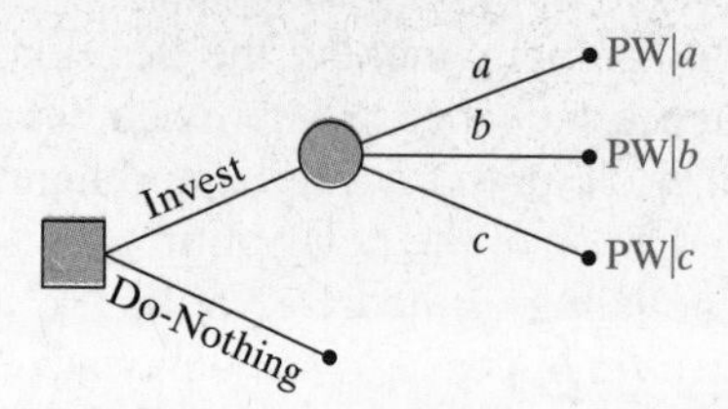

Figure 13.5
Generic representation of a single-stage decision tree.

one period in which a decision is made (hence the name "single-stage"), so the only viable decisions at time zero are either to invest or not to invest. These decisions can be represented in general by the decision tree in Figure 13.5.

The decision not to invest is equivalent to rejecting a project. The arc associated with this decision terminates at the value of $0, assuming that the do-nothing option is valued as such. The decision to invest leads to a chance node. Each arc emanating from the chance node represents possible outcomes for the variable or parameters in question, which may correspond to the pessimistic, optimistic, or average scenarios defined in Section 10.7. The value associated with each of these arcs is the present worth of the decision *if* the outcome represented by that arc were to occur.

While it is possible to analyze a decision tree with a variety of measures, we restrict our discussion to expected present worth, as that is the most straightforward and most commonly used measure. One could obviously compute additional measures of worth, such as future worth, or compute additional data, such as the variance of the present worth, but we will not address these issues here.

Solving a single-stage decision tree merely requires computing the expected present worth at each chance node and then choosing the largest present-worth option at each decision node. The expected present worth is computed backwards in time (from right to left in the tree), because the only data available at time zero are the present-worth values for each possible outcome and the probability of each outcome. This approach is termed a *rollback procedure* for the more complicated trees in the next section. We illustrate the procedure in the next example.

EXAMPLE 13.2 *Single-Stage Decision Tree*

In 2004, Cairn Energy PLC of the United Kingdom announced that it had discovered oil in its N-B-1 exploration well in northwestern India. Initial estimates placed the find between 50 million and 200 million barrels of recoverable oil.[4,5]

Make the following assumptions: Oil generates revenues of $20 per barrel in the first year of operation and is assumed to increase in price 3% per year against a

[4] Long, M., "Oil Find in India Boosts Prospects for Cairn Energy," *Dow Jones Newswires*, January 20, 2004.

[5] "Cairn Energy Saw 6,000 Barrels Cumulative Flow Rate," *Dow Jones Newswires*, February 2, 2004.

constant cost of \$8 a barrel, regardless of the amount of oil extracted. Oil is to be extracted at the rate of 10,000 barrels per day after a \$400 million investment in the field. With 360 days of production per year, we assume a continuous flow rate of 3.6 million barrels per year. The length of the project is dependent on the amount of oil in the well. The following three scenarios with respect to the size of the field have been identified and assigned probabilities based on initial testing and experience:

1. LOW: 50 million barrels with probability 0.20.
2. MEDIUM: 125 million barrels with probability 0.70.
3. HIGH: 200 million barrels with probability 0.10.

Finally, assume \$50 million in remediation costs at the end of the project. If the interest rate is 14% per year, compounded continuously, should the investment be made? Use expected present worth to make your decision.

Solution. The difference in the three possible outcomes is the amount of recoverable oil in the field. Assuming a fixed rate of extraction (3.6 million barrels per year), the length of the project varies according to the outcome. We convert the continuous cash flow from the oil revenues to an annual, discrete cash flow (see the website for converting continuous flows to discrete flows) as follows:

$$A = \bar{A}(^{F/\bar{A},14\%,1}) = \left(\frac{3.6\text{M barrels}}{\text{year}} \times \frac{\$20}{\text{barrel}}\right)\overset{F/\bar{A},14\%,1}{(1.0734)} = \$77.28 \text{ million}.$$

This is assumed to be the amount in the first year, increasing at a rate of 3% per year thereafter. For simplicity, we assume that the discrete cash flow, and not the continuous flow, increases each year. The costs associated with this flow are

$$A = \left(\frac{3.6\text{M barrels}}{\text{year}} \times \frac{\$8}{\text{barrel}}\right)\overset{F/\bar{A},14\%,1}{(1.0734)} = \$30.91 \text{ million},$$

which are assumed to remain constant in each year over the horizon.

As previously noted, the difference between the scenarios is the length of time that the well will be operated. For the LOW estimate, the length of the project is 14 years, while it lasts for 35 and 56 years, for the MEDIUM and HIGH estimates, respectively. We convert the 14% nominal rate compounded continuously to 15% per year so as to avoid confusing the factors ($e^{0.14} - 1 = .150$). Thus, we define the present worth of each possible outcome as follows:

LOW:

$$\text{PW}(15\%) = -\$400\text{M} + \$77.28\text{M}\overset{P/A_1,3\%,15\%,14}{(6.5420)} - \$30.91\text{M}\overset{P/A,15\%,14}{(5.7165)}$$
$$- \$50\text{M}\overset{P/F,15\%,14}{(0.1408)} = -\$77.68\text{M};$$

MEDIUM:

$$\text{PW}(15\%) = -\$400\text{M} + \$77.28\text{M}\overset{P/A_1,3\%,15\%,35}{(8.1385)} - \$30.91\text{M}\overset{P/A,15\%,35}{(6.6039)}$$
$$- \$50\text{M}\overset{P/F,15\%,35}{(0.0074)} = \$25.51\text{M};$$

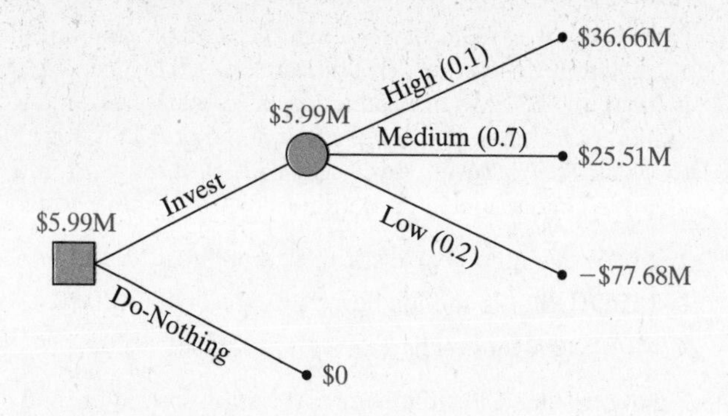

Figure 13.6 Single-stage decision tree for oil-field investment.

HIGH:

$$PW(15\%) = -\$400M + \$77.28M(\overset{P/A_1,3\%,15\%,56}{8.2954}) - \$30.91M(\overset{P/A,15\%,56}{6.6504}) - \$50M(\overset{P/F,15\%,56}{0.0004}) = \$36.66M.$$

With this information, we can construct our single-stage decision tree as depicted in Figure 13.6.

As shown in the figure, the decisions are whether to invest in the project or not to invest in it. (We ignore the delay option here, as we are only analyzing a single-stage decision tree.) The decision not to invest leads to a present worth of zero, while the decision to invest is defined by the expected value of the three possible outcomes, or

$$E(PW(15\%)) = 0.20(-\$77.68M) + 0.70(\$25.51M) + 0.10(\$36.66M) = \$5.99 \text{ million}.$$

Thus, the decision is to invest in the oil field, since it carries an expected present worth of $5.99 million, which is greater than the $0 value associated with the do-nothing alternative.

The analysis of the previous example is no different than the scenario analysis presented in Section 10.7. We have merely added a visual image to the procedure. However, this analysis is critical, as it sets the stage for analyzing more complicated decisions.

An interesting question to ask when analyzing a single-stage decision tree is "What would happen if we had perfect information?" In this situation, perfect information is the ability to see the future: You could predict the future outcome with probability one. If the uncertain future is defined by LOW, MEDIUM, and HIGH oil reserves, then perfect information would tell you with certainty which outcome will occur. We can place a value on this information, because if we know the outcome with certainty, we can choose the best decision (invest or don't invest) without apprehension. We illustrate by returning to our previous example.

EXAMPLE 13.3

Value of Perfect Information

Determine the value of perfect information for Example 13.2. Assume the same data and parameters as before.

Solution. Assuming that we know with certainty the amount of oil in the well, our decisions would be as follows:

Outcome	Decision
LOW	Reject the project.
MEDIUM	Accept the project.
HIGH	Accept the project.

These decisions each maximize present worth, given that the outcome is known with certainty. Note that the only decision which has changed from our expected-value decision (to invest) is if the LOW outcome occurs. The expected value of a decision with perfect information is

$$E\,(\text{PW}(15\%)) = \underbrace{0.20(\$0)}_{\text{LOW: Do Not Invest}} + \underbrace{0.70(\$25.51\text{M})}_{\text{MEDIUM: Invest}} + \underbrace{0.10(\$36.66\text{M})}_{\text{HIGH: Invest}}$$
$$= \$21.52 \text{ million.}$$

Thus, with an expected value of \$5.99 million, the value of perfect information is

$$\text{VPI} = \$21.52\text{M} - \$5.99\text{M} = \$15.53 \text{ million.}$$

This means that we would not be willing to pay more than \$15.53 million for perfect information.

What does the preceding discussion tell us? Since we will never have perfect information (i.e., we will never know outcomes with certainty), its value tells us the maximum we would be willing to pay if it were possible. This also caps how much we would be willing to pay for *imperfect* information, which is clearly the kind of information we are likely to receive. This knowledge will come in handy as we now move towards more complicated decisions.

13.2.2 Two-Stage Decision Tree

We noted earlier that time (delay) may allow the decision maker to reduce the risk of an investment decision. A natural question is "How can one lower the risk associated with a decision?" Recall that the definition of decision making under risk is "making a decision when the outcomes are probabilistic." In other words, the possible outcomes and their probabilities are known. In order to reduce the risk associated with these decisions, one must "move" the associated probabilities with outcomes to zero or one, or at least as close as possible to those values. That is, if the probability associated with a given outcome is 60 percent, we can reduce the risk associated with making the decision if the probability can be driven up to 80 percent or down to, say, 20 percent, because we are much more confident that these probabilities identify the true outcome.

In order to change the probability of a given outcome, we must take the time and possibly money to gather more information. This strategy is often called *mitigating risk* and is the reason that we are discussing sequential decisions, which are multistage decisions. The strategy works as follows: A decision is made to gather information. The information is gathered and assessed. From this assessment, the probabilities of future outcomes can be revised and, hopefully, "improved"; that is, the probability of events get closer to zero or one. From this point, a decision about whether to execute the decision to invest can be made with more confidence.

A two-stage decision tree is nothing more than two single-stage decision trees placed back to back. However, the probabilities associated with certain outcomes are more complicated, since they are path dependent—or conditional—on a previous set of outcomes.

A typical two-stage decision tree is given in Figure 13.7. The first-stage decision is whether to wait (and possibly gather information or test) or go ahead with the decision (invest or not invest). If we go ahead and decide to invest, our path merely leads to the single-stage decision tree defined in the previous section.

The decision to wait (test) leads to a chance node with a set of outcomes. If the decision is truly just to wait, as in letting time pass, these outcomes may represent changes in the market, such as "Market Improves," "Market Continues," or "Market Deteriorates." If the decision is to gather information, the outcomes may represent possible results, such as "Good," "Moderate," or "Negative."

Unlike the arcs in a single-stage tree, the arcs emanating from the chance node in a multistage tree do not lead directly to monetary values. Rather, the arcs lead to another set of decision nodes in which the original decision of investing or not investing is revisited. These decision arcs lead to the final set of chance nodes associated with the outcomes that ultimately define the present worth of each decision. Note that the present worth for an outcome is calculated in the same way as it was in the single-stage tree, since its value is dependent only on the outcome occurring, not its probability. If the two stages are spaced one discounting period apart, then the present-worth values at the end of the second stage will be discounted one more period than the single-stage solutions.

The difference between the single- and two-stage tree is most easily seen in the last stage. Although the present-worth values of the outcomes are the same (possibly discounted an extra period for the additional stage), the probabilities of a certain outcome occurring have changed. In the single-stage tree, there is a probability $\Pr(A)$ of event A occurring. In the two-stage tree, there are three probabilities: the probability of A occurring, conditional on the outcome

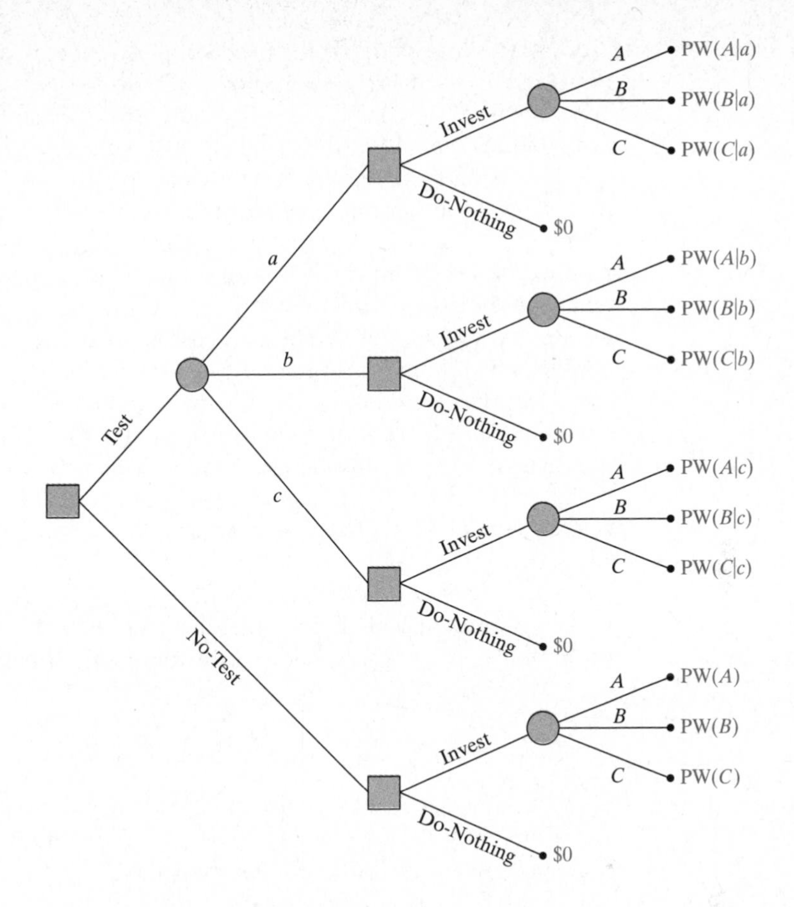

Figure 13.7
Generic representation of a two-stage decision tree.

of each of the chance nodes, or $\Pr(A|a)$, $\Pr(A|b)$, and $\Pr(A|c)$.[6] We denote first-stage outcomes with lowercase letters and second-stage outcomes with uppercase letters. In the single-stage tree, the probabilities are $\Pr(A)$, $\Pr(B)$, and $\Pr(C)$, but in the two-stage tree, each of these probabilities is conditional in that $\Pr(A)$ has been supplanted by $\Pr(A|a)$, $\Pr(A|b)$, and $\Pr(A|c)$. The hope is that the conditional probabilities provide more information than the original probabilities. If the conditional probabilities are closer to the values of 0 or 1, then the information is more certain than before. This is what is meant by reducing the risk inherent in a project. We describe how to calculate these probabilities and solve a decision tree in the next two subsections.

13.2.3 Calculating Conditional Probabilities

Before delving into the issue of computing conditional probabilities, it is helpful to take note of what is known at the start of the decision-making

[6] The term $\Pr(A|a)$ is the probability of A occurring, given that a has already occurred. It is read as "the probability of A, given a."

process. It is assumed that the probabilities of outcomes A, B, and C, namely, $\Pr(A)$, $\Pr(B)$, and $\Pr(C)$, are known. These are the same probabilities that are known in the single-stage decision tree. They also represent the *only* probabilities that are needed in a single-stage decision tree.

The second source of uncertainty comes in the first-stage of the two-stage tree, when awaiting the results of the first decision, such as the outcome of a survey or a sample or just the passage of time. If these results are to be helpful, we must be provided with some information as to how the outputs from the first stage affect the outputs of the second stage. For example, if we are taking a survey of expected customers in order to better gauge future demand, then some information must be made available that can connect these two uncertainties (survey outcome and expected demand). Clearly, we would like to know the probability of demand, given the outcome of a survey. Unfortunately, the *opposite* information is generally provided by the survey maker, who generally defines the probability of a survey outcome, given a certain demand level. That is, we require $\Pr(A|a)$, while we are generally given $\Pr(a|A)$.

Thus, we are often required to make use of Bayes' Theorem, which states that a conditional probability is merely the joint probability (the probability of both events occurring), divided by the marginal probability of an outcome, or

$$\Pr(A|a) = \frac{\Pr(A, a)}{\Pr(a)}.$$

Unfortunately, the only information that we have available at time zero is the probability of outcome A, $\Pr(A)$, and the conditional probability of the first-stage outcome, $\Pr(a|A)$. Thus, we make use of Bayes' Theorem as follows:

$$\Pr(A, a) = \Pr(A|a)\Pr(a).$$

Equivalently, we can write the joint probability as

$$\Pr(A, a) = \Pr(a, A) = \Pr(a|A)\Pr(A).$$

The values of $\Pr(a|A)$ and $\Pr(A)$ are known; thus, the joint probability $\Pr(a, A)$ can be calculated. This leaves only the calculation of the marginal probability $\Pr(a)$ to derive our desired value of $\Pr(A|a)$. That value can be gleaned from the fact that the marginal probability is merely the sum of all joint probabilities that include the outcome a. For our small example, this is

$$\Pr(a) = \Pr(A, a) + \Pr(B, a) + \Pr(C, a).$$

Given the joint probability and the marginal probability for each possible survey outcome, we can now populate the decision tree with probabilities on each outcome arc. We illustrate the calculation of these probabilities in the next example.

EXAMPLE 13.4

Two-Stage Decision-Tree Conditional Probabilities

We continue the analysis of the oil-well investment from the previous examples. After the initial discovery, Cairn Energy announced that it would dig an appraisal well and perform additional seismic testing to better understand the amount of reserves that the well holds. Assume that the additional testing can be characterized as either "Favorable," "Unfavorable," or "Inconclusive." (We occasionally refer to these as "Fav," "Unf," and "Inc," respectively, to conserve space.) More importantly, we assume that we know the conditional probabilities as defined in Table 13.1. The table is meant to be read as the probability of a testing result given the amount of oil in the field. For example, $\Pr(\text{Fav}|\text{LOW}) = 0.05$ and $\Pr(\text{Fav}|\text{HIGH}) = 0.80$. Note that the sum down each of the columns in the table is one, as it encompasses all of the possible test results for a given outcome.

Solution. Given the conditional probabilities in Table 13.1 and recalling that $\Pr(\text{LOW}) = 0.20$, $\Pr(\text{MED}) = 0.70$, and $\Pr(\text{HIGH}) = 0.10$, we first calculate the joint probabilities of each test result and outcome. For a Favorable test and LOW oil reserves,

$$\Pr(\text{Fav,LOW}) = \Pr(\text{Fav}|\text{LOW})\Pr(\text{LOW}) = (0.05)(0.20) = 0.01.$$

This approach can be used to determine all of the joint probabilities as given in Table 13.2.

The marginal probabilities of each possible test result can now be derived as follows:

$$\begin{aligned}\Pr(\text{Fav}) &= \Pr(\text{Fav,LOW}) + \Pr(\text{Fav,MEDIUM}) + \Pr(\text{Fav,HIGH}) \\ &= 0.01 + 0.175 + 0.08 = 0.265.\end{aligned}$$

TABLE 13.1

Conditional probabilities of test results, given the size of the oil field.

	Oil Reserves		
Test Result	LOW	MEDIUM	HIGH
Favorable	0.05	0.25	0.80
Inconclusive	0.25	0.50	0.15
Unfavorable	0.70	0.25	0.05

TABLE 13.2

Joint and marginal probabilities for testing size of oil field.

	Oil Reserves			
Test Result	LOW	MEDIUM	HIGH	Pr(Test Result)
Favorable	0.01	0.175	0.08	0.265
Inconclusive	0.05	0.35	0.015	0.415
Unfavorable	0.14	0.175	0.005	0.32

TABLE 13.3 **Conditional probabilities of oil field reserves, given test results.**

Oil Reserves	Test Result Favorable	Inconclusive	Unfavorable
LOW	0.0377	0.1205	0.4375
MEDIUM	0.6604	0.8434	0.5469
HIGH	0.3019	0.0361	0.0156

The final conditional probabilities needed are those for the size of the oil reserves, given a test result—for example,

$$\Pr(\text{LOW}|\text{Fav}) = \frac{\Pr(\text{LOW, Fav})}{\Pr(\text{Fav})} = \frac{0.01}{0.265} = 0.0377.$$

The complete list of conditional probabilities is given in Table 13.3. Note that the columns again sum to one as the probabilities are conditioned on the test outcome. With this information, we are prepared to populate the two-stage decision tree and perform the rollback procedure as outlined in the next section.

13.2.4 Rollback Solution Procedure

Once the tree has been populated with probabilities on each arc emanating from a chance node and the present worth of each possible outcome has been calculated, we can begin our rollback procedure for determining the best path of decisions. The procedure is actually the same as that for the single-stage decision tree, only it is repeated for each stage.

The expected value for a given set of outcomes is calculated at each chance node, beginning with the chance nodes at the far right of the decision tree. Once these are computed, we determine the best decision by finding the maximum expected present worth from the decision nodes connected to

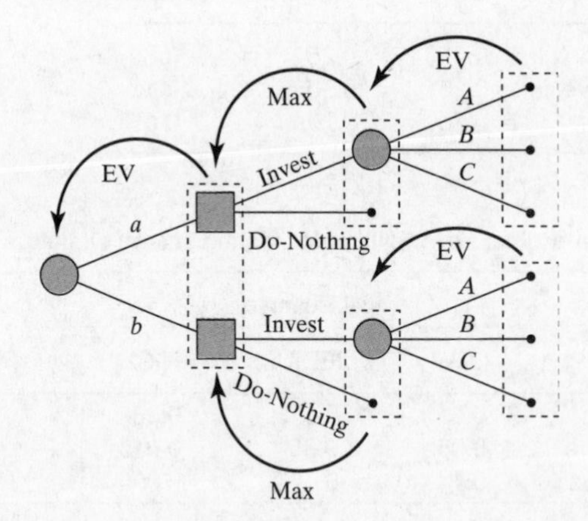

Figure 13.8 Rollback solution procedure for two-stage decision tree.

each of the chance nodes previously analyzed. This procedure is depicted in Figure 13.8. We then continue to roll back through time, calculating expected values at chance nodes and selecting the decision with the maximum present worth at each decision node until we reach the original decision node in the first stage. The choice at this node defines the initial decision of our optimal policy. The expected worth of that decision is the expected value of the present worth that has been "dragged" through the tree in the rollback procedure. We can follow the tree, depending on how nature takes its course, to determine the best course of action over time. In the next example, we illustrate this procedure by continuing the analysis from the previous examples.

EXAMPLE 13.5 *Two-Stage Decision-Tree Rollback Procedure*

We combine the results from our last two examples to complete our two-stage decision tree. Assume that it will cost $5 million to complete testing procedures and that testing and analysis takes one year.

Solution. The complete two-stage decision tree is given in Figure 13.9. The initial decision is regarding whether to perform the tests. If the tests are not performed, the option arc leads to a subtree that was analyzed in our single-stage analysis. The marginal probabilities of the LOW, MEDIUM, and HIGH oil reserves are given here, together with the associated present worth of an invest or do-not-invest decision. Note that the values in this part of the tree have not changed from what they were in the single-stage analysis.

If the decision to perform the tests, at the cost of $5 million, is taken, then the tree greatly expands from the single-stage tree. The first chance node encountered defines the outcome of the test, as either Favorable (Fav), Inconclusive (Inc), or Unfavorable (Unf). Let us assume that we follow the Favorable path, which leads to our decision of whether to invest or not to invest in the field, just as in the single-stage decision tree. If we choose not to invest, a present worth of zero is returned. If we choose to invest, the outcomes of LOW (L), MEDIUM (M), and HIGH (H) are now dependent on the Favorable outcome of the test. Thus, the conditional probabilities are placed on these arcs. The present worth for each outcome is as it was in the original example, except that it is discounted an additional period for the time to test.

To compute the expected value of this decision, we merely take the present worth for each outcome, multiplied by its conditional probability. Continuing to use the Favorable branch as an example, we obtain

$$E\,(\text{PW}(15\%)|\text{Fav}) = \Pr(\text{LOW}|\text{Fav})\text{PW}_{\text{LOW}}(15\%) + \Pr(\text{MED}|\text{Fav})\text{PW}_{\text{MED}}(15\%) + \Pr(\text{HIGH}|\text{Fav})\text{PW}_{\text{HIGH}}(15\%).$$

Plugging in our conditional probabilities and present-worth values (assuming the 15% discount rate), we get

$$\begin{aligned} E\,(\text{PW}|\text{Fav}) &= \max(0.0377(-\$67.55\text{M}) + 0.6604(\$22.18\text{M}) + 0.3019(\$31.88\text{M}), \$0) \\ &= \$21.73\text{M}, \end{aligned}$$

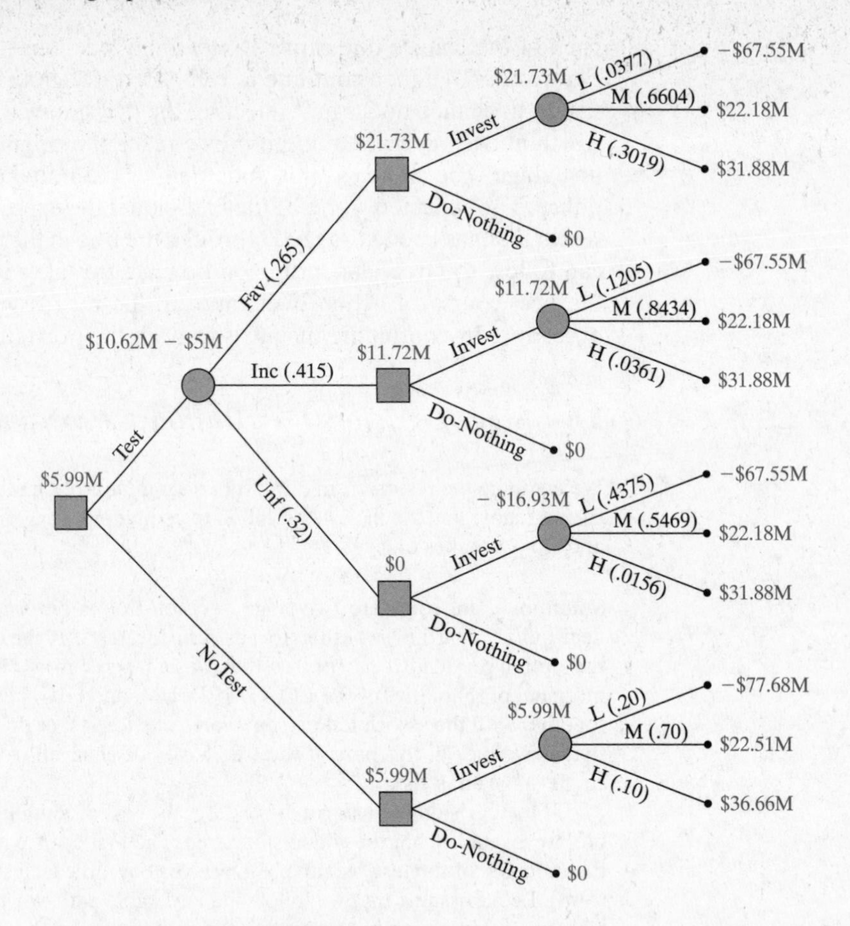

Figure 13.9 Complete two-stage decision tree for oil-field investment.

$$E\,(\text{PW}|\text{Inc}) = \max\,(0.1205(-\$67.55\text{M}) + 0.8434(\$22.18\text{M}) + 0.0361(\$31.88\text{M}),\ \$0)$$
$$= \$11.72\text{M},$$
$$E\,(\text{PW}|\text{Unf}) = \max\,(0.4375(-\$67.55\text{M}) + 0.5469(\$22.18\text{M}) + 0.0156(\$31.88\text{M}),\ \$0)$$
$$= \$0.$$

Note that the previous present-worth values for LOW, MED, and HIGH have been discounted at 15% for the testing period (one year). The respective expected values are marked on each node in the decision tree. Once the expected values have been computed for each possible test result, the expected value of the decision to perform the test can be computed as follows:

$$E\,(\text{PW}(15\%)) = \Pr(\text{Fav})\text{PW}_{\text{Fav}}(15\%) + \Pr(\text{Inc})\text{PW}_{\text{Inc}}(15\%) + \Pr(\text{Unf})\text{PW}_{\text{Unf}}(15\%).$$

Again, substituting the appropriate values results in

$$E\,(\text{PW}(15\%)) = 0.265(\$21.73\text{M}) + 0.415(\$11.72\text{M}) + 0.32(\$0\text{M}) = \$10.62 \text{ million}.$$

Note that our decision was not to invest in the project if the test result was unfavorable. Because the $5 million testing fee must also be subtracted from this option, the value of the decision to test is only $5.62 million.

The optimal decision at time zero is thus

$$E\,(\text{PW}(15\%)) = \max\,(\$5.62\text{M}, \$5.99\text{M}) = \$5.99 \text{ million.}$$

The decision associated with the $5.99 million is not to test, which was calculated in the single-stage decision tree. Thus, it is best to explore the field without further testing, as the testing does not provide sufficient information to improve our decision making for its cost (although it is a close decision).

We have now defined a two-stage decision tree in which we may take some action, such as a survey or test, in order to gather more information about the future. Previously, we assumed that we knew the cost of this first stage in the process. A question that may be of interest is the value of the information. Put another way, what would we be willing to pay for the information, given its possible influence on future decisions? This is essentially a break-even analysis, because we clearly do not want to pay more than the expected benefit of gathering information.

The value of the project without the testing was determined to be $5.99 million, and the value with the test (excluding the price of the test) was $10.62 million. Thus, the most we would be willing to pay for testing is $10.62 million – $5.99 million = $4.63 million. As expected, this is less than the value of perfect information defined earlier ($15.53 million).

13.2.5 Multistage Decision Trees

There is no limit to the number of stages that can be evaluated in decision-tree analysis. For the oil-well example that we analyzed in the previous section, it would seem quite likely that further testing is needed. If this testing led to the conclusion that there was a large amount (200 million barrels) of oil in the ground, that might be enough incentive to increase the speed of production such that the oil can be extracted in 10 to 20 years, as opposed to the 56 years calculated in the analysis. This decision to speed up mineral extraction projects with additional wells or mines was discussed in Section 9.2.1. These consecutive decisions can also be modeled in multistage trees. Figure 13.10 illustrates the decisions involved in a three-stage tree.

A number of other examples fit into this sequential decision-making framework:

- *Research and development.* A company will often fund research in stages (called R&D). If the R&D progresses, more money is released for future testing. Stages are often defined by the testing and evaluation of prototypes.

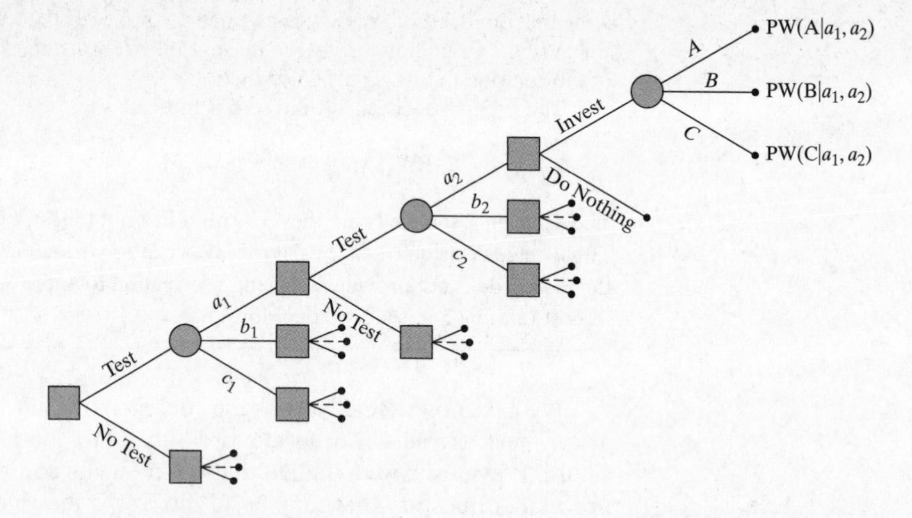

Figure 13.10 Three-stage mineral extraction decision tree.

- *Pharmaceutical testing.* An example of R&D is the development of new drugs. In addition to undergoing rigorous laboratory testing and design, all drugs must be approved by the Food and Drug Administration.
- *Regulated development.* The development of drugs is but one example of a regulated industry that requires testing. The development of aircraft also requires a rigorous sequence of testing before full-scale investment is pursued.
- *License procurement.* Often, an entity must secure the rights to something before it progresses on an investment. For example, a mining company must secure land for development and a cellular phone company must purchase a license to operate in a given area. The respective items are often secured through bidding processes. Multistage decision-tree analyses often aid in determining bidding prices for these initial investments, which are frequently followed up with additional testing (seismic or market survey testing for our examples) before full-scale investment proceeds.
- *Phased expansions.* A company may decide to procure land for a large facility, but bring capacity online in phases, as in our introductory real decision problems or the example of expanding oil-field production. Phased expansion reduces the risk of building too much capacity. Realizing demand and prices over time aids in making the decision on timing the next expansion.

The method described for the two-stage decision tree is the same for any multiple-stage problem and can be applied in those situations as well. It should be clear that the number of stages, the number of decisions at each stage, and the number of possible outcomes at each chance node contribute greatly to the difficulty in solving a decision tree. The difficulty lies in the calculation of the conditional probabilities, as they can be based on a number of factors over many periods.

13.3 Real Options

There has been a growing body of research in both engineering economy and finance concerning the application of financial option pricing theory to the analysis of capital investments. A financial option is a contract that can be purchased. If purchased, the contract gives the owner an *option* to take some later action defined by the contract. The arrangement is termed an "option" because the owner has the right, but is not obliged, to make the subsequent investment.

For example, a call option on a stock gives the owner the right, but not the obligation, to purchase a number of shares of a given stock for a given price by (or on) a given date. Assume that we buy a call option which allows for the purchase of stock at $20 per share by some date. If the stock price rises above $20, termed the **strike price**, by the due date, then we would exercise the option and purchase the stock for $20 per share. If nothing else, the stock could be sold immediately for a profit, as the current price would be higher than $20 per share (ignoring, of course, the fact that we paid for the option.) In this situation, the option is said to be "in the money." If the stock price dips below $20 per share by the due date, the option is not to be exercised, as we could purchase the stock cheaper directly from the market. In this situation, the option is "out of the money." There is a variety of options, which vary according to parameters such as when the option can be exercised. Options can also be used for other types of transactions; for example, "put" options give the holder the right to *sell* shares of a stock at a given price. Obviously, the put option is appealing if it is believed that the stock price will go down with time.

Option pricing theory determines the price, or value, of the option—essentially, the cost of the contract. The price of the option is dependent on the current stock price, maturity date, and expected movement in the stock price over the duration of the contract. The application of financial option theory to capital investments is generally termed *real options* analysis.

Although techniques for pricing options can be quite sophisticated, we can model the value of the option in a simple manner by using our two-stage decision tree, as shown in Figure 13.11. The decision at time zero is whether to purchase the option. After some period of time, the stock price is revealed. If the stock price is greater than the strike price, then a positive return is generated. Otherwise, no return is generated. Given possible stock prices with associated probabilities, a value for the option can be computed just as we analyzed the value of a test in our two-stage decision tree.

There are many applications where the notion of an option is direct, such as the following:

- Black Stone Minerals Company LP purchased royalty interests and subsurface mineral rights on about 3.3 million acres in the southern United States from Pure Resources, a subsidiary of Unocal, for $190

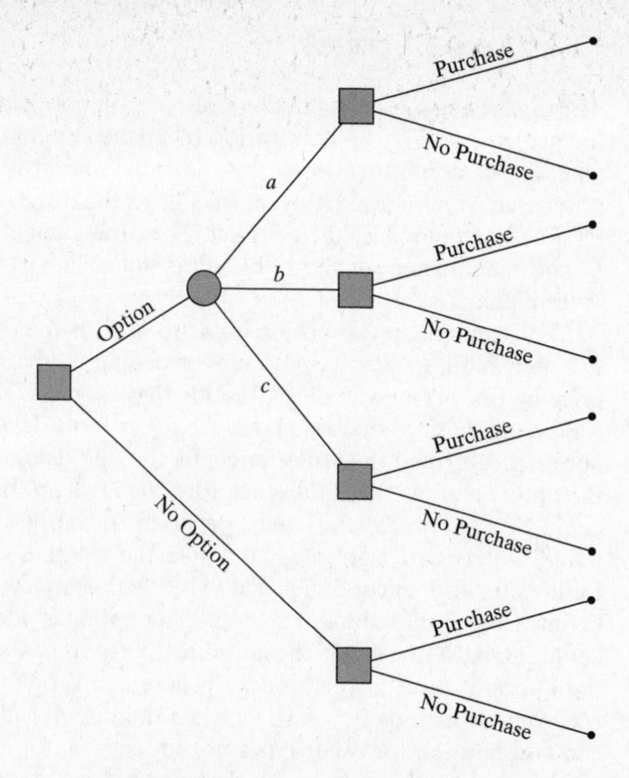

Figure 13.11 Two-stage decision tree for pricing an option.

million in spring of 2004.[7] The purchase gives Black Stone the right (option) to invest and harvest the resources underneath the property.

- Thirty-three telecommunications companies showed interest in bidding for Pakistan's cellular licenses in early 2004.[8] The licenses give the operators the right (option) to develop networks and sell services in the area.
- BHP Billiton, Ltd., was the highest bidder for exploration leases of 32 blocks (areas) in the Gulf of Mexico. The leases were valued at over $500,000 per block and give BHP the right to explore for natural resources in the block.[9]

In these situations, a company can purchase (or has purchased) the right to carry out some investment, including drilling for oil and offering cellular phone services. In these cases, the licenses, leases, patents, or rights can be viewed as options, as they provide the ability to make a subsequent investment.

[7] Derpinghaus, T., "Unocal Unit to Sell Fee Mineral Interests to Black Stone Minerals," *Dow Jones Newswires*, March 11, 2004.

[8] Anwar, H., "Big Players to Bid for Pakistan's Two Cellular Licenses," *Dow Jones Newswires*, January 22, 2004.

[9] Johnston, E., "BHP Ramps Up Gulf of Mexico Push with New Leases," *Dow Jones Newswires*, March 19, 2004.

Other applications are not quite as direct. Consider, for example, the situation in which there is value in delaying an investment. If we have an estimate of the investment cost and also an estimate of the return on the project (all cash flows after the initial investment), then we can apply options theory. The investment cost is considered the exercise (strike) price, because it is known and represents the cost of executing the investment. The value of the returns is what changes with time (we noted that the delay option should be considered if parameters change with time) and thus is analogous to the actual stock price. If the actual price is greater than the exercise price, then the option is in the money and should be pursued. If we have a measure of the project's volatility (uncertainty in cash flows), we can use the options methodology to price the option, a strategy that is equivalent to determining the value in delaying the investment. Again, we can apply this method of thinking in any situation where we can delay an investment, such as a phased expansion and abandoning an asset. (Of course, there is debate regarding using this methodology in this context, as it is unclear whether all of the required assumptions hold.)

It should also be noted that it has become quite common for firms to secure "options" on orders, especially with purchases of large items such as aircraft. Consider the following examples:

- Canada's Seaspan placed a $688 million order with South Korea's Samsung Heavy Industries Co. to build eight 9600 twenty-foot equivalent unit (TEU) container ships (a ship that can hold 9600, 20-foot long containers). The order also included an option to build four additional 9600 TEU vessels.[10]
- Korean Air Co. signed an agreement to purchase five A380 superjumbo jets from Airbus for $270 million each, with the option to purchase an additional three at a later time.[11]
- Bombardier Transportation received orders for an additional 120 M-7 electric commuter rail cars from Metropolitan Transportation Authority/Metro-North Railroad in New York. The order was the result of exercising options valued at C$280 million. After this order, placed in April of 2004, 288 options remained open, while a total of 978 firm orders for railcars had been placed.[12]

In these examples, the purchaser has negotiated the ability to purchase additional assets at a later date at a given price, presumably the same as in the original order. An interesting issue is the price of this option, which can

[10] Kim, Y.-H., "S Korea Samsung Heavy Gets $688M Order from Canada," *Dow Jones Newswires,* November 17, 2003.

[11] Chang, S., "Korean Air in Airbus Deal to Buy up to 8 Aircraft," *Dow Jones Newswires,* October 23, 2003.

[12] Li, J., "Bombardier to Produce 120 Additional M-7 Commuter Rail Cars for MTA/Metro-North Railroad in New York," *Dow Jones Newswires,* April 22, 2004.

be modeled as a two-stage decision tree. The first stage is the purchase of the option for an unknown price. In the second stage, the market price of the asset is revealed. The decision is whether to invest (purchase the option). Clearly, the option will be worth the change in the market price over time, *assuming that it rises*, discounted for the one period of interest. If the price falls, then the firm will not want to exercise the option. This analysis is similar to determining the value of a test or experiment in the two-stage decision tree. However, we are not performing a test in this instance, but merely allowing time to progress, much as we did in our delay decision networks. If we can estimate the possible movements in the market price of the asset (with probabilities), then the value of the option can be computed.

13.4 Examining the Real Decision Problems

We return to our introductory real decision problems of Bayer expanding production capacity for a high-tech plastic and the questions posed there:

1. What expansion options should be considered at time zero? (Assume capacity increments of 50,000 tons.)

   As demand grows from zero to 200,000 tons per year, it seems reasonable to analyze investments of size zero through 200,000 tons in increments of 50,000. This results in five decisions to analyze at time zero.

2. As demand grows to 200,000 tons per year, what expansion decisions should be considered over the next four years? Draw the cash flow diagram for one of these paths, assuming a 10-year horizon with no salvage value.

   A path of expansions can be taken such that the total capacity built does not exceed 200,000 pounds per year. For example, one could build 200,000 tons of capacity at time zero, or one could build 50,000 tons of capacity in each of the first four periods. The cash flow diagram for delaying one year and then incrementally adding 50,000 tons of capacity in each ensuing year is given in Figure 13.12. Note that capacity trails demand in the first four periods.

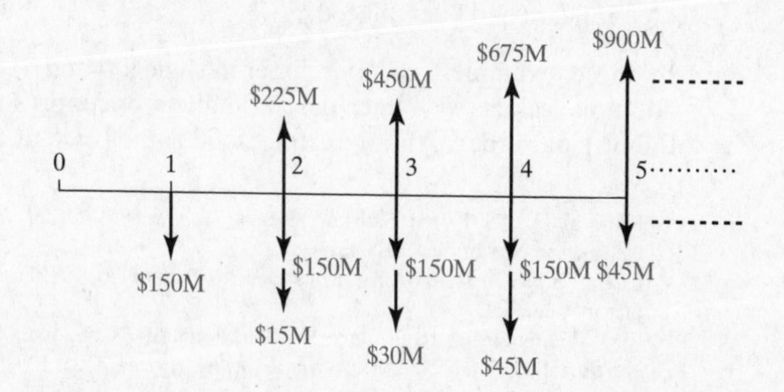

Figure 13.12 Delayed and phased expansion of plastic production.

3. What probabilistic information should be incorporated to improve decision making? How does this information alter the decision network?

It is clear that expansions should take place if demand is realized. Thus, demand should be treated as probabilistic. Furthermore, because demand in one period is likely to be dependent on demand in the preceding period, conditional probabilities defining demand in one period, given last period's demand, would be extremely helpful in guiding the decisions. Unfortunately, incorporating this information can lead to a large network that must be analyzed.

13.5 Key Points

- The decision to delay should be considered when external factors, such as demand or prices, may change with time.
- A decision network represents possible dynamic decisions made over time, considering the case of certainty. Decision nodes represent periods when a decision must be made from the set of choices, modeled as option arcs.
- The value of perfect information is the difference between the return that would have been generated with perfect information about the future and the return generated with expected values for all parameters.
- Decision trees are similar to decision networks, but they allow probabilistic information to be included. Two-stage decision trees allow for the modeling of sequential decisions, such as whether an investment should be made after a survey has been taken or a test has been executed.
- Real options analyze investments for which delay is a viable alternative. The underlying theory comes from the field of financial options pricing.

13.6 Further Reading

Park, C.S., and H.S.B. Herath, "Exploiting Uncertainty—Investment Opportunities as Real Options: A New Way of Thinking in Engineering Economics," *The Engineering Economist,* 45(1):1–36, 2000.

Park, C.S., and G.P. Sharp-Bette, *Advanced Engineering Economics.* John Wiley and Sons, New York, 1990.

Marshall, K.T., and R.M. Oliver, *Decision Making and Forecasting.* McGraw-Hill, New York, 1995.

Canada, J.R., W.G. Sullivan, and J.A. White, *Capital Investment Analysis for Engineering and Management.* Prentice Hall, Upper Saddle River, New Jersey, 1996.

Thuesen, G.J., and W.J. Fabrycky, *Engineering Economy,* 9th ed. Prentice Hall, Upper Saddle River, New Jersey, 2001.

Trigeorgis, L. (Ed.) *Real Options in Capital Investments: Models, Strategies, and Applications.* Praeger, Westport, Connecticut, 1995.

13.7 Questions and Problems

13.7.1 Concept Questions

1. Why do we distinguish between decision networks and decision trees? What is the difference?
2. When should dynamic options, including the delay option, be considered?
3. If it is unlikely that external or internal factors will change over time, should the delay option be considered? Why?
4. What is the value in delaying an investment?
5. Are there cases when we should evaluate accelerating (as opposed to delaying) an investment? How would accelerating a project be related (mathematically) to delaying a project?
6. What is the relationship between scenario analysis and a single-stage decision tree?
7. Define the value of perfect information. What does this mean? What does it bound?
8. What is the purpose of most two-stage decision analyses? Provide two examples.
9. What complicates two-stage decision-tree solutions?
10. Can the two-stage methodology be extended to multiple stages? What data must be available?
11. Why is the term "real options" used in studying dynamic investment options? What is the relationship of real options analysis to financial options analysis?
12. Why would a company purchase an option to buy equipment at a given price by a given time? Devise a decision-tree analysis to "price" or value the option.

13.7.2 Drill Problems

1. The price of a barrel of oil is currently $45 and is expected to rise 4% per period for the foreseeable future. Examine the option to delay the following investment for up to two years: The company wishes to invest $100 million to explore an oil field with expected reserves of 20 million barrels. The cost to extract a barrel is assumed to be $25 at a rate of 1 million barrels per year. Fixed O&M costs are $5 million per year, there is no salvage value, the horizon is 20 years, and the MARR is 25% per year.
2. Does the answer to the previous problem remain the same if the price is expected to decline 4% per year?
3. An investment of $2.5 million has three possible outcomes based on demand:

- Good: Annual returns of $700,000 over 12 years with no salvage value.
- Moderate: Annual returns of $500,000 over 10 years with $100,000 salvage value.
- Poor: Annual returns of $400,000 over 8 years with $300,000 salvage value.

Assuming that the probabilities of the good, moderate, and poor returns are 0.25, 0.60, and 0.15, respectively, should the investment be considered? Draw a single-stage decision tree, and compute the expected present worth (at an 18% annual interest rate) to make a decision.

4. Consider again the previous problem, but assume that the returns are dependent on the selling price of a commodity. (Assume that the prices are $0.08 (good), $0.06 (moderate), and $0.05 (poor) per unit, against costs of $0.01 per unit on the sale of 10 million units.) Should a delay option be considered? Explain.

5. Draw a decision tree for the previous problem, assuming the investment can be delayed for zero, one, or two periods at time zero. After one period, the probabilities are 0.30, 0.60, and 0.10, for respective prices of $0.08, $0.06, and $0.05 per unit. After another period, they are 0.35, 0.60, and 0.05, respectively. (Note that the prices in each period are independent of those in other periods.) Should the delay option be exercised? If so, for how long?

6. In the previous problem, how much are you willing to pay for the probabilistic information about the future (in each period)?

7. A company is considering launching a new product. The possible outcomes that have been identified are as follows:

- Successful: Annual sales of 25 million units resulting in a present worth of $250 million.
- Moderate: Annual sales of 10 million units resulting in a present worth of $10 million.
- Dismal: Annual sales of 2 million units resulting in a present worth of –$10 million.

On the basis of information about other products, probabilities of 0.20, 0.40, and 0.30, respectively, have been estimated for the three outcomes. Should the project be pursued?

8. Consider the previous problem again. What is the value of perfect information? Assume that a market survey can be taken in order to better gauge future demand. The following information is given by the marketing department:

	Product Launch		
Test Result	Successful	Moderate	Dismal
Favorable	0.65	0.30	0.10
Inconclusive	0.30	0.50	0.10
Unfavorable	0.05	0.20	0.80

What is the best decision path, given this information? How much are you willing to pay for the information?

9. Bids are being taken on plots of land that are believed to contain oil. The following estimates of the size of the field have been provided to all companies:

- Large (0.20): 150 million barrels.
- Medium (0.40): 75 million barrels.
- Small (0.20): 25 million barrels.

Assume that an investment (excluding the bid) would require $300 million and extract 5 million barrels of oil per year for the length of the life of the field. Remediation costs would be $200 million at the end of the project, regardless of its length. Revenues are expected to be $30 per barrel against costs of $14 per barrel and annual fixed costs of $5 million per year. Assume an interest rate of 18%. If bidding were free, should you make the investment? What would you be willing to pay for the bid?

10. Consider the previous problem again. What is the value of perfect information? Construct a two-stage decision tree under the assumption of unknown conditional probabilities. Given this information, continue your analysis and construct a test (use the definition of conditional probabilities) that eases your worries about the investment. Are the probabilities feasible?

13.7.3 Application Problems

1. The California Energy Commission allowed Valero Energy Corp. to postpone bringing its 51-MW power plant at its refinery online from December 31, 2002, until November 1, 2005, due to depressed electricity prices. The plant is expected to cost $100 million to construct.[13]

Make the following assumptions: The cost of building the plant is $100 million and it takes one year to construct. The plant can produce 51 MW of power, which it supplies to the refinery for $15 million per year in operating and maintenance costs. The plant has a 10-year life with no salvage value, and the interest rate is 14% per year. However, the plant can purchase the same energy from the state's electrical grid for $25 million a year.

(a) Should the plant be constructed, or should electricity be purchased from the grid? Assume that the cost of using the grid is constant with time.

(b) Given your answer in part (a), should the investment ever be considered? What must change in order for the investment to be considered? Would this motivate a delay option?

(c) Assume that the price of acquiring electricity from the grid is expected to rise $5 million per year after the first year. Evaluate the options of building the plant right away or delaying for up to three periods. What is the best decision? What is the present worth of the decision?

(d) Assume that the price can move probabilistically and either increase $3 million or $5 million in each period, with 0.50 probability for either move. Evaluate the same options as in the previous problem. What is the minimum-expected-cost decision?

2. Renesas Technology, Corp., increased its large-capacity flash memory production from 9000 wafers per month to 12,000 wafers in the fall of 2004 at the cost of ¥33

[13] Berthold, J., "Valero Can Move Online Date of 51MW Calif Unit to Nov '05," *Dow Jones Newswires,* October 22, 2003.

billion. The expansion allowed for the annual production of 3 million flash memory chips used in digital consumer products such as handheld camcorders.[14]

Make the following assumptions: The expansion from 9000 wafers to 12,000 takes one quarter and that one wafer generates revenues of $500 against costs of $200. Assume further that periodic (quarterly) costs of $1 million increase to $1.25 million after the expansion. If time zero is the end of the fall of 2003 and current plans are to expand during the summer of 2004 (after the winter and spring of that year), with capacity coming online at the beginning of the fall of 2004, evaluate accelerating the expansion to occur either at time zero or after the winter of 2004. Regardless of when the expansion occurs, production is expected to continue at capacity through the end of 2006, with a salvage value of the additional capacity of $5 million at that time. The cost of the expansion is $25 million. The quarterly MARR is 5%.

(a) Assume that the $500 revenue per wafer is for each quarter in years 2005 and beyond. Revenues of $525, $520, $515, and $510 are available in the four consecutive quarters of 2004, as competition also increases capacity. Should the expansion be moved up (accelerated)? If so, when?

(b) Assume that revenues are highly dependent on the competition, such that they have a 0.60 probability of being $520 in the first period and settling to $510 for the remainder of the first year, or a 0.40 probability of being $510 per period for the first year (and then returning to $500 as previously assumed). Does this information change the decision you arrived at in part (a)?

3. Infineon Technologies North America Corp. plans to spend $9 million to expand its microchip design center in Vermont. The cost will cover adding 10,000 square feet to its plant and hiring 30 additional workers to research memory in cell phones, personal digital assistants (PDAs), and digital cameras.[15]

(a) If the annual cost per employee is $150,000 (salary plus burden) and all other costs are $10 per square foot per year, what must the annual contribution (in dollars) of the group be in order to facilitate the expansion, assuming a 15% annual interest rate and a five-year horizon?

(b) Assume the answer to the previous problem is $7.5 million, such that each employee contributes about $250,000 in revenue. Using this number and the previous data, what is the present worth of adding 10 people in each of the first three periods of the horizon. Note that the $9 million spent on the building is still spent at time zero. What is the new break-even revenue? When might staggering hiring be a viable strategy to consider?

4. A 500 million-cubic-meter coalfield was discovered in Inner Mongolia, with seismographic and geological data showing a 20-km-long, 5-km-wide coal seam. A mine was constructed in 2004 at the cost of YUAN18 million to extract 250,000 tons of coal a year.[16]

---

[14] Yamada, M., "Japan's Renesas to Invest Y33B in Flash Memory Unit," *Dow Jones Newswires*, December 25, 2003.

[15] Associated Press, "Infineon Plans Expansion in Vermont," *Dow Jones Newswires*, December 14, 2003.

[16] Gan, T. (ed.), "Massive Coal Deposit Found in Northern China—Report," *Dow Jones Newswires,* February 6, 2004.

Make the following assumptions: Initial studies showed that there was a 0.50 probability of a coal deposit of 300 million tons, a 0.20 probability of 100 million tons, and a 0.30 probability of 500 million tons. Assume that expected revenues are YUAN250 per ton of coal, with per unit extraction costs of YUAN200 per ton. Fixed costs of YUAN5 million per year are expected through the life of the mine. Remediation costs are assumed to be YUAN20 million. The interest rate is 25% per year.

(a) Using a single-stage analysis, should the mine be built?

(b) What is the value of perfect information?

Suppose the investment had been YUAN1.7 billion to produce 25 million tons of coal per year, with annual costs of YUAN500 million. The remediation costs are YUAN200 million in this situation.

(c) Using the same scenarios from (a) and a single-stage analysis, should the mine be built?

(d) What is the value of perfect information?

Consider again the larger investment. Seismographic and geological testing can be performed to learn more about the mine. Assume that the information in Table 13.4 is known before testing. Given the 0.50 probability of medium, 0.20 probability of low, and 0.30 probability of high amounts of coal, and given the testing, answer the following questions:

(e) If the testing is free, should it be done? Explain.

(f) What is the expected value of this decision?

(g) What is the maximum amount that you would be willing to pay for the seismic tests? How does this amount compare with the value of perfect information?

(h) If the tests come back as unfavorable, what should be done? What if the tests come back as inconclusive?

5. Pulp and paper producer Celulosa Argentina SA was considering spending $30 million over three years to increase production at its Santa Fé plant by 60%, to 180,000 tons of cellulose and 140,000 tons of paper per year.[17]

Make the following assumptions: Assume that the time-zero capacity is 120,000 tons of cellulose and 80,000 tons of paper per year. Assume further that capacity

TABLE 13.4 **Conditional probabilities for coal mine size testing.**

	Coal Deposit		
Test Result	LOW	MEDIUM	HIGH
Favorable	0.05	0.25	0.85
Inconclusive	0.15	0.60	0.14
Unfavorable	0.80	0.15	0.01

[17] Wong, W., "Argentina's Celulosa to Invest $30M to Boost Production," *Dow Jones Newswires*, November 12, 2003.

expansions must be in increments of 20,000 tons and that cellulose and paper expansions occur at the same time (and in the same amount).

(a) Assume a five-year horizon and assume that the expansion(s) can occur in any of the first three periods. If capacity cannot exceed 180,000 tons per year for cellulose and 140,000 tons per year for paper, what is the total number of possible expansion plans over the three years?

(b) Restrict the options to expanding small (20,000 tons), medium (40,000 tons), and large (60,000 tons) at either time 0 or time 1. How many possible expansions are their now? Draw a decision network to verify your solution.

(c) Consider part (b) again. Assume that the cost is $12 million to install the small-capacity plant, $22 million for the medium one, and $30 million for the large one. If revenues are expected to be $850 per ton for cellulose and $425 per ton for paper, against per unit costs of $750 and $375, respectively, in addition to annual costs of $1 million per year per 20,000 tons of expansion, what expansion strategy (if any) should be chosen. Assume no salvage value after ten years and an interest rate of 11%.

(d) Assume that the revenue per ton of paper and cellulose is $400 and $800, respectively, during the first period and grows by $20 per pound over the horizon time. Does this change the solution you found in part (c)?

(e) Restrict the decisions to be studied to either a large expansion or small expansion at time zero and, if the small expansion is taken, a medium expansion at time one. Assume that the revenue per ton of paper and cellulose is $400 and $800, respectively, during the first period and grows by either $20 per pound (with a probability of 0.50) or stays the same in each ensuing period. What decision maximizes expected present worth?

6. Consider the introductory real decision problems and data from this chapter. Draw a decision network in which one can expand with 100,000 or 200,000 tons of capacity in each period. Should both of these decisions be considered over the horizon? Redraw the network, assuming the probability of demand growing low (25,000 tons), medium (50,000), or high (75,000) in each period.

7. Progress Energy Corporation started its 516-MW natural-gas-fired, combined-cycle unit power plant in December of 2003. At the cost of $230 million, the plant adds to the 482-MW unit already in place at the Hines Energy Complex in Polk County, Florida. Another 516-MW plant is to become operational in December of 2005 at a cost of $230 million. The investment is in response to a state mandate requiring that utilities maintain reserves of at least 20%.[18]

Make the following assumptions: The demand for power is growing at a rate of 250MW per year over 10 years. The cost of expansion can be approximated by the power law and sizing model with an exponent of 0.75. Assume that the investment cost of a 500-MW unit is $230 million, the cost operating the new plant is $5 million per 100MW, and an interest rate of 13% per year. What feasible expansion decisions should be examined over time? Draw the network for the first three periods.

---

[18] Kamp, J., "Progress Energy/Hines Unit-2: Adds 516MW of New Power," *Dow Jones Newswires*, December 9, 2003.

8. In 2003, ChevronTexaco Corp. and Total SA paid $40 million to Socar, the State Oil Company of Azerbaijan, for not drilling a well in the Absheron project. The contract called for two wells to be drilled, but the first drilling showed no commercial reserves of oil or gas.[19]

Make the following assumptions: Before the first well was drilled, it was estimated that there was a 0.50 probability that there were no reserves, a 0.30 probability that there were 50 million barrels of reserves, and a 0.20 probability that there were 100 million barrels of reserves in the well. Assume also that oil generates revenue of $35 per barrel in the first year of operation and will increase in price 3% per year against a constant cost of $13.50 a barrel, regardless of the amount of oil extracted. Oil is to be extracted at the rate of 10,000 barrels per day after a $400 million investment in the field. With 360 days of production per year, we assume production of 3.6 million barrels per year. Assume further that there will be $100 million in remediation costs at the end of the project. The interest rate is 18% per year.

(a) Should the investment be made? Use expected present worth to make your decision.

(b) Now assume that a contract is signed such that two wells are going to be drilled for testing. Draw the decision tree that allows for the following choices:

- *If the first well is drilled.* On the basis of the results, (1) the investment can be made and production begins; (2) the project may be terminated (do nothing); or (3) another well can be drilled.
- *If the second well is drilled.* On the basis of the results, (1) the investment can be made and production begins; or (2) the project may be terminated (do nothing).
- *If neither well is drilled.* (1) The investment can be made and production begins; or (2) the project may be terminated (do nothing).

Assume that we have the data in Table 13.5 for the first well. These data only return a result of favorable or unfavorable.

Given the results of the first test, the data for a second test (if undertaken) are listed in Table 13.6. Note that our outcomes are much more complicated now, since they are dependent on the results of two tests. For example, if the first test is favorable, then the second test can be either favorable or unfavorable. This outcome is read "Pr(Fav,Fav|LOW) is 0.03."

(c) What is the value of perfect information?

(d) What is the best course of action?

TABLE 13.5 **Conditional probabilities for first oil-field size test.**

	Coal Deposit		
Test Result	LOW	MEDIUM	HIGH
Favorable	0.15	0.40	0.70
Unfavorable	0.85	0.60	0.30

[19] Sultanova, A., and A. Ivanova-Galitsina, "ChevronTexaco, Total Compensate Socar for Undrilled Well," *Dow Jones Newswires*, November 18, 2003.

TABLE 13.6 **Conditional probabilities for second oil-field size test.**

Test Result 1	Test Result 2	Coal Deposit LOW	MEDIUM	HIGH
Favorable	Favorable	0.03	0.20	0.38
Favorable	Unfavorable	0.25	0.30	0.30
Unfavorable	Favorable	0.25	0.30	0.30
Unfavorable	Unfavorable	0.47	0.20	0.02

9. Nucor purchased a steel mill from North Star Steel in Kingman, Arizona, for $35 million in the spring of 2003. As of the spring of 2004, Nucor had not operated the mill, due to high electricity costs.[20]

Make the following assumptions: Assume that the plant is not in operation and it costs $100,000 per quarter to maintain it. If the plant is in operation, it generates $5 million per quarter in revenue against $4 million in costs, excluding electricity. Annual electricity costs are expected to decline from $4.1 million in the first year by $100,000 per year for the next two years (and then level off). If the plant becomes operational, it will have a salvage value of $10 million at the end of five years. The cost to begin operations is $1 million, regardless of when it occurs. If the plant never becomes operational, it has a salvage value of $30 million after five years. Assume a MARR of 16% per year.

(a) Does the $35 million purchase price for the plant influence the decision? Explain.

(b) Model and solve the problem with a decision network over 5 years such that the plant may delay operations for zero, one, two, three, four or five years at time zero.

(c) Consider part (b) again, but this time assume that the price of electricity is probabilistic. Explain how you would approach this problem. Draw a decision tree in your answer.

13.7.4 Fundamentals of Engineering Exam Prep

1. Decisions to accept, reject, or delay an investment are

(a) Independent.

(b) Mutually exclusive.

(c) Both (a) and (b).

(d) None of the above.

2. The decision to invest later is

(a) Contingent on delaying earlier.

(b) Mutually exclusive from investing earlier.

[20] "Nucor Steel's Ariz Plant Unlikely to Reopen Soon - Owner," *Dow Jones Newswires,* March 29, 2004.

(c) Mutually exclusive from investing even later.

(d) All of the above.

3. A decision tree

(a) Allows one to map out sequential decisions.

(b) Allows one to incorporate probabilistic information into investment decisions.

(c) Both (a) and (b).

(d) None of the above.

4. An example of a sequential investment decision is

(a) Purchasing a machine.

(b) Getting results from seismic testing before drilling for oil in a field.

(c) Leasing an asset for three years.

(d) None of the above.

5. An example of a sequential investment decision is

(a) Designing a prototype for market testing before mass-producing a product.

(b) Researching the effectiveness of a drug before mass-producing it.

(c) Building a facility shell, installing two production lines, and then installing two more production lines (at the same facility) two years later.

(d) All of the above.

14 Multicriteria Evaluation

(*Courtesy of DaimlerChrysler AG.*)

Real Decisions: The Hybrids are Coming!

The city of New York will take delivery of 825 hybrid buses, as opposed to traditional diesel buses, through 2007. The $500,000 Orion VII buses by DaimlerChrysler get 3.6 miles per gallon (mpg) which betters the 2.5 mpg of traditional diesels, although they cost 30% more. In addition to getting better gas mileage, the hybrids require no transmission overhauls (since they have no transmission) and fewer brake pad replacements due to smoother acceleration and stopping.[1]

Make the following assumptions: A diesel bus costs $375,000 and lasts 12 years, at which time it has a $10,000 salvage value. Assume further that maintenance costs total $1000 in year 1 and increase 8% per year over the life of the bus. Maintenance costs for a hybrid start at $500

[1] Howard, B., "New York's New Hybrid Buses; Cleaner air and 1 mpg better (and that's a whole lot better)," *PC Magazine,* December 15, 2005.

and increase 5% per year. The hybrid bus has a salvage value of $25,000 after 15 years of service.

Finally, assume that a bus operates 30,000 miles per year and fuel costs $2.25 per gallon, increasing 3% per year. The interest rate for discounting is 10%. All this information leads to a number of interesting questions:

1. Which bus provides service at minimum cost? What assumptions did you make?
2. What is the break-even annual operating mileage, gas price, interest rate, and difference in fuel efficiency between the two types of bus?
3. What noneconomic factors are relevant? How should they be incorporated into the analysis?

In addition to answering these questions, after studying this chapter you will be able to:

- Perform break-even and sensitivity analyses for multiple alternatives. (Section 14.1)
- Compare multiple alternatives with the use of simulation. (Section 14.2)
- Define the "efficient frontier" and identify dominated projects when comparing multiple alternatives with multicriteria. (Section 14.3)
- Eliminate projects from consideration through minimum-threshold analysis or ranking methods. (Sections 14.3–14.4)
- Incorporate noneconomic criteria into the selection process with multiple projects. (Section 14.5)

In our discussions of the risk involved with investments, we offered a number of analyses of a single project in previous chapters. For each individual project we consider in a multiple-project situation, we can perform the same types of analyses as we did in Chapter 10. The added information imposes another layer of difficulty on making our decision, as we must now compare multiple projects with multiple criteria. In the current chapter, we explicitly consider sensitivity analysis, break-even analysis, and simulation analysis before discussing some general methods for dealing with multicriteria.

14.1 Sensitivity and Break-Even Analysis

We defined sensitivity and break-even analysis for a single project in Sections 10.4 and 10.5. In those analyses, input parameters such as cash flows, the interest rate, and the project horizon, were perturbed to glean more information about the investment alternative. In the case of multiple projects, we can perform these same analyses in order to determine ranges of estimates in which a given project is preferred to other alternatives.

Performing this analysis does not differ from performing those introduced earlier. Rather, the interpretation of the results is what differs now, as multiple projects are being evaluated. We rely more heavily on graphical analysis when considering multiple projects, because a graph allows one to "see" the different projects more clearly.

We have two options with respect to viewing the additional information. In the case of two projects, we can graph the difference in our performance measure, such as annual equivalent costs or present worth, between the two projects against the error. This allows us to use spider plots for various input parameters.

However, we may also plot the results of the individual projects separately against the actual input data, as opposed to percentage errors. This information can be useful if we are in a situation where budgets are limited and we want to see the different project outcomes with respect to the critical input parameter of investment cost. Note that the method allows one to plot more than two projects simultaneously, which is not possible when looking only at differences. We illustrate in the next example.

EXAMPLE 14.1 *Sensitivity and Break-Even Analysis of Multiple Alternatives*

Xcel Energy, Inc., decided to build a 750-MW power plant at its Comanche Generating Station outside of Pueblo, Colorado, at the estimated cost of $1.3 billion. The plant is expected to be operational by late 2009, with construction beginning in 2005. The plant will burn low-sulfur coal from the nearby area.[2] The site was chosen over Xcel's

[2] "Xcel Energy selects preferred site for new coal generation," *News Release*, www.xcelenergy.com, February 17, 2004.

Pawnee Generating Station site outside of Brush, Colorado. The city of Pueblo will annex the plant, generating additional tax revenues for the city and providing a reliable, cheaper water source for the plant.[3]

Make the following assumptions in considering the decision of where to locate the plant: Since both possible plant locations (Pueblo and Brush) are within close proximity to each other and to the customer base, the revenues can be ignored, as they are assumed to be the same. Thus, only the construction, operating, and maintenance costs need to be considered. Both facilities are to be depreciated under MACRS GDS over a 20-year recovery period (150% declining balance switching to straightline with the half-year convention). Compare the AEC of each facility, assuming a tax rate of 35%, an after-tax MARR of 6.5%, and 10 years of operations (altogether, a 13-year horizon). Both facilities will have salvage values of \$300 million at the end of the horizon.

For the Pueblo location, assume \$1.3 billion in construction costs spread evenly over four years. O&M costs are assumed to be \$3 million per year, and water costs are \$750,000 per year (due to arrangements with the city).

The Brush facility costs \$1.275 billion to construct (more space is available for work) and is depreciated accordingly, with a similar salvage value. Annual O&M costs are expected to be \$5 million per year (due to further distribution costs), and higher water costs are estimated at \$2.5 million in the first year of operation, increasing 10% each year thereafter.

Solution. Assuming deterministic costs, the spreadsheet in Figure 14.1 gives the after-tax cash flows for both the Pueblo and Brush locations. Note that these flows assume that costs are positive and revenues are negative. The AEC values are also shown, assuming the 6.5% interest rate.

The after-tax cash flow with positive costs for year 6 for the Pueblo facility is

$$\begin{aligned} A_6 &= (1-t)E_n - t(D_n) \\ &= (1-0.35)(\$3\text{M} + \$.075\text{M}) - (0.35)(0.0688)(\$1.3\text{B}) = -\$27.96 \text{ million.} \end{aligned}$$

For the Brush facility, the cash flow in year 10 is

$$\begin{aligned} A_{10} &= (1-t)E_n - t(D_n) \\ &= (1-0.35)(\$5\text{M} + \$2.5\text{M}(1+0.05)^6) - (0.35)(0.0489)(\$1.275\text{B}) \\ &= -\$15.69 \text{ million.} \end{aligned}$$

The Pueblo facility has an AEC of \$96.26 million, which is \$160,000 per year lower than the Brush facility's AEC.

Two inputs appear to be critical to this decision: (1) the initial investment cost, because it affects cash flows in the first four periods, depreciation values, and the gain at the time of the sale; and (2) water costs, since they are fixed for the Pueblo facility and increasing for the Brush facility. For (1), we assume that the investment cost is variable, but still spread evenly over four years. For (2), the initial and increasing water costs are perturbed (separately). Figure 14.2 illustrates the resulting

---

[3] "NEWS WRAP: Xcel Picks Pueblo for New Coal-Fired Pwr Plant," *Dow Jones Newswires,* February 17, 2004.

	A	B	C	D	E	F	G	H	I	J
1	Example 14.1: Power Plant Location									
2										
3	Pueblo Facility							**Input**		
4	**Period**	**ATCF**	**Purchase/Sale**	**Depreciation**	**O&M**	**Water**		Investment	$1,300.00	million
5	0	$325.00	$325.00					O&M	$3.00	million
6	1	$325.00	$325.00	=(1-$I$12)*(E10+F10)				Water	$0.75	million
7	2	$325.00	$325.00	-$I$12*D10				Depreciation	MACRS	1.5
8	3	$325.00	$325.00					Rec Period	20	years
9	4	-$14.63		$48.75	$3.00	$0.75		Salvage	$300.00	million
10	5	-$30.41		$93.86	$3.00	$0.75		MARR	6.5%	per year
11	6	-$27.96		$86.84	$3.00	$0.75		Periods	10	years
12	7	-$25.68		$80.34	$3.00	$0.75		Tax Rate	35%	
13	8	-$23.54		$74.23	$3.00	$0.75				
14	9	-$21.59		$68.64	$3.00	$0.75		**Output**		
15	10	-$19.81		$63.57	$3.00	$0.75		AEC	$96.26	million
16	11	-$18.13		$58.76	$3.00	$0.75				
17	12	-$17.86		$57.98	$3.00	$0.75				
18	13	-$531.02	-$300.00	$28.99	$3.00	$0.75				
19										
20	Brush Facility							**Input**		
21	**Period**	**ATCF**	**Purchase/Sale**	**Depreciation**	**O&M**	**Water**		Investment	$1,275.00	million
22	0	$318.75	$318.75					O&M	$5.00	million
23	1	$318.75	$318.75					Water	$2.50	million
24	2	$318.75	$318.75					g (Water)	10%	per year
25	3	$318.75	$318.75					Depreciation	MACRS	1.5
26	4	-$11.86		$47.81	$5.00	$2.50		Rec Period	20	years
27	5	-$27.18		$92.06	$5.00	$2.75		Salvage	$300.00	million
28	6	-$24.59		$85.17	$5.00	$3.03		MARR	6.5%	per year
29	7	-$22.17		$78.80	$5.00	$3.33		Periods	10	years
30	8	-$19.85		$72.80	$5.00	$3.66		Tax Rate	35%	
31	9	-$17.69		$67.32	$5.00	$4.03				
32	10	-$15.69		$62.35	$5.00	$4.43		**Output**		
33	11	-$13.75		$57.63	$5.00	$4.87		AEC	$96.42	million
34	12	-$13.17		$56.87	$5.00	$5.36				
35	13	-$521.89	-$300.00	$28.43	$5.00	$5.89				

Figure 14.1 After-tax cash flows with positive costs in millions, based on power plant location.

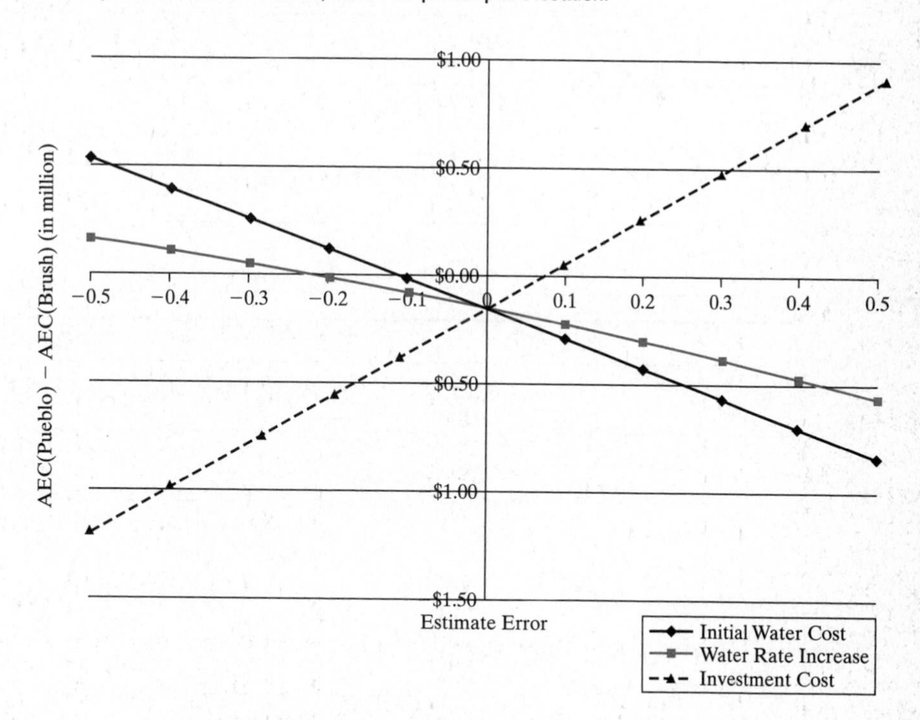

Figure 14.2 Spider plot examining differences in AEC values for errors in estimates.

spider plot of the differences in the AEC values (Pueblo–Brush) for the given errors in the estimates.

Note that negative values in the difference in AEC errors mean that the Pueblo facility is preferred, and positive values promote the Brush decision. The spider plot shows that, for errors of −10% or lower in the initial water costs, or of −20% or lower in the increase in water costs (at the Brush facility) such that the water costs turn out to be much cheaper than estimated, the Brush facility is preferred. For the investment cost, the decision changes with errors near 10% such that investment costs are a bit higher than anticipated. If desired, we could convert the *y*-axis values to percentages in order to see the relative effects of the errors on the difference in AEC values.

From what we have learned, we can perform a break-even analysis on the water costs to determine the cost that changes our decision with respect to the Pueblo and Brush facilities. For the water rate increase, we can compute this value directly by equating our AEC equations and solving for a variable increase. Using Goal Seek in Excel, we find this to be 7.7% for the water rate increase for the Brush facility. We can view this information graphically, as in Figure 14.3. Similar analyses can be performed for the initial water cost or the investment cost.

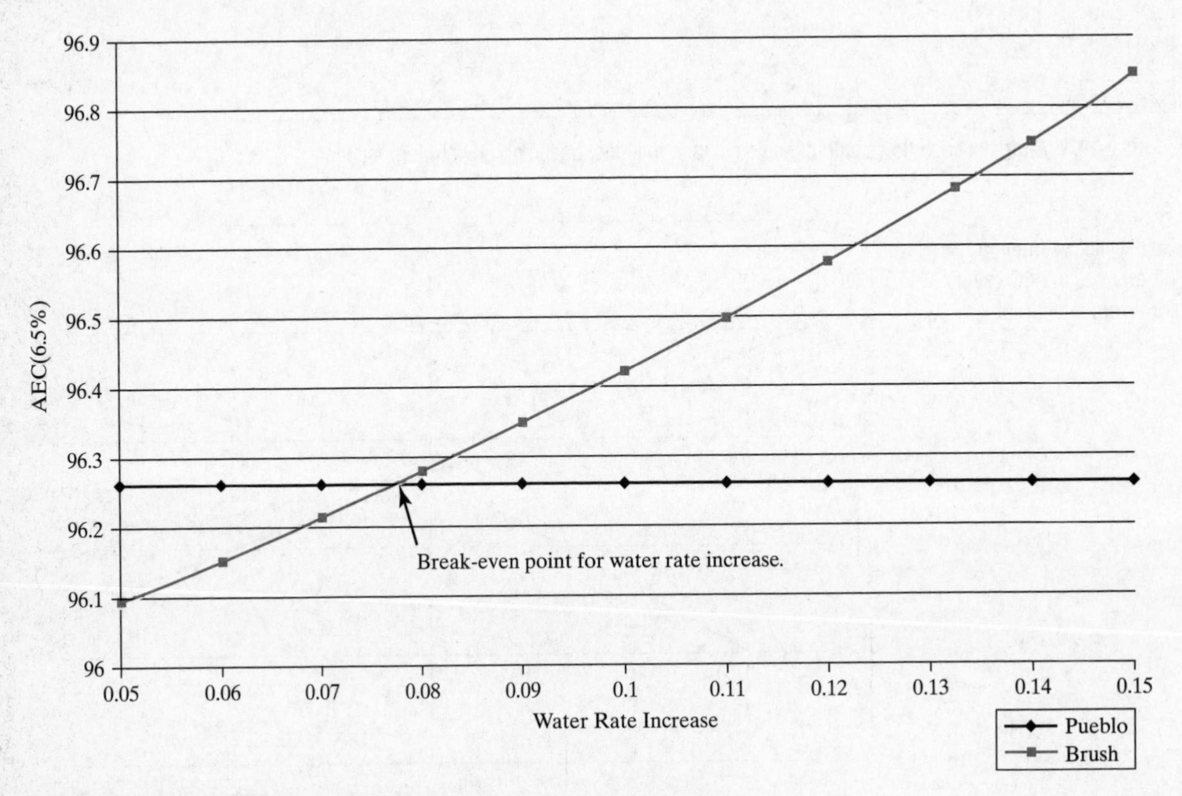

Figure 14.3 Break-even analysis concerning water rate increase parameter.

14.2 Simulation Analysis

We continue to apply analysis tools introduced when considering a single project. In Chapter 10, we illustrated how Monte Carlo simulation could be used to analyze projects that had a variety of probabilistic attributes. For a given simulation, we could compute the sample mean and variance to glean more information about the present worth of our project. In addition, we computed the probability that the project would not be successful (indicated by a negative present worth) by determining the ratio of the number of simulations that resulted in a negative present worth to the total number of simulation runs.

We can easily extend this analysis to the case of multiple projects. Specifically, we can simulate the outcome of each project and calculate the *difference* in the present worth for each simulation run. After performing a number of simulations, we can determine the sample mean and variance on the difference of the cash flows. Further, we can directly compute the probability that project A will have a higher present worth than project B by calculating the ratio of the number of simulation runs in which the present worth of project A exceeds that of B to the total number of simulation runs. We illustrate in the next example.

EXAMPLE 14.2

Simulation Analysis of Multiple Projects

We return to our previous example of choosing between two locations, Pueblo and Brush, for building a power plant. We make the following assumptions: The respective investment costs are probabilistic, with the Pueblo facility cost following a Normal distribution with mean $1.3 billion and standard deviation $100 million. The Brush facility has a similar standard deviation with a mean of $1.275 billion. Further, the costs associated with water for the Brush facility are random, so that the initial cost follows a Uniform distribution, U[$1.5M,$3.5M], and the annual rate increase follows a Normal distribution, N(10%,2.5%). Which location is preferred?

Solution. We simulated the cash flows of each facility and calculated the statistics in Table 14.1 on the basis of 100 simulation runs. In addition to the mean and standard deviation of the annual equivalent costs, the number of times an option resulted in a lower AEC was recorded.

TABLE 14.1

Statistical results of 100 simulation runs based on power-plant location.

AEC(6.5%) Statistics	Pueblo	Brush
Mean (Sample)	$96.12M	$95.69M
Standard Deviation (Sample)	$8.44M	$8.31M
Lower AEC	48	52

The results of the simulation are interesting because they favor the Brush facility, deviating from our deterministic analysis. The table shows that the Brush facility has a lower AEC mean and standard deviation. Furthermore, we estimate that

$$\Pr\left(\text{AEC}(6.5\%)_{\text{Brush}} \leq \text{AEC}(6.5\%)_{\text{Pueblo}}\right) = 0.52.$$

While the data tilts in favor of the Brush facility, it is clear that there is not a large difference in costs between the two locations. There may be noneconomic factors that could sway the decision in either direction.

If we had multiple projects (more than two), then we could perform this analysis repeatedly and compare each pair of potential projects. This pairwise comparison will provide valuable information to the decision maker despite the additional work.

14.3 Project Elimination Methods

In Chapter 11, we noted that we have the ability to gather an enormous amount of information about each of our projects, at least in the case of a single project. For example, we could compute some measure, or measures, of worth, its mean and variance, the payback period, and ranges of estimates for which the project is acceptable. The question that we consider now is how to choose between projects when there are *multicriteria* available. For example, in considering two projects, we could have computed the mean and variance of the present worth of each project and the payback period with interest. The choice becomes difficult when there is no clear-cut winner, such as a project with a higher present-worth mean, but also a higher standard deviation compared with another project. The question is, How do we choose between these projects?

Our approach in this section is to eliminate inferior projects through a variety of techniques. Clearly, if we have a choice between two projects and one project dominates in all aspects, then the decision is easy. Unfortunately, that will generally not be the case, and it will fall upon the decision maker to make use of subjective information and good judgment.

14.3.1 Dominance and Efficient Frontier Analysis

A project **dominates** another project if it is superior in all performance measures being evaluated. Thus, if two projects are being compared with respect to, say, present worth and payback period, the project with the higher present worth and lower payback period is said to dominate the other project. However, if one project has a higher present worth and the other project has a lower payback period, then neither project dominates the other one.

With multiple projects, it is easy to tell which projects are dominated by constructing an **efficiency frontier**. In two dimensions, an efficiency frontier is a plot with the x-axis representing one project measure such as present worth, and the y-axis representing the other project measures such as the payback period. A project is plotted according to its two measures resulting in a point located at coordinate (x, y). If the objective is to maximize both x and y, then any project defined by the coordinate (x', y'), where $x > x'$ and $y > y'$, is dominated. Further, the nondominated projects are said to form the efficiency frontier. Projects that lie along the efficiency frontier are the only projects that should be considered for selection, as they are not dominated in all measures being evaluated. The points on this frontier are often referred to as *Pareto optimal* solutions (after Vilfredo Pareto, the famed Italian economist and sociologist). We illustrate constructing this frontier in the following example.

EXAMPLE 14.3

Present-Worth and Payback Period Analysis

The Sunrise gas field, which lies in the Timor Sea between East Timor and northern Australia (see Chapter 12), is believed to hold 9 trillion cubic feet of natural gas, in addition to other liquids and oils. Woodside Energy, Royal Dutch/Shell, ConocoPhillips, and Osaka Gas are the four major partners developing the project, with Woodside being the project leader. Three projects have been proposed for developing the field:[4,5,6,7,8]

1. Pipe the natural gas to Darwin in northern Australia (a distance of about 500 km). ConocoPhillips has infrastructure in Darwin due to other exploration activities. The gas could be sold to local markets in Australia or, with the construction of a facility, converted to LNG for shipping to other markets, both in Asia and the United States.
2. Pipe the natural gas to East Timor (a distance of about 150 km). With the construction of a facility, the LNG could be shipped to Asia and the United States.
3. Construct a floating liquefied-natural-gas (FLNG) terminal at the drilling site. The gas would be transformed into its liquid form on a converted barge for subsequent transport to markets.

Because these are revenue projects, the do-nothing alternative is also a viable option.

Make the following assumptions: The pipeline cost to Darwin is $730 million, while the pipeline cost to East Timor is $317 million. The cost to construct an LNG-processing facility would be $1 billion at Darwin and $1.3 billion at East Timor, due

---

[4] Irwin, J., "E Timor Interested in LNG Plant for Sunrise Gas," *International Oil Daily,* www.etan.org, January 26, 2004.

[5] Counsel, J., "Send gas to East Timor—Expert," *Herald,* www.smh.com.au, September 26, 2002.

[6] "Sunrise Gas Project," *Our Business,* Woodside Australian Energy, www.woodside.com.au.

[7] "Sunrise Gas Field, Timor Sea, Australia," *The Website for the HydroCarbons Industry,* www.hydrocarbons-technology.com.

[8] Symon, A., "Timor Sea Natural Gas Development: Still in Embryo," *Economic Issues,* No. 2, South Australian Centre for Economic Studies, August 2001.

mainly to differences in infrastructure. The cost to build the FLNG-processing facility is $1.9 billion. These costs also include those required to extract the gas. It would take five years for production to begin once the projects commence, and all investment costs are spread evenly over the first five years.

The FLNG is expected to cost more to maintain, due to being continuously offshore, but is expected to reach its maximum production capacity sooner (since no pipeline work is required). Assume O&M costs of $125 million in the first year of operation, growing 3% each year thereafter. Further, assume constant transportation costs of $25 million over the life of the project. (Transportation costs are lower in the FLNG option than in the other options because the FLNG is accessible to all markets.) Due to accessibility, this option is expected to "fetch" $3.45 per 1000 cubic feet of natural gas delivered. The amount of gas extracted is expected to be constant at 300 billion cubic feet per year for 30 years.

The Darwin project is expected to cost $75 million to maintain the first year, with costs growing 1.5% per year. Expected revenues are $3.25 per 1000 cubic feet, based on production of 180 billion cubic feet in year 1 growing at 36% per year until reaching a capacity of 333 billion cubic feet in year 3 of production. Note that the gas reserves are depleted in year 26 of production with this option. Transportation costs are expected to be $45 million per year.

Finally, production at East Timor is to grow from 180 billion cubic feet by 20% per year to a capacity of 311 billion cubic feet, also generating revenues of $3.25 per 1000 cubic feet. Production lasts for 30 years. Transportation costs are expected to be $55 million per year, with O&M costs growing by 2% per year from $100 million in year 1.

Assume that each option ends with zero salvage value. Which project should be selected, using present worth and payback period with interest as the two financial measures? Assume a MARR of 25%.

Solution. Constructing the expected cash flows for each option leads to the results listed in Table 14.2. Although the do-nothing option is still available, we ignore it for now because defining its other properties, such as a payback period, is problematic. Rather, we will evaluate the other three proposals and, once a choice is made, consider the do-nothing option before moving forward.

TABLE 14.2 **PW(25%) and payback period with interest (years) for natural-gas-delivery projects.**

Project	PW(25%) (in M)	Payback Period
East Timor	$59.77	18
Darwin	$148.73	15
FLNG	$143.30	14

To compare these projects and illustrate the concepts of dominance and the efficiency frontier, we plotted the three options with the present worth as the x-coordinate and the payback period with interest as the y-coordinate. The graph is given in Figure 14.4.

It is clear from the figure that the East Timor project is dominated by both the Darwin and FLNG projects. This is because the East Timor project has a smaller

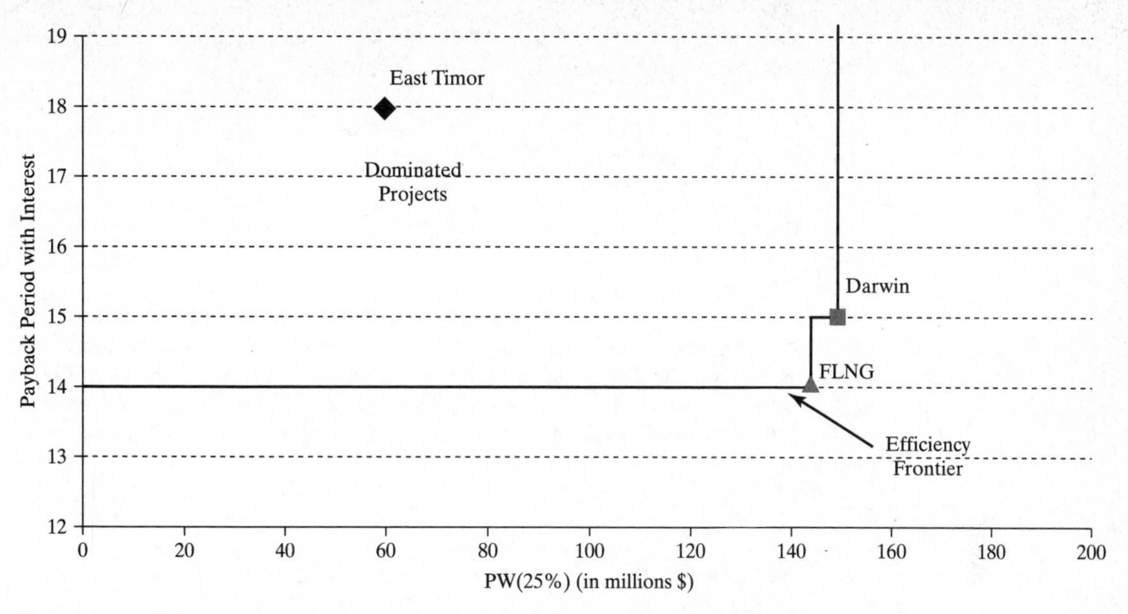

Figure 14.4 PW(25%) and payback period for three mutually exclusive natural-gas-delivery options.

present worth and longer payback period with interest compared with the other projects.

Now consider the FLNG project. If we draw both a horizontal line (to the left) and a vertical line (upwards) from its $(x, y) = (143.30, 14)$ coordinate, we define all other (x', y') coordinates that the FLNG project dominates, such that $x > x'$ and $y < y'$. This is because all of these projects would have lower present-worth values *and* longer payback periods with interest. The East Timor project lies in the zone created by these two lines. If we consider two similar lines from the (148.73,15) coordinate representing the Darwin project, we find that the East Timor project is again dominated. In fact, the two zones created by these two projects overlap considerably, with Figure 14.4 outlining their union.

From this information, we eliminate East Timor from further consideration and concentrate on the Darwin and FLNG projects, as they define the efficiency frontier. The decision will come down to whether a higher reward (Darwin project) or lower risk (FLNG) is desired. Of course, the do-nothing option is still available if both of these projects are deemed too risky.

The method is actually quite general, in that it can be used for any number of projects and any number of project measures. The only difficulty with using multiple measures is that they often cannot be graphed easily. However, a graphical analysis is not required and the project measures themselves can be examined. If a graphical analysis is preferred (because it does make it easy to spot dominated projects), we can construct multiple two-dimensional graphs.

Points that do not lie on the efficiency frontier in any graph are dominated, while the other points define the efficiency frontier.

EXAMPLE 14.4

Multicriteria Analysis

We continue our previous example, but expand upon the data available. Specifically, we perform project balance and break-even analyses.

Solution. The data available for our decision are contained in Table 14.3. Note that we continue to ignore the do-nothing option in this analysis. We have added (1) the ratio of the area of positive balance to the area of negative balance from the project-balance curve and (2) the minimum natural-gas price that allows for a positive present worth (break-even analysis). These data are in addition to our previous measures of present worth and payback period with interest.

TABLE 14.3 **Multicriteria for three natural-gas-delivery project proposals.**

Project	PW(25%) (in M)	Payback Period	APB/ANB	Price per 1000 ft^3
East Timor	$59.77	18	14.47	$3.111
Darwin	$148.73	15	47.73	$2.934
FLNG	$143.30	14	49.98	$3.158

Examining the new data, we see that the prospects for East Timor improve because the break-even analysis has resulted in a lower natural-gas price than that of the FLNG project. Despite this information, the Darwin project still dominates East Timor in all four measures; thus, the East Timor project is to be rejected, leaving the decision between Darwin and FLNG. Darwin performs better in terms of present worth and the price of natural gas, while FLNG has a better payback period and ratio of positive to negative balance in the project-balance curve.

14.3.2 Minimum-Threshold Analysis

We introduced the concept of minimum threshold analysis in Chapter 11 for a single project. A project was rejected if it did not meet the minimum threshold for all relevant measures. For the case of multiple projects, we can apply a similar decision methodology in either a static or dynamic setting.

In the static application, we would compare each alternative with a standard or threshold. Common rules for eliminating projects are as follows:

1. The project does not meet the thresholds of any of the measures under consideration.
2. The project does not meet the thresholds of all of the measures under consideration.

It should be clear that the first rule does not eliminate many projects, while the second can eliminate a lot of projects.

In the dynamic application, the criteria (measures) are ordered. Common rules include the following:

1. Eliminate all projects that do not meet the minimum threshold of the first criterion. Move to the second criterion and repeat. Continue until one project remains or all criteria are exhausted.
2. Accept the project with the highest value of the first criterion. Consider the second (and further) criteria to break ties.

As with our static application, these dynamic applications are a bit extreme in that the first application may eliminate few projects while the second application may select one project, but only using one (or few) criterion.

As an alternative, we can choose to visualize the data through use of a "decision evaluation display," which brings perspective to the thresholds and data from multiple projects. We illustrate by reexamining our previous example, but this time with minimum-threshold analysis.

EXAMPLE 14.5

Minimum Threshold Analysis

Consider the four criteria and three projects from the previous example. Assume that the order of importance for our project ranking, with dynamic analysis, is (1) present worth, (2) natural-gas price, (3) payback period with interest, and (4) ratio of area of positive balance to area of negative balance in the project balance curve. Assume further that the minimum thresholds for these measures are, respectively, \$0, \$3.15, 17 years, and 10.0. From the preceding methods, determine the project(s) to be considered further.

Solution. We first address the static application of minimum-threshold analysis. The results are as follows:

1. No project is eliminated, because all projects have positive present-worth values.
2. The East Timor project is eliminated because its payback period is too high. The FLNG project is eliminated because its break-even natural gas price is too high. The Darwin project is the only project remaining for further consideration.

The dynamic application results in the following decisions:

1. No project is eliminated when considering present worth. The FLNG project is eliminated when considering the price of natural gas. The East Timor project is eliminated when considering the payback period. The Darwin project again remains.
2. The Darwin project is accepted, as it has the highest present worth.

Thus, both applications of threshold analysis result in the Darwin project being preferred.

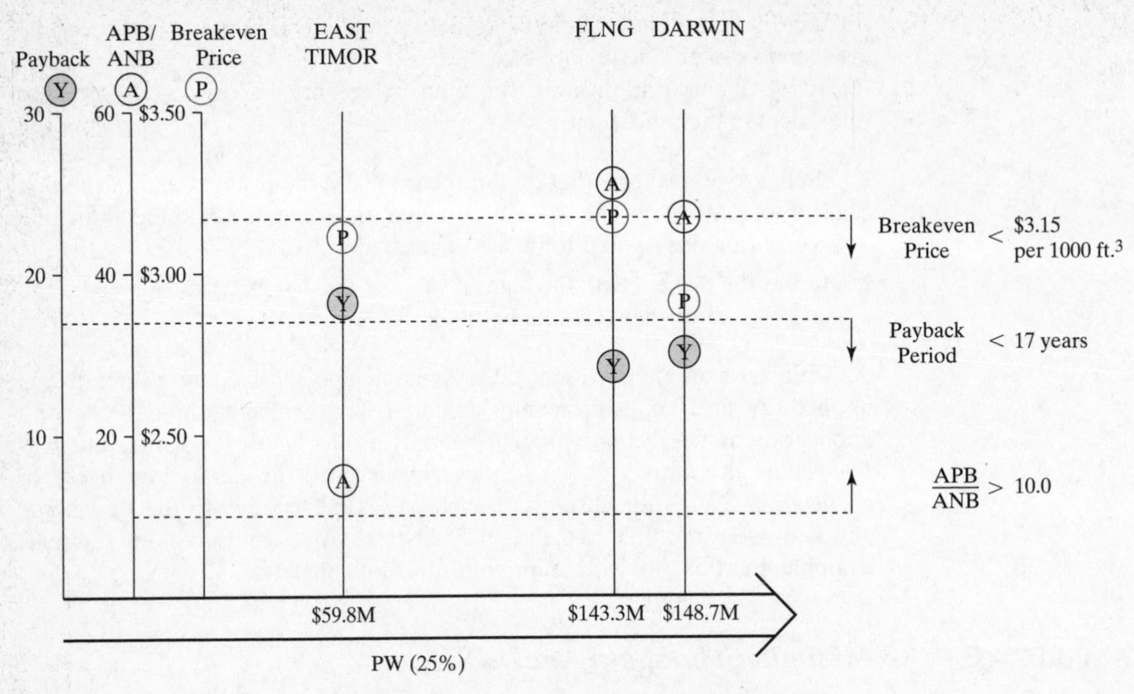

Figure 14.5 Decision evaluation display for natural-gas extraction with multiple criteria.

As noted earlier, the use of multiple criteria can be especially difficult because it is hard to visualize. Figure 14.5 provides a decision evaluation display of all four data parameters for the three projects being evaluated.

The x-axis illustrates the measure of worth (present worth), with the projects ordered according to increasing worth. Three y-axes are defined, one for each of the remaining criteria. These axes show a scale of possible values, with the threshold values providing breakpoints in the quadrant to the right. In this area, each alternative is given, again ordered by present worth, with the value of every other criterion presented on an axis, labeled with a Y, A, or P. From this figure, it is readily seen which projects are feasible for all criteria. More importantly, unlike our methods for eliminating projects, this graph provides a "feel" for the magnitudes that make or break a project. While such a display will not make the decision any easier, it does provide a way in which to succinctly capture all relevant data in a single picture.

We now turn our attention to ways in which to aggregate the information, as opposed to examining each data point individually, as in this section. This approach of keeping the data separate leads to a variety of situations in which the decision remains among a number of alternatives. The next section generally rectifies that conflict.

14.4 Project-Ranking Methods

In Chapter 11, we introduced the concept of consolidating multiple criteria into a single value with either scoring methods or utility theory. In the case of multiple projects, a value is computed for each project, and the project with the highest value among all of the projects is selected. We illustrate by revisiting our previous example.

EXAMPLE 14.6

Ranking Projects

Use the data from the previous example and compute a score for each project. The project weights are $\frac{4}{10}$ for present worth, $\frac{3}{10}$ for natural-gas price, $\frac{2}{10}$ for payback period, and $\frac{1}{10}$ for area ratio. In addition, ideal measures for each criterion are (1) \$200 million for present worth; (2) \$2.5 price for natural gas; (3) 10 for payback period; and (4) 50 for area ratio.

Solution. The score for each project is

$$\text{SCORE}_E = \frac{4}{10}\left(\frac{\$59.77}{\$200}\right) + \frac{3}{10}\left(\frac{\$2.5}{\$3.111}\right) + \frac{2}{10}\left(\frac{10}{18}\right) + \frac{1}{10}\left(\frac{14.47}{50}\right) = 0.501$$

$$\text{SCORE}_D = \frac{4}{10}\left(\frac{\$148.73}{\$200}\right) + \frac{3}{10}\left(\frac{\$2.5}{\$2.934}\right) + \frac{2}{10}\left(\frac{10}{15}\right) + \frac{1}{10}\left(\frac{47.73}{50}\right) = 0.782$$

$$\text{SCORE}_F = \frac{4}{10}\left(\frac{\$143.30}{\$200}\right) + \frac{3}{10}\left(\frac{\$2.5}{\$3.158}\right) + \frac{2}{10}\left(\frac{10}{14}\right) + \frac{1}{10}\left(\frac{49.98}{50}\right) = 0.767$$

As expected, the decision is close between Darwin and the FLNG, with Darwin being preferred, given the evaluation parameters.

We do not give an example illustrating utility theory here, but it would be similar to our scoring method here, in that some subset of the measures would be converted into a single number through a utility function (as in Chapter 11 for a single project). From these numbers, the project with the highest utility would be chosen.

14.5 Considering Noneconomic Factors

We now come to the final issue regarding multiple projects: noneconomic factors. In the case of a single project, we emphasized that noneconomic factors should not influence the economic analysis of a project. Rather, if the noneconomic factors were positive and the economic analysis was negative, or if the noneconomic factors were negative and the economic analysis was positive, then the noneconomic factors should be examined. Specifically, the question to answer is whether the noneconomic factor is "worth" the gap in value defined by either moving a project from an accept to a reject decision or vice versa.

In the case of multiple projects, the application is similar. Incremental analysis defines a gap between two projects that can be used to question the influence of noneconomic factors. To illustrate, consider the time when, in 2003, Airbus announced that it had awarded the EuroProp consortium, composed of Rolls-Royce PLC (of Great Britain), Snecma Moteurs (France), MTU Aero Engines (Germany), and Industria de Turbo Propulsores (Spain), a $3.4 billion contract to supply engines for the new A400M military transport aircraft over Pratt & Whitney, a division of United Technologies of the United States. This was despite the fact that Pratt & Whitney's bid was 20% (about $400 million) lower than that of Europrop. Reasons for going with the more expensive bid included pressure from governments in Europe (potential customers) desiring the plane to have "European engines."[9] In this application, the noneconomic factor of government influence was deemed to be more valuable than the 20% gap in price. Clearly, this noneconomic factor has some economic influence, since sales could be affected by the decision. But Airbus said that if the contract were based only on price, it would have gone to Pratt & Whitney. As this example illustrates, judgment will always play a major role in decision making if noneconomic factors are highly influential.

14.6 Examining the Real Decision Problems

We revisit our introductory real decision problems concerning the city of New York purchasing hybrid buses and the questions posed:

1. Which bus provides service at minimum cost? What assumptions did you make?

Since these are service projects with unequal lives, we must either estimate costs over a common life or compare AEC values over the respective service lives under the assumption of an infinite horizon and repeatability. We make this final assumption because it is reasonable to assume that buses will be required indefinitely and technology will not evolve too rapidly. Note that we are not asking the question of *how long* each bus will be retained, as we assume that it will be for the service life. (That question is addressed in Chapter 16.)

The spreadsheet in Figure 14.6 provides the cash flows and AEC(10%) calculations. The analysis shows that it costs $742 (cell F16) more per year to operate the hybrid bus.

---

[9] Lunsford, J.L., "Airbus Bypasses Pratt & Whitney, Keeps Engine Contract in Europe," *The Wall Street Journal,* May 7, 2003.

	A	B	C	D	E	F	G	H
1	Chapter 14: Bus Choice				**Input**	**Hybrid**	**Diesel**	
2					Purchase	$500,000	$375,000	
3	**Age**	**Hybrid**	**Diesel**		Annual Miles	30000	30000	
4	0	$500,000	$375,000		Gallon Gas	$2.25	$2.25	
5	1	$19,250	$28,000		MPG	3.6	2.5	
6	2	$19,838	$28,885		Op Cost per Mile	$0.63	$0.90	
7	3	$20,443	$29,800		g (Op Cost)	3%	3%	per year
8	4	$21,067	$30,746		Maintenance	$500.00	$1,000.00	
9	5	$21,711	$31,724		g (Maintenance)	5%	8%	per year
10	6	$22,375	$32,736		Salvage	$25,000.00	$10,000.00	
11	7	$23,059	$33,783		Life	15	12	
12	8	$23,764	$34,866		Interest	10%	10%	per year
13	9	$24,491	$35,986		=PMT(G12,G11,-NPV(G12,C4:C16))			
14	10	$25,240	$37,146		**Output**			
15	11	$26,013	$38,347		AEC	$79,720	$78,978	
16	12	$26,810	$29,590		Difference	$742		
17	13	$27,631			Breakeven			
18	14	$28,478			Op Miles	32780		
19	15	$4,351			Gallon Gas	$2.46		
20		=$F$3*$F$6*(1+$F$7)^(A18-1)+			Interest	9.24%		
21		$F$8*(1+$F$9)^(A18-1)			MPG Differential	1.24		

Figure 14.6 Cash flow diagrams and analysis for bus technology selection.

2. What is the break-even annual operating mileage, gas price, interest rate, and difference in fuel efficiency between the two types of bus?

   The break-even values were determined by driving the difference in the AEC values (cell F16) to zero with Goal Seek. This occurs if the annual mileage increases to 32,780, fuel costs rise to $2.46, the interest rate drops to 9.24%, or the difference in fuel efficiency is realized as 1.24 mpg (as opposed to the estimated 1.1 mpg). Thus, small changes in the operating environment will lead to the hybrid bus being chosen. It may be prudent to perform some analyses with respect to fuel prices, as these appear critical.

3. What noneconomic factors are relevant? How should they be incorporated into the analysis?

   According to Howard,[1] hybrid buses emit less than half the nitrous oxides and one-quarter the carbon monoxide compared with low-emission diesel transit buses or buses powered by compressed natural gas. The question must be asked as to whether this reduction is worth $742 per year. Assuming that the other data are correct, it appears that city of New York officials value the reduction in pollution by at least that much.

14.7 Key Points

- The methods for dealing with risk introduced for single projects are generally applicable to analyzing multiple projects.
- Break-even analysis and sensitivity analysis determine when certain projects are preferred over ranges of parameter values.

- Simulation analysis can be used to determine the probability that one project achieves a greater present worth (or some other applicable measure of worth) than another project.
- When considering multiple criteria for multiple projects, a project is dominated if another project performs better for all criteria under consideration.
- Projects that are not dominated define the efficiency frontier. These are the only projects that should be considered for acceptance.
- Minimum threshold analysis provides a way in which to eliminate projects by rejecting those which do not meet minimum standards. These projects may be viewed in a decision evaluation display.
- Projects can be ranked with either scoring methods or utility theory; in either case, the project with the highest score is accepted.
- When comparing two mutually exclusive projects, noneconomic factors should be considered if they can reverse the decision that results from rigorous economic analysis. Judgment plays a role in these decisions.

14.8 Further Reading

Boucher, T.O., O. Gogus, and E.M. Wicks, "A Comparison between Two Multiattribute Decision Methodologies Used in Capital Investment Decision Analysis," *The Engineering Economist,* 42(3):179–202, 1997.

Canada, J.R., and W.G. Sullivan, *Economic and Multiattribute Evaluation of Advanced Manufacturing Systems.* Prentice Hall, Englewood Cliffs, New Jersey, 1989.

Eschenbach, T., "Sensitivity Analysis," Chapter 3 in *Case Studies in Engineering Economy,* John Wiley and Sons, New York, 1989.

Fabrycky, W.J., G.J. Thuesen, and D. Verma, *Economic Decision Analysis,* 3d ed. Prentice Hall, Upper Saddle River, New Jersey, 1998.

Hartman, J.C., "Replacement Analysis with Advanced Technologies: A Traditional Framework," in *Economic Evaluation of Advanced Technologies: Techniques and Case Studies,* edited by J.P. Lavelle, H.R. Liggett, and H.R. Parsaei, Taylor and Francis, New York, 2002.

Law, A.M., and W.D. Kelton, *Simulation Modeling and Analysis,* 2d ed. McGraw Hill, New York, 1991.

Merino, D., "Developing Economic and Non-Economic Incentives to Select Among Technical Alternatives," *The Engineering Economist,* 34(4):275–290, 1989.

14.9 Questions and Problems

14.9.1 Concept Questions

1. What differs between sensitivity analysis and break-even analysis when considering multiple projects?
2. How can data be displayed in sensitivity analysis when considering multiple projects? Is there always a choice?
3. How can simulation be used to describe the differences between two projects?
4. What is the meaning of dominance and how does it apply to capital investment analysis?
5. What is the efficiency frontier? How is it used in decision making?
6. What is threshold analysis? How can it be applied with multiple projects? Would you expect all applications to lead to similar results? Explain.
7. How are project-ranking (scoring) methods and utility theory similar? What is their purpose in multiple-project analysis?
8. If you must choose and implement only one project from a list of alternatives, should you apply threshold analysis or a scoring method? Explain.
9. How does one incorporate noneconomic factors into an analysis of multiple (more than two) alternatives?
10. Why is judgment so critical when one is making the final decision about an investment?

14.9.2 Drill Problems

1. Consider the following list of projects (data in millions):

Alternative	Investment	PW(15%)	σ_{PW}	Payback
1	\$297	\$241	\$98	2
2	\$118	\$221	\$207	6
3	\$297	\$377	\$239	4

Assume that Projects 1 and 3 are mutually exclusive and that there is a \$500 million budget.

(a) Use total investment analysis and present worth to identify the best portfolio under certainty.

(b) What is the variance of the portfolio? (Merely sum the variances, not the standard deviations, of the individual projects, as they are independent.)

(c) Find the mean and variance of the present worth for all feasible portfolios. Construct an efficiency frontier. What does it tell you? Does this conflict with any answers in (a)?

(d) Use threshold analysis with a minimum present-worth standard of \$0, a maximum standard deviation of \$500 million, and a maximum payback period of 4 (assume that the payback period for a portfolio is the maximum payback period of any project in the portfolio). What projects should be considered?

(e) Assume that the criteria of present worth, standard deviation, and payback period are ordered as listed, with ideal values of $500 million, $75 million, and 2, respectively. Provide a score for each portfolio and select the best one.

2. Consider the following list of projects (data in millions):

Alternative	Investment	PW(15%)	σ_{PW}	Payback
1	$471	$716	$544	3
2	$465	$757	$616	4
3	$405	$179	$40	2

Assume that all three projects are mutually exclusive.

(a) Are any projects dominated? Explain.

(b) Use threshold analysis with a minimum present-worth standard of $0, a maximum standard deviation of $600 million, and a maximum payback period of 4 (assume that the payback period for a portfolio is the maximum payback period of any project in the portfolio). What projects should be considered?

(c) Assume that the criteria of present worth, standard deviation, and payback period are ordered as listed, with ideal values of $1 billion, $20 million, and 2, respectively. Provide a score for each portfolio and select the best one.

(d) Apply the Freund utility model to the portfolios to identify the best one ($B = 0.05$).

3. Consider the following list of projects (data in millions):

Alternative	Investment	PW(15%)	σ_{PW}	Payback
1	$100	$573	$1642	12
2	$303	$283	$132	7
3	$317	$647	$660	5
4	$235	$665	$941	4

Assume that Projects 2 and 3 are mutually exclusive and that there is an $600 million budget.

(a) Find the mean and variance of the present worth for all feasible portfolios. Construct an efficiency frontier. What does it tell you?

(b) Use threshold analysis with a minimum present-worth standard of $0, a maximum standard deviation of $2 billion, and a maximum payback period of 10 (assume that the payback period for a portfolio is the maximum payback period of any project in the portfolio). What projects should be considered?

(c) Assume that the criteria of present worth, standard deviation, and payback period are ordered as listed, with ideal values of $2 billion, $125 million, and 4, respectively. Provide a score for each portfolio and select the best one.

4. Consider the 10 projects given in Problems 1, 2, and 3. Write the integer programming formulation, assuming project (listed in order) 1 is contingent on 2 being accepted; projects 4, 5, and 6 are mutually exclusive; and projects 8 and 9 are mutually exclusive. Solve with Excel under the assumption of no budget constraint, a $1.5 billion budget, a $750 million budget, and a $300 million budget. Also, examine adding constraints that limit the payback period (values of 8 and 12) and the standard deviation of the portfolio ($1 billion, $2 billion, and $3 billion). How do these solutions compare with those from Drill Problem 6 in Chapter 12.

5. Two assets are being considered for installation to perform a job. (The options are mutually exclusive.) Asset A costs $500,000 to purchase and $400,000 per year to maintain, lasts for eight years, and has a salvage value of $50,000. Asset B costs $350,000 to purchase and $450,000 in the first year to maintain, increasing by 3% each year thereafter. Asset B lasts for six years with no discernible salvage value. Assume a 9% annual rate of interest and a long enough horizon to assume repeatability.

(a) Analyze the sensitivity of the choice with respect to the investment cost, the initial O&M cost, and the rate of increase. Present the analysis as succinctly as possible.

(b) Assume that the maintenance costs can be broken down further such that for Asset A, costs are $300,000 per year plus $5000 per production batch, while Asset B has costs of $250,000 per year and per batch costs of $4000, which increase at a rate of 15% per year. Develop a break-even analysis on the number of batches produced each year.

(c) Assume that the number of batches is fixed at 1000 per year, and perform a break-even analysis on the increasing rate of batch costs for B.

6. Consider the previous problem again, but this time assume that the maintenance costs are probabilistic. For Asset A, assume that costs follow a Normal distribution with mean $380,000 and standard deviation $40,000 in each period. For Asset B, assume that the mean is $400,000 in year 1 and increases each year by $5000, while the standard deviation is $60,000 per year. Perform a simulation, and determine the mean and standard deviation of each asset's AEC values. Also, determine the probability that one asset will have lower annual costs than the other.

14.9.3 Application Problems

1. Locus Pharmaceuticals of Blue Bell, Pennsylvania, signed up for IBM's "supercomputing-on-demand service" in order to speed research on cures for AIDS, cancer, and inflammation. Currently, it takes two weeks to run certain algorithms on Locus's in-house supercomputer, whereas the time is reduced to one day with the service. In addition, the need to purchase a supercomputer is eliminated.[10]

Make the following assumptions: The cost for Locus to purchase a supercomputer to meet the company's current needs is $2.5 million with $200,000 in annual maintenance. The supercomputer needs to be replaced every three years, at which time it has a salvage value of $10,000. The machine supplies adequate service for 15,000 tests per year. However, for every test above 15,000, a cost of $1000 per test

[10] Fuscaldo, D., "IBM Gets Pharma Customer for Supercomputing on Demand," *Dow Jones Newswires,* December 1, 2003.

is added due to the excessive solution time. Assume that the supercomputing cost is $400 per test and the interest rate is 20% per year.

(a) If Locus requires 10,000 tests per year, what is the preferred choice?

(b) What is the break-even value for the number of tests per year? Provide a graph of the break-even analysis.

(c) Assume that 20,000 tests are carried out in a year. Perform a sensitivity analysis for the per test price for both options. (The fee is charged only on 5000 tests for the in-house option.)

(d) Perform a simulation of testing over a three-year period. Assume that a test can last U[3,5] days on the in-house supercomputer and U[1,2] days through outsourcing. The charge is $400 per day of testing for outsourcing. The first 45,000 days of testing are free on the in-house computer, but any additional testing costs Locus $1000 per day. The number of tests in a year follows a Normal distribution with mean 13,000 and standard deviation 1200. (It should be clear that multiple tests run simultaneously; hence, there can be more tests than days in a year.) All other costs are assumed as before. Compute the annual equivalent cost of each technology choice and its standard deviation.

(e) Using all the information in part (d), compute the probability that the in-house solution will have a lower cost than the on-demand service.

2. In 2003, a review panel from Valley Metro Rail, of Phoenix, Arizona, selected Kinki Sharyo Co. of Japan to supply cars for the Phoenix area's light rail system. The contract is for $115 million and includes 36 railcars priced at $2.92 million each, with options on an additional 39 vehicles for the 20-mile system, planned to start operations in December 2006. Kinki Sharyo was selected over Bombardier of Canada, Siemens of Germany, and CAF USA of Spain. The panel considered "technical criteria, company capabilities, and price" in selecting the bidder.[11]

Make the following assumptions: Assume that the scores for each supplier for each category are defined as in Table 14.4. The criteria are listed in the table according to their order of importance. For the given data,

TABLE 14.4 Scores and data for various attributes of potential railcar suppliers.

	Category				
Supplier	Price	AEC	Reliability	Capability	Technical Specifications
Supplier A	$104.60	$10.83	0.835	6	9
Supplier B	$105.58	$12.48	0.915	10	7
Supplier C	$100.64	$13.62	0.943	6	10
Supplier D	$106.69	$14.03	0.899	7	7
Threshold	$106.00	$14.00	0.90	7	7
Ideal	$100.00	$10.00	0.980	10	10

[11] "Kinki Sharyo Gets Prelim Deal for Phoenix Rail Cars," *Dow Jones Newswires,* November 21, 2003.

(a) Are there any suppliers that are dominated? Which suppliers make up the efficient frontier?

(b) Perform both types of static threshold analyses, using the given standards. How do the results differ?

(c) Perform both types of dynamic threshold analyses, using the given standards. How do the results differ?

(d) Draw a decision evaluation display for the suppliers and parameters.

(e) Provide a ranking of criteria by computing a score (weighted average of attribute scores that have been scaled) for each supplier.

3. At the beginning of the year, Unocal Corporation announced that 2004 capital expenditures would total $1.93 billion. Planned investments included deepwater exploration in the Gulf of Mexico and in the waters off Indonesia, oil and gas development in Thailand, oil development in the Caspian Sea, natural-gas development in China, and developments of the Mad Dog field in the United States and the K2 field in the Gulf of Mexico.[12]

We seek to evaluate Unocal's numerous projects. Because individual project data are not available here, assume that information on the various projects is as given in Table 14.5. Assume further that Unocal's capital budget is capped at $1 billion and that the projects are independent.

TABLE 14.5

Investment alternatives for energy company.

Option	Investment Cost	PW(14%)	σ_{PW}
1	$345.00	$655.44	$331.83
2	$322.00	$2371.17	$970.44
3	$999.00	$1984.61	$787.05
4	$659.00	$2195.41	$1506.51
5	$335.00	$2004.82	$1385.18
6	$764.00	$7634.59	$4491.35
7	$947.00	$6237.34	$2724.90
8	$192.00	$437.28	$281.84
9	$935.00	$2719.88	$1644.79
10	$788.00	$320.72	$68.14

(a) Establish the set of feasible portfolios.

(b) For each portfolio, the expected present worth is the sum of the present-worth values of the individual projects, while the variance is the sum of the variances of the individual projects. (Note that the table gives the standard deviation.) Determine the expected present worth and variance for each feasible portfolio, and generate an efficiency frontier. Which portfolios are dominated?

(c) Use the Freund utility function described earlier ($B = 0.01$) to determine the utility of each portfolio for ranking. Are the results as expected, given the information about the efficiency frontier? How influential is B?

[12] Unocal Corporation, "Unocal Sees 2004 Capital Expenditures at $1.93 Billion; Focus on Major Development Projects," *Press Release,* www.unocal.com, January 27, 2004.

4. Consider again Application Problem 5 of Chapter 12. The amount of data available has expanded, as given in Table 14.6. The criteria are listed in their order of importance.

TABLE 14.6

Multicriteria for media company expansion options.

Option	PW(16%) (M)	IRR	Payback Period
1	$15.78	40%	3
2	$11.96	27%	4
3	$10.04	35%	3
4	$36.12	74%	2
Threshold	$10.00	16%	3
Ideal	$55.00	80%	2

(a) Assume that only one of the four projects will be chosen. Using only present worth and payback analysis, construct an efficiency frontier. Which projects are dominated? From which set should the winner be chosen? Why?

(b) Return to the original Application Problem 5 of Chapter 12 and consider part (b). Are the data in Table 14.6 sufficient? If not, develop a table with the correct attribute values and options.

(c) Given the correct options, perform both types of static threshold analyses, using the given standards. How do the results differ?

(d) Perform both types of dynamic threshold analyses, using the given standards. Are the results different?

(e) Provide a ranking of criteria by computing a score (weighted average of attribute scores that have been scaled) for each option. Does your optimal choice coincide with that from Application Problem 5 of Chapter 12?

5. Revisit Application Problem 4 in Chapter 12 concerning the purchase of CNG-fueled garbage trucks. What is the break-even value of cleaner air for the CNG trucks versus the diesel trucks? Ignoring the clean air, what is the break-even increase in cost for the CNG trucks, given the other data?

14.9.4 Fundamentals of Engineering Exam Prep

1. A company is considering outsourcing a process. To purchase a machine would cost $200,000, with $15,000 in annual fixed costs and $3.00 per part. If outsourced, the parts could be purchased for $4.00 each. If the horizon is five years with no salvage value and the interest rate is 10%, then the number of parts at which the options break even is most nearly

(a) 68,000 parts per year.

(b) 42,000 parts per year.

(c) 50,000 parts per year.

(d) 77,000 parts per year.

2. A machine can be purchased for \$20,000 and used for three years with a salvage value of \$0. It would cost \$500 per year to maintain the machine during that time. If the interest rate is 6%, then an annual lease that has equivalent costs would be computed as

(a) $\$20{,}000(A/F, 6\%, 3) + \$500$.

(b) $\$20{,}000 + \$500(P/A, 6\%, 3)$.

(c) $\$20{,}000(A/P, 6\%, 3) + \$500$.

(d) $\$20{,}000 + \$500$.

3. Engineering economy studies should

(a) Consider only intangibles.

(b) Ignore intangibles.

(c) Consider economic factors and intangibles.

(d) Both (b) and (c).

4. A choice between two machines must be made for implementation in a production process with costs given in the following table:

Parameter	Machine A	Machine B
Initial Cost	\$150,000	\$250,000
Annual Cost	\$10,000	\$2000
Salvage Value	\$0	\$5000
Service Life	5	10

Use the data in the table and a 10% annual rate of interest to compute the break-even annual cost for Machine A. Then

(a) Machine A costs more for any nonnegative annual cost.

(b) The break-even annual cost is \$2800.

(c) The break-even annual cost is \$7200.

(d) The break-even annual cost is \$11,200.

5. Three mutually exclusive investments with the following data (in millions) are being examined:

Parameter	A	B	C
PW(15%)	\$20	\$25	\$18
Standard Deviation	\$4	\$5	\$7
Payback Period	5	6	10

When making the final investment decision,

(a) Alternative C should be rejected.

(b) Alternatives A and B could both be considered.

(c) Both (a) and (b).

(d) None of the above.

Part V
Postimplementation Analysis

15 Postimplementation and Evaluation

(*Courtesy of Nissan North America.*)

Real Decisions: No Used Cars Here!

The motivation for this example comes from Nissan and its automobile production plant in Canton, Mississippi. The plant, which opened in May of 2004, was built at the cost of $1.4 billion, with a capacity to produce 400,000 cars annually. Before bringing the plant to its full capacity, Nissan announced that it was considering expanding the plant due to brisk sales and double-digit growth.[1]

Make the following assumptions: Define a baseline investment of $1.4 billion, spread equally over 2 years. Upon completion of the plant, sales are expected to be 200,000 vehicles in the first year, increasing by 100,000 per year until reaching the capacity of 400,000 vehicles per year. The plant is expected to be operational for 10 years, with a $50

[1] "Nissan CEO Ponders Building Another North American Plant," *Dow Jones Newswires*, January 6, 2004.

million salvage value at the end of that time. Revenues are expected to be $35,000 per vehicle, against costs of $33,000 per vehicle and annual fixed costs of $200 million (including overhead costs). The interest rate is 20% per year.

Assume that it is the end of 2005 and that the first 4 years of cash flows have been realized. The one difference from the expected numbers was that per vehicle revenues of $35,500 were realized on output of 300,000 vehicles in the first year and 400,000 vehicles in the second year. All other costs and revenues were as expected. This scenario leads to a number of interesting questions:

1. How is the investment performing with respect to expectations?
2. Should the project continue in its present state, or should some changes be made?
3. If changes should be made, how should they be analyzed?

In addition to answering these questions, after studying this chapter you will be able to:

- Explain why it is important to track an investment's progress over time. (Section 15.2)
- Analyze errors in the estimation process. (Section 15.2)
- Track a project by charting its expected and actual project balance. (Section 15.2)
- Link results from tracking a project to further investment options, such as abandonment. (Section 15.2)
- Update forecasts with data gathered when tracking a project. (Section 15.2)
- Populate a database with data gathered for use in future decisions. (Section 15.3)
- Explain a company's balance sheet, income statement, and cash flow statement and how the company is affected by a project's outcomes. (Section 15.4)
- Explain how activity-based costing allocates costs to different activities involved in the production and delivery of a product or service. (Section 15.5)

Through our evaluations, we have stressed that there is tremendous risk when undertaking an investment in an engineering project. Possible solution alternatives are determined early in the decision-making process, and from these choices, forecasts and predictions are made in order to provide information that may be used in analysis. Considering that the estimates are of cash flows that may not occur until far in the future, one must accept the fact that errors will occur.

Thus, it is of utmost importance that when the decision is made to invest in a project, our analysis does not stop. This is because, after the decision has been made, its outcomes become reality in the form of actual out-of-pocket costs and in-pocket revenues. These outcomes must be tracked in order to accumulate information and learn from mistakes made in forecasting and in the decision-making process. This learning can be used both (1) to "correct" one's decisions, if possible, regarding the current project—analogous to a pilot steering an airplane back on the correct course—and (2) to aid in future decision-making processes, as history often repeats itself.

As was highlighted in Chapter 13, investments are not static, and they must be monitored to determine further courses of action. After a project is underway, the options available to the decision maker may include the following:

1. Continue the project as expected.
2. Delay some phase of the project.
3. Accelerate some phase of the project.
4. Abandon the project.
5. Expand the project.
6. Continue the project after replacing processes or equipment.

The key to making these decisions about a project that is moving forward is to understand the current standing of the project. In this chapter, we focus on postimplementation analysis from the perspective of information gathering and illustrate how it may be useful in making subsequent decisions.

15.1 Sunk Costs Revisited

We raised the issue of sunk costs in Chapter 7, but it is worth repeating here briefly in order to motivate our discussion of postimplementation issues. As noted earlier, there are many options available to the decision maker after an investment has been made. To fully appreciate these options, sunk costs must be both understood and, more importantly, ignored when evaluating the options. Recall that sunk costs are expenses that *have been* paid (i.e., in the past). Since they have occurred in the past, they are truly irrelevant in current and future decision making. Thus, they must be *ignored* when one is considering future options for an investment project—especially when the project has been implemented.

This point is raised again because the options available to the decision maker in the postimplementation phase must be made going forward. These new options are generally the result of changes in the expected plan or in the expected outcome of a project. That is, if things are not going well, delaying or abandoning an investment may be the best course of action. Replacing equipment may be warranted if things are going well, but not as well as expected. Expansion may be desired if things are going very well. Changing the course of a project (delay, abandon, expand, or replace) is an acknowledgment that something in the analysis may have been wrong, such as a bad forecast or a bad decision. The only way to improve the situation is to look forward—a strategy that requires the ability to examine future options with an open mind and forget about the sunk costs of an investment.

15.2 Tracking an Investment

In our discussion of techniques used to analyze a project, we may have forgotten that the data, including all cash flows, their timing, and other relevant parameters, such as the interest rate and project horizon, are forecasted data. Consequently, they are prone to error, and it is imperative that one chart the actual progress of the solution that has been implemented. This information is valuable for a number of reasons, including its use in:

1. **Updating current project estimates, as uncertainty is resolved over time.** Updating provides information for further courses of action, which include, as noted earlier, continuing, delaying, abandoning, or expanding the project.
2. **Providing information and data for future, related projects.** In engineering, history truly does repeat itself, as designs and projects are often related to previous designs. The cash flows of future projects will be easier to estimate with a database of relevant information collected from current projects.

We present a number of techniques for tracking and evaluating the progress of an investment. These techniques also provide the opportunity to evaluate the decision-making process.

15.2.1 Error Analysis

In order to get a measure of the progress of a project, it is helpful to gauge the accuracy of its estimates. In forecasting, this is known as *error analysis*. We present three different methods by which to measure errors in forecasting, or, simply, forecast error. The first is the *mean absolute deviation*, or MAD, computed as the absolute value of the difference between an estimate and its realization, averaged over all estimates. We discuss the MAD in terms of cash flows, but it can be used for any parameter. We define A_n as our realized, or true, value and $\tilde{A}_n$ as our original estimate. Then the MAD of error for n periods of cash flows that have occurred is

$$\text{MAD} = \frac{\sum_{t=0}^{n} |A_n - \tilde{A}_n|}{n+1}.$$

Note that we divide by $n + 1$, as this assumes that a cash flow exists at time zero. Equivalently, in terms of information, the *mean squared error*, or MSE, replaces the absolute value of the error with the squared error:

$$\text{MSE} = \frac{\sum_{t=0}^{n} (A_n - \tilde{A}_n)^2}{n+1}.$$

The values of the MAD and the MSE give similar indications of the error of the estimates. The MSE is usually easier to deal with mathematically and thus is generally preferred.

Unfortunately, the data generated from MSE or MAD analysis can be hard to conceptualize if one is dealing with extremely large numbers, such as the typical multimillion-dollar investments that we have analyzed thus far in the text. If this is perceived to be a problem, the *mean absolute percentage error*, or MAPE, may capture the intended data better. With the MAPE, the error is scaled by the actual cash flow, so that

$$\text{MAPE} = \frac{\sum_{t=0}^{n} \left| \frac{A_n - \tilde{A}_n}{A_n} \right|}{n+1} \times 100,$$

which is an average of the percentage of absolute error. We illustrate all three error measures in the next example.

EXAMPLE 15.1

Error Analysis

Distributors of Sony Corp. DVDs built a new manufacturing plant in Russia in early 2004 at the cost of $1.5 million. The plant produces DVDs from Columbia and Tristar films, and the DVDs are expected to retail for about 750 rubles in Russia.[2]

Make the following assumptions: The plant was built at the end of 2003 at the cost of $1.5 million. It produces one million DVDs with an average retail price of $10.06 per unit, against costs of $9.25 per unit and $200,000 in fixed annual costs. The plant was expected to be operational for 10 years with no salvage value. Assume that 10 years have passed and that the expected and realized cash flows are as given in the spreadsheet in Figure 15.1. Compute the MAD, the MSE, and the MAPE for the realized cash flows.

[2] Moscow Bureau, "Sony DVD Distributors to Build Plant in Russia—*Vedomosti*," *Dow Jones Newswires,* November 13, 2003.

	A	B	C	D	E	F	G	H	I	J	K
1	Example 15.1: DVD Production Facility Tracking								**Input**		
2									Investment	$1,500,000.00	
3									DVD Sales	1000000	per year
4	**Time**	**Estimate**	**Actual**	**Error**	**Sqr Err**	**Abs Err**	**Abs Rel Err**		Revenue	$10.06	per DVD
5	0	($1,500,000)	($1,521,121)	($21,121)	$446,109,823	$21,121	1.39%		Cost	$9.25	per DVD
6	1	$610,000	$623,597	$13,597	$184,871,987	$13,597	2.18%		Operating Cost	$200,000.00	per year
7	2	$610,000	$688,733	$78,733	$6,198,937,880	$78,733	11.43%		MARR	12%	per year
8	3	$610,000	$594,989	($15,011)	$225,342,597	$15,011	2.52%		Periods	10	years
9	4	$610,000	$527,803	($82,197)	$6,756,281,320	$82,197	15.57%				
10	5	$610,000	$702,091	$92,091	$8,480,731,565	$92,091	13.12%		**Output**		
11	6	$610,000	$579,522	($30,478)	$928,916,530	$30,478	5.26%		Avg Error	$997.77	
12	7	$610,000	$629,181	$19,181	$367,918,125	$19,181	3.05%		MSE	$3,299,944,789	
13	8	$610,000	$533,402	($76,598)	$5,867,306,913	$76,598	14.36%		MAD	$49,210	
14	9	$610,000	$682,540	$72,540	$5,262,051,358	$72,540	10.63%		MAPE	7.86%	
15	10	$610,000	$570,239	($39,761)	$1,580,924,579	$39,761	6.97%		Present Worth	$1,946,636.05	
16											
17					=D14^2			=ABS(D14/C14)		=AVERAGE(E5:E15)	
18		=$J$3*($J$4-$J$5)-$J$6		=C15-B15		=ABS(C15)					

Figure 15.1 Expected net cash flows and realizations with errors over time.

Solution. The cash flow errors are provided in Figure 15.1, along with the squared errors, absolute errors, and scaled absolute errors. The averages of these respective errors lead to the calculations of the MSE (cell J12), MAD (J13), and MAPE (J14) as $3.3 billion, $49,210, and 7.86%, respectively. Because the MAPE errors are less than 15%, we, as decision makers, should be happy with the estimates, given our bounds for errors in Chapter 7.

Note that the use of the MAD, MSE, or MAPE defines all errors as absolute, regardless of whether they are over- or underestimates. This is to ensure that positive and negative errors do not cancel each other out, leaving the decision maker with a false sense of security with regard to forecasting. Unfortunately, tracking errors with MAD, MSE, or MAPE does not correspond directly with the guidelines from the American Association of Cost Engineers International, as discussed in Chapter 7. For example, the guidelines suggest that definitive forecast errors should lie in the range from −5% to 15%. The MAD, MSE, and MAPE, by contrast, produce single-valued (positive) errors, due to their treatment of all errors as absolute.

To circumvent this problem, one could just apply the values of MAPE (since it defines a percentage) to the error criteria (that definite estimates should fall within −5% and 15% of true values) and use one's judgment. That is, if the MAPE produces errors around 15% for a definitive estimate, then the method is probably working fairly well, regardless of the direction of the error.

Alternatively, we can track individual errors by considering their direction (i.e., whether we have positive or negative errors). This may provide more specific information about the current project and help analyze the options available, such as abandonment or expansion. When a project is underway, it is generally good news to have revenues that are greater than expected and costs that are lower than expected. This is related to our discussion of risk in which we pointed out the difference between the variance and the semivariance, noting that returns that vary from the expected in a positive direction are welcomed. This information is useful for tracking purposes.

Thus, we can divide our errors into positive and negative and compute separate measures. For the MAPE measure, forecast errors that are detrimental to an investment (overstated revenues and understated costs) can be measured as

$$\text{MAPE}^{-} = \frac{\sum_{n'} \left| \frac{(A_n - \tilde{A}_n)^{-}}{A_n} \right|}{n'} \times 100.$$

The value of $(A_n - \tilde{A}_n)^{-}$ refers to the fact that the value of the error, $A_n - \tilde{A}_n$, is included in the calculation only if it is negative. We assume that there are n' of these errors, and we average over only those values. This happens when a positive cash flow that is realized is less than the expected cash flow. Similarly, if a negative cash flow that is realized is more negative than its expected value, it is included in the calculation.

We can define a value of MAPE^{+} analogously that calculates the average percentage error of the positive errors. Given these two pieces of information, one can make comparisons with the printed guidelines of cost estimates more easily and determine whether corrective actions are needed, either with the project or the process. We reexamine the previous example to illustrate these calculations.

EXAMPLE 15.2

Error Analysis Revisited

Continue the previous example, but compute MAPE for both positive and negative errors.

Solution. The spreadsheet in Figure 15.2 breaks out the scaled absolute errors, with use of the IF function in Excel, depending on whether the error occurred in a positive or negative direction with respect to the estimate. These values are used to compute

	A	B	C	D	E	F	G	H	I	J
1	Example 15.2: DVD Production Facility Tracking							**Input**		
2								Investment	$1,500,000.00	
3								DVD Sales	1000000	per year
4	**Time**	**Estimate**	**Actual**	**Error**	**+ Rel Err**	**- Rel Err**		Revenue	$10.06	per DVD
5	0	-$1,500,000	-$1,521,121	-$21,121	1.39%	--		Cost	$9.25	per DVD
6	1	$610,000	$623,597	$13,597	--	2.18%		Operating Cost	$200,000.00	per year
7	2	$610,000	$688,733	$78,733	--	11.43%		MARR	12%	
8	3	$610,000	$594,989	-$15,011	2.52%	--		Periods	10	years
9	4	$610,000	$527,803	-$82,197	15.57%	--				
10	5	$610,000	$702,091	$92,091	--	13.12%		**Output**		
11	6	$610,000	$579,522	-$30,478	5.26%	--		MAPE+	7.68%	
12	7	$610,000	$629,181	$19,181	--	3.05%		MAPE-	8.08%	
13	8	$610,000	$533,402	-$76,598	14.36%	--				
14	9	$610,000	$682,540	$72,540	--	10.63%			=AVERAGE(F5:F15)	
15	10	$610,000	$570,239	-$39,761	6.97%	--	=IF(D12>0,ABS(D12/C12),0)			
16										
17					=IF(D15<0,ABS(D15/C15),0)					

Figure 15.2 Scaled absolute errors with respect to their direction from the estimate.

the $MAPE^+$ and $MAPE^-$ measures of error. The cash flows in periods 0, 3, 4, 6, 8, and 10 result in negative errors, while the remaining periods have positive errors.

We see that the MAPE values for negative and positive errors are quite similar, so there does not appear to be a bias towards over- or underestimating cash flows. Of course, there are a limited number of samples to evaluate, which is typical in capital investment analysis. Since both of the MAPE values of 7.68% and 8.08% fall under 10%, we conclude that the estimation process is working well. The negative error is slightly higher than desired according to our rule for definitive estimates, but it is within reason.

The absolute measures of error, including the MAD, MSE, and MAPE, will help in improving the estimation process, in which we clearly are interested in errors in both directions. We noted that we, as investors, are happy when returns tend to be better than expected. However, if returns are continually underestimated, it is possible that potential projects will be rejected due to an expected lack of return, when, in actuality, the projects perform well. Thus, these absolute measures provide a balanced examination of both understating and overstating possible returns—information that can be used to improve estimation processes.

The MAPE, $MAPE^-$, and $MAPE^+$ are quite different and should be used for different purposes. The MAPE, as noted earlier, should be used to revise future forecasting techniques and data collection. The values of $MAPE^-$ or $MAPE^+$ should highlight whether the forecasts tend to overstate positive or understate negative cash flows. However, they should also be used to signal potential problems with the current project that may be able to be corrected.

15.2.2 Cash Flow Charting

Control charts are often used in manufacturing applications in order to signal when a process is deviating from its expected behavior (when it is "out of control.") The method is often associated with the field of quality control, as deviations in a process can ultimately lead to problems with the quality of the final product. The concept is simple: An expected behavior with allowable deviations is provided as a baseline, and the actual behavior is plotted against the expected behavior. When the actual behavior exceeds the allowable range of operation, the process is said to be out of control and the decision maker should try to rectify the situation.

This concept is not unique to cash flow analysis; indeed, researchers have generally applied the theory to tracking an investment over time. The difficulty with its application in investment analysis is that not much data are available for analysis. In a manufacturing setting, a number of similar machines may be in use, each providing data that can be averaged such that anomalies attributed to a single machine can be lessened. Or repeated readings from a machine can be taken over time and averaged to give a more accurate description of the process. This is not possible with an investment, because only one cash flow is realized in each period and the cash flows are often expected to be different in each period.

Using the concept of project balance and borrowing a method from quality control known as the cumulative-sum quality-control (CUSUM) chart, we can track our investment and evaluate its progress. CUSUM charts differ from typical quality-control charting methods (e.g., X-bar and Shewhart charts) in that they examine the *cumulative* errors in a process over time, not merely the individual samples (which may be averages of numerous readings) taken over time. That is, measures are taken periodically and compared with target values. A running total, or cumulative sum, of the errors is compiled in order to provide information to the decision maker. The method allows one to more easily see drifts in the process, as the accumulation of errors, not single error samples, is analyzed. The process is ideal for low sample sizes—even a size of one, as is typical with capital investments.

By definition, the project balance (Chapter 10) is a cumulative sum of the cash flows from an investment, considering the time value of money. Thus, if we use the expected project balances from our estimated cash flows over time to define the "expected" process, we can measure the deviations from what is expected over time. Since we are using project balances, our errors are naturally summed, much as errors are summed in a CUSUM chart.

Unfortunately, we cannot apply the ideas from CUSUM charts directly to our cash flows. Both the interest rate used in calculating the project balance and the fact that our data are not independent from period to period (most likely) prohibit this. However, we can use the concept of a CUSUM chart to visualize our project over time. Further, we can gather simple statistics from this process in order to help guide which analyses should be performed in the future.

To be more specific, we again define cash flows that occur as A_n and the estimates of these cash flows as $\tilde{A}_n$. The error in the project balance at time zero, e_0, is merely the difference between the realized and estimated cash flow at that time, or

$$e_0 = A_0 - \tilde{A}_0 = \text{PB}_0(i) - \tilde{\text{PB}}_0(i).$$

For period 1, we carry forward the error from time zero and add our new error, just as we did with the definition of project balance:

$$e_1 = \left(A_0 - \tilde{A}_0\right)(1+i) + A_1 - \tilde{A}_1 = A_0(1+i) + A_1 - \tilde{A}_0(1+i) - \tilde{A}_1,$$

$$e_1 = \text{PB}_1(i) - \tilde{\text{PB}}_1(i).$$

Generalizing, we have, at any time period n,

$$e_n = \text{PB}_n(i) - \tilde{\text{PB}}_n(i).$$

Thus, in period N:

$$e_N = \text{PB}_N(i) - \tilde{\text{PB}}_N(i) = \text{FW}(i) - \tilde{\text{FW}}(i),$$

where $\tilde{\text{FW}}(i)$ is our predicted future worth and $\text{FW}(i)$ is our realized future worth.

With these definitions, the value of the error is cumulative, as with CUSUM charting, but it also incorporates the time value of money. This is important, as errors incurred early in the process have a greater effect on the worth of the project. We illustrate with an example and later discuss how this additional information can be used for decision-making purposes.

EXAMPLE 15.3 *Project Tracking*

In 2004, Novartis AG announced that it was building a production site for generic pharmaceutical drugs in Strykow, Poland, at the cost of EUR70 million. The facility is expected to produce 1.5 billion tablets and capsules annually.[3]

Make the following assumptions: The facility opens at time zero at the cost of EUR70 million and generates revenues of EUR1 per capsule or tablet at the cost of EUR0.95 and EUR10 million in annual operating costs. The plant has a six-year life, with a EUR5 million salvage value. Track the project under the assumption of a 30% annual interest rate with three scenarios representing three possible outcomes, defined as "good," "average," and "poor." The scenarios are defined as follows according to Uniform distributions:

Scenario	Per Tablet or Per Capsule Revenue in EUR
Good	U[1,1.05]
Average	U[0.95,1.05]
Poor	U[0.95,1]

Solution. Table 15.1 shows the cash flows generated from the descriptions of the three scenarios for the first four years of revenues. Also given are the resulting project balances for the three scenarios, in addition to the estimated data.

To mimic reality, only one simulation was run for each scenario (good, average, or poor). For the good scenario, the average revenue per tablet over four years was EUR1.04. For the average and poor scenarios, the average price was EUR1.01 and EUR0.97, respectively. This difference led to the discrepancies in the cash flows and resulting project balances over time.

Figure 15.3 charts the error in the project balances over time with respect to each of the scenarios. As may be expected, the poor scenario diverges negatively from the expected project balance, while the good scenario shows a much better project balance

[3] Weitz, S., "Novartis to Open New Generics Facility in Poland," *Dow Jones Newswires,* March 12, 2004.

TABLE 15.1 Expected project balances (in M) and three possible realizations of cash flows over time.

Year	Expected		Good		Average		Poor	
n	A_n	PB_n	A_n	PB_n	A_n	PB_n	A_n	PB_n
0	−€70.00	−€70.00	−€70.00	−€70.00	−€70.00	−€70.00	−€70.00	−€70.00
1	€65.00	−€26.00	€120.00	€29.00	€23.64	−€67.36	€8.80	−€82.20
2	€65.00	€31.20	€138.03	€175.72	€117.63	€30.06	€29.11	−€77.74
3	€65.00	€105.56	€127.15	€355.59	€67.90	€106.98	€53.95	−€47.12
4	€65.00	€202.23	€139.47	€601.74	€64.45	€203.53	€6.40	−€54.85
5	€65.00	€327.90	—	—	—	—	—	—
6	€70.00	€496.27	—	—	—	—	—	—

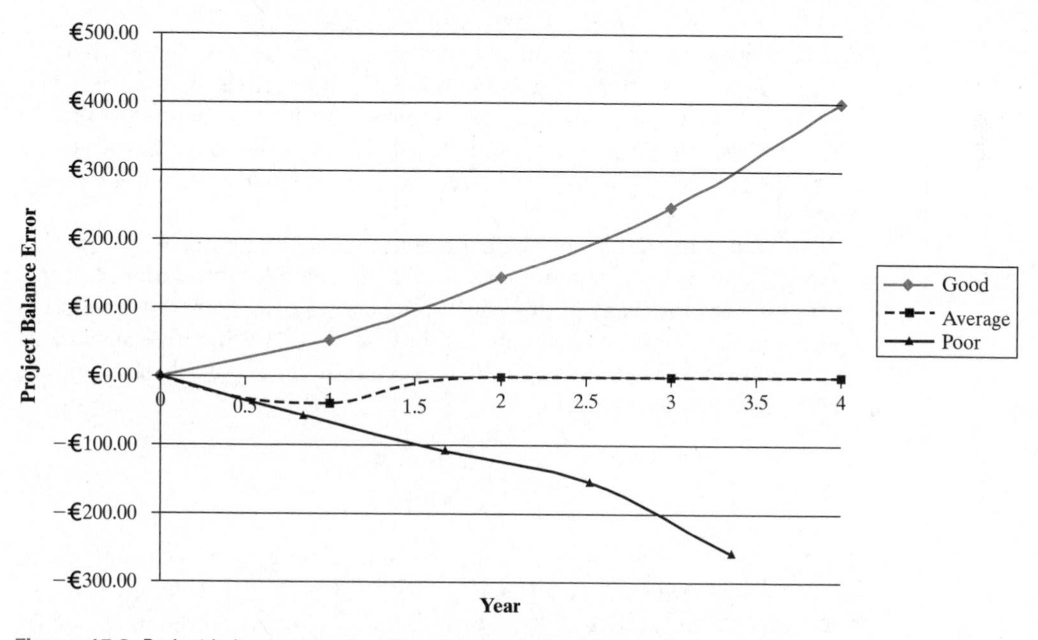

Figure 15.3 Project balance errors (in millions) for each scenario over time.

over the life of the project. The average project deviates at the beginning (due to randomness in the price), but achieves a final project balance that is only negligibly different from the original project estimates.

It should be clear that positive errors in the project balance over time are good, because it means that the returns are greater than expected. Negative errors signal potential problems that may need to be addressed. We now examine how to utilize this additional information.

15.2.3 Considering Dynamic Options

Since we are now charting our project over time, we can put the new information mentioned in the previous subsection to use. We have discussed the fact that investment decisions are not static. It is quite rare that one can make an investment decision, pay the investment cost, and then sit back to receive the benefits of the investment. It is likely that the investment will require nurturing, possibly in the form of additional funding, to take advantage of new opportunities that arise with time. Or it may be that the project is not progressing well and needs to be terminated.

We can use the information that we have charted to help start the process of evaluating our dynamic options. The power of a CUSUM chart lies in its visualization of possible trouble. We will use a similar approach. By taking the difference of the realized cash flow and the estimated cash flow, any positive deviations will become evident as positive slopes in the chart of project balance deviations over time. Similarly, a negative slope signals that revenues are below expectations or expenses are higher than expected. In sum, a positive slope, as shown in Figure 15.3, indicates that the project is running above its target and a negative slope means that it is running below its target.

The values in the project balance control chart may become unwieldy (too large), thus impairing our ability to read and gather information. For this reason, we scale the data. In this matter, the choice of a reference, or base, is important. We choose the estimated future worth, or $\tilde{\mathrm{FW}}(i)$, for our base, as it is the ultimate expected value to be achieved by the project balance. Thus, we now redefine our error as

$$e_n = \frac{\mathrm{PB}_n(i) - \tilde{\mathrm{PB}}_n(i)}{\tilde{\mathrm{FW}}(i)}.$$

Note that by choosing $\tilde{\mathrm{FW}}(i)$ as our denominator value, we have guaranteed that the project will not achieve a positive future worth (our measure of project acceptability) if $e_N \leq -1$. This follows from the fact that, in period N, $\mathrm{PB}_N(i) = \mathrm{FW}(i)$, such that

$$e_N = \frac{\mathrm{PB}_N(i) - \tilde{\mathrm{PB}}_N(i)}{\tilde{\mathrm{FW}}(i)} = \frac{\mathrm{FW}(i) - \tilde{\mathrm{FW}}(i)}{\tilde{\mathrm{FW}}(i)} = \frac{\mathrm{FW}(i)}{\tilde{\mathrm{FW}}(i)} - 1.$$

Thus, the value of e_N defines the success of a project compared with the original projections. If the project is unsuccessful, then the realized future worth, $\mathrm{FW}(i)$, will be negative. $\tilde{\mathrm{FW}}(i)$ is positive by definition if the project is accepted, so $e_N < -1$ for an unsuccessful project.

If the project generates a positive return ($\mathrm{FW}(i) > 0$), then $e_N > -1$, since the ratio of the realized future worth to the expected future worth is greater than zero. If the project performs exactly as expected, the error, as one would expect, goes to zero. Finally, if the project performs better than expected, then the realized future worth is greater than the expected future worth, defining $e_N > 0$. In sum, we have the following set of conditions:

Project Balance Error	Project Result
$e_N < -1$	Project loses money (negative future worth realized).
$-1 \leq e_N < 0$	Project does not meet expectations.
$e_N = 0$	Project performs as expected.
$e_N > 0$	Project exceeds expectations.

This analysis focuses on the error at time N. However, the error can be calculated at any time $n \leq N$. With each calculation, there are two valuable pieces of information available for our use: (1) the positive or negative error and (2) the trajectory of the project-balance error. Let us consider (1) first.

Since we have defined the error as the actual project balance minus the expected project balance, a positive value is welcomed. Thus, if the value of e_n is positive, then we are above our target and the project is showing promise. If however, the value of e_n is negative, then we are below target. (A value of zero means that the project is on target.) It is reasonable to ask what magnitude of error is acceptable. There is no definitive answer to this question other than that $e_N < -1$ signals that a project will not achieve a positive return. For any period $n < N$, a negative value of e_n merely warns the decision maker that the project may be in trouble.

To glean further information from e_n, we next consider (2), or the trajectory of the error. When tracking the error in the project balance, we are essentially graphing a curve with our project-balance control chart. Thus, the slope at any point on the curve provides the expected path of the curve at that point in time. We can *crudely* estimate the slope of the curve at a point by drawing a line through the last two data points (periods n and $n-1$). This line gives a general idea of the path of the project balance. If we project the line to period N, the project's horizon, we gather more information concerning the potential future worth of the project. As there is only one period of time between $n-1$ and n, we define the slope of the error at time n as

$$s_n = e_n - e_{n-1}.$$

We can use this value of the slope to project an estimate of the final project-balance error, or

$$\tilde{e_N} = e_n + s_n(N - n). \tag{15.1}$$

This trajectory can be used to point the decision maker towards future decisions. For example, we may use the following guidelines to start further investigations:

Error Value	Project Prediction	Action(s) to Consider
$\tilde{e_N} < -1$	Risk of loss	Delay or abandon project.
$-1 \le \tilde{e_N} < 0$	Below expectations	Delay or replace project.
$\tilde{e_N} = 0$	Expected returns	Continue project.
$\tilde{e_N} > 0$	Above expectations	Accelerate or expand project.

Note that we have specifically said that the estimates of e_N may be used as *guidelines to start further investigations*. These values are not to be taken as decision-making rules. Rather, they are to be taken as indicators alerting the decision maker to the fact that he or she should consider further analysis, perhaps including delaying, abandoning, accelerating, expanding, or replacing the project. (We consider abandonment and replacement in Chapter 16; delay was considered in Chapter 13 and expansion was examined in Chapters 9 and 13.)

Further, the estimates of e_N should not be taken as absolute givens. For example, if the value of $\tilde{e_N}$ slips to -0.05, then one should not immediately assume that the project should be delayed or abandoned. Rather, the estimates are just guidelines alerting the decision maker to potential problems. They are intended to help the decision maker identify problems and take corrective action before losses are incurred. We illustrate in the next example.

EXAMPLE 15.4

Project-Balance Trajectories

Continue the previous example, and compute the scaled project balance errors and the trajectories after each new period of data.

Solution. Table 15.2 provides the scaled errors from the three scenarios. Using the scaled project-balance errors and the trajectories provides a much quicker read of the situation. For the "good" scenario, the higher revenues lead to positive errors, and after only two periods it is projected that the project will achieve higher returns than estimated. The "poor" scenario illustrates negative errors, with a project error approaching -1 after the fourth period, signaling that the project is in jeopardy of not succeeding. (Recall that $N = 6$.)

TABLE 15.2

Project-balance errors, scaled and projected, for generic drug facility.

Year	Good		Average		Poor	
n	e_n	$\tilde{e_N}$	e_n	$\tilde{e_N}$	e_n	$\tilde{e_N}$
0	0.0000	0.0000	0.0000	0.0000	0.0000	0.0000
1	0.1108	0.6649	−0.0833	−0.5000	−0.1132	−0.6794
2	0.2912	1.0128	−0.0023	0.3219	−0.2195	−0.6447
3	0.5038	1.1417	0.0029	0.0183	−0.3076	−0.5720
4	0.8050	1.4075	0.0026	0.0021	−0.5180	−0.9388

The "average" case bounces about in the first few periods, but settles to predict returns in line with the original estimates. The conclusions to be made are that, for the good scenario, expansion of the project may be warranted, while delay or possible abandonment should be examined in the "poor" estimate. The "average" case seems to be moving as expected, so we should likely leave matters alone.

Trajectories alert the decision maker to consider corrective actions, which may be as simple as keeping a closer eye on the project or as drastic as abandoning the investment. Before making a drastic decision, however, the decision maker must perform an extensive analysis.

15.2.4 Updating the Forecast

Cash flow charting or error analysis will lead the decision maker to conclusions that the project is performing (a) as expected, (b) better than expected, or (c) worse than expected. Given one of these scenarios, a rational question is "What do we do now?" First, it makes sense to revise forecasts of the cash flows, because the information gained thereby can be used for subsequent analyses.

Unfortunately, the proper method to update forecasts is highly dependent on how the forecasts were initially acquired and what data are available. If we only have an original estimate and the realized value, then we do not have a lot of information with which to work. Essentially, there are three approaches that we can take:

1. **Revisit scenario estimates from the decision-making process.** In outlining our approach to estimating costs for a project, we noted that it was wise to generate more than one forecast. Accordingly, we formulated optimistic, pessimistic, and average outcomes for the project. With these projections, we can move from one scenario to another, such as moving from the average estimate to the pessimistic estimate if charting reveals that the project is not performing well or moving from the average estimate to the optimistic estimate if charting reveals that the project is performing better than expected. We can then use the information we obtain to update expected outcomes.
2. **Smoothing new data with old data.** Another popular method that is used extensively in demand forecasting is smoothing of data. This method essentially takes an old estimate and revises it with new information. Often, a percentage, α, of the old estimate is added to a percentage, $1 - \alpha$, of the new estimate (obtained from updated information).
3. **Reinitiate the estimation process.** Another approach is to use the new data and start the cost estimation process again. For example, curve-fitting techniques described in Chapter 7 can be used to "refit" curves and project future costs or revenues.

The fact is, updating forecasts takes a bit of judgment mixed with the appropriate technique that is dependent on the situation. At the least, information used to develop initial forecasts should be revisited with the new information at hand such that they can be reforecasted in a new light. As noted earlier, once the forecasts are updated, the decision maker must ascertain whether any corrective actions to the projects are necessary.

15.3 Maintaining a Database

It is in the final stage of the decision-making process, when a project is tracked, that the estimation process is "put to the test," because cash flows are realized. But tracking is not enough: We must retain the information we garner for future decisions.

Recall some of the techniques used to forecast future costs and revenues, including the factor method, the power law and sizing model, and curve fitting. One commonality among these techniques is that they require input data. Clearly, the input data can be estimated, too, but here is where historical data serve better. Thus, the information that is being tracked to guide the current investment should be recorded in a database for future use.

We will not discuss the intricacies of database development here. Rather, we note information that would be useful to track over time. For the project itself, the following data should be gathered:

- *Costs*. Investment, operating, and maintenance costs, including the costs of labor, materials, and energy usage.
- *Revenues*. Operating revenues and those from the sale of assets.
- *Impact data*. Data on product demand and pricing.

Also, data that are not under the direct control of the decision maker, but that affect the decision, should be maintained in the database. Often, these data can be retrieved at a later time, but it is better to include them with the data gathered in tracking a project, so that the next decision maker does not have to treat the process as a new investigation. "Outside" information may include any or all of the following:

- Data on competition, including pricing, warranties, sales strategies, supply, and demand.
- Commodity prices.
- Currency exchange rates.
- Market interest rates.
- Abnormal conditions, including extreme weather conditions, the political climate, etc.

Ideally, the database will display the forecasts, actual outcomes, and error analysis together with the external data. In that way, one can examine various cause-and-effect scenarios simultaneously.

15.4 Cost Accounting

In studying whether a single project or a set of projects should be accepted for investment, it is easy to lose sight of the bigger picture for the company or government agency. This is why companies employ numerous accountants. Accountants are responsible for compiling financial data for a company at given intervals, generally quarterly and annually. The reports are often shared with the public in the case of publicly owned companies. Private companies perform this same function in order to keep track of costs and prepare for the possibility of becoming a publicly traded company.

While we will not go into the intricacies of cost accounting, we will briefly highlight the major financial and accounting statements. It is important for the engineer to understand that the costs and revenues experienced from an engineering project contribute to the financial well-being of the company, which is often made public in order to entice future investment and growth in the firm. In the rest of this section, we present the major financial statements and highlight where engineering projects tend to have an impact.

15.4.1 Balance Sheet

The balance sheet provides a snapshot of a company's financial standing on a given date. This report is usually generated quarterly to let shareholders know the standing of the company. Figure 15.4 gives an example of a balance sheet, according to which

$$\text{Assets} = \text{Liabilities} + \text{Owners' Equity}.$$

Assets comprise "current assets," which are liquid (i.e., they generally can or will be converted to cash within a year), such as cash or short-term bonds, and other assets, such as equipment, which is not liquid. Liabilities consist of all items that require payment, including debt, which is generally broken out into short-term and long-term debt. The difference between current assets and current liabilities is working capital. Engineering projects may affect both assets and liabilities through the acquisition of equipment and sales. Working capital is often critical to fund purchases of materials and inventory.

The remaining part of the equation is owners' equity, which is an accumulation of earnings over time (retained earnings), and invested capital, which is money provided by investors for growth, generally through selling shares of stock. Both of these funds may be used to fuel growth.

TEXAS INSTRUMENTS INCORPORATED AND SUBSIDIARIES
Consolidated Statements of Income
(In millions of dollars, except per-share amounts)

	For Years Ended Dec. 31 2005	Dec. 31 2004
Net revenue	$ 13392	$ 12580
Cost of revenue (COR)	7029	6954
Gross profit	6363	5626
Gross profit % of revenue	47.5 %	44.7 %
Research and development (R&D)	2015	1978
R&D% of revenue	15 %	15.7 %
Selling, general and administrative (SG&A)	1557	1441
SG&A% of revenue	11.6 %	11.5 %
Total operating expenses	3572	3419
Profit from operations	2791	2207
Operating profit % of revenue	20.8 %	17.5 %
Other Income (expense) net	206	235
Interest on loans	9	21
Income before income taxes	2988	2421
Provision for income taxes	664	560
Net Income	$ 2324	$ 1861
Basic earnings per common share	$ 1.42	$ 1.08
Diluted earnings per common share	$ 1.39	$ 1.05
Average shares outstanding, basic	1640	1730
Average shares outstanding, diluted	1671	1768
Cash dividends declared per share of common stock	$ .105	$ 0.89

(a)

TEXAS INSTRUMENTS INCORPORATED AND SUBSIDIARIES
Statement of Cash Flows
(In millions of dollars)

	For Years Ended Dec. 31 2005	Dec. 31 2004
Cash flows from operating activities:		
Net income	$ 2324	$ 1861
Adjustments to reconcile net income to cash provided by operating activities:		
Depreciation	1375	1479
Stock-based compensation	178	18
Amortization of acquisition-related costs	56	70
(Gains)/losses on investments	–	1
(Gains)/losses on sales of assets	(26)	–
Deferred income taxes	(194)	68
(Increase) decrease fom change in:		
Accounts receivable	(139)	(238)
Inventories	(25)	(272)
Prepaid expenses and other current assets	117	148
Accounts payable and accrued expenses	264	(71)
Income taxes payable	35	59
Accrued profit sharing and retirement	(145)	235
Noncurrent accrued retirement costs	(166)	(248)
Other	118	36
Net cash provided by operating activities	3772	3146
Cash flows from investing activities:		
Additions to property, plant and equipment	(1330)	(1298)
Sales of assets	47	–
Purchases of short-term investments	(5851)	(3674)
Sales and maturities of cash investments	5430	3809
Purchases of equity investments	(17)	(22)
Sales of equity and debt investments	53	32
Acquisition of businesses, net of cash acquired	(19)	(8)
Net cash provided by (used in) investing activities	(1687)	(1161)
Cash flows from financing activities:		
Additions to loans payable and long-term debt	275	–
Payments on loans payable and long-term debt	(11)	(435)
Dividends paid on common stock	(173)	(154)
Sales and other common stock transactions	461	192
Excess tax benefit from stock-option exercises	59	–
Common stock repurchases	(4151)	(753)
Net cash used in financing activities	(3540)	(1150)
Effect of exchange rate changes on cash	6	15
Net increase (decrease) in cash and cash equivalents	(1449)	850
Cash and cash equivalents at beginning of period	2668	1818
Cash and cash equivalents at end of period	$ 1219	$ 2668

(b)

TEXAS INSTRUMENTS INCORPORATED AND SUBSIDIARIES
Consolidated Balance Sheets
(In millions of dollars)

	Dec. 31 2005	Dec. 31 2004
Assets		
Cash and cash equivalents	$ 1219	$ 2668
Short-term investments	4116	3690
Accounts receivable, net of allowances for customer adjustments and doubtful accounts	1812	1696
Raw materials	122	117
Work in process	827	756
Finished goods	324	383
Inventories	1273	1256
Deferred income taxes	619	554
Prepaid expenses and other current assets	146	272
Total current assets	9185	10136
Property, plant and equipment at cost	8921	9573
Less accumulated depreciation	(5022)	(5655)
Property, plant and equipment, net	3899	3918
Equity and debt investments	236	264
Goodwill	713	701
Acquisition-related intangibles	64	111
Deferred income taxes	393	449
Capitalized software licenses, net	245	307
Prepaid retirement costs	210	277
Other assets	118	136
Total assets	$ 15063	$ 16299
Liabilities and Stockholders' Equity		
Loans payable and current portion long-term debt	$ 301	$ 11
Accounts payable	750	552
Accrued expenses and other liablities	998	892
Income taxes payable	163	203
Profit sharing contributions and accrued retirement	134	267
Total current liabilities	2346	1925
Long term debt	360	368
Accrued retirement costs	136	589
Deferred income taxes	23	40
Deferred credits and other liabilities	261	314
Stockholders' equity:		
Common stock, $1 par value. Authorized -- 2.4 million shares.		
Shares issued: December 31, 2005 -- 1,738,780,512;		
September 30, 2005 -- 1,738,650,318;		
December 31, 2004 -- 1,738,156,615;	1739	1738
Paid-in capital	742	750
Retained earnings	13394	11242
Less treasury common stock at cost:		
Shares: December 31, 2005 -- 142,190,707		
September 30, 2005 -- 120, 597,527;		
December 31, 2004 -- 20,041,497	(3856)	(480)
Accumulated other comprehensive income (loss):		
Minimum pension liability	(65)	(168)
Unrealized holding gains (losses) on investments	(16)	(15)
Deferred compensation	(1)	(4)
Total stockholders' equity	11937	13063
Total liabilities and stockholders' equity	$ 15063	$ 16299

(c)

Figure 15.4 Income Statement (a) feeds cash flow statement (b) which feeds balance sheet (c). Data is from Texas Instruments, 2004–2005. Source: www.ti.com/corp/docs/investor/quarterly/4q05.html.

15.4.2 Income Statement

An income statement is a historical record of accounts over some stated period. This is in stark contrast to the balance sheet, which is truly a snapshot in time. The income statement generally reflects the previous quarter's, half-year's, or year's data.

The income statement reports the income from sales, less expenses incurred, calculating the profit at numerous levels (gross profit, operating profit, and profit before and after taxes). This is essentially what is done in creating an after-tax cash flow when analyzing an engineering project. The difference here is that all projects are aggregated together for the company. A sample of an income statement is given in Figure 15.4.

15.4.3 Cash Flow Statement

The final statement used to clarify the financial picture of a company is the cash flow statement. Unlike the income statement, which focuses on profits, the cash flow statement addresses the *movement* of cash, including any financing, such as debt repayment or dividend payments. That is why the cash flow statement is broken out according to operating activities (such as engineering projects generating sales, depreciation, and working-capital adjustments), investing activities (including capital expenditures and investments), and financing activities (purchases of stock or sales and repayment of debt). A sample of a cash flow statement is given in Figure 15.4.

These statements are shown here to reiterate that investment projects contribute to the overall financial standing of a company. Further, gathering this accounting data is analogous to our project tracking—albeit at a much more aggregate level.

15.5 Activity-Based Costing

Activity-based costing, or ABC, is an approach designed to accurately attribute costs of operations, which are generally the result of capital investments, to their cause, through the use of *cost drivers*. We have illustrated a number of examples in this text in which either "fixed" costs or "overhead" costs have been assigned to an investment. Traditionally, overhead costs are assigned to an activity by using a single overhead rate. The proponents of ABC argue that these costs should be more accurately assigned to their source, in order to help reduce waste and cost products more accurately. We mention ABC here because its use is directly attributable to tracking investments and assigning all costs incurred to their origin.

To implement an activity-based cost system is not trivial. First, the appropriate cost drivers need to be identified. That is, what are traditionally defined as indirect cost activities, such as engineering, material handling, or quality assurance, must be assigned cost drivers that are appropriate to the activity. For example, engineering might have its level of activity defined according to the number of design changes, and quality assurance might count the number of inspections performed. The key is that the driver is matched to the activity. Once these drivers have been identified, the costs for each activity can be tracked according to the driver. The ultimate goal is to assign these indirect costs more accurately to the proper activities.

Why is activity-based costing related to engineering economy and investment analysis? We already noted that we generally include overhead costs in our analyses. The fact is, if these overhead costs are not properly, or fairly, allocated to projects and activities, then certain projects may be "overcosted" while others are "undercosted." This means that we would be evaluating projects with the wrong data. Activity-based costing attempts to correct this error by assigning indirect costs to their appropriate activities. With activity-based costing, not only do indirect cost allocations mirror reality, but actual cost estimation procedures improve because developing an ABC system requires a true understanding of the costs incurred by an enterprise.

15.6 Examining the Real Decision Problems

We return to our introductory real decision problems concerning Nissan's desire to expand and the questions posed:

1. How is the investment performing with respect to expectations?

   The expected cash flows and project balances of the original investment are given side by side with the actual cash flow and project-balance realizations in Table 15.3. The increased sales and higher selling price in the first two years contribute to positive errors in the project balances.

   The errors and their projections to the final period are graphed in Figure 15.5. In addition, the project balance is shown on the graph, scaled so that each periodic project balance is divided by the expected future worth such that the graph moves to 1.0 over time.

   As the figure shows, the project performs as expected in the first two periods, but outperforms expectations in the third period, noted by the error term being greater than zero. In the fourth year, the error is calculated as

$$e_{2005} = \frac{\text{PB}_{2005}(i) - \tilde{\text{PB}}_{2005}(i)}{\tilde{\text{FW}}(i)} = \frac{-\$1578\text{M} - -\$758\text{M}}{\$3166\text{M}} = 0.259.$$

   Since the error is positive, the project is performing better than expected.

2. Should the project continue in its present state, or should some changes be made?

TABLE 15.3 **Expected and realized cash flows and project balances (in M) for automobile production.**

Year	Estimated		Actual		Error	
n	A_n	PB_n	A_n	PB_n	e_n	$\tilde{e_N}$
2002	–\$700	–\$700	–\$700	–\$700	0.0000	0.0000
2003	–\$700	–\$1540	–\$700	–\$1540	0.0000	0.0000
2004	\$200	–\$1648	\$550	–\$1298	0.1105	1.1055
2005	\$400	–\$1578	\$800	–\$758	0.2590	1.4466
2006	\$600	–\$1293	–	–	–	–
2007	\$600	–\$952	–	–	–	–
2008	\$600	–\$542	–	–	–	–
2009	\$600	–\$51	–	–	–	–
2010	\$600	\$539	–	–	–	–
2011	\$600	\$1247	–	–	–	–
2012	\$600	\$2097	–	–	–	–
2013	\$650	\$3166	–	–	–	–

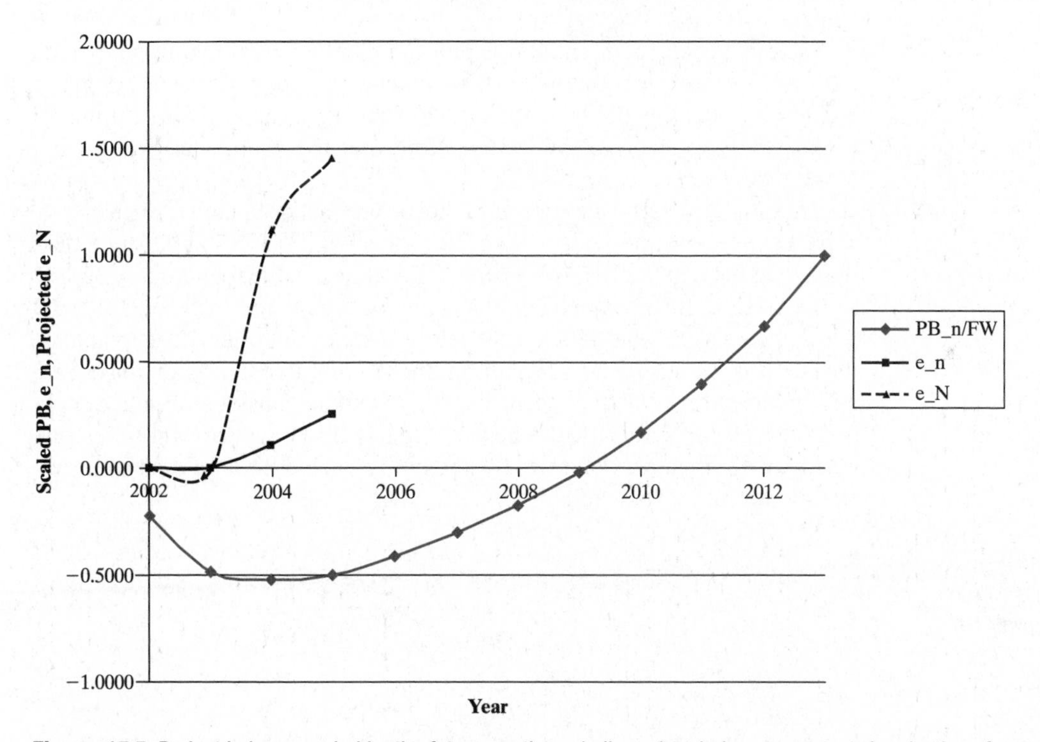

Figure 15.5 Project balance scaled by the future worth, periodic project-balance errors, and projection of errors to period N for automobile plant investment.

We can continue the analysis by projecting the slope of the error from the third and fourth periods to the final period, resulting in our estimated value of e_N ($\tilde{e_N}$) as:

$$\tilde{e_{2013}} = 0.259 + (0.259 - 0.1105)(2013 - 2005) = 1.4466.$$

TABLE 15.4 **Probabilities of annual sales levels, given expansion time.**

Expansion	Annual Sales		
Time	100,000	150,000	200,000
0	0.50	0.25	0.25
1	0.25	0.50	0.25
2	0.25	0.25	0.50

Since the projected error in the final period is greater than one, examining an expansion option is justified.

3. If changes should be made, how should they be analyzed?

Because an expansion may be justified, our analysis should follow that typical of Chapters 9 through 11 and 13, paying specific attention to dynamic options, since there is flexibility as to when to start an expansion. Consider the following example:

Assume that a $500 million expansion can be completed in one year's time to increase capacity by 200,000 vehicles per year. However, demand increases either by 100,000, 150,000, or 200,000 vehicles, with estimated probabilities as given in Table 15.4. Note that the probabilities of having better sales increase with time.

Assume further that per unit costs are as predicted ($33,000 per vehicle), but revenues from *all* vehicle sales will be $35,000 per unit (less than is realized in the first two years of operations), due to increased supply to the system. If the expansion is not pursued, revenues will total $35,500 per vehicle. All of the data are as before, except that the salvage value of the plant is increased by $25 million and annual fixed costs increase $7.5 million per year after the expansion, which can be considered immediately or at the end of either of the next two periods. The expansion takes one period, with no impact on sales for the entire period. Should the expansion commence and if so, when?

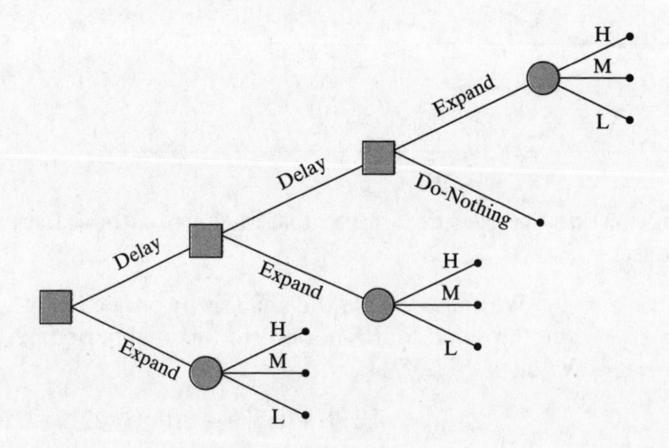

Figure 15.6 Decision tree for expansion of automobile production capacity.

The decision tree that describes the problem being addressed here is given in Figure 15.6. The decision to expand leads to a chance node with three arcs defining the level of sales (H for high, M for medium, and L for low).

Since this is a decision tree with probabilistic arcs, we must traverse it backwards. We begin with the final decision node of whether to expand at period 2 or continue with operations. (Either option assumes that no expansions occurred earlier.) If the expansion is to be undertaken, then the present worth (valued in period 2) *of all sales costs* is

$$E\,(\text{PW}(20\%)) = \$800\text{M} - \$500\text{M} + 0.25\left(\$793\text{M}(\overset{P/A,20\%,5}{2.9906})\right) + 0.25\left(\$893\text{M}(\overset{P/A,20\%,5}{2.9906})\right) + 0.50\left(\$993\text{M}(\overset{P/A,20\%,5}{2.9906})\right) + \$75\text{M}(\overset{P/F,20\%,5}{0.4019}) = \$3104 \text{ million.}$$

The present worth of the do-nothing alternative (in period 2) is computed as

$$\text{PW}(20\%) = \$800\text{M} + \$800\text{M}(\overset{P/A,20\%,5}{2.9906}) + \$50\text{M}(\overset{P/F,20\%,5}{0.4019}) = \$3213 \text{ million.}$$

At this stage in the tree, the best decision is to do nothing, as it carries a larger present worth. Peeling back this value through the tree to period 1, we evaluate the decisions as follows:

$$E\,(\text{PW}(20\%)) = \$800\text{M} - \$500\text{M} + 0.25\left(\$793\text{M}(\overset{P/A,20\%,6}{3.3255})\right) + 0.50\left(\$893\text{M}(\overset{P/A,20\%,6}{3.3255})\right) + 0.25\left(\$993\text{M}(\overset{P/A,20\%,6}{3.3255})\right) + \$75\text{M}(\overset{P/F,20\%,6}{0.3349}) = \$3318 \text{ million.}$$

For the do-nothing decision at time 1, which is essentially the delay option

$$\text{PW}(20\%) = \$800\text{M} + \frac{\$3213\text{M}}{1 + 0.20} = \$3477 \text{ million.}$$

Again, the decision is to do nothing (delay), as it leads to a higher present worth. Finally, the decision to expand at time zero carries an expected present worth of

$$E\,(\text{PW}(20\%)) = \$800\text{M} - \$500\text{M} + 0.50\left(\$793\text{M}(\overset{P/A,20\%,7}{3.6046})\right) + 0.25\left(\$893\text{M}(\overset{P/A,20\%,7}{3.6046})\right) + 0.25\left(\$993\text{M}(\overset{P/A,20\%,7}{3.6046})\right) + \$75\text{M}(\overset{P/F,20\%,7}{0.2791}) = \$3469 \text{ million.}$$

For the delay alternative,

$$PW(20\%) = \$800M + \frac{\$3477M}{1 + 0.20} = \$3698 \text{ million.}$$

Thus, the decision at time zero, and in each ensuing period, is to delay the expansion. Consequently, the plant would be expected to continue to work at full capacity because an expansion is not justified at this time.

The preceding analysis illustrates that just because things are going well does not necessarily justify new investment. A rigorous analysis must be performed to make the right decision.

15.7 Key Points

1. A project should be tracked in order to (1) motivate future options, including abandonment, expansion, or replacement, and (2) improve the decision-making process.

2. Errors in estimates can be measured with the statistics MAD, MSE, and MAPE. These measures may also separate negative from positive errors.

3. A project can be tracked by comparing its estimated project balance with the project balance that actually occurs.

4. The trajectory of the project-balance error may provide motivation to examine future options, including expansion or abandonment of the project.

5. Tracking a project may motivate updating forecasts.

6. A database of all relevant information about a project should be maintained to help with future estimation processes.

7. Tracking data is a natural company function that is required in financial reporting. Typical data to be shared include income statements, cash flow statements, and balance sheets.

8. Activity-based costing is an alternative way of tracking costs that differs from traditional cost accounting in that overhead costs are more accurately assigned to their relevant activities or products.

15.8 Further Reading

Canada, J.R., W.G. Sullivan, and J.A. White, *Capital Investment Analysis for Engineering and Management,* 2d ed. Prentice Hall, Upper Saddle River, New Jersey, 1996.

DeVor, R.E., T. Chang, and J.W. Sutherland, *Statistical Quality Design and Control: Contemporary Concepts and Methods*. Macmillan Publishing Company, New York, 1992.

Mitra, A., *Fundamentals of Quality Control and Improvement,* 2d ed. Prentice Hall, Upper Saddle River, New Jersey, 1998.

Herath, H.S.B., C.S. Park, and G.C. Prueitt, "Monitoring Projects Using Cash Flow Control Charts," *The Engineering Economist,* 41(1):27–52, 1995.

Newbold, P., and T. Bos, *Introductory Business and Economic Forecasting,* 2d ed. South-Western Publishing Co., Cincinnati, 1994.

Park, C.S., and G.P. Sharp-Bette, *Advanced Engineering Economics*. John Wiley and Sons, New York, 1990.

Prueitt, G.C., and C.S. Park, "Monitoring Projects Performance with Post-Audit Information: Cash Flow Control Charts," *The Engineering Economist,* 36(4):307–335, 1991.

Riggs, H.E., *Financial and Cost Analysis for Engineering and Technology Management*. John Wiley and Sons, New York, 1994.

15.9 Questions and Problems

15.9.1 Concept Questions

1. Give two reasons for tracking an investment.
2. What does computing the MAPE of an investment tell us? How should this information be used by a decision maker? How does the information differ compared with calculating MAPE^+ or MAPE^-? Explain.
3. Why does it make sense to track a project with its project balance?
4. How should the information from tracking a project be used? Should decisions be made directly from that information? Why or why not?
5. What might be some other ways in which to forecast e_N? What are the advantages and disadvantages of your proposed methods?
6. What is the relationship between tracking a project and maintaining a database? What items should be kept in a database for future estimation purposes?
7. How are cost accounting and financial reporting related to the postimplementation step in the decision-making process?
8. What are the three major accounting statements, and what does each of them reveal to an investor?
9. How do project returns affect the accounting statements?
10. What is ABC, and why has it received attention for both product costing and cost estimation?

15.9.2 Drill Problems

1. Consider the following data from a recently completed project, and answer the questions that follow:

Period n	$\tilde{A}_n$	$\tilde{PB}_n(12\%)$	A_n	$PB_n(12\%)$
0	−$100.00	−$100.00	−$95.00	−$95.00
1	$25.00	−$87.00	$26.47	−$79.93
2	$30.00	−$67.44	$30.14	−$59.38
3	$35.00	−$40.53	$31.89	−$34.62
4	$40.00	−$5.40	$33.65	−$5.12
5	$45.00	$38.96	$37.01	$31.28
6	$50.00	$93.63	$40.32	$75.35
7	$55.00	$159.87	$41.03	$125.43
8	$60.00	$239.05	$44.13	$184.61

(a) Compute the MAPE of the estimate errors. Are your results acceptable?

(b) Use the project-balance errors to assess the project over its lifetime. Was the correct decision made to leave this project alone? If not, what other options should have been examined and when? Support your answer.

2. Consider the following, different outcome for the project in Drill Problem 1, and answer the questions that follow:

Period n	$\tilde{A}_n$	$\tilde{PB}_n(12\%)$	A_n	$PB_n(12\%)$
0	−$100.00	−$100.00	−$95.00	−$95.00
1	$25.00	−$87.00	$20.14	−$86.26
2	$30.00	−$67.44	$32.27	−$64.34
3	$35.00	−$40.53	$60.34	−$11.71
4	$40.00	−$5.40	$72.02	$58.90
5	$45.00	$38.96	$97.18	$163.15
6	$50.00	$93.63	$108.30	$291.02
7	$55.00	$159.87	$109.72	$435.66
8	$60.00	$239.05	$128.12	$616.06

(a) Compute the MAPE of the estimate errors. Are your results acceptable?

(b) Use the project-balance errors to assess the project over its lifetime. Was the correct decision made to leave this project alone? If not, what other options should have been examined and when? Support your answer.

3. Consider the project from Drill Problem 1 one last time, but assume that it is only half complete:

Period n	$\tilde{A}_n$	$\tilde{PB}_n(12\%)$	A_n	$PB_n(12\%)$
0	−$100.00	−$100.00	−$95.00	−$95.00
1	$25.00	−$87.00	$26.67	−$79.73
2	$30.00	−$67.44	$9.10	−$80.19
3	$35.00	−$40.53	$0.82	−$88.99
4	$40.00	−$5.40	−$1.20	−$100.87
5	$45.00	$38.96	—	—
6	$50.00	$93.63	—	—
7	$55.00	$159.87	—	—
8	$60.00	$239.05	—	—

Given the status of the project, what, if anything, should be done? Justify your answer and suggest the next steps.

4. Suggest another method in which to project the error to period N. Illustrate the approach with the data from the previous three problems. What are the benefits of your approach compared with the one provided in this chapter? What are the disbenefits?

5. Compute the $MAPE^{+}$ and $MAPE^{-}$ for Drill Problems 1 and 2. Does there appear to be a problem with the estimation process? Explain.

15.9.3 Application Problems

1. In May of 2004, Reliant Energy mothballed an 822-MW power plant in Choctaw County, Mississippi. The natural-gas-fired, combined-cycle facility opened in July of 2003, but wholesale energy prices were too low to cover cash costs to operate the facility. The plant would be mothballed until the wholesale market improved.[4]

   Make the following assumptions: Assume that the plant cost $340 million to construct. The cost was evenly split over the eight quarters preceding July of 2003. At the time the facility was approved, it was assumed that it would generate $350 million per year in revenues against costs of $200 million per year over a 10-year horizon with a salvage value of $10 million. Assume that actual quarterly revenues for the 2 years during which the plant operated were $75 million, due to depressed wholesale markets, with costs totalling $65 million per quarter because of drastic increases in natural-gas prices. The interest rate is assumed to be 2.5% per quarter.

   (a) Compute the MAD, MSE, and MAPE of the estimation errors for the data that have been realized, given that the construction costs followed the estimated costs.

   (b) Chart the actual project balance against the expected project balance. Compute the errors in each period.

   (c) Project the error in the project-balance estimate through the end of the horizon. Do the data motivate examining the delay or abandonment option? Explain.

2. Consider, again, the previous problem. Assume that it costs $100,000 per quarter to mothball the facility (i.e., delay operations), but nothing to restart operations.

   (a) Assume that it is July of 2003. Revenues are expected to drop to $70 million, $65 million, and $60 million over the next three quarters before improving by $10 million each quarter thereafter (as other energy sources are retired), until peaking at $120 million per quarter. O&M costs are expected to remain at $70 million per quarter (if the plant is operating) for the foreseeable future. To limit the analysis, assume that a delay must occur immediately, but can last up to two

[4] McNamara, K., "Reliant Energy Mothballs 822-Megawatt Mississippi Power Plant; Plans for Mothballing PJM Assets Change," *Dow Jones Newswires*, May 4, 2004.

years. Should the facility be mothballed? If so, for how long? Assume a 5-year horizon and a $25 million salvage value.

(b) Repeat the analysis of part (a) if the operating costs are expected to increase $200,000 per quarter for five quarters after the initial $70 million for the first quarter.

(c) Repeat part (a), but this time assuming that the electricity prices are probabilistic. This will result in either a solid revenue stream of $75 million, $70 million, $80 million, $95 million, and $120 million over the first five quarters, continuing at $120 million for the remainder of the horizon; an average revenue stream of $70 million, $70 million, $80 million, $90 million, and $110 million over the first five quarters, continuing at $110 million for the remainder of the horizon; or a poor revenue stream of $70 million, $70 million, $70 million, $90 million, and $100 million over the first five quarters, continuing at $100 million for the remainder of the horizon. Consider delays of 0, 1, or 2 quarters. If the solid stream of revenues has a 20% probability of occurring, 50% for the average, and 30% for the poor, what is the best strategy?

3. At the end of 2003, Toyota announced that it would discontinue production of the sport-utility vehicle Voltz, jointly manufactured with General Motors, in the spring of 2004. The vehicle, produced in Fremont, California, was released exclusively for the Japanese market in August of 2002. Total sales for the first year were 8802, far short of the 1500 cars per month hoped.[5]

Make the following assumptions: Assume that Toyota and General Motors invested $400 million to produce the Voltz, with expected sales of 1500 per month generating profits of $6500 per vehicle over a five-year period with a $100 million salvage value. Assume further an interest rate of 0.75% per month.

(a) Assume that monthly sales were 400, 400, 400, 500, 600, 700, 800, 900, 1000, 1000, 1000, and 1000 and that profits averaged $5000 per vehicle. Calculate the error between the estimates and the cash flow realizations. (Assume that the estimated investment cost was paid.) Calculate also the MAD, MSE, and MAPE for the flows. Is it worth calculating $MAPE^+$ or $MAPE^-$? Explain.

(b) Chart the actual project balance against the expected project balance. Compute the errors in each period.

(c) Project the error in the project-balance estimate through the end of the horizon. Do the data motivate examining the delay or abandonment option? Explain.

4. General Shales contemplated expanding its brick plant in Brickhaven, North Carolina, to meet strong market demand. The expansion would cost $5.6 million to increase capacity from 60 million to 90 million units per year.[6]

Make the following assumptions: The original 60 million unit annual capacity facility was constructed two years prior to the expansion at the cost of $16 million. The plant was expected to produce profits of $0.09 per unit, with fixed annual costs

---

[5] "Toyota to Halt Voltz SUV Production at US Plant in Spring," *Dow Jones Newswires*, December 24, 2003.

[6] Maya, B., "Wienerberger Expands Capacity in US Brick Plants," *Dow Jones Newswires*, April 30, 2004.

of $1 million per year. Sales were expected to grow from 10 million units in the first quarter at a rate of 5% per quarter until reaching the capacity of 60 million units. However, sales totaled 12 million in the first period, grew to 13.5 million in the second quarter, and reached capacity in the third quarter. In addition, profits of $0.11 were realized for the first two years of operations. The project was approved with a MARR of 18% compounded quarterly and an assumed eight-year plant life with no salvage value.

(a) Calculate the error between the estimates and the cash flow realizations. (Assume that the estimated investment cost was paid.) Calculate the MAD, MSE, and MAPE for the flows.

(b) Chart the actual project balance against the expected project balance. Compute the errors in each period.

(c) Project the error in the project-balance estimate through the end of the horizon. Do the data motivate examining the delay, abandonment, or expansion option? Explain.

(d) Reevaluate the chart, assuming that sales were realized as expected (12 million in the first period, etc.), but profits were $0.08 per unit and costs were $300,000 per quarter. Compute the project-balance errors in each period.

(e) Again, project the error to the end of the project. How should this information be read? Explain.

5. Bema Gold Corporation and Kinross Gold Corporation placed their Refugio, Chile, mine on care and maintenance in May of 2001 due to declining gold prices and decreasing gold production. The two companies recommenced mining operations in December of 2003.[7]

Make the following assumptions: In order to define a baseline to motivate the decision to delay mining operations (i.e, place the mine on care and maintenance), assume that an investment of $130 million was made at the end of 1995, with gold production commencing in 1996. Annual production was expected to be 230,000 ounces over 10 years, with a gold price of $350 per ounce against operating costs of $225 an ounce. Fixed operating and maintenance costs total $500,000 per year, and $5 million in remediation costs are expected at the end of the mine's life.

Assume further that production began at 150,000 ounces in 1996, increasing by 10,000 ounces per year through 2000, and that the operating cost was $240 per ounce over this period. The price of gold in 1996 through 2000 was $390, $330, $290, $280, and $280 per ounce, respectively.

Given the preceding information, chart the project, using the project balance, under the assumption that it is the end of 2000 and the interest rate is 14% per year. Is the decision to suspend operations justified? Support your answer.

6. Select a company and download its annual report. Determine the amount of the company's capital expenditure in the last fiscal year, and determine where it affects the balance sheet, income statement, and cash flow statement.

---

[7] "Bema Gold Corp Starts Gold Mining at Refugio Mine, Chile," *Dow Jones Newswires,* December 8, 2003.

15.9.4 Fundamentals of Engineering Exam Prep

1. Sunk costs

(a) Are important in determining when to abandon a project.

(b) Should influence future decisions.

(c) Should be ignored when one is making an abandonment decision.

(d) None of the above.

2. Activity-based costing is

(a) A method of allocating costs to activities.

(b) A method of costing products.

(c) A method of allocating indirect or overhead costs to actual products or processes.

(d) All of the above.

3. Tracking an investment after it has started is important to

(a) Verify estimated quantities and costs.

(b) Provide data for future estimation tasks and investment decisions.

(c) Provide input to determine when a project should be terminated.

(d) All of the above.

4. If a project is progressing well, it may be prudent to

(a) Ignore the project altogether.

(b) Examine the option to expand.

(c) Abandon the project.

(d) Replace technologies used in the project.

5. If a project is progressing below expected standards, it may be prudent to

(a) Consider replacing technologies or processes.

(b) Examine the option to abandon.

(c) Examine the option to expand.

(d) Both (a) and (b).

16 Abandonment and Replacement Analysis

(*Courtesy of American Eurocopter LLC.*)

Real Decisions: Speed May be Monitored from Aircraft Above

In 2006, the city of San Diego approved the purchase of four new helicopters from American Eurocopter to replace a 38-, a 31-, and two 12-year-old helicopters. The old fleet was appraised at $1.75 million, while the new machines cost $12.2 million, including training and accessories. Although the oldest helicopter was retrofit at the cost of $393,000 five years earlier, the aging helicopters have corrosion problems and increasing unexpected maintenance costs. The city is financing the purchase over 7 years at a total cost (with interest) of $14.2 million. The helicopters are expected to last for 15 years. The helicopters fly about 3000 hours per year.[1]

[1] Hall, M.T., "City to replace 4 helicopters by borrowing up to $13 million," *The San Diego Union-Tribune,* p. B-4, November 15, 2005.

Make the following assumptions: The 31-year-old helicopter has a $90K market value, the 12-year-old helicopters are worth $670K each, and the 38-year-old helicopter is worth $320K, because the retrofitting left it currently valued and operating as a 19-year-old machine. A new helicopter costs $3.05 million. The value of a helicopter drops 30% after its first year of use and 10% each year thereafter.

Routine operating and maintenance costs are $175 per operating hour (3000 hours per year, split evenly among the fleet), increasing at a rate of 8% per year. Unexpected maintenance costs are $1000 per failure. The failure rate is increasing with age such that the expected number of failures for a helicopter of age n is estimated by

$$\text{Failures}_n = \lfloor (0.25)(1 + 0.30)^{n-1} \rfloor.$$

Thus, the expected number of failures is growing at a rate of 30% from 0.25 in the first year, but this value is rounded down to the nearest integer, such that a new helicopter is not expected to experience a failure. The interest rate is 5% per year. These data lead to a number of interesting questions:

1. What is the economic life of a new helicopter?
2. Should the 31-year-old helicopter be replaced?
3. How can the financing be included in the analysis?

In addition to answering these questions, after studying this chapter you will be able to:

- Determine when a project should be terminated, or abandoned. (Section 16.1)
- Define the economic life of an asset. (Section 16.2)
- Compute the best time to replace an asset under the assumption of deterioration or technological change (or both). (Section 16.2)
- Compute an after-tax cash flow for a given replacement strategy. (Section 16.2)

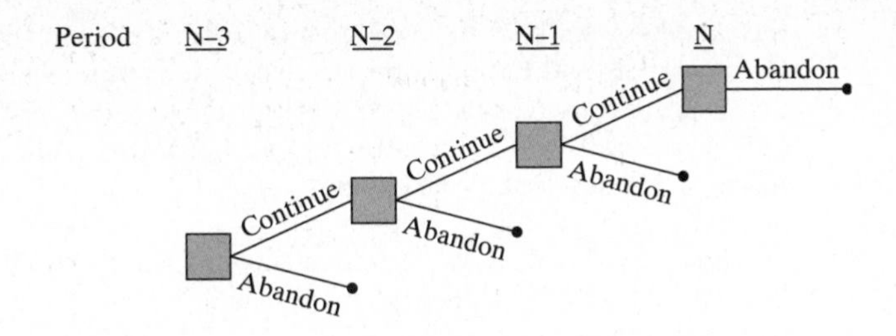

Figure 16.1 Network of possible abandonment decisions.

When a project is evaluated for possible investment, the horizon N of the project is generally defined. However, the estimate is subject to change as information used to make an initial decision becomes available with time. In the previous chapter, we illustrated how one could track an investment through time. If deviations from the actual project outcomes are significantly different from those predicted, the decision maker must contemplate the possibility that a project could be doing unexpectedly better or worse. If the former is true, then one may consider accelerating or expanding the project. The techniques used to analyze this option were given in Chapters 9 and 13.

If the latter is true, then one may consider delaying some portion of the project. Techniques for this dynamic analysis were given in Chapter 13. Replacing equipment may help improve a project, as analyzed later in the current chapter. However, if the project does not have any hope of improving, even with a delay, then it may be wise to abandon it altogether and reinvest any remaining proceeds into other projects with better potential.

16.1 Abandonment Analysis

We now turn our attention to the abandonment option. Generally, if we are in period n of a project that is expected to last for N periods, then we have $N - n + 1$ possible decisions: (1) abandon now; (2) abandon after one period; (3) abandon after two periods; and finally, option $(N - n + 1)$, abandon at the end of period N, as originally planned.

These decisions are depicted in the network in Figure 16.1. Note that the network is completely general in that we can visit the abandon decision in each period.

We motivate our discussion through a detailed set of examples, the next one of which sets the stage by tracking a project over time.

EXAMPLE 16.1

Tracking a Project

In March of 2004, General Motors (GM) announced that it would cease production of the Saturn L-300 sedan ahead of schedule due to low sales.[2] GM spent $550 million on

[2] "GM to Close Delaware Plant for Almost a Year," *Dow Jones Newswires*, March 1, 2004.

the plant between 1996 and 1999[3] to produce the sedan in hopes of competing with the Honda Accord and Toyota Camry, each of which sold about 200,000 vehicles annually.[4] Sales lagged considerably, near 24,000 (in 1999), 94,000 (2000), 98,000 (2001), 81,000 (2002), and 65,000 (2003) over time.[5,6] GM redesigned the 2003 model and increased advertising, but sales did not budge.

We make the following assumptions to define our baseline: The $550 million investment in the plant was spread evenly in 1996 through 1999. Expected sales were 100,000 vehicles in 1999, 150,000 in 2000, and 200,000 in years 2001 through 2006. The plant had a resale value of $50 million at the end of 2006, as it could be used for other vehicles or for a replacement vehicle for the L-series. Since the price tag of the L-series is between $16,000 and $24,000, we assume a $20,000-per-car revenue at the cost of $18,500 per car to produce and $10 million per year in fixed costs. Finally, assume that the redesign and additional advertising cost $50 million in 2002. The MARR is 12%.

Solution. Table 16.1 lists the expected and realized cash flows and project balances for the automobile assembly plant. The expected values run from 1996 through 2006, while the realized flows are from 1996 through 2003. The future worth of the original investment in 2006 is $1.68 billion. This amount represents the baseline to be used in our analyses for tracking the investment. The estimated and actual project balance over time are graphed in Figure 16.2.

The final two columns in Table 16.1 illustrate the error. Recall from the previous chapter that e_n is the difference between the expected and actual project balance in

TABLE 16.1 Expected project balances and realizations (in millions) over time.

Year	Estimated		Actual		Error	
n	A_n	PB_n	A_n	PB_n	e_n	$\tilde{e}_N$
1996	–$137.50	–$137.50	–$137.50	–$137.50	0	—
1997	–$137.50	–$291.50	–$137.50	–$291.50	0	0
1998	–$137.50	–$463.98	–$137.50	–$463.98	0	0
1999	$2.50	–$517.16	–$111.50	–$631.16	–0.068	–0.541
2000	$215.00	–$364.22	$131.00	–$575.90	–0.126	–0.474
2001	$290.00	–$117.92	$137.00	–$508.00	–0.232	–0.761
2002	$290.00	$157.93	$61.50	–$507.46	–0.395	–1.049
2003	$290.00	$466.88	$87.50	–$480.86	–0.563	–1.065
2004	$290.00	$812.90				
2005	$290.00	$1,200.45				
2006	$340.00	$1,684.51				

[3] Tadesse, L. B., "Boxwood Plant's Fate is on the Line," *The News Journal*, www.delawareonline.com, June 22, 2003.

[4] McCracken, J., "GM Plans to Nix Saturn L-300 and Replace It with a Sportsback," *Knight Ridder Newspapers*, CTCarsAndTrucks.com, December 6, 2003.

[5] "Saturn: Production and Sales Figures," www.autointell.net.

[6] Garsten , E., "Saturn May Drop L-Series," *The Detroit News*, www.detnews.com, February 29, 2004.

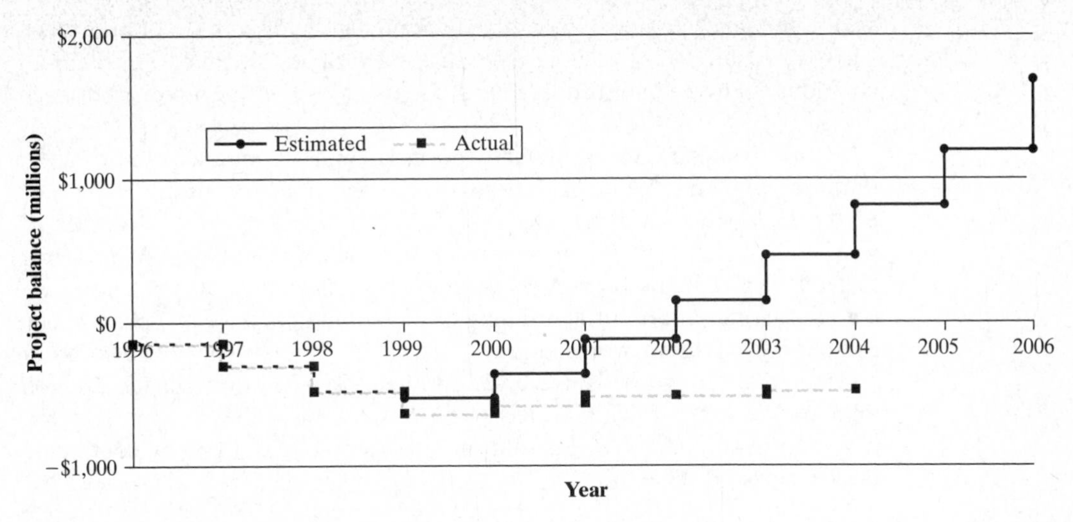

Figure 16.2 Expected and actual project balances for car plant investment.

year n, normalized by the expected future worth. For example, in year 2000,

$$e_{2000} = \frac{PB_{2000}(i) - \tilde{PB}_{2000}(i)}{\tilde{FW}(i)} = \frac{-\$575.90M - -\$364.22M}{\$1684.51M} = -0.126.$$

The estimated value of e_N ($\tilde{e_N}$) is computed by projecting the slope of the errors from the previous two readings, such that, at year 2000,

$$\tilde{e_{2006}} = -0.126 + (-0.126 - -0.068)(2006 - 2000) = -0.474.$$

The negative reading implies that the project is not realizing its potential due to sluggish sales. GM invested in the line with a redesign and increased advertising, but the predicted error in 2002 was −1.049, signaling that the project was in jeopardy of losing money (in terms of future worth). The fact that $\tilde{e_N}$ is less than −1 signals that examining the abandonment option for the project is warranted.

The preceding example indicates that we may want to consider abandoning the project, not only because it is not generating the expected returns, but because it may not achieve a positive return on the investment—meaning that the future worth of the project may turn out to be negative. This information provides an impetus to examine the abandonment option in detail.

16.1.1 Dynamic Deterministic Evaluation

We return to our decision network of Figure 16.1 to consider the abandonment option over time. Each path defines a cash flow diagram for a specific decision as to when to abandon the project. A decision to abandon is associated with a salvage value, or possibly a remediation cost, to discontinue the project, while a decision to continue is generally met with additional revenues for an additional period. As illustrated in the figure, the decision is revisited each period.

Note that each abandonment option (timing) is mutually exclusive in that only one abandonment option is to be chosen. Note further that each of these decisions has an unequal life, corresponding to the number of years the project remains in operation until being abandoned. We assume that this is a revenue project, as we must be generating revenues if we are considering the option of continuing or abandoning a project. Thus, we can compute the present worth of each option for comparison. (This strategy assumes that all proceeds are invested at the MARR through the horizon time of the longest-lived project—the longest time that the project can remain operational.) We illustrate the abandonment analysis by continuing our previous example.

EXAMPLE 16.2

Dynamic Abandonment Analysis

In the previous example, the historical data were used merely to illustrate how the abandonment option was motivated. We now assume that it is the end of 2003. The options available are to abandon the project now or continue through 2004, 2005, or 2006. Assume that sales in 2004 will total 30,000 vehicles, decreasing 5000 each year through 2006. Assume further that the profit per vehicle declines to $900 per unit in 2004 and an additional $100 each ensuing year, due to underutilization of the factory and increased advertising to promote sales. Finally, assume that the salvage value of the plant declines $12 million per period due to machinery neglect (since the factory is underutilized) and it costs $25 million to close the plant (costs include employee severance, etc.).

Solution. The decision network is illustrated in Figure 16.3. Each node represents the period when a decision can be made, and each arc in the network is populated with the one-period cost or revenue associated with each decision.

The figure defines the four options to abandon the project either in 2003 (time 0), 2004 (time 1), 2005 (time 2), or 2006 (time 3). Note that revenues are expected to be $27 million, $20 million, and $14 million in years 2004 through 2006, respectively, with

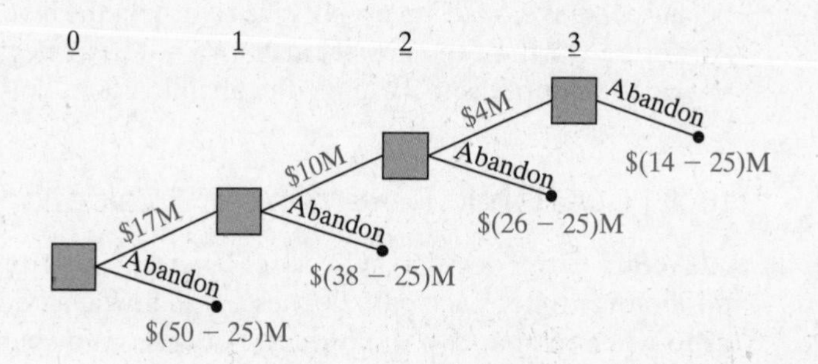

Figure 16.3 Abandonment options in years 2003 through 2006.

annual O&M costs of $10 million. The present worth of each option is as follows:

$$PW_0(12\%) = \$50M - \$25M = \$25 \text{ million};$$

$$PW_1(12\%) = (\$17M + \$38M - \$25M)\overset{P/F,12\%,1}{(0.8929)} = \$26.79 \text{ million};$$

$$PW_2(12\%) = \$17M\overset{P/F,12\%,1}{(0.8929)} + (\$10M + \$26M - \$25M)\overset{P/F,12\%,2}{(0.7972)} = \$23.95 \text{ million};$$

$$PW_3(12\%) = \$17M\overset{P/F,12\%,1}{(0.8929)} + \$10M\overset{P/F,12\%,2}{(0.7972)} + (\$4M + \$14M - \$25M)\overset{P/F,12\%,3}{(0.7118)}$$
$$= \$18.17 \text{ million}.$$

From this analysis, the best option is to operate the plant for one more period and then abandon the project.

16.1.2 Probabilistic Evaluation

We can expand the abandonment analysis to include probabilistic information. Our previous decision network is expanded into the decision tree in Figure 16.4. After each decision, a chance node signals the possibility of multiple outcomes, which are identified on the ensuing arcs.

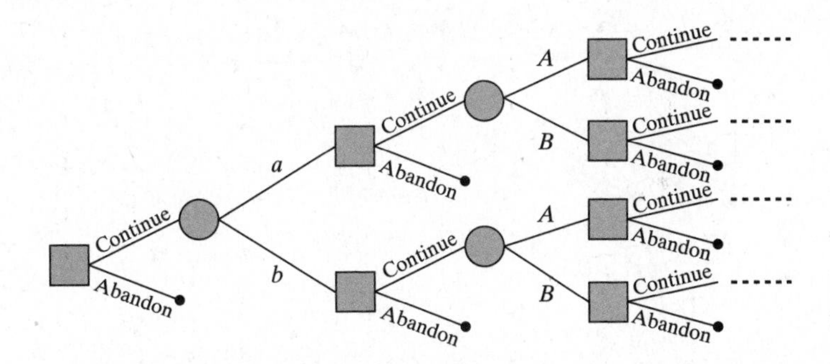

Figure 16.4 Decision tree for abandonment option.

The decisions have not changed, but the best option must be determined by finding the path with the minimum expected cost or maximum present worth from the start node through all possible terminal nodes. This requires that the solution be backtracked through the network. We illustrate the procedure in the next example, which revisits the investment in the new automobile design.

EXAMPLE 16.3 *Probabilistic Abandonment Analysis*

We continue our analysis started in the previous example. It is assumed that sales can reach between 10,000 and 60,000 over time and the level in each ensuing year is dependent on the previous year's sales. The levels of sales and the associated probabilities are given in the decision tree in Figure 16.5. All other data are the same as before.

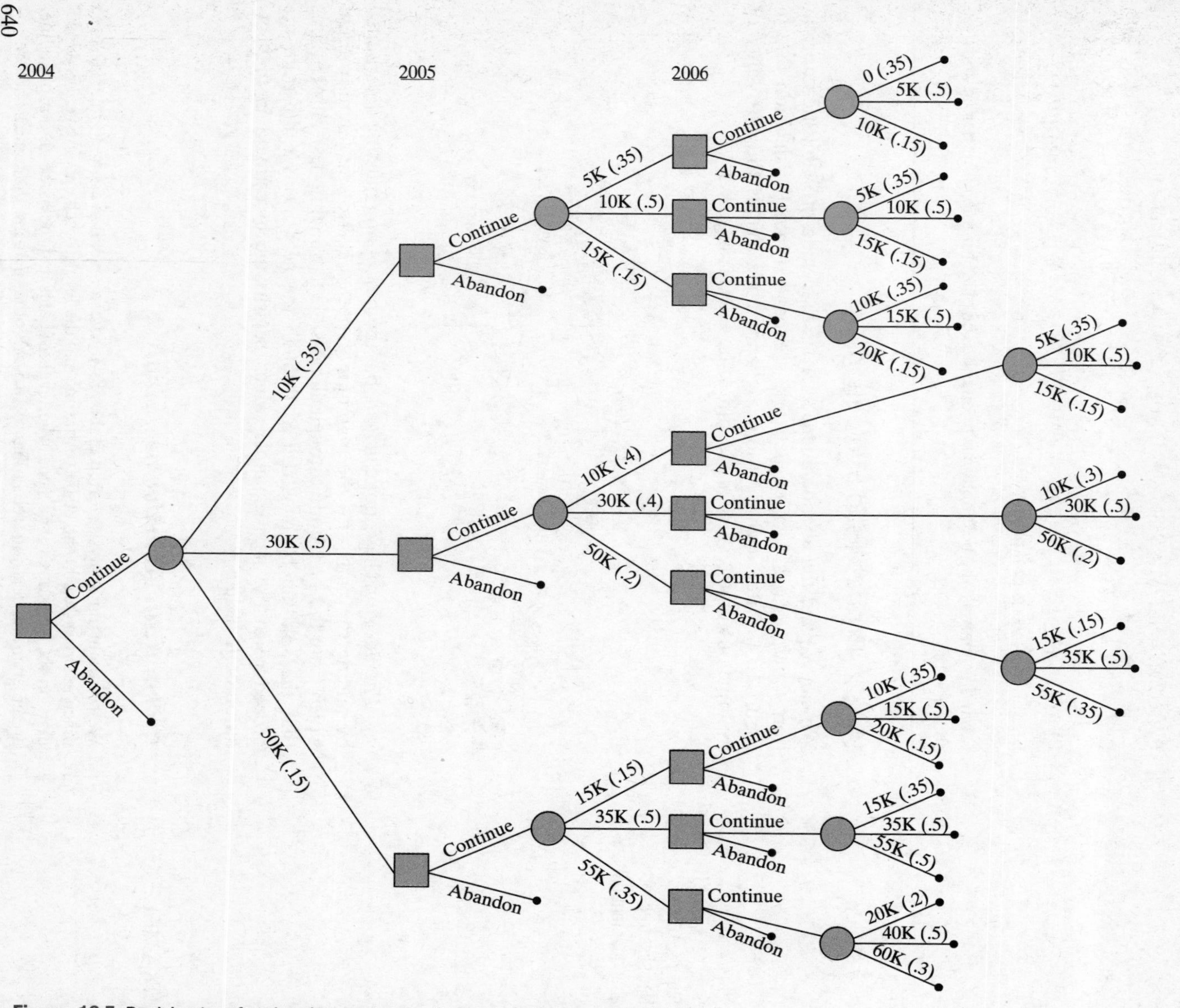

Figure 16.5 Decision tree for abandonment option for automobile assembly plant.

Solution. Figure 16.5 illustrates the decisions at each period. The present worth of each option is dependent on the level of sales in each period. It should be clear from the figure that these problems can become quite complicated if there are numerous possible outcomes at each chance node and numerous decisions at each decision node.

Consider the top right of the decision tree. This is the path followed if we decide to continue operations in each period, yet experience the lowest possible demand in each period. If demand is 10,000 vehicles in 2004 and 5000 vehicles in 2005, we would be faced with a decision in 2006 to continue operations or abandon the project. If we abandon the project at the end of 2005, we receive \$1 million from the salvage value of \$26 million, less \$25 million in closing costs. If we decide to continue operations at this point, the expected revenues are

$$(0.35)\$0 + (0.50)(5000 \text{ vehicles})\left(\frac{\$700}{\text{vehicle}}\right) + (0.15)(10{,}000 \text{ vehicles})\left(\frac{\$700}{\text{vehicle}}\right)$$
$$= \$2.8 \text{ million}.$$

With \$10 million in operating costs, a \$14 million salvage value, and a \$25 million cost to close the facility, the net cash flow for this decision is

$$\$2.8\text{M} - \$10\text{M} + \$14\text{M} - \$25\text{M} = -\$18.2 \text{ million}.$$

The decision is between the \$1 million net salvage value and $-\$18.2$ million/1.12 = $-\$16.25$ million, as the two options occur one period apart. The decision with the greatest present worth—to abandon the project—is chosen for this node. The analysis can be repeated for each of the eight remaining scenarios in 2006 (end 2005). The maximum of the expected values between operating and abandoning is given on the decision node in Figure 16.6.

Continuing back in the tree to 2004, we analyze a similar decision to illustrate the rollback procedure. Assume that we have just experienced demand of 50,000 vehicles in 2004. Our decision is whether to abandon the project and receive \$13 million in net salvage value (\$38 – \$25 million) or continue. If we continue, the expected value of the decision at the end of 2005 is

$$(0.15)\left[(15{,}000 \text{ vehicles})\left(\frac{\$800}{\text{vehicle}}\right) + \$1\text{M}\right]$$
$$+ (0.50)\left[(35{,}000 \text{ vehicles})\left(\frac{\$800}{\text{vehicle}}\right) + \$1\text{M}\right]$$
$$+ (0.35)\left[(55{,}000 \text{ vehicles})\left(\frac{\$800}{\text{vehicle}}\right) + \$7.5\text{M}\right] - \$10\text{M} = \$24.5 \text{ million}.$$

Note that the \$1 million, \$1 million, and \$7.5 million on the end of the revenue streams come from the value of the optimal decision in the ensuing period, which we have already calculated in the previous step of this procedure. We choose to continue operations here, as \$21.9 million (\$24.5 million/1.12) is greater than the \$13 million abandon option.

Completing the calculations for the tree leaves us with the final decision between abandoning the project in 2003 for \$25 million (\$50 million salvage value, less \$25

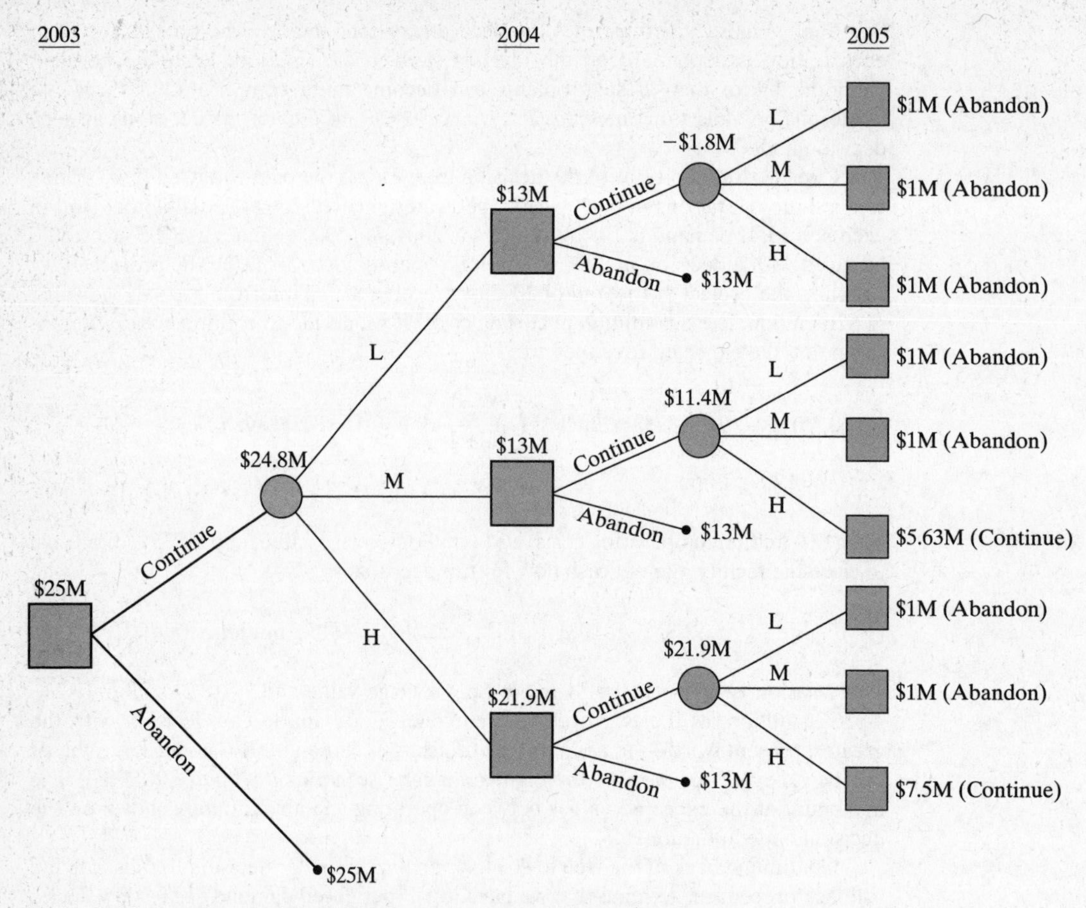

Figure 16.6 Complete decision tree solution for automotive plant example.

million closing costs) and continuing. At time 2004, the expected present worth of continuing the project is

$$(0.35)\left[(10{,}000 \text{ vehicles})\left(\frac{\$900}{\text{vehicle}}\right) + \$13\text{M}\right]$$

$$+ (0.50)\left[(30{,}000 \text{ vehicles}\left(\frac{\$900}{\text{vehicle}}\right) + \$13\text{M}\right]$$

$$+ (0.15)\left[(50{,}000 \text{ vehicles})\left(\frac{\$900}{\text{vehicle}}\right) + \$21.9\text{M}\right] - \$10\text{M} = \$27.7 \text{ million.}$$

Again, the \$13 million, \$13 million, and \$21.9 million values come from ensuing decisions. This results in a present worth of \$24.8 million (\$27.7 million/1.12) in 2003. Thus, the decision to abandon operations is preferred, albeit barely, as it carries a present worth of \$25 million at the end of 2003.

GM decided to shut down the plant after experiencing extremely low sales in January of 2004 (78% lower than January of 2003[7]). Thus, it appears that a strategy similar to the one we calculated was taken by GM.

Using the decision tree in this manner is quite flexible, as we have allowed for multiple stages and three possible outcomes for each decision. Again, a trade-off must generally be made between the complexity required to ensure a good solution and the complexity with which to solve the problem. It is clear that it can be used in abandonment analysis.

16.2 Replacement Analysis

Our final decision that can arise during the course of tracking a project falls between the extremes of expanding and abandoning the project: replacing some of the equipment used by the project. The idea behind a replacement decision is that the project may continue into the future, but not as it is. There may be opportunities to improve, either by reducing costs or increasing revenues, both made possible by examining whether assets should be replaced over time. Equipment replacement decisions are generally motivated by the following:

- **Deterioration.** An asset wears with use over time. This is especially true in heavy engineering industries, such as the mining and energy sectors, where assets can be exposed to harsh conditions and continual use. As the asset deteriorates, its operating and maintenance (O&M) costs will generally rise, while its salvage value will decline. Thus, it may be economical to replace the equipment periodically with new equipment.
- **Obsolescence.** Many assets, such as a computer, do not deteriorate, but become obsolete. That is, the asset may perform as it once did, but compared with other assets available on the market, it may not perform its functions as efficiently. Thus, replacing the asset can lead to gains in productivity.

The replacement decision involves two steps: (1) The timing of the replacement must be determined, and (2) the new equipment (or process) must be chosen. The replacement decision can be thought of as the combination of an abandonment decision followed by an investment decision.

Before proceeding with our analysis, we define some terms to streamline our discussion. An asset that is currently owned is termed a **defender**. If an entity operates multiple assets, it may have multiple defenders. An asset that is available to replace the defender is termed a **challenger**. It is possible

[7] Carty, S.S., "GM May Stop Producing the Saturn L-300—Detroit News," *Dow Jones Newswires,* March 1, 2004.

to have more than one challenger available in a given period, and there is no limit to the kind of challenger involved. It may be an asset of the same make and model, but possibly released (new or used) in a different year, or it may be a competitor's product. In general, the economic aspects of any potential challenger—including the machine's purchase price, operating and maintenance costs, and salvage values over its maximum service life—must be known for analysis.

We assume that all assets are defined by a maximum **service life**—the maximum age that an asset can be retained such that if that age is reached, replacement is mandatory. The service life is generally the maximum number of periods that the asset can maintain normal operations. Note that this can be a considerable amount of time, although it may be at a considerable cost. A number of external factors contribute to the definition of the maximum service life, including the length of time that spare parts or maintenance expertise is available.

Finally, the **economic life** of an asset is the optimal time to keep the asset. The term has special meaning under certain cost assumptions, but we refer to it in general as the best age at which to replace an asset.

Note that we often refer to these problems as "equipment replacement" problems. This is merely the term used in the literature, but the methods are not relegated to single-asset analyses; they can be used to evaluate the replacement of an entire process or plant.

16.2.1 Considering Deterioration

We first consider the simplified situation in which an asset deteriorates as it ages, leading to an increase in periodic O&M costs and a decline in salvage value. We assume that the asset can perform its duties over its service life, but at these additional costs. We analyze the case of technologically advanced challengers later.

Infinite-Horizon Analysis

We first examine a restricted scenario for replacement analysis, assuming an extremely long (indeed, infinite) time horizon and no technological change. Further, we assume that there is only one challenger available in each period; in terms of economics, the challenger is defined as a new version of the defender. Given this scenario, the first five periods of two possible replacement solutions are depicted as cash flow diagrams in Figure 16.7.

In the first solution (a) shown in the figure, an asset is purchased at time zero and is retained for two periods, at which time it is replaced with an asset that is retained for three periods. In the second solution (b), an asset is replaced after one period of use for each of the first three periods. The subsequent asset is retained longer than the two remaining periods in the cash flow diagram. These two simple solutions should illustrate the difficulty with replacement problems: Even when the possible challengers are restricted

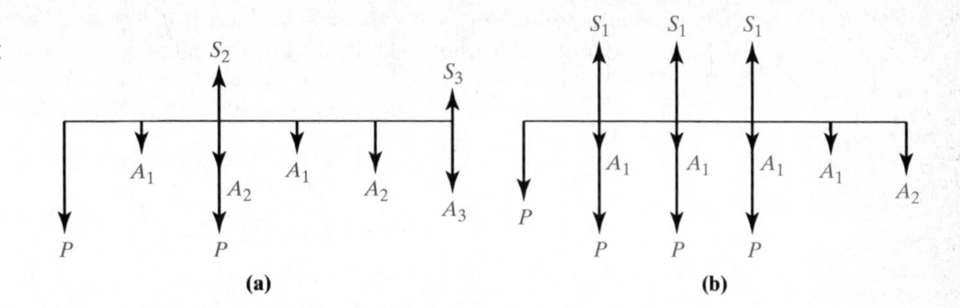

Figure 16.7
Two possible replacement solutions over time.

(e.g., to one challenger and no technological change in this example), there is an enormous number of possible solutions, since the asset can be kept or replaced after every period.

We seek the *lowest-cost* replacement policy, given this situation of an infinite horizon and no technological change. Note that our definition of the maximum service life of an asset stated that the asset could perform its duties over the given time frame. Thus, it is reasonable to assume that revenues are unaffected by the choice of a challenger or defender. This defines the case of a "service project" with mutually exclusive alternatives, as defined in Chapter 12. In addition, the do-nothing option is not feasible in this case, as an asset must be installed for service. (Alternatively, one can view keeping the defender as the do-nothing option.) The assumption of an infinite horizon allows us to easily apply the assumption of repeatability, because both the defender and challenger will have equal horizons for analysis.

Examining Figure 16.7, we may find that some solutions are clearly better than others. In fact, if an asset is replaced with a new asset that is identical in terms of economics, then it would seem reasonable that whatever we do with the first asset we should also do with the second asset, since operations will continue indefinitely. This is precisely true and defines a *stationary policy*, in that we want to repeat our decision over time. In this case, the decision is the number of periods to keep an asset. Once an asset reaches this limit, we replace the asset and start again.

To illustrate this stationary concept, consider the typical use of an asset as it ages. O&M costs rise as the asset becomes more expensive to operate. However, the capital costs of the asset tend to decline. Capital costs include the purchase cost of an asset and its salvage value. If we consider retaining an asset for n periods, where n is less than or equal to our maximum life of N periods, we could determine the annual equivalent operating and maintenance costs for each possible lifetime, $n = 1, 2, \ldots, N$. If A_n represents the cost of using an n-period-old asset (at the end of the period) for one period, then the annual equivalent of the O&M costs is

$$\text{AEC}_{\text{O\&M}}(i) = A_1(^{P/F,i,1})(^{A/P,i,n}) + A_2(^{P/F,i,2})(^{A/P,i,n}) + \cdots + A_n(^{P/F,i,n})(^{A/P,i,n}).$$

Assuming a purchase price P and a salvage value S_n for an asset of age n, we find that the annual equivalent of the capital costs would be

$$\mathrm{AEC_{Cap}}(i) = P(^{A/P,i,n}) - S_n(^{A/F,i,n}).$$

Note that this value of $\mathrm{AEC_{Cap}}(i)$ is often defined equivalently as

$$\mathrm{AEC_{Cap}}(i) = (P - S_n)(^{A/P,i,n}) + S_n i, \tag{16.1}$$

which is called the *capital recovery with return* of an asset. This definition comes from the fact that $(^{A/P,i,n}) - (^{A/F,i,n}) = i$. The first term of the capital recovery can be described as the cost, with interest, of using an asset over time as its value drops from the purchase price to the salvage value, while the second term adds back the return obtained from the salvage value. In addition to having this economic interpretation, Equation (16.1) is computationally efficient for calculating annual equivalent capital costs.

If we graph the *total* annual equivalent costs of owning and operating an asset for n periods, or $\mathrm{AEC}_n(i)$, which is the sum of $\mathrm{AEC_{O\&M}}(i) + \mathrm{AEC_{Cap}}(i)$, over all possible ages $n = 1, 2, \ldots, N$, in general, the curve looks like the one depicted in Figure 16.8. The graph also illustrates the individual annual equivalent O&M and capital costs.

The figure takes on the attributes we would expect. On the one hand, as O&M costs rise with the age of the asset, the associated annual equivalent costs rise with age. On the other hand, the high purchase cost P is spread over n periods such that the annual equivalent capital costs decline with time. These attributes lead to a total annual equivalent cost curve that declines before rising after meeting some minimum value. The minimum value corresponds to the **economic life** of the asset, or n^*, as it is the age at which costs are minimized.

The economic life of an asset is a critical value, given that we have assumed that the challenger is identical to the defender and the horizon is infinite. In this situation, we are going to be faced with an infinite stream of cash flows representing the costs of owning and operating our asset (and its replacements) over time. This scenario seems daunting, but it is a frequent assumption if entities expect to continue operations well into the future. The

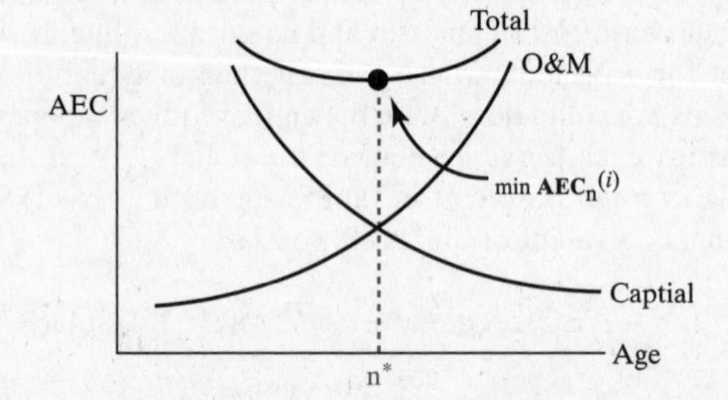

Figure 16.8 Graph of AEC for operating and owning an asset over all possible lifetimes.

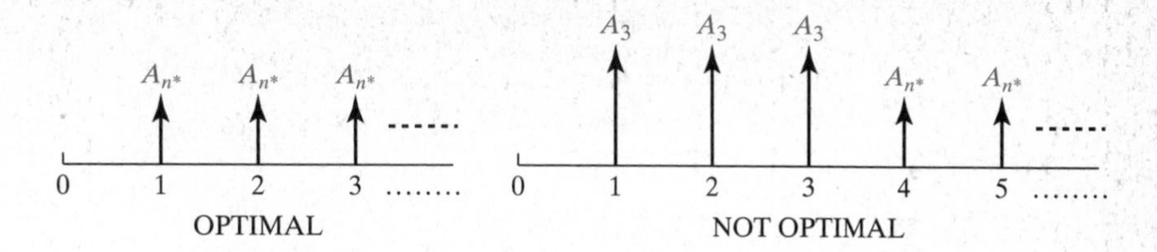

Figure 16.9 The lowest-cost replacement schedule replaces the asset repeatedly at the end of its economic life.

question is, "What would be the lowest-cost stream?" The answer should be clear from an examination of Figure 16.8. If n^* provides the lowest annual equivalent cost among all possible asset ages, then we have the lowest-cost stream over the infinite horizon if we repeatedly purchase and sell our asset when it reaches age n^*. If, at any time over the horizon, we keep the asset for a time longer or shorter than n^*, we have annual equivalent costs higher than those produced when the asset is kept for n^* periods. Thus, the minimum-cost decision is to repeatedly replace an asset at age n^*. This is shown pictorially in Figure 16.9. If any asset in the stream of assets over the infinite horizon is retained for a period not equal to n^*, the annual equivalent costs are higher for that period. Thus, a better (and thus optimal) answer to our question is to always replace the asset at n^*.

The economic life of the asset is determined by computing the annual equivalent cost of retaining the asset for $1, 2, \ldots, N$ periods. The minimum cost from all these costs is then chosen. The next example illustrates and provides further intuition concerning the economic life of an asset.

EXAMPLE 16.4

Economic Life

In early 2004, Statoil ASA closed a deal to pay Proffice SEK30 million to replace cranes on two of the company's oil rigs: the Snorre A and Draupner S. The price included the procurement, manufacturing, and installation of the cranes.[8]

Make the following assumptions: The purchase price of a new crane is SEK15 million, there is no technological change, and a crane is needed to operate over an infinite horizon. A crane can remain in operation for seven years. O&M costs are expected to be SEK500,000 in the first year and grow at the rate of 65% due to heavy wear in the oil industry. The crane is expected to lose 20% of its value annually. The O&M costs and salvage values for each age of operation are given in Table 16.2. The interest rate is 18%.

Solution. Solving this problem is a bit tedious, but we will be thorough in order to illustrate it properly. (In contrast, we use a spreadsheet to find the economic life of an asset in solving the introductory real decision problems at the end of the chapter.) For the given costs, we must determine the annual equivalent costs of ownership for each possible length (life) of operation of the asset. Note that we suspend the units (SEK) in intermittent calculations for clarity.

[8] "Proffice Gets SEK30M Crane Deal from Statoil," *Dow Jones Newswires,* January 22, 2004.

TABLE 16.2 **O&M cost and salvage value data for crane over seven-year service life.**

Age	O&M Cost (SEK)	Salvage Value (SEK)
0	–	15,000,000
1	500,000	12,000,000
2	825,000	9,600,000
3	1,361,250	7,680,000
4	2,246,063	6,144,000
5	3,706,003	4,915,200
6	6,114,905	3,932,160
7	10,089,594	3,145,728

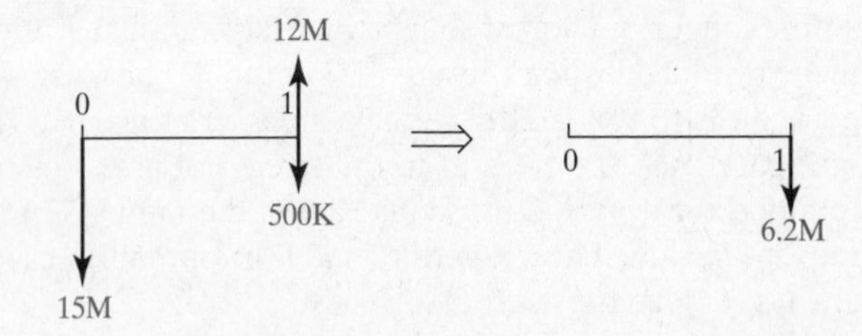

Figure 16.10 Costs realized when keeping the asset for one year.

The relevant costs of retaining the crane for *one year* are depicted in the cash flow diagram in Figure 16.10. The figure also depicts the annual equivalent of these costs, which is merely a single cash flow at time 1, calculated as follows:

$$\text{AEC}_1(18\%) = 15\text{M}\overset{F/P,18\%,1}{(1.1800)} + 500\text{K} - 12\text{M} = \text{SEK}6.2\text{M}.$$

For an asset life of *two years*, the relevant costs are given in Figure 16.11. Again, the annual equivalent costs are depicted in the figure. The AEC can be calculated in a number of ways. Here, we move all the cash flows to time zero and then spread them evenly over the two years:

$$\text{AEC}_2(18\%) = \left(15\text{M} + 500\text{K}\overset{P/F,18\%,1}{(0.8475)} + (825\text{K} - 9.6\text{M})\overset{P/F,18\%,2}{(0.7182)}\right)\overset{A/P,18\%,2}{(0.6387)} = \text{SEK}5.83\text{M}.$$

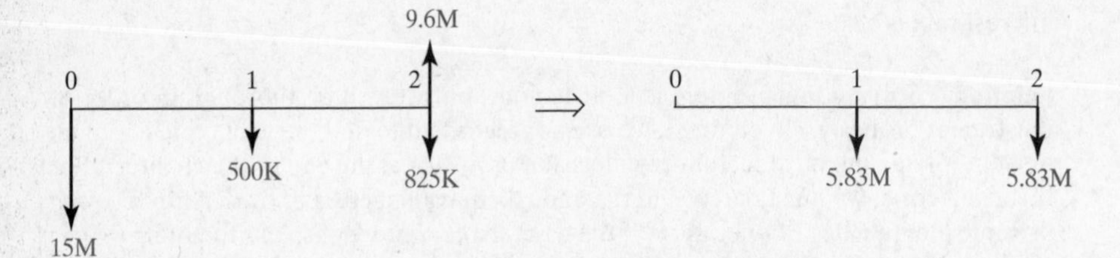

Figure 16.11 Costs realized when keeping the asset for two years.

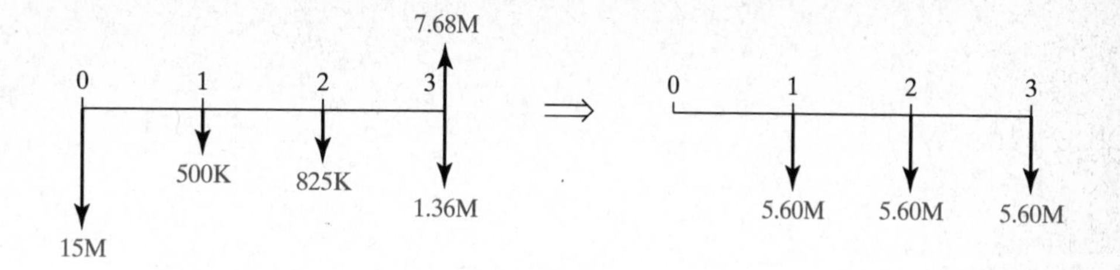

Figure 16.12 Costs realized when keeping the asset for three years.

Figure 16.12 gives the relevant costs and the annual equivalent of retaining the asset for *three years*. The annual equivalent cost is calculated as follows:

$$\text{AEC}_3(18\%) = \Big(15\text{M} + 500\text{K}\overset{P/F,18\%,1}{(0.8475)} + 825\text{K}\overset{P/F,18\%,2}{(0.7182)}$$

$$+ (1.36\text{M} - 7.68\text{M})\overset{P/F,18\%,3}{(0.6086)}\Big)\overset{A/P,18\%,3}{(0.4599)} = \text{SEK}5.60\text{M}.$$

We are not relegated to finding the present worth and then converting it to an annual equivalent cost. For *four years*, we illustrate the use of a mixture of interest factors:

$$\text{AEC}_4(18\%) = \Big(15\text{M} + 500\text{K}\overset{P/F,18\%,1}{(0.8475)} + 825\text{K}\overset{P/F,18\%,2}{(0.7182)}$$

$$+ 1.36\text{M}\overset{P/F,18\%,3}{(0.6086)}\Big)\overset{A/P,18\%,4}{(0.3717)} + (2.25\text{M} - 6.14\text{M})\overset{A/F,18\%,4}{(0.1917)}$$

$$= \text{SEK}5.50\text{M}.$$

The relevant cash flow diagram is given in Figure 16.13.

The calculations for years five, six, and seven are as follow:

$$\text{AEC}_5(18\%) = \Big(15\text{M} + 500\text{K}\overset{P/F,18\%,1}{(0.8475)} + 825\text{K}\overset{P/F,18\%,2}{(0.7182)} + 1.36\text{M}\overset{P/F,18\%,3}{(0.6086)}$$

$$+ 2.25\text{M}\overset{P/F,18\%,4}{(0.5158)}\Big)\overset{A/P,18\%,5}{(0.3198)} + (3.71\text{M} - 4.92\text{M})\overset{A/F,18\%,5}{(0.1398)}$$

$$= \text{SEK}5.60\text{M};$$

$$\text{AEC}_6(18\%) = \Big(15\text{M} + 500\text{K}\overset{P/F,18\%,1}{(0.8475)} + 825\text{K}\overset{P/F,18\%,2}{(0.7182)} + 1.36\text{M}\overset{P/F,18\%,3}{(0.6086)}$$

$$+ 2.25\text{M}\overset{P/F,18\%,4}{(0.5158)} + 3.71\text{M}\overset{P/F,18\%,5}{(0.4371)}\Big)\overset{A/P,18\%,6}{(0.2859)}$$

$$+ (6.11\text{M} - 3.93\text{M})\overset{A/F,18\%,6}{(0.1059)} = \text{SEK}5.84\text{M};$$

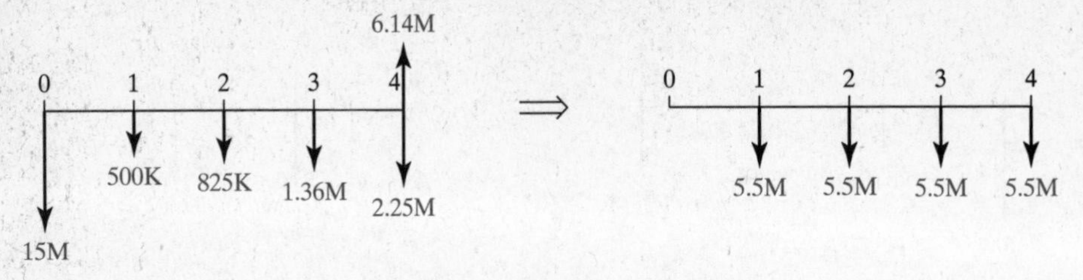

Figure 16.13 Costs realized when keeping the asset for four years.

TABLE 16.3 Annual equivalent costs (in SEK) for each age over service life for crane.

Age	AEC(18%)
1	6,200,000
2	5,826,147
3	5,597,478
4	**5,514,452**
5	5,588,013
6	5,841,633
7	6,314,570

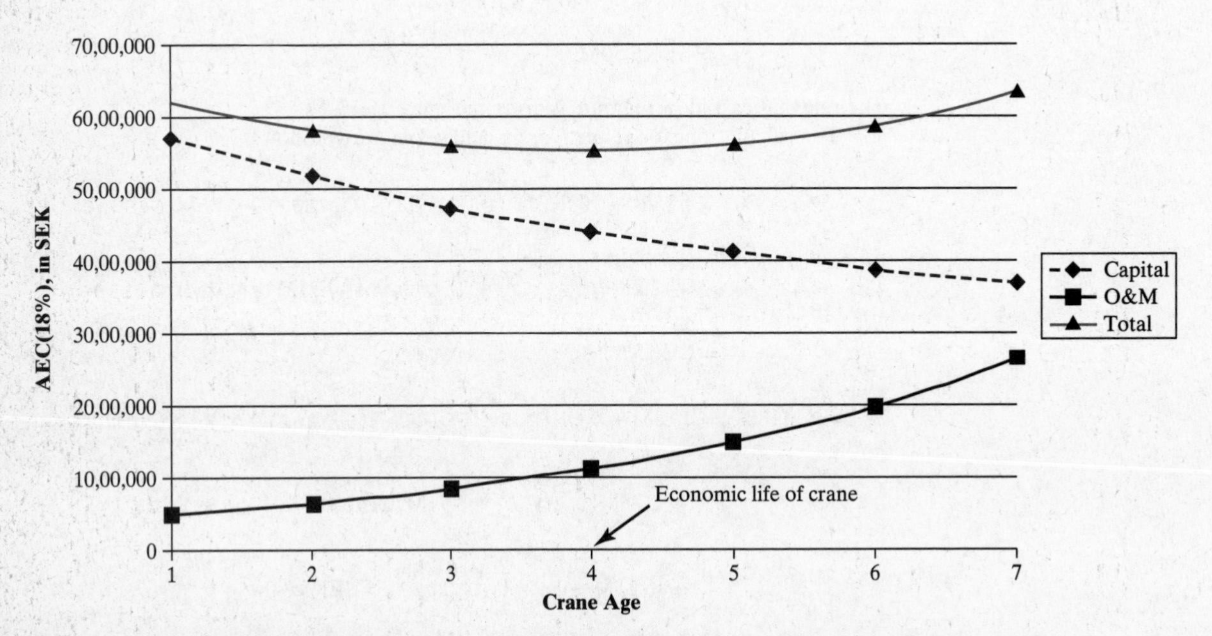

Figure 16.14 Annual equivalent costs of owning and operating a crane over its possible service lives.

$$\text{AEC}_7(18\%) = \Bigg(15\text{M} + 500\text{K}(\overset{P/F,18\%,1}{0.8475}) + 825\text{K}(\overset{P/F,18\%,2}{0.7182}) + 1.36\text{M}(\overset{P/F,18\%,3}{0.6086})$$

$$+ 2.25\text{M}(\overset{P/F,18\%,4}{0.5158}) + 3.71\text{M}(\overset{P/F,18\%,5}{0.4371}) + 6.15\text{M}(\overset{P/F,18\%,6}{0.3704})\Bigg)(\overset{A/P,18\%,7}{0.2624})$$

$$+ (10.1\text{M} - 3.15\text{M})(\overset{A/F,18\%,7}{0.0824}) = \text{SEK}6.31\text{M}.$$

The annual equivalent costs are tabulated in Table 16.3. The economic life is defined as the age at which the annual equivalent cost is minimized. This occurs at age four, so that $n^* = 4$, and is also depicted graphically in Figure 16.14.

Table 16.3 shows that, under the assumptions of repeatability and an infinite horizon, the best replacement strategy is to continually replace the crane when it reaches age four. This results in an annual equivalent cost stream of SEK5.5 million each year, indefinitely. Keeping the asset for any different length of time results in a higher cost solution.

Note that in computing the economic life, we have assumed that we start our analysis with a new asset. A fundamental assumption in replacement analysis, and in our postimplementation analysis, is that the defender being analyzed has been in use for some period of time. But then, how does one take advantage of the economic life information if the asset in use is not new? This is an important question because the economics of a new asset are different from those of a used asset. Consider again Figure 16.14, which illustrated the annual equivalent costs of owning (capital) and operating (O&M) the crane. This figure was derived under the assumption that the crane was purchased new. The figure would not be the same if the crane had been a *used* asset, because the decisions would then no longer be whether to keep the crane for one, two, or three periods through N periods, but rather, whether to keep the crane for just one or two *more* periods.

Thus, if we have an asset that is currently n periods old, we would like to determine when to begin the cycle of replacing the asset at its economic life. The answer is not as simple as comparing the value of n with n^* and keeping the asset if it is younger than n^* or replacing it if it is n^* or older. This is because the value of n^* has been determined under different circumstances (namely, the assumption of starting with a new asset).

To ensure that we get the minimum-cost solution, we turn to the dynamic analysis discussed in Chapter 13 and examine each possible decision. If the asset is n periods old, then we could (1) replace it immediately, (2) replace it after one more period of use, (3) replace it after two more periods of use, etc., until we reach our final possible decision to keep the asset until it reaches age N, at which time it must be replaced. Once the asset is replaced, we begin our cycle of replacing the asset at its economic life, since this is the optimal replacement strategy given that a new asset has been purchased. These decisions are depicted in the decision network in Figure 16.15.

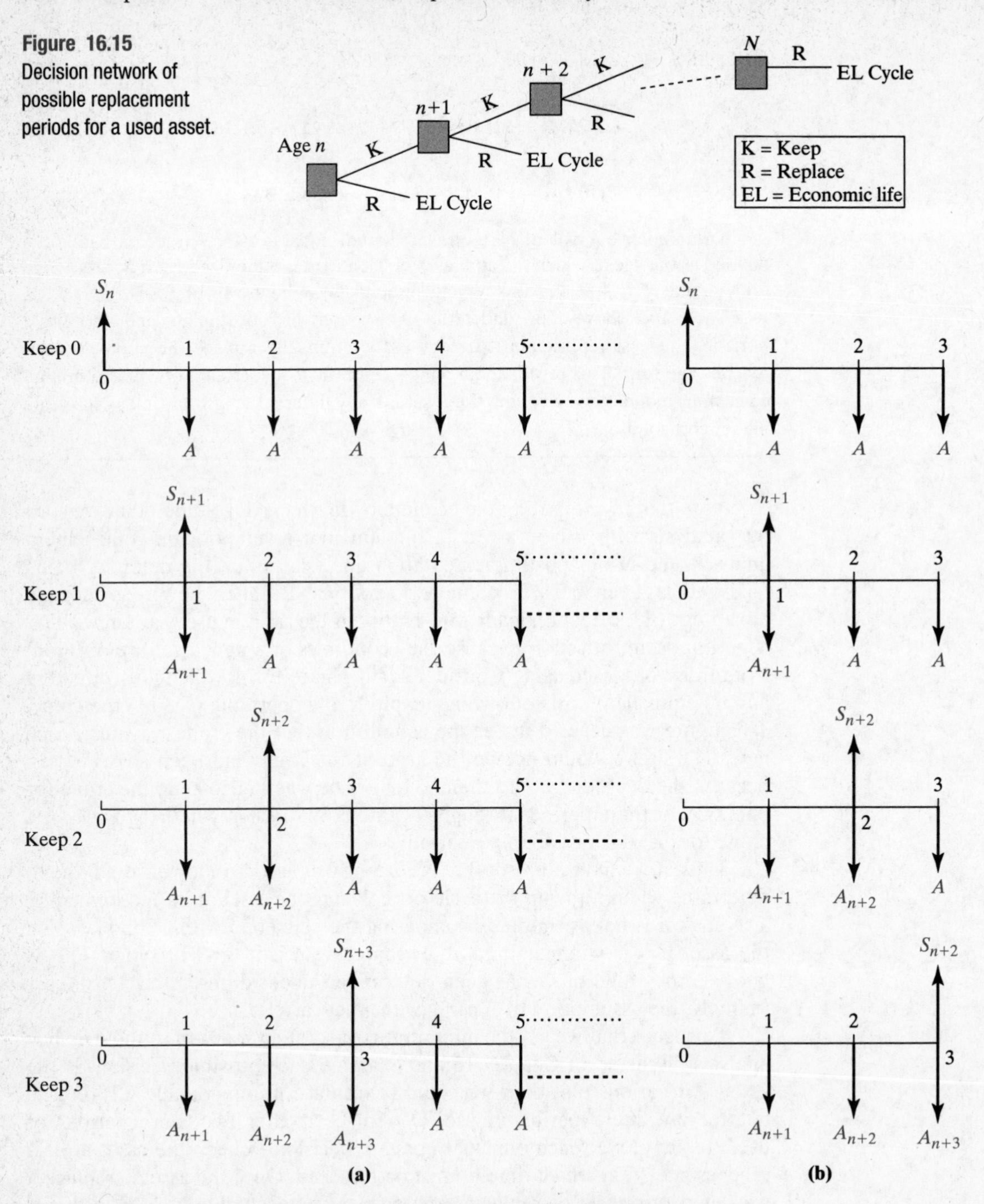

Figure 16.15 Decision network of possible replacement periods for a used asset.

Figure 16.16 (a) Cash flows from decision network (b) reduced to a common horizon.

We have drawn many decision networks and noted that each individual path defines a cash flow diagram that can be analyzed. Because these paths represent mutually exclusive decisions, we want to identify the path that maximizes some measure of worth or, in this case, minimizes cost. What must be noted in this replacement example is that a path does not define a cash flow diagram that ends at time N, because we are assuming an infinite horizon. Once the end of a decision path has been reached, the remaining cash flows (over the infinite horizon) are the annual equivalent costs associated with the economic life of the new asset (defined as A). Thus, our mutually exclusive decisions each have equivalent (infinite) horizons and can be analyzed. This fact is important; recall from Chapter 12 that service projects require equal horizons for a fair comparison. The cash flows for the respective decisions to keep the defender for one, two, and three more periods are given in Figure 16.16(a).

To analyze the options, we can simplify our work with our knowledge of incremental investment analysis introduced in Chapter 12. Incremental analysis examines the difference between two competing cash flow diagrams in order to determine which one is preferred. In Figure 16.16(a), the cash flows are aligned such that all of the options have the same cash flows beyond period N. If we look at the incremental difference between retaining the asset for one year and retaining it for two years, we note that the cash flows for periods three through infinity are the same. In analyzing the two options, these similar cash flows can be eliminated because they are equivalent after period two. Applying this logic to our specific situation, we can reduce the cash flows in Figure 16.16(a) to those in (b). In general, if we have an asset of age $n < N$ that can be retained until age N, then we must analyze cash flows over $N - n$ periods, as the cash flows are the same for all periods thereafter for all possible scenarios. The choice amongst these various cash flows can be made with incremental or total investment analysis. We illustrate by revisiting our previous example, but this time assuming that our defender is age $n > 0$.

EXAMPLE 16.5

Starting an Economic Life Cycle

We return to our previous example, but now assume that the crane has just completed its fifth year of operation. When should the cycle of replacing the crane at age four commence?

Solution. The decisions available at time zero (when the crane is age five) are depicted in Figure 16.17. The crane can be replaced immediately, resulting in the owner receiving the salvage value of SEK4.92 million immediately. Or the crane may be retained for up to two more periods of service. The cash flows for each of these three possible decisions are drawn at the respective nodes in the decision network.

After period two, all of the cash flows are equivalent and equal to SEK5.51 million per year, which is the periodic cost of retaining a newly purchased crane for

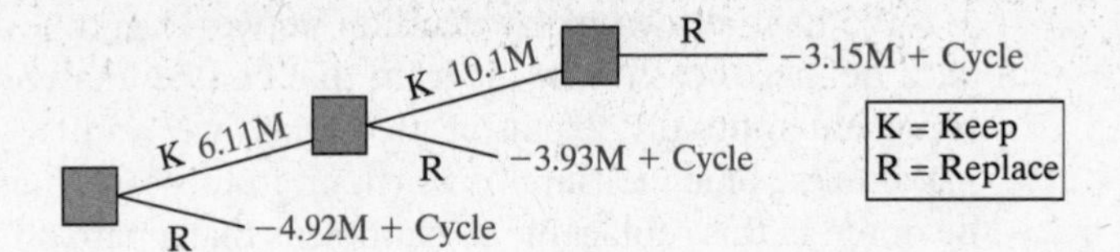

Figure 16.17 Decisions and associated cash flows for five-period-old crane.

its economic life. Thus, they may be ignored. Furthermore, because the three service options now have an equivalent horizon of two periods, we may analyze them with present worth or annual equivalent worth. The annual equivalent cost of each decision is as follows:

$$AEC_0(18\%) = -4.92M\overset{A/P,18\%,2}{(0.6387)} + 5.51M = SEK2.38M;$$

$$AEC_1(18\%) = (6.11M - 3.93M)\overset{P/F,18\%,1}{(0.8475)}\overset{A/P,18\%,2}{(0.6387)} + 5.51M\overset{A/F,18\%,2}{(0.4587)}$$
$$= SEK3.71M;$$

$$AEC_2(18\%) = 6.11M\overset{P/F,18\%,1}{(0.8475)}\overset{A/P,18\%,2}{(0.6387)} + (10.1M - 3.15M)\overset{A/F,18\%,2}{(0.4587)}$$
$$= SEK6.5M.$$

Since we seek the lowest-cost solution, the decision is to replace the crane immediately and begin the economic life cycle with a new crane. We should be confident about this decision because we have evaluated all of the possible alternatives.

Before we move to our next complication in replacement analysis, it is worthy to note how the previous analyses compare with the analyses of mutually exclusive service projects in Chapter 12. Because we ignored revenues, we ensured that we had equal horizons for all of our investment alternatives through the assumptions of repeatability and an infinite horizon. However, the glaring difference in replacement analysis was that we did not select a service life for the assets *before* the analysis. In Chapter 12, we stated the length of time that an asset or project would last and then assumed repeatability if necessary. In this chapter, we *determine* the minimum-cost time to keep the project or asset. As we evaluate all possible replacement ages, we guarantee that the minimum cost solution is found.

Finite-Horizon Analysis

What if we require our asset for only a finite number of periods? The preceding analyses assumed that we would need the services of the asset and its replacements indefinitely. However, there are a number of situations in which that would not be the case. For example, a contractor may require assets for a particular job, but not need their services beyond the life of the contract. Or a company may temporarily expand into a market for a given length of time. In these situations, assets are required only for a finite period of time, but it may still be economical to consider replacing the asset during some period or periods over the finite horizon. Unfortunately, the assumption of a finite horizon greatly complicates our analysis. However, our assumption of no technological change is probably more justified in this situation rather than under an infinite horizon.

The difficulty with a finite horizon is that we cannot apply our economic-life solution defined in the previous section to this situation. For example, if we require an asset to be in service for 12 periods and the economic life n^* of the asset is five periods, we have no solution that "fits" the required period of analysis. We could keep the asset for two cycles of five periods each, totaling 10 periods of service, but this leaves many questions for the final two periods.

To determine the best course of action, we need to once again apply our knowledge with respect to dynamic decisions. For the 12-period example described in the previous paragraph, the most general situation would allow us to replace the asset after each period over the 12 periods of analysis.

It may seem natural to define our decision network in terms of our decision at each period: Either keep or replace the asset. Such a network is illustrated in Figure 16.18. This is a viable approach that will surely produce the best answer, but is tedious to solve because there are two decisions in each period. If we have T periods of analysis, we will have to define 2^T paths, each representing a different replacement schedule and subsequent cash flow diagram for analysis.

Instead of modeling a keep–replace decision at each node (which represents a period), we define the decision as the length of time to keep the asset, such as $1, 2, \ldots, N$ periods. We use this model because it saves us work and allows us to take advantage of any data that we may already have calculated in solving for the economic life of the asset. The network using this decision-making structure is given in Figure 16.19.

A decision node coincides with each period t over the horizon T. The decisions are the number of periods n to keep an asset, represented by N arcs

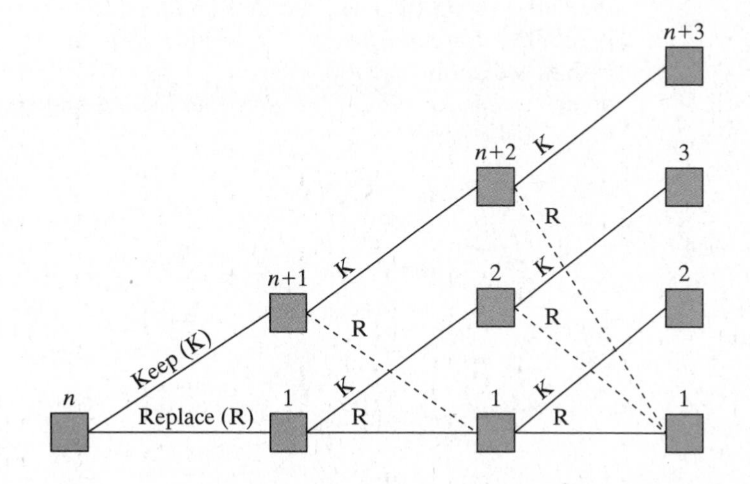

Figure 16.18 Replacement-problem decision network with keep–replace decision each period. Each node is labeled with the age of the asset.

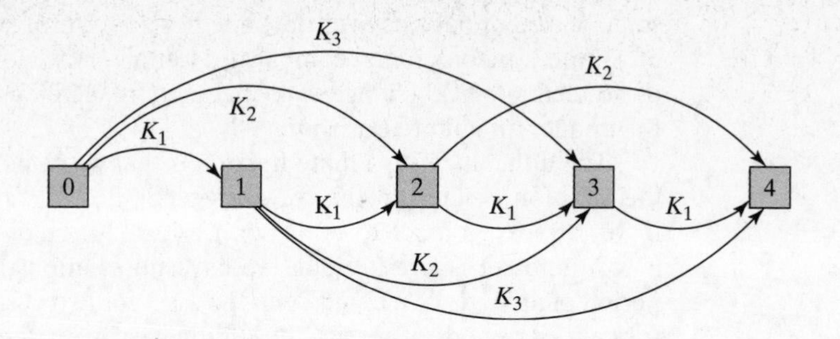

Figure 16.19 Replacement-problem network with decisions to keep the asset $1, 2, \ldots,$ or N periods with $N = 3$ and $T = 4$.

emanating from a node, where N is the maximum service life of the asset. This assumes that the arc ends at a node $n + t \leq T$, as T is the end of the horizon. (Arcs are labeled K_1 through K_N.)

We must provide one modification to the decision network described in Figure 16.19 in order to solve our replacement problem. Note that at time zero, the decisions are whether to keep the asset for $1, 2, \ldots,$ or N periods. If we own an asset at time zero (which is our assumption in this chapter), then we actually have the ability to keep the defender for up to its maximum service life or replace it with a challenger that can be retained up to its maximum service life. These additional arcs are considered only at time zero, because once we replace the defender, it has been assumed that there is only one challenger available in each period over the finite horizon. These additional decisions are illustrated in Figure 16.20 with the letter D.

Given our complete network, we can trace each possible path that represents our sequence of decisions over the horizon. Each path defines a replacement schedule with a corresponding cash flow diagram. Note that each path is defined by the same length of time, such that each mutually exclusive service alternative (a sequence of service lives) once again has the same horizon for analysis. The minimum-cost decision among these mutually exclusive alternatives is then chosen. For example, we may choose to retain the defender for two more periods and then replace it with a challenger that is retained for three periods. This path is highlighted in Figure 16.20. The cash flow diagram corresponding to the set of decisions leading to that path is depicted in Figure 16.21. We illustrate the method by which it was obtained in the next example.

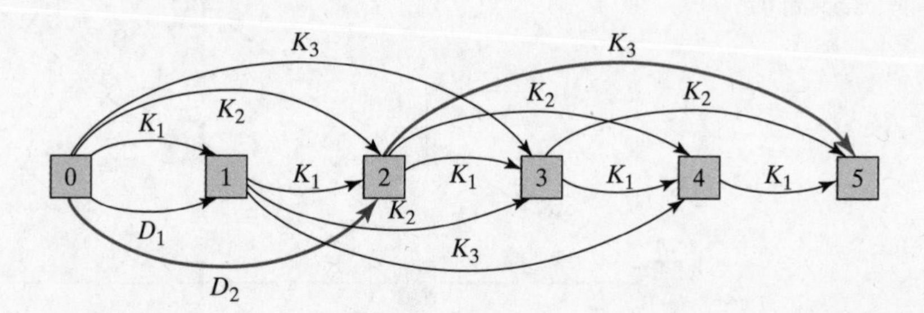

Figure 16.20 Incorporation of defender decisions into replacement-problem decision network with $N = 3$ and $T = 5$.

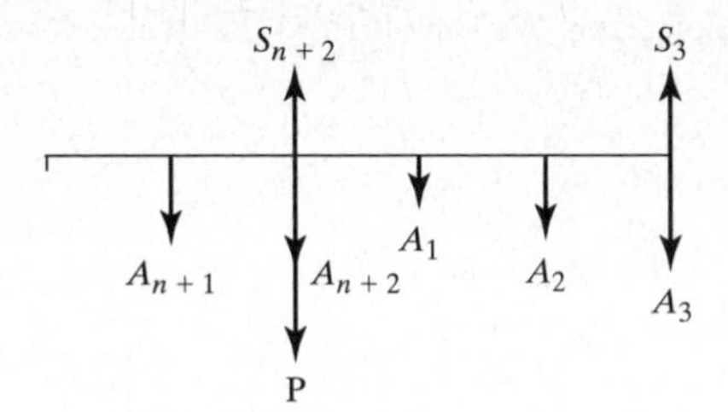

Figure 16.21 Cash flow for keeping the defender for two periods and its replacement for three periods.

EXAMPLE 16.6

Finite-Horizon Analysis

Consider again the crane replacement problem that we studied in the previous two examples, but now assume that the current crane is five years old and assume that a crane needs to be in service for the next five years. All of the other data are as before.

Solution. The decision network for our problem is given in Figure 16.22. Note that, at time zero, the crane (defender) is five years old. The time-zero decisions regarding the defender are (1) to replace it immediately; (2) to keep it for one more year (D_1); and (3) to keep it for two more years (D_2). If alternative (1) is chosen, then five possible choices remain: The challenger can be retained for one (K_1) through five (K_5) periods. If the decision is to keep the challenger for fewer than five periods, then the replacement decision is revisited during the appropriate period.

Our earlier Table 16.3 shows the annual equivalent costs for retaining a new asset between one and seven years. These costs are relevant for the challenger in any period, as we are assuming no technological change. The costs for retaining the defender are computed from Table 16.2. For example, retaining the defender for 1 period costs SEK6.11M with a salvage value of SEK3.93M. This results in a present worth cost of SEK1.85M. The present worth of these costs have been placed on the arcs in the decision network in Figure 16.22.

Before examining each path, we can reduce our workload by realizing that there are two ways to move from node 0 to node 1. These represent the two decisions of either keeping the defender for one period or replacing the defender with the challenger and keeping the challenger for one period. Since these are the only two paths from node 0 to node 1 and we know the cost of each of these paths, we only need to consider the cheaper of the two options when examining all of the paths in the network from node 0 to node 5. Thus, we can reduce the network in Figure 16.22 by eliminating the higher cost parallel arcs between any two nodes. This reduction eliminates the arcs associated with the defender (dashed arcs) such that the optimal decision at time zero

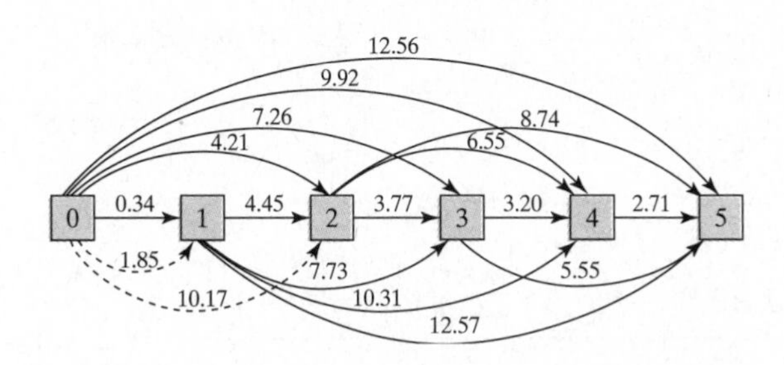

Figure 16.22 Decisions with present-worth costs (in M of SEK) over five periods for crane replacement.

is to replace the defender. We now turn to the decision concerning the length of time to keep a challenger purchased at time zero.

We can enumerate each decision and evaluate its cost. Consider the decision to keep the crane purchased at time zero for one period, repeatedly, over the horizon. We know that the annual equivalent cost to keep the crane for one year is SEK6.2 million, as computed in Example 16.4. Thus, the decision to keep it for one year five successive times is defined by the cash flow diagram in Figure 16.23(a). Note that there is a cash inflow of SEK4.92 million at time zero corresponding to the sale of the five-year-old crane. The present worth of the costs associated with this sequence of decisions is computed as

$$\text{PW}(18\%) = -\text{SEK}4.92\text{M} + \text{SEK}6.2\text{M}(\overset{P/A,18\%,5}{3.1272}) = \text{SEK}14.47 \text{ million.}$$

Another possible decision would be to keep the challenger for two periods and replace it with a new crane that is retained for the final three periods over the horizon. The cash flow diagram associated with this decision is given in Figure 16.23(b). The present

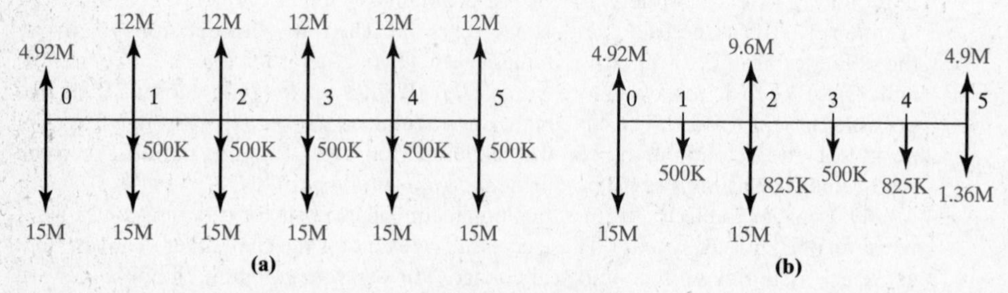

Figure 16.23 Cash flow diagrams in (SEK) for (a) keeping each challenger for one period and (b) keeping the initial challenger for two periods and a new challenger for the final three periods.

TABLE 16.4 **Present worth of costs for each possible replacement schedule over five-year horizon.**

Decision	PW(18%) Costs (M of SEK)
0–1–2–3–4–5	14.47
0–1–2–3–5	14.12
0–1–2–4–5	14.05
0–1–2–5	13.53
0–1–3–4–5	13.98
0–1–3–5	13.62
0–1–4–5	13.36
0–1–5	12.91
0–2–3–4–5	13.89
0–2–3–5	13.53
0–2–4–5	13.47
0–2–5	12.95
0–3–4–5	13.16
0–3–5	12.81
0–4–5	12.63
0–5	**12.56**

worth of the costs is calculated as

$$\text{PW}(18\%) = -\text{SEK}4.92\text{M} + \text{SEK}5.83\text{M}\overset{P/A,18\%,2}{(1.5656)} + \text{SEK}5.6\text{M}\overset{P/A,18\%,3}{(2.1743)}\overset{P/F,18\%,2}{(0.7182)}$$

$$= \text{SEK}12.95\text{M}.$$

We can continue to evaluate each of the possible decisions in this manner. At time zero, there are five such decisions. In period 1, there are four possible decisions, etc. This leads to a total of 16 possible paths whose costs can be summed from Figure 16.22.

All of the possible solutions are enumerated in Table 16.4, along with the present worth of the costs associated with each of the decisions. Choosing the minimum-cost decision from this set, we see that the optimal decision is to replace the defender immediately and keep the challenger for the duration of the horizon (five years). This answer was not unexpected, given our economic-life solution earlier.

Despite introducing a decision network that is slightly different from our previous decision networks, we may not be overly excited about seeking the solution, as there are numerous paths to evaluate. This should be somewhat expected, since we have extreme flexibility to manage the situation (we can replace the asset each period) and we desire the *best* decision. (It all comes at a price.)

We can lighten our workload somewhat by utilizing a technique known as **dynamic programming** to find the best answer. With this technique, the amount of work is considerably less than enumeration, which is what we are doing when we evaluate each individual path. (We will not address the theory of dynamic programming here; rather, we illustrate its use in the next section.)

Note that our modeling approach has once again generalized our analysis of mutually exclusive service options, since we are seeking the minimum-cost decision from a set of options with the same service lives. However, using our approach, we find the optimal length to retain each possible asset and do not make any assumptions about repeatability. Unfortunately, our workload has increased, as we are forced to look at a number of possible solutions.

16.2.2 Considering Technological Change

The term "technological change" is used to refer to the improvement in future challengers available for replacement. There are many methods with which to model technological change. Because technological improvements are generally accompanied by improvements in efficiency, it is commonly assumed that assets will cost less to operate and will retain their value longer compared with previously purchased assets. Specifically, initial O&M costs are generally lower and the increase with age may be slower. In addition, drops in initial and successive salvage values may be lower over time.

Estimating individual cash flows and their changes with technology over time is not an easy task, as we noted in Chapter 7. If data are available from previous models, one may be able to use curve-fitting techniques to define functions that describe the technological change. Individual cash flows can be modeled, such as estimating the initial period of O&M costs for successive challengers, as can life-cycle costs, such as estimating the total cost of retaining an asset for n periods. Typical functions used to model technological change include linear, exponential, and bounded geometric functions of increasing efficiency. Again, these functions can be used to model individual cash flow changes or the total cost over some lifetime.

For example, suppose we define the annual equivalent cost of purchasing an asset at time t, use the asset for n periods, and then sell it as $\text{AEC}_n(t)$. We could model the cost of the challenger that is available at time $t+1$ as

$$\text{AEC}_n(t+1) = f(\text{AEC}_n(t)).$$

Without technological change, $\text{AEC}_n(t)$ would equal $\text{AEC}_n(t')$ for any $t' \neq t$. With technological change, $\text{AEC}_n(t') < \text{AEC}_n(t)$, for any $t' > t$, since we would expect the next challenger to be more efficient. Again, a common assumption for the function describing technological change is that it is linear, exponential, or geometrically bounded.

If we continue to assume that there is only one challenger available in a given period, then the only change in the solution procedure is in the definition of the costs on the arcs in Figure 16.19: Presumably, the costs on arcs emanating from node t in the network will be cheaper than those emanating from $t-1$ for similar decisions.

Finite-Horizon Analysis

The method described to solve the finite-horizon equipment replacement problem in the previous section can be used to solve *any* replacement problem over a finite horizon. We only need to populate the decision network with the appropriate costs to model technological change.

We present two examples concerning technological change. The first assumes that we forecast individual cash flows for all replacement assets over the horizon. We solve this problem in the next example by using dynamic programming.

EXAMPLE 16.7 *Technological Change in Replacement Analysis*

Anglo American invested $37 million to replace oil-fired boilers with gas boilers at the company's Merebank paper mill in KwaZulu-Natal, South Africa. The conversion was aimed at improving efficiency and reducing emissions.[9]

Make the following assumptions: Assume a 10-year-old oil-fired boiler is owned that has a maximum service life of 12 years. The salvage values (S_n) and O&M costs

[9] "Anglo American to Expand at Merebank Paper Mill," *Dow Jones Newswires*, September 29, 2003.

(A_n) associated with the current boiler (the defender) are given in the spreadsheet in Figure 16.24. Note that S_0 is the purchase price. (Due to space limits, the data center is not shown; thus, the 11% interest rate is hard coded into the formula illustrated.) The table also gives the costs of a new gas-fired boiler available on the market today that can be purchased either at time zero or time 1. Furthermore, a newer boiler is expected to be available on the market two years from now (and available for the foreseeable future thereafter). The costs for this advanced boiler are also given in the table. It carries a higher purchase and first-year O&M cost, but a slower decline in salvage values and a lower increase in O&M costs. Assuming an 11% annual rate of interest, determine the optimal replacement policy for the boilers over a five-year horizon.

Solution. We solve this example with dynamic programming. First, we populate our decision network with the costs associated with all three boilers. Note that the defender (the old boiler) is available only at time zero and, if immediately salvaged, lowers the cost of purchasing the new boiler. The boiler available at time zero may only be purchased at time zero or one. Both of the new boilers can be retained through the horizon. Figure 16.25 illustrates the costs involved over the five-year horizon. Note that all costs are in terms of present worth. The spreadsheet in Figure 16.24 gives costs for the old boiler and boiler available at time zero in terms of present worth (time zero). The present worth costs for the boiler available at time 2 are in time-2 dollars, and thus discounted further on the arcs in Figure 16.25.

We have two pairs of arcs at node 0: one pair connecting nodes 0 and 1 and the other connecting nodes 0 and 2. This is from the choice of whether to keep the defender or replace it with a challenger. The costs associated with the defender are lower; thus, we eliminate the $0 \rightarrow 1$ and $0 \rightarrow 2$ arcs of the challenger at time zero.

The idea behind dynamic programming is to solve the problem in stages. By solving the problem backwards, we define the cost of going from one stage (node) to the end. Thus, we only have to compute the cost of getting from the current stage to all of the stages that follow, since the best solutions from those nodes are already known. Let us illustrate.

In our problem, each stage is a period represented by one of the nodes in Figure 16.25. We begin at the final stage (stage 5) and work our way to the initial stage

	A	B	C	D	E	F	G	H	I	J
1	Example 16.7: Boiler Replacement									
2										
3	Age		Old Boiler			New Boiler Time Zero			New Boiler Time Two	
4		SV	OM	PW(11%)	SV	OM	PW(11%)	SV	OM	PW(11%)
5	0	--	--	--	$20.00	--	--	$21.00	--	--
6	1	--	--	--	$18.20	$0.0102	$3.61	$19.25	$0.0111	$3.67
7	2	--	--	--	$16.40	$0.0106	$6.71	$17.50	$0.0113	$6.82
8	3	=NPV(0.11,$F$6:F9)+$E$5-E9/(1.11)^A9			$14.60	$0.0113	$9.35	$15.75	$0.0117	$9.51
9	4	--	--	--	$12.80	$0.0122	$11.60	$14.00	$0.0122	$11.81
10	5	--	--	--	$11.00	$0.0135	$13.51	$12.25	$0.0128	$13.77
11	6	--	--	--	$9.20	$0.0152	$15.13	$10.50	$0.0136	$15.44
12	7	--	--	--	$7.40	$0.0174	$16.49	$8.75	$0.0145	$16.84
13	8	--	--	--	$5.60	$0.0204	$17.64	$7.00	$0.0157	$18.03
14	9	--	--	--	$3.80	$0.0244	$18.59	$5.25	$0.0172	$19.02
15	10	$1.00	--	--	$2.00	$0.0297	$19.38	$3.50	$0.0190	$19.85
16	11	$0.50	$0.43	$0.94	$1.00	$0.0369	$19.78	$1.75	$0.0212	$20.53
17	12	$0.25	$0.46	$1.56	$0.50	$0.0469	$19.97	$0.75	$0.0239	$20.88

Figure 16.24 Salvage Values (S_n), O&M costs (A_n), and PW (at time zero and time 2) of ownership of boilers.

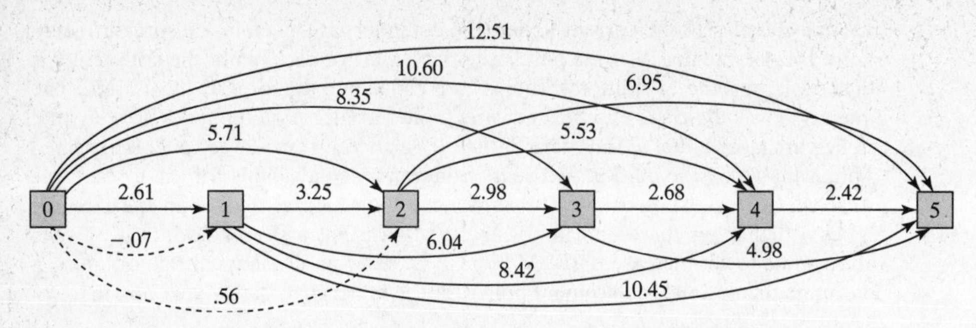

Figure 16.25 Decision network with present-worth costs in millions of dollars.

(0). Stage 5 is the terminal node, and we assign it a value of zero as an end condition because the problem ends at that point.

We then move back to stage 4. From that point, there is only one way to node 5 and that is on the arc (4,5). This arc represents purchasing a new boiler at time 4 and using it for one year before retiring it. The cost (present worth) for this decision is \$2.42 million. This is the minimum cost of reaching node 5 from node 4. We label it as

$$\min_{4} = \$2.42\text{M}.$$

We now step back to period 3 and wish to determine the best way to get to node 5 from node 3. There are two ways: (1) via stage 4 and (2) directly. As we have just computed the cost to traverse from node 4 to node 5 in the most efficient manner, we need only add the cost to get to node 4 in order to determine the best way to get to node 5 from node 3 via node 4. The direct path from 3 to 5 is merely along the arc that connects the two nodes directly. The minimum of these two paths is

$$\min_{3} = \min\{\underbrace{(\$2.68 + \$2.42)}_{\text{Via 4}}, \underbrace{\$4.98}_{\text{Direct}}\} = \$4.98\text{M}.$$

Thus, from node 3, the minimum-cost path to node 5 is the direct one, which corresponds to purchasing a new boiler and retaining it for two periods.

From stage 2, the paths to node 5 are either via node 3 or node 4 or directly to 5. The minimum cost is

$$\min_{2} = \min\{\underbrace{(\$2.98 + \$4.98)}_{\text{Via 3}}, \underbrace{(\$5.53 + \$2.42)}_{\text{Via 4}}, \underbrace{\$6.95}_{\text{Direct}}\} = \$6.95\text{M}.$$

We repeat this step at nodes 1 and 0. Table 16.5 gives a summary of the results, with the minimum-cost decisions in bold.

Now we must convert our solution across the network into a sequence of decisions for the replacement problem. According to the solution, we traverse the network from node 0 to node 5 via node 2. Thus, our decision is to retain the defender for its final two periods and purchase the advanced challenger at time 2. We then retain the challenger for the remainder of the horizon. The total present-worth cost of this sequence of decisions is \$7.51 million.

TABLE 16.5

Dynamic programming solution to boiler replacement example.

Stage	Route	PW(11%) Cost (in M)
5	End	**$0.00**
4	Direct	**$2.42**
3	Via 4	$5.10
	Direct	**$4.98**
2	Via 3	$7.96
	Via 4	$7.95
	Direct	**$6.95**
1	Via 2	**$10.21**
	Via 3	$11.03
	Via 4	$10.84
	Direct	$10.45
0	Via 1	$10.15
	Via 2	**$7.51**
	Via 3	$13.33
	Via 4	$13.02
	Direct	$12.51

The dynamic programming approach allows us to model a variety of costs in the replacement decision network. Note that we can easily expand the number of challengers in a period, as these can be reduced in a manner similar to the way we reduce defender and challenger arcs at time zero.

The dynamic programming approach is also flexible in that it can maximize present worth. If technological change defines a challenger that can provide additional revenues, not possible with the defender, then the revenues must be included in the analysis and present worth should be maximized. This includes the case where an asset (challenger) provides more capacity than the defender.

The previous example utilized forecasted data for all of the relevant cash flows for the defender and challengers over time. We now assume that technological change can be modeled according to some function, as noted earlier. We will use the expected life-cycle costs in this situation.

EXAMPLE 16.8

Technological Change in Replacement Analysis Revisited

A certain Middle Eastern petrochemical company procured a refrigeration system for a styrene plant from Toromont Process Systems. The C$9 million contract calls for the engineering, design, and manufacture of a system that includes two compressors requiring a total of 5500 horsepower.[10]

Make the following assumptions: The system purchased at time zero incurs the O&M costs and salvage values given in Table 16.6. Assume further that the compressors

[10] Tsau, W., "Toromont Receives C$9M Intl Order," *Dow Jones Newswires,* January 23, 2004.

TABLE 16.6 **Costs for owning and operating a compressor purchased at time zero.**

Age n	Salvage Value	O&M
0	C$4,000,000	–
1	C$2,800,000	C$120,000
2	C$1,960,000	C$172,800
3	C$1,372,000	C$298,598
4	C$960,400	C$619,174
5	C$0	C$1,540,702

available on the market for replacement improve 5 percent each year with respect to annual equivalent costs, such that

$$\text{AEC}_n(t) = \left(\frac{1}{1+0.05}\right)^t \text{AEC}_n(0).$$

Assuming a 17% annual interest rate, determine the optimal replacement schedule over a six-year horizon.

Solution. To illustrate how technological change is modeled, Table 16.7 shows the annual equivalent costs for keeping the compressor for up to five years if the asset is purchased in one of the first three periods. Note that AEC(t) follows the assumption of technological change, with costs discounted at 5 percent per year.

We convert the AEC values shown into equivalent present-worth values at the time the equipment is procured and populate the decision network given in Figure 16.26. The same rate of technological change is assumed throughout the horizon. For example, if a system is purchased at time 1 and retained for 2 periods, represented by the arc from node 1 to node 3, the present worth of that decision, at time 1, is:

$$\text{PW}(17\%) = \text{C\$1.680M}(\overset{P/A,17\%,2}{1.5852}) = \text{C\$2.66M},$$

which is the cost identified in Figure 16.26.

Reading through Table 16.8, we see that the optimal decision at time zero is via node 4, which means that we should purchase a challenger at time zero and keep it for four periods. The optimal decision at stage 4 is direct to the horizon time, meaning that

TABLE 16.7 **$\text{AEC}_n(t)$ for compressor available at times t = 0, 1, and 2.**

Age (n)	$\text{AEC}_n(0)$	$\text{AEC}_n(1)$	$\text{AEC}_n(2)$
1	C$2,000,000	C$1,904,762	C$1,814,059
2	C$1,764,424	C$1,680,404	C$1,600,385
3	C$1,610,527	C$1,533,835	C$1,460,796
4	C$1,543,119	C$1,469,637	C$1,399,654
5	C$1,702,969	C$1,621,875	C$1,544,643

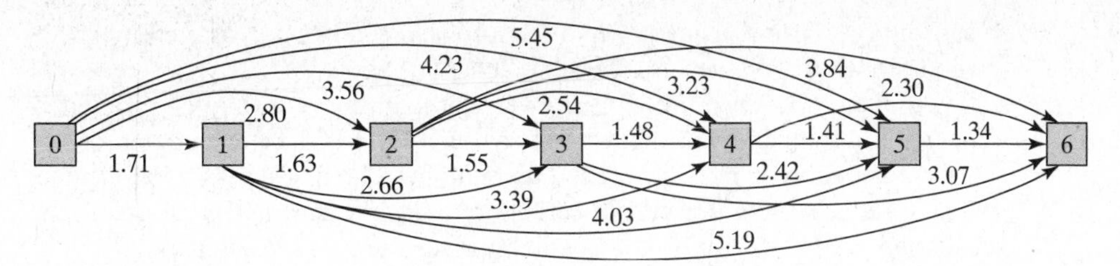

Figure 16.26 Network with costs (in millions of Canadian dollars) for compressor replacement over six-year horizon.

TABLE 16.8

Dynamic programming solution to compressor replacement example.

Stage	Route	PW(17%) Cost (in M)
5	Direct	**C$610,897**
4	Via 5	C$1,361,385
	Direct	**C$1,227,977**
3	Via 4	C$2,149,951
	Via 5	C$2,119,468
	Direct	**C$1,919,346**
2	Via 3	C$3,051,991
	Via 4	C$3,081,256
	Via 5	C$2,968,814
	Direct	**C$2,804,866**
1	Via 2	C$4,196,320
	Via 3	C$4,196,098
	Via 4	C$4,124,678
	Via 5	C$4,056,675
	Direct	**C$4,434,991**
0	Via 1	C$5,766,076
	Via 2	C$5,601,856
	Via 3	C$5,477,942
	Via 4	**C$5,461,115**
	Via 5	C$6,059,284
	Direct	Infeasible

a challenger is purchased at the end of period 4 and is retained for the final two periods. The total present worth of these costs is C$5.46 million. This path was identified with the use of dynamic programming, after converting the costs on the arcs in Figure 16.26 to time zero present worth costs.

Infinite-Horizon Analysis

Dealing with technological change over an infinite horizon is another complication in replacement analysis. Unlike the case without technological change, where we took advantage of an infinite horizon in order to derive our economic-life solution, with technological change we can no longer guarantee a stationary solution. That is, with technological change, we would not expect to continually keep assets for the same length of time.

To overcome these difficulties, we will examine two simpler problems. The first problem assumes that we have a defender, but the challenger is different—presumably, a more advanced technology. The decision is to determine when to replace the n-period-old defender, under the assumption that the new challenger will be the only asset available after the first replacement.

In Section 16.2.1, we illustrated how to compute the economic life of an asset and how to implement this solution when the defender is not new. We follow the same approach here, except that the replacement asset is defined by the minimum AEC of a technologically advanced challenger. Thus, the question is to determine how many periods, $0, 1, \ldots, N - n$, to retain the defender before replacing it with the challenger. After the defender is replaced, the challenger is to be continually replaced at its economic life, as it is assumed that this is the only asset available over the remaining (infinite) horizon.

The second problem we address is an indirect solution to the infinite-horizon problem. Instead of determining the best replacement schedule (of keep and replace decisions over time) for the defender and each successive replacement asset, we determine the best time-zero decision for the defender over an infinite horizon. That is, we only concern ourselves with what we have to do *now*, and we do not worry about the future.

Fortunately, we have the tools to solve this problem, as it follows from our finite-horizon procedure illustrated in the previous section. The only difference is that we do not know exactly what length-of-horizon problem to solve in order to guarantee that we have found the best solution. Algorithmically, we solve a number of problems until we are convinced that the time-zero decision will not change for any horizon that is longer. This can be guaranteed if we solve problems with horizons of $T, T+1, \ldots, T+N$ and the time-zero decision does not change. To facilitate the process, in the next example we will solve our decision network problem with a *forward recursion*, as opposed to the backward recursion that we used previously.

EXAMPLE 16.9 *Optimal Time-Zero Decisions for Infinite Horizon*

Consider again Example 16.8, and determine the optimal time-zero decision under the assumption of an infinite horizon. All other data remain as before, but now with technological change continuing indefinitely into the future.

Solution. We again solve this problem with dynamic programming, although this time we utilize a forward recursion. This facilitates our algorithm for finding the best time-zero decision, because we can "add" nodes into our network while moving forward in time until we are assured that we have attained that decision.

The network with costs is as before (Figure 16.26), but this time we do not have a definite horizon. The plan is to solve a number of successive problems until N consecutive problems produce the same time-zero decision. As we move forward with this recursion, we note first that the initial condition (time zero) has a cost of zero. Then we move to period 1 and its representative node. In our previous implementation of dynamic programming, we were interested in the minimum-cost path from a node to the end. Now we want to determine the minimum-cost path from the initial node to the node under consideration.

There is only one way to reach node 1 from node 0, corresponding to the decision to keep an asset purchased at time zero for one period. Thus,

$$\min_{1} = \text{C\$1.71M}.$$

TABLE 16.9

Forward dynamic programming solution to compressor replacement example.

Stage	Route	PW(17%) Cost	Time Zero Decision
1	Via 0	**C$1,709,402**	Keep 1 Period
2	Via 0	**C$2,796,990**	
	Via 1	C$3,100,856	Keep 2 Periods
3	Via 0	**C$3,558,596**	
	Via 1	C$3,986,154	
	Via 2	C$3,929,635	Keep 3 Periods
4	Via 0	**C$4,233,137**	
	Via 1	C$4,606,102	
	Via 2	C$4,650,269	
	Via 3	C$4,480,570	Keep 4 Periods
5	Via 0	C$5,448,386	
	Via 1	C$5,155,179	
	Via 2	C$5,154,907	
	Via 3	C$5,067,167	
	Via 4	**C$4,983,625**	Keep 4 Periods
6	Via 1	C$6,144,393	
	Via 2	C$5,601,856	
	Via 3	C$5,477,942	
	Via 4	**C$5,461,115**	
	Via 5	C$5,594,522	Keep 4 Periods
7	Via 2	C$6,407,077	
	Via 3	C$5,841,759	
	Via 4	**C$5,795,487**	
	Via 5	C$5,983,199	
	Via 6	C$5,958,386	Keep 4 Periods
8	Via 3	C$6,497,210	
	Via 4	**C$6,091,634**	
	Via 5	C$6,255,378	
	Via 6	C$6,274,769	
	Via 7	C$6,200,266	Keep 4 Periods

We now step forward in time to period 2 and note that there are two ways to reach node 2: (1) directly from node 0, meaning that the initial challenger is purchased and retained for two periods, and (2) from node 1, which was previously solved. Recall that the costs beyond time zero in Figure 16.26 must be discounted. Hence,

$$\min_{2} = \min\{\underbrace{(C\$1.39 + C\$1.71)}_{\text{Via 1}}, \underbrace{C\$2.80}_{\text{Direct}}\} = C\$2.80M.$$

Therefore, the best way to reach node 2 is directly from node 0. At period 3,

$$\min_{3} = \min\{\underbrace{(C\$2.28 + C\$1.71)}_{\text{Via 1}}, \underbrace{(C\$1.13 + C\$2.80)}_{\text{Via 2}}, \underbrace{C\$3.56}_{\text{Direct}}\} = C\$3.56M.$$

We continue to move forward in the network in this manner. As $N = 5$, we stop when we have five consecutive solutions in which the time-zero decision does not change. The complete dynamic-programming results are given in Table 16.9.

The table illustrates that in solving for the first four horizons of $T = 1$ through $T = 4$, respectively, the time-zero decision changes accordingly with the horizon (i.e., keep 1 period, keep 2 periods, etc.). However, under the assumption of a five-year horizon (i.e., the problem ends at stage 5), the optimal time-zero decision is to retain the asset for four periods, followed by keeping the new challenger for one more period. The six-year horizon solution corresponds to our solution over a six-year horizon with the backward recursion.

As we solve through stages 4, 5, 6, 7, and 8, the time-zero decision remains the same: Keep the initial asset for four periods. Since we have now gone through five consecutive horizons in which there is no change in the time-zero decision, we have found the optimal time-zero decision for an infinite horizon.

Although we have not determined a replacement schedule over the entire infinite horizon, we have identified the best option at time zero. If desired, we could continue the method forward in order to determine the best period-one decisions. Or we could wait a period and solve with updated data.

16.2.3 After-Tax Issues in Replacement Analysis

Taxes were introduced in Chapter 8, and they play a prominent role in replacement analysis in the form of depreciation and gains or losses from the sale of assets. As noted in Section 8.5.2, the difference when replacing an asset, as opposed to retiring an asset, is that a gain or loss is not realized on the sale. Rather, any residual book value of the asset being replaced (the defender) is transferred to the new asset being purchased (the challenger). This feature can complicate things considerably in after-tax replacement analysis.

The difficulty with properly incorporating taxes into our analysis is that the cash flows for the defenders and challengers are no longer independent, because of the transfer in residual book value. To illustrate with this difficulty, consider the next example.

EXAMPLE 16.10

After-Tax Replacement Analysis

In early 2004, Verizon Wireless announced that it would spend $1 billion over two years to deploy a high-speed wireless Internet data network nationwide.[11] Clearly, a lot of technology, including routers and switches, will have to be replaced over the course of this investment.

Make the following assumptions: A new switch costs $500,000, has a life of five years (due to technological change), and costs $100,000 to maintain the first year, increasing by $25,000 each year thereafter. Its salvage value remains relatively high, decreasing by $50,000 each of the first two years, after which the switch loses $150,000 in each successive year (due to obsolescence). Assume also that the switches currently in the network are worth only $50,000 and have a remaining life of two years. They will be worthless after the current period and will cost $250,000 and $300,000 to operate and maintain for their final two years of service, respectively. Assume further that the assets are depreciated according to MACRS (200% declining balance switching to straight line) with a five-year recovery period and the half-year convention. The current defender is three years old and was purchased for $425,000. Consider the replacement options at time zero and note the complications introduced by after-tax analysis.

Solution. Consider our typical decision network for this three-period problem, as in Figure 16.27. The five decisions at time zero are as follows: Keep the defender for (1) one or (2) two periods; and replace the defender with a challenger and keep it for (3) one, (4) two, or (5) three periods. These options are fairly straightforward in that if the challenger is selected, the residual book value of the defender would be transferred to the challenger. Under the MACRS depreciation schedule (Table 8.3), the book value of the defender *if sold at time zero* is

$$B_3 = (1 - (.20 + .32 + .192/2))\,\$425{,}000 = \$163{,}200.$$

Note that the amount of depreciation in year 3 is halved due to the early disposal. If this were a retirement, then a loss of $163,200 – $50,000 = $113,200 would be incurred, reducing the amount of taxes to be paid elsewhere in the project or company. However, in the replacement situation, the book value of the new switch would now be $500,000 + $113,200 = $613,200.

Incorporating this data into the arcs from time zero is straightforward. The $613,200 book value would be used to compute depreciation charges on the challenger

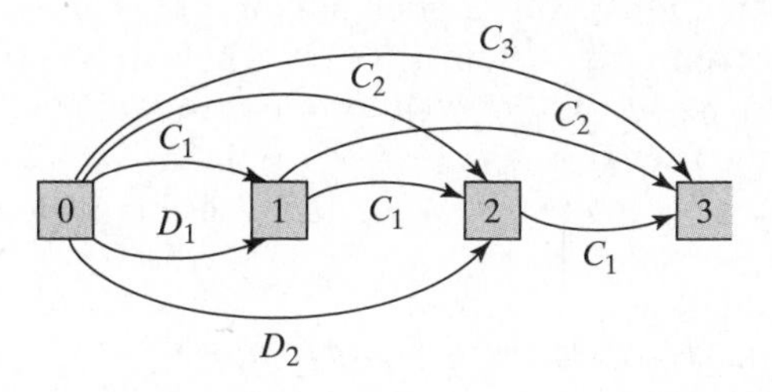

Figure 16.27 Network of decisions for telephone switch replacement.

[11] Latour, A., and J. Drucker, "Verizon Wireless to Invest $1 Billion," *The Wall Street Journal Online,* January 8, 2004.

arcs. If the defender is to be kept, then the original book value of $425,000 would be used to compute after-tax cash flows.

The difficulty with this concept comes at the time of the second replacement. Consider period 1. There are two ways in which to reach node 1 in the network. Either (1) the defender is retained for one period and replaced with a challenger, or (2) the defender is replaced at time zero and the acquired challenger is retained for only one period.

(1) If the defender is retained for one more year and then sold, the book value of the defender at the time of the sale is

$$B_4 = (1 - (.20 + .32 + .192 + .1152/2))\,\$425{,}000 = \$97{,}920.$$

Since the defender is sold for $0 after one additional year of service, the $97,920 is transferred to the challenger, so that its book value is now $597,920.

(2) In this situation, the defender is sold at time zero and a challenger is purchased. As noted earlier, the challenger has a book value of $613,200. Now if the challenger is kept for only one year, then the book value at time 1 (when it is replaced) is

$$B_1 = (1 - .20)\,\$613{,}200 = \$490{,}560.$$

Since the asset is sold for $450,000 after the first period of service, we have a loss of $40,560 on the transaction.

It may not be readily obvious, but we now have a dilemma. If we follow path (1) of keeping the defender, then we are purchasing a new challenger at time 1 and wish to transfer $97,920 in residual book value to the new purchase. However, if we have discarded the defender and retained a challenger from time zero for one period, then we are purchasing a new challenger at time 1 and wish to transfer $40,560 in residual book value to the new purchase. The dilemma is that the book value of the asset purchased at time 1 is different, depending on the path taken to get to the associated node.

To be correct, we would need to add a number of nodes to the network in order to represent challengers being purchased at different times with different book values. This, however, is clearly a complication that we wish to avoid.

We can *approximate* a good solution by treating replacements as retirements according to current tax law. Thus, at the time of a sale, a gain or loss is recognized and the appropriate tax credits or payments are made. We continue the previous example by populating the network with after-tax cash flows. Solving the network would follow as in the previous sections.

EXAMPLE 16.11

After-Tax Replacement Analysis

Continue the previous example by populating the network with the appropriate costs (at their present worths), assuming a 20% interest rate and a 39% effective tax rate.

Assume also that we are ignoring revenues (minimizing cost) and that all losses receive tax credits.

Solution. In Chapter 8, we defined an after-tax cash flow in period n as

$$A_n = -P + S_n - t(S_n - B_n) - (1 - t)E_n + tD_n.$$

This equation assumes that we are maximizing present worth (revenues minus expenses). We now focus on minimizing the present worth of costs and thus will "flip" the signs. For example, if we sell the defender immediately, the after-tax cash flow is

$$-S_n + t(S_n - B_n) = -\$50{,}000 + (0.39)(\$50{,}000 - \$163{,}200) = -\$94{,}148.$$

This dollar amount is negative because it is a net revenue and we are minimizing costs. It is added to the time-zero cash flow of purchasing the challenger at time zero (at a $500,000 purchase price).

Now consider the decision to purchase a challenger at time zero and keep it for two periods. The cash flows associated with this transaction are as follows:

1. *Time 0.* $500,000 purchase, less net revenue of $94,148, as just noted.
2. *Time 1.* (1 – 0.39)$100,000 – (0.39)(0.20)($500,000) = $22,000.
3. *Time 2.* (1 – 0.39)$125,000 – (0.39)(0.32/2)($500,000) – $400,000 + (0.39)($400,000 – $320,000) = –$323,750.

Table 16.10 shows the after-tax cash flows for each of the decisions and the associated present worths. These are the numbers that can be used to populate the decision network.

Note that if the defender is kept, then the eventual loss from the sale of the asset and its tax implications are defined in the cash flows for the defender. This avoids the problem of having different nodes representing the same asset with differing initial book values. Again, the network is solved as before.

TABLE 16.10 After-tax cash flows for telecommunications switch replacement.

Period	Decision	Time: 0	1	2	3	PW(20%)
	Defender					
0	Keep 1	$0	$104,764	—	—	$87,303
	Keep 2	$0	$133,406	$154,358	—	$218,365
	Challenger					
0	Keep 1	$405,852	–$408,500	—	—	$65,435
	Keep 2	$405,852	$22,000	–$323,750	—	$199,359
	Keep 3	$405,852	$22,000	$13,850	–$154,600	$344,336
1	Keep 1	—	$500,000	–$408,500	—	$132,986
	Keep 2	—	$500,000	$22,000	–$323,750	$244,589
2	Keep 1	—	—	$500,000	–$408,500	$110,822

We reiterate that this solution to the after-tax problem is an approximation. A dynamic programming formulation can be developed to solve it exactly, but it is computationally burdensome.

16.3 Examining the Real Decision Problems

We return to our introductory real decision problems concerning the San Diego police department's desire to replace its helicopter fleet and the questions posed there:

1. What is the economic life of a new helicopter?

The spreadsheet in Figure 16.28 presents the economic-life calculations for the helicopter. The calculations in column D make use of the capital-recovery factor with return, Equation (16.1), using the PMT function. The annual O&M costs are converted to a present value with the NPV function and are then spread evenly over the asset's life with the use of the PMT function.

	A	B	C	D	E	F	G	H
1	Chapter 16: Helicopter Replacement					**Input**		
2						Purchase	$3,050,000	
3	**Age**	**O&M**	**SV**	**AEC**		Op Hours	750	
4	0		$3,050,000			Op Cost	$175.00	
5	1	$131,250	$2,135,000	$1,198,750		g (Op Cost)	8%	per year
6	2	$141,750	$1,921,500	$839,360		Maintenance	$1,000.00	
7	3	$153,090	$1,729,350	$713,097		Incident	0.25	
8	4	$165,337	$1,556,415	$646,194		g (Incident)	30%	per year
9	5	$178,564	$1,400,774	$603,816		g (Sal 1)	-30%	per year
10	6	$192,849	$1,260,696	$574,287		g (Sal)	-10%	per year
11	7	$209,277	$1,134,627	$552,683		Interest	5%	per year
12	8	$225,939	$1,021,164	$536,289				
13	9	$244,935	$919,047	$523,757		**Output**		
14	10	$264,369	$827,143	$514,095		MIN AEC	$494,634	
15	11	$286,359	$744,428	$506,798				
16	12	$310,028	$669,986	$501,451				
17	13	$335,510	$602,987	$497,756				
18	14	$363,951	$542,688	$495,544				
19	15	$394,507	$488,420	$494,634				
20	16	$428,347	$439,578	$494,929				
21	17	$465,655	$395,620	$496,348				
22	18	$506,627	$356,058	$498,823				
23	19	$552,478	$320,452	$502,329				
24	20	$602,436	$288,407	$506,810				
25	21	$658,751	$259,566	$512,275				
26	22	$721,691	$233,610	$518,724				
27	23	$793,546	$210,249	$526,204				
28	24	$874,630	$189,224	$534,742				
29	25	$967,280	$170,301	$544,399				
30	26	$1,074,862	$153,271	$555,277				
31	27	$1,199,771	$137,944	$567,487				
32	28	$1,346,433	$124,150	$581,178				
33	29	$1,519,308	$111,735	$596,530				
34	30	$1,725,892	$100,561	$613,781				
35	31	$1,974,724	$90,505	$633,227				
36	32	$2,277,382	$81,455	$655,242				
37	33	$2,646,492	$73,309	$680,266				
38	34	$3,102,732	$65,978	$708,872				
39	35	$3,666,830	$59,380	$741,731				

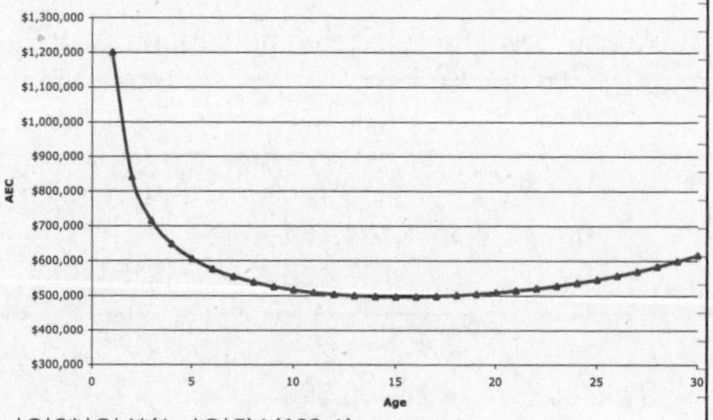

```
=PMT($G$11,A12,-($C$4-C12))+C12*$G$11
+PMT($G$11,A12,-NPV($G$11,$B$5:B12))
```

```
=$G$3*$G$4*(1+$G$5)^(A32-1)+
$G$6*(ROUNDDOWN($G$7*(1+$G$8)^(A32-1),0))
```

Figure 16.28 Economic-life calculations of helicopter over 35 years.

The minimum-cost AEC(5%) value is $494,634, which occurs at age 15. The graph shows the typical economic-life cost curve, which is extremely steep for the first few years.

2. Should the 31-year-old helicopter be replaced?

The cost to operate the helicopter for another year is nearly $2M, and it will lose roughly another $10K in salvage value. These costs are clearly greater than those associated with the economic life of a new helicopter, and they do not improve if the 31-year-old asset is retained longer. Thus, it should be replaced immediately.

3. How can the financing be included in the analysis?

Instead of using a single purchase price that is to be spread over the life being analyzed, the analysis can use the interest and principal payments directly. These payments can merely be added into the O&M costs in column B, and the purchase price can be removed from calculations in column D, to arrive at the new economic life. Note that the interest rate used for calculations may have to be adjusted to at least the cost of capital (cost of the loan).

16.4 Key Points

- If a project is underperforming, delay or abandonment options should be examined.
- The abandonment option should be examined if delaying a project will not improve its economic viability.
- Replacement decisions are often motivated by deterioration, whereby an asset becomes more expensive to operate and loses value with time, or obsolescence, such that assets available on the market are more efficient.
- An asset that is owned is referred to as a defender, while a potential replacement asset is a challenger.
- The economic life of an asset is the length of time the asset should be retained in order to minimize cost.
- The maximum service life of an asset is the maximum length of time the asset can be kept in operation.
- The economic life of an asset, assuming repeatability (no technological change) and an infinite horizon, is defined as the age that minimizes the annual equivalent costs of owning and operating the asset.
- The economic-life solution for the infinite-horizon problem does not translate to the case of the finite horizon, which must be modeled with a decision network and solved through enumeration or dynamic programming.

- Technological change is generally modeled through decreased costs of operation for future challengers and can be incorporated into finite-horizon replacement models or infinite-horizon replacement models under specific assumptions.
- Multiple challengers can easily be incorporated into the decision network analysis as pre-processing eliminates inferior challengers.
- Taxes greatly complicate replacement analysis because the residual book value is transferred from the defender to the challenger at the time of replacement. A computationally efficient approximation is to ignore the transfer and pay (credit) taxes on any gains (losses) during the time of the exchange.
- When a replacement is being considered, but the challenger possesses additional capability or capacity, present worth should be maximized rather than costs minimized.

16.5 Further Reading

Bean, J.C., J.R. Lohmann, and R.L. Smith, "A Dynamic Infinite Horizon Replacement Economy Decision Model," *The Engineering Economist,* 30:99–120, 1985.

Bellman, R.E., "Equipment Replacement Policy," *Journal of the Society for the Industrial Applications of Mathematics,* 3:133–136, 1955.

Dobson, J., and R. Dorsey, "Reputation, Information, and Project Termination in Capital Budgeting," *The Engineering Economist,* 38(2):143–152, 1993.

Dreyfus, S.E., and A.M. Law, *Art and Theory of Dynamic Programming.* Academic Press, New York, 1977.

Fleischer, G.A., *Introduction to Engineering Economy.* PWS Publishing Co., Boston, 1994.

Hartman, J.C., and R.V. Hartman, "After-Tax Economic Replacement Analysis," *The Engineering Economist,* 46(3):181–204, 2001.

Eschenbach, T.G., *Engineering Economy, Applying Theory to Practice,* Irwin, Chicago, 1995.

Oakford, R.V., J.R. Lohmann, and A. Salazar, "A Dynamic Replacement Economy Decision Model," *IIE Transactions,* 16:65–72, 1984.

Oakford, R.V., *Capital Budgeting: A Quantitative Evaluation of Investment Alternatives.* The Ronald Press Co., New York, 1970.

Park, C.S., and G.P. Sharp-Bette, *Advanced Engineering Economics.* John Wiley and Sons, New York, 1990.

Thuesen, G.J., and W.J. Fabrycky, *Engineering Economy,* 9th ed. Prentice Hall, Upper Saddle River, New Jersey, 2001.

Wagner, H.M., *Principles of Operations Research.* Prentice-Hall, Englewood Cliffs, New Jersey, 1975.

16.6 Questions and Problems

16.6.1 Concept Questions

1. When should the abandonment option be taken into consideration?
2. If the expected project lifetime has four periods remaining, how many possible abandonment decisions are there?
3. What measure of worth should be used when evaluating abandonment options with regard to different periods? Why?
4. Is the abandonment option computationally easier or harder than the delay option? Why?
5. What motivates replacement decisions?
6. What is the economic life of an asset? When is it constant over time?
7. How does one model deterioration, obsolescence, and technological change in replacement problems?
8. Compare the decision networks for the finite-horizon replacement models discussed, one with keep–replace decisions and one with decisions involving keeping an asset for n periods.
9. Why must we simplify infinite-horizon problems under technological change?
10. Is it reasonable to only require an optimal solution at time zero for a replacement problem? Explain.
11. Are multiple challengers a great complication in our analysis?
12. How do taxes affect replacement analysis problems and why are they difficult to incorporate?
13. Describe two situations in which replacement analysis must include revenues.

16.6.2 Drill Problems

1. An asset that can last for seven years is purchased for $200,000. It costs $10,000 to operate in the first year, growing by 25% per year, and loses 10% of its value every period. What is the economic life of the asset, assuming a 20% annual interest rate and $N = 8$?
2. Recompute the economic life in Problem 1 if the asset has no salvage value after the first year of operation (as customized equipment often does not have) and $N = 20$.
3. Recompute the economic life in Problem 1 if O&M costs do not rise with time.
4. Consider an asset with the following costs:

Age	O&M	SV
0		\$550,000
1	\$90,000	\$467,500
2	\$108,000	\$397,375
3	\$129,600	\$337,769
4	\$155,520	\$287,103

Using the preceding data and a 12% rate of interest,

(a) What are the annual equivalent capital costs over time?

(b) What are the annual equivalent operating costs over time?

(c) Graph both costs and determine the economic life of the asset graphically.

5. An asset is purchased for \$350,000. Its O&M costs are \$50,000 in the first year and are expected to grow 25% each year through its maximum service life of five years. The salvage value of the asset drops 30% after the first year and an additional 10% each year thereafter. If the interest rate is 15%,

(a) What is the economic life of the asset?

(b) If an asset currently owned is 3-years-old, should it be kept or replaced? If kept, for how long?

(c) Re-solve the previous problem, assuming that the asset is needed to be in service for only four more periods.

6. Consider Drill Problem 5 again, but assume that a superior challenger is available which costs \$360,000. However, O&M costs for the challenger start at \$40,000 and increase 15% each year, while the salvage value declines 10% each period. The challenger also has a maximum service life of five years.

(a) If this asset is the only challenger for the foreseeable future, when should the three-year-old asset from Drill Problem 5 be replaced, assuming an infinite horizon.

(b) Re-solve part (a) if the horizon is four years.

7. Consider Drill Problem 5 yet once more, and assume that a new challenger is available each period and shows 7% improvement in annual equivalent costs for each age over the previous period's model. Re-solve this problem over the four-year horizon.

8. Re-address Drill Problem 7 in the context of an infinite horizon. How would you find an optimal time-zero solution?

16.6.3 Application Problems

1. Let us continue the analysis of the abandonment of the Toyota Voltz in Application Problem 3 of Chapter 15.

(a) Assume that sales are expected to continue at 750 per month with profits of \$4500 per vehicle. Assume further that it costs \$10 million to close the operation against a $150 million salvage value (at time zero) which declines $2.5 million each month. When should the project be abandoned?

(b) What is the break-even value for cars sales in order for the plant to remain operational?

(c) Repeat part (a), assuming that the profit per vehicle declines $1000 per year.

(d) Repeat part (a), assuming that sales drop 200 units per month with probability 0.5, stay the same with probability 0.15, or increase with probability 0.10.

2. Apple Computer, Inc., said it expected to record $10 million in charges for closing a manufacturing facility in Sacramento, California, in addition to $1.9 million in severance charges. The move is expected to generate $6 million in quarterly savings.[12]

This example is a bit different from our typical abandonment analysis, because closing the facility results in revenues (savings). Make the following assumptions: The facility was to remain open only for an additional five quarters. Closing the facility at any time costs $11.9 million, resulting in quarterly revenues of $6 million for the remaining periods of the horizon (through period 5). Further, if the facility remains open through period 5, then only $10 million is paid for closing the facility at that time.

(a) Model this problem as an abandonment problem with a decision network. If the interest rate is 5% per quarter, when should the facility be closed?

(b) Abandonment analysis usually maximizes present worth, since the different options are revenue projects with different study horizons. Is the analysis in part (a) justified? Explain.

3. In 2003, Boeing announced that it would stop producing the 757. The move was motivated by Continental Airlines stating that it would purchase only 5 of the aircraft in 2004, as opposed to its original order of 11. The line had 18 orders to complete at the time of the announcement.[13]

Assume that a total profit of $1.5 million is made on the sale of each plane. Assume further that it costs $10 million per quarter to keep the facility open. The plant, which has a capacity of 30 planes per quarter, can be closed at any time for $20 million. It takes one full quarter to shut the facility down (and it cannot be reopened). The 18 orders on the books for 2004 may be dropped for a payment of $100,000 each. For an interest rate of 4.5% per quarter and assuming that it is the end of 2003, answer the following questions:

(a) Given the data, determine whether the facility should be closed immediately or after one more period of production. Draw the cash flow diagrams for each option.

(b) How big must an order be for the second quarter in order to keep the facility open for an additional period?

---

[12] Enrich, D., "Apple Sees $10M 3Q Chg from Closing Sacramento Facility," *Dow Jones Newswires,* May 6, 2004.

[13] Souder, E., "UPDATE: Boeing Parks 757 as Airlines want Smaller Planes," *Dow Jones Newswires,* October 17, 2003.

TABLE 16.11 **O&M costs and salvage values for defender jet.**

Age	Salvage Value (M)	O&M (M)
0	$140.00	
1	$134.01	$25.75
2	$127.92	$26.50
3	$121.73	$27.26
4	$115.44	$28.02
5	$109.06	$28.78
6	$102.57	$29.54
7	$95.98	$30.30
8	$89.29	$31.07
9	$82.50	$31.84
10	$75.50	$32.62
11	$69.10	$33.39
12	$62.60	$34.17
13	$56.00	$34.95
14	$49.30	$35.74
15	$42.50	$36.52

(c) If there is a 0.33 probability of an order being received at the end of the first quarter for 20 more planes, a 0.33 probability of an order for 10 more planes, and a 0.34 probability of no more orders, should the facility be kept open another period? Note that the decision to close the facility takes one period to execute.

4. To begin replacement of its 747-300 fleet, Boeing delivered 10 new 777-200ER jets to KLM Royal Dutch Airlines. The new jets are known to be technologically advanced and environmentally friendly.[14] Perform the following analyses under the assumption of a MARR of 22%:

(a) Assume that a 747-300 has the costs and salvage values as given in Table 16.11. What is the economic life of the jet, assuming an interest rate of 22%, no technological change, and an infinite horizon?

(b) If a 10-year-old 747-300 is owned at time zero, when should the economic life cycle defined in part (a) start?

(c) Assume that the 777-200ER is the only replacement available for the 10-year-old 747. The costs for the 777 are as shown in Table 16.12. If the economic life of the 777-200ER is 15 years (annual equivalent cost of $47.63 million), when should the 747 be replaced? Verify the economic life of the 777.

(d) Consider part (c) again, but this time over a five-year horizon. Solve with either enumeration or dynamic programming.

(e) Consider part (d) again, but assume that either the 777-200ER or a new 747-300 can be purchased. If they provide the same revenues, what is the optimal replacement schedule over the five periods?

[14] "Boeing Delivers First 10 777-200ER Jets to KLM," *Dow Jones Newswires,* October 24, 2003.

TABLE 16.12 **O&M costs and salvage values for challenger jet.**

Age	Salvage Value (M)	O&M (M)
0	$165.00	
1	$154.77	$10.19
2	$154.30	$10.38
3	$153.61	$10.57
4	$152.69	$10.76
5	$151.54	$10.96
6	$150.18	$11.15
7	$148.60	$11.35
8	$146.82	$11.54
9	$144.84	$11.74
10	$142.66	$11.94
11	$140.31	$12.14
12	$137.78	$12.34
13	$135.10	$12.54
14	$132.26	$12.74
15	$129.28	$12.94

(f) Consider part (e) again, but now assume that the revenues are not the same. Assume the 777-200ER is configured to hold 400 passengers while the 747-300 holds 496 passengers. Per passenger revenues average $225 per flight for the 777 and $200 for the 747, with 1000 flights per year. How sensitive is your answer to capacity and revenue differences?

5. In the fall of 2003, L.M. Ericsson signed an agreement with Oman Telecommunications Co. worth OMR4.5 million to upgrade digital exchanges to allow for ISDN and ADSL features.[15] Assume that this upgrade requires the replacement of five telecommunications switches with new ones. Consider one switch and assume that the replacement does not affect revenues. Using the costs and data assumptions (but ignoring taxes) in Example 16.10, determine the optimal replacement time for a current three-year-old switch in the network, assuming a MARR of 12% and the following information:

(a) The challenger switch is to be used for an infinite horizon and is the only challenger available over that time.

(b) The horizon is five years, and the same challenger that is available at time zero is available in each period.

(c) The horizon is five years, and the available challenger improves each period such that its AEC values decline 5% from the previous model for the same service life.

(d) Draw the correct *after-tax* cash flow diagram of the optimal decision to part (b), assuming an effective tax rate of 39%. (Depreciation assumptions are given in

[15] Fardan, A. "Motorola, Ericsson Sign Telecom Upgrade Contracts in Oman," *Dow Jones Newswires*, October 27, 2003.

the example.) What is the present worth of this cash flow (be sure to adjust the MARR to an after-tax MARR) compared with the present worth of the before-tax cash flow.

(e) The horizon is infinite, and the technological change assumption from part (c) holds. Ignore taxes and determine the optimal time-zero decision.

6. In early 2004, the U.S. government announced that it would retire the space shuttle program and reallocate funding to develop a new space vehicle (with $1 billion earmarked for the next five years).[16] Explain how this type of analysis with benefits and costs could be undertaken, and explicitly define the data that would be needed.

7. Given the police helicopter fleet replacement data in the chapter's introductory real decision problems, should the 12-year-old helicopters be replaced? If not, how big of a fleet discount (off the purchase price) is required to change the decision?

8. Consider again Application Problem 1 of Chapter 12. Assume that you want to purchase the $225 million satellite. What is the economic life of the asset? Re-solve the problem under the assumption that revenues start at $75 million per period, but decline 20% per year due to technological obsolescence, since older satellites cannot offer the latest offerings (such as high-definition radio or TV). Examine 8% declines too.

9. Consider again Application Problem 3 of Chapter 12. Treat this as a replacement problem in which "modernizing" the machine is equivalent to purchasing a challenger. Assume that the salvage value of the defender is EUR10 million at time zero and declines EUR1 million each year (until reaching zero). The salvage value of the modernized machine is assumed to decline 20% per year of use. As in the original problem, analyze replacement decisions over horizons of three, five, eight years, and over an infinite horizon. Assume that the modernized machine will always be available.

16.6.4 Fundamentals of Engineering Exam Prep

1. Consider an asset that costs $500,000 to purchase and whose salvage value decreases 20% each year. The first year, operating and maintenance costs are $100,000, increasing by 30% each year thereafter. The asset can be retained up to and including seven years. Use the following data for this questions and the next four (i.e., through Question 5):

Age	O&M	SV	AEC(15%)
0		$500,000	
1	$100,000	$400,000	$275,000
2	$130,000	$320,000	$272,674
3	$169,000	$256,000	$275,072
4	$219,700	$204,800	$281,927
5	$285,610	$163,840	$293,104
6	$371,293	$131,072	$308,587
7	$482,681	$104,858	$328,464

[16] Souder, E., "Boeing to Develop Space Vehicles as it Retires Shuttle," *Dow Jones Newswires,* January 15, 2004.

The annual equivalent cost values have been computed for retaining the asset between one and seven years, inclusive. The annual equivalent cost for retaining the asset for two years is computed as

(a) $500,000$(A/P, 15\%, 2)$ + $100,000 – $320,000$(A/F, 15\%, 2)$.

(b) $500,000$(A/P, 15\%, 2)$ + $130,000 – $320,000$(A/F, 15\%, 2)$.

(c) ($500,000 + $100,000/1.15)$(A/P, 15\%, 2)$ + ($130,000 – $320,000)$(A/F, 15\%, 2)$.

(d) $500,000$(A/P, 15\%, 7)$ + $100,000$(P/A_1, 30\%, 15\%, 7)$ – $104,858$(A/F, 15\%, 7)$.

2. The capitalized cost for purchasing the asset and retaining it for three years is computed as

   (a) $500,000$(A/P, 15\%, 3)$ + $100,000 – $256,000$(A/F, 15\%, 3)$.

   (b) $500,000$(A/P, 15\%, 3)$ – $256,000$(A/F, 15\%, 3)$.

   (c) ($500,000 – $256,000)$(A/P, 15\%, 3)$ + $256,000(0.15).

   (d) Both (b) and (c).

3. What is the economic life of the asset?

   (a) Two years.

   (b) Three years.

   (c) Four years.

   (d) Five years.

4. Suppose you own an asset of age three. You should

   (a) Retain it at a marginal cost of $219,700 for the next period.

   (b) Salvage it due to the marginal cost of retaining it for $277,940 in the next period.

   (c) Retain it through age seven.

   (d) None of the above.

5. It is decided to replace the asset from Question 1, but there are two choices for replacement: a new version of the same asset or a more technologically advanced version that costs more ($550,000), but costs less to operate (starting O&M costs are $90,000 and increase at a rate of 20%). Furthermore, the salvage value declines only 15% each period. The costs and annual equivalents are as follows:

Age	O&M	SV	AEC(15%)
0		$550,000	
1	$90,000	$467,500	$255,000
2	$108,000	$397,375	$251,860
3	$129,600	$337,769	$250,983
4	$155,520	$287,103	$252,158

The current asset should be

(a) Replaced with a new version of itself.

(b) Replaced with the technologically advanced asset, which should be retained for four years.

(c) Replaced with the technologically advanced asset, which should be retained for two years.

(d) Replaced with the technologically advanced asset, which should be retained for three years.

6. The annualized operating costs for using the technologically advanced asset for three years is calculated as

(a) $\$550,000(A/P, 15\%, 3)$.

(b) $\$550{,}000(A/P, 15\%, 3) - \$250{,}983(A/F, 15\%, 3)$.

(c) $(\$90{,}000(P/F, 15\%, 1) + \$108{,}000(P/F, 15\%, 2) + \$129{,}600(P/F, 15\%, 3))$ $(A/P, 15\%, 3)$

(d) $129,600.

7. An asset should be replaced at the end of its economic life if

(a) The asset is required to be in service only for a few years.

(b) Repeatability and an infinite horizon are assumed.

(c) New assets available on the market are much more technologically advanced.

(d) None of the above.

8. Reasons to replace an asset include the fact that

(a) It does not perform in its intended capacity.

(b) It has deteriorated and operates at very high cost.

(c) Assets available on the marketplace are technologically superior.

(d) All of the above.

9. Technologically advanced assets are generally modeled as

(a) Having higher costs of operation.

(b) Having fewer capabilities.

(c) Having lower costs of operation and higher salvage values.

(d) None of the above.

10. The economic life of an asset is generally defined as

(a) The age that minimizes annualized capital and operating costs.

(b) The maximum service life of the asset.

(c) The age that minimizes the annualized capital costs.

(d) The age at which operating costs exceed depreciation charges.

Part VI
Appendix

Solutions to Fundamentals of Engineering Exam Prep Questions

Chapter 1

1(d); 2(b); 3(c); 4(c); 5(b); 6(d).

Chapter 2

1(c); 2(b); 3(d); 4(a); 5(b); 6(a); 7(d); 8(b); 9(a); 10(b); 11(a); 12(c); 13(c); 14(d); 15(d); 16(b); 17(a); 18(a); 19(c); 20(d); 21(a); 22(b); 23(b); 24(c); 25(b); 26(b); 27(c); 28(d); 29(d); 30(a).

Chapter 3

1(b); 2(d); 3(a); 4(b); 5(c); 6(d); 7(a); 8(d); 9(b); 10(b); 11(d); 12(c); 13(a); 14(c); 15(b).

Chapter 4

1(b); 2(a); 3(d); 4(c); 5(c); 6(a); 7(b); 8(c); 9(d); 10(a); 11(c); 12(d); 13(c); 14(b); 15(a); 16(b); 17(b); 18(a); 19(c); 20(a); 21(d); 22(b); 23(b); 24(b); 25(c); 26(d); 27(d); 28(b); 29(a); 30(c).

Chapter 5

1(d); 2(d); 3(d); 4(c); 5(a).

Chapter 6

1(a); 2(b); 3(d); 4(d); 5(c).

Chapter 7

1(c); 2(c); 3(a); 4(b); 5(d); 6(c); 7(b); 8(d); 9(b); 10(d); 11(a); 12(c); 13(c); 14(d); 15(a); 16(b); 17(c); 18(d); 19(a); 20(d).

Chapter 8

1(c); 2(d); 3(a); 4(b); 5(d); 6(b); 7(c); 8(b); 9(a); 10(b); 11(c); 12(a); 13(d); 14(a); 15(d); 16(c); 17(b); 18(a); 19(d); 20(d).

Chapter 9

1(c); 2(a); 3(b); 4(c); 5(d); 6(b); 7(a); 8(a); 9(d); 10(c);

Chapter 10

1(b); 2(d); 3(a); 4(b); 5(b); 6(a); 7(c); 8(a); 9(a); 10(b).

Chapter 11

1(d); 2(a); 3(d); 4(b); 5(c); 6(c).

Chapter 12

1(b); 2(b); 3(c); 4(d); 5(a); 6(b); 7(c); 8(a); 9(a); 10(b); 11(b); 12(c); 13(c); 14(d); 15(c).

Chapter 13

1(b); 2(d); 3(c); 4(b); 5(d).

Chapter 14

1(a); 2(c); 3(c); 4(b); 5(c).

Chapter 15

1(c); 2(d); 3(d); 4(b); 5(d).

Chapter 16

1(c); 2(d); 3(a); 4(b); 5(d); 6(d); 7(c); 8(b); 9(d); 10(c).

Interest Rate Factors for Discrete Compounding

This appendix presents the tables for all interest factors assuming discrete compounding

TABLE A.1 **$i = 0.25\%$**

N	F/P	F/A	F/G	P/F	P/A	P/G	A/F	A/P	A/G
1	1.0025	1.0000	0.0000	0.9975	0.9975	0.0000	1.0000	1.0025	0.0000
2	1.0050	2.0025	1.0000	0.9950	1.9925	0.9950	0.4994	0.5019	0.4994
3	1.0075	3.0075	3.0025	0.9925	2.9851	2.9801	0.3325	0.3350	0.9983
4	1.0100	4.0150	6.0100	0.9901	3.9751	5.9503	0.2491	0.2516	1.4969
5	1.0126	5.0251	10.0250	0.9876	4.9627	9.9007	0.1990	0.2015	1.9950
6	1.0151	6.0376	15.0501	0.9851	5.9478	14.8263	0.1656	0.1681	2.4927
7	1.0176	7.0527	21.0877	0.9827	6.9305	20.7223	0.1418	0.1443	2.9900
8	1.0202	8.0704	28.1404	0.9802	7.9107	27.5839	0.1239	0.1264	3.4869
9	1.0227	9.0905	36.2108	0.9778	8.8885	35.4061	0.1100	0.1125	3.9834
10	1.0253	10.1133	45.3013	0.9753	9.8639	44.1842	0.0989	0.1014	4.4794
11	1.0278	11.1385	55.4146	0.9729	10.8368	53.9133	0.0898	0.0923	4.9750
12	1.0304	12.1664	66.5531	0.9705	11.8073	64.5886	0.0822	0.0847	5.4702
13	1.0330	13.1968	78.7195	0.9681	12.7753	76.2053	0.0758	0.0783	5.9650
14	1.0356	14.2298	91.9163	0.9656	13.7410	88.7587	0.0703	0.0728	6.4594
15	1.0382	15.2654	106.1461	0.9632	14.7042	102.2441	0.0655	0.0680	6.9534
16	1.0408	16.3035	121.4114	0.9608	15.6650	116.6567	0.0613	0.0638	7.4469
17	1.0434	17.3443	137.7150	0.9584	16.6235	131.9917	0.0577	0.0602	7.9401
18	1.0460	18.3876	155.0593	0.9561	17.5795	148.2446	0.0544	0.0569	8.4328
19	1.0486	19.4336	173.4469	0.9537	18.5332	165.4106	0.0515	0.0540	8.9251
20	1.0512	20.4822	192.8805	0.9513	19.4845	183.4851	0.0488	0.0513	9.4170
21	1.0538	21.5334	213.3627	0.9489	20.4334	202.4634	0.0464	0.0489	9.9085
22	1.0565	22.5872	234.8961	0.9466	21.3800	222.3410	0.0443	0.0468	10.3995
23	1.0591	23.6437	257.4834	0.9442	22.3241	243.1131	0.0423	0.0448	10.8901
24	1.0618	24.7028	281.1271	0.9418	23.2660	264.7753	0.0405	0.0430	11.3804
25	1.0644	25.7646	305.8299	0.9395	24.2055	287.3230	0.0388	0.0413	11.8702
26	1.0671	26.8290	331.5945	0.9371	25.1426	310.7516	0.0373	0.0398	12.3596
27	1.0697	27.8961	358.4235	0.9348	26.0774	335.0566	0.0358	0.0383	12.8485
28	1.0724	28.9658	386.3195	0.9325	27.0099	360.2334	0.0345	0.0370	13.3371
29	1.0751	30.0382	415.2853	0.9301	27.9400	386.2776	0.0333	0.0358	13.8252
30	1.0778	31.1133	445.3235	0.9278	28.8679	413.1847	0.0321	0.0346	14.3130
35	1.0913	36.5292	611.6949	0.9163	33.4724	560.5076	0.0274	0.0299	16.7454
40	1.1050	42.0132	805.2816	0.9050	38.0199	728.7399	0.0238	0.0263	19.1673
45	1.1189	47.5661	1026.4256	0.8937	42.5109	917.3400	0.0210	0.0235	21.5789
50	1.1330	53.1887	1275.4731	0.8826	46.9462	1125.7767	0.0188	0.0213	23.9802
55	1.1472	58.8819	1552.7746	0.8717	51.3264	1353.5286	0.0170	0.0195	26.3710
60	1.1616	64.6467	1858.6850	0.8609	55.6524	1600.0845	0.0155	0.0180	28.7514
65	1.1762	70.4839	2193.5639	0.8502	59.9246	1864.9427	0.0142	0.0167	31.1215
70	1.1910	76.3944	2557.7749	0.8396	64.1439	2147.6111	0.0131	0.0156	33.4812
75	1.2059	82.3792	2951.6868	0.8292	68.3108	2447.6069	0.0121	0.0146	35.8305
80	1.2211	88.4392	3375.6726	0.8189	72.4260	2764.4568	0.0113	0.0138	38.1694
85	1.2364	94.5753	3830.1100	0.8088	76.4901	3097.6963	0.0106	0.0131	40.4980
90	1.2520	100.7885	4315.3817	0.7987	80.5038	3446.8700	0.0099	0.0124	42.8162
95	1.2677	107.0797	4831.8750	0.7888	84.4677	3811.5311	0.0093	0.0118	45.1241
100	1.2836	113.4500	5379.9822	0.7790	88.3825	4191.2417	0.0088	0.0113	47.4216

TABLE A.2 $i = 0.50\%$

N	F/P	F/A	F/G	P/F	P/A	P/G	A/F	A/P	A/G
1	1.0050	1.0000	0.0000	0.9950	0.9950	0.0000	1.0000	1.0050	0.0000
2	1.0100	2.0050	1.0000	0.9901	1.9851	0.9901	0.4988	0.5038	0.4988
3	1.0151	3.0150	3.0050	0.9851	2.9702	2.9604	0.3317	0.3367	0.9967
4	1.0202	4.0301	6.0200	0.9802	3.9505	5.9011	0.2481	0.2531	1.4938
5	1.0253	5.0503	10.0501	0.9754	4.9259	9.8026	0.1980	0.2030	1.9900
6	1.0304	6.0755	15.1004	0.9705	5.8964	14.6552	0.1646	0.1696	2.4855
7	1.0355	7.1059	21.1759	0.9657	6.8621	20.4493	0.1407	0.1457	2.9801
8	1.0407	8.1414	28.2818	0.9609	7.8230	27.1755	0.1228	0.1278	3.4738
9	1.0459	9.1821	36.4232	0.9561	8.7791	34.8244	0.1089	0.1139	3.9668
10	1.0511	10.2280	45.6053	0.9513	9.7304	43.3865	0.0978	0.1028	4.4589
11	1.0564	11.2792	55.8333	0.9466	10.6770	52.8526	0.0887	0.0937	4.9501
12	1.0617	12.3356	67.1125	0.9419	11.6189	63.2136	0.0811	0.0861	5.4406
13	1.0670	13.3972	79.4480	0.9372	12.5562	74.4602	0.0746	0.0796	5.9302
14	1.0723	14.4642	92.8453	0.9326	13.4887	86.5835	0.0691	0.0741	6.4190
15	1.0777	15.5365	107.3095	0.9279	14.4166	99.5743	0.0644	0.0694	6.9069
16	1.0831	16.6142	122.8461	0.9233	15.3399	113.4238	0.0602	0.0652	7.3940
17	1.0885	17.6973	139.4603	0.9187	16.2586	128.1231	0.0565	0.0615	7.8803
18	1.0939	18.7858	157.1576	0.9141	17.1728	143.6634	0.0532	0.0582	8.3658
19	1.0994	19.8797	175.9434	0.9096	18.0824	160.0360	0.0503	0.0553	8.8504
20	1.1049	20.9791	195.8231	0.9051	18.9874	177.2322	0.0477	0.0527	9.3342
21	1.1104	22.0840	216.8022	0.9006	19.8880	195.2434	0.0453	0.0503	9.8172
22	1.1160	23.1944	238.8862	0.8961	20.7841	214.0611	0.0431	0.0481	10.2993
23	1.1216	24.3104	262.0806	0.8916	21.6757	233.6768	0.0411	0.0461	10.7806
24	1.1272	25.4320	286.3910	0.8872	22.5629	254.0820	0.0393	0.0443	11.2611
25	1.1328	26.5591	311.8230	0.8828	23.4456	275.2686	0.0377	0.0427	11.7407
26	1.1385	27.6919	338.3821	0.8784	24.3240	297.2281	0.0361	0.0411	12.2195
27	1.1442	28.8304	366.0740	0.8740	25.1980	319.9523	0.0347	0.0397	12.6975
28	1.1499	29.9745	394.9044	0.8697	26.0677	343.4332	0.0334	0.0384	13.1747
29	1.1556	31.1244	424.8789	0.8653	26.9330	367.6625	0.0321	0.0371	13.6510
30	1.1614	32.2800	456.0033	0.8610	27.7941	392.6324	0.0310	0.0360	14.1265
35	1.1907	38.1454	629.0756	0.8398	32.0354	528.3123	0.0262	0.0312	16.4915
40	1.2208	44.1588	831.7695	0.8191	36.1722	681.3347	0.0226	0.0276	18.8359
45	1.2516	50.3242	1064.8328	0.7990	40.2072	850.7631	0.0199	0.0249	21.1595
50	1.2832	56.6452	1329.0326	0.7793	44.1428	1035.6966	0.0177	0.0227	23.4624
55	1.3156	63.1258	1625.1550	0.7601	47.9814	1235.2686	0.0158	0.0208	25.7447
60	1.3489	69.7700	1954.0061	0.7414	51.7256	1448.6458	0.0143	0.0193	28.0064
65	1.3829	76.5821	2316.4124	0.7231	55.3775	1675.0272	0.0131	0.0181	30.2475
70	1.4178	83.5661	2713.2211	0.7053	58.9394	1913.6427	0.0120	0.0170	32.4680
75	1.4536	90.7265	3145.3010	0.6879	62.4136	2163.7525	0.0110	0.0160	34.6679
80	1.4903	98.0677	3613.5427	0.6710	65.8023	2424.6455	0.0102	0.0152	36.8474
85	1.5280	105.5943	4118.8594	0.6545	69.1075	2695.6389	0.0095	0.0145	39.0065
90	1.5666	113.3109	4662.1872	0.6383	72.3313	2976.0769	0.0088	0.0138	41.1451
95	1.6061	121.2224	5244.4859	0.6226	75.4757	3265.3298	0.0082	0.0132	43.2633
100	1.6467	129.3337	5866.7397	0.6073	78.5426	3562.7934	0.0077	0.0127	45.3613

TABLE A.3 $i = 0.75\%$

N	F/P	F/A	F/G	P/F	P/A	P/G	A/F	A/P	A/G
1	1.0075	1.0000	0.0000	0.9926	0.9926	0.0000	1.0000	1.0075	0.0000
2	1.0151	2.0075	1.0000	0.9852	1.9777	0.9852	0.4981	0.5056	0.4981
3	1.0227	3.0226	3.0075	0.9778	2.9556	2.9408	0.3308	0.3383	0.9950
4	1.0303	4.0452	6.0301	0.9706	3.9261	5.8525	0.2472	0.2547	1.4907
5	1.0381	5.0756	10.0753	0.9633	4.8894	9.7058	0.1970	0.2045	1.9851
6	1.0459	6.1136	15.1508	0.9562	5.8456	14.4866	0.1636	0.1711	2.4782
7	1.0537	7.1595	21.2645	0.9490	6.7946	20.1808	0.1397	0.1472	2.9701
8	1.0616	8.2132	28.4240	0.9420	7.7366	26.7747	0.1218	0.1293	3.4608
9	1.0696	9.2748	36.6371	0.9350	8.6716	34.2544	0.1078	0.1153	3.9502
10	1.0776	10.3443	45.9119	0.9280	9.5996	42.6064	0.0967	0.1042	4.4384
11	1.0857	11.4219	56.2563	0.9211	10.5207	51.8174	0.0876	0.0951	4.9253
12	1.0938	12.5076	67.6782	0.9142	11.4349	61.8740	0.0800	0.0875	5.4110
13	1.1020	13.6014	80.1858	0.9074	12.3423	72.7632	0.0735	0.0810	5.8954
14	1.1103	14.7034	93.7872	0.9007	13.2430	84.4720	0.0680	0.0755	6.3786
15	1.1186	15.8137	108.4906	0.8940	14.1370	96.9876	0.0632	0.0707	6.8606
16	1.1270	16.9323	124.3042	0.8873	15.0243	110.2973	0.0591	0.0666	7.3413
17	1.1354	18.0593	141.2365	0.8807	15.9050	124.3887	0.0554	0.0629	7.8207
18	1.1440	19.1947	159.2958	0.8742	16.7792	139.2494	0.0521	0.0596	8.2989
19	1.1525	20.3387	178.4905	0.8676	17.6468	154.8671	0.0492	0.0567	8.7759
20	1.1612	21.4912	198.8292	0.8612	18.5080	171.2297	0.0465	0.0540	9.2516
21	1.1699	22.6524	220.3204	0.8548	19.3628	188.3253	0.0441	0.0516	9.7261
22	1.1787	23.8223	242.9728	0.8484	20.2112	206.1420	0.0420	0.0495	10.1994
23	1.1875	25.0010	266.7951	0.8421	21.0533	224.6682	0.0400	0.0475	10.6714
24	1.1964	26.1885	291.7961	0.8358	21.8891	243.8923	0.0382	0.0457	11.1422
25	1.2054	27.3849	317.9845	0.8296	22.7188	263.8029	0.0365	0.0440	11.6117
26	1.2144	28.5903	345.3694	0.8234	23.5422	284.3888	0.0350	0.0425	12.0800
27	1.2235	29.8047	373.9597	0.8173	24.3595	305.6387	0.0336	0.0411	12.5470
28	1.2327	31.0282	403.7644	0.8112	25.1707	327.5416	0.0322	0.0397	13.0128
29	1.2420	32.2609	434.7926	0.8052	25.9759	350.0867	0.0310	0.0385	13.4774
30	1.2513	33.5029	467.0536	0.7992	26.7751	373.2631	0.0298	0.0373	13.9407
35	1.2989	39.8538	647.1750	0.7699	30.6827	498.2471	0.0251	0.0326	16.2387
40	1.3483	46.4465	859.5309	0.7416	34.4469	637.4693	0.0215	0.0290	18.5058
45	1.3997	53.2901	1105.3483	0.7145	38.0732	789.7173	0.0188	0.0263	20.7421
50	1.4530	60.3943	1385.9010	0.6883	41.5664	953.8486	0.0166	0.0241	22.9476
55	1.5083	67.7688	1702.5112	0.6630	44.9316	1128.7869	0.0148	0.0223	25.1223
60	1.5657	75.4241	2056.5516	0.6387	48.1734	1313.5189	0.0133	0.0208	27.2665
65	1.6253	83.3709	2449.4470	0.6153	51.2963	1507.0910	0.0120	0.0195	29.3801
70	1.6872	91.6201	2882.6764	0.5927	54.3046	1708.6065	0.0109	0.0184	31.4634
75	1.7514	100.1833	3357.7753	0.5710	57.2027	1917.2225	0.0100	0.0175	33.5163
80	1.8180	109.0725	3876.3374	0.5500	59.9944	2132.1472	0.0092	0.0167	35.5391
85	1.8873	118.3001	4440.0174	0.5299	62.6838	2352.6375	0.0085	0.0160	37.5318
90	1.9591	127.8790	5050.5326	0.5104	65.2746	2577.9961	0.0078	0.0153	39.4946
95	2.0337	137.8225	5709.6660	0.4917	67.7704	2807.5694	0.0073	0.0148	41.4277
100	2.1111	148.1445	6419.2683	0.4737	70.1746	3040.7453	0.0068	0.0143	43.3311

TABLE A.4 $i = 1\%$

N	F/P	F/A	F/G	P/F	P/A	P/G	A/F	A/P	A/G
1	1.0100	1.0000	0.0000	0.9901	0.9901	0.0000	1.0000	1.0100	0.0000
2	1.0201	2.0100	1.0000	0.9803	1.9704	0.9803	0.4975	0.5075	0.4975
3	1.0303	3.0301	3.0100	0.9706	2.9410	2.9215	0.3300	0.3400	0.9934
4	1.0406	4.0604	6.0401	0.9610	3.9020	5.8044	0.2463	0.2563	1.4876
5	1.0510	5.1010	10.1005	0.9515	4.8534	9.6103	0.1960	0.2060	1.9801
6	1.0615	6.1520	15.2015	0.9420	5.7955	14.3205	0.1625	0.1725	2.4710
7	1.0721	7.2135	21.3535	0.9327	6.7282	19.9168	0.1386	0.1486	2.9602
8	1.0829	8.2857	28.5671	0.9235	7.6517	26.3812	0.1207	0.1307	3.4478
9	1.0937	9.3685	36.8527	0.9143	8.5660	33.6959	0.1067	0.1167	3.9337
10	1.1046	10.4622	46.2213	0.9053	9.4713	41.8435	0.0956	0.1056	4.4179
11	1.1157	11.5668	56.6835	0.8963	10.3676	50.8067	0.0865	0.0965	4.9005
12	1.1268	12.6825	68.2503	0.8874	11.2551	60.5687	0.0788	0.0888	5.3815
13	1.1381	13.8093	80.9328	0.8787	12.1337	71.1126	0.0724	0.0824	5.8607
14	1.1495	14.9474	94.7421	0.8700	13.0037	82.4221	0.0669	0.0769	6.3384
15	1.1610	16.0969	109.6896	0.8613	13.8651	94.4810	0.0621	0.0721	6.8143
16	1.1726	17.2579	125.7864	0.8528	14.7179	107.2734	0.0579	0.0679	7.2886
17	1.1843	18.4304	143.0443	0.8444	15.5623	120.7834	0.0543	0.0643	7.7613
18	1.1961	19.6147	161.4748	0.8360	16.3983	134.9957	0.0510	0.0610	8.2323
19	1.2081	20.8109	181.0895	0.8277	17.2260	149.8950	0.0481	0.0581	8.7017
20	1.2202	22.0190	201.9004	0.8195	18.0456	165.4664	0.0454	0.0554	9.1694
21	1.2324	23.2392	223.9194	0.8114	18.8570	181.6950	0.0430	0.0530	9.6354
22	1.2447	24.4716	247.1586	0.8034	19.6604	198.5663	0.0409	0.0509	10.0998
23	1.2572	25.7163	271.6302	0.7954	20.4558	216.0660	0.0389	0.0489	10.5626
24	1.2697	26.9735	297.3465	0.7876	21.2434	234.1800	0.0371	0.0471	11.0237
25	1.2824	28.2432	324.3200	0.7798	22.0232	252.8945	0.0354	0.0454	11.4831
26	1.2953	29.5256	352.5631	0.7720	22.7952	272.1957	0.0339	0.0439	11.9409
27	1.3082	30.8209	382.0888	0.7644	23.5596	292.0702	0.0324	0.0424	12.3971
28	1.3213	32.1291	412.9097	0.7568	24.3164	312.5047	0.0311	0.0411	12.8516
29	1.3345	33.4504	445.0388	0.7493	25.0658	333.4863	0.0299	0.0399	13.3044
30	1.3478	34.7849	478.4892	0.7419	25.8077	355.0021	0.0287	0.0387	13.7557
35	1.4166	41.6603	666.0276	0.7059	29.4086	470.1583	0.0240	0.0340	15.9871
40	1.4889	48.8864	888.6373	0.6717	32.8347	596.8561	0.0205	0.0305	18.1776
45	1.5648	56.4811	1148.1075	0.6391	36.0945	733.7037	0.0177	0.0277	20.3273
50	1.6446	64.4632	1446.3182	0.6080	39.1961	879.4176	0.0155	0.0255	22.4363
55	1.7285	72.8525	1785.2457	0.5785	42.1472	1032.8148	0.0137	0.0237	24.5049
60	1.8167	81.6697	2166.9670	0.5504	44.9550	1192.8061	0.0122	0.0222	26.5333
65	1.9094	90.9366	2593.6649	0.5237	47.6266	1358.3903	0.0110	0.0210	28.5217
70	2.0068	100.6763	3067.6337	0.4983	50.1685	1528.6474	0.0099	0.0199	30.4703
75	2.1091	110.9128	3591.2847	0.4741	52.5871	1702.7340	0.0090	0.0190	32.3793
80	2.2167	121.6715	4167.1522	0.4511	54.8882	1879.8771	0.0082	0.0182	34.2492
85	2.3298	132.9790	4797.8997	0.4292	57.0777	2059.3701	0.0075	0.0175	36.0801
90	2.4486	144.8633	5486.3267	0.4084	59.1609	2240.5675	0.0069	0.0169	37.8724
95	2.5735	157.3538	6235.3755	0.3886	61.1430	2422.8811	0.0064	0.0164	39.6265
100	2.7048	170.4814	7048.1383	0.3697	63.0289	2605.7758	0.0059	0.0159	41.3426

TABLE A.5 $i = 1.25\%$

N	F/P	F/A	F/G	P/F	P/A	P/G	A/F	A/P	A/G
1	1.0125	1.0000	0.0000	0.9877	0.9877	0.0000	1.0000	1.0125	0.0000
2	1.0252	2.0125	1.0000	0.9755	1.9631	0.9755	0.4969	0.5094	0.4969
3	1.0380	3.0377	3.0125	0.9634	2.9265	2.9023	0.3292	0.3417	0.9917
4	1.0509	4.0756	6.0502	0.9515	3.8781	5.7569	0.2454	0.2579	1.4845
5	1.0641	5.1266	10.1258	0.9398	4.8178	9.5160	0.1951	0.2076	1.9752
6	1.0774	6.1907	15.2524	0.9282	5.7460	14.1569	0.1615	0.1740	2.4638
7	1.0909	7.2680	21.4430	0.9167	6.6627	19.6571	0.1376	0.1501	2.9503
8	1.1045	8.3589	28.7110	0.9054	7.5681	25.9949	0.1196	0.1321	3.4348
9	1.1183	9.4634	37.0699	0.8942	8.4623	33.1487	0.1057	0.1182	3.9172
10	1.1323	10.5817	46.5333	0.8832	9.3455	41.0973	0.0945	0.1070	4.3975
11	1.1464	11.7139	57.1150	0.8723	10.2178	49.8201	0.0854	0.0979	4.8758
12	1.1608	12.8604	68.8289	0.8615	11.0793	59.2967	0.0778	0.0903	5.3520
13	1.1753	14.0211	81.6893	0.8509	11.9302	69.5072	0.0713	0.0838	5.8262
14	1.1900	15.1964	95.7104	0.8404	12.7706	80.4320	0.0658	0.0783	6.2982
15	1.2048	16.3863	110.9068	0.8300	13.6005	92.0519	0.0610	0.0735	6.7682
16	1.2199	17.5912	127.2931	0.8197	14.4203	104.3481	0.0568	0.0693	7.2362
17	1.2351	18.8111	144.8843	0.8096	15.2299	117.3021	0.0532	0.0657	7.7021
18	1.2506	20.0462	163.6953	0.7996	16.0295	130.8958	0.0499	0.0624	8.1659
19	1.2662	21.2968	183.7415	0.7898	16.8193	145.1115	0.0470	0.0595	8.6277
20	1.2820	22.5630	205.0383	0.7800	17.5993	159.9316	0.0443	0.0568	9.0874
21	1.2981	23.8450	227.6013	0.7704	18.3697	175.3392	0.0419	0.0544	9.5450
22	1.3143	25.1431	251.4463	0.7609	19.1306	191.3174	0.0398	0.0523	10.0006
23	1.3307	26.4574	276.5894	0.7515	19.8820	207.8499	0.0378	0.0503	10.4542
24	1.3474	27.7881	303.0467	0.7422	20.6242	224.9204	0.0360	0.0485	10.9056
25	1.3642	29.1354	330.8348	0.7330	21.3573	242.5132	0.0343	0.0468	11.3551
26	1.3812	30.4996	359.9702	0.7240	22.0813	260.6128	0.0328	0.0453	11.8024
27	1.3985	31.8809	390.4699	0.7150	22.7963	279.2040	0.0314	0.0439	12.2478
28	1.4160	33.2794	422.3507	0.7062	23.5025	298.2719	0.0300	0.0425	12.6911
29	1.4337	34.6954	455.6301	0.6975	24.2000	317.8019	0.0288	0.0413	13.1323
30	1.4516	36.1291	490.3255	0.6889	24.889	337.7797	0.0277	0.0402	13.5715
35	1.5446	43.5709	685.6696	0.6474	28.2079	443.9037	0.0230	0.0355	15.7369
40	1.6436	51.4896	919.1646	0.6084	31.3269	559.2320	0.0194	0.0319	17.8515
45	1.7489	59.9157	1193.2553	0.5718	34.2582	682.2710	0.0167	0.0292	19.9156
50	1.8610	68.8818	1510.5432	0.5373	37.0129	811.6738	0.0145	0.0270	21.9295
55	1.9803	78.4225	1873.7964	0.5050	39.6017	946.2277	0.0128	0.0253	23.8936
60	2.1072	88.5745	2285.9606	0.4746	42.0346	1084.8429	0.0113	0.0238	25.8083
65	2.2422	99.3771	2750.1700	0.4460	44.3210	1226.5421	0.0101	0.0226	27.6741
70	2.3859	110.8720	3269.7598	0.4191	46.4697	1370.4513	0.0090	0.0215	29.4913
75	2.5388	123.1035	3848.2789	0.3939	48.4890	1515.7904	0.0081	0.0206	31.2605
80	2.7015	136.1188	4489.5036	0.3702	50.3867	1661.8651	0.0073	0.0198	32.9822
85	2.8746	149.9682	5197.4522	0.3479	52.1701	1808.0598	0.0067	0.0192	34.6570
90	3.0588	164.7050	5976.4006	0.3269	53.8461	1953.8303	0.0061	0.0186	36.2855
95	3.2548	180.3862	6830.8985	0.3072	55.4211	2098.6973	0.0055	0.0180	37.8682
100	3.4634	197.0723	7765.7874	0.2887	56.9013	2242.2411	0.0051	0.0176	39.4058

TABLE A.6 $i = 1.5\%$

N	F/P	F/A	F/G	P/F	P/A	P/G	A/F	A/P	A/G
1	1.0150	1.0000	0.0000	0.9852	0.9852	0.0000	1.0000	1.0150	0.0000
2	1.0302	2.0150	1.0000	0.9707	1.9559	0.9707	0.4963	0.5113	0.4963
3	1.0457	3.0452	3.0150	0.9563	2.9122	2.8833	0.3284	0.3434	0.9901
4	1.0614	4.0909	6.0602	0.9422	3.8544	5.7098	0.2444	0.2594	1.4814
5	1.0773	5.1523	10.1511	0.9283	4.7826	9.4229	0.1941	0.2091	1.9702
6	1.0934	6.2296	15.3034	0.9145	5.6972	13.9956	0.1605	0.1755	2.4566
7	1.1098	7.3230	21.5329	0.9010	6.5982	19.4018	0.1366	0.1516	2.9405
8	1.1265	8.4328	28.8559	0.8877	7.4859	25.6157	0.1186	0.1336	3.4219
9	1.1434	9.5593	37.2888	0.8746	8.3605	32.6125	0.1046	0.1196	3.9008
10	1.1605	10.7027	46.8481	0.8617	9.2222	40.3675	0.0934	0.1084	4.3772
11	1.1779	11.8633	57.5508	0.8489	10.0711	48.8568	0.0843	0.0993	4.8512
12	1.1956	13.0412	69.4141	0.8364	10.9075	58.0571	0.0767	0.0917	5.3227
13	1.2136	14.2368	82.4553	0.8240	11.7315	67.9454	0.0702	0.0852	5.7917
14	1.2318	15.4504	96.6921	0.8118	12.5434	78.4994	0.0647	0.0797	6.2582
15	1.2502	16.6821	112.1425	0.7999	13.3432	89.6974	0.0599	0.0749	6.7223
16	1.2690	17.9324	128.8247	0.7880	14.1313	101.5178	0.0558	0.0708	7.1839
17	1.2880	19.2014	146.7570	0.7764	14.9076	113.9400	0.0521	0.0671	7.6431
18	1.3073	20.4894	165.9584	0.7649	15.6726	126.9435	0.0488	0.0638	8.0997
19	1.3270	21.7967	186.4478	0.7536	16.4262	140.5084	0.0459	0.0609	8.5539
20	1.3469	23.1237	208.2445	0.7425	17.1686	154.6154	0.0432	0.0582	9.0057
21	1.3671	24.4705	231.3681	0.7315	17.9001	169.2453	0.0409	0.0559	9.4550
22	1.3876	25.8376	255.8387	0.7207	18.6208	184.3798	0.0387	0.0537	9.9018
23	1.4084	27.2251	281.6762	0.7100	19.3309	200.0006	0.0367	0.0517	10.3462
24	1.4295	28.6335	308.9014	0.6995	20.0304	216.0901	0.0349	0.0499	10.7881
25	1.4509	30.0630	337.5349	0.6892	20.7196	232.6310	0.0333	0.0483	11.2276
26	1.4727	31.5140	367.5979	0.6790	21.3986	249.6065	0.0317	0.0467	11.6646
27	1.4948	32.9867	399.1119	0.6690	22.0676	267.0002	0.0303	0.0453	12.0992
28	1.5172	34.4815	432.0986	0.6591	22.7267	284.7958	0.0290	0.0440	12.5313
29	1.5400	35.9987	466.5801	0.6494	23.3761	302.9779	0.0278	0.0428	12.9610
30	1.5631	37.5387	502.5788	0.6398	24.0158	321.5310	0.0266	0.0416	13.3883
35	1.6839	45.5921	706.1392	0.5939	27.0756	419.3521	0.0219	0.0369	15.4882
40	1.8140	54.2679	951.1929	0.5513	29.9158	524.3568	0.0184	0.0334	17.5277
45	1.9542	63.6142	1240.9467	0.5117	32.5523	635.0110	0.0157	0.0307	19.5074
50	2.1052	73.6828	1578.8552	0.4750	34.9997	749.9636	0.0136	0.0286	21.4277
55	2.2679	84.5296	1968.6399	0.4409	37.2715	868.0285	0.0118	0.0268	23.2894
60	2.4432	96.2147	2414.3101	0.4093	39.3803	988.1674	0.0104	0.0254	25.0930
65	2.6320	108.8028	2920.1848	0.3799	41.3378	1109.4752	0.0092	0.0242	26.8393
70	2.8355	122.3638	3490.9169	0.3527	43.1549	1231.1658	0.0082	0.0232	28.5290
75	3.0546	136.9728	4131.5187	0.3274	44.8416	1352.5600	0.0073	0.0223	30.1631
80	3.2907	152.7109	4847.3902	0.3039	46.4073	1473.0741	0.0065	0.0215	31.7423
85	3.5450	169.6652	5644.3484	0.2821	47.8607	1592.2095	0.0059	0.0209	33.2676
90	3.8189	187.9299	6528.6600	0.2619	49.2099	1709.5439	0.0053	0.0203	34.7399
95	4.1141	207.6061	7507.0762	0.2431	50.4622	1824.7224	0.0048	0.0198	36.1602
100	4.4320	228.8030	8586.8696	0.2256	51.6247	1937.4506	0.0044	0.0194	37.5295

TABLE A.7 $i = 1.75\%$

N	F/P	F/A	F/G	P/F	P/A	P/G	A/F	A/P	A/G
1	1.0175	1.0000	0.0000	0.9828	0.9828	0.0000	1.0000	1.0175	0.0000
2	1.0353	2.0175	1.0000	0.9659	1.9487	0.9659	0.4957	0.5132	0.4957
3	1.0534	3.0528	3.0175	0.9493	2.8980	2.8645	0.3276	0.3451	0.9884
4	1.0719	4.1062	6.0703	0.9330	3.8309	5.6633	0.2435	0.2610	1.4783
5	1.0906	5.1781	10.1765	0.9169	4.7479	9.3310	0.1931	0.2106	1.9653
6	1.1097	6.2687	15.3546	0.9011	5.6490	13.8367	0.1595	0.1770	2.4494
7	1.1291	7.3784	21.6233	0.8856	6.5346	19.1506	0.1355	0.1530	2.9306
8	1.1489	8.5075	29.0017	0.8704	7.4051	25.2435	0.1175	0.1350	3.4089
9	1.1690	9.6564	37.5093	0.8554	8.2605	32.0870	0.1036	0.1211	3.8844
10	1.1894	10.8254	47.1657	0.8407	9.1012	39.6535	0.0924	0.1099	4.3569
11	1.2103	12.0148	57.9911	0.8263	9.9275	47.9162	0.0832	0.1007	4.8266
12	1.2314	13.2251	70.0059	0.8121	10.7395	56.8489	0.0756	0.0931	5.2934
13	1.2530	14.4565	83.2310	0.7981	11.5376	66.4260	0.0692	0.0867	5.7573
14	1.2749	15.7095	97.6876	0.7844	12.3220	76.6227	0.0637	0.0812	6.2184
15	1.2972	16.9844	113.3971	0.7709	13.0929	87.4149	0.0589	0.0764	6.6765
16	1.3199	18.2817	130.3816	0.7576	13.8505	98.7792	0.0547	0.0722	7.1318
17	1.3430	19.6016	148.6632	0.7446	14.5951	110.6926	0.0510	0.0685	7.5842
18	1.3665	20.9446	168.2648	0.7318	15.3269	123.1328	0.0477	0.0652	8.0338
19	1.3904	22.3112	189.2095	0.7192	16.0461	136.0783	0.0448	0.0623	8.4805
20	1.4148	23.7016	211.5206	0.7068	16.7529	149.5080	0.0422	0.0597	8.9243
21	1.4395	25.1164	235.2223	0.6947	17.4475	163.4013	0.0398	0.0573	9.3653
22	1.4647	26.5559	260.3386	0.6827	18.1303	177.7385	0.0377	0.0552	9.8034
23	1.4904	28.0207	286.8946	0.6710	18.8012	192.5000	0.0357	0.0532	10.2387
24	1.5164	29.5110	314.9152	0.6594	19.4607	207.6671	0.0339	0.0514	10.6711
25	1.5430	31.0275	344.4262	0.6481	20.1088	223.2214	0.0322	0.0497	11.1007
26	1.5700	32.5704	375.4537	0.6369	20.7457	239.1451	0.0307	0.0482	11.5274
27	1.5975	34.1404	408.0241	0.6260	21.3717	255.4210	0.0293	0.0468	11.9513
28	1.6254	35.7379	442.1646	0.6152	21.9870	272.0321	0.0280	0.0455	12.3724
29	1.6539	37.3633	477.9024	0.6046	22.5916	288.9623	0.0268	0.0443	12.7907
30	1.6828	39.0172	515.2657	0.5942	23.1858	306.1954	0.0256	0.0431	13.2061
35	1.8353	47.7308	727.4766	0.5449	26.0073	396.3824	0.0210	0.0385	15.2412
40	2.0016	57.2341	984.8077	0.4996	28.5942	492.0109	0.0175	0.0350	17.2066
45	2.1830	67.5986	1291.3476	0.4581	30.9663	591.5540	0.0148	0.0323	19.1032
50	2.3808	78.9022	1651.5557	0.4200	33.1412	693.7010	0.0127	0.0302	20.9317
55	2.5965	91.2302	2070.2950	0.3851	35.1354	797.3321	0.0110	0.0285	22.6931
60	2.8318	104.6752	2552.8695	0.3531	36.9640	901.4954	0.0096	0.0271	24.3885
65	3.0884	119.3386	3105.0636	0.3238	38.6406	1005.3872	0.0084	0.0259	26.0189
70	3.3683	135.3308	3733.1862	0.2969	40.1779	1108.3333	0.0074	0.0249	27.5856
75	3.6735	152.7721	4444.1175	0.2722	41.5875	1209.7738	0.0065	0.0240	29.0899
80	4.0064	171.7938	5245.3614	0.2496	42.8799	1309.2482	0.0058	0.0233	30.5329
85	4.3694	192.5393	6145.1017	0.2289	44.0650	1406.3828	0.0052	0.0227	31.9161
90	4.7654	215.1646	7152.2638	0.2098	45.1516	1500.8798	0.0046	0.0221	33.2409
95	5.1972	239.8402	8276.5820	0.1924	46.1479	1592.5069	0.0042	0.0217	34.5087
100	5.6682	266.7518	9528.6725	0.1764	47.0615	1681.0886	0.0037	0.0212	35.7211

TABLE A.8 $i = 2\%$

N	F/P	F/A	F/G	P/F	P/A	P/G	A/F	A/P	A/G
1	1.0200	1.0000	0.0000	0.9804	0.9804	0.0000	1.0000	1.0200	0.0000
2	1.0404	2.0200	1.0000	0.9612	1.9416	0.9612	0.4950	0.5150	0.4950
3	1.0612	3.0604	3.0200	0.9423	2.8839	2.8458	0.3268	0.3468	0.9868
4	1.0824	4.1216	6.0804	0.9238	3.8077	5.6173	0.2426	0.2626	1.4752
5	1.1041	5.2040	10.2020	0.9057	4.7135	9.2403	0.1922	0.2122	1.9604
6	1.1262	6.3081	15.4060	0.8880	5.6014	13.6801	0.1585	0.1785	2.4423
7	1.1487	7.4343	21.7142	0.8706	6.4720	18.9035	0.1345	0.1545	2.9208
8	1.1717	8.5830	29.1485	0.8535	7.3255	24.8779	0.1165	0.1365	3.3961
9	1.1951	9.7546	37.7314	0.8368	8.1622	31.5720	0.1025	0.1225	3.8681
10	1.2190	10.9497	47.4860	0.8203	8.9826	38.9551	0.0913	0.1113	4.3367
11	1.2434	12.1687	58.4358	0.8043	9.7868	46.9977	0.0822	0.1022	4.8021
12	1.2682	13.4121	70.6045	0.7885	10.5753	55.6712	0.0746	0.0946	5.2642
13	1.2936	14.6803	84.0166	0.7730	11.3484	64.9475	0.0681	0.0881	5.7231
14	1.3195	15.9739	98.6969	0.7579	12.1062	74.7999	0.0626	0.0826	6.1786
15	1.3459	17.2934	114.6708	0.7430	12.8493	85.2021	0.0578	0.0778	6.6309
16	1.3728	18.6393	131.9643	0.7284	13.5777	96.1288	0.0537	0.0737	7.0799
17	1.4002	20.0121	150.6035	0.7142	14.2919	107.5554	0.0500	0.0700	7.5256
18	1.4282	21.4123	170.6156	0.7002	14.9920	119.4581	0.0467	0.0667	7.9681
19	1.4568	22.8406	192.0279	0.6864	15.6785	131.8139	0.0438	0.0638	8.4073
20	1.4859	24.2974	214.8685	0.6730	16.3514	144.6003	0.0412	0.0612	8.8433
21	1.5157	25.7833	239.1659	0.6598	17.0112	157.7959	0.0388	0.0588	9.2760
22	1.5460	27.2990	264.9492	0.6468	17.6580	171.3795	0.0366	0.0566	9.7055
23	1.5769	28.8450	292.2482	0.6342	18.2922	185.3309	0.0347	0.0547	10.1317
24	1.6084	30.4219	321.0931	0.6217	18.9139	199.6305	0.0329	0.0529	10.5547
25	1.6406	32.0303	351.5150	0.6095	19.5235	214.2592	0.0312	0.0512	10.9745
26	1.6734	33.6709	383.5453	0.5976	20.1210	229.1987	0.0297	0.0497	11.3910
27	1.7069	35.3443	417.2162	0.5859	20.7069	244.4311	0.0283	0.0483	11.8043
28	1.7410	37.0512	452.5605	0.5744	21.2813	259.9392	0.0270	0.0470	12.2145
29	1.7758	38.7922	489.6117	0.5631	21.8444	275.7064	0.0258	0.0458	12.6214
30	1.8114	40.5681	528.4040	0.5521	22.3965	291.7164	0.0246	0.0446	13.0251
35	1.9999	49.9945	749.7239	0.5000	24.9986	374.8826	0.0200	0.0400	14.9961
40	2.2080	60.4020	1020.0992	0.4529	27.3555	461.9931	0.0166	0.0366	16.8885
45	2.4379	71.8927	1344.6355	0.4102	29.4902	551.5652	0.0139	0.0339	18.7034
50	2.6916	84.5794	1728.9701	0.3715	31.4236	642.3606	0.0118	0.0318	20.4420
55	2.9717	98.5865	2179.3267	0.3365	33.1748	733.3527	0.0101	0.0301	22.1057
60	3.2810	114.0515	2702.5770	0.3048	34.7609	823.6975	0.0088	0.0288	23.6961
65	3.6225	131.1262	3306.3078	0.2761	36.1975	912.7085	0.0076	0.0276	25.2147
70	3.9996	149.9779	3998.8956	0.2500	37.4986	999.8343	0.0067	0.0267	26.6632
75	4.4158	170.7918	4789.5886	0.2265	38.6771	1084.6393	0.0059	0.0259	28.0434
80	4.8754	193.7720	5688.5979	0.2051	39.7445	1166.7868	0.0052	0.0252	29.3572
85	5.3829	219.1439	6707.1969	0.1858	40.7113	1246.0241	0.0046	0.0246	30.6064
90	5.9431	247.1567	7857.8328	0.1683	41.5869	1322.1701	0.0040	0.0240	31.7929
95	6.5617	278.0850	9154.2480	0.1524	42.3800	1395.1033	0.0036	0.0236	32.9189
100	7.2446	312.2323	10611.6153	0.1380	43.0984	1464.7527	0.0032	0.0232	33.9863

TABLE A.9 $i = 3\%$

N	F/P	F/A	F/G	P/F	P/A	P/G	A/F	A/P	A/G
1	1.0300	1.0000	0.0000	0.9709	0.9709	0.0000	1.0000	1.0300	0.0000
2	1.0609	2.0300	1.0000	0.9426	1.9135	0.9426	0.4926	0.5226	0.4926
3	1.0927	3.0909	3.0300	0.9151	2.8286	2.7729	0.3235	0.3535	0.9803
4	1.1255	4.1836	6.1209	0.8885	3.7171	5.4383	0.2390	0.2690	1.4631
5	1.1593	5.3091	10.3045	0.8626	4.5797	8.8888	0.1884	0.2184	1.9409
6	1.1941	6.4684	15.6137	0.8375	5.4172	13.0762	0.1546	0.1846	2.4138
7	1.2299	7.6625	22.0821	0.8131	6.2303	17.9547	0.1305	0.1605	2.8819
8	1.2668	8.8923	29.7445	0.7894	7.0197	23.4806	0.1125	0.1425	3.3450
9	1.3048	10.1591	38.6369	0.7664	7.7861	29.6119	0.0984	0.1284	3.8032
10	1.3439	11.4639	48.7960	0.7441	8.5302	36.3088	0.0872	0.1172	4.2565
11	1.3842	12.8078	60.2599	0.7224	9.2526	43.5330	0.0781	0.1081	4.7049
12	1.4258	14.1920	73.0677	0.7014	9.9540	51.2482	0.0705	0.1005	5.1485
13	1.4685	15.6178	87.2597	0.6810	10.6350	59.4196	0.0640	0.0940	5.5872
14	1.5126	17.0863	102.8775	0.6611	11.2961	68.0141	0.0585	0.0885	6.0210
15	1.5580	18.5989	119.9638	0.6419	11.9379	77.0002	0.0538	0.0838	6.4500
16	1.6047	20.1569	138.5627	0.6232	12.5611	86.3477	0.0496	0.0796	6.8742
17	1.6528	21.7616	158.7196	0.6050	13.1661	96.0280	0.0460	0.0760	7.2936
18	1.7024	23.4144	180.4812	0.5874	13.7535	106.0137	0.0427	0.0727	7.7081
19	1.7535	25.1169	203.8956	0.5703	14.3238	116.2788	0.0398	0.0698	8.1179
20	1.8061	26.8704	229.0125	0.5537	14.8775	126.7987	0.0372	0.0672	8.5229
21	1.8603	28.6765	255.8829	0.5375	15.4150	137.5496	0.0349	0.0649	8.9231
22	1.9161	30.5368	284.5593	0.5219	15.9369	148.5094	0.0327	0.0627	9.3186
23	1.9736	32.4529	315.0961	0.5067	16.4436	159.6566	0.0308	0.0608	9.7093
24	2.0328	34.4265	347.5490	0.4919	16.9355	170.9711	0.0290	0.0590	10.0954
25	2.0938	36.4593	381.9755	0.4776	17.4131	182.4336	0.0274	0.0574	10.4768
26	2.1566	38.5530	418.4347	0.4637	17.8768	194.0260	0.0259	0.0559	10.8535
27	2.2213	40.7096	456.9878	0.4502	18.3270	205.7309	0.0246	0.0546	11.2255
28	2.2879	42.9309	497.6974	0.4371	18.7641	217.5320	0.0233	0.0533	11.5930
29	2.3566	45.2189	540.6283	0.4243	19.1885	229.4137	0.0221	0.0521	11.9558
30	2.4273	47.5754	585.8472	0.4120	19.6004	241.3613	0.0210	0.0510	12.3141
35	2.8139	60.4621	848.7361	0.3554	21.4872	301.6267	0.0165	0.0465	14.0375
40	3.2620	75.4013	1180.0420	0.3066	23.1148	361.7499	0.0133	0.0433	15.6502
45	3.7816	92.7199	1590.6620	0.2644	24.5187	420.6325	0.0108	0.0408	17.1556
50	4.3839	112.7969	2093.2289	0.2281	25.7298	477.4803	0.0089	0.0389	18.5575
55	5.0821	136.0716	2702.3873	0.1968	26.7744	531.7411	0.0073	0.0373	19.8600
60	5.8916	163.0534	3435.1146	0.1697	27.6756	583.0526	0.0061	0.0361	21.0674
65	6.8300	194.3328	4311.0919	0.1464	28.4529	631.2010	0.0051	0.0351	22.1841
70	7.9178	230.5941	5353.1355	0.1263	29.1234	676.0869	0.0043	0.0343	23.2145
75	9.1789	272.6309	6587.6952	0.1089	29.7018	717.6978	0.0037	0.0337	24.1634
80	10.6409	321.3630	8045.4340	0.0940	30.2008	756.0865	0.0031	0.0331	25.0353
85	12.3357	377.8570	9761.8984	0.0811	30.6312	791.3529	0.0026	0.0326	25.8349
90	14.3005	443.3489	11778.2968	0.0699	31.0024	823.6302	0.0023	0.0323	26.5667
95	16.5782	519.2720	14142.4009	0.0603	31.3227	853.0742	0.0019	0.0319	27.2351
100	19.2186	607.2877	16909.5911	0.0520	31.5989	879.8540	0.0016	0.0316	27.8444

TABLE A.10 $i = 4\%$

N	F/P	F/A	F/G	P/F	P/A	P/G	A/F	A/P	A/G
1	1.0400	1.0000	0.0000	0.9615	0.9615	0.0000	1.0000	1.0400	0.0000
2	1.0816	2.0400	1.0000	0.9246	1.8861	0.9246	0.4902	0.5302	0.4902
3	1.1249	3.1216	3.0400	0.8890	2.7751	2.7025	0.3203	0.3603	0.9739
4	1.1699	4.2465	6.1616	0.8548	3.6299	5.2670	0.2355	0.2755	1.4510
5	1.2167	5.4163	10.4081	0.8219	4.4518	8.5547	0.1846	0.2246	1.9216
6	1.2653	6.6330	15.8244	0.7903	5.2421	12.5062	0.1508	0.1908	2.3857
7	1.3159	7.8983	22.4574	0.7599	6.0021	17.0657	0.1266	0.1666	2.8433
8	1.3686	9.2142	30.3557	0.7307	6.7327	22.1806	0.1085	0.1485	3.2944
9	1.4233	10.5828	39.5699	0.7026	7.4353	27.8013	0.0945	0.1345	3.7391
10	1.4802	12.0061	50.1527	0.6756	8.1109	33.8814	0.0833	0.1233	4.1773
11	1.5395	13.4864	62.1588	0.6496	8.7605	40.3772	0.0741	0.1141	4.6090
12	1.6010	15.0258	75.6451	0.6246	9.3851	47.2477	0.0666	0.1066	5.0343
13	1.6651	16.6268	90.6709	0.6006	9.9856	54.4546	0.0601	0.1001	5.4533
14	1.7317	18.2919	107.2978	0.5775	10.5631	61.9618	0.0547	0.0947	5.8659
15	1.8009	20.0236	125.5897	0.5553	11.1184	69.7355	0.0499	0.0899	6.2721
16	1.8730	21.8245	145.6133	0.5339	11.6523	77.7441	0.0458	0.0858	6.6720
17	1.9479	23.6975	167.4378	0.5134	12.1657	85.9581	0.0422	0.0822	7.0656
18	2.0258	25.6454	191.1353	0.4936	12.6593	94.3498	0.0390	0.0790	7.4530
19	2.1068	27.6712	216.7807	0.4746	13.1339	102.8933	0.0361	0.0761	7.8342
20	2.1911	29.7781	244.4520	0.4564	13.5903	111.5647	0.0336	0.0736	8.2091
21	2.2788	31.9692	274.2300	0.4388	14.0292	120.3414	0.0313	0.0713	8.5779
22	2.3699	34.2480	306.1992	0.4220	14.4511	129.2024	0.0292	0.0692	8.9407
23	2.4647	36.6179	340.4472	0.4057	14.8568	138.1284	0.0273	0.0673	9.2973
24	2.5633	39.0826	377.0651	0.3901	15.2470	147.1012	0.0256	0.0656	9.6479
25	2.6658	41.6459	416.1477	0.3751	15.6221	156.1040	0.0240	0.0640	9.9925
26	2.7725	44.3117	457.7936	0.3607	15.9828	165.1212	0.0226	0.0626	10.3312
27	2.8834	47.0842	502.1054	0.3468	16.3296	174.1385	0.0212	0.0612	10.6640
28	2.9987	49.9676	549.1896	0.3335	16.6631	183.1424	0.0200	0.0600	10.9909
29	3.1187	52.9663	599.1572	0.3207	16.9837	192.1206	0.0189	0.0589	11.3120
30	3.2434	56.0849	652.1234	0.3083	17.2920	201.0618	0.0178	0.0578	11.6274
35	3.9461	73.6522	966.3056	0.2534	18.6646	244.8768	0.0136	0.0536	13.1198
40	4.8010	95.0255	1375.6379	0.2083	19.7928	286.5303	0.0105	0.0505	14.4765
45	5.8412	121.0294	1900.7348	0.1712	20.7200	325.4028	0.0083	0.0483	15.7047
50	7.1067	152.6671	2566.6771	0.1407	21.4822	361.1638	0.0066	0.0466	16.8122
55	8.6464	191.1592	3403.9793	0.1157	22.1086	393.6890	0.0052	0.0452	17.8070
60	10.5196	237.9907	4449.7671	0.0951	22.6235	422.9966	0.0042	0.0442	18.6972
65	12.7987	294.9684	5749.2095	0.0781	23.0467	449.2014	0.0034	0.0434	19.4909
70	15.5716	364.2905	7357.2615	0.0642	23.3945	472.4789	0.0027	0.0427	20.1961
75	18.9453	448.6314	9340.7842	0.0528	23.6804	493.0408	0.0022	0.0422	20.8206
80	23.0498	551.2450	11781.1244	0.0434	23.9154	511.1161	0.0018	0.0418	21.3718
85	28.0436	676.0901	14777.2531	0.0357	24.1085	526.9384	0.0015	0.0415	21.8569
90	34.1193	827.9833	18449.5833	0.0293	24.2673	540.7369	0.0012	0.0412	22.2826
95	41.5114	1012.7846	22944.6162	0.0241	24.3978	552.7307	0.0010	0.0410	22.6550
100	50.5049	1237.6237	28440.5926	0.0198	24.5050	563.1249	0.0008	0.0408	22.9800

TABLE A.11 $i = 5\%$

N	F/P	F/A	F/G	P/F	P/A	P/G	A/F	A/P	A/G
1	1.0500	1.0000	0.0000	0.9524	0.9524	0.0000	1.0000	1.0500	0.0000
2	1.1025	2.0500	1.0000	0.9070	1.8594	0.9070	0.4878	0.5378	0.4878
3	1.1576	3.1525	3.0500	0.8638	2.7232	2.6347	0.3172	0.3672	0.9675
4	1.2155	4.3101	6.2025	0.8227	3.5460	5.1028	0.2320	0.2820	1.4391
5	1.2763	5.5256	10.5126	0.7835	4.3295	8.2369	0.1810	0.2310	1.9025
6	1.3401	6.8019	16.0383	0.7462	5.0757	11.9680	0.1470	0.1970	2.3579
7	1.4071	8.1420	22.8402	0.7107	5.7864	16.2321	0.1228	0.1728	2.8052
8	1.4775	9.5491	30.9822	0.6768	6.4632	20.9700	0.1047	0.1547	3.2445
9	1.5513	11.0266	40.5313	0.6446	7.1078	26.1268	0.0907	0.1407	3.6758
10	1.6289	12.5779	51.5579	0.6139	7.7217	31.6520	0.0795	0.1295	4.0991
11	1.7103	14.2068	64.1357	0.5847	8.3064	37.4988	0.0704	0.1204	4.5144
12	1.7959	15.9171	78.3425	0.5568	8.8633	43.6241	0.0628	0.1128	4.9219
13	1.8856	17.7130	94.2597	0.5303	9.3936	49.9879	0.0565	0.1065	5.3215
14	1.9799	19.5986	111.9726	0.5051	9.8986	56.5538	0.0510	0.1010	5.7133
15	2.0789	21.5786	131.5713	0.4810	10.3797	63.2880	0.0463	0.0963	6.0973
16	2.1829	23.6575	153.1498	0.4581	10.8378	70.1597	0.0423	0.0923	6.4736
17	2.2920	25.8404	176.8073	0.4363	11.2741	77.1405	0.0387	0.0887	6.8423
18	2.4066	28.1324	202.6477	0.4155	11.6896	84.2043	0.0355	0.0855	7.2034
19	2.5270	30.5390	230.7801	0.3957	12.0853	91.3275	0.0327	0.0827	7.5569
20	2.6533	33.0660	261.3191	0.3769	12.4622	98.4884	0.0302	0.0802	7.9030
21	2.7860	35.7193	294.3850	0.3589	12.8212	105.6673	0.0280	0.0780	8.2416
22	2.9253	38.5052	330.1043	0.3418	13.1630	112.8461	0.0260	0.0760	8.5730
23	3.0715	41.4305	368.6095	0.3256	13.4886	120.0087	0.0241	0.0741	8.8971
24	3.2251	44.5020	410.0400	0.3101	13.7986	127.1402	0.0225	0.0725	9.2140
25	3.3864	47.7271	454.5420	0.2953	14.0939	134.2275	0.0210	0.0710	9.5238
26	3.5557	51.1135	502.2691	0.2812	14.3752	141.2585	0.0196	0.0696	9.8266
27	3.7335	54.6691	553.3825	0.2678	14.6430	148.2226	0.0183	0.0683	10.1224
28	3.9201	58.4026	608.0517	0.2551	14.8981	155.1101	0.0171	0.0671	10.4114
29	4.1161	62.3227	666.4542	0.2429	15.1411	161.9126	0.0160	0.0660	10.6936
30	4.3219	66.4388	728.7770	0.2314	15.3725	168.6226	0.0151	0.0651	10.9691
35	5.5160	90.3203	1106.4061	0.1813	16.3742	200.5807	0.0111	0.0611	12.2498
40	7.0400	120.7998	1615.9955	0.1420	17.1591	229.5452	0.0083	0.0583	13.3775
45	8.9850	159.7002	2294.0031	0.1113	17.7741	255.3145	0.0063	0.0563	14.3644
50	11.4674	209.3480	3186.9599	0.0872	18.2559	277.9148	0.0048	0.0548	15.2233
55	14.6356	272.7126	4354.2524	0.0683	18.6335	297.5104	0.0037	0.0537	15.9664
60	18.6792	353.5837	5871.6744	0.0535	18.9293	314.3432	0.0028	0.0528	16.6062
65	23.8399	456.7980	7835.9602	0.0419	19.1611	328.6910	0.0022	0.0522	17.1541
70	30.4264	588.5285	10370.5702	0.0329	19.3427	340.8409	0.0017	0.0517	17.6212
75	38.8327	756.6537	13633.0744	0.0258	19.4850	351.0721	0.0013	0.0513	18.0176
80	49.5614	971.2288	17824.5764	0.0202	19.5965	359.6460	0.0010	0.0510	18.3526
85	63.2544	1245.0871	23201.7414	0.0158	19.6838	366.8007	0.0008	0.0508	18.6346
90	80.7304	1594.6073	30092.1460	0.0124	19.7523	372.7488	0.0006	0.0506	18.8712
95	103.0347	2040.6935	38913.8706	0.0097	19.8059	377.6774	0.0005	0.0505	19.0689
100	131.5013	2610.0252	50200.5031	0.0076	19.8479	381.7492	0.0004	0.0504	19.2337

TABLE A.12 $i = 6\%$

N	F/P	F/A	F/G	P/F	P/A	P/G	A/F	A/P	A/G
1	1.0600	1.0000	0.0000	0.9434	0.9434	0.0000	1.0000	1.0600	0.0000
2	1.1236	2.0600	1.0000	0.8900	1.8334	0.8900	0.4854	0.5454	0.4854
3	1.1910	3.1836	3.0600	0.8396	2.6730	2.5692	0.3141	0.3741	0.9612
4	1.2625	4.3746	6.2436	0.7921	3.4651	4.9455	0.2286	0.2886	1.4272
5	1.3382	5.6371	10.6182	0.7473	4.2124	7.9345	0.1774	0.2374	1.8836
6	1.4185	6.9753	16.2553	0.7050	4.9173	11.4594	0.1434	0.2034	2.3304
7	1.5036	8.3938	23.2306	0.6651	5.5824	15.4497	0.1191	0.1791	2.7676
8	1.5938	9.8975	31.6245	0.6274	6.2098	19.8416	0.1010	0.1610	3.1952
9	1.6895	11.4913	41.5219	0.5919	6.8017	24.5768	0.0870	0.1470	3.6133
10	1.7908	13.1808	53.0132	0.5584	7.3601	29.6023	0.0759	0.1359	4.0220
11	1.8983	14.9716	66.1940	0.5268	7.8869	34.8702	0.0668	0.1268	4.4213
12	2.0122	16.8699	81.1657	0.4970	8.3838	40.3369	0.0593	0.1193	4.8113
13	2.1329	18.8821	98.0356	0.4688	8.8527	45.9629	0.0530	0.1130	5.1920
14	2.2609	21.0151	116.9178	0.4423	9.2950	51.7128	0.0476	0.1076	5.5635
15	2.3966	23.2760	137.9328	0.4173	9.7122	57.5546	0.0430	0.1030	5.9260
16	2.5404	25.6725	161.2088	0.3936	10.1059	63.4592	0.0390	0.0990	6.2794
17	2.6928	28.2129	186.8813	0.3714	10.4773	69.4011	0.0354	0.0954	6.6240
18	2.8543	30.9057	215.0942	0.3503	10.8276	75.3569	0.0324	0.0924	6.9597
19	3.0256	33.7600	245.9999	0.3305	11.1581	81.3062	0.0296	0.0896	7.2867
20	3.2071	36.7856	279.7599	0.3118	11.4699	87.2304	0.0272	0.0872	7.6051
21	3.3996	39.9927	316.5454	0.2942	11.7641	93.1136	0.0250	0.0850	7.9151
22	3.6035	43.3923	356.5382	0.2775	12.0416	98.9412	0.0230	0.0830	8.2166
23	3.8197	46.9958	399.9305	0.2618	12.3034	104.7007	0.0213	0.0813	8.5099
24	4.0489	50.8156	446.9263	0.2470	12.5504	110.3812	0.0197	0.0797	8.7951
25	4.2919	54.8645	497.7419	0.2330	12.7834	115.9732	0.0182	0.0782	9.0722
26	4.5494	59.1564	552.6064	0.2198	13.0032	121.4684	0.0169	0.0769	9.3414
27	4.8223	63.7058	611.7628	0.2074	13.2105	126.8600	0.0157	0.0757	9.6029
28	5.1117	68.5281	675.4685	0.1956	13.4062	132.1420	0.0146	0.0746	9.8568
29	5.4184	73.6398	743.9966	0.1846	13.5907	137.3096	0.0136	0.0736	10.1032
30	5.7435	79.0582	817.6364	0.1741	13.7648	142.3588	0.0126	0.0726	10.3422
35	7.6861	111.4348	1273.9130	0.1301	14.4982	165.7427	0.0090	0.0690	11.4319
40	10.2857	154.7620	1912.6994	0.0972	15.0463	185.9568	0.0065	0.0665	12.3590
45	13.7646	212.7435	2795.7252	0.0727	15.4558	203.1096	0.0047	0.0647	13.1413
50	18.4202	290.3359	4005.5984	0.0543	15.7619	217.4574	0.0034	0.0634	13.7964
55	24.6503	394.1720	5652.8671	0.0406	15.9905	229.3222	0.0025	0.0625	14.3411
60	32.9877	533.1282	7885.4697	0.0303	16.1614	239.0428	0.0019	0.0619	14.7909
65	44.1450	719.0829	10901.3810	0.0227	16.2891	246.9450	0.0014	0.0614	15.1601
70	59.0759	967.9322	14965.5362	0.0169	16.3845	253.3271	0.0010	0.0610	15.4613
75	79.0569	1300.9487	20432.4780	0.0126	16.4558	258.4527	0.0008	0.0608	15.7058
80	105.7960	1746.5999	2776.6649	0.0095	16.5091	262.5493	0.0006	0.0606	15.9033
85	141.5789	2342.9817	37633.0290	0.0071	16.5489	265.8096	0.0004	0.0604	16.0620
90	189.4645	3141.0752	50851.2531	0.0053	16.5787	268.3946	0.0003	0.0603	16.1891
95	253.5463	4209.1042	68568.4042	0.0039	16.6009	270.4375	0.0002	0.0602	16.2905
100	339.3021	5638.3681	92306.1343	0.0029	16.6175	272.0471	0.0002	0.0602	16.3711

TABLE A.13 $i = 7\%$

N	F/P	F/A	F/G	P/F	P/A	P/G	A/F	A/P	A/G
1	1.0700	1.0000	0.0000	0.9346	0.9346	0.0000	1.0000	1.0700	0.0000
2	1.1449	2.0700	1.0000	0.8734	1.8080	0.8734	0.4831	0.5531	0.4831
3	1.2250	3.2149	3.0700	0.8163	2.6243	2.5060	0.3111	0.3811	0.9549
4	1.3108	4.4399	6.2849	0.7629	3.3872	4.7947	0.2252	0.2952	1.4155
5	1.4026	5.7507	10.7248	0.7130	4.1002	7.6467	0.1739	0.2439	1.8650
6	1.5007	7.1533	16.4756	0.6663	4.7665	10.9784	0.1398	0.2098	2.3032
7	1.6058	8.6540	23.6289	0.6227	5.3893	14.7149	0.1156	0.1856	2.7304
8	1.7182	10.2598	32.2829	0.5820	5.9713	18.7889	0.0975	0.1675	3.1465
9	1.8385	11.9780	42.5427	0.5439	6.5152	23.1404	0.0835	0.1535	3.5517
10	1.9672	13.8164	54.5207	0.5083	7.0236	27.7156	0.0724	0.1424	3.9461
11	2.1049	15.7836	68.3371	0.4751	7.4987	32.4665	0.0634	0.1334	4.3296
12	2.2522	17.8885	84.1207	0.4440	7.9427	37.3506	0.0559	0.1259	4.7025
13	2.4098	20.1406	102.0092	0.4150	8.3577	42.3302	0.0497	0.1197	5.0648
14	2.5785	22.5505	122.1498	0.3878	8.7455	47.3718	0.0443	0.1143	5.4167
15	2.7590	25.1290	144.7003	0.3624	9.1079	52.4461	0.0398	0.1098	5.7583
16	2.9522	27.8881	169.8293	0.3387	9.4466	57.5271	0.0359	0.1059	6.0897
17	3.1588	30.8402	197.7174	0.3166	9.7632	62.5923	0.0324	0.1024	6.4110
18	3.3799	33.9990	228.5576	0.2959	10.0591	67.6219	0.0294	0.0994	6.7225
19	3.6165	37.3790	262.5566	0.2765	10.3356	72.5991	0.0268	0.0968	7.0242
20	3.8697	40.9955	299.9356	0.2584	10.5940	77.5091	0.0244	0.0944	7.3163
21	4.1406	44.8652	340.9311	0.2415	10.8355	82.3393	0.0223	0.0923	7.5990
22	4.4304	49.0057	385.7963	0.2257	11.0612	87.0793	0.0204	0.0904	7.8725
23	4.7405	53.4361	434.8020	0.2109	11.2722	91.7201	0.0187	0.0887	8.1369
24	5.0724	58.1767	488.2382	0.1971	11.4693	96.2545	0.0172	0.0872	8.3923
25	5.4274	63.2490	546.4148	0.1842	11.6536	100.6765	0.0158	0.0858	8.6391
26	5.8074	68.6765	609.6639	0.1722	11.8258	104.9814	0.0146	0.0846	8.8773
27	6.2139	74.4838	678.3403	0.1609	11.9867	109.1656	0.0134	0.0834	9.1072
28	6.6488	80.6977	752.8242	0.1504	12.1371	113.2264	0.0124	0.0824	9.3289
29	7.1143	87.3465	833.5218	0.1406	12.2777	117.1622	0.0114	0.0814	9.5427
30	7.6123	94.4608	920.8684	0.1314	12.4090	120.9718	0.0106	0.0806	9.7487
35	10.6766	138.2369	1474.8125	0.0937	12.9477	138.1353	0.0072	0.0772	10.6687
40	14.9745	199.6351	2280.5016	0.0668	13.3317	152.2928	0.0050	0.0750	11.4233
45	21.0025	285.7493	3439.2759	0.0476	13.6055	163.7559	0.0035	0.0735	12.0360
50	29.4570	406.5289	5093.2704	0.0339	13.8007	172.9051	0.0025	0.0725	12.5287
55	41.3150	575.9286	7441.8370	0.0242	13.9399	180.1243	0.0017	0.0717	12.9215
60	57.9464	813.5204	10764.5769	0.0173	14.0392	185.7677	0.0012	0.0712	13.2321
65	81.2729	1146.7552	15453.6452	0.0123	14.1099	190.1452	0.0009	0.0709	13.4760
70	113.9894	1614.1342	22059.0596	0.0088	14.1604	193.5185	0.0006	0.0706	13.6662
75	159.8760	2269.6574	31352.2488	0.0063	14.1964	196.1035	0.0004	0.0704	13.8136
80	224.2344	3189.0627	44415.1811	0.0045	14.2220	198.0748	0.0003	0.0703	13.9273
85	314.5003	4478.5761	62765.3731	0.0032	14.2403	199.5717	0.0002	0.0702	14.0146
90	441.1030	6287.1854	88531.2204	0.0023	14.2533	200.7042	0.0002	0.0702	14.0812
95	618.6697	8823.8535	124697.9077	0.0016	14.2626	201.5581	0.0001	0.0701	14.1319
100	867.7163	12381.6618	175452.3113	0.0012	14.2693	202.2001	0.0001	0.0701	14.1703

TABLE A.14 $i = 8\%$

N	F/P	F/A	F/G	P/F	P/A	P/G	A/F	A/P	A/G
1	1.0800	1.0000	0.0000	0.9259	0.9259	0.0000	1.0000	1.0800	0.0000
2	1.1664	2.0800	1.0000	0.8573	1.7833	0.8573	0.4808	0.5608	0.4808
3	1.2597	3.2464	3.0800	0.7938	2.5771	2.4450	0.3080	0.3880	0.9487
4	1.3605	4.5061	6.3264	0.7350	3.3121	4.6501	0.2219	0.3019	1.4040
5	1.4693	5.8666	10.8325	0.6806	3.9927	7.3724	0.1705	0.2505	1.8465
6	1.5869	7.3359	16.6991	0.6302	4.6229	10.5233	0.1363	0.2163	2.2763
7	1.7138	8.9228	24.0350	0.5835	5.2064	14.0242	0.1121	0.1921	2.6937
8	1.8509	10.6366	32.9578	0.5403	5.7466	17.8061	0.0940	0.1740	3.0985
9	1.9990	12.4876	43.5945	0.5002	6.2469	21.8081	0.0801	0.1601	3.4910
10	2.1589	14.4866	56.0820	0.4632	6.7101	25.9768	0.0690	0.1490	3.8713
11	2.3316	16.6455	70.5686	0.4289	7.1390	30.2657	0.0601	0.1401	4.2395
12	2.5182	18.9771	87.2141	0.3971	7.5361	34.6339	0.0527	0.1327	4.5957
13	2.7196	21.4953	106.1912	0.3677	7.9038	39.0463	0.0465	0.1265	4.9402
14	2.9372	24.2149	127.6865	0.3405	8.2442	43.4723	0.0413	0.1213	5.2731
15	3.1722	27.1521	151.9014	0.3152	8.5595	47.8857	0.0368	0.1168	5.5945
16	3.4259	30.3243	179.0535	0.2919	8.8514	52.2640	0.0330	0.1130	5.9046
17	3.7000	33.7502	209.3778	0.2703	9.1216	56.5883	0.0296	0.1096	6.2037
18	3.9960	37.4502	243.1280	0.2502	9.3719	60.8426	0.0267	0.1067	6.4920
19	4.3157	41.4463	280.5783	0.2317	9.6036	65.0134	0.0241	0.1041	6.7697
20	4.6610	45.7620	322.0246	0.2145	9.8181	69.0898	0.0219	0.1019	7.0369
21	5.0338	50.4229	367.7865	0.1987	10.0168	73.0629	0.0198	0.0998	7.2940
22	5.4365	55.4568	418.2094	0.1839	10.2007	76.9257	0.0180	0.0980	7.5412
23	5.8715	60.8933	473.6662	0.1703	10.3711	80.6726	0.0164	0.0964	7.7786
24	6.3412	66.7648	534.5595	0.1577	10.5288	84.2997	0.0150	0.0950	8.0066
25	6.8485	73.1059	601.3242	0.1460	10.6748	87.8041	0.0137	0.0937	8.2254
26	7.3964	79.9544	674.4302	0.1352	10.8100	91.1842	0.0125	0.0925	8.4352
27	7.9881	87.3508	754.3846	0.1252	10.9352	94.4390	0.0114	0.0914	8.6363
28	8.6271	95.3388	841.7354	0.1159	11.0511	97.5687	0.0105	0.0905	8.8289
29	9.3173	103.9659	937.0742	0.1073	11.1584	100.5738	0.0096	0.0896	9.0133
30	10.0627	113.2832	1041.0401	0.0994	11.2578	103.4558	0.0088	0.0888	9.1897
35	14.7853	172.3168	1716.4600	0.0676	11.6546	116.0920	0.0058	0.0858	9.9611
40	21.7245	259.0565	2738.2065	0.0460	11.9246	126.0422	0.0039	0.0839	10.5699
45	31.9204	386.5056	4268.8202	0.0313	12.1084	133.7331	0.0026	0.0826	11.0447
50	46.9016	573.7702	6547.1270	0.0213	12.2335	139.5928	0.0017	0.0817	11.4107
55	68.9139	848.9232	9924.0400	0.0145	12.3186	144.0065	0.0012	0.0812	11.6902
60	101.2571	1253.2133	14915.1662	0.0099	12.3766	147.3000	0.0008	0.0808	11.9015
65	148.7798	1847.2481	22278.1010	0.0067	12.4160	149.7387	0.0005	0.0805	12.0602
70	218.6064	2720.0801	33126.0009	0.0046	12.4428	151.5326	0.0004	0.0804	12.1783
75	321.2045	4002.5566	49094.4578	0.0031	12.4611	152.8448	0.0002	0.0802	12.2658
80	471.9548	5886.9354	72586.6929	0.0021	12.4735	153.8001	0.0002	0.0802	12.3301
85	693.4565	8655.7061	107133.8264	0.0014	12.4820	154.4925	0.0001	0.0801	12.3772
90	1018.9151	12723.9386	157924.2327	0.0010	12.4877	154.9925	0.0001	0.0801	12.4116
95	1497.1205	18701.5069	232581.3357	0.0007	12.4917	155.3524	0.0001	0.0801	12.4365
100	2199.7613	27484.5157	342306.4463	0.0005	12.4943	155.6107	0.0000	0.0800	12.4545

TABLE A.15 $i = 9\%$

N	F/P	F/A	F/G	P/F	P/A	P/G	A/F	A/P	A/G
1	1.0900	1.0000	0.0000	0.9174	0.9174	0.0000	1.0000	1.0900	0.0000
2	1.1881	2.0900	1.0000	0.8417	1.7591	0.8417	0.4785	0.5685	0.4785
3	1.2950	3.2781	3.0900	0.7722	2.5313	2.3860	0.3051	0.3951	0.9426
4	1.4116	4.5731	6.3681	0.7084	3.2397	4.5113	0.2187	0.3087	1.3925
5	1.5386	5.9847	10.9412	0.6499	3.8897	7.1110	0.1671	0.2571	1.8282
6	1.6771	7.5233	16.9259	0.5963	4.4859	10.0924	0.1329	0.2229	2.2498
7	1.8280	9.2004	24.4493	0.5470	5.0330	13.3746	0.1087	0.1987	2.6574
8	1.9926	11.0285	33.6497	0.5019	5.5348	16.8877	0.0907	0.1807	3.0512
9	2.1719	13.0210	44.6782	0.4604	5.9952	20.5711	0.0768	0.1668	3.4312
10	2.3674	15.1929	57.6992	0.4224	6.4177	24.3728	0.0658	0.1558	3.7978
11	2.5804	17.5603	72.8921	0.3875	6.8052	28.2481	0.0569	0.1469	4.1510
12	2.8127	20.1407	90.4524	0.3555	7.1607	32.1590	0.0497	0.1397	4.4910
13	3.0658	22.9534	110.5932	0.3262	7.4869	36.0731	0.0436	0.1336	4.8182
14	3.3417	26.0192	133.5465	0.2992	7.7862	39.9633	0.0384	0.1284	5.1326
15	3.6425	29.3609	159.5657	0.2745	8.0607	43.8069	0.0341	0.1241	5.4346
16	3.9703	33.0034	188.9267	0.2519	8.3126	47.5849	0.0303	0.1203	5.7245
17	4.3276	36.9737	221.9301	0.2311	8.5436	51.2821	0.0270	0.1170	6.0024
18	4.7171	41.3013	258.9038	0.2120	8.7556	54.8860	0.0242	0.1142	6.2687
19	5.1417	46.0185	300.2051	0.1945	8.9501	58.3868	0.0217	0.1117	6.5236
20	5.6044	51.1601	346.2236	0.1784	9.1285	61.7770	0.0195	0.1095	6.7674
21	6.1088	56.7645	397.3837	0.1637	9.2922	65.0509	0.0176	0.1076	7.0006
22	6.6586	62.8733	454.1482	0.1502	9.4424	68.2048	0.0159	0.1059	7.2232
23	7.2579	69.5319	517.0215	0.1378	9.5802	71.2359	0.0144	0.1044	7.4357
24	7.9111	76.7898	586.5535	0.1264	9.7066	74.1433	0.0130	0.1030	7.6384
25	8.6231	84.7009	663.3433	0.1160	9.8226	76.9265	0.0118	0.1018	7.8316
26	9.3992	93.3240	748.0442	0.1064	9.9290	79.5863	0.0107	0.1007	8.0156
27	10.2451	102.7231	841.3682	0.0976	10.0266	82.1241	0.0097	0.0997	8.1906
28	11.1671	112.9682	944.0913	0.0895	10.1161	84.5419	0.0089	0.0989	8.3571
29	12.1722	124.1354	1057.0595	0.0822	10.1983	86.8422	0.0081	0.0981	8.5154
30	13.2677	136.3075	1181.1949	0.0754	10.2737	89.0280	0.0073	0.0973	8.6657
35	20.4140	215.7108	2007.8973	0.0490	10.5668	98.3590	0.0046	0.0946	9.3083
40	31.4094	337.8824	3309.8049	0.0318	10.7574	105.3762	0.0030	0.0930	9.7957
45	48.3273	525.8587	5342.8748	0.0207	10.8812	110.5561	0.0019	0.0919	10.1603
50	74.3575	815.0836	8500.9284	0.0134	10.9617	114.3251	0.0012	0.0912	10.4295
55	114.4083	1260.0918	13389.9088	0.0087	11.0140	117.0362	0.0008	0.0908	10.6261
60	176.0313	1944.7921	20942.1348	0.0057	11.0480	118.9683	0.0005	0.0905	10.7683
65	270.8460	2998.2885	32592.0942	0.0037	11.0701	120.3344	0.0003	0.0903	10.8702
70	416.7301	4619.2232	50546.9242	0.0024	11.0844	121.2942	0.0002	0.0902	10.9427
75	641.1909	7113.2321	78202.5794	0.0016	11.0938	121.9646	0.0001	0.0901	10.9940
80	986.5517	10950.5741	120784.1566	0.0010	11.0998	122.4306	0.0001	0.0901	11.0299
85	1517.9320	16854.8003	186331.1147	0.0007	11.1038	122.7533	0.0001	0.0901	11.0551
90	2335.5266	25939.1842	287213.1583	0.0004	11.1064	122.9758	0.0000	0.0900	11.0726
95	3593.4971	39916.6350	442462.6107	0.0003	11.1080	123.1287	0.0000	0.0900	11.0847
100	5529.0408	61422.6755	681363.0607	0.0002	11.1091	123.2335	0.0000	0.0900	11.0930

TABLE A.16 $i = 10\%$

N	F/P	F/A	F/G	P/F	P/A	P/G	A/F	A/P	A/G
1	1.1000	1.0000	0.0000	0.9091	0.9091	0.0000	1.0000	1.1000	0.0000
2	1.2100	2.1000	1.0000	0.8264	1.7355	0.8264	0.4762	0.5762	0.4762
3	1.3310	3.3100	3.1000	0.7513	2.4869	2.3291	0.3021	0.4021	0.9366
4	1.4641	4.6410	6.4100	0.6830	3.1699	4.3781	0.2155	0.3155	1.3812
5	1.6105	6.1051	11.0510	0.6209	3.7908	6.8618	0.1638	0.2638	1.8101
6	1.7716	7.7156	17.1561	0.5645	4.3553	9.6842	0.1296	0.2296	2.2236
7	1.9487	9.4872	24.8717	0.5132	4.8684	12.7631	0.1054	0.2054	2.6216
8	2.1436	11.4359	34.3589	0.4665	5.3349	16.0287	0.0874	0.1874	3.0045
9	2.3579	13.5795	45.7948	0.4241	5.7590	19.4215	0.0736	0.1736	3.3724
10	2.5937	15.9374	59.3742	0.3855	6.1446	22.8913	0.0627	0.1627	3.7255
11	2.8531	18.5312	75.3117	0.3505	6.4951	26.3963	0.0540	0.1540	4.0641
12	3.1384	21.3843	93.8428	0.3186	6.8137	29.9012	0.0468	0.1468	4.3884
13	3.4523	24.5227	115.2271	0.2897	7.1034	33.3772	0.0408	0.1408	4.6988
14	3.7975	27.9750	139.7498	0.2633	7.3667	36.8005	0.0357	0.1357	4.9955
15	4.1772	31.7725	167.7248	0.2394	7.6061	40.1520	0.0315	0.1315	5.2789
16	4.5950	35.9497	199.4973	0.2176	7.8237	43.4164	0.0278	0.1278	5.5493
17	5.0545	40.5447	235.4470	0.1978	8.0216	46.5819	0.0247	0.1247	5.8071
18	5.5599	45.5992	275.9917	0.1799	8.2014	49.6395	0.0219	0.1219	6.0526
19	6.1159	51.1591	321.5909	0.1635	8.3649	52.5827	0.0195	0.1195	6.2861
20	6.7275	57.2750	372.7500	0.1486	8.5136	55.4069	0.0175	0.1175	6.5081
21	7.4002	64.0025	430.0250	0.1351	8.6487	58.1095	0.0156	0.1156	6.7189
22	8.1403	71.4027	494.0275	0.1228	8.7715	60.6893	0.0140	0.1140	6.9189
23	8.9543	79.5430	565.4302	0.1117	8.8832	63.1462	0.0126	0.1126	7.1085
24	9.8497	88.4973	644.9733	0.1015	8.9847	65.4813	0.0113	0.1113	7.2881
25	10.8347	98.3471	733.4706	0.0923	9.0770	67.6964	0.0102	0.1102	7.4580
26	11.9182	109.1818	831.8177	0.0839	9.1609	69.7940	0.0092	0.1092	7.6186
27	13.1100	121.0999	940.9994	0.0763	9.2372	71.7773	0.0083	0.1083	7.7704
28	14.4210	134.2099	1062.0994	0.0693	9.3066	73.6495	0.0075	0.1075	7.9137
29	15.8631	148.6309	1196.3093	0.0630	9.3696	75.4146	0.0067	0.1067	8.0489
30	17.4494	164.4940	1344.9402	0.0573	9.4269	77.0766	0.0061	0.1061	8.1762
31	19.1943	181.9434	1509.4342	0.0521	9.4790	78.6395	0.0055	0.1055	8.2962
32	21.1138	201.1378	1691.3777	0.0474	9.5264	80.1078	0.0050	0.1050	8.4091
33	23.2252	222.2515	1892.5154	0.0431	9.5694	81.4856	0.0045	0.1045	8.5152
34	25.5477	245.4767	2114.7670	0.0391	9.6086	82.7773	0.0041	0.1041	8.6149
35	28.1024	271.0244	2360.2437	0.0356	9.6442	83.9872	0.0037	0.1037	8.7086
36	30.9127	299.1268	2631.2681	0.0323	9.6765	85.1194	0.0033	0.1033	8.7965
37	34.0039	330.0395	2930.3949	0.0294	9.7059	86.1781	0.0030	0.1030	8.8789
38	37.4043	364.0434	3260.4343	0.0267	9.7327	87.1673	0.0027	0.1027	8.9562
39	41.1448	401.4478	3624.4778	0.0243	9.7570	88.0908	0.0025	0.1025	9.0285
40	45.2593	442.5926	4025.9256	0.0221	9.7791	88.9525	0.0023	0.1023	9.0962

TABLE A.17 $i = 11\%$

N	F/P	F/A	F/G	P/F	P/A	P/G	A/F	A/P	A/G
1	1.1100	1.0000	0.0000	0.9009	0.9009	0.0000	1.0000	1.1100	0.0000
2	1.2321	2.1100	1.0000	0.8116	1.7125	0.8116	0.4739	0.5839	0.4739
3	1.3676	3.3421	3.1100	0.7312	2.4437	2.2740	0.2992	0.4092	0.9306
4	1.5181	4.7097	6.4521	0.6587	3.1024	4.2502	0.2123	0.3223	1.3700
5	1.6851	6.2278	11.1618	0.5935	3.6959	6.6240	0.1606	0.2706	1.7923
6	1.8704	7.9129	17.3896	0.5346	4.2305	9.2972	0.1264	0.2364	2.1976
7	2.0762	9.7833	25.3025	0.4817	4.7122	12.1872	0.1022	0.2122	2.5863
8	2.3045	11.8594	35.0858	0.4339	5.1461	15.2246	0.0843	0.1943	2.9585
9	2.5580	14.1640	46.9452	0.3909	5.5370	18.3520	0.0706	0.1806	3.3144
10	2.8394	16.7220	61.1092	0.3522	5.8892	21.5217	0.0598	0.1698	3.6544
11	3.1518	19.5614	77.8312	0.3173	6.2065	24.6945	0.0511	0.1611	3.9788
12	3.4985	22.7132	97.3926	0.2858	6.4924	27.8388	0.0440	0.1540	4.2879
13	3.8833	26.2116	120.1058	0.2575	6.7499	30.9290	0.0382	0.1482	4.5822
14	4.3104	30.0949	146.3174	0.2320	6.9819	33.9449	0.0332	0.1432	4.8619
15	4.7846	34.4054	176.4124	0.2090	7.1909	36.8709	0.0291	0.1391	5.1275
16	5.3109	39.1899	210.8177	0.1883	7.3792	39.6953	0.0255	0.1355	5.3794
17	5.8951	44.5008	250.0077	0.1696	7.5488	42.4095	0.0225	0.1325	5.6180
18	6.5436	50.3959	294.5085	0.1528	7.7016	45.0074	0.0198	0.1298	5.8439
19	7.2633	56.9395	344.9044	0.1377	7.8393	47.4856	0.0176	0.1276	6.0574
20	8.0623	64.2028	401.8439	0.1240	7.9633	49.8423	0.0156	0.1256	6.2590
21	8.9492	72.2651	466.0468	0.1117	8.0751	52.0771	0.0138	0.1238	6.4491
22	9.9336	81.2143	538.3119	0.1007	8.1757	54.1912	0.0123	0.1223	6.6283
23	11.0263	91.1479	619.5262	0.0907	8.2664	56.1864	0.0110	0.1210	6.7969
24	12.2392	102.1742	710.6741	0.0817	8.3481	58.0656	0.0098	0.1198	6.9555
25	13.5855	114.4133	812.8482	0.0736	8.4217	59.8322	0.0087	0.1187	7.1045
26	15.0799	127.9988	927.2616	0.0663	8.4881	61.4900	0.0078	0.1178	7.2443
27	16.7386	143.0786	1055.2603	0.0597	8.5478	63.0433	0.0070	0.1170	7.3754
28	18.5799	159.8173	1198.3390	0.0538	8.6016	64.4965	0.0063	0.1163	7.4982
29	20.6237	178.3972	1358.1562	0.0485	8.6501	65.8542	0.0056	0.1156	7.6131
30	22.8923	199.0209	1536.5534	0.0437	8.6938	67.1210	0.0050	0.1150	7.7206
31	25.4104	221.9132	1735.5743	0.0394	8.7331	68.3016	0.0045	0.1145	7.8210
32	28.2056	247.3236	1957.4875	0.0355	8.7686	69.4007	0.0040	0.1140	7.9147
33	31.3082	275.5292	2204.8111	0.0319	8.8005	70.4228	0.0036	0.1136	8.0021
34	34.7521	306.8374	2480.3403	0.0288	8.8293	71.3724	0.0033	0.1133	8.0836
35	38.5749	341.5896	2787.1778	0.0259	8.8552	72.2538	0.0029	0.1129	8.1594
36	42.8181	380.1644	3128.7673	0.0234	8.8786	73.0712	0.0026	0.1126	8.2300
37	47.5281	422.9825	3508.9317	0.0210	8.8996	73.8286	0.0024	0.1124	8.2957
38	52.7562	470.5106	3931.9142	0.0190	8.9186	74.5300	0.0021	0.1121	8.3567
39	58.5593	523.2667	4402.4248	0.0171	8.9357	75.1789	0.0019	0.1119	8.4133
40	65.0009	581.8261	4925.6915	0.0154	8.9511	75.7789	0.0017	0.1117	8.4659

TABLE A.18 $i = 12\%$

N	F/P	F/A	F/G	P/F	P/A	P/G	A/F	A/P	A/G
1	1.1200	1.0000	0.0000	0.8929	0.8929	0.0000	1.0000	1.1200	0.0000
2	1.2544	2.1200	1.0000	0.7972	1.6901	0.7972	0.4717	0.5917	0.4717
3	1.4049	3.3744	3.1200	0.7118	2.4018	2.2208	0.2963	0.4163	0.9246
4	1.5735	4.7793	6.4944	0.6355	3.0373	4.1273	0.2092	0.3292	1.3589
5	1.7623	6.3528	11.2737	0.5674	3.6048	6.3970	0.1574	0.2774	1.7746
6	1.9738	8.1152	17.6266	0.5066	4.1114	8.9302	0.1232	0.2432	2.1720
7	2.2107	10.0890	25.7418	0.4523	4.5638	11.6443	0.0991	0.2191	2.5515
8	2.4760	12.2997	35.8308	0.4039	4.9676	14.4714	0.0813	0.2013	2.9131
9	2.7731	14.7757	48.1305	0.3606	5.3282	17.3563	0.0677	0.1877	3.2574
10	3.1058	17.5487	62.9061	0.3220	5.6502	20.2541	0.0570	0.1770	3.5847
11	3.4785	20.6546	80.4549	0.2875	5.9377	23.1288	0.0484	0.1684	3.8953
12	3.8960	24.1331	101.1094	0.2567	6.1944	25.9523	0.0414	0.1614	4.1897
13	4.3635	28.0291	125.2426	0.2292	6.4235	28.7024	0.0357	0.1557	4.4683
14	4.8871	32.3926	153.2717	0.2046	6.6282	31.3624	0.0309	0.1509	4.7317
15	5.4736	37.2797	185.6643	0.1827	6.8109	33.9202	0.0268	0.1468	4.9803
16	6.1304	42.7533	222.9440	0.1631	6.9740	36.3670	0.0234	0.1434	5.2147
17	6.8660	48.8837	265.6973	0.1456	7.1196	38.6973	0.0205	0.1405	5.4353
18	7.6900	55.7497	314.5810	0.1300	7.2497	40.9080	0.0179	0.1379	5.6427
19	8.6128	63.4397	370.3307	0.1161	7.3658	42.9979	0.0158	0.1358	5.8375
20	9.6463	72.0524	433.7704	0.1037	7.4694	44.9676	0.0139	0.1339	6.0202
21	10.8038	81.6987	505.8228	0.0926	7.5620	46.8188	0.0122	0.1322	6.1913
22	12.1003	92.5026	587.5215	0.0826	7.6446	48.5543	0.0108	0.1308	6.3514
23	13.5523	104.6029	680.0241	0.0738	7.7184	50.1776	0.0096	0.1296	6.5010
24	15.1786	118.1552	784.6270	0.0659	7.7843	51.6929	0.0085	0.1285	6.6406
25	17.0001	133.3339	902.7823	0.0588	7.8431	53.1046	0.0075	0.1275	6.7708
26	19.0401	150.3339	1036.1161	0.0525	7.8957	54.4177	0.0067	0.1267	6.8921
27	21.3249	169.3740	1186.4501	0.0469	7.9426	55.6369	0.0059	0.1259	7.0049
28	23.8839	190.6989	1355.8241	0.0419	7.9844	56.7674	0.0052	0.1252	7.1098
29	26.7499	214.5828	1546.5229	0.0374	8.0218	57.8141	0.0047	0.1247	7.2071
30	29.9599	241.3327	1761.1057	0.0334	8.0552	58.7821	0.0041	0.1241	7.2974
31	33.5551	271.2926	2002.4384	0.0298	8.0850	59.6761	0.0037	0.1237	7.3811
32	37.5817	304.8477	2273.7310	0.0266	8.1116	60.5010	0.0033	0.1233	7.4586
33	42.0915	342.4294	2578.5787	0.0238	8.1354	61.2612	0.0029	0.1229	7.5302
34	47.1425	384.5210	2921.0082	0.0212	8.1566	61.9612	0.0026	0.1226	7.5965
35	52.7996	431.6635	3305.5291	0.0189	8.1755	62.6052	0.0023	0.1223	7.6577
36	59.1356	484.4631	3737.1926	0.0169	8.1924	63.1970	0.0021	0.1221	7.7141
37	66.2318	543.5987	4221.6558	0.0151	8.2075	63.7406	0.0018	0.1218	7.7661
38	74.1797	609.8305	4765.2544	0.0135	8.2210	64.2394	0.0016	0.1216	7.8141
39	83.0812	684.0102	5375.0850	0.0120	8.2330	64.6967	0.0015	0.1215	7.8582
40	93.0510	767.0914	6059.0952	0.0107	8.2438	65.1159	0.0013	0.1213	7.8988

TABLE A.19 $i = 13\%$

N	F/P	F/A	F/G	P/F	P/A	P/G	A/F	A/P	A/G
1	1.1300	1.0000	0.0000	0.8850	0.8850	0.0000	1.0000	1.1300	0.0000
2	1.2769	2.1300	1.0000	0.7831	1.6681	0.7831	0.4695	0.5995	0.4695
3	1.4429	3.4069	3.1300	0.6931	2.3612	2.1692	0.2935	0.4235	0.9187
4	1.6305	4.8498	6.5369	0.6133	2.9745	4.0092	0.2062	0.3362	1.3479
5	1.8424	6.4803	11.3867	0.5428	3.5172	6.1802	0.1543	0.2843	1.7571
6	2.0820	8.3227	17.8670	0.4803	3.9975	8.5818	0.1202	0.2502	2.1468
7	2.3526	10.4047	26.1897	0.4251	4.4226	11.1322	0.0961	0.2261	2.5171
8	2.6584	12.7573	36.5943	0.3762	4.7988	13.7653	0.0784	0.2084	2.8685
9	3.0040	15.4157	49.3516	0.3329	5.1317	16.4284	0.0649	0.1949	3.2014
10	3.3946	18.4197	64.7673	0.2946	5.4262	19.0797	0.0543	0.1843	3.5162
11	3.8359	21.8143	83.1871	0.2607	5.6869	21.6867	0.0458	0.1758	3.8134
12	4.3345	25.6502	105.0014	0.2307	5.9176	24.2244	0.0390	0.1690	4.0936
13	4.8980	29.9847	130.6515	0.2042	6.1218	26.6744	0.0334	0.1634	4.3573
14	5.5348	34.8827	160.6362	0.1807	6.3025	29.0232	0.0287	0.1587	4.6050
15	6.2543	40.4175	195.5190	0.1599	6.4624	31.2617	0.0247	0.1547	4.8375
16	7.0673	46.6717	235.9364	0.1415	6.6039	33.3841	0.0214	0.1514	5.0552
17	7.9861	53.7391	282.6082	0.1252	6.7291	35.3876	0.0186	0.1486	5.2589
18	9.0243	61.7251	336.3472	0.1108	6.8399	37.2714	0.0162	0.1462	5.4491
19	10.1974	70.7494	398.0724	0.0981	6.9380	39.0366	0.0141	0.1441	5.6265
20	11.5231	80.9468	468.8218	0.0868	7.0248	40.6854	0.0124	0.1424	5.7917
21	13.0211	92.4699	549.7686	0.0768	7.1016	42.2214	0.0108	0.1408	5.9454
22	14.7138	105.4910	642.2385	0.0680	7.1695	43.6486	0.0095	0.1395	6.0881
23	16.6266	120.2048	747.7295	0.0601	7.2297	44.9718	0.0083	0.1383	6.2205
24	18.7881	136.8315	867.9343	0.0532	7.2829	46.1960	0.0073	0.1373	6.3431
25	21.2305	155.6196	1004.7658	0.0471	7.3300	47.3264	0.0064	0.1364	6.4566
26	23.9905	176.8501	1160.3854	0.0417	7.3717	48.3685	0.0057	0.1357	6.5614
27	27.1093	200.8406	1337.2355	0.0369	7.4086	49.3276	0.0050	0.1350	6.6582
28	30.6335	227.9499	1538.0761	0.0326	7.4412	50.2090	0.0044	0.1344	6.7474
29	34.6158	258.5834	1766.0260	0.0289	7.4701	51.0179	0.0039	0.1339	6.8296
30	39.1159	293.1992	2024.6093	0.0256	7.4957	51.7592	0.0034	0.1334	6.9052
31	44.2010	332.3151	2317.8086	0.0226	7.5183	52.4380	0.0030	0.1330	6.9747
32	49.9471	376.5161	2650.1237	0.0200	7.5383	53.0586	0.0027	0.1327	7.0385
33	56.4402	426.4632	3026.6398	0.0177	7.5560	53.6256	0.0023	0.1323	7.0971
34	63.7774	482.9034	3453.1029	0.0157	7.5717	54.1430	0.0021	0.1321	7.1507
35	72.0685	546.6808	3936.0063	0.0139	7.5856	54.6148	0.0018	0.1318	7.1998
36	81.4374	618.7493	4482.6871	0.0123	7.5979	55.0446	0.0016	0.1316	7.2448
37	92.0243	700.1867	5101.4364	0.0109	7.6087	55.4358	0.0014	0.1314	7.2858
38	103.9874	792.2110	5801.6232	0.0096	7.6183	55.7916	0.0013	0.1313	7.3233
39	117.5058	896.1984	6593.8342	0.0085	7.6268	56.1150	0.0011	0.1311	7.3576
40	132.7816	1013.7042	7490.0326	0.0075	7.6344	56.4087	0.0010	0.1310	7.3888

TABLE A.20 $i = 14\%$

N	F/P	F/A	F/G	P/F	P/A	P/G	A/F	A/P	A/G
1	1.1400	1.0000	0.0000	0.8772	0.8772	0.0000	1.0000	1.1400	0.0000
2	1.2996	2.1400	1.0000	0.7695	1.6467	0.7695	0.4673	0.6073	0.4673
3	1.4815	3.4396	3.1400	0.6750	2.3216	2.1194	0.2907	0.4307	0.9129
4	1.6890	4.9211	6.5796	0.5921	2.9137	3.8957	0.2032	0.3432	1.3370
5	1.9254	6.6101	11.5007	0.5194	3.4331	5.9731	0.1513	0.2913	1.7399
6	2.1950	8.5355	18.1108	0.4556	3.8887	8.2511	0.1172	0.2572	2.1218
7	2.5023	10.7305	26.6464	0.3996	4.2883	10.6489	0.0932	0.2332	2.4832
8	2.8526	13.2328	37.3769	0.3506	4.6389	13.1028	0.0756	0.2156	2.8246
9	3.2519	16.0853	50.6096	0.3075	4.9464	15.5629	0.0622	0.2022	3.1463
10	3.7072	19.3373	66.6950	0.2697	5.2161	17.9906	0.0517	0.1917	3.4490
11	4.2262	23.0445	86.0323	0.2366	5.4527	20.3567	0.0434	0.1834	3.7333
12	4.8179	27.2707	109.0768	0.2076	5.6603	22.6399	0.0367	0.1767	3.9998
13	5.4924	32.0887	136.3475	0.1821	5.8424	24.8247	0.0312	0.1712	4.2491
14	6.2613	37.5811	168.4362	0.1597	6.0021	26.9009	0.0266	0.1666	4.4819
15	7.1379	43.8424	206.0172	0.1401	6.1422	28.8623	0.0228	0.1628	4.6990
16	8.1372	50.9804	249.8597	0.1229	6.2651	30.7057	0.0196	0.1596	4.9011
17	9.2765	59.1176	300.8400	0.1078	6.3729	32.4305	0.0169	0.1569	5.0888
18	10.5752	68.3941	359.9576	0.0946	6.4674	34.0380	0.0146	0.1546	5.2630
19	12.0557	78.9692	428.3517	0.0829	6.5504	35.5311	0.0127	0.1527	5.4243
20	13.7435	91.0249	507.3209	0.0728	6.6231	36.9135	0.0110	0.1510	5.5734
21	15.6676	104.7684	598.3458	0.0638	6.6870	38.1901	0.0095	0.1495	5.7111
22	17.8610	120.4360	703.1143	0.0560	6.7429	39.3658	0.0083	0.1483	5.8381
23	20.3616	138.2970	823.5503	0.0491	6.7921	40.4463	0.0072	0.1472	5.9549
24	23.2122	158.6586	961.8473	0.0431	6.8351	41.4371	0.0063	0.1463	6.0624
25	26.4619	181.8708	1120.5059	0.0378	6.8729	42.3441	0.0055	0.1455	6.1610
26	30.1666	208.3327	1302.3767	0.0331	6.9061	43.1728	0.0048	0.1448	6.2514
27	34.3899	238.4993	1510.7095	0.0291	6.9352	43.9289	0.0042	0.1442	6.3342
28	39.2045	272.8892	1749.2088	0.0255	6.9607	44.6176	0.0037	0.1437	6.4100
29	44.6931	312.0937	2022.0980	0.0224	6.9830	45.2441	0.0032	0.1432	6.4791
30	50.9502	356.7868	2334.1918	0.0196	7.0027	45.8132	0.0028	0.1428	6.5423
31	58.0832	407.7370	2690.9786	0.0172	7.0199	46.3297	0.0025	0.1425	6.5998
32	66.2148	465.8202	3098.7156	0.0151	7.0350	46.7979	0.0021	0.1421	6.6522
33	75.4849	532.0350	3564.5358	0.0132	7.0482	47.2218	0.0019	0.1419	6.6998
34	86.0528	607.5199	4096.5708	0.0116	7.0599	47.6053	0.0016	0.1416	6.7431
35	98.1002	693.5727	4704.0907	0.0102	7.0700	47.9519	0.0014	0.1414	6.7824
36	111.8342	791.6729	5397.6634	0.0089	7.0790	48.2649	0.0013	0.1413	6.8180
37	127.4910	903.5071	6189.3363	0.0078	7.0868	48.5472	0.0011	0.1411	6.8503
38	145.3397	1030.9981	7092.8434	0.0069	7.0937	48.8018	0.0010	0.1410	6.8796
39	165.6873	1176.3378	8123.8415	0.0060	7.0997	49.0312	0.0009	0.1409	6.9060
40	188.8835	1342.0251	9300.1793	0.0053	7.1050	49.2376	0.0007	0.1407	6.9300

TABLE A.21 $i = 15\%$

N	F/P	F/A	F/G	P/F	P/A	P/G	A/F	A/P	A/G
1	1.1500	1.0000	0.0000	0.8696	0.8696	0.0000	1.0000	1.1500	0.0000
2	1.3225	2.1500	1.0000	0.7561	1.6257	0.7561	0.4651	0.6151	0.4651
3	1.5209	3.4725	3.1500	0.6575	2.2832	2.0712	0.2880	0.4380	0.9071
4	1.7490	4.9934	6.6225	0.5718	2.8550	3.7864	0.2003	0.3503	1.3263
5	2.0114	6.7424	11.6159	0.4972	3.3522	5.7751	0.1483	0.2983	1.7228
6	2.3131	8.7537	18.3583	0.4323	3.7845	7.9368	0.1142	0.2642	2.0972
7	2.6600	11.0668	27.1120	0.3759	4.1604	10.1924	0.0904	0.2404	2.4498
8	3.0590	13.7268	38.1788	0.3269	4.4873	12.4807	0.0729	0.2229	2.7813
9	3.5179	16.7858	51.9056	0.2843	4.7716	14.7548	0.0596	0.2096	3.0922
10	4.0456	20.3037	68.6915	0.2472	5.0188	16.9795	0.0493	0.1993	3.3832
11	4.6524	24.3493	88.9952	0.2149	5.2337	19.1289	0.0411	0.1911	3.6549
12	5.3503	29.0017	113.3444	0.1869	5.4206	21.1849	0.0345	0.1845	3.9082
13	6.1528	34.3519	142.3461	0.1625	5.5831	23.1352	0.0291	0.1791	4.1438
14	7.0757	40.5047	176.6980	0.1413	5.7245	24.9725	0.0247	0.1747	4.3624
15	8.1371	47.5804	217.2027	0.1229	5.8474	26.6930	0.0210	0.1710	4.5650
16	9.3576	55.7175	264.7831	0.1069	5.9542	28.2960	0.0179	0.1679	4.7522
17	10.7613	65.0751	320.5006	0.0929	6.0472	29.7828	0.0154	0.1654	4.9251
18	12.3755	75.8364	385.5757	0.0808	6.1280	31.1565	0.0132	0.1632	5.0843
19	14.2318	88.2118	461.4121	0.0703	6.1982	32.4213	0.0113	0.1613	5.2307
20	16.3665	102.4436	549.6239	0.0611	6.2593	33.5822	0.0098	0.1598	5.3651
21	18.8215	118.8101	652.0675	0.0531	6.3125	34.6448	0.0084	0.1584	5.4883
22	21.6447	137.6316	770.8776	0.0462	6.3587	35.6150	0.0073	0.1573	5.6010
23	24.8915	159.2764	908.5092	0.0402	6.3988	36.4988	0.0063	0.1563	5.7040
24	28.6252	184.1678	1067.7856	0.0349	6.4338	37.3023	0.0054	0.1554	5.7979
25	32.9190	212.7930	1251.9534	0.0304	6.4641	38.0314	0.0047	0.1547	5.8834
26	37.8568	245.7120	1464.7465	0.0264	6.4906	38.6918	0.0041	0.1541	5.9612
27	43.5353	283.5688	1710.4584	0.0230	6.5135	39.2890	0.0035	0.1535	6.0319
28	50.0656	327.1041	1994.0272	0.0200	6.5335	39.8283	0.0031	0.1531	6.0960
29	57.5755	377.1697	2321.1313	0.0174	6.5509	40.3146	0.0027	0.1527	6.1541
30	66.2118	434.7451	2698.3010	0.0151	6.5660	40.7526	0.0023	0.1523	6.2066
31	76.1435	500.9569	3133.0461	0.0131	6.5791	41.1466	0.0020	0.1520	6.2541
32	87.5651	577.1005	3634.0030	0.0114	6.5905	41.5006	0.0017	0.1517	6.2970
33	100.6998	664.6655	4211.1035	0.0099	6.6005	41.8184	0.0015	0.1515	6.3357
34	115.8048	765.3654	4875.7690	0.0086	6.6091	42.1033	0.0013	0.1513	6.3705
35	133.1755	881.1702	5641.1344	0.0075	6.6166	42.3586	0.0011	0.1511	6.4019
36	153.1519	1014.3457	6522.3045	0.0065	6.6231	42.5872	0.0010	0.1510	6.4301
37	176.1246	1167.4975	7536.6502	0.0057	6.6288	42.7916	0.0009	0.1509	6.4554
38	202.5433	1343.6222	8704.1477	0.0049	6.6338	42.9743	0.0007	0.1507	6.4781
39	232.9248	1546.1655	10047.7699	0.0043	6.6380	43.1374	0.0006	0.1506	6.4985
40	267.8635	1779.0903	11593.9354	0.0037	6.6418	43.2830	0.0006	0.1506	6.5168

TABLE A.22 $i = 16\%$

N	F/P	F/A	F/G	P/F	P/A	P/G	A/F	A/P	A/G
1	1.1600	1.0000	0.0000	0.8621	0.8621	0.0000	1.0000	1.1600	0.0000
2	1.3456	2.1600	1.0000	0.7432	1.6052	0.7432	0.4630	0.6230	0.4630
3	1.5609	3.5056	3.1600	0.6407	2.2459	2.0245	0.2853	0.4453	0.9014
4	1.8106	5.0665	6.6656	0.5523	2.7982	3.6814	0.1974	0.3574	1.3156
5	2.1003	6.8771	11.7321	0.4761	3.2743	5.5858	0.1454	0.3054	1.7060
6	2.4364	8.9775	18.6092	0.4104	3.6847	7.6380	0.1114	0.2714	2.0729
7	2.8262	11.4139	27.5867	0.3538	4.0386	9.7610	0.0876	0.2476	2.4169
8	3.2784	14.2401	39.0006	0.3050	4.3436	11.8962	0.0702	0.2302	2.7388
9	3.8030	17.5185	53.2407	0.2630	4.6065	13.9998	0.0571	0.2171	3.0391
10	4.4114	21.3215	70.7592	0.2267	4.8332	16.0399	0.0469	0.2069	3.3187
11	5.1173	25.7329	92.0807	0.1954	5.0286	17.9941	0.0389	0.1989	3.5783
12	5.9360	30.8502	117.8136	0.1685	5.1971	19.8472	0.0324	0.1924	3.8189
13	6.8858	36.7862	148.6637	0.1452	5.3423	21.5899	0.0272	0.1872	4.0413
14	7.9875	43.6720	185.4499	0.1252	5.4675	23.2175	0.0229	0.1829	4.2464
15	9.2655	51.6595	229.1219	0.1079	5.5755	24.7284	0.0194	0.1794	4.4352
16	10.7480	60.9250	280.7814	0.0930	5.6685	26.1241	0.0164	0.1764	4.6086
17	12.4677	71.6730	341.7064	0.0802	5.7487	27.4074	0.0140	0.1740	4.7676
18	14.4625	84.1407	413.3795	0.0691	5.8178	28.5828	0.0119	0.1719	4.9130
19	16.7765	98.6032	497.5202	0.0596	5.8775	29.6557	0.0101	0.1701	5.0457
20	19.4608	115.3797	596.1234	0.0514	5.9288	30.6321	0.0087	0.1687	5.1666
21	22.5745	134.8405	711.5032	0.0443	5.9731	31.5180	0.0074	0.1674	5.2766
22	26.1864	157.4150	846.3437	0.0382	6.0113	32.3200	0.0064	0.1664	5.3765
23	30.3762	183.6014	1003.7587	0.0329	6.0442	33.0442	0.0054	0.1654	5.4671
24	35.2364	213.9776	1187.3600	0.0284	6.0726	33.6970	0.0047	0.1647	5.5490
25	40.8742	249.2140	1401.3376	0.0245	6.0971	34.2841	0.0040	0.1640	5.6230
26	47.4141	290.0883	1650.5517	0.0211	6.1182	34.8114	0.0034	0.1634	5.6898
27	55.0004	337.5024	1940.6399	0.0182	6.1364	35.2841	0.0030	0.1630	5.7500
28	63.8004	392.5028	2278.1423	0.0157	6.1520	35.7073	0.0025	0.1625	5.8041
29	74.0085	456.3032	2670.6451	0.0135	6.1656	36.0856	0.0022	0.1622	5.8528
30	85.8499	530.3117	3126.9483	0.0116	6.1772	36.4234	0.0019	0.1619	5.8964
31	99.5859	616.1616	3657.2600	0.0100	6.1872	36.7247	0.0016	0.1616	5.9356
32	115.5196	715.7475	4273.4217	0.0087	6.1959	36.9930	0.0014	0.1614	5.9706
33	134.0027	831.2671	4989.1691	0.0075	6.2034	37.2318	0.0012	0.1612	6.0019
34	155.4432	965.2698	5820.4362	0.0064	6.2098	37.4441	0.0010	0.1610	6.0299
35	180.3141	1120.7130	6785.7060	0.0055	6.2153	37.6327	0.0009	0.1609	6.0548
36	209.1643	1301.0270	7906.4189	0.0048	6.2201	37.8000	0.0008	0.1608	6.0771
37	242.6306	1510.1914	9207.4460	0.0041	6.2242	37.9484	0.0007	0.1607	6.0969
38	281.4515	1752.8220	10717.6373	0.0036	6.2278	38.0799	0.0006	0.1606	6.1145
39	326.4838	2034.2735	12470.4593	0.0031	6.2309	38.1963	0.0005	0.1605	6.1302
40	378.7212	2360.7572	14504.7328	0.0026	6.2335	38.2992	0.0004	0.1604	6.1441

TABLE A.23 $i = 17\%$

N	F/P	F/A	F/G	P/F	P/A	P/G	A/F	A/P	A/G
1	1.1700	1.0000	0.0000	0.8547	0.8547	0.0000	1.0000	1.1700	0.0000
2	1.3689	2.1700	1.0000	0.7305	1.5852	0.7305	0.4608	0.6308	0.4608
3	1.6016	3.5389	3.1700	0.6244	2.2096	1.9793	0.2826	0.4526	0.8958
4	1.8739	5.1405	6.7089	0.5337	2.7432	3.5802	0.1945	0.3645	1.3051
5	2.1924	7.0144	11.8494	0.4561	3.1993	5.4046	0.1426	0.3126	1.6893
6	2.5652	9.2068	18.8638	0.3898	3.5892	7.3538	0.1086	0.2786	2.0489
7	3.0012	11.7720	28.0707	0.3332	3.9224	9.3530	0.0849	0.2549	2.3845
8	3.5115	14.7733	39.8427	0.2848	4.2072	11.3465	0.0677	0.2377	2.6969
9	4.1084	18.2847	54.6159	0.2434	4.4506	13.2937	0.0547	0.2247	2.9870
10	4.8068	22.3931	72.9006	0.2080	4.6586	15.1661	0.0447	0.2147	3.2555
11	5.6240	27.1999	95.2937	0.1778	4.8364	16.9442	0.0368	0.2068	3.5035
12	6.5801	32.8239	122.4937	0.1520	4.9884	18.6159	0.0305	0.2005	3.7318
13	7.6987	39.4040	155.3176	0.1299	5.1183	20.1746	0.0254	0.1954	3.9417
14	9.0075	47.1027	194.7216	0.1110	5.2293	21.6178	0.0212	0.1912	4.1340
15	10.5387	56.1101	241.8243	0.0949	5.3242	22.9463	0.0178	0.1878	4.3098
16	12.3303	66.6488	297.9344	0.0811	5.4053	24.1628	0.0150	0.1850	4.4702
17	14.4265	78.9792	364.5832	0.0693	5.4746	25.2719	0.0127	0.1827	4.6162
18	16.8790	93.4056	443.5624	0.0592	5.5339	26.2790	0.0107	0.1807	4.7488
19	19.7484	110.2846	536.9680	0.0506	5.5845	27.1905	0.0091	0.1791	4.8689
20	23.1056	130.0329	647.2526	0.0433	5.6278	28.0128	0.0077	0.1777	4.9776
21	27.0336	153.1385	777.2855	0.0370	5.6648	28.7526	0.0065	0.1765	5.0757
22	31.6293	180.1721	930.4240	0.0316	5.6964	29.4166	0.0056	0.1756	5.1641
23	37.0062	211.8013	1110.5961	0.0270	5.7234	30.0111	0.0047	0.1747	5.2436
24	43.2973	248.8076	1322.3975	0.0231	5.7465	30.5423	0.0040	0.1740	5.3149
25	50.6578	292.1049	1571.2050	0.0197	5.7662	31.0160	0.0034	0.1734	5.3789
26	59.2697	342.7627	1863.3099	0.0169	5.7831	31.4378	0.0029	0.1729	5.4362
27	69.3455	402.0323	2206.0726	0.0144	5.7975	31.8128	0.0025	0.1725	5.4873
28	81.1342	471.3778	2608.1049	0.0123	5.8099	32.1456	0.0021	0.1721	5.5329
29	94.9271	552.5121	3079.4827	0.0105	5.8204	32.4405	0.0018	0.1718	5.5736
30	111.0647	647.4391	3631.9948	0.0090	5.8294	32.7016	0.0015	0.1715	5.6098
31	129.9456	758.5038	4279.4339	0.0077	5.8371	32.9325	0.0013	0.1713	5.6419
32	152.0364	888.4494	5037.9377	0.0066	5.8437	33.1364	0.0011	0.1711	5.6705
33	177.8826	1040.4858	5926.3871	0.0056	5.8493	33.3163	0.0010	0.1710	5.6958
34	208.1226	1218.3684	6966.8729	0.0048	5.8541	33.4748	0.0008	0.1708	5.7182
35	243.5035	1426.4910	8185.2413	0.0041	5.8582	33.6145	0.0007	0.1707	5.7380
36	284.8991	1669.9945	9611.7323	0.0035	5.8617	33.7373	0.0006	0.1706	5.7555
37	333.3319	1954.8936	11281.7268	0.0030	5.8647	33.8453	0.0005	0.1705	5.7710
38	389.9983	2288.2255	13236.6204	0.0026	5.8673	33.9402	0.0004	0.1704	5.7847
39	456.2980	2678.2238	15524.8458	0.0022	5.8695	34.0235	0.0004	0.1704	5.7967
40	533.8687	3134.5218	18203.0696	0.0019	5.8713	34.0965	0.0003	0.1703	5.8073

TABLE A.24 $i = 18\%$

N	F/P	F/A	F/G	P/F	P/A	P/G	A/F	A/P	A/G
1	1.1800	1.0000	0.0000	0.8475	0.8475	0.0000	1.0000	1.1800	0.0000
2	1.3924	2.1800	1.0000	0.7182	1.5656	0.7182	0.4587	0.6387	0.4587
3	1.6430	3.5724	3.1800	0.6086	2.1743	1.9354	0.2799	0.4599	0.8902
4	1.9388	5.2154	6.7524	0.5158	2.6901	3.4828	0.1917	0.3717	1.2947
5	2.2878	7.1542	11.9678	0.4371	3.1272	5.2312	0.1398	0.3198	1.6728
6	2.6996	9.4420	19.1220	0.3704	3.4976	7.0834	0.1059	0.2859	2.0252
7	3.1855	12.1415	28.5640	0.3139	3.8115	8.9670	0.0824	0.2624	2.3526
8	3.7589	15.3270	40.7055	0.2660	4.0776	10.8292	0.0652	0.2452	2.6558
9	4.4355	19.0859	56.0325	0.2255	4.3030	12.6329	0.0524	0.2324	2.9358
10	5.2338	23.5213	75.1184	0.1911	4.4941	14.3525	0.0425	0.2225	3.1936
11	6.1759	28.7551	98.6397	0.1619	4.6560	15.9716	0.0348	0.2148	3.4303
12	7.2876	34.9311	127.3948	0.1372	4.7932	17.4811	0.0286	0.2086	3.6470
13	8.5994	42.2187	162.3259	0.1163	4.9095	18.8765	0.0237	0.2037	3.8449
14	10.1472	50.8180	204.5446	0.0985	5.0081	20.1576	0.0197	0.1997	4.0250
15	11.9737	60.9653	255.3626	0.0835	5.0916	21.3269	0.0164	0.1964	4.1887
16	14.1290	72.9390	316.3279	0.0708	5.1624	22.3885	0.0137	0.1937	4.3369
17	16.6722	87.0680	389.2669	0.0600	5.2223	23.3482	0.0115	0.1915	4.4708
18	19.6733	103.7403	476.3349	0.0508	5.2732	24.2123	0.0096	0.1896	4.5916
19	23.2144	123.4135	580.0752	0.0431	5.3162	24.9877	0.0081	0.1881	4.7003
20	27.3930	146.6280	703.4887	0.0365	5.3527	25.6813	0.0068	0.1868	4.7978
21	32.3238	174.0210	850.1167	0.0309	5.3837	26.3000	0.0057	0.1857	4.8851
22	38.1421	206.3448	1024.1377	0.0262	5.4099	26.8506	0.0048	0.1848	4.9632
23	45.0076	244.4868	1230.4825	0.0222	5.4321	27.3394	0.0041	0.1841	5.0329
24	53.1090	289.4945	1474.9693	0.0188	5.4509	27.7725	0.0035	0.1835	5.0950
25	62.6686	342.6035	1764.4638	0.0160	5.4669	28.1555	0.0029	0.1829	5.1502
26	73.9490	405.2721	2107.0673	0.0135	5.4804	28.4935	0.0025	0.1825	5.1991
27	87.2598	479.2211	2512.3394	0.0115	5.4919	28.7915	0.0021	0.1821	5.2425
28	102.9666	566.4809	2991.5605	0.0097	5.5016	29.0537	0.0018	0.1818	5.2810
29	121.5005	669.4475	3558.0414	0.0082	5.5098	29.2842	0.0015	0.1815	5.3149
30	143.3706	790.9480	4227.4888	0.0070	5.5168	29.4864	0.0013	0.1813	5.3448
31	169.1774	934.3186	5018.4368	0.0059	5.5227	29.6638	0.0011	0.1811	5.3712
32	199.6293	1103.4960	5952.7555	0.0050	5.5277	29.8191	0.0009	0.1809	5.3945
33	235.5625	1303.1253	7056.2514	0.0042	5.5320	29.9549	0.0008	0.1808	5.4149
34	277.9638	1538.6878	8359.3767	0.0036	5.5356	30.0736	0.0006	0.1806	5.4328
35	327.9973	1816.6516	9898.0645	0.0030	5.5386	30.1773	0.0006	0.1806	5.4485
36	387.0368	2144.6489	11714.7161	0.0026	5.5412	30.2677	0.0005	0.1805	5.4623
37	456.7034	2531.6857	13859.3650	0.0022	5.5434	30.3465	0.0004	0.1804	5.4744
38	538.9100	2988.3891	16391.0507	0.0019	5.5452	30.4152	0.0003	0.1803	5.4849
39	635.9139	3527.2992	19379.4399	0.0016	5.5468	30.4749	0.0003	0.1803	5.4941
40	750.3783	4163.2130	22906.7390	0.0013	5.5482	30.5269	0.0002	0.1802	5.5022

TABLE A.25 **$i = 19\%$**

N	F/P	F/A	F/G	P/F	P/A	P/G	A/F	A/P	A/G
1	1.1900	1.0000	0.0000	0.8403	0.8403	0.0000	1.0000	1.1900	0.0000
2	1.4161	2.1900	1.0000	0.7062	1.5465	0.7062	0.4566	0.6466	0.4566
3	1.6852	3.6061	3.1900	0.5934	2.1399	1.8930	0.2773	0.4673	0.8846
4	2.0053	5.2913	6.7961	0.4987	2.6386	3.3890	0.1890	0.3790	1.2844
5	2.3864	7.2966	12.0874	0.4190	3.0576	5.0652	0.1371	0.3271	1.6566
6	2.8398	9.6830	19.3840	0.3521	3.4098	6.8259	0.1033	0.2933	2.0019
7	3.3793	12.5227	29.0669	0.2959	3.7057	8.6014	0.0799	0.2699	2.3211
8	4.0214	15.9020	41.5896	0.2487	3.9544	10.3421	0.0629	0.2529	2.6154
9	4.7854	19.9234	57.4916	0.2090	4.1633	12.0138	0.0502	0.2402	2.8856
10	5.6947	24.7089	77.4151	0.1756	4.3389	13.5943	0.0405	0.2305	3.1331
11	6.7767	30.4035	102.1239	0.1476	4.4865	15.0699	0.0329	0.2229	3.3589
12	8.0642	37.1802	132.5275	0.1240	4.6105	16.4340	0.0269	0.2169	3.5645
13	9.5964	45.2445	169.7077	0.1042	4.7147	17.6844	0.0221	0.2121	3.7509
14	11.4198	54.8409	214.9522	0.0876	4.8023	18.8228	0.0182	0.2082	3.9196
15	13.5895	66.2607	269.7931	0.0736	4.8759	19.8530	0.0151	0.2051	4.0717
16	16.1715	79.8502	336.0537	0.0618	4.9377	20.7806	0.0125	0.2025	4.2086
17	19.2441	96.0218	415.9040	0.0520	4.9897	21.6120	0.0104	0.2004	4.3314
18	22.9005	115.2659	511.9257	0.0437	5.0333	22.3543	0.0087	0.1987	4.4413
19	27.2516	138.1664	627.1916	0.0367	5.0700	23.0148	0.0072	0.1972	4.5394
20	32.4294	165.4180	765.3580	0.0308	5.1009	23.6007	0.0060	0.1960	4.6268
21	38.5910	197.8474	930.7760	0.0259	5.1268	24.1190	0.0051	0.1951	4.7045
22	45.9233	236.4385	1128.6235	0.0218	5.1486	24.5763	0.0042	0.1942	4.7734
23	54.6487	282.3618	1365.0619	0.0183	5.1668	24.9788	0.0035	0.1935	4.8344
24	65.0320	337.0105	1647.4237	0.0154	5.1822	25.3325	0.0030	0.1930	4.8883
25	77.3881	402.0425	1984.4342	0.0129	5.1951	25.6426	0.0025	0.1925	4.9359
26	92.0918	479.4306	2386.4767	0.0109	5.2060	25.9141	0.0021	0.1921	4.9777
27	109.5893	571.5224	2865.9072	0.0091	5.2151	26.1514	0.0017	0.1917	5.0145
28	130.4112	681.1116	3437.4296	0.0077	5.2228	26.3584	0.0015	0.1915	5.0468
29	155.1893	811.5228	4118.5412	0.0064	5.2292	26.5388	0.0012	0.1912	5.0751
30	184.6753	966.7122	4930.0640	0.0054	5.2347	26.6958	0.0010	0.1910	5.0998
31	219.7636	1151.3875	5896.7762	0.0046	5.2392	26.8324	0.0009	0.1909	5.1215
32	261.5187	1371.1511	7048.1637	0.0038	5.2430	26.9509	0.0007	0.1907	5.1403
33	311.2073	1632.6698	8419.3148	0.0032	5.2462	27.0537	0.0006	0.1906	5.1568
34	370.3366	1943.8771	10051.9846	0.0027	5.2489	27.1428	0.0005	0.1905	5.1711
35	440.7006	2314.2137	11995.8617	0.0023	5.2512	27.2200	0.0004	0.1904	5.1836
36	524.4337	2754.9143	14310.0754	0.0019	5.2531	27.2867	0.0004	0.1904	5.1944
37	624.0761	3279.3481	17064.9897	0.0016	5.2547	27.3444	0.0003	0.1903	5.2038
38	742.6506	3903.4242	20344.3378	0.0013	5.2561	27.3942	0.0003	0.1903	5.2119
39	883.7542	4646.0748	24247.7620	0.0011	5.2572	27.4372	0.0002	0.1902	5.2190
40	1051.6675	5529.8290	28893.8367	0.0010	5.2582	27.4743	0.0002	0.1902	5.2251

TABLE A.26 *i* = 20%

N	F/P	F/A	F/G	P/F	P/A	P/G	A/F	A/P	A/G
1	1.2000	1.0000	0.0000	0.8333	0.8333	0.0000	1.0000	1.2000	0.0000
2	1.4400	2.2000	1.0000	0.6944	1.5278	0.6944	0.4545	0.6545	0.4545
3	1.7280	3.6400	3.2000	0.5787	2.1065	1.8519	0.2747	0.4747	0.8791
4	2.0736	5.3680	6.8400	0.4823	2.5887	3.2986	0.1863	0.3863	1.2742
5	2.4883	7.4416	12.2080	0.4019	2.9906	4.9061	0.1344	0.3344	1.6405
6	2.9860	9.9299	19.6496	0.3349	3.3255	6.5806	0.1007	0.3007	1.9788
7	3.5832	12.9159	29.5795	0.2791	3.6046	8.2551	0.0774	0.2774	2.2902
8	4.2998	16.4991	42.4954	0.2326	3.8372	9.8831	0.0606	0.2606	2.5756
9	5.1598	20.7989	58.9945	0.1938	4.0310	11.4335	0.0481	0.2481	2.8364
10	6.1917	25.9587	79.7934	0.1615	4.1925	12.8871	0.0385	0.2385	3.0739
11	7.4301	32.1504	105.7521	0.1346	4.3271	14.2330	0.0311	0.2311	3.2893
12	8.9161	39.5805	137.9025	0.1122	4.4392	15.4667	0.0253	0.2253	3.4841
13	10.6993	48.4966	177.4830	0.0935	4.5327	16.5883	0.0206	0.2206	3.6597
14	12.8392	59.1959	225.9796	0.0779	4.6106	17.6008	0.0169	0.2169	3.8175
15	15.4070	72.0351	285.1755	0.0649	4.6755	18.5095	0.0139	0.2139	3.9588
16	18.4884	87.4421	357.2106	0.0541	4.7296	19.3208	0.0114	0.2114	4.0851
17	22.1861	105.9306	444.6528	0.0451	4.7746	20.0419	0.0094	0.2094	4.1976
18	26.6233	128.1167	550.5833	0.0376	4.8122	20.6805	0.0078	0.2078	4.2975
19	31.9480	154.7400	678.7000	0.0313	4.8435	21.2439	0.0065	0.2065	4.3861
20	38.3376	186.6880	833.4400	0.0261	4.8696	21.7395	0.0054	0.2054	4.4643
21	46.0051	225.0256	1020.1280	0.0217	4.8913	22.1742	0.0044	0.2044	4.5334
22	55.2061	271.0307	1245.1536	0.0181	4.9094	22.5546	0.0037	0.2037	4.5941
23	66.2474	326.2369	1516.1843	0.0151	4.9245	22.8867	0.0031	0.2031	4.6475
24	79.4968	392.4842	1842.4212	0.0126	4.9371	23.1760	0.0025	0.2025	4.6943
25	95.3962	471.9811	2234.9054	0.0105	4.9476	23.4276	0.0021	0.2021	4.7352
26	114.4755	567.3773	2706.8865	0.0087	4.9563	23.6460	0.0018	0.2018	4.7709
27	137.3706	681.8528	3274.2638	0.0073	4.9636	23.8353	0.0015	0.2015	4.8020
28	164.8447	819.2233	3956.1166	0.0061	4.9697	23.9991	0.0012	0.2012	4.8291
29	197.8136	984.0680	4775.3399	0.0051	4.9747	24.1406	0.0010	0.2010	4.8527
30	237.3763	1181.8816	5759.4078	0.0042	4.9789	24.2628	0.0008	0.2008	4.8731
31	284.8516	1419.2579	6941.2894	0.0035	4.9824	24.3681	0.0007	0.2007	4.8908
32	341.8219	1704.1095	8360.5473	0.0029	4.9854	24.4588	0.0006	0.2006	4.9061
33	410.1863	2045.9314	10064.6568	0.0024	4.9878	24.5368	0.0005	0.2005	4.9194
34	492.2235	2456.1176	12110.5881	0.0020	4.9898	24.6038	0.0004	0.2004	4.9308
35	590.6682	2948.3411	14566.7057	0.0017	4.9915	24.6614	0.0003	0.2003	4.9406
36	708.8019	3539.0094	17515.0469	0.0014	4.9929	24.7108	0.0003	0.2003	4.9491
37	850.5622	4247.8112	21054.0562	0.0012	4.9941	24.7531	0.0002	0.2002	4.9564
38	1020.6747	5098.3735	25301.8675	0.0010	4.9951	24.7894	0.0002	0.2002	4.9627
39	1224.8096	6119.0482	30400.2410	0.0008	4.9959	24.8204	0.0002	0.2002	4.9681
40	1469.7716	7343.8578	36519.2892	0.0007	4.9966	24.8469	0.0001	0.2001	4.9728

TABLE A.27 ***i* = 21%**

N	F/P	F/A	F/G	P/F	P/A	P/G	A/F	A/P	A/G
1	1.2100	1.0000	0.0000	0.8264	0.8264	0.0000	1.0000	1.2100	0.0000
2	1.4641	2.2100	1.0000	0.6830	1.5095	0.6830	0.4525	0.6625	0.4525
3	1.7716	3.6741	3.2100	0.5645	2.0739	1.8120	0.2722	0.4822	0.8737
4	2.1436	5.4457	6.8841	0.4665	2.5404	3.2115	0.1836	0.3936	1.2641
5	2.5937	7.5892	12.3298	0.3855	2.9260	4.7537	0.1318	0.3418	1.6246
6	3.1384	10.1830	19.9190	0.3186	3.2446	6.3468	0.0982	0.3082	1.9561
7	3.7975	13.3214	30.1020	0.2633	3.5079	7.9268	0.0751	0.2851	2.2597
8	4.5950	17.1189	43.4234	0.2176	3.7256	9.4502	0.0584	0.2684	2.5366
9	5.5599	21.7139	60.5423	0.1799	3.9054	10.8891	0.0461	0.2561	2.7882
10	6.7275	27.2738	82.2562	0.1486	4.0541	12.2269	0.0367	0.2467	3.0159
11	8.1403	34.0013	109.5300	0.1228	4.1769	13.4553	0.0294	0.2394	3.2213
12	9.8497	42.1416	143.5314	0.1015	4.2784	14.5721	0.0237	0.2337	3.4059
13	11.9182	51.9913	185.6729	0.0839	4.3624	15.5790	0.0192	0.2292	3.5712
14	14.4210	63.9095	237.6643	0.0693	4.4317	16.4804	0.0156	0.2256	3.7188
15	17.4494	78.3305	301.5737	0.0573	4.4890	17.2828	0.0128	0.2228	3.8500
16	21.1138	95.7799	379.9042	0.0474	4.5364	17.9932	0.0104	0.2204	3.9664
17	25.5477	116.8937	475.6841	0.0391	4.5755	18.6195	0.0086	0.2186	4.0694
18	30.9127	142.4413	592.5778	0.0323	4.6079	19.1694	0.0070	0.2170	4.1602
19	37.4043	173.3540	735.0191	0.0267	4.6346	19.6506	0.0058	0.2158	4.2400
20	45.2593	210.7584	908.3731	0.0221	4.6567	20.0704	0.0047	0.2147	4.3100
21	54.7637	256.0176	1119.1315	0.0183	4.6750	20.4356	0.0039	0.2139	4.3713
22	66.2641	310.7813	1375.1491	0.0151	4.6900	20.7526	0.0032	0.2132	4.4248
23	80.1795	377.0454	1685.9304	0.0125	4.7025	21.0269	0.0027	0.2127	4.4714
24	97.0172	457.2249	2062.9758	0.0103	4.7128	21.2640	0.0022	0.2122	4.5119
25	117.3909	554.2422	2520.2007	0.0085	4.7213	21.4685	0.0018	0.2118	4.5471
26	142.0429	671.6330	3074.4429	0.0070	4.7284	21.6445	0.0015	0.2115	4.5776
27	171.8719	813.6759	3746.0759	0.0058	4.7342	21.7957	0.0012	0.2112	4.6039
28	207.9651	985.5479	4559.7519	0.0048	4.7390	21.9256	0.0010	0.2110	4.6266
29	251.6377	1193.5129	5545.2997	0.0040	4.7430	22.0368	0.0008	0.2108	4.6462
30	304.4816	1445.1507	6738.8127	0.0033	4.7463	22.1321	0.0007	0.2107	4.6631
31	368.4228	1749.6323	8183.9634	0.0027	4.7490	22.2135	0.0006	0.2106	4.6775
32	445.7916	2118.0551	9933.5957	0.0022	4.7512	22.2830	0.0005	0.2105	4.6900
33	539.4078	2563.8467	12051.6507	0.0019	4.7531	22.3424	0.0004	0.2104	4.7006
34	652.6834	3103.2545	14615.4974	0.0015	4.7546	22.3929	0.0003	0.2103	4.7097
35	789.7470	3755.9379	17718.7519	0.0013	4.7559	22.4360	0.0003	0.2103	4.7175
36	955.5938	4545.6848	21474.6897	0.0010	4.7569	22.4726	0.0002	0.2102	4.7242
37	1156.2685	5501.2787	26020.3746	0.0009	4.7578	22.5037	0.0002	0.2102	4.7299
38	1399.0849	6657.5472	31521.6533	0.0007	4.7585	22.5302	0.0002	0.2102	4.7347
39	1692.8927	8056.6321	38179.2004	0.0006	4.7591	22.5526	0.0001	0.2101	4.7389
40	2048.4002	9749.5248	46235.8325	0.0005	4.7596	22.5717	0.0001	0.2101	4.7424

TABLE A.28 $i = 22\%$

N	F/P	F/A	F/G	P/F	P/A	P/G	A/F	A/P	A/G
1	1.2200	1.0000	0.0000	0.8197	0.8197	0.0000	1.0000	1.2200	0.0000
2	1.4884	2.2200	1.0000	0.6719	1.4915	0.6719	0.4505	0.6705	0.4505
3	1.8158	3.7084	3.2200	0.5507	2.0422	1.7733	0.2697	0.4897	0.8683
4	2.2153	5.5242	6.9284	0.4514	2.4936	3.1275	0.1810	0.4010	1.2542
5	2.7027	7.7396	12.4526	0.3700	2.8636	4.6075	0.1292	0.3492	1.6090
6	3.2973	10.4423	20.1922	0.3033	3.1669	6.1239	0.0958	0.3158	1.9337
7	4.0227	13.7396	30.6345	0.2486	3.4155	7.6154	0.0728	0.2928	2.2297
8	4.9077	17.7623	44.3741	0.2038	3.6193	9.0417	0.0563	0.2763	2.4982
9	5.9874	22.6700	62.1364	0.1670	3.7863	10.3779	0.0441	0.2641	2.7409
10	7.3046	28.6574	84.8064	0.1369	3.9232	11.6100	0.0349	0.2549	2.9593
11	8.9117	35.9620	113.4638	0.1122	4.0354	12.7321	0.0278	0.2478	3.1551
12	10.8722	44.8737	149.4259	0.0920	4.1274	13.7438	0.0223	0.2423	3.3299
13	13.2641	55.7459	194.2996	0.0754	4.2028	14.6485	0.0179	0.2379	3.4855
14	16.1822	69.0100	250.0455	0.0618	4.2646	15.4519	0.0145	0.2345	3.6233
15	19.7423	85.1922	319.0555	0.0507	4.3152	16.1610	0.0117	0.2317	3.7451
16	24.0856	104.9345	404.2477	0.0415	4.3567	16.7838	0.0095	0.2295	3.8524
17	29.3844	129.0201	509.1822	0.0340	4.3908	17.3283	0.0078	0.2278	3.9465
18	35.8490	158.4045	638.2023	0.0279	4.4187	17.8025	0.0063	0.2263	4.0289
19	43.7358	194.2535	796.6068	0.0229	4.4415	18.2141	0.0051	0.2251	4.1009
20	53.3576	237.9893	990.8603	0.0187	4.4603	18.5702	0.0042	0.2242	4.1635
21	65.0963	291.3469	1228.8496	0.0154	4.4756	18.8774	0.0034	0.2234	4.2178
22	79.4175	356.4432	1520.1965	0.0126	4.4882	19.1418	0.0028	0.2228	4.2649
23	96.8894	435.8607	1876.6398	0.0103	4.4985	19.3689	0.0023	0.2223	4.3056
24	118.2050	532.7501	2312.5005	0.0085	4.5070	19.5635	0.0019	0.2219	4.3407
25	144.2101	650.9551	2845.2506	0.0069	4.5139	19.7299	0.0015	0.2215	4.3709
26	175.9364	795.1653	3496.2057	0.0057	4.5196	19.8720	0.0013	0.2213	4.3968
27	214.6424	971.1016	4291.3710	0.0047	4.5243	19.9931	0.0010	0.2210	4.4191
28	261.8637	1185.7440	5262.4726	0.0038	4.5281	20.0962	0.0008	0.2208	4.4381
29	319.4737	1447.6077	6448.2166	0.0031	4.5312	20.1839	0.0007	0.2207	4.4544
30	389.7579	1767.0813	7895.8243	0.0026	4.5338	20.2583	0.0006	0.2206	4.4683
31	475.5046	2156.8392	9662.9056	0.0021	4.5359	20.3214	0.0005	0.2205	4.4801
32	580.1156	2632.3439	11819.7448	0.0017	4.5376	20.3748	0.0004	0.2204	4.4902
33	707.7411	3212.4595	14452.0887	0.0014	4.5390	20.4200	0.0003	0.2203	4.4988
34	863.4441	3920.2006	17664.5482	0.0012	4.5402	20.4582	0.0003	0.2203	4.5060
35	1053.4018	4783.6447	21584.7488	0.0009	4.5411	20.4905	0.0002	0.2202	4.5122
36	1285.1502	5837.0466	26368.3935	0.0008	4.5419	20.5178	0.0002	0.2202	4.5174
37	1567.8833	7122.1968	32205.4401	0.0006	4.5426	20.5407	0.0001	0.2201	4.5218
38	1912.8176	8690.0801	39327.6370	0.0005	4.5431	20.5601	0.0001	0.2201	4.5256
39	2333.6375	10602.8978	48017.7171	0.0004	4.5435	20.5763	0.0001	0.2201	4.5287
40	2847.0378	12936.5353	58620.6148	0.0004	4.5439	20.5900	0.0001	0.2201	4.5314

TABLE A.29 $i = 23\%$

N	F/P	F/A	F/G	P/F	P/A	P/G	A/F	A/P	A/G
1	1.2300	1.0000	0.0000	0.8130	0.8130	0.0000	1.0000	1.2300	0.0000
2	1.5129	2.2300	1.0000	0.6610	1.4740	0.6610	0.4484	0.6784	0.4484
3	1.8609	3.7429	3.2300	0.5374	2.0114	1.7358	0.2672	0.4972	0.8630
4	2.2889	5.6038	6.9729	0.4369	2.4483	3.0464	0.1785	0.4085	1.2443
5	2.8153	7.8926	12.5767	0.3552	2.8035	4.4672	0.1267	0.3567	1.5935
6	3.4628	10.7079	20.4693	0.2888	3.0923	5.9112	0.0934	0.3234	1.9116
7	4.2593	14.1708	31.1772	0.2348	3.3270	7.3198	0.0706	0.3006	2.2001
8	5.2389	18.4300	45.3480	0.1909	3.5179	8.6560	0.0543	0.2843	2.4605
9	6.4439	23.6690	63.7780	0.1552	3.6731	9.8975	0.0422	0.2722	2.6946
10	7.9259	30.1128	87.4470	0.1262	3.7993	11.0330	0.0332	0.2632	2.9040
11	9.7489	38.0388	117.5598	0.1026	3.9018	12.0588	0.0263	0.2563	3.0905
12	11.9912	47.7877	155.5986	0.0834	3.9852	12.9761	0.0209	0.2509	3.2560
13	14.7491	59.7788	203.3862	0.0678	4.0530	13.7897	0.0167	0.2467	3.4023
14	18.1414	74.5280	263.1651	0.0551	4.1082	14.5063	0.0134	0.2434	3.5311
15	22.3140	92.6694	337.6930	0.0448	4.1530	15.1337	0.0108	0.2408	3.6441
16	27.4462	114.9834	430.3624	0.0364	4.1894	15.6802	0.0087	0.2387	3.7428
17	33.7588	142.4295	545.3458	0.0296	4.2190	16.1542	0.0070	0.2370	3.8289
18	41.5233	176.1883	687.7753	0.0241	4.2431	16.5636	0.0057	0.2357	3.9036
19	51.0737	217.7116	863.9636	0.0196	4.2627	16.9160	0.0046	0.2346	3.9684
20	62.8206	268.7853	1081.6753	0.0159	4.2786	17.2185	0.0037	0.2337	4.0243
21	77.2694	331.6059	1350.4606	0.0129	4.2916	17.4773	0.0030	0.2330	4.0725
22	95.0413	408.8753	1682.0665	0.0105	4.3021	17.6983	0.0024	0.2324	4.1139
23	116.9008	503.9166	2090.9418	0.0086	4.3106	17.8865	0.0020	0.2320	4.1494
24	143.7880	620.8174	2594.8584	0.0070	4.3176	18.0464	0.0016	0.2316	4.1797
25	176.8593	764.6054	3215.6759	0.0057	4.3232	18.1821	0.0013	0.2313	4.2057
26	217.5369	941.4647	3980.2813	0.0046	4.3278	18.2970	0.0011	0.2311	4.2278
27	267.5704	1159.0016	4921.7460	0.0037	4.3316	18.3942	0.0009	0.2309	4.2465
28	329.1115	1426.5719	6080.7476	0.0030	4.3346	18.4763	0.0007	0.2307	4.2625
29	404.8072	1755.6835	7507.3195	0.0025	4.3371	18.5454	0.0006	0.2306	4.2760
30	497.9129	2160.4907	9263.0030	0.0020	4.3391	18.6037	0.0005	0.2305	4.2875
31	612.4328	2658.4036	11423.4937	0.0016	4.3407	18.6526	0.0004	0.2304	4.2971
32	753.2924	3270.8364	14081.8973	0.0013	4.3421	18.6938	0.0003	0.2303	4.3053
33	926.5496	4024.1287	17352.7336	0.0011	4.3431	18.7283	0.0002	0.2302	4.3122
34	1139.6560	4950.6783	21376.8624	0.0009	4.3440	18.7573	0.0002	0.2302	4.3180
35	1401.7769	6090.3344	26327.5407	0.0007	4.3447	18.7815	0.0002	0.2302	4.3228
36	1724.1856	7492.1113	32417.8751	0.0006	4.3453	18.8018	0.0001	0.2301	4.3269
37	2120.7483	9216.2969	39909.9864	0.0005	4.3458	18.8188	0.0001	0.2301	4.3304
38	2608.5204	11337.0451	49126.2832	0.0004	4.3462	18.8330	0.0001	0.2301	4.3333
39	3208.4801	13945.5655	60463.3284	0.0003	4.3465	18.8449	0.0001	0.2301	4.3357
40	3946.4305	17154.0456	74408.8939	0.0003	4.3467	18.8547	0.0001	0.2301	4.3377

TABLE A.30 $i = 24\%$

N	F/P	F/A	F/G	P/F	P/A	P/G	A/F	A/P	A/G
1	1.2400	1.0000	0.0000	0.8065	0.8065	0.0000	1.0000	1.2400	0.0000
2	1.5376	2.2400	1.0000	0.6504	1.4568	0.6504	0.4464	0.6864	0.4464
3	1.9066	3.7776	3.2400	0.5245	1.9813	1.6993	0.2647	0.5047	0.8577
4	2.3642	5.6842	7.0176	0.4230	2.4043	2.9683	0.1759	0.4159	1.2346
5	2.9316	8.0484	12.7018	0.3411	2.7454	4.3327	0.1242	0.3642	1.5782
6	3.6352	10.9801	20.7503	0.2751	3.0205	5.7081	0.0911	0.3311	1.8898
7	4.5077	14.6153	31.7303	0.2218	3.2423	7.0392	0.0684	0.3084	2.1710
8	5.5895	19.1229	46.3456	0.1789	3.4212	8.2915	0.0523	0.2923	2.4236
9	6.9310	24.7125	65.4685	0.1443	3.5655	9.4458	0.0405	0.2805	2.6492
10	8.5944	31.6434	90.1810	0.1164	3.6819	10.4930	0.0316	0.2716	2.8499
11	10.6571	40.2379	121.8244	0.0938	3.7757	11.4313	0.0249	0.2649	3.0276
12	13.2148	50.8950	162.0623	0.0757	3.8514	12.2637	0.0196	0.2596	3.1843
13	16.3863	64.1097	212.9573	0.0610	3.9124	12.9960	0.0156	0.2556	3.3218
14	20.3191	80.4961	277.0670	0.0492	3.9616	13.6358	0.0124	0.2524	3.4420
15	25.1956	100.8151	357.5631	0.0397	4.0013	14.1915	0.0099	0.2499	3.5467
16	31.2426	126.0108	458.3782	0.0320	4.0333	14.6716	0.0079	0.2479	3.6376
17	38.7408	157.2534	584.3890	0.0258	4.0591	15.0846	0.0064	0.2464	3.7162
18	48.0386	195.9942	741.6423	0.0208	4.0799	15.4385	0.0051	0.2451	3.7840
19	59.5679	244.0328	937.6365	0.0168	4.0967	15.7406	0.0041	0.2441	3.8423
20	73.8641	303.6006	1181.6693	0.0135	4.1103	15.9979	0.0033	0.2433	3.8922
21	91.5915	377.4648	1485.2699	0.0109	4.1212	16.2162	0.0026	0.2426	3.9349
22	113.5735	469.0563	1862.7347	0.0088	4.1300	16.4011	0.0021	0.2421	3.9712
23	140.8312	582.6298	2331.7910	0.0071	4.1371	16.5574	0.0017	0.2417	4.0022
24	174.6306	723.4610	2914.4208	0.0057	4.1428	16.6891	0.0014	0.2414	4.0284
25	216.5420	898.0916	3637.8818	0.0046	4.1474	16.7999	0.0011	0.2411	4.0507
26	268.5121	1114.6336	4535.9735	0.0037	4.1511	16.8930	0.0009	0.2409	4.0695
27	332.9550	1383.1457	5650.6071	0.0030	4.1542	16.9711	0.0007	0.2407	4.0853
28	412.8642	1716.1007	7033.7528	0.0024	4.1566	17.0365	0.0006	0.2406	4.0987
29	511.9516	2128.9648	8749.8535	0.0020	4.1585	17.0912	0.0005	0.2405	4.1099
30	634.8199	2640.9164	10878.8183	0.0016	4.1601	17.1369	0.0004	0.2404	4.1193
31	787.1767	3275.7363	13519.7347	0.0013	4.1614	17.1750	0.0003	0.2403	4.1272
32	976.0991	4062.9130	16795.4710	0.0010	4.1624	17.2067	0.0002	0.2402	4.1338
33	1210.3629	5039.0122	20858.3840	0.0008	4.1632	17.2332	0.0002	0.2402	4.1394
34	1500.8500	6249.3751	25897.3962	0.0007	4.1639	17.2552	0.0002	0.2402	4.1440
35	1861.0540	7750.2251	32146.7713	0.0005	4.1644	17.2734	0.0001	0.2401	4.1479
36	2307.7070	9611.2791	39896.9964	0.0004	4.1649	17.2886	0.0001	0.2401	4.1511
37	2861.5567	11918.9861	49508.2755	0.0003	4.1652	17.3012	0.0001	0.2401	4.1537
38	3548.3303	14780.5428	61427.2616	0.0003	4.1655	17.3116	0.0001	0.2401	4.1560
39	4399.9295	18328.8731	76207.8044	0.0002	4.1657	17.3202	0.0001	0.2401	4.1578
40	5455.9126	22728.8026	94536.6775	0.0002	4.1659	17.3274	0.0000	0.2400	4.1593

TABLE A.31 $i = 25\%$

N	F/P	F/A	F/G	P/F	P/A	P/G	A/F	A/P	A/G
1	1.2500	1.0000	0.0000	0.8000	0.8000	0.0000	1.0000	1.2500	0.0000
2	1.5625	2.2500	1.0000	0.6400	1.4400	0.6400	0.4444	0.6944	0.4444
3	1.9531	3.8125	3.2500	0.5120	1.9520	1.6640	0.2623	0.5123	0.8525
4	2.4414	5.7656	7.0625	0.4096	2.3616	2.8928	0.1734	0.4234	1.2249
5	3.0518	8.2070	12.8281	0.3277	2.6893	4.2035	0.1218	0.3718	1.5631
6	3.8147	11.2588	21.0352	0.2621	2.9514	5.5142	0.0888	0.3388	1.8683
7	4.7684	15.0735	32.2939	0.2097	3.1611	6.7725	0.0663	0.3163	2.1424
8	5.9605	19.8419	47.3674	0.1678	3.3289	7.9469	0.0504	0.3004	2.3872
9	7.4506	25.8023	67.2093	0.1342	3.4631	9.0207	0.0388	0.2888	2.6048
10	9.3132	33.2529	93.0116	0.1074	3.5705	9.9870	0.0301	0.2801	2.7971
11	11.6415	42.5661	126.2645	0.0859	3.6564	10.8460	0.0235	0.2735	2.9663
12	14.5519	54.2077	168.8306	0.0687	3.7251	11.6020	0.0184	0.2684	3.1145
13	18.1899	68.7596	223.0383	0.0550	3.7801	12.2617	0.0145	0.2645	3.2437
14	22.7374	86.9495	291.7979	0.0440	3.8241	12.8334	0.0115	0.2615	3.3559
15	28.4217	109.6868	378.7474	0.0352	3.8593	13.3260	0.0091	0.2591	3.4530
16	35.5271	138.1085	488.4342	0.0281	3.8874	13.7482	0.0072	0.2572	3.5366
17	44.4089	173.6357	626.5427	0.0225	3.9099	14.1085	0.0058	0.2558	3.6084
18	55.5112	218.0446	800.1784	0.0180	3.9279	14.4147	0.0046	0.2546	3.6698
19	69.3889	273.5558	1018.2230	0.0144	3.9424	14.6741	0.0037	0.2537	3.7222
20	86.7362	342.9447	1291.7788	0.0115	3.9539	14.8932	0.0029	0.2529	3.7667
21	108.4202	429.6809	1634.7235	0.0092	3.9631	15.0777	0.0023	0.2523	3.8045
22	135.5253	538.1011	2064.4043	0.0074	3.9705	15.2326	0.0019	0.2519	3.8365
23	169.4066	673.6264	2602.5054	0.0059	3.9764	15.3625	0.0015	0.2515	3.8634
24	211.7582	843.0329	3276.1318	0.0047	3.9811	15.4711	0.0012	0.2512	3.8861
25	264.6978	1054.7912	4119.1647	0.0038	3.9849	15.5618	0.0009	0.2509	3.9052
26	330.8722	1319.4890	5173.9559	0.0030	3.9879	15.6373	0.0008	0.2508	3.9212
27	413.5903	1650.3612	6493.4449	0.0024	3.9903	15.7002	0.0006	0.2506	3.9346
28	516.9879	2063.9515	8143.8061	0.0019	3.9923	15.7524	0.0005	0.2505	3.9457
29	646.2349	2580.9394	10207.7577	0.0015	3.9938	15.7957	0.0004	0.2504	3.9551
30	807.7936	3227.1743	12788.6971	0.0012	3.9950	15.8316	0.0003	0.2503	3.9628
31	1009.7420	4034.9678	16015.8713	0.0010	3.9960	15.8614	0.0002	0.2502	3.9693
32	1262.1774	5044.7098	20050.8392	0.0008	3.9968	15.8859	0.0002	0.2502	3.9746
33	1577.7218	6306.8872	25095.5490	0.0006	3.9975	15.9062	0.0002	0.2502	3.9791
34	1972.1523	7884.6091	31402.4362	0.0005	3.9980	15.9229	0.0001	0.2501	3.9828
35	2465.1903	9856.7613	39287.0453	0.0004	3.9984	15.9367	0.0001	0.2501	3.9858
36	3081.4879	12321.9516	49143.8066	0.0003	3.9987	15.9481	0.0001	0.2501	3.9883
37	3851.8599	15403.4396	61465.7582	0.0003	3.9990	15.9574	0.0001	0.2501	3.9904
38	4814.8249	19255.2994	76869.1978	0.0002	3.9992	15.9651	0.0001	0.2501	3.9921
39	6018.5311	24070.1243	96124.4972	0.0002	3.9993	15.9714	0.0000	0.2500	3.9935
40	7523.1638	30088.6554	120194.6215	0.0001	3.9995	15.9766	0.0000	0.2500	3.9947

TABLE A.32 **$i = 30\%$**

N	F/P	F/A	F/G	P/F	P/A	P/G	A/F	A/P	A/G
1	1.3000	1.0000	0.0000	0.7692	0.7692	0.0000	1.0000	1.3000	0.0000
2	1.6900	2.3000	1.0000	0.5917	1.3609	0.5917	0.4348	0.7348	0.4348
3	2.1970	3.9900	3.3000	0.4552	1.8161	1.5020	0.2506	0.5506	0.8271
4	2.8561	6.1870	7.2900	0.3501	2.1662	2.5524	0.1616	0.4616	1.1783
5	3.7129	9.0431	13.4770	0.2693	2.4356	3.6297	0.1106	0.4106	1.4903
6	4.8268	12.7560	22.5201	0.2072	2.6427	4.6656	0.0784	0.3784	1.7654
7	6.2749	17.5828	35.2761	0.1594	2.8021	5.6218	0.0569	0.3569	2.0063
8	8.1573	23.8577	52.8590	0.1226	2.9247	6.4800	0.0419	0.3419	2.2156
9	10.6045	32.0150	76.7167	0.0943	3.0190	7.2343	0.0312	0.3312	2.3963
10	13.7858	42.6195	108.7317	0.0725	3.0915	7.8872	0.0235	0.3235	2.5512
11	17.9216	56.4053	151.3512	0.0558	3.1473	8.4452	0.0177	0.3177	2.6833
12	23.2981	74.3270	207.7565	0.0429	3.1903	8.9173	0.0135	0.3135	2.7952
13	30.2875	97.6250	282.0835	0.0330	3.2233	9.3135	0.0102	0.3102	2.8895
14	39.3738	127.9125	379.7085	0.0254	3.2487	9.6437	0.0078	0.3078	2.9685
15	51.1859	167.2863	507.6210	0.0195	3.2682	9.9172	0.0060	0.3060	3.0344
16	66.5417	218.4722	674.9073	0.0150	3.2832	10.1426	0.0046	0.3046	3.0892
17	86.5042	285.0139	893.3795	0.0116	3.2948	10.3276	0.0035	0.3035	3.1345
18	112.4554	371.5180	1178.3934	0.0089	3.3037	10.4788	0.0027	0.3027	3.1718
19	146.1920	483.9734	1549.9114	0.0068	3.3105	10.6019	0.0021	0.3021	3.2025
20	190.0496	630.1655	2033.8849	0.0053	3.3158	10.7019	0.0016	0.3016	3.2275
21	247.0645	820.2151	2664.0503	0.0040	3.3198	10.7828	0.0012	0.3012	3.2480
22	321.1839	1067.2796	3484.2654	0.0031	3.3230	10.8482	0.0009	0.3009	3.2646
23	417.5391	1388.4635	4551.5450	0.0024	3.3254	10.9009	0.0007	0.3007	3.2781
24	542.8008	1806.0026	5940.0086	0.0018	3.3272	10.9433	0.0006	0.3006	3.2890
25	705.6410	2348.8033	7746.0111	0.0014	3.3286	10.9773	0.0004	0.3004	3.2979
26	917.3333	3054.4443	10094.8145	0.0011	3.3297	11.0045	0.0003	0.3003	3.3050
27	1192.5333	3971.7776	13149.2588	0.0008	3.3305	11.0263	0.0003	0.3003	3.3107
28	1550.2933	5164.3109	17121.0364	0.0006	3.3312	11.0437	0.0002	0.3002	3.3153
29	2015.3813	6714.6042	22285.3474	0.0005	3.3317	11.0576	0.0001	0.3001	3.3189
30	2619.9956	8729.9855	28999.9516	0.0004	3.3321	11.0687	0.0001	0.3001	3.3219

TABLE A.33 *i* = 35%

N	F/P	F/A	F/G	P/F	P/A	P/G	A/F	A/P	A/G
1	1.3500	1.0000	0.0000	0.7407	0.7407	0.0000	1.0000	1.3500	0.0000
2	1.8225	2.3500	1.0000	0.5487	1.2894	0.5487	0.4255	0.7755	0.4255
3	2.4604	4.1725	3.3500	0.4064	1.6959	1.3616	0.2397	0.5897	0.8029
4	3.3215	6.6329	7.5225	0.3011	1.9969	2.2648	0.1508	0.5008	1.1341
5	4.4840	9.9544	14.1554	0.2230	2.2200	3.1568	0.1005	0.4505	1.4220
6	6.0534	14.4384	24.1098	0.1652	2.3852	3.9828	0.0693	0.4193	1.6698
7	8.1722	20.4919	38.5482	0.1224	2.5075	4.7170	0.0488	0.3988	1.8811
8	11.0324	28.6640	59.0400	0.0906	2.5982	5.3515	0.0349	0.3849	2.0597
9	14.8937	39.6964	87.7040	0.0671	2.6653	5.8886	0.0252	0.3752	2.2094
10	20.1066	54.5902	127.4005	0.0497	2.7150	6.3363	0.0183	0.3683	2.3338
11	27.1439	74.6967	181.9906	0.0368	2.7519	6.7047	0.0134	0.3634	2.4364
12	36.6442	101.8406	256.6873	0.0273	2.7792	7.0049	0.0098	0.3598	2.5205
13	49.4697	138.4848	358.5279	0.0202	2.7994	7.2474	0.0072	0.3572	2.5889
14	66.7841	187.9544	497.0127	0.0150	2.8144	7.4421	0.0053	0.3553	2.6443
15	90.1585	254.7385	684.9671	0.0111	2.8255	7.5974	0.0039	0.3539	2.6889
16	121.7139	344.8970	939.7056	0.0082	2.8337	7.7206	0.0029	0.3529	2.7246
17	164.3138	466.6109	1284.6025	0.0061	2.8398	7.8180	0.0021	0.3521	2.7530
18	221.8236	630.9247	1751.2134	0.0045	2.8443	7.8946	0.0016	0.3516	2.7756
19	299.4619	852.7483	2382.1381	0.0033	2.8476	7.9547	0.0012	0.3512	2.7935
20	404.2736	1152.2103	3234.8864	0.0025	2.8501	8.0017	0.0009	0.3509	2.8075
21	545.7693	1556.4838	4387.0967	0.0018	2.8519	8.0384	0.0006	0.3506	2.8186
22	736.7886	2102.2532	5943.5805	0.0014	2.8533	8.0669	0.0005	0.3505	2.8272
23	994.6646	2839.0418	8045.8337	0.0010	2.8543	8.0890	0.0004	0.3504	2.8340
24	1342.7973	3833.7064	10884.8755	0.0007	2.8550	8.1061	0.0003	0.3503	2.8393
25	1812.7763	5176.5037	14718.5820	0.0006	2.8556	8.1194	0.0002	0.3502	2.8433
26	2447.2480	6989.2800	19895.0857	0.0004	2.8560	8.1296	0.0001	0.3501	2.8465
27	3303.7848	9436.5280	26884.3656	0.0003	2.8563	8.1374	0.0001	0.3501	2.8490
28	4460.1095	12740.3128	36320.8936	0.0002	2.8565	8.1435	0.0001	0.3501	2.8509
29	6021.1478	17200.4222	49061.2064	0.0002	2.8567	8.1481	0.0001	0.3501	2.8523
30	8128.5495	23221.5700	66261.6286	0.0001	2.8568	8.1517	0.0000	0.3500	2.8535

TABLE A.34 $i = 40\%$

N	F/P	F/A	F/G	P/F	P/A	P/G	A/F	A/P	A/G
1	1.4000	1.0000	0.0000	0.7143	0.7143	0.0000	1.0000	1.4000	0.0000
2	1.9600	2.4000	1.0000	0.5102	1.2245	0.5102	0.4167	0.8167	0.4167
3	2.7440	4.3600	3.4000	0.3644	1.5889	1.2391	0.2294	0.6294	0.7798
4	3.8416	7.1040	7.7600	0.2603	1.8492	2.0200	0.1408	0.5408	1.0923
5	5.3782	10.9456	14.8640	0.1859	2.0352	2.7637	0.0914	0.4914	1.3580
6	7.5295	16.3238	25.8096	0.1328	2.1680	3.4278	0.0613	0.4613	1.5811
7	10.5414	23.8534	42.1334	0.0949	2.2628	3.9970	0.0419	0.4419	1.7664
8	14.7579	34.3947	65.9868	0.0678	2.3306	4.4713	0.0291	0.4291	1.9185
9	20.6610	49.1526	100.3815	0.0484	2.3790	4.8585	0.0203	0.4203	2.0422
10	28.9255	69.8137	149.5342	0.0346	2.4136	5.1696	0.0143	0.4143	2.1419
11	40.4957	98.7391	219.3478	0.0247	2.4383	5.4166	0.0101	0.4101	2.2215
12	56.6939	139.2348	318.0870	0.0176	2.4559	5.6106	0.0072	0.4072	2.2845
13	79.3715	195.9287	457.3217	0.0126	2.4685	5.7618	0.0051	0.4051	2.3341
14	111.1201	275.3002	653.2504	0.0090	2.4775	5.8788	0.0036	0.4036	2.3729
15	155.5681	386.4202	928.5506	0.0064	2.4839	5.9688	0.0026	0.4026	2.4030
16	217.7953	541.9883	1314.9708	0.0046	2.4885	6.0376	0.0018	0.4018	2.4262
17	304.9135	759.7837	1856.9592	0.0033	2.4918	6.0901	0.0013	0.4013	2.4441
18	426.8789	1064.6971	2616.7428	0.0023	2.4941	6.1299	0.0009	0.4009	2.4577
19	597.6304	1491.5760	3681.4400	0.0017	2.4958	6.1601	0.0007	0.4007	2.4682
20	836.6826	2089.2064	5173.0160	0.0012	2.4970	6.1828	0.0005	0.4005	2.4761
21	1171.3556	2925.8889	7262.2223	0.0009	2.4979	6.1998	0.0003	0.4003	2.4821
22	1639.8978	4097.2445	10188.1113	0.0006	2.4985	6.2127	0.0002	0.4002	2.4866
23	2295.8569	5737.1423	14285.3558	0.0004	2.4989	6.2222	0.0002	0.4002	2.4900
24	3214.1997	8032.9993	20022.4981	0.0003	2.4992	6.2294	0.0001	0.4001	2.4925
25	4499.8796	11247.1990	28055.4974	0.0002	2.4994	6.2347	0.0001	0.4001	2.4944
26	6299.8314	15747.0785	39302.6963	0.0002	2.4996	6.2387	0.0001	0.4001	2.4959
27	8819.7640	22046.9099	55049.7749	0.0001	2.4997	6.2416	0.0000	0.4000	2.4969
28	12347.6696	30866.6739	77096.6848	0.0001	2.4998	6.2438	0.0000	0.4000	2.4977
29	17286.7374	43214.3435	107963.3587	0.0001	2.4999	6.2454	0.0000	0.4000	2.4983
30	24201.4324	60501.0809	151177.7022	0.0000	2.4999	6.2466	0.0000	0.4000	2.4988

TABLE A.35 $i = 45\%$

N	F/P	F/A	F/G	P/F	P/A	P/G	A/F	A/P	A/G
1	1.4500	1.0000	0.0000	0.6897	0.6897	0.0000	1.0000	1.4500	0.0000
2	2.1025	2.4500	1.0000	0.4756	1.1653	0.4756	0.4082	0.8582	0.4082
3	3.0486	4.5525	3.4500	0.3280	1.4933	1.1317	0.2197	0.6697	0.7578
4	4.4205	7.6011	8.0025	0.2262	1.7195	1.8103	0.1316	0.5816	1.0528
5	6.4097	12.0216	15.6036	0.1560	1.8755	2.4344	0.0832	0.5332	1.2980
6	9.2941	18.4314	27.6253	0.1076	1.9831	2.9723	0.0543	0.5043	1.4988
7	13.4765	27.7255	46.0566	0.0742	2.0573	3.4176	0.0361	0.4861	1.6612
8	19.5409	41.2019	73.7821	0.0512	2.1085	3.7758	0.0243	0.4743	1.7907
9	28.3343	60.7428	114.9840	0.0353	2.1438	4.0581	0.0165	0.4665	1.8930
10	41.0847	89.0771	175.7269	0.0243	2.1681	4.2772	0.0112	0.4612	1.9728
11	59.5728	130.1618	264.8040	0.0168	2.1849	4.4450	0.0077	0.4577	2.0344
12	86.3806	189.7346	394.9657	0.0116	2.1965	4.5724	0.0053	0.4553	2.0817
13	125.2518	276.1151	584.7003	0.0080	2.2045	4.6682	0.0036	0.4536	2.1176
14	181.6151	401.3670	860.8155	0.0055	2.2100	4.7398	0.0025	0.4525	2.1447
15	263.3419	582.9821	1262.1824	0.0038	2.2138	4.7929	0.0017	0.4517	2.1650
16	381.8458	846.3240	1845.1645	0.0026	2.2164	4.8322	0.0012	0.4512	2.1802
17	553.6764	1228.1699	2691.4886	0.0018	2.2182	4.8611	0.0008	0.4508	2.1915
18	802.8308	1781.8463	3919.6584	0.0012	2.2195	4.8823	0.0006	0.4506	2.1998
19	1164.1047	2584.6771	5701.5047	0.0009	2.2203	4.8978	0.0004	0.4504	2.2059
20	1687.9518	3748.7818	8286.1818	0.0006	2.2209	4.9090	0.0003	0.4503	2.2104
21	2447.5301	5436.7336	12034.9636	0.0004	2.2213	4.9172	0.0002	0.4502	2.2136
22	3548.9187	7884.2638	17471.6972	0.0003	2.2216	4.9231	0.0001	0.4501	2.2160
23	5145.9321	11433.1824	25355.9610	0.0002	2.2218	4.9274	0.0001	0.4501	2.2178
24	7461.6015	16579.1145	36789.1434	0.0001	2.2219	4.9305	0.0001	0.4501	2.2190
25	10819.3222	24040.7161	53368.2580	0.0001	2.2220	4.9327	0.0000	0.4500	2.2199
26	15688.0172	34860.0383	77408.9741	0.0001	2.2221	4.9343	0.0000	0.4500	2.2206
27	22747.6250	50548.0556	112269.0124	0.0000	2.2221	4.9354	0.0000	0.4500	2.2210
28	32984.0563	73295.6806	162817.0680	0.0000	2.2222	4.9362	0.0000	0.4500	2.2214
29	47826.8816	106279.7368	236112.7485	0.0000	2.2222	4.9368	0.0000	0.4500	2.2216
30	69348.9783	154106.6184	342392.4854	0.0000	2.2222	4.9372	0.0000	0.4500	2.2218

TABLE A.36 $i = 50\%$

N	F/P	F/A	F/G	P/F	P/A	P/G	A/F	A/P	A/G
1	1.5000	1.0000	0.0000	0.6667	0.6667	0.0000	1.0000	1.5000	0.0000
2	2.2500	2.5000	1.0000	0.4444	1.1111	0.4444	0.4000	0.9000	0.4000
3	3.3750	4.7500	3.5000	0.2963	1.4074	1.0370	0.2105	0.7105	0.7368
4	5.0625	8.1250	8.2500	0.1975	1.6049	1.6296	0.1231	0.6231	1.0154
5	7.5938	13.1875	16.3750	0.1317	1.7366	2.1564	0.0758	0.5758	1.2417
6	11.3906	20.7813	29.5625	0.0878	1.8244	2.5953	0.0481	0.5481	1.4226
7	17.0859	32.1719	50.3438	0.0585	1.8829	2.9465	0.0311	0.5311	1.5648
8	25.6289	49.2578	82.5156	0.0390	1.9220	3.2196	0.0203	0.5203	1.6752
9	38.4434	74.8867	131.7734	0.0260	1.9480	3.4277	0.0134	0.5134	1.7596
10	57.6650	113.3301	206.6602	0.0173	1.9653	3.5838	0.0088	0.5088	1.8235
11	86.4976	170.9951	319.9902	0.0116	1.9769	3.6994	0.0058	0.5058	1.8713
12	129.7463	257.4927	490.9854	0.0077	1.9846	3.7842	0.0039	0.5039	1.9068
13	194.6195	387.2390	748.4780	0.0051	1.9897	3.8459	0.0026	0.5026	1.9329
14	291.9293	581.8585	1135.7170	0.0034	1.9931	3.8904	0.0017	0.5017	1.9519
15	437.8939	873.7878	1717.5756	0.0023	1.9954	3.9224	0.0011	0.5011	1.9657
16	656.8408	1311.6817	2591.3633	0.0015	1.9970	3.9452	0.0008	0.5008	1.9756
17	985.2613	1968.5225	3903.0450	0.0010	1.9980	3.9614	0.0005	0.5005	1.9827
18	1477.8919	2953.7838	5871.5675	0.0007	1.9986	3.9729	0.0003	0.5003	1.9878
19	2216.8378	4431.6756	8825.3513	0.0005	1.9991	3.9811	0.0002	0.5002	1.9914
20	3325.2567	6648.5135	13257.0269	0.0003	1.9994	3.9868	0.0002	0.5002	1.9940
21	4987.8851	9973.7702	19905.5404	0.0002	1.9996	3.9908	0.0001	0.5001	1.9958
22	7481.8276	14961.6553	29879.3106	0.0001	1.9997	3.9936	0.0001	0.5001	1.9971
23	11222.7415	22443.4829	44840.9659	0.0001	1.9998	3.9955	0.0000	0.5000	1.9980
24	16834.1122	33666.2244	67284.4488	0.0001	1.9999	3.9969	0.0000	0.5000	1.9986
25	25251.1683	50500.3366	100950.6732	0.0000	1.9999	3.9979	0.0000	0.5000	1.9990
26	37876.7524	75751.5049	151451.0098	0.0000	1.9999	3.9985	0.0000	0.5000	1.9993
27	56815.1287	113628.2573	227202.5146	0.0000	2.0000	3.9990	0.0000	0.5000	1.9995
28	85222.6930	170443.3860	340830.7720	0.0000	2.0000	3.9993	0.0000	0.5000	1.9997
29	127834.0395	255666.0790	511274.1580	0.0000	2.0000	3.9995	0.0000	0.5000	1.9998
30	191751.0592	383500.1185	766940.2369	0.0000	2.0000	3.9997	0.0000	0.5000	1.9998

Index